EVERYMAN'S LIBRARY

EVERYMAN,
I WILL GO WITH THEE,
AND BE THY GUIDE,
IN THY MOST NEED
TO GO BY THY SIDE

CHARLES DARWIN

The Origin of Species and
The Voyage of the Beagle

with an Introduction by Richard Dawkins

EVERYMAN'S LIBRARY

Alfred A. Knopf New York London Toronto

258

THIS IS A BORZOI BOOK

PUBLISHED BY ALFRED A. KNOPF

The Voyage of the Beagle first included in Everyman's Library, 1906
The Origin of Species first included in Everyman's Library, 1928

Introduction, Bibliography and Chronology Copyright © 2003 by
Everyman's Library
Typography by Peter B. Willberg

US website: www.randomhouse/everymans

ISBN: 1-4000-4127-9 (US)
1-85715-258-1 (UK)

A CIP catalogue reference for this book is available from the
British Library

Library of Congress Cataloging-in-Publication Data
Darwin, Charles, 1809–1882.
The origin of species; and, The voyage of the Beagle/Charles Darwin; with an
introduction by Richard Dawkins.
p.   cm.—(Everyman's library)
Reprint of 2 works: First work originally published: On the origin of species by
means of natural selection. London: J. Murray, 1859; 2nd work originally
published: Journal of researches into the natural history and geology of the
various countries visited by H.M.S. Beagle, London: H. Colburn, 1839.
Includes bibliographical references (p.).
ISBN 1-4000-4127-9 (hc.: alk. paper)
1. Evolution (Biology) 2. Natural selection. 3. Beagle Expedition
(1831–1836) 4. Natural history. 5. Geology. 6. South-America—Description
and travel. 7. Voyages around the world. I. Darwin, Charles, 1809–1882.
Voyage of the Beagle. II. Title.
QH365.O2   2003
576.8'2—dc21   2003049220

Book design by Barbara de Wilde and Carol Devine Carson
Printed and bound in Germany by GGP Media, Germany

CHARLES DARWIN

———

CONTENTS

INTRODUCTION

Suppose we measure the power of a scientific theory as a ratio: how much it explains divided by how much it needs to assume in order to do that explaining. By this criterion, Darwin's theory of evolution by natural selection is second to none. Think of what it explains – and I really mean explains in the fullest sense of the word: your existence and mine; the form, diversity, and apparently designed complexity and elegance of all living things, not only on this planet but probably wherever in the universe organized complexity may be found. The explanatory work that the theory does, then – the numerator of the ratio – is immense. But the theory itself – the denominator – could hardly be smaller or more simple; you can write it out in a phrase: 'Non-random survival of randomly varying hereditary elements.' That isn't quite how Darwin himself would have put it (he phrased it in terms of the survival and reproduction of individual organisms). But it captures the essence of his idea in a way that – I am convinced – he would recognize and, indeed, enjoy if he were alive today.

We can probably pare the denominator down even further, to a single essential: heredity. Given that there exists a system of high fidelity copying, two things will follow immediately: there will be a population of the entities copied; and, since no copying process is perfect (there is mutation, in other words), they won't all be exactly the same. It then follows that successful hereditary types will tend to increase in number at the expense of unsuccessful hereditary types. That is probably all that is needed for life to get going and, given enough time, to evolve living machinery whose complexity and diversity progresses without obvious limit.

Consider where we would be without Darwin's idea. We'd presumably have some sort of science of biology. Students would take degrees in the subject, books would be written, Nobel prizes in physiology and medicine would be won. We might know in some detail how living organisms work; we'd know that a human body is a teeming army of cells, a thousand times more numerous than the people in the world; we'd know that every one of these cells is a mass-production molecule-factory packed with the membranous equivalent of miles of sophisticated conveyor belt. We'd know a great deal about how our bodies work, and how the bodies of shrimps, elephants and redwood trees work. We'd be compelled to recognize how complicatedly organisms are fitted to survive in their particular worlds. But we wouldn't have the foggiest idea why. We'd read volumes about living things but we wouldn't have a clue about where they came from originally nor why they work so efficiently and purposefully. It would undoubtedly be the most baffling problem in biology – probably in the whole of science and the whole of philosophy. This was the problem that Darwin decisively solved. We are now as certain as one can ever be in science that a version of Darwin's solution is the correct one.

The Darwinian solution to the riddle of existence is so powerfully simple, so felicitous to the modern mind, that it is hard for us to understand why it had to wait until the mid-nineteenth century before anyone thought of it. It is further surprising – and probably telling – that this great inspiration, which looks so elegant from the modern armchair, eluded centuries of philosophers, mathematicians and polymaths. Plato and Aristotle never even got close, fooling about with 'essences' and 'ideal forms'; Leibnitz and Newton gave us calculus (which might seem a more exacting intellectual achievement) but not an inkling of the reason for our own existence; Hume would surely have recognized the idea as a great one if it had occurred to him, but it never did. Having eluded this galaxy of all-time talent, the answer finally came, almost simultaneously, to two English naturalists. Darwin and Wallace, they alone, thought of it.

For we must not forget Alfred Russel Wallace. This happens to Wallace quite often, I am afraid. He tends to get a poor deal at the hands of posterity, partly through his own generous nature. It was Wallace himself who coined the word 'Darwin-ism', he regularly referred to it as Darwin's theory and he referred to himself as 'more Darwinian than Darwin'. The main reason we know Darwin's name more than Wallace's is that Darwin went on, a year later, to publish *The Origin of Species*. The *Origin* not only explained and advocated the Darwin/Wallace theory of natural selection as the mechanism of evolution. It also – and this had to be done at book length – set out the multifarious evidence for the *fact* of evolution itself.

Given the failure of great philosophers and other thinkers throughout history to tumble to the breathtakingly powerful yet simple principle of natural selection, it is worth attending to the story of Darwin and Wallace, in case the similarities between them are revealing. The drama of how Wallace's letter arrived at Downe House on 18 June 1858, casting Darwin into an agony of indecision and worry, is too well known for me to retell it. In my view the whole episode is one of the more creditable and agreeable in the history of scientific priority disputes – precisely because it wasn't a dispute, although it so very easily could have become one. It was resolved amicably, and with heartwarming generosity on both sides, especially, I must say, on Wallace's. As Darwin later wrote:

Early in 1856 Lyell advised me to write out my views pretty fully, and I began at once to do so on a scale three or four times as extensive as that which was afterwards followed in my *Origin of Species*; yet it was only an abstract of the materials which I had collected, and I got through about half the work on this scale. But my plans were overthrown, for early in the summer of 1858 Mr Wallace, who was then in the Malay archipelago, sent me an essay *On the Tendency of Varieties to depart indefinitely from the Original Type*; and this essay contained exactly the same theory as mine. Mr Wallace expressed the wish that if I thought well of his essay, I should send it to Lyell for perusal.

The circumstances under which I consented at the request of Lyell and Hooker to allow of an extract from my MS., together with a letter to Asa Gray, dated September 5, 1857, to be published at the same time with Wallace's Essay, are given in the *Journal of the Proceedings of the Linnean Society*, 1858, p. 45. I was at first very unwilling to consent, as I thought Mr Wallace might consider my doing so unjustifiable, for I did not then know how generous and noble was his disposition. The extract from my MS. and the letter to Asa Gray ... had neither been intended for publication, and were badly written. Mr Wallace's essay, on the other hand, was admirably expressed and quite clear. Nevertheless our joint productions excited very little attention, and the only published notice of them which I can remember was by Professor Haughton of Dublin, whose verdict was that all that was new in them was false, and what was true was old. This shows how necessary it is that any new view should be explained at considerable length in order to arouse public attention.

Darwin was over-modest about his own two papers which were read at the Linnean Society. Both are models of the explainer's art. Wallace's paper is also very clearly argued. His ideas were, indeed, remarkably similar to Darwin's own, and there is no doubt that Wallace arrived at them independently. In my opinion the Wallace paper needs to be read in conjunction with his earlier paper, published in 1855, in the *Annals and Magazine of Natural History*. Darwin read this paper when it came out. Indeed, it led to Wallace joining his large circle of correspondents, and to his engaging Wallace's services as a collector. But, oddly, Darwin did not see in the 1855 paper any warning that Wallace was by then a convinced evolutionist of a very Darwinian stamp. I mean Darwinian as opposed to the Lamarckian view of evolution which saw modern species as all on a ladder, changing into one another as they moved up the ladder. By contrast Wallace, in 1855, had a clear view of evolution as a branching tree, exactly like Darwin's famous diagram which became the only illustration in *The Origin of Species*. The 1855 paper, however, made no mention of natural selection or the struggle for existence.

That was left to Wallace's 1858 paper, the one that hit Darwin like a lightning bolt. Here, Wallace even used the phrase 'Struggle for Existence', and he devoted considerable attention to the exponential increase in numbers (another key Darwinian point). Wallace wrote:

The greater or less fecundity of an animal is often considered to be one of the chief causes of its abundance or scarcity; but a consideration of the facts will show us that it really has little or nothing to do with the matter. Even the least prolific of animals would increase rapidly if unchecked, whereas it is evident that the animal population of the globe must be stationary, or perhaps ... decreasing.

Wallace deduced from this that

The numbers that die annually must be immense; and as the individual existence of each animal depends upon itself, those that die must be the weakest...

Wallace's peroration could have been Darwin himself writing:

The powerful retractile talons of the falcon- and the cat-tribes have not been produced or increased by the volition of those animals; but among the different varieties which occurred in the earlier and less highly organized forms of these groups, those always survived longest which had the greatest facilities for seizing their prey. Neither did the giraffe acquire its long neck by desiring to reach the foliage of the more lofty shrubs, and constantly stretching its neck for the purpose, but because any varieties which occurred among its anti-types with a longer neck than usual at once secured a fresh range of pasture over the same ground as their shorter-necked companions, and on the first scarcity of food were thereby enabled to outlive them. Even the peculiar colours of many animals, especially insects, so closely resembling the soil or the leaves or the trunks on which they habitually reside, are explained on the same principle; for though in the course of ages varieties of many tints may have occurred, yet those races having colours best adapted to concealment from their enemies would inevitably survive the longest. We have also here an acting cause to account for that balance so often observed in nature, – a deficiency in one set of organs always being compensated by an increased development of some others – powerful wings accompanying weak feet, or great velocity making up for the absence of defensive weapons; for it has been shown that all varieties in which an unbalanced deficiency occurred could not long continue their existence. The action of this principle is exactly like that of the centrifugal governor of the steam engine, which checks and corrects any irregularities almost before they become evident.

The image of the steam governor is a powerful one which, I can't help feeling, Darwin might have envied.

Historians of science have raised the suggestion that Wallace's version of natural selection was not quite so Darwinian as Darwin himself believed. Wallace persistently used the word 'variety' as the level of entity at which natural selection acts, for example in the long passage I have just quoted. And some have suggested that Wallace, unlike Darwin who clearly saw selection as choosing among *individuals*, was proposing what modern theorists rightly denigrate as 'group selection'. This would be true if, by 'varieties', Wallace meant geographically separated groups or races of individuals. But I believe a careful reading of Wallace's paper rules it out. I think that by 'variety' Wallace meant what we would nowadays call 'genetic type', even what a modern population geneticist (as opposed to molecular geneticist) might mean by a gene. I think that, to Wallace in this paper, variety meant not local race of eagles, for example, but 'that set of individual eagles whose talons were hereditarily sharper than usual'.

If I am right, it is a similar misunderstanding to the one suffered by Darwin, whose use of the word 'race' in the subtitle of *The Origin of Species* is sometimes misread in support of racialism. That subtitle, or alternative title rather, is *The Preservation of Favoured Races in the Struggle for Life*. Once again, Darwin was using 'race' to mean 'that set of individuals who share a particular hereditary characteristic, such as sharp talons', *not* a geographically distinct race such as the Hoodie Crow. If he had meant that, Darwin too would have been guilty of the group selection fallacy. I believe that neither Darwin nor Wallace were. And, by the same token, I do not believe that Wallace's conception of natural selection was different from Darwin's.

As for the calumny that Darwin plagiarized Wallace, that is rubbish. The evidence is very clear that Darwin did think of natural selection before the younger man, although he did not publish it. We have his abstract of 1842 and his longer essay of 1844, both of which establish his priority clearly, as did his letter to Asa Gray of 1857 which was also read out at the Linnean Society in 1858. Why he delayed so long before publishing is one of the great mysteries of the history of science. Some historians have suggested that he was afraid of the religious implications, others the political ones. Maybe he was just a perfectionist.

When Wallace's letter arrived, Darwin was more surprised than we moderns might think he had any right to be. He wrote to Lyell:

I never saw a more striking coincidence; if Wallace had had my manuscript sketch, written out in 1842, he could not have made a better short abstract of it. Even his terms now stand as Heads of my Chapters.

INTRODUCTION

There were other similarities. Both men took Charles Lyell's *Principles of Geology* with them on their voyages, and surely were influenced by Lyell's 'uniformitarian' (anti-catastrophist) view of world history: the view that the forces governing the modern earth are the same as those governing the ancient earth, that change is gradual, and that the earth is very old. The coincidence extended to both Darwin and Wallace being inspired by Malthus on population. In Wallace's case, the recollection of an earlier reading of Malthus dropped into his overheated brain in the midst of a malarial fever, on the island of Ternate in the Moluccas archipelago:

I was suffering from a sharp attack of intermittent fever, and every day during the cold and succeeding hot fits had to lie down for several hours, during which time I had nothing to do but to think over any subjects then particularly interesting me ... One day something brought to my recollection Malthus's 'Principles of Population'. I thought of his clear exposition of 'the positive checks to increase' – disease, accidents, war, and famine – which keep down the population of savage races to so much lower an average than that of more civilized peoples. It then occurred to me ...

(from Wallace's autobiography)

It is easy to imagine the impact on Wallace's fevered brain, teeming with the diversity of the tropics, of Malthus's demonstration that populations tend to outstrip resources, from which it followed that life must be a continued struggle. The principle of natural selection jumped out at him, and it is tempting to imagine it as one of those cool lucid respites in the malarial cycle, when the jungle's clamour seems suddenly hushed.

Darwin, by his own account (although modern historians have disputed it), was immediately inspired by Malthus's emphasis on overpopulation and competition, the moment he read it. He wrote in his autobiography:

In October, 1838, that is, fifteen months after I had begun my systematic inquiry, I happened to read for amusement Malthus on population, and being well prepared to appreciate the struggle for existence which everywhere goes on from long continuous observation of the habits of animals and plants, it at once struck me that under these circumstances favourable variations would tend to be preserved and unfavourable ones to be destroyed. The result of this would be the formation of new species. Here, then, I had at last got a theory by which to work; but I was so anxious to avoid prejudice, that I determined not for some time to write even the briefest sketch of it. In June 1842, I first allowed myself the satisfaction of writing a very brief abstract of my theory in pencil in 35 pages; and this was enlarged during the summer of 1844 into one of 230 pages, which I had fairly copied out and still possess.

Darwin's determination not to write his hypothesis down, 'to avoid prejudice', seems odd until one remembers that, in his time, scientists were supposed to proceed by pure induction, letting facts mount up and speak for themselves to a mind unbiased by theory. In fact, as Michael Ghiselin showed in his excellent *The Triumph of the Darwinian Method*, Darwin, like any good scientist, worked by dreaming up hypotheses and then testing them against the facts, not only in the case of the theory of natural selection itself, but in all his major works (and, by the way, even if he had not given us the *Origin*, Darwin would still come down to us as a considerable figure in Victorian science).

There are other candidates for priority, apart from Darwin and Wallace. I'm not talking about the idea of evolution itself, of course: there are numerous precedents there, including Erasmus Darwin, his remarkable grandfather. But as discoverers of natural selection, other Victorians have been championed with something like the zeal of Baconians disputing the authorship of Shakespeare. These include Patrick Matthew, whose cause is vigorously argued by W. J. Dempster in *Patrick Matthew and Natural Selection*, and Edward Blyth who is similarly advocated by Loren Eiseley. Darwin himself mentions an earlier candidate, W. C. Wells. Matthew complained that Darwin had overlooked him, and Darwin subsequently did mention him in later editions of the *Origin*.

In 1831 Mr Patrick Matthew published his work on 'Naval Timber and Arboriculture', in which he gives precisely the same view of the origin of species as that ... propounded by Mr Wallace and myself in the 'Linnean Journal', and as that enlarged in the present volume. Unfortunately the view was given by Mr Matthew very briefly in scattered passages in an Appendix to a work on a different subject, so that it remained unnoticed until Mr Matthew himself drew attention to it in the 'Gardener's Chronicle' ... (Quoted in Darwin's 'Historical Sketch', first added in the third edition of the *Origin*.)

I think it is by no means clear that these authors understood the importance of natural selection. Matthew himself didn't seem to regard it as a major discovery at all:

To me the conception of this law of Nature came intuitively as a self-evident fact, almost without an effort of concentrated thought ... Mr Darwin here seems to have more merit in the discovery than I have had; to me it did not appear a discovery. He seems to have worked it out by inductive reason, slowly and with due caution to have made his way synthetically from fact to fact onwards; while with me it was by a general glance at the scheme of Nature that I estimated this select production of species as an *a priori* recognisable fact – an axiom requiring only to be pointed out to be admitted by unprejudiced minds of sufficient grasp.

The evidence is compatible with the view that these purported predecessors of Darwin and Wallace saw natural selection as a purely negative force, weeding out misfits rather than building up the whole evolution of life (this, indeed, is a misconception under which modern creationists can be found labouring).

What is the present status of Darwinism? Should we today read Darwin's works for their historic interest only, or are they still regarded as scientifically true? We can now see that, to a remarkable extent, Darwin got it right. Above all, we still hold to his solution of the really big question, the question of how the laws of blind physics could generate ever increasing prodigies of statistically improbable, organized complexity (a feat that, as we shall see, has even – erroneously – been thought to fly in the face of one of the most important of those laws, the second law of thermodynamics). Quite apart from the correctness of his fundamental point, Darwin showed astonishing prescience when it came to detail. His passages on ecology, on the tangled webs of interaction between animals and plants, sound uncannily modern. It is his ideas on the detailed nature of heredity that we can now see to have been most seriously wrong. Indeed, reforms in ideas on heredity have led, in two successive episodes, to announcements of the births of 'neo'-Darwinisms.

The first neo-Darwinism dates from the 1890s and was mainly associated with the name of the great German biologist August Weismann. Weismann's most important contribution was the idea of the genetic material as an autonomous river flowing through time, with bodies – you and me – as temporary conduits for it. Weismannism is diametrically opposed to Lamarckism. Weismann's genetic river flows on inviolate, and acquired characteristics are never inherited. The second neo-Darwinism, the one to which the name is applied today, is the population genetic version of Darwinism initiated by R.A. Fisher, J.B.S. Haldane and Sewall Wright. This is conventionally dated from about 1930, when Fisher published *The Genetical Theory of Natural Selection*, although his writings on the subject go back at least a decade earlier.

Darwin's view of heredity was that of his time. There seemed no particular reason to doubt that environmentally induced changes in the body impressed themselves on the hereditary substance and so were passed on to future generations. Darwin did not doubt it, and when it suited him he freely made use of the idea. Indeed, in *Variation under Domestication*, he developed his own theory of acquired heredity – 'Pangenesis'. Tiny particles called gemmules circulated in the blood, becoming imprinted with information from all parts of the body and conveying it to the sex cells and hence into the next generation. This is one of the very few of Darwin's ideas that we can say with confidence is utterly wrong, not just in detail but in spirit.

Darwin also accepted the assumption of blending inheritance prevalent at that period. Children clearly resembled both their parents, but some seemed to take more after the father, others more after the mother. Presumably some substance from each parent was blended with that of the other, though not necessarily in equal proportions. The incompatibility of all forms of blending inheritance with natural selection was pointed out by Fleeming Jenkin, Professor of Engineering at Glasgow, in 1867 in the *North British Review*. Jenkin, who was hostile to Darwin's ideas, produced an ingenious and apparently knock-down argument against the theory of natural selection. He used a human example to express his point, and couched it in the racialist terms that were almost universal then, so I'll put it in a less offensive way. If children are intermediate between their parents – a blend like a mixture of two pots of paint – new variations will inevitably tend to disappear as the generations go by. No matter how strongly natural selection favours a characteristic, that characteristic is bound to become successively diluted as time passes. Nothing can save the beneficial characteristic from being swamped out of existence.

It is tempting to say – and frequently said – that it was Jenkin's challenge that goaded Darwin into importing a large measure of Lamarckian heredity into the fifth and sixth editions of the *Origin*. After all, it is known that Darwin was worried by Jenkin's argument, and the assumption of blending inheritance is indeed well-nigh fatal to natural selection. Although this feels right, it has been pointed out that in fact most of the fifth edition revisions had already been made before Darwin could have read Jenkin's article. Whatever the stimulus for these revisions, they are among the most important reasons for preferring the first edition for this Everyman reprinting. The first edition was written without benefit of criticism from Darwin's lesser contemporaries, and it was the better for it.

Incidentally, the *Origin* is not the only one of Darwin's books of which this might be said. In the second edition of *The Descent of Man*, Darwin raised the question of the 1:1 sex ratio, only to drop it again:

I formerly thought that when a tendency to produce the two sexes in equal numbers was advantageous to the species, it would follow from natural selection, but I now see that the whole problem is so intricate that it is safer to leave its solution to the future.

(*The Descent of Man*, second edition)

It was the splendidly clear-thinking Ronald Fisher who came to the rescue in 1930.* He introduced his discussion by quoting the above passage from Darwin, and then went on to his own elegant solution:

*R. A. Fisher, *The Genetical Theory of Natural Selection*, Oxford University Press, 1930.

In organisms of all kinds the young are launched upon their careers endowed with a certain amount of biological capital derived from their parents ... Let us consider the reproductive value of these offspring at the moment when this parental expenditure on their behalf has just ceased. If we consider the aggregate of an entire generation of such offspring it is clear that the total reproductive value of the males in this group is exactly equal to the total value of all the females, because each sex must supply half the ancestry of all future generations of the species. From this it follows that the sex ratio will so adjust itself, under the influence of Natural Selection, that the total parental expenditure incurred in respect of children of each sex, shall be equal; for if this were not so and the total expenditure incurred in producing males, for instance, were less than the total expenditure incurred in producing females, then since the total reproductive value of the males is equal to that of the females, it would follow that those parents, the innate tendencies of which caused them to produce males in excess, would, the for same expenditure, produce a greater amount of reproductive value; and in consequence would be the progenitors of a larger fraction of future generations than would parents having a congenital bias towards the production of females. Selection would thus raise the sex-ratio until the expenditure upon males became equal to that upon females.

But what Fisher and many others overlooked is that Darwin himself came much closer to the correct, Fisherian solution in the *first* edition of *Descent* than in the second edition which is the one Fisher quoted.

Let us now take the case of a species producing, from the unknown causes just alluded to, an excess of one sex – we will say of males – these being superfluous and useless, or nearly useless. Could the sexes be equalized through natural selection? We may feel sure, from all characters being variable, that certain pairs would produce a some-what less excess of males over females than other pairs. The former, supposing the actual number of the offspring to remain constant, would necessarily produce more females, and would therefore be more productive. On the doctrine of chances a greater number of the offspring of the more productive pairs would survive; and these would inherit a tendency to procreate fewer males and more females. Thus a tendency toward equalization of the sexes would be brought about.

This really is a remarkably close approximation to Fisher's

own formulation. If only Darwin had had the courage of his first edition convictions! The whole question of Fisher's predecessors in sex ratio theory is admirably discussed by one of Fisher's most distinguished pupils, A. W. F. Edwards.*

Returning to the *Origin*, and Fleeming Jenkin's riddle of the apparently disappearing variance, the solution eventually turned out to lie in the fact that heredity is not blending at all, as Jenkin and Darwin thought, but 'Mendelian' – not a matter of blending substances but of reshuffling indivisible particles. One of the famous ironies of the history of science is that Mendel's experiments had already been published in 1865, two years before Jenkin's troubling assault. Properly interpreted, Mendel would eventually turn out to be Darwin's saviour from the erroneous criticism of Jenkin. But, far from being properly interpreted, Mendel was not even read, by Darwin or apparently anyone else, until 1900. And even after the rise of Mendelism its leading exponents, the 'Mendelians' William Bateson, Hugo De Vries and others, so misunderstood its significance as to think it positively opposed to Darwinism. This was largely because the most visible mutations were too large and dramatic to do the job that Darwinism would demand of them.

The matter was not definitively resolved until it was finally taken in hand, again by R. A. Fisher. Incidentally, there is a pleasing continuity in the fact that Fisher was backed and supported in his young days by one of Charles Darwin's sons, the eugenicist Major Leonard Darwin. Fisher's daughter, Joan Fisher Box was later to write, in her biography of her father:†

Major Darwin, by his faith and his personal generosity, stood god-father to Fisher's scientific career, and in the 30 years of deepening friendship until Darwin's death in 1942, Fisher honored him as a father.

W. Weinberg in Germany and the eccentric English math-ematician G. H. Hardy had shown by elementary algebra that

*A. W. F. Edwards, 'Natural Selection and the Sex Ratio: Fisher's Sources', *American Naturalist*, **151**, 564–9, 1998.
†Joan Fisher Box, *R. A. Fisher: the Life of a Scientist*, Wiley, New York, 1978.

there would be no inherent tendency for genes to change in frequency. In the same way as Newton argued that bodies in motion would stay in motion unless positively resisted, Hardy and Weinberg showed, with almost embarrassingly simple algebra, that gene frequencies in populations would remain the same unless positively perturbed by some force such as selection. Starting from the so-called Hardy/Weinberg Law, Fisher, Haldane and Wright built up a new way of thinking about Darwinism which is pretty much the way we think about it today. Evolution became seen as changes in frequencies of rival ('allelic') particles in a pool of particles, the gene pool. As some genes become more frequent in the population and others less frequent, so the shapes and sizes of organisms change. As I would put it today, geological time is the stage for a continually renewed dance of immortal genes, changing partners and casting off an unending string of mortal bodies. So the spectre of Fleeming Jenkin was finally laid.

What might have been pointed out in Jenkin's own time was that his argument was only incidentally anti-Darwinian. It was also an argument against manifest fact. If Jenkin had been right, variation should not be conserved and should decrease with dramatic rapidity. Any cohort of grandchildren should be visibly less variable than their grandparents were. Mix black paint and white, and the result is grey paint. Mix grey with grey and you'll get more grey: all the king's horses and all the king's men will never reconstitute the original black and white. Yet with living things, as the generations go by, it is perfectly obvious that this is not happening. The descendants of variable ancestors do not merge into grey uniformity. They remain, on the whole, just as variable as earlier generations ever were. Jenkin thought his argument anti-Darwinian. It was actually anti-known-fact and therefore clearly wrong.

Fisher pointed out that Darwin, as early as 1857, had a dim perception of some sort of particulate inheritance. In a letter to Huxley, Darwin had written:

I have lately been inclined to speculate, very crudely and indistinctly, that propagation by true fertilization will turn out to be a sort of mixture, and not true fusion, of two distinct individuals, as each parent has its parents and ancestors. I can understand on no other view the way in which crossed forms go back to so large an extent to ancestral forms. But all this, of course, is infinitely crude.

Being wise after the event, one can say that the principle of particulate or Mendelian inheritance, and its significance in refuting Jenkin, is staring us in the face every time we contemplate sex itself. We all have a mother and a father but we are not intermediate hermaphrodites. If maleness and femaleness do not blend, Jenkin-style, why should we assume that anything else does? Some things are so obvious that it requires a Fisher to see beyond the obviousness.

But, yet again, what even Fisher did not realize is how tantalizingly close Darwin himself came. In an 1866 letter to Wallace,* which Fisher would surely have quoted had he known of it, Darwin said:

My dear Wallace
. . . I do not think you understand what I mean by the non-blending of certain varieties. It does not refer to fertility; an instance will explain. I crossed the Painted Lady and Purple sweetpeas, which are very differently coloured varieties, and got, even out of the same pod, both varieties perfect but none intermediate. Something of this kind I should think must occur at least with your butterflies & the three forms of Lythrum; tho' these cases are in appearance so wonderful, I do not know that they are really more so than every female in the world producing distinct male and female offspring. . .
Believe me, yours very sincerely
Ch. Darwin

Here Darwin comes closer to anticipating Mendel than in the passage quoted by Fisher, and he even mentions his own Mendel-like experiments on sweet peas.

Fleeming Jenkin was not the only physical scientist to give

*Letter dated 'Tuesday, February, 1866'. Published in James Marchant, *Alfred Russel Wallace: Letters and Reminiscences*, Vol. 1, Cassell, London, 1916. I am extremely grateful to Dr Seymour J. Garte of New York University, who found this letter by chance in a volume of correspondence between Darwin and Wallace in the British Library in London, immediately recognized its significance and sent a copy to me. I have previously published this letter in *A Devil's Chaplain*, Weidenfeld and Nicolson, London, 2003.

Darwin a hard time. Sir William Thomson, Lord Kelvin, incidentally a former collaborator of Fleeming Jenkin in laying the transatlantic cable, was seen by many as the leading physicist of his generation. He pronounced, with the lofty certainty of a physicist talking down to a mere biologist, that the earth was probably only a few tens of millions of years old, and certainly far too young to allow evolution to achieve what Darwin supposed. Darwin quailed a little before such massively authoritative pontification, but he remained true to his own, more deeply reasoned convictions.

In our century, of course, he was triumphantly vindicated. Radioactive dating methods show the earth's age to be measured in thousands of millions of years, ample – indeed, almost too ample – for Darwin's purposes. Even by the end of the nineteenth century Kelvin's conclusions were already being questioned by physical scientists, among them Darwin's son George. It is easy to forgive Kelvin his error; harder to forgive his assumption that, if the best physicist of the day says one thing and the best biologist the opposite, it is axiomatic that the physicist must be right. In our own time the eminent astrophysicist Sir Fred Hoyle ran an idiosyncratic anti-Darwinian campaign which led him, together with another astronomer, C. Wickramasinghe, even to question the authenticity of one of our best-known fossils, the primitive bird *Archaeopteryx*, and to claim that 'Any *physicist* looking at it would have worries' (quoted in *New Scientist*, 14 April 1985, p. 3, my emphasis). Hoyle and Wickramasinghe have also written:

At one stage we received a suggestion by telephone that a panel of four sedimentary geologists be appointed, two by the Museum, two by us, to pronounce on the authenticity of the fossil. Our reply was typical of all that is objectionable in our way of expressing things.

The remarks about sedimentary geologists that follow do indeed bear out this judgment, though why they are so proud of being objectionable is less clear. They conclude (emphasis mine):

Besides which, controversies are not settled in the *quantitative sciences* in such a ludicrous way.

The same assumption about the hierarchy of sciences, with physics at the top, can be detected in the following letter to *The Times* of 19 December 1981 by a professional physicist (and a Fellow of the Royal Society):

As a physicist, I cannot accept [natural selection] ... It seems to me to be impossible that chance variation should have produced the remarkable machine that is the human body. Take only one example – the eye ...

The objection to Darwinism that it is a theory of 'chance' is one of the most popular and most foolish of all. Mutation is indeed random, although only in the sense that it is not directed towards improvement. Natural selection is quintessentially the opposite of random. Any fool can see that random chance cannot put together living complexity. That is precisely why Darwinism is necessary. It is a preposterous irony that the statistical improbability of living organization is regularly advanced as though it counted *against* Darwinism, rather than in favour of it.

Other alleged objections were mostly disposed of by Darwin himself in his 'Difficulties on Theory' chapter of the *Origin*, and the three subsequent chapters on 'Instinct', 'Hybridism' and 'On the Imperfection of the Geological Record'. Some more modern objections Darwin would have rumbled with ease. I'll mention two, 'thermodynamics' and 'tautology', which are based upon half-understood smatterings of science and philosophy.

The second law of thermodynamics states that any closed system tends to become more disordered. This is fundamentally because there are more ways of being disordered than ways of being in a state that we would recognize as ordered. Evolution represents an increase in the ordering of matter. Therefore, the argument triumphantly concludes, evolution violates the second law and evolution cannot occur! The physicist's simple reply to this argument is that we are not dealing with a closed system: external energy is fed in. But a more cutting retort than this is called for. It is the very existence of life – whether it comes about by evolution, divine creation or by any other means – that constitutes, if anything does, an increase in the ordering of matter. If the thermo-dynamic argument could be used to prove the non-existence of evolution, it could equally be used to prove the non-existence of life. The idea that evolution violates the second law is not merely mistaken – it turns out to be actively foolish.

Half-understood philosophy is no less pernicious than half-understood physics. Most of us nowadays have picked up the philosopher's nostrum that a good scientific theory should be 'falsifiable'. Any theory that turns out to be a necessary truth, a tautology, is of no value according to this philosophy. Darwinian natural selection is sometimes summed up in Herbert Spencer's phrase 'survival of the fittest'. The 'fittest' are defined as those that survive. So the phrase, and by triumphant implication the theory, is circular, tautological and empty. Once again, we can give a scholarly reply and also a more scathing one. The scholarly reply is that it is only recently that 'fittest' has come to be defined in such a way as to make 'survival of the fittest' circular. For Darwin the fittest were the strongest, the fastest, the sharpest-eyed. As part of the neo-Darwinian revolution, theorists needed a name for the quantity that natural selection tends to maximize, a mathematical quantity sometimes symbolized as W. They lit upon 'fitness'. If fitness is *defined* in this way, it is obvious that the phrase 'survival of the fittest' has got to become a tautology.

To come to the more scathing reply, 'survival of the fittest' is a tautology in the same sense as it is tautological to say: 'If a horse runs at twice the speed of a man, the horse will cover a given distance in half the time as a man'. This follows from the definition of 'speed'. Only someone 'educated far beyond his capacity to undertake analytical thought'* would conclude that because the statement is tautological neither horse nor man will complete the course! Darwin himself would perhaps have appealed, as he so often did, to the facts of domestication. Nobody denies that selective breeding has been extremely effective in making racehorses run as fast as they do, or in improving the milk yield of cows. If we define the 'fittest'

*P. B. Medawar, *The Art of the Soluble*, Oxford University Press, 1982 (p. 249).

racehorse as the type most likely to be selected for breeding, then what is going on in a stud farm is a process of (artificial) 'survival of the fittest' which is every bit as tautological as survival of the fittest in wild nature. But nobody ever tried to persuade a horse breeder or a cattle breeder: 'You're wasting your time trying to improve the breed; your whole enterprise is a circular tautology and therefore doomed to failure.'

On the Origin of Species (the 'On' was dropped from the sixth edition) is not the only important book that Darwin wrote. I have already mentioned *The Descent of Man*. But the one that Everyman have chosen to reprint together with the *Origin* is Darwin's first and most charming book, commonly known as *The Voyage of the Beagle* or *A Naturalist's Voyage*, which was based upon his diaries of H.M.S. *Beagle*. I say 'commonly known as', because the Victorians had a more fluid view of book titles than we do; it is even possible to be unsure which title applies to which book. The full title of the first edition, published in 1839 by Henry Colburn, London, was *Voyages of the Adventure and Beagle. Narrative of the Surveying Voyages of His Majesty's Ships Adventure and Beagle between the years 1826 and 1836 describing their Examination of the Southern Shores of South America and the Beagle's circumnavigation of the globe. In Three Volumes*. Volumes I and II were by Captains King and Fitzroy. Darwin's was Volume III, entitled *Journal and Remarks, 1832–1836*. Soon afterwards, Colburn reissued Darwin's volume separately, under a new title. For his second edition, in 1845, Darwin went alone to a new publisher, John Murray, with whom he was to stay for the rest of his writing life. This second edition's title, slightly modified from Colburn's second title, was *Journal of Researches into the Natural History and Geology of the Countries Visited during the Voyage of H.M.S. Beagle round the World*. It is that second edition, originally published in 1845, which is reprinted here.

Several further editions have appeared, mostly versions of the second, 1845 edition. Many of these have an abbreviated title on the spine, and sometimes on the title page as well: *A Naturalist's Voyage* or *A Naturalist's Voyage Around the World*. Others, including my own copy published in 1891 by Ward Lock, has a different abbreviated title on the cover: *Darwin's Journal During the Voyage of H.M.S. Beagle round the World*. To add one more layer to the complication, in 1933 Darwin's granddaughter Nora Barlow edited *Charles Darwin's Diary of the Voyage of the Beagle* for Cambridge University Press. This is a reprinting of the original diaries on which the earlier book was based.

By whatever name, it is a delightful and beautifully written travel book which always makes me think of my earlier hero, Hugh Lofting's 'Doctor Dolittle' (now, alas, banned by righteous librarians for 'racism' – regrettable, no doubt, but surely a peccadillo in comparision to the rampant speciesism of our society which Dr Dolittle could do so much to remove from childish minds).

Darwin's travel notes offer tantalizing glimpses of how his mind was working towards the great truth that he was later to unveil for us. On every page, the book gives us hints of the great sage that the young Darwin was to become. We are left with a picture of a clever, sensitive and humane observer, a restlessly curious intellect, and a tireless collector whose youthful energy makes a poignant contrast with his invalided maturity and with the brooding melancholy that stares across to us from the portraits of his old age. There is adventure here. And humour, as in the story of how Darwin tried to use the bolas, the South American hunting weapon consisting of a weight on the end of a rope, but succeeded only in capturing the legs of his own horse – to the huge amusement of the gauchos.

We also can appreciate the liberal decency of the young Darwin, his passionate opposition to slavery (which caused a falling-out with Fitzroy), and his bewildered disgust at, for example, the indiscriminate hunting of South American Indians by the Spanish:

This is a dark picture; but how much more shocking is the unquestionable fact, that all the women who appear above twenty years old are massacred in cold blood! When I exclaimed that this appeared rather inhuman, he answered, 'Why, what can be done? They breed so!'

Everyone is convinced that this is the most just war, because it is against barbarians. Who would believe in this age that such atrocities could be committed in a Christian civilized country?

But most of his attention is focused upon his beloved natural history, and geology. *The Voyage of the Beagle* is a travel book by one of the most intelligently observant and thoughtful travellers of all time. What a privilege to see the world through such eyes. What a bonus to glimpse the mind of a developing genius in the flower of his robust and energetic youth.

Charles Darwin should be read because, give or take a few minor details, the evolutionary world view that he has given us, that world view which is so utterly different from anything that went before, is, according to all that we have learnt since his time, literally true and unlikely ever to be superseded. Evolution as a fact has the same status as the fact that the earth is round and not flat (though people can be found to call either 'mere theories'). It is Darwin who conclusively showed to the world that evolution is true.

He also gave us by far the most plausible theory for *how* evolution has taken place, the theory of natural selection. I would go further. I have argued elsewhere that, at least as far as adaptation – the strong illusion of design that pervades all living things – is concerned, not only is natural selection the correct explanation for adaptive evolution on this planet. It is the only explanation ever suggested that *in principle* is capable of explaining adaptation. We therefore have grounds for believing that Darwinism has a universality that otherwise we expect to find only in the great theories of physics. If there is life – if there is any form of organized complexity – it will have evolved by some form of Darwinian natural selection. If that sounds like a rash or an arrogant thing to say – who, after all, would have thought that Newton would be superseded by Einstein and Planck? – reflect that there are truths that everybody would accept are simply true and will never be superseded.

It is sometimes suggested that Darwin is one of a trio of nineteenth-century giants, along with Marx and Freud. But are Marx and Freud really in his league? If we are ever contacted by alien beings from anywhere else in the universe sufficiently advanced technologically to reach us, will we have anything to say to them, when they step from their celestial *Beagle*? We shall surely share with them at least parts of mathematics and physics. They will have computed the same value of π; they will have the geometrical theorem that we attribute to Pythagoras; they will revere their equivalents of Einstein and of Planck. But there is no reason to suppose that they will have a Marx or a Freud. Why should these men's discoveries have any applicability outside the narrow confines of one species of animal on one planet in one galaxy? It is only the anthropologists on the alien expedition who will have any time for Marx and Freud. But if I am right about the universal significance of Darwinian natural selection, our aliens *will* revere the immortal memory of their Charles Darwin.

Richard Dawkins

RICHARD DAWKINS FRS is a Fellow of New College, Oxford, and holder of the Charles Simonyi Chair of Public Understanding of Science. He has written extensively on Darwin and the theory of evolution. His books include *The Selfish Gene*, *The Extended Phenotype*, *The Blind Watchmaker*, *River Out of Eden*, *Climbing Mount Improbable*, *Unweaving the Rainbow* and, most recently, *A Devil's Chaplain*.

NOTE ON THE TEXT

This edition reprints the text of the second edition of *The Voyage of the Beagle* and the first edition of *The Origin of Species*. The Historical Sketch and the Glossary, however, did not appear with the first edition but were added by Darwin later.

SELECT BIBLIOGRAPHY

BIOGRAPHY

JANET BROWNE, *Darwin: Voyaging*, Jonathan Cape, 1995.

JANET BROWNE, *Darwin: The Power of Place*, Jonathan Cape, 2002.

CHARLES DARWIN, *The Correspondence of Charles Darwin*, ed. Frederick Burkhardt et al., 13 vols, further vols forthcoming, Cambridge University Press, 1985–.

FRANCIS DARWIN, *Life and Letters of Charles Darwin*, John Murray, 1887.

FRANCIS DARWIN, *More Letters of Charles Darwin*, John Murray, 1903.

WORKS

There are various annotated editions of *The Voyage of the Beagle* and *The Origin of Species*. Definitive texts and their variants can be found in *The Works of Charles Darwin*, ed. P. H. Barratt and R. B. Freeman, 29 vols, Pickering & Chatto, 1986–.

COMMENTARY

MEA ALLAN, *Darwin and His Flowers: The Key to Natural Selection*, Faber, 1977.

BERNARD GRANT CAMPBELL, *Sexual Selection and the Descent of Man*, Heinemann, 1972.

RICHARD DAWKINS, ed., *The Descent of Man*, Gibson Square Books, 2003.

RICHARD DAWKINS, *River Out of Eden*, Weidenfeld, 1993.

GAVIN DE BEER, *Charles Darwin: Evolution by Natural Selection*, Nelson, 1963.

R. A. FISHER, *The Genetical Theory of Natural Selection*, Oxford University Press, 1930.

M. T. GHISELIN, *The Triumph of the Darwinian Method*, University of California Press, 1969.

S. J. GOULD, *Ever Since Darwin*, Burnett Books, 1978.

H. E. GRUBER, *Darwin on Man*, Wildwood House, 1974.

GERTRUDE HIMMELFARB, *Darwin and the Darwinian Revolution*, Doubleday, 1959.

ADRIAN DESMOND and JAMES MOORE, *Darwin*, Penguin, 1992.

LAURIE ROHDE GODFREY, ed., *What Darwin Began: Modern Darwinian and Non-Darwinian Perspectives on Evolution*, Allyn and Bacon, 1984.

MICHAEL RUSE, *The Darwinian Revolution*, University of Chicago Press, 1979.

G. G. SIMPSON, *The Meaning of Evolution*, Yale University Press, 1949.

RELATED TOPICS

C. BIBBY, *T. H. Huxley: Scientist, Humanist and Educator*, Watts, 1959.

W. GEORGE, *Biologist Philosopher: A Study of the Life and Writing of Alfred Russel Wallace*, Abelard-Schuman, 1964.

W. GEORGE, *Gregor Mendel and Heredity*, Priory Press, 1975.

K. LYELL, ed., *Life, Letters and Journals of Sir Charles Lyell*, John Murray, 1881.

W. B. TURRILL, *Joseph Dalton Hooker*, Nelson, 1963.

L. WILSON, *Charles Lyell: The Years to 1841*, Yale University Press.

CHRONOLOGY

―――――

DATE	AUTHOR'S LIFE	LITERARY CONTEXT
1789		Erasmus Darwin: *Botanic Garden*.
1793		
1794		Erasmus Darwin: *Zoonomia* (to 1796).
1798		Malthus: *An Essay on the Principle of Population*.
1799–1804		
1802		Paley: *Natural Theology*.
1807		Humboldt: *Voyages aux régions equinoxiales du nouveau continent*.
1809	Born 9 February in Shrewsbury, son of Dr Robert Darwin and Susannah Wedgwood, grandson of Erasmus Darwin the scientist, doctor, poet and polymath, and Josiah Wedgwood the potter.	Lamarck: *Philosophie zoologique*. Goethe: *Elective Affinities*.
1811		Jane Austen: *Sense and Sensibility*. Shelley: *The Necessity of Atheism*.
1812		Byron: *Childe Harold* Cantos 1 & 2. Hegel: *Logic* (to 1816).
1814		Lamarck: *Histoire des animaux sans vertèbres* (to 1822). Wordsworth: *The Excuisian*.
1815		
1816		Jane Austen: *Emma*. Scott: *Old Mortality*.
1817	Mother died.	Georges Cuvier: *Le Règne animal distribué d'après son organisation*. Keats: *Poems*. Coleridge: *Biographia Literaria*.
1818	Sent to Shrewsbury School.	Scott: *Heart of Midlothian*. Etienne Geoffroy Saint-Hilaire: *Philosophie anatomique* (to 1820). Mary Shelley: *Frankenstein*. Jane Austen: *Northanger Abbey*, *Persuasion*.

HISTORICAL EVENTS

French Revolution.
Britain declares war on France.

Naturalist Baron von Humboldt travels in South America.

Sir John Moore dies at Corunna; allied victories in the Peninsula; Napoleon captures Vienna.

Prince of Wales made Regent (George III recognized as insane). Bread riots in Nottingham; Luddite uprisings begin.

Napoleon invades Russia. Tory government under Lord Liverpool in Britain (to 1827).
Napoleon abdicates. Congress of Vienna (to 1815).

Napoleon's Hundred Days. Battle of Waterloo. Second Bourbon restoration. Corn Laws in Britain.
Spa Field riot. Economic depression. Holy Alliance between Russia, Prussia and Austria to preserve European status quo.

Congress of Aix-la-Chapelle.

DATE	AUTHOR'S LIFE	LITERARY CONTEXT
1819		Scott: *Ivanhoe*; *The Bride of Lammermoor*.
1820		Hegel: *Philosophy of Right*.
1824		
1825	Enrolled at Edinburgh University to read medicine. Didn't like it, and spent much of his time studying natural history instead, especially seashore invertebrates.	Saint-Simon: *Nouveau Christianisme*.
1828	Having given up all hope of being a doctor, entered Christ's College, Cambridge, to train for the clergy. Read Paley's *Evidences* which made a great impression on him. Came under the influence of the botanist Henslow, and the geologist Sedgwick.	
1829		Balzac: *La Comédie humaine* (17 vols, to 1848).
1830		Stendhal: *Le Rouge et le Noir*. Charles Lyell: *Principles of Geology* (to 1833). Cuvier and Saint-Hilaire: *Philosophie zoologique*.
1831	Overcoming father's initial opposition with aid of uncle Wedgwood, sailed on the surveying ship H.M.S. *Beagle*, having been recommended by Henslow for the post of gentleman companion to Captain Robert Fitzroy.	Hugo: *Notre-Dame de Paris*.
1832		Death of Goethe and Sir Walter Scott.
1832–5	Various overland expeditions in South America, each time returning to the *Beagle*, which was charting South American waters.	
1833		George Sand: *Lélia*. Carlyle: *Sartor Resartus*.
1834		
1835	Visit to the Galapagos Islands, 16 September–20 October.	Humboldt: *Géographie du nouveau continent* (to 1838).

CHRONOLOGY

HISTORICAL EVENTS

Peterloo massacre; Six Acts. Birth of Queen Victoria. Factory Act restricts child labour.
Death of George III; accession of George IV. Trial of Queen Caroline.
Repeal of Combination Acts. *Westminster Review* founded.

Catholic Emancipation. Stephenson's 'Rocket'.

Death of George IV; accession of William IV. The 'Bourgeois Revolution' in France: flight of Charles X and accession of Louis-Philippe.
The first great railway, between Liverpool and Manchester, opened.

First Parliamentary Reform Act.

New Factory Act introduces further reforms. Start of Oxford Movement; Newman launches 'Tracts for the Times'. Abolition of slavery in the British Empire.
New Poor Law abolishes outdoor relief. Tolpuddle Martyrs.
Peel's Tamworth Manifesto. Fox Talbot: first photographic negative.

DATE	AUTHOR'S LIFE	LITERARY CONTEXT
1836	Return of the *Beagle* to England, via Australia and South Africa (2 October).	
1837		Dickens: *Pickwick Papers*. Carlyle: *History of the French Revolution*.
1838	Darwin read Malthus's essay on population, which he later said gave him the inspiration for natural selection (28 September).	Charles Hennell: *An Inquiry into the Origins of Christianity*.
1839	Married his cousin Emma Wedgwood (29 January). Publication of *Journal of Researches* (*The Voyage of the Beagle*).	Carlyle: *Chartism*.
1840		
1841		Dickens: *The Old Curiosity Shop*. Isidore Geoffroy Saint-Hilaire: *Essais de zoologie générale*.
1842	First written sketch of evolution theory, unpublished. *Coral Reefs*. Moved to Down House, Kent, where he was to live for the rest of his life.	
1843	*Volcanic Islands*.	Ruskin: *Modern Painters* (to 1860).
1844	Wrote longer essay on evolution by natural selection, again unpublished but with instructions to Emma to publish it in the event of his death.	Robert Chambers: *Vestiges of Creation* (published anonymously).
1845		Humboldt: *Cosmos* (to 1862). Disraeli: *Sybil, or The Two Nations*.
1846	Began long study of the taxonomy of barnacles.	Strauss: *Das Leben Jesu* (1835) trans. George Eliot as *The Life of Jesus*.
1847		J. L. R. Agassiz: *Système glaciarre*. C. Brontë: *Jane Eyre*. E. Brontë: *Wuthering Heights*. Thackeray: *Vanity Fair* (to 1848).
1848	Death of father, Dr Robert Darwin.	Dickens: *Dombey and Son*. Marx and Engels: *The Communist Manifesto*. Elizabeth Gaskell: *Mary Barton*. J. S. Mill: *Political Economy*. Macaulay: *History of England* (to 1859).

CHRONOLOGY

HISTORICAL EVENTS

Queen Victoria comes to the throne. Electric telegraph.

Chartists demand reform, including universal suffrage. Brunel's 'Great Western'.

Daguerrotype photograph. Faraday's theory of electromagnetism.

Marriage of the Queen to Prince Albert.
Newman's Tract XC. Income tax introduced. 'Young England' Movement.

Joule's theory of thermodynamics.

Irish potato famine.

Repeal of the Corn Laws under Sir Robert Peel. Railway expansion.

Revolutions in Europe; Chartist unrest in England. Pre-Raphaelite Brotherhood founded.

CHARLES DARWIN

DATE	AUTHOR'S LIFE	LITERARY CONTEXT
1849		Richard Wagner: *Die Kunst und die Revolution*.
1850		First publication of Wordsworth's *Prelude*. Tennyson: *In Memoriam*. Death of Wordsworth and Balzac. Herbert Spencer: *Social Statics*. Dickens: *David Copperfield*.
1851	Death of beloved daughter Annie, which some see as a formative moment in his loss of religious faith.	
1852		Thackeray: *Henry Esmond*.
1853		Comte: *Cours de philosophie positive* (1830–42) trans. Harriet Martineau as *Positive Philosophy*.
1854	Barnacle work completed.	Dickens: *Hard Times*. Feuerbach: *Das Wesen des Christenthums* (1841) trans. *George Eliot as The Essence of Christianity*.
1855	Publication of Wallace's first paper on evolution (but not natural selection) in *Annals and Magazine of Natural History*.	Herbert Spencer: *Principles of Psychology*. Lewes: *Life of Goethe*. Elizabeth Gaskell: *North and South*.
1857		Flaubert: *Madame Bovary*. Dickens: *Little Dorrit*. Trollope: *Barchester Towers*. Lewes: *Sea-side Studies*.
1858	Darwin receives Wallace's unpublished paper clearly setting out the theory of evolution by natural selection (18 June). Darwin's and Wallace's unpublished papers read, in the absence of both authors, at the Linnean Society, London (1 July).	
1859	*The Origin of Species* (22 November).	J. S. Mill: *On Liberty*.
1860		Collins: *The Woman in White*. George Eliot: *The Mill on the Floss*. Ruskin: *Unto this Last*.

HISTORICAL EVENTS

Mazzini's short-lived Roman republic falls to French army. Other European revolutions suppressed and period of reaction follows.
Don Pacifico incident. Public Libraries Act.

Great Exhibition at the Crystal Palace. Second census records a population of 17,927,609 in England and Wales as compared with 8,872,980 in 1801. Foundation of Royal School of Mines – first state institution for scientific research and teaching.
Louis Napoleon proclaimed Emperor of France.

Crimean War (to 1856). British Medical Association founded. Dissenters allowed to matriculate at Oxford.

Palmerston becomes Prime Minister (to 1858).

Indian Mutiny. Matrimonial Causes Act.

First transatlantic cable laid between Ireland and Newfoundland.
Jews admitted to Parliament.

Palmerston's second ministry (to 1865).

DATE	AUTHOR'S LIFE	LITERARY CONTEXT
1861		Dickens: *Great Expectations*. J. S. Mill: *Utilitarianism*.
1862	*Fertilization of Orchids*.	Spencer: *First Principles*. Turgenev: *Fathers and Children*. Hugo: *Les Misérables*.
1863		Charles Lyell: *The Geological Evidences of the Antiquity of Man*. T. H. Huxley: *Man's Place in Nature*.
1864		Spencer: *Principles of Biology*.
1865		Gregor Mendel: 'Experiments on Plant Hybridization'. Tolstoy: *War and Peace* (to 1869).
1867		Marx: *Das Kapital* (vol. 1).
1868	*Variation of Animals and Plants*.	Collins: *The Moonstone*.
1869		Francis Galton (Darwin's cousin): *Hereditary Genius*. Trollope: *Phineas Finn*. J. S. Mill: *The Subjection of Women*. Arnold: *Culture and Anarchy*.
1870		Alfred Russel Wallace: *Contributions to the Theory of Natural Selection*. Death of Dickens.
1871	*Descent of Man*.	George Eliot: *Middlemarch* (to 1872).
1872	*Expression of the Emotions*.	
1874		Hardy: *Far from the Madding Crowd*.
1875	*Insectivorous Plants*.	Tolstoy: *Anna Karenina* (to 1878). Trollope: *The Way We Live Now*.
1876	*Effects of Cross- and Self-fertilization*.	Alfred Russel Wallace: *Geographical Distribution of Animals*. George Eliot: *Daniel Deronda*.

CHRONOLOGY

HISTORICAL EVENTS

Garibaldi and 'The Thousand' conquer Sicily. Death of Prince Albert.
Victor Emmanuel becomes first king of a united Italy. Emancipation of the
serfs in Russia. American Civil War (to 1865).
Maxwell's theory of electromagnetism.

Pasteur publishes theory of germs causing disease. First Women's Suffrage
Committee formed in Manchester.

Second Reform Bill brought in by Disraeli. First antiseptic operation
performed by Joseph Lister.
First annual Trades Union Congress held in Manchester. Gladstone
becomes Prime Minister (December).
Suez Canal opened. Emily Davies founds a women's college at Hitchin,
which moves to Girton, Cambridge in 1872.

Married Women's Property Act. Education Act. Civil service in England
thrown open to competitive examination. Franco-Prussian War (to 1871).

Bismarck becomes Chancellor of German Empire. Abolition of religious
Tests in the Universities.
Vote by secret ballot at Parliamentary elections imposed by law.

Disraeli becomes Prime Minister. Remington typewriter.

Queen assumes title of Empress of India. Bell patents telephone.

DATE	AUTHOR'S LIFE	LITERARY CONTEXT
1877	*Forms of Flowers*. Honorary degree from Cambridge.	Mallock: *The New Republic*.
1878		Hardy: *The Return of the Native*. James: *Daisy Miller*.
1879		Ernst Haeckel: *Evolution of Man*. Spencer: *Principles of Ethics* (to 1893). Meredith: *The Egoist*.
1880	*Power of Movement in Plants*.	
1881	*Formation of Vegetable Mould* (Worm book).	James: *The Portrait of a Lady*.
1882	Death of Darwin at Down House. Buried in Westminster Abbey following campaign led by Huxley (19 April).	
1889		Alfred Russel Wallace: *Darwinism*. Francis Galton: *Natural Inheritance*.

HISTORICAL EVENTS

Russo-Turkish War (to 1878).

Congress of Berlin settles Eastern Question.

Zulu War. First women students at Oxford (Lady Margaret Hall, Somerville College).

Gladstone again Prime Minister. Parnell demands Home Rule for Ireland.

JOURNAL OF RESEARCHES
INTO THE NATURAL HISTORY
AND GEOLOGY
OF THE COUNTRIES VISITED
DURING THE VOYAGE
OF H.M.S. BEAGLE
ROUND THE WORLD, 1832–36

CONTENTS

Capybara, or River Hog – Tucutuco – Molothrus, cuckoo-
like habits – Tyrant-flycatcher – Mocking-bird – Carrion
Hawks – Tubes formed by Lightning – House struck

Monte – Effects of Cattle on the Vegetation – Cardoon –
Buenos Ayres – Corral where Cattle are slaughtered

Excursion to St. Fé – Thistle-Beds – Habits of the Bizcacha
– Little Owl – Saline Streams – Level Plains – Mastodon –
St. Fé – Change in Landscape – Geology – Tooth of extinct
Horse – Relation of the Fossil and recent Quadrupeds of
North and South America – Effects of a great Drought –
Parana – Habits of the Jaguar – Scissor-beak – Kingfisher,
Parrot, and Scissor-tail – Revolution – Buenos Ayres –
State of Government

Excursion to Colonia del Sacramiento – Value of an Estan-
cia – Cattle, how counted – Singular Breed of Oxen –
Perforated Pebbles – Shepherd-dogs – Horses Broken-in,
Gauchos Riding – Character of Inhabitants – Rio Plata –
Flocks of Butterflies – Aëronaut Spiders – Phosphores-
cence of the Sea – Port Desire – Guanaco – Port St. Julian
– Geology of Patagonia – Fossil gigantic Animal – Types of
Organization constant – Change in the Zoology of Amer-
ica – Causes of Extinction

Santa Cruz – Expedition up the River – Indians – Immense
Streams of Basaltic Lava – Fragments not transported by the

Origin and Formation – Bathurst, general civility of the lower orders – State of Society – Van Diemen's Land – Hobart Town – Aborigines all banished – Mount Wellington – King George's Sound – Cheerless aspect of the Country – Bald Head, calcareous Casts of Branches of Trees – Party of Natives – Leave Australia

Keeling Island – Singular appearance – Scanty Flora – Transport of Seeds – Birds and Insects – Ebbing and flowing Springs – Fields of dead Coral – Stones transported in the Roots of Trees – Great Crab – Stinging Corals – Coral-eating Fish – Coral Formations – Lagoon Islands, or Atolls – Depth at which reef-building Corals can live – Vast Areas interspersed with low Coral Islands – Subsidence of their Foundations – Barrier Reefs – Fringing Reefs – Conversion of Fringing Reefs into Barrier Reefs, and into Atolls – Evidence of Changes in Level – Breaches in Barrier Reefs – Maldiva Atolls; their peculiar structure – Dead and submerged Reefs – Areas of Subsidence and Elevation – Distribution of Volcanoes – Subsidence slow, and vast in amount

Mauritius, beautiful appearance of – Great crateriform Ring of Mountains – Hindoos – St. Helena – History of the Changes in the Vegetation – Cause of the Extinction of Land-shells – Ascension – Variation in the Imported

Rats – Volcanic Bombs – Beds of Infusoria – Bahia – Brazil – Splendour of Tropical Scenery – Pernambuco – Singular Reef – Slavery – Return to England – Retrospect on our Voyage

PREFACE

I HAVE stated in the preface to the first Edition of this work, and in the Zoology of the Voyage of the *Beagle*, that it was in consequence of a wish expressed by Captain Fitz Roy, of having some scientific person on board, accompanied by an offer from him of giving up part of his own accommodations, that I volunteered my services, which received, through the kindness of the hydrographer, Captain Beaufort, the sanction of the Lords of the Admiralty. As I feel that the opportunities which I enjoyed of studying the Natural History of the different countries we visited, have been wholly due to Captain Fitz Roy, I hope I may here be permitted to repeat my expression of gratitude to him; and to add that, during the five years we were together, I received from him the most cordial friendship and steady assistance. Both to Captain Fitz Roy and to all the Officers of the *Beagle*[1] I shall ever feel most thankful for the undeviating kindness with which I was treated during our long voyage.

This volume contains, in the form of a Journal, a history of our voyage, and a sketch of those observations in Natural History and Geology, which I think will possess some interest for the general reader. I have in this edition largely condensed and corrected some parts, and have added a little to others, in order to render the volume more fitted for popular reading; but I trust that naturalists will remember, that they must refer for details to the larger publications, which comprise the scientific results of the Expedition. The Zoology of the Voyage of the *Beagle* includes an account of the Fossil Mammalia, by Professor Owen; of the Living Mammalia, by Mr. Waterhouse; of the Birds, by Mr. Gould; of the Fish, by the Rev. L. Jenyns; and of the Reptiles, by Mr. Bell. I have appended to the descriptions of each species an account of its habits and range. These works, which I owe to the high talents and disinterested zeal of the above

---

1 I must take this opportunity of returning my sincere thanks to Mr. Bynoe, the surgeon of the *Beagle*, for his very kind attention to me when I was ill at Valparaiso.

distinguished authors, could not have been undertaken, had it not been for the liberality of the Lords Commissioners of Her Majesty's Treasury, who, through the representation of the Right Honourable the Chancellor of the Exchequer, have been pleased to grant a sum of one thousand pounds towards defraying part of the expenses of publication.

I have myself published separate volumes on the 'Structure and Distribution of Coral Reefs;' on the 'Volcanic Islands visited during the Voyage of the *Beagle*;' and on the 'Geology of South America.' The sixth volume of the 'Geological Transactions' contains two papers of mine on the Erratic Boulders and Volcanic Phenomena of South America. Messrs. Waterhouse, Walker, Newman, and White, have published several able papers on the Insects which were collected, and I trust that many others will hereafter follow. The plants from the southern parts of America will be given by Dr. J. Hooker, in his great work on the Botany of the Southern Hemisphere. The Flora of the Galapagos Archipelago is the subject of a separate memoir by him, in the 'Linnean Transactions.' The Reverend Professor Henslow has published a list of the plants collected by me at the Keeling Islands; and the Reverend J. M. Berkeley has described my cryptogamic plants.

I shall have the pleasure of acknowledging the great assistance which I have received from several other naturalists, in the course of this and my other works; but I must be here allowed to return my most sincere thanks to the Reverend Professor Henslow, who, when I was an under-graduate at Cambridge, was one chief means of giving me a taste for Natural History, – who, during my absence, took charge of the collections I sent home, and by his correspondence directed my endeavours, – and who, since my return, has constantly rendered me every assistance which the kindest friend could offer.

Down, Bromley, Kent.
June, 1845.

POSTSCRIPT

I TAKE the opportunity of a new edition of my Journal to correct a few errors. At pages 97–8 I have stated that the majority of the shells which were embedded with the extinct mammals at Punta Alta, in Bahia Blanca, were still living species. These shells have since been examined (see 'Geological Observations in South America,' p. 83) by M. Aleide d'Orbigny, and he pronounces them all to be recent. M. Aug. Bravard has lately described, in a Spanish work ('Observaciones Geologicas,' 1857), this district, and he believes that the bones of the extinct mammals were washed out of the underlying Pampean deposit, and subsequently became embedded with the still existing shells; but I am not convinced by his remarks. M. Bravard believes that the whole enormous Pampean deposit is a sub-aërial formation, like sand-dunes: this seems to me to be an untenable doctrine.

At page 390 I give a list of the birds inhabiting the Galapagos Archipelago. The progress of research has shown that some of these birds, which were then thought to be confined to the islands, occur on the American continent. The eminent ornithologist, Mr. Sclater, informs me that this is the case with the Strix punctatissima and Pyrocephalus nanus; and probably with the Otus galapagoensis and Zenaida galapagoensis: so that the number of endemic birds is reduced to twenty-three, or probably to twenty-one. Mr. Sclater thinks that one or two of these endemic forms should be ranked rather as varieties than species, which always seemed to me probable.

The snake mentioned at page 393, as being, on the authority of M. Bibron, the same with a Chilian species, is stated by Dr. Günter (Zoolog. Soc., Jan. 24th, 1859) to be a peculiar species, not known to inhabit any other country.

Feb. 1st, 1860.

CHAPTER I

ST. JAGO – CAPE DE
VERD ISLANDS

Porto Praya – Ribeira Grande – Atmospheric Dust with
Infusoria – Habits of a Sea-slug and Cuttle-fish – St. Paul's
Rocks, non-volcanic – Singular Incrustations – Insects the
first Colonists of Islands – Fernando Noronha – Bahia –
Burnished Rocks – Habits of a Diodon – Pelagic Confervæ
and Infusoria – Causes of discoloured Sea.

AFTER having been twice driven back by heavy south-western gales,
Her Majesty's ship *Beagle*, a ten-gun brig, under the command of
Captain Fitz Roy, R.N., sailed from Devonport on the 27th of
December, 1831. The object of the expedition was to complete the
survey of Patagonia and Tierra del Fuego, commenced under Captain
King in 1826 to 1830 – to survey the shores of Chile, Peru, and of
some islands in the Pacific – and to carry a chain of chronometrical
measurements round the World. On the 6th of January we reached
Teneriffe, but were prevented landing, by fears of our bringing the
cholera: the next morning we saw the sun rise behind the rugged
outline of the Grand Canary island, and suddenly illumine the Peak of
Teneriffe, whilst the lower parts were veiled in fleecy clouds. This
was the first of many delightful days never to be forgotten. On the
16th of January, 1832, we anchored at Porto Praya, in St. Jago, the
chief island of the Cape de Verd archipelago.

The neighbourhood of Porto Praya, viewed from the sea, wears
a desolate aspect. The volcanic fires of a past age, and the scorching
heat of a tropical sun, have in most places rendered the soil unfit
for vegetation. The country rises in successive steps of table-land,
interspersed with some truncate conical hills, and the horizon
is bounded by an irregular chain of more lofty mountains. The
scene, as beheld through the hazy atmosphere of this climate, is one
of great interest; if, indeed, a person, fresh from sea, and who has just
walked, for the first time, in a grove of cocoa-nut trees, can be a judge

of anything but his own happiness. The island would generally be considered as very uninteresting; but to any one accustomed only to an English landscape, the novel aspect of an utterly sterile land possesses a grandeur which more vegetation might spoil. A single green leaf can scarcely be discovered over wide tracts of the lava plains; yet flocks of goats, together with a few cows, contrive to exist. It rains very seldom, but during a short portion of the year heavy torrents fall, and immediately afterwards a light vegetation springs out of every crevice. This soon withers; and upon such naturally formed hay the animals live. It had not now rained for an entire year. When the island was discovered, the immediate neighbourhood of Porto Praya was clothed with trees,[1] the reckless destruction of which has caused here, as at St. Helena, and at some of the Canary islands, almost entire sterility. The broad, flat-bottomed valleys, many of which serve during a few days only in the season as watercourses, are clothed with thickets of leafless bushes. Few living creatures inhabit these valleys. The commonest bird is a kingfisher (Dacelo Iagoensis), which tamely sits on the branches of the castor-oil plant, and thence darts on grasshoppers and lizards. It is brightly coloured, but not so beautiful as the European species: in its flight, manners, and place of habitation, which is generally in the driest valley, there is also a wide difference.

One day, two of the officers and myself rode to Ribeira Grande, a village a few miles eastward of Porto Praya. Until we reached the valley of St. Martin, the country presented its usual dull brown appearance; but here, a very small rill of water produces a most refreshing margin of luxuriant vegetation. In the course of an hour we arrived at Ribeira Grande, and were surprised at the sight of a large ruined fort and cathedral. This little town, before its harbour was filled up, was the principal place in the island: it now presents a melancholy, but very picturesque appearance. Having procured a black Padre for a guide, and a Spaniard who had served in the Peninsular war as an interpreter, we visited a collection of buildings, of which an ancient church formed the principal part. It is here the

---

1 I state this on the authority of Dr. E. Dieffenbach, in his German translation of the first edition of this Journal.

governors and captain-generals of the islands have been buried. Some of the tombstones recorded dates of the sixteenth century.[1] The heraldic ornaments were the only things in this retired place that reminded us of Europe. The church or chapel formed one side of a quadrangle, in the middle of which a large clump of bananas were growing. On another side was a hospital, containing about a dozen miserable-looking inmates.

We returned to the Vênda to eat our dinners. A considerable number of men, women, and children, all as black as jet, collected to watch us. Our companions were extremely merry; and everything we said or did was followed by their hearty laughter. Before leaving the town we visited the cathedral. It does not appear so rich as the smaller church, but boasts of a little organ, which sent forth singularly inharmonious cries. We presented the black priest with a few shillings, and the Spaniard, patting him on the head, said, with much candour, he thought his colour made no great difference. We then returned, as fast as the ponies would go, to Porto Praya.

Another day we rode to the village of St. Domingo, situated near the centre of the island. On a small plain which we crossed, a few stunted acacias were growing; their tops had been bent by the steady trade-wind, in a singular manner – some of them even at right angles to their trunks. The direction of the branches was exactly N.E. by N., and S.W. by S., and these natural vanes must indicate the prevailing direction of the force of the trade-wind. The travelling had made so little impression on the barren soil, that we here missed our track, and took that to Fuentes. This we did not find out till we arrived there; and we were afterwards glad of our mistake. Fuentes is a pretty village, with a small stream; and everything appeared to prosper well, excepting, indeed, that which ought to do so most – its inhabitants. The black children, completely naked, and looking very wretched, were carrying bundles of firewood half as big as their own bodies.

Near Fuentes we saw a large flock of guinea-fowl – probably fifty or sixty in number. They were extremely wary, and could not

1 The Cape de Verd Islands were discovered in 1449. There was a tombstone of a bishop with the date of 1571; and a crest of a hand and dagger, dated 1497.

be approached. They avoided us, like partridges on a rainy day in September, running with their heads cocked up; and if pursued, they readily took to the wing.

The scenery of St. Domingo possesses a beauty totally unexpected, from the prevalent gloomy character of the rest of the island. The village is situated at the bottom of a valley, bounded by lofty and jagged walls of stratified lava. The black rocks afford a most striking contrast with the bright green vegetation, which follows the banks of a little stream of clear water. It happened to be a grand feast-day, and the village was full of people. On our return we overtook a party of about twenty young black girls, dressed in excellent taste; their black skins and snow-white linen being set off by coloured turbans and large shawls. As soon as we approached near, they suddenly all turned round, and covering the path with their shawls, sung with great energy a wild song, beating time with their hands upon their legs. We threw them some vintéms, which were received with screams of laughter, and we left them redoubling the noise of their song.

One morning the view was singularly clear; the distant mountains being projected with the sharpest outline, on a heavy bank of dark blue clouds. Judging from the appearance, and from similar cases in England, I supposed that the air was saturated with moisture. The fact, however, turned out quite the contrary. The hygrometer gave a difference of 29.6 degrees, between the temperature of the air, and the point at which dew was precipitated. This difference was nearly double that which I had observed on the previous mornings. This unusual degree of atmospheric dryness was accompanied by continual flashes of lightning. Is it not an uncommon case, thus to find a remarkable degree of aerial transparency with such a state of weather?

Generally the atmosphere is hazy; and this is caused by the falling of impalpably fine dust, which was found to have slightly injured the astronomical instruments. The morning before we anchored at Porto Praya, I collected a little packet of this brown-coloured fine dust, which appeared to have been filtered from the wind by the gauze of the vane at the mast-head. Mr. Lyell has also given me four packets of dust which fell on a vessel a few hundred miles northward of these islands. Professor Ehrenberg[1] finds that this dust consists in great part of infusoria with siliceous shields, and of the siliceous tissue of plants. In five little packets which I sent him, he has ascertained no less than sixty-seven different organic forms! The infusoria, with the exception of two marine species, are all inhabitants of fresh-water. I have found no less than fifteen different accounts of dust having fallen on vessels when far out in the Atlantic. From the direction of the wind whenever it has fallen, and from its having always fallen during those months when the harmattan is known to raise clouds of dust high into the atmosphere, we may feel sure that it all comes from Africa. It is, however, a very singular fact, that, although Professor Ehrenberg knows many species of infusoria peculiar to Africa, he finds none of these in the dust which I sent him: on the other hand, he finds in it two species which hitherto he knows as living only in South America. The dust falls in such quantities as to dirty everything on board, and to hurt people's eyes; vessels even have run on shore owing to the obscurity of the atmosphere. It has often fallen on ships when several hundred, and even more than a thousand miles from the coast of Africa, and at points sixteen hundred miles distant in a north and south direction. In some dust which was collected on a vessel three hundred miles from the land, I was much surprised to find particles of stone above the thousandth of an inch square, mixed with finer matter. After this fact one need not be surprised at the diffusion of the far lighter and smaller sporules of cryptogamic plants.

The geology of this island is the most interesting part of its natural history. On entering the harbour, a perfectly horizontal white band in the face of the sea cliff, may be seen running for some miles along the coast, and at the height of about forty-five feet above the water. Upon examination, this white stratum is found to consist of calcareous matter, with numerous shells embedded, most or all of which now exist on the neighbouring coast. It rests on ancient volcanic rocks, and has been covered by a stream of basalt, which must have entered the sea when the white shelly bed was lying at the bottom. It is interesting to trace the changes, produced by the heat of the overlying lava, on

1 I must take this opportunity of acknowledging the great kindness with which this illustrious naturalist has examined many of my specimens. I have sent (June, 1845) a full account of the falling of this dust to the Geological Society.

the friable mass, which in parts has been converted into a crystalline limestone, and in other parts into a compact spotted stone. Where the lime has been caught up by the scoriaceous fragments of the lower surface of the stream, it is converted into groups of beautifully radiated fibres resembling arragonite. The beds of lava rise in successive gently-slopping plains, towards the interior, whence the deluges of melted stone have originally proceeded. Within historical times, no signs of volcanic activity have, I believe, been manifested in any part of St. Jago. Even the form of a crater can but rarely be discovered on the summits of the many red cindery hills; yet the more recent streams can be distinguished on the coast, forming lines of cliffs of less height, but stretching out in advance of those belonging to an older series: the height of the cliffs thus affording a rude measure of the age of the streams.

During our stay, I observed the habits of some marine animals. A large Aplysia is very common. This sea-slug is about five inches long; and is of a dirty-yellowish colour, veined with purple. On each side of the lower surface, or foot, there is a broad membrane, which appears sometimes to act as a ventilator, in causing a current of water to flow over the dorsal branchiæ or lungs. It feeds on the delicate sea-weeds which grow among the stones in muddy and shallow water; and I found in its stomach several small pebbles, as in the gizzard of a bird. This slug, when disturbed, emits a very fine purplish-red fluid, which stains the water for the space of a foot around. Besides this means of defence, an acrid secretion, which is spread over its body, causes a sharp, stinging sensation, similar to that produced by the Physalia, or Portuguese man-of-war.

I was much interested, on several occasions, by watching the habits of an Octopus, or cuttle-fish. Although common in the pools of water left by the retiring tide, these animals were not easily caught. By means of their long arms and suckers, they could drag their bodies into very narrow crevices; and when thus fixed, it required great force to remove them. At other times they darted tail first, with the rapidity of an arrow, from one side of the pool to the other, at the same instant discolouring the water with a dark chestnut-brown ink. These animals also escape detection by a very extraordinary, chameleon-like power of changing their colour. They appear to vary their tints according to the nature of the ground over which they pass: when in deep water, their general shade was brownish purple, but when placed on the land, or in shallow water, this dark tint changed into one of a yellowish green. The colour, examined more carefully, was a French grey, with numerous minute spots of bright yellow: the former of these varied in intensity; the latter entirely disappeared and appeared again by turns. These changes were effected in such a manner, that clouds, varying in tint between a hyacinth red and a chestnut-brown,[1] were continually passing over the body. Any part, being subjected to a slight shock of galvanism, became almost black: a similar effect, but in a less degree, was produced by scratching the skin with a needle. These clouds, or blushes as they may be called, are said to be produced by the alternate expansion and contraction of minute vesicles containing variously coloured fluids.[2]

This cuttle-fish displayed its chameleon-like power both during the act of swimming and whilst remaining stationary at the bottom. I was much amused by the various arts to escape detection used by one individual, which seemed fully aware that I was watching it. Remaining for a time motionless, it would then stealthily advance an inch or two, like a cat after a mouse; sometimes changing its colour: it thus proceeded, till having gained a deeper part, it darted away, leaving a dusky train of ink to hide the hole into which it had crawled.

While looking for marine animals, with my head about two feet above the rocky shore, I was more than once saluted by a jet of water, accompanied by a slight grating noise. At first I could not think what it was, but afterwards I found out that it was this cuttle-fish, which, though concealed in a hole, thus often led me to its discovery. That it possesses the power of ejecting water there is no doubt, and it appeared to me that it could certainly take good aim by directing the tube or siphon on the under side of its body. From the difficulty which these animals have in carrying their heads, they cannot crawl with ease when placed on the ground. I observed that one which I kept in the cabin was slightly phosphorescent in the dark.

1 So named according to Patrick Symes's nomenclature.
2 See Encyclop. of Anat. and Physiol., article *Cephalopoda*.

ST. PAUL'S ROCKS. – In crossing the Atlantic we hove-to, during the morning of February 16th, close to the island of St. Paul's. This cluster of rocks is situated in 0° 58′ north latitude, and 29° 15′ west longitude. It is 540 miles distant from the coast of America, and 350 from the island of Fernando Noronha. The highest point is only fifty feet above the level of the sea, and the entire circumference is under three-quarters of a mile. This small point rises abruptly out of the depths of the ocean. Its mineralogical constitution is not simple; in some parts the rock is of a cherty, in others of a felspathic nature, including thin veins of serpentine. It is a remarkable fact, that all the many small islands, lying far from any continent, in the Pacific, Indian, and Atlantic Oceans, with the exception of the Seychelles and this little point of rock, are, I believe, composed either of coral or of erupted matter. The volcanic nature of these oceanic islands is evidently an extension of that law, and the effect of those same causes, whether chemical or mechanical, from which it results that a vast majority of the volcanoes now in action stand either near sea-coasts or as islands in the midst of the sea.

The rocks of St. Paul appear from a distance of a brilliantly white colour. This is partly owing to the dung of a vast multitude of seafowl, and partly to a coating of a hard glossy substance with a pearly lustre, which is intimately united to the surface of the rocks. This, when examined with a lens, is found to consist of numerous exceedingly thin layers, its total thickness being about the tenth of an inch. It contains much animal matter, and its origin, no doubt, is due to the action of the rain or spray on the birds' dung. Below some small masses of guano at Ascension, and on the Abrolhos Islets, I found certain stalactitic branching bodies, formed apparently in the same manner as the thin white coating on these rocks. The branching bodies so closely resembled in general appearance certain nulliporæ (a family of hard calcareous sea-plants), that in lately looking hastily over my collection I did not perceive the difference. The globular extremities of the branches are of a pearly texture, like the enamel of teeth, but so hard as just to scratch plate-glass. I may here mention, that on a part of the coast of Ascension, where there is a vast accumulation of shelly sand, an incrustation is deposited on the tidal rocks, by the water of the sea, resembling, as represented in the woodcut, certain cryptogamic plants (Marchantiæ) often seen on damp walls. The surface of the fronds is beautifully glossy; and those parts formed where fully exposed to the light, are of a jet black colour, but those shaded under ledges are only grey. I have shown specimens of this incrustation to several geologists, and they all thought that they were of volcanic or igneous origin! In its hardness and translucency – in its polish, equal to that of the finest oliva-shell – in the bad smell given out, and loss of colour under the blow-pipe – it shows a close similarity with living sea-shells. Moreover, in sea-shells, it is known that the parts habitually covered and shaded by the mantle of the animal, are of a paler colour than those fully exposed to the light, just as is the case with this incrustation. When we remember that lime, either as a phosphate or carbonate, enters into the composition of the hard parts, such as bones and shells, of all living animals, it is an interesting physiological fact[1] to find substances harder than the

1 Mr. Horner and Sir David Brewster have described (Philosophical Transactions, 1836, p. 65) a singular 'artificial substance resembling shell.' It is deposited in fine, transparent, highly polished, brown-coloured laminæ, possessing peculiar optical prop-erties, on the inside of a vessel, in which cloth, first prepared with glue and then with lime, is made to revolve rapidly in water. It is much softer, more transparent, and contains more animal matter, than the natural incrustation at Ascension; but we here again see the strong tendency which carbonate of lime and animal matter evince to form a solid substance allied to shell.

enamel of teeth, and coloured surfaces as well polished as those of a fresh shell, reformed through inorganic means from dead organic matter – mocking, also, in shape some of the lower vegetable productions.

We found on St. Paul's only two kinds of birds – the booby and the noddy. The former is a species of gannet, and the latter a tern. Both are of a tame and stupid disposition, and are so unaccustomed to visitors, that I could have killed any number of them with my geological hammer. The booby lays her eggs on the bare rock; but the tern makes a very simple nest with seaweed. By the side of many of these nests a small flying-fish was placed; which, I suppose, had been brought by the male bird for its partner. It was amusing to watch how quickly a large and active crab (Graspus), which inhabits the crevices of the rock, stole the fish from the side of the nest, as soon as we had disturbed the parent birds. Sir W. Symonds, one of the few persons who have landed here, informs me that he saw the crabs dragging even the young birds out of their nests, and devouring them. Not a single plant, not even a lichen, grows on this islet; yet it is inhabited by several insects and spiders. The following list completes, I believe, the terrestrial fauna: a fly (Olfersia) living on the booby, and a tick which must have come here as a parasite on the birds; a small brown moth, belonging to a genus that feeds on feathers; a beetle (Quedius) and a woodlouse from beneath the dung; and lastly, numerous spiders, which I suppose prey on these small attendants and scavengers of the waterfowl. The often repeated description of the stately palm and other noble tropical plants, then birds, and lastly man, taking possession of the coral islets as soon as formed, in the Pacific, is probably not quite correct; I fear it destroys the poetry of this story, that feather and dirt-feeding and parasitic insects and spiders should be the first inhabitants of newly-formed oceanic land.

The smallest rock in the tropical seas, by giving a foundation for the growth of innumerable kinds of seaweed and compound animals, supports likewise a large number of fish. The sharks and the seamen in the boats maintained a constant struggle which should secure the greater share of the prey caught by the fishing-lines. I have heard that a rock near the Bermudas, lying many miles out at sea, and at a considerable depth, was first discovered by the circumstance of fish having been observed in the neighbourhood.

FERNANDO NORONHA, *Feb. 20th.* – As far as I was enabled to observe, during the few hours we stayed at this place, the constitution of the island is volcanic, but probably not of a recent date. The most remarkable feature is a conical hill, about one thousand feet high, the upper part of which is exceedingly steep, and on one side overhangs its base. The rock is phonolite, and is divided into irregular columns. On viewing one of these isolated masses, at first one is inclined to believe that it has been suddenly pushed up in a semi-fluid state. At St. Helena, however, I ascertained that some pinnacles, of a nearly similar figure and constitution, had been formed by the injection of melted rock into yielding strata, which thus had formed the moulds for these gigantic obelisks. The whole island is covered with wood; but from the dryness of the climate there is no appearance of luxuriance. Half-way up the mountain, some great masses of the columnar rock, shaded by laurel-like trees, and ornamented by others covered with fine pink flowers but without a single leaf, gave a pleasing effect to the nearer parts of the scenery.

BAHIA, OR SAN SALVADOR. BRAZIL, *Feb. 29th.* – The day has passed delightfully. Delight itself, however, is a weak term to express the feelings of a naturalist who, for the first time, has wandered by himself in a Brazilian forest. The elegance of the grasses, the novelty of the parasitical plants, the beauty of the flowers, the glossy green of the foliage, but above all the general luxuriance of the vegetation, filled me with admiration. A most paradoxical mixture of sound and silence pervades the shady parts of the wood. The noise from the insects is so loud, that it may be heard even in a vessel anchored several hundred yards from the shore; yet within the recesses of the forest a universal silence appears to reign. To a person fond of natural history, such a day as this brings with it a deeper pleasure than he can ever hope to experience again. After wandering about for some hours, I returned to the landing-place; but, before reaching it, I was overtaken by a tropical storm. I tried to find shelter under a tree, which was so thick that it would never have been penetrated by common English rain; but here, in a couple of minutes, a little torrent flowed down the trunk. It is to this violence of the rain that we must attribute the verdure at the bottom of the thickest woods: if the showers were like those of a colder clime, the greater part would be absorbed or evaporated before it reached the ground. I will not at present attempt to describe the gaudy scenery of this noble bay, because, in our homeward voyage, we called here a second time, and I shall then have occasion to remark on it.

Along the whole coast of Brazil, for a length of at least 2000 miles, and certainly for a considerable space inland, wherever solid rock occurs, it belongs to a granitic formation. The circumstance of this enormous area being constituted of materials which most geologists believe to have been crystallized when heated under pressure, gives rise to many curious reflections. Was this effect produced beneath the depths of a profound ocean? or did a covering of strata formerly extend over it, which has since been removed? Can we believe that any power, acting for a time short of infinity, could have denuded the granite over so many thousand square leagues?

On a point not far from the city, where a rivulet entered the sea, I observed a fact connected with a subject discussed by Humboldt.[1] At the cataracts of the great rivers Orinoco, Nile, and Congo, the syenitic rocks are coated by a black substance, appearing as if they had been polished with plumbago. The layer is of extreme thinness; and on analysis by Berzelius it was found to consist of the oxides of manganese and iron. In the Orinoco it occurs on the rocks periodically washed by the floods, and in those parts alone where the stream is rapid; or, as the Indians say, 'the rocks are black where the waters are white.' Here the coating is of a rich brown instead of a black colour, and seems to be composed of ferruginous matter alone. Hand specimens fail to give a just idea of these brown burnished stones which glitter in the sun's rays. They occur only within the limits of the tidal waves; and as the rivulet slowly trickles down, the surf must supply the polishing power of the cataracts in the great rivers. In like manner, the rise and fall of the tide probably answer to the periodical inundations; and thus the same effects are produced under apparently

1 Pers. Narr., vol. v. pt. i. p. 18.

different but really similar circumstances. The origin, however, of these coatings of metallic oxides, which seem as if cemented to the rocks, is not understood; and no reason, I believe, can be assigned for their thickness remaining the same.

One day I was amused by watching the habits of the Diodon antennatus, which was caught swimming near the shore. This fish, with its flabby skin, is well known to possess the singular power of distending itself into a nearly spherical form. After having been taken out of the water for a short time, and then again immersed in it, a considerable quantity both of water and air is absorbed by the mouth, and perhaps likewise by the branchial orifices. This process is effected by two methods: the air is swallowed, and is then forced into the cavity of the body, its return being prevented by a muscular contraction which is externally visible: but the water enters in a gentle stream through the mouth, which is kept wide open and motionless; this latter action must, therefore, depend on suction. The skin about the abdomen is much looser than that on the back; hence, during the inflation, the lower surface becomes far more distended than the upper; and the fish, in consequence, floats with its back downwards. Cuvier doubts whether the Diodon in this position is able to swim; but not only can it thus move forward in a straight line, but it can turn round to either side. This latter movement is effected solely by the aid of the pectoral fins; the tail being collapsed, and not used. From the body being buoyed up with so much air, the branchial openings are out of water, but a stream drawn in by the mouth constantly flows through them.

The fish, having remained in this distended state for a short time, generally expelled the air and water with considerable force from the branchial apertures and mouth. It could emit, at will, a certain portion of the water; and it appears, therefore, probable that this fluid is taken in partly for the sake of regulating its specific gravity. This Diodon possessed several means of defence. It could give a severe bite, and could eject water from its mouth to some distance, at the same time making a curious noise by the movement of its jaws. By the inflation of its body, the papillæ, with which the skin is covered, become erect and pointed. But the most curious circumstance is, that it secretes from the skin of its belly, when handled, a most beautiful carmine-red fibrous matter, which stains ivory and paper in so permanent a manner, that the tint is retained with all its brightness to the present day: I am quite ignorant of the nature and use of this secretion. I have heard from Dr. Allan of Forres, that he has frequently found a Diodon, floating alive and distended, in the stomach of the shark; and that on several occasions he has known it eat its way, not only through the coats of the stomach, but through the sides of the monster, which has thus been killed. Who would ever have imagined that a little soft fish could have destroyed the great and savage shark?

March 18th. – We sailed from Bahia. A few days afterwards, when not far distant from the Abrolhos Islets, my attention was called to a reddish-brown appearance in the sea. The whole surface of the water, as it appeared under a weak lens, seemed as if covered by chopped bits of hay, with their ends jagged. These are minute cylindrical confervæ, in bundles or rafts of from twenty to sixty in each. Mr. Berkeley informs me that they are the same species (Trichodesmium ery-thræum) with that found over large spaces in the Red Sea, and whence its name of Red Sea is derived.[1] Their numbers must be infinite: the ship passed through several bands of them, one of which was about ten yards wide, and, judging from the mud-like colour of the water, at least two and a half miles long. In almost every long voyage some account is given of these confervæ. They appear especially common in the sea near Australia; and off Cape Leeuwin I found an allied, but smaller and apparently different species. Captain Cook, in his third voyage, remarks, that the sailors gave to this appearance the name of sea-sawdust.

Near Keeling Atoll, in the Indian Ocean, I observed many little masses of confervæ a few inches square, consisting of long cylindrical threads of excessive thinness, so as to be barely visible to the naked eye, mingled with other rather larger bodies, finely conical at both ends. Two of these are shown in the woodcut united together. They vary in length from .04 to .06, and even to .08 of an

1 M. Montagne, in Comptes Rendus, &c., Juillet, 1844; and Annal. des Scienc. Nat., Dec. 1844.

inch in length; and in diameter from .006 to .008 of an inch. Near one extremity of the cylindrical part, a green septum, formed of granular matter, and thickest in the middle, may generally be seen. This, I believe, is the bottom of a most delicate, colourless sac, composed of a pulpy substance, which lines the exterior case, but does not extend within the extreme conical points. In some specimens, small but perfect spheres of brownish granular matter supplied the places of the septa; and I observed the curious process by which they were produced. The pulpy matter of the internal coating suddenly grouped itself into lines, some of which assumed a form radiating from a common centre; it then continued, with an irregular and rapid movement, to contract itself, so that in the course of a second the whole was united into a perfect little sphere, which occupied the position of the septum at one end of the now quite hollow case. The formation of the granular sphere was hastened by any accidental injury. I may add, that frequently a pair of these bodies were attached to each other, as represented above, cone beside cone, at that end where the septum occurs.

I will here add a few other observations connected with the discoloration of the sea from organic causes. On the coast of Chile, a few leagues north of Concepcion, the *Beagle* one day passed through great bands of muddy water, exactly like that of a swollen river; and again, a degree south of Valparaiso, when fifty miles from the land, the same appearance was still more extensive. Some of the water placed in a glass was of a pale reddish tint; and, examined under a microscope, was seen to swarm with minute animalcula darting about, and often exploding. Their shape is oval, and contracted in the middle by a ring of vibrating curved ciliæ. It was, however, very difficult to examine them with care, for almost the instant motion ceased, even while crossing the field of vision, their bodies burst. Sometimes both ends burst at once, sometimes only one, and a quantity of coarse, brownish, granular matter was ejected. The animal an instant before bursting expanded to half again its natural size; and the explosion took place about fifteen seconds after the rapid progressive motion had ceased: in a few cases it was preceded for a short interval by a rotatory movement on the longer axis. About two minutes after any number were isolated in a drop of water, they thus perished. The animals move with the narrow apex forwards, by the aid of their vibratory ciliæ, and generally by rapid starts. They are exceedingly minute, and quite invisible to the naked eye, only covering a space equal to the square of the thousandth of an inch. Their numbers were infinite; for the smallest drop of water which I could remove contained very many. In one day we passed through two spaces of water thus stained, one of which alone must have extended over several square miles. What incalculable numbers of these microscopical animals! The colour of the water, as seen at some distance, was like that of a river which has flowed through a red clay district; but under the shade of the vessel's side it was quite as dark as chocolate. The line where the red and blue water joined was distinctly defined. The weather for some days previously had been calm, and the ocean abounded, to an unusual degree, with living creatures.[1]

In the sea around Tierra del Fuego, and at no great distance from the land, I have seen narrow lines of water of a bright red colour, from the number of crustacea, which somewhat resemble in form large prawns. The sealers call them whale-food. Whether whales feed on them I do not know; but terns, cormorants, and immense herds of great unwieldy seals derive, on some parts of the coast, their chief sustenance from these swimming crabs. Seamen invariably attribute the discoloration of the water to spawn; but I found this to be the case only on one occasion. At the distance of several leagues from the Archipelago of the Galapagos, the ship sailed through three strips of a dark yellowish, or mud-like water; these strips were some miles long, but only a few yards wide, and they were separated from the surrounding water by a sinuous yet distinct margin. The colour was caused by little gelatinous balls, about the fifth of an inch in diameter, in which numerous minute spherical ovules

1 M. Lesson (Voyage de la *Coquille*, tom. i. p. 255) mentions red water off Lima, apparently produced by the same cause. Peron, the distinguished naturalist, in the Voyage aux Terres Australes, gives no less than twelve references to voyagers who have alluded to the discoloured waters of the sea (vol. ii. p. 239). To the references given by Peron may be added, Humboldt's Pers. Narr., vol. vi. p. 804; Flinders' Voyage, vol. i. p. 92; Labillardière, vol. i. p. 287; Ulloa's Voyage; Voyage of the *Astrolabe* and of the *Coquilla*; Captain King's Survey of Australia, &c.

were embedded: they were of two distinct kinds, one being of a reddish colour and of a different shape from the other. I cannot form a conjecture as to what two kinds of animals these belonged. Captain Colnett remarks, that this appearance is very common among the Galapagos Islands, and that the direction of the bands indicates that of the currents; in the described case, however, the line was caused by the wind. The only other appearance which I have to notice, is a thin oily coat on the water which displays iridescent colours. I saw a considerable tract of the ocean thus covered on the coast of Brazil; the seamen attributed it to the putrefying carcass of some whale, which probably was floating at no great distance. I do not here mention the minute gelatinous particles, hereafter to be referred to, which are frequently dispersed throughout the water, for they are not sufficiently abundant to create any change of colour.

There are two circumstances in the above accounts which appear remarkable: first, how do the various bodies which form the bands with defined edges keep together? In the case of the prawn-like crabs, their movements were as coinstantaneous as in a regiment of soldiers; but this cannot happen from any thing like voluntary action with the ovules, or the confervæ, nor is it probable among the infusoria. Secondly, what causes the length and narrowness of the bands? The appearance so much resembles that which may be seen in every torrent, where the stream uncoils into long streaks the froth collected in the eddies, that I must attribute the effect to a similar action either of the currents of the air or sea. Under this supposition we must believe that the various organized bodies are produced in certain favourable places, and are thence removed by the set of either wind or water. I confess, however, there is a very great difficulty in imagining any one spot to be the birthplace of the millions of millions of animalcula and confervæ: for whence come the germs at such points? – the parent bodies having been distributed by the winds and waves over the immense ocean. But on no other hypothesis can I understand their linear grouping. I may add that Scoresby remarks, that green water abounding with pelagic animals is invariably found in a certain part of the Arctic Sea.

CHAPTER II

RIO DE JANEIRO

Rio de Janeiro – Excursion north of Cape Frio – Great
Evaporation – Slavery – Botofogo Bay – Terrestrial Planariæ – Clouds on the Corcovado – Heavy Rain – Musical
Frogs – Phosphorescent Insects – Elater, springing powers
of – Blue Haze – Noise made by a Butterfly – Entomology
– Ants – Wasp killing a Spider – Parasitical Spider – Arti-
fices of an Epeira – Gregarious Spider – Spider with an
unsymmetrical Web.

APRIL 4th to July 5th, 1832. – A few days after our arrival I became
acquainted with an Englishman who was going to visit his estate,
situated, rather more than a hundred miles from the capital, to the
northward of Cape Frio. I gladly accepted his kind offer of allowing
me to accompany him.

April 8th. – Our party amounted to seven. The first stage was very
interesting. The day was powerfully hot, and as we passed through the
woods, every thing was motionless, excepting the large and brilliant
butterflies, which lazily fluttered about. The view seen when crossing
the hills behind Praia Grande was most beautiful; the colours were
intense, and the prevailing tint a dark blue; the sky and the calm waters
of the bay vied with each other in splendour. After passing through
some cultivated country, we entered a forest, which in the grandeur
of all its parts could not be exceeded. We arrived by midday at
Ithacaia; this small village is situated on a plain, and round the central
house are the huts of the negroes. These, from their regular form and
position, reminded me of the drawings of the Hottentot habitations in
Southern Africa. As the moon rose early, we determined to start the
same evening for our sleeping-place at the Lagoa Marica. As it was
growing dark we passed under one of the massive, bare, and steep hills
of granite which are so common in this country. This spot is notorious
from having been, for a long time, the residence of some runaway
slaves, who, by cultivating a little ground near the top, contrived to

eke out a subsistence. At length they were discovered, and a party of soldiers being sent, the whole were seized with the exception of one old woman, who, sooner than again be led into slavery, dashed herself to pieces from the summit of the mountain. In a Roman matron this would have been called the noble love of freedom: in a poor negress it is mere brutal obstinacy. We continued riding for some hours. For the few last miles the road was intricate, and it passed through a desert waste of marshes and lagoons. The scene by the dimmed light of the moon was most desolate. A few fireflies flitted by us; and the solitary snipe, as it rose, uttered its plaintive cry. The distant and sullen roar of the sea scarcely broke the stillness of the night.

April 9th. We left our miserable sleeping-place before sunrise. The road passed through a narrow sandy plain, lying between the sea and the interior salt lagoons. The number of beautiful fishing birds, such as egrets and cranes, and the succulent plants assuming most fantastical forms, gave to the scene an interest which it would not otherwise have possessed. The few stunted trees were loaded with parasitical plants, among which the beauty and delicious fragrance of some of the orchidæ were most to be admired. As the sun rose, the day became extremely hot, and the reflection of the light and heat from the white sand was very distressing. We dined at Mandetiba; the thermometer in the shade being 84°. The beautiful view of the distant wooded hills, reflected in the perfectly calm water of an extensive lagoon, quite refreshed us. As the vênda[1] here was a very good one, and I have the pleasant, but rare remembrance, of an excellent dinner, I will be grateful and presently describe it, as the type of its class. These houses are often large, and are built of thick upright posts, with boughs interwoven, and afterwards plastered. They seldom have floors, and never glazed windows; but are generally pretty well roofed. Universally the front part is open, forming a kind of verandah, in which tables and benches are placed. The bed-rooms join on each side, and here the passenger may sleep as comfortably as he can, on a wooden platform, covered by a thin straw mat. The vênda stands in a courtyard, where the horses are fed. On first arriving, it was our custom to unsaddle the horses and give them their Indian corn;

1 Vênda, the Portuguese name for an inn.

then, with a low bow, to ask the senhôr to do us the favour to give us something to eat. 'Any thing you choose, sir,' was his usual answer. For the few first times, vainly I thanked providence for having guided us to so good a man. The conversation proceeding, the case universally became deplorable. 'Any fish can you do us the favour of giving?' – 'Oh! no, sir.' – 'Any soup?' – 'No, sir.' – 'Any bread?' – 'Oh! no, sir.' – 'Any dried meat?' – 'Oh! no, sir.' If we were lucky, by waiting a couple of hours, we obtained fowls, rice, and farinha. It not unfrequently happened, that we were obliged to kill, with stones, the poultry for our own supper. When, thoroughly exhausted by fatigue and hunger, we timorously hinted that we should be glad of our meal, the pompous, and (though true) most unsatisfactory answer was, 'It will be ready when it is ready.' If we had dared to remonstrate any further, we should have been told to proceed on our journey, as being too impertinent. The hosts are most ungracious and disagreeable in their manners; their houses and their persons are often filthily dirty; the want of the accommodation of forks, knives, and spoons is common; and I am sure no cottage or hovel in England could be found in a state so utterly destitute of every comfort. At Campos Novos, however, we fared sumptuously; having rice and fowls, biscuit, wine, and spirits, for dinner; coffee in the evening, and fish with coffee for breakfast. All this, with good food for the horses, only cost 2s. 6d. per head. Yet the host of this vênda, being asked if he knew any thing of a whip which one of the party had lost, gruffly answered, 'How should I know? why did you not take care of it? – I suppose the dogs have eaten it.'

Leaving Mandetiba, we continued to pass through an intricate wilderness of lakes; in some of which were fresh, in others salt water shells. Of the former kind, I found a Limnæa in great numbers in a lake, into which, the inhabitants assured me that the sea enters once a year, and sometimes oftener, and makes the water quite salt. I have no doubt many interesting facts, in relation to marine and fresh water animals, might be observed in this chain of lagoons, which skirt the coast of Brazil. M. Gay[1] has stated that he found in the neighbourhood of Rio, shells of the marine genera solen and mytilus, and fresh

1 Annales des Sciences Naturelles for 1833.

water ampullariæ, living together in brackish water. I also frequently observed in the lagoon near the Botanic Garden, where the water is only a little less salt than in the sea, a species of hydrophilus, very similar to a water-beetle common in the ditches of England: in the same lake the only shell belonged to a genus generally found in estuaries.

Leaving the coast for a time, we again entered the forest. The trees were very lofty, and remarkable, compared with those of Europe, from the whiteness of their trunks. I see by my note-book, 'wonderful and beautiful, flowering parasites,' invariably struck me as the most novel object in these grand scenes. Travelling onwards we passed through tracts of pasturage, much injured by the enormous conical ants' nests, which were nearly twelve feet high. They gave to the plain exactly the appearance of the mud volcanos at Jorullo, as figured by Humboldt. We arrived at Engenhodo after it was dark, having been ten hours on horseback. I never ceased, during the whole journey, to be surprised at the amount of labour which the horses were capable of enduring; they appeared also to recover from any injury much sooner than those of our English breed. The Vampire bat is often the cause of much trouble, by biting the horses on their withers. The injury is generally not so much owing to the loss of blood, as to the inflammation which the pressure of the saddle afterwards produces. The whole circumstance has lately been doubted in England; I was therefore fortunate in being present when one (Desmodus d'orbignyi, Wat.) was actually caught on a horse's back. We were bivouacking late one evening near Coquimbo, in Chile, when my servant, noticing that one of the horses was very restive, went to see what was the matter, and fancying he could distinguish something, suddenly put his hand on the beast's withers, and secured the vampire. In the morning the spot where the bite had been inflicted was easily distinguished from being slightly swollen and bloody. The third day afterwards we rode the horse, without any ill effects.

April 13th. – After three days' travelling we arrived at Socêgo, the estate of Senhôr Manuel Figuireda, a relation of one of our party. The house was simple, and, though like a barn in form, was well suited to the climate. In the sitting-room gilded chairs and sofas were oddly contrasted with the whitewashed walls, thatched roof, and windows without glass. The house, together with the granaries, the stables, and workshops for the blacks, who had been taught various trades, formed a rude kind of quadrangle; in the centre of which a large pile of coffee was drying. These buildings stand on a little hill, overlooking the cultivated ground, and surrounded on every side by a wall of dark green luxuriant forest. The chief produce of this part of the country is coffee. Each tree is supposed to yield annually, on an average, two pounds; but some give as much as eight. Mandioca or cassada is likewise cultivated in great quantity. Every part of this plant is useful: the leaves and stalks are eaten by the horses, and the roots are ground into a pulp, which, when pressed dry and baked, forms the farinha, the principal article of sustenance in the Brazils. It is a curious, though well-known fact, that the juice of this most nutritious plant is highly poisonous. A few years ago a cow died at this Fazênda, in consequence of having drunk some of it. Senhôr Figuireda told me that he had planted, the year before, one bag of feijaô or beans, and three of rice; the former of which produced eighty, and the latter three hundred and twenty fold. The pasturage supports a fine stock of cattle, and the woods are so full of game, that a deer had been killed on each of the three previous days. This profusion of food showed itself at dinner, where, if the tables did not groan, the guests surely did: for each person is expected to eat of every dish. One day, having, as I thought, nicely calculated so that nothing should go away untasted, to my utter dismay a roast turkey and a pig appeared in all their substantial reality. During the meals, it was the employment of a man to drive out of the room sundry old hounds, and dozens of little black children, which crawled in together, at every opportunity. As long as the idea of slavery could be banished, there was something exceedingly fascinating in this simple and patriarchal style of living: it was such a perfect retirement and independence from the rest of the world. As soon as any stranger is seen arriving, a large bell is set tolling, and generally some small cannon are fired. The event is thus announced to the rocks and woods, but to nothing else. One morning I walked out an hour before daylight to admire the solemn stillness of the scene; at last, the silence was broken by the morning hymn, raised on high by the whole body of the blacks; and in this manner their daily work is generally begun. On such fazêndas as these, I have no doubt the slaves pass happy and contented lives. On Saturday and Sunday they work for themselves, and in this fertile climate the labour of two days is sufficient to support a man and his family for the whole week.

April 14th. – Leaving Socêgo, we rode to another estate on the Rio Macâe, which was the last patch of cultivated ground in that direction. The estate was two and a half miles long, and the owner had forgotten how many broad. Only a very small piece had been cleared, yet almost every acre was capable of yielding all the various rich productions of a tropical land. Considering the enormous area of Brazil, the proportion of cultivated ground can scarcely be considered as any thing, compared to that which is left in the state of nature: at some future age, how vast a population it will support! During the second day's journey we found the road so shut up, that it was necessary that a man should go ahead with a sword to cut away the creepers. The forest abounded with beautiful objects; among which the tree ferns, though not large, were, from their bright green foliage, and the elegant curvature of their fronds, most worthy of admiration. In the evening it rained very heavily, and although the thermometer stood at $65°$, I felt very cold. As soon as the rain ceased, it was curious to observe the extraordinary evaporation which commenced over the whole extent of the forest. At the height of a hundred feet the hills were buried in a dense white vapour, which rose like columns of smoke from the most thickly-wooded parts, and especially from the valleys. I observed this phenomenon on several occasions: I suppose it is owing to the large surface of foliage, previously heated by the sun's rays.

While staying at this estate, I was very nearly being an eye-witness to one of those atrocious acts which can only take place in a slave country. Owing to a quarrel and a law-suit, the owner was on the point of taking all the women and children from the male slaves, and selling them separately at the public auction at Rio. Interest, and not any feeling of compassion, prevented this act. Indeed, I do not believe the inhumanity of separating thirty families, who had lived together for many years, even occurred to the owner. Yet I will pledge myself, that in humanity and good feeling he was superior to the common run of men. It may be said there exists no limit to the blindness of interest and selfish habit. I may mention one very trifling anecdote, which at the time struck me more forcibly than any story of cruelty. I was crossing a ferry with a negro, who was uncommonly stupid. In endeavouring to make him understand, I talked loud, and made signs, in doing which I passed my hand near his face. He, I suppose, thought I was in a passion, and was going to strike him; for instantly, with a frightened look and half-shut eyes, he dropped his hands. I shall never forget my feelings of surprise, disgust, and shame, at seeing a great powerful man afraid even to ward off a blow, directed, as he thought, at his face. This man had been trained to a degradation lower than the slavery of the most helpless animal.

April 18*th*. – In returning we spent two days at Socêgo, and I employed them in collecting insects in the forest. The greater number of trees, although so lofty, are not more than three or four feet in circumference. There are, of course, a few of much greater dimension. Senhôr Manuel was then making a canoe 70 feet in length from a solid trunk, which had originally been 110 feet long, and of great thickness. The contrast of palm trees, growing amidst the common branching kinds, never fails to give the scene an inter-tropical character. Here the woods were ornamented by the Cabbage Palm – one of the most beautiful of its family. With a stem so narrow that it might be clasped with the two hands, it waves its elegant head at the height of forty or fifty feet above the ground. The woody creepers, themselves covered by other creepers, were of great thickness: some which I measured were two feet in circumference. Many of the older trees presented a very curious appearance from the tresses of a liana hanging from their boughs, and resembling bundles of hay. If the eye was turned from the world of foliage above, to the ground beneath, it was attracted by the extreme elegance of the leaves of the ferns and mimosæ. The latter, in some parts, covered the surface with a brushwood only a few inches high. In walking across these thick beds of mimosæ, a broad track was marked by the change of shade, produced by the drooping of their sensitive petioles. It is easy to specify the individual objects of admiration in these grand scenes; but it is not possible to give an adequate idea of the higher feelings of wonder, astonishment, and devotion, which fill and ele-vate the mind.

April 19*th.* – Leaving Socêgo, during the two first days, we retraced our steps. It was very wearisome work, as the road generally ran across a glaring hot sandy plain, not far from the coast. I noticed that each time the horse put its foot on the fine siliceous sand, a gentle chirping noise was produced. On the third day we took a different line, and passed through the gay little village of Madre de Deôs. This is one of the principal lines of road in Brazil; yet it was in so bad a state that no wheel vehicle, excepting the clumsy bullock-waggon, could pass along. In our whole journey we did not cross a single bridge built of stone; and those made of logs of wood were frequently so much out of repair, that it was necessary to go on one side to avoid them. All distances are inaccurately known. The road is often marked by crosses, in the place of milestones, to signify where human blood has been spilled. On the evening of the 23rd we arrived at Rio, having finished our pleasant little excursion.

During the remainder of my stay at Rio, I resided in a cottage at Botofogo Bay. It was impossible to wish for anything more delightful than thus to spend some weeks in so magnificent a country. In England any person fond of natural history enjoys in his walks a great advantage, by always having something to attract his attention; but in these fertile climates, teeming with life, the attractions are so numerous, that he is scarcely able to walk at all.

The few observations which I was enabled to make were almost exclusively confined to the invertebrate animals. The existence of a division of the genus Planaria, which inhabits the dry land, interested me much. These animals are of so simple a structure, that Cuvier has arranged them with the intestinal worms, though never found within the bodies of other animals. Numerous species inhabit both salt and fresh water; but those to which I allude were found, even in the drier parts of the forest, beneath logs of rotten wood, on which I believe they feed. In general form they resemble little slugs, but are very much narrower in proportion, and several of the species are beautifully coloured with longitudinal stripes. Their structure is very simple: near the middle of the under or crawling surface there are two small transverse slits, from the anterior one of which a funnel-shaped and highly irritable mouth can be protruded. For some time after the rest of the animal was completely dead from the effects of salt water or any other cause, this organ still retained its vitality.

I found no less than twelve different species of terrestrial Planariæ in different parts of the southern hemisphere.[1] Some specimens which I obtained at Van Diemen's Land, I kept alive for nearly two months, feeding them on rotten wood. Having cut one of them transversely into two nearly equal parts, in the course of a fortnight both had the shape of perfect animals. I had, however, so divided the body, that one of the halves contained both the inferior orifices, and the other, in consequence, none. In the course of twenty-five days from the operation, the more perfect half could not have been distinguished from any other specimen. The other had increased much in size; and towards its posterior end, a clear space was formed in the parenchymatous mass, in which a rudimentary cup-shaped mouth could clearly be distinguished; on the under surface, however, no corresponding slit was yet open. If the increased heat of the weather, as we approached the equator, had not destroyed all the individuals, there can be no doubt that this last step would have completed its structure. Although so well-known an experiment, it was interesting to watch the gradual production of every essential organ, out of the simple extremity of another animal. It is extremely difficult to preserve these Planariæ; as soon as the cessation of life allows the ordinary laws of change to act, their entire bodies become soft and fluid, with a rapidity which I have never seen equalled.

I first visited the forest in which these Planariæ were found, in company with an old Portuguese priest who took me out to hunt with him. The sport consisted in turning into the cover a few dogs, and then patiently waiting to fire at any animal which might appear. We were accompanied by the son of a neighbouring farmer – a good specimen of a wild Brazilian youth. He was dressed in a tattered old shirt and trousers, and had his head uncovered: he carried an old-fashioned gun and a large knife. The habit of carrying the knife is universal; and in traversing a thick wood it is almost necessary, on account of the creeping plants. The frequent occurrence of murder

1 I have described and named these species in the 'Annals of Nat. Hist.,' vol. xiv. p. 241.

may be partly attributed to this habit. The Brazilians are so dexterous with the knife, that they can throw it to some distance with precision, and with sufficient force to cause a fatal wound. I have seen a number of little boys practising this art as a game of play, and from their skill in hitting an upright stick, they promised well for more earnest attempts. My companion, the day before, had shot two large bearded monkeys. These animals have prehensile tails, the extremity of which, even after death, can support the whole weight of the body. One of them thus remained fast to a branch, and it was necessary to cut down a large tree to procure it. This was soon effected, and down came tree and monkey with an awful crash. Our day's sport, besides the monkey, was confined to sundry small green parrots and a few toucans. I profited, however, by my acquaintance with the Portuguese padre, for on another occasion he gave me a fine specimen of the Yagouaroundi cat.

Every one has heard of the beauty of the scenery near Botofogo. The house in which I lived was seated close beneath the well-known mountain of the Corcovado. It has been remarked, with much truth, that abruptly conical hills are characteristic of the formation which Humboldt designates as gneiss-granite. Nothing can be more striking than the effect of these huge rounded masses of naked rock rising out of the most luxuriant vegetation.

I was often interested by watching the clouds, which, rolling in from seaward, formed a bank just beneath the highest point of the Corcovado. This mountain, like most others, when thus partly veiled, appeared to rise to a far prouder elevation than its real height of 2300 feet. Mr. Daniell has observed, in his meteorological essays, that a cloud sometimes appears fixed on a mountain summit, while the wind continues to blow over it. The same phenomenon here presented a slightly different appearance. In this case the cloud was clearly seen to curl over, and rapidly pass by the summit, and yet was neither diminished nor increased in size. The sun was setting, and a gentle southerly breeze, striking against the southern side of the rock, mingled its current with the colder air above; and the vapour was thus condensed: but as the light wreaths of cloud passed over the ridge, and came within the influence of the warmer atmosphere of the northern sloping bank, they were immediately re-dissolved.

The climate, during the months of May and June, or the beginning of winter, was delightful. The mean temperature, from observations taken at nine o'clock, both morning and evening, was only 72°. It often rained heavily, but the drying southerly winds soon again rendered the walks pleasant. One morning, in the course of six hours, 1.6 inches of rain fell. As this storm passed over the forests which surround the Corcovado, the sound produced by the drops pattering on the countless multitude of leaves was very remarkable; it could be heard at the distance of a quarter of a mile, and was like the rushing of a great body of water. After the hotter days, it was delicious to sit quietly in the garden and watch the evening pass into night. Nature, in these climes, chooses her vocalists from more humble performers than in Europe. A small frog, of the genus Hyla, sits on a blade of grass about an inch above the surface of the water, and sends forth a pleasing chirp: when several are together they sing in harmony on different notes. I had some difficulty in catching a specimen of this frog. The genus Hyla has its toes terminated by small suckers; and I found this animal could crawl up a pane of glass, when placed absolutely perpendicular. Various cicadæ and crickets, at the same time, keep up a ceaseless shrill cry, but which, softened by the distance, is not unpleasant. Every evening after dark this great concert commenced; and often have I sat listening to it, until my attention has been drawn away by some curious passing insect.

At these times the fireflies are seen flitting about from hedge to hedge. On a dark night the light can be seen at about two hundred paces distant. It is remarkable that in all the different kinds of glowworms, shining elaters, and various marine animals (such as the crustacea, medusæ, nereidæ, a coralline of the genus Clytia, and Pyrosoma), which I have observed, the light has been of a well-marked green colour. All the fireflies, which I caught here, belonged to the Lampyridæ (in which family the English glowworm is included), and the greater number of specimens were of Lampyris occidentalis.[1] I found that this insect emitted the most brilliant flashes when irritated: in the intervals, the abdominal rings were obscured.

1 I am greatly indebted to Mr. Waterhouse for his kindness in naming for me this and many other insects, and in giving me much valuable assistance.

The flash was almost coinstantaneous in the two rings, but it was just perceptible first in the anterior one. The shining matter was fluid and very adhesive: little spots, where the skin had been torn, continued bright with a slight scintillation, whilst the uninjured parts were obscured. When the insect was decapitated the rings remained un-interruptedly bright, but not so brilliant as before: local irritation with a needle always increased the vividness of the light. The rings in one instance retained their luminous property nearly twenty-four hours after the death of the insect. From these facts it would appear prob-able, that the animal has only the power of concealing or extinguish-ing the light for short intervals, and that at other times the display is involuntary. On the muddy and wet gravel-walks I found the larvæ of this lampyris in great numbers: they resembled in general form the female of the English glowworm. These larvæ possessed but feeble luminous powers; very differently from their parents, on the slightest touch they feigned death, and ceased to shine; nor did irritation excite any fresh display. I kept several of them alive for some time: their tails are very singular organs, for they act, by a well-fitted contrivance, as suckers or organs of attachment, and likewise as reservoirs for saliva, or some such fluid. I repeatedly fed them on raw meat; and I invari-ably observed, that every now and then the extremity of the tail was applied to the mouth, and a drop of fluid exuded on the meat, which was then in the act of being consumed. The tail, notwithstanding so much practice, does not seem to be able to find its way to the mouth; at least the neck was always touched first, and apparently as a guide.

When we were at Bahia, an elater or beetle (Pyrophorus lumino-sus, Illig.) seemed the most common luminous insect. The light in this case was also rendered more brilliant by irritation. I amused myself one day by observing the springing powers of this insect, which have not, as it appears to me, been properly described.[1] The elater, when placed on its back and preparing to spring, moved its head and thorax backwards, so that the pectoral spine was drawn out, and rested on the edge of its sheath. The same backward movement being continued, the spine, by the full action of the muscles, was bent like a spring; and the insect at this moment rested on the extremity of

1 Kirby's Entomology, vol. ii. p. 317.

its head and wing-cases. The effort being suddenly relaxed, the head and thorax flew up, and in consequence, the base of the wing-cases struck the supporting surface with such force, that the insect by the reaction was jerked upwards to the height of one or two inches. The projecting points of the thorax, and the sheath of the spine, served to steady the whole body during the spring. In the descriptions which I have read, sufficient stress does not appear to have been laid on the elasticity of the spine: so sudden a spring could not be the result of simple muscular contraction, without the aid of some mechanical contrivance.

On several occasions I enjoyed some short but most pleasant excursions in the neighbouring country. One day I went to the Botanic Garden, where many plants, well known for their great utility, might be seen growing. The leaves of the camphor, pepper, cinnamon, and clove trees were delightfully aromatic; and the bread-fruit, the jaca, and the mango, vied with each other in the magnificence of their foliage. The landscape in the neighbourhood of Bahia almost takes its character from the two latter trees. Before seeing them, I had no idea that any trees could cast so black a shade on the ground. Both of them bear to the evergreen vegetation of these climates the same kind of relation which laurels and hollies in England do to the lighter green of the deciduous trees. It may be observed, that the houses within the tropics are surrounded by the most beautiful forms of vegetation, because many of them are at the same time most useful to man. Who can doubt that these qualities are united in the banana, the cocoa-nut, the many kinds of palm, the orange, and the bread-fruit tree?

During this day I was particularly struck with a remark of Humboldt's, who often alludes to 'the thin vapour which, without changing the transparency of the air, renders its tints more harmonious, and softens its effects.' This is an appearance which I have never observed in the temperate zones. The atmosphere, seen through a short space of half or three quarters of a mile, was perfectly lucid, but at a greater distance all colours were blended into a most beautiful haze, of a pale French grey, mingled with a little blue. The condition of the atmosphere between the morning and about noon, when the effect was most evident, had undergone little change, excepting in its dryness. In the interval, the difference between the dew point and temperature had increased from $7°.5$ to $17°$.

On another occasion I started early and walked to the Gavia, or topsail mountain. The air was delightfully cool and fragrant; and the drops of dew still glittered on the leaves of the large liliaceous plants, which shaded the streamlets of clear water. Sitting down on a block of granite, it was delightful to watch the various insects and birds as they flew past. The humming-bird seems particularly fond of such shady retired spots. Whenever I saw these little creatures buzzing round a flower, with their wings vibrating so rapidly as to be scarcely visible, I was reminded of the sphinx moths: their movements and habits are indeed in many respects very similar.

Following a pathway I entered a noble forest, and from a height of five or six hundred feet, one of those splendid views was presented, which are so common on every side of Rio. At this elevation the landscape attains its most brilliant tint; and every form, every shade, so completely surpasses in magnificence all that the European has ever beheld in his own country, that he knows not how to express his feelings. The general effect frequently recalled to my mind the gayest scenery of the Opera-house or the great theatres. I never returned from these excursions empty-handed. This day I found a specimen of a curious fungus, called Hymenophallus. Most people know the English Phallus, which in autumn taints the air with its odious smell: this, however, as the entomologist is aware, is to some of our beetles a delightful fragrance. So was it here; for a Strongylus, attracted by the odour, alighted on the fungus as I carried it in my hand. We here see in two distant countries a similar relation between plants and insects of the same families, though the species of both are different. When man is the agent in introducing into a country a new species, this relation is often broken: as one instance of this I may mention, that the leaves of the cabbages and lettuces, which in England afford food to such a multitude of slugs and caterpillars, in the gardens near Rio are untouched.

During our stay at Brazil I made a large collection of insects. A few general observations on the comparative importance of the different orders may be interesting to the English entomologist. The large and brilliantly-coloured Lepidoptera bespeak the zone they inhabit, far more plainly than any other race of animals. I allude only to the butterflies; for the moths, contrary to what might have been expected from the rankness of the vegetation, certainly appeared in much fewer numbers than in our own temperate regions. I was much surprised at the habits of Papilio feronia. This butterfly is not uncommon, and generally frequents the orange-groves. Although a high flier, yet it very frequently alights on the trunks of trees. On these occasions its head is invariably placed downwards; and its wings are expanded in a horizontal plane, instead of being folded vertically, as is commonly the case. This is the only butterfly which I have ever seen, that uses its legs for running. Not being aware of this fact, the insect, more than once, as I cautiously approached with my forceps, shuffled on one side just as the instrument was on the point of closing, and thus escaped. But a far more singular fact is the power which this species possesses of making a noise.[1] Several times when a pair, probably male and female, were chasing each other in an irregular course, they passed within a few yards of me; and I distinctly heard a clicking noise, similar to that produced by a toothed wheel passing under a spring catch. The noise was continued at short intervals, and could be distinguished at about twenty yards' distance: I am certain there is no error in the observation.

I was disappointed in the general aspect of the Coleoptera. The number of minute and obscurely-coloured beetles is exceedingly great.[2] The cabinets of Europe can, as yet, boast only of the larger

1 Mr. Doubleday has lately described (before the Entomological Society, March 3rd, 1845) a peculiar structure in the wings of this butterfly, which seems to be the means of its making its noise. He says, 'It is remarkable for having a sort of drum at the base of the fore wings, between the costal nervure and the subcostal. These two nervures, more-over, have a peculiar screw-like diaphragm or vessel in the interior.' I find in Langs-dorff's travels (in the years 1803–7, p. 74) it is said, that in the island of St. Catherine's on the coast of Brazil a butterfly called Februa Hoffmanseggi, makes a noise, when flying away, like a rattle.

2 I may mention, as a common instance of one day's (June 23rd) collecting, when I was not attending particularly to the Coleoptera, that I caught sixty-eight species of that order. Among these, there were only two of the Carabidæ, four Brachelytra, fifteen Rhyncophora, and fourteen of the Chrysomelidæ. Thirty-seven species of Arachnidæ, which I brought home, will be sufficient to prove that I was not paying overmuch attention to the generally favoured order of Coleoptera.

species from tropical climates. It is sufficient to disturb the composure of an entomologist's mind, to look forward to the future dimensions of a complete catalogue. The carnivorous beetles, or Carabidæ, appear in extremely few numbers within the tropics: this is the more remarkable when compared to the case of the carnivorous quadrupeds, which are so abundant in hot countries. I was struck with this observation both on entering Brazil, and when I saw the many elegant and active forms of the Harpalidæ re-appearing on the temperate plains of La Plata. Do the very numerous spiders and rapacious Hymenoptera supply the place of the carnivorous beetles? The carrion-feeders and Brachelytera are very uncommon; on the other hand, the Rhyncophora and Chrysomelidæ, all of which depend on the vegetable world for subsistence, are present in astonishing numbers. I do not here refer to the number of different species, but to that of the individual insects; for on this it is that the most striking character in the entomology of different countries depends. The orders Orthoptera and Hemiptera are particularly numerous; as likewise is the stinging division of the Hymenoptera; the bees, perhaps, being excepted. A person, on first entering a tropical forest, is astonished at the labours of the ants: well-beaten paths branch off in every direction, on which an army of never-failing foragers may be seen, some going forth, and others returning, burdened with pieces of green leaf, often larger than their own bodies.

A small dark-coloured ant sometimes migrates in countless numbers. One day, at Bahia, my attention was drawn by observing many spiders, cockroaches, and other insects, and some lizards, rushing in the greatest agitation across a bare piece of ground. A little way behind, every stalk and leaf was blackened by a small ant. The swarm having crossed the bare space, divided itself, and descended an old wall. By this means many insects were fairly enclosed; and the efforts which the poor little creatures made to extricate themselves from such a death were wonderful. When the ants came to the road they changed their course, and in narrow files re-ascended the wall. Having placed a small stone so as to intercept one of the lines, the whole body attacked it, and then immediately retired. Shortly afterwards another body came to the charge, and again having failed to make any impression, this line of march was entirely given up.

By going an inch round, the file might have avoided the stone, and this doubtless would have happened, if it had been originally there: but having been attacked, the lion-hearted little warriors scorned the idea of yielding.

Certain wasp-like insects, which construct in the corners of the verandahs clay cells for their larvæ, are very numerous in the neighbourhood of Rio. These cells they stuff full of half-dead spiders and caterpillars, which they seem wonderfully to know how to sting to that degree as to leave them paralysed but alive, until their eggs are hatched; and the larvæ feed on the horrid mass of powerless, half-killed victims – a sight which has been described by an enthusiastic naturalist[1] as curious and pleasing! I was much interested one day by watching a deadly contest between a Pepsis and a large spider of the genus Lycosa. The wasp made a sudden dash at its prey, and then flew away: the spider was evidently wounded, for, trying to escape, it rolled down a little slope, but had still strength sufficient to crawl into a thick tuft of grass. The wasp soon returned, and seemed surprised at not immediately finding its victim. It then commenced as regular a hunt as ever hound did after fox; making short semicircular casts, and all the time rapidly vibrating its wings and antennæ. The spider, though well concealed, was soon discovered; and the wasp, evidently still afraid of its adversary's jaws, after much manœuvring, inflicted two stings on the under side of its thorax. At last, carefully examining with its antennæ the now motionless spider, it proceeded to drag away the body. But I stopped both tyrant and prey.[2]

The number of spiders, in proportion to other insects, is here compared with England very much larger; perhaps more so than with any other division of the articulate animals. The variety of species among the jumping spiders appears almost infinite. The genus, or

1 In a MS. in the British Museum by Mr. Abbott, who made his observations in Georgia; see Mr. A. White's paper in the 'Annals of Nat. Hist.,' vol. vii. p. 472. Lieut. Hutton has described a sphex with similar habits in India, in the 'Journal of the Asiatic Society,' vol. i. p. 555.

2 Don Felix Azara (vol. i. p. 175), mentioning a hymenopterous insect, probably of the same genus, says, he saw it dragging a dead spider through tall grass, in a straight line to its nest, which was one hundred and sixty-three paces distant. He adds that the wasp, in order to find the road, every now and then made 'demi-tours d'environ trois palmes.'

rather family of Epeira, is here characterized by many singular forms; some species have pointed coriaceous shells, others enlarged and spiny tibiæ. Every path in the forest is barricaded with the strong yellow web of a species, belonging to the same division with the Epeira clavipes of Fabricius, which was formerly said by Sloane to make, in the West Indies, webs so strong as to catch birds. A small and pretty kind of spider, with very long fore-legs, and which appears to belong to an undescribed genus, lives as a parasite on almost every one of these webs. I suppose it is too insignificant to be noticed by the great Epeira, and is therefore allowed to prey on the minute insects, which, adhering to the lines, would otherwise be wasted. When frightened, this little spider either feigns death by extending its front legs, or suddenly drops from the web. A large Epeira of the same division with Epeira tuberculata and conica is extremely common, especially in dry situations. Its web, which is generally placed among the great leaves of the common agave, is sometimes strengthened near the centre by a pair or even four zigzag ribbons, which connect two adjoining rays. When any large insect, as a grasshopper or wasp, is caught, the spider, by a dexterous movement, makes it revolve very rapidly, and at the same time emitting a band of threads from its spinners, soon envelops its prey in a case like the cocoon of a silkworm. The spider now examines the powerless victim, and gives the fatal bite on the hinder part of its thorax; then retreating, patiently waits till the poison has taken effect. The virulence of this poison may be judged of from the fact that in half a minute I opened the mesh, and found a large wasp quite lifeless. This Epeira always stands with its head downwards near the centre of the web. When disturbed, it acts differently according to circumstances: if there is a thicket below, it suddenly falls down; and I have distinctly seen the thread from the spinners lengthened by the animal while yet station-ary, as preparatory to its fall. If the ground is clear beneath, the Epeira seldom falls, but moves quickly through a central passage from one to the other side. When still further disturbed, it practises a most curious manœuvre: standing in the middle, it violently jerks the web, which is attached to elastic twigs, till at last the whole acquires such a rapid vibratory movement, that even the outline of the spider's body becomes indistinct.

It is well known that most of the British spiders, when a large insect is caught in their webs, endeavour to cut the lines and liberate their prey, to save their nets from being entirely spoiled. I once, however, saw in a hot-house in Shropshire a large female wasp caught in the irregular web of a quite small spider; and this spider, instead of cutting the web, most perseveringly continued to entangle the body, and especially the wings, of its prey. The wasp at first aimed in vain repeated thrusts with its sting at its little antagonist. Pitying the wasp, after allowing it to struggle for more than an hour, I killed it and put it back into the web. The spider soon returned; and an hour afterwards I was much surprised to find it with its jaws buried in the orifice, through which the sting is protruded by the living wasp. I drove the spider away two or three times, but for the next twenty-four hours I always found it again sucking at the same place. The spider became much distended by the juices of its prey, which was many times larger than itself.

I may here just mention, that I found, near St. Fé Bajada, many large black spiders, with ruby-coloured marks on their backs, having gregarious habits. The webs were placed vertically, as is invariably the case with the genus Epeira: they were separated from each other by a space of about two feet, but were all attached to certain common lines, which were of great length, and extended to all parts of the community. In this manner the tops of some large bushes were encompassed by the united nets. Azara[1] has described a gregarious spider in Paraguay, which Walckenaer thinks must be a Theridion, but probably it is an Epeira, and perhaps even the same species with mine. I cannot, however, recollect seeing a central nest as large as a hat, in which, during autumn, when the spiders die, Azara says the eggs are deposited. As all the spiders which I saw were of the same size, they must have been nearly of the same age. This gregarious habit, in so typical a genus as Epeira, among insects, which are so bloodthirsty and solitary that even the two sexes attack each other, is a very singular fact.

In a lofty valley of the Cordillera, near Mendoza, I found another spider with a singularly-formed web. Strong lines radiated in a

1 Azara's Voyage, vol. i. p. 213.

vertical plane from a common centre, where the insect had its station; but only two of the rays were connected by a symmetrical mesh-work; so that the net, instead of being, as is generally the case, circular, consisted of a wedge-shaped segment. All the webs were similarly constructed.

CHAPTER III

MALDONADO

Monte Video – Maldonado – Excursion to R. Polanco –
Lazo and Bolas – Partridges – Absence of Trees – Deer –
Capybara, or River Hog – Tucutuco – Molothrus,
cuckoo-like habits – Tyrant-flycatcher – Mocking-bird –
Carrion Hawks – Tubes formed by Lightning – House
struck.

JULY 5th, 1832. – In the morning we got under way, and stood out of
the splendid harbour of Rio de Janeiro. In our passage to the Plata, we
saw nothing particular, excepting on one day a great shoal of porpoises, many hundreds in number. The whole sea was in places
furrowed by them; and a most extraordinary spectacle was presented,
as hundreds, proceeding together by jumps, in which their whole
bodies were exposed, thus cut the water. When the ship was running
nine knots an hour, these animals could cross and recross the bows
with the greatest ease, and then dash away right ahead. As soon as we
entered the estuary of the Plata, the weather was very unsettled. One
dark night we were surrounded by numerous seals and penguins,
which made such strange noises, that the officer on watch reported he
could hear the cattle bellowing on shore. On a second night we
witnessed a splendid scene of natural fireworks; the masthead and
yard-arm-ends shone with St. Elmo's light; and the form of the vane
could almost be traced, as if it had been rubbed with phosphorus. The
sea was so highly luminous, that the tracks of the penguins were
marked by a fiery wake, and the darkness of the sky was momentarily
illuminated by the most vivid lightning.

When within the mouth of the river, I was interested by observing
how slowly the waters of the sea and river mixed. The latter, muddy
and discoloured, from its less specific gravity, floated on the surface of
the salt water. This was curiously exhibited in the wake of the vessel,
where a line of blue water was seen mingling in little eddies, with the
adjoining fluid.

July 26th. – We anchored at Monte Video. The *Beagle* was employed in surveying the extreme southern and eastern coasts of America, south of the Plata, during the two succeeding years. To prevent useless repetitions, I will extract those parts of my journal which refer to the same districts, without always attending to the order in which we visited them.

MALDONADO is situated on the northern bank of the Plata, and not very far from the mouth of the estuary. It is a most quiet, forlorn, little town; built, as is universally the case in these countries, with the streets running at right angles to each other, and having in the middle a large plaza or square, which, from its size, renders the scantiness of the population more evident. It possesses scarcely any trade; the exports being confined to a few hides and living cattle. The inhabitants are chiefly land-owners, together with a few shop-keepers and the necessary tradesmen, such as blacksmiths and carpenters, who do nearly all the business for a circuit of fifty miles round. The town is separated from the river by a band of sand-hillocks, about a mile broad: it is surrounded, on all other sides, by an open slightly-undulating country, covered by one uniform layer of fine green turf, on which countless herds of cattle, sheep, and horses graze. There is very little land cultivated even close to the town. A few hedges, made of cacti and agave, mark out where some wheat or Indian corn has been planted. The features of the country are very similar along the whole northern bank of the Plata. The only difference is, that here the granitic hills are a little bolder. The scenery is very uninteresting; there is scarcely a house, an enclosed piece of ground, or even a tree, to give it an air of cheerfulness. Yet, after being imprisoned for some time in a ship, there is a charm in the unconfined feeling of walking over boundless plains of turf. Moreover, if your view is limited to a small space, many objects possess beauty. Some of the smaller birds are brilliantly coloured; and the bright green sward, browsed short by the cattle, is ornamented by dwarf flowers, among which a plant, looking like the daisy, claimed the place of an old friend. What would a florist say to whole tracts so thickly covered by the Verbena melindres, as, even at a distance, to appear of the most gaudy scarlet?

I staid ten weeks at Maldonado, in which time a nearly perfect collection of the animals, birds, and reptiles, was procured. Before making any observations respecting them, I will give an account of a little excursion I made as far as the river Polanco, which is about seventy miles distant, in a northerly direction. I may mention, as a proof how cheap everything is in this country, that I paid only two dollars a day, or eight shillings, for two men, together with a troop of about a dozen riding-horses. My companions were well armed with pistols and sabres; a precaution which I thought rather unnecessary; but the first piece of news we heard was, that, the day before, a traveller from Monte Video had been found dead on the road, with his throat cut. This happened close to a cross, the record of a former murder.

On the first night we slept at a retired little country-house; and there I soon found out that I possessed two or three articles, especially a pocket compass, which created unbounded astonishment. In every house I was asked to show the compass, and by its aid, together with a map, to point out the direction of various places. It excited the liveliest admiration that I, a perfect stranger, should know the road (for direction and road are synonymous in this open country) to places where I had never been. At one house a young woman, who was ill in bed, sent to entreat me to come and show her the compass. If their surprise was great, mine was greater, to find such ignorance among people who possessed their thousands of cattle, and 'estancias' of great extent. It can only be accounted for by the circumstance that this retired part of the country is seldom visited by foreigners. I was asked whether the earth or sun moved; whether it was hotter or colder to the north; where Spain was, and many other such questions. The greater number of the inhabitants had an indistinct idea that England, London, and North America, were different names for the same place; but the better informed well knew that London and North America were separate countries close together, and that England was a large town in London! I carried with me some promethean matches, which I ignited by biting; it was thought so wonderful that a man should strike fire with his teeth, that it was usual to collect the whole family to see it: I was once offered a dollar for a single one. Washing my face in the morning caused much speculation at the village of Las Minas; a superior tradesman closely cross-questioned me about so singular a practice; and likewise why on board we wore our beards; for he had heard from my guide that we did so. He eyed me with much suspicion; perhaps he had heard of ablutions in the Mahomedan religion, and knowing me to be a heretick, probably he came to the conclusion that all hereticks were Turks. It is the general custom in this country to ask for a night's lodging at the first convenient house. The astonishment at the compass, and my other feats in jugglery, was to a certain degree advantageous, as with that, and the long stories my guides told of my breaking stones, knowing venomous from harmless snakes, collecting insects, etc., I repaid them for their hospitality. I am writing as if I had been among the inhabitants of central Africa: Banda Oriental would not be flattered by the comparison; but such were my feelings at the time.

The next day we rode to the village of Las Minas. The country was rather more hilly, but otherwise continued the same; an inhabitant of the Pampas no doubt would have considered it as truly Alpine. The country is so thinly inhabited, that during the whole day we scarcely met a single person. Las Minas is much smaller even than Maldonado. It is seated on a little plain, and is surrounded by low rocky mountains. It is of the usual symmetrical form; and with its whitewashed church standing in the centre, had rather a pretty appearance. The outskirting houses rose out of the plain like isolated beings, without the accompaniment of gardens or courtyards. This is generally the case in the country, and all the houses have, in consequence, an uncomfortable aspect. At night we stopped at a pulperia, or drinking-shop. During the evening a great number of Gauchos came in to drink spirits and smoke cigars: their appearance is very striking; they are generally tall and handsome; but with a proud and dissolute expression of countenance. They frequently wear their moustaches, and long black hair curling down their backs. With their brightly-coloured garments, great spurs clanking about their heels, and knives stuck as daggers (and often so used) at their waists, they look a very different race of men from what might be expected from their name of Gauchos, or simple countrymen. Their politeness is excessive; they never drink their spirits without expecting you to taste it; but whilst making their exceedingly graceful bow, they seem quite as ready, if occasion offered, to cut your throat.

On the third day we pursued rather an irregular course, as I was employed in examining some beds of marble. On the fine plains of turf we saw many ostriches (Struthio rhea). Some of the flocks contained as many as twenty or thirty birds. These, when standing on any little eminence, and seen against the clear sky, presented a very noble appearance. I never met with such tame ostriches in any other part of the country: it was easy to gallop up within a short distance of them; but then, expanding their wings, they made all sail right before the wind, and soon left the horse astern.

At night we came to the house of Don Juan Fuentes, a rich landed proprietor, but not personally known to either of my companions. On approaching the house of a stranger, it is usual to follow several little points of etiquette: riding up slowly to the door, the salutation of Ave Maria is given, and until somebody comes out and asks you to alight, it is not customary even to get off your horse: the formal answer of the owner is, 'sin pecado concebida' – that is, conceived without sin. Having entered the house, some general conversation is kept up for a few minutes, till permission is asked to pass the night there. This is granted as a matter of course. The stranger then takes his meals with the family, and a room is assigned him, where with the horsecloths belonging to his recado (or saddle of the Pampas) he makes his bed. It is curious how similar circumstances produce such similar results in manners. At the Cape of Good Hope the same hospitality, and very nearly the same points of etiquette, are universally observed. The difference, however, between the character of the Spaniard and that of the Dutch boor is shown, by the former never asking his guest a single question beyond the strictest rules of politeness, whilst the honest Dutchman demands where he has been, where he is going, what is his business, and even how many brothers, sisters, or children he may happen to have.

Shortly after our arrival at Don Juan's, one of the large herds of cattle was driven in towards the house, and three beasts were picked out to be slaughtered for the supply of the establishment. These half-wild cattle are very active; and knowing full well the fatal lazo, they led the horses a long and laborious chase. After witnessing the rude wealth displayed in the number of cattle, men, and horses, Don Juan's miserable house was quite curious. The floor consisted of hardened mud, and the windows were without glass; the sitting-room boasted only of a few of the roughest chairs and stools, with a couple of tables. The supper, although several strangers were present, consisted of two huge piles, one of roast beef, the other of boiled, with some pieces of pumpkin: besides this latter there was no other vegetable, and not even a morsel of bread. For drinking, a large earthenware jug of water served the whole party. Yet this man was the owner of several square miles of land, of which nearly every acre would produce corn, and, with a little trouble, all the common vegetables. The evening was spent in smoking, with a little impromptu singing, accompanied by the guitar. The signoritas all sat together in one corner of the room, and did not sup with the men.

So many works have been written about these countries, that it is almost superfluous to describe either the lazo or the bolas. The lazo consists of a very strong, but thin, well-plaited rope, made of raw hide. One end is attached to the broad surcingle, which fastens together the complicated gear of the recado, or saddle used in the Pampas; the other is terminated by a small ring of iron or brass, by which a noose can be formed. The Gaucho, when he is going to use the lazo, keeps a small coil in his bridle-hand, and in the other holds the running noose, which is made very large, generally having a diameter of about eight feet. This he whirls round his head, and by the dexterous movement of his wrist keeps the noose open; then, throwing it, he causes it to fall on any particular spot he chooses. The lazo, when not used, is tied up in a small coil to the after part of the recado. The bolas, or balls, are of two kinds: the simplest, which is chiefly used for catching ostriches, consists of two round stones, covered with leather, and united by a thin plaited thong, about eight feet long. The other kind differs only in having three balls united by the thongs to a common centre. The Gaucho holds the smallest of the three in his hand, and whirls the other two round and round his head; then, taking aim, sends them like chain shot revolving through the air. The balls no sooner strike any object, than, winding round it, they cross each other, and become firmly hitched. The size and weight of the balls varies, according to the purpose for which they are made: when of stone, although not larger than an apple, they are sent with such force as sometimes to break the leg even of a horse. I have seen the balls made of wood, and as large as a turnip, for the sake of catching these animals without injuring them. The balls are sometimes made of iron, and these can be hurled to the greatest distance. The main difficulty in using either lazo or bolas is to ride so well as to be able at full speed, and while suddenly turning about, to whirl them so steadily round the head, as to take aim: on foot any person would soon learn the art. One day, as I was amusing myself by galloping and whirling the balls round my head, by accident the free one struck a bush; and its revolving motion being thus destroyed, it immediately fell to the ground, and like magic caught one hind leg of my horse; the other ball was then jerked out of my hand, and the horse fairly secured. Luckily he was an old practised animal, and knew what it meant; otherwise he would probably have kicked till he had thrown himself down. The Gauchos roared with laughter; they cried out that they had seen every sort of animal caught, but had never before seen a man caught by himself.

During the two succeeding days, I reached the furthest point which I was anxious to examine. The country wore the same aspect, till at last the fine green turf became more wearisome than a dusty turnpike road. We everywhere saw great numbers of part-ridges (Nothura major). These birds do not go in coveys, nor do they conceal themselves like the English kind. It appears a very silly bird. A man on horseback by riding round and round in a circle, or rather in a spire, so as to approach closer each time, may knock on the head as many as he pleases. The more common method is to catch them with a running noose, or little lazo, made of the stem of an ostrich's feather, fastened to the end of a long stick. A boy on a quiet old horse will frequently thus catch thirty or forty in a day. In Arctic North America[1] the Indians catch the Varying Hare by walking spirally round and round it, when on its form: the middle of the day is reckoned the best time, when the sun is high, and the shadow of the hunter not very long.

On our return to Maldonado, we followed rather a different line of road. Near Pan de Azucar, a landmark well known to

1 Hearne's Journey, p. 383.

all those who have sailed up the Plata, I stayed a day at the house of a most hospitable old Spaniard. Early in the morning we ascended the Sierra de las Animas. By the aid of the rising sun the scenery was almost picturesque. To the westward the view extended over an immense level plain as far as the Mount, at Monte Video, and to the eastward, over the mammillated country of Maldonado. On the summit of the mountain there were several small heaps of stones, which evidently had lain there for many years. My companion assured me that they were the work of the Indians in the old time. The heaps were similar, but on a much smaller scale, to those so commonly found on the mountains of Wales. The desire to signalize any event, on the highest point of the neighbouring land, seems an universal passion with mankind. At the present day, not a single Indian, either civilized or wild, exists in this part of the province; nor am I aware that the former inhabitants have left behind them any more permanent records than these insignificant piles on the summit of the Sierra de las Animas.

The general, and almost entire absence of trees in Banda Oriental is remarkable. Some of the rocky hills are partly covered by thickets, and on the banks of the larger streams, especially to the north of Las Minas, willow-trees are not uncommon. Near the Arroyo Tapes I heard of a wood of palms; and one of these trees, of considerable size, I saw near the Pan de Azucar, in lat. 35°. These, and the trees planted by the Spaniards, offer the only exceptions to the general scarcity of wood. Among the introduced kinds may be enumerated poplars, olives, peach, and other fruit trees: the peaches succeed so well, that they afford the main supply of firewood to the city of Buenos Ayres. Extremely level countries, such as the Pampas, seldom appear favourable to the growth of trees. This may possibly be attributed either to the force of the winds, or the kind of drainage. In the nature of the land, however, around Maldonado, no such reason is apparent; the rocky mountains afford protected situations, enjoying various kinds of soil; streamlets of water are common at the bottoms of nearly every valley; and the clayey nature of the earth seems adapted to retain moisture. It has been inferred with much probability, that the presence of woodland is generally determined[1] by the annual amount of moisture; yet in this province abundant and heavy rain falls during the winter; and the summer, though dry, is not so in any excessive degree.[2] We see nearly the whole of Australia covered by lofty trees, yet that country possesses a far more arid climate. Hence we must look to some other and unknown cause.

Confining our view to South America, we should certainly be tempted to believe that trees flourished only under a very humid climate; for the limit of the forest-land follows, in a most remarkable manner, that of the damp winds. In the southern part of the continent, where the western gales, charged with moisture from the Pacific, prevail, every island on the broken west coast, from lat. 38° to the extreme point of Tierra del Fuego, is densely covered by impenetrable forests. On the eastern side of the Cordillera, over the same extent of latitude, where a blue sky and a fine climate prove that the atmosphere has been deprived of its moisture by passing over the mountains, the arid plains of Patagonia support a most scanty vegetation. In the more northern parts of the continent, within the limits of the constant south-eastern trade-wind, the eastern side is ornamented by magnificent forests; whilst the western coast, from lat. 4° S. to lat. 32° S., may be described as a desert: on this western coast, northward of lat. 4° S., where the trade-wind loses its regularity, and heavy torrents of rain fall periodically, the shores of the Pacific, so utterly desert in Peru, assume near Cape Blanco the character of luxuriance so celebrated at Guyaquil and Panama. Hence in the southern and northern parts of the continent, the forest and desert lands occupy reversed positions with respect to the Cordillera, and these positions are apparently determined by the direction of the prevalent winds. In the middle of the continent there is a broad intermediate band, including central Chile and the provinces of La Plata, where the rain-bringing winds have not to pass over lofty mountains, and where the land is neither a desert nor covered by forests. But even the rule, if confined to South America, of trees flourishing only in a

1 Maclaren, art. 'America,' Encyclop. Britann.

2 Azara says, 'Je crois que la quantité annuelle des pluies est, dans toutes ces contrées, plus considérable qu'en Espagne.' – Vol. i. p. 36.

climate rendered humid by rain-bearing winds, has a strongly marked exception in the case of the Falkland Islands. These islands, situated in the same latitude with Tierra del Fuego and only between two and three hundred miles distant from it, having a nearly similar climate, with a geological formation almost identical, with favourable situations and the same kind of peaty soil, yet can boast of few plants deserving even the title of bushes; whilst in Tierra del Fuego it is impossible to find an acre of land not covered by the densest forest. In this case, both the direction of the heavy gales of wind and of the currents of the sea are favourable to the transport of seeds from Tierra del Fuego, as is shown by the canoes and trunks of trees drifted from that country, and frequently thrown on the shores of the Western Falkland. Hence perhaps it is, that there are many plants in common to the two countries: but with respect to the trees of Tierra del Fuego, even attempts made to transplant them have failed.

During our stay at Maldonado I collected several quadrupeds, eighty kinds of birds, and many reptiles, including nine species of snakes. Of the indigenous mammalia, the only one now left of any size, which is common, is the Cervus campestris. This deer is exceedingly abundant, often in small herds, throughout the countries bordering the Plata and in Northern Patagonia. If a person crawling close along the ground, slowly advances towards a herd, the deer frequently, out of curiosity, approach to reconnoitre him. I have by this means killed, from one spot, three out of the same herd. Although so tame and inquisitive, yet when approached on horseback, they are exceedingly wary. In this country nobody goes on foot, and the deer knows man as its enemy only when he is mounted and armed with the bolas. At Bahia Blanca, a recent establishment in Northern Patagonia, I was surprised to find how little the deer cared for the noise of a gun: one day I fired ten times from within eighty yards at one animal; and it was much more startled at the ball cutting up the ground than at the report of the rifle. My powder being exhausted, I was obliged to get up (to my shame as a sportsman be it spoken, though well able to kill birds on the wing) and halloo till the deer ran away.

The most curious fact with respect to this animal, is the overpoweringly strong and offensive odour which proceeds from the buck. It is quite indescribable: several times whilst skinning the specimen which is now mounted at the Zoological Museum, I was almost overcome by nausea. I tied up the skin in a silk pocket-handkerchief, and so carried it home: this handkerchief, after being well washed, I continually used, and it was of course as repeatedly washed; yet every time, for a space of one year and seven months, when first unfolded, I distinctly perceived the odour. This appears an astonishing instance of the permanence of some matter, which nevertheless in its rature must be most subtile and volatile. Frequently, when passing at the distance of half a mile to leeward of a herd, I have perceived the whole air tainted with the effluvium. I believe the smell from the buck is most powerful at the period when its horns are perfect, or free from the hairy skin. When in this state the meat is, of course, quite uneatable; but the Gauchos assert, that if buried for some time in fresh earth, the taint is removed. I have somewhere read that the islanders in the north of Scotland treat the rank carcasses of the fish-eating birds in the same manner.

The order Rodentia is here very numerous in species: of mice alone I obtained no less than eight kinds.[1] The largest gnawing animal in the world, the Hydrochærus capybara (the water-hog), is here also common. One which I shot at Monte Video weighed ninety-eight pounds: its length, from the end of the snout to the stump-like tail, was three feet two inches; and its girth three feet eight. These great Rodents occasionally frequent the islands in the mouth of the Plata, where the water is quite salt, but are far more abundant on the borders of fresh-water lakes and rivers. Near Maldonado three or four generally live together. In the daytime they either lie among the aquatic plants, or openly feed on the turf plain.[2] When viewed at a distance,

1 In South America I collected altogether twenty-seven species of mice; and thirteen more are known from the works of Azara and other authors. Those collected by myself have been named and described by Mr. Waterhouse at the meetings of the Zoological Society. I must be allowed to take this opportunity of returning my cordial thanks to Mr. Waterhouse, and to the other gentlemen attached to that Society, for their kind and most liberal assistance on all occasions.

2 In the stomach and duodenum of a capybara which I opened, I found a very large quantity of a thin yellowish fluid, in which scarcely a fibre could be distinguished. Mr. Owen informs me that a part of the œsophagus is so constructed that nothing much larger than a crowquill can be passed down. Certainly the broad teeth and strong jaws of this animal are well fitted to grind into pulp the aquatic plants on which it feeds.

from their manner of walking and colour they resemble pigs: but when seated on their haunches, and attentively watching any object with one eye, they re-assume the appearance of their congeners, cavies and rabbits. Both the front and side view of their head has quite a ludicrous aspect, from the great depth of their jaw. These animals, at Maldonado, were very tame; by cautiously walking, I approached within three yards of four old ones. This tameness may probably be accounted for, by the Jaguar having been banished for some years, and by the Gaucho not thinking it worth his while to hunt them. As I approached nearer and nearer they frequently made their peculiar noise, which is a low abrupt grunt, not having much actual sound, but rather arising from the sudden expulsion of air: the only noise I know at all like it, is the first hoarse bark of a large dog. Having watched the four from almost within arm's length (and they me) for several minutes, they rushed into the water at full gallop with the greatest impetuosity, and emitted at the same time their bark. After diving a short distance they came again to the surface, but only just showed the upper part of their heads. When the female is swimming in the water, and has young ones, they are said to sit on her back. These animals are easily killed in numbers; but their skins are of trifling value, and the meat is very indifferent. On the islands in the Rio Parana they are exceedingly abundant, and afford the ordinary prey to the Jaguar.

The Tucutuco (Ctenomys Brasiliensis) is a curious small animal, which may be briefly described as a Gnawer, with the habits of a mole. It is extremely numerous in some parts of the country, but is difficult to be procured, and never, I believe, comes out of the ground. It throws up at the mouth of its burrows hillocks of earth like those of the mole, but smaller. Considerable tracts of country are so completely undermined by these animals, that horses in passing over, sink above their fetlocks. The tucutucos appear, to a certain degree, to be gregarious: the man who procured the specimens for me had caught six together, and he said this was a common occurrence. They are nocturnal in their habits; and their principal food is the roots of plants, which are the object of their extensive and superficial burrows. This animal is universally known by a very peculiar noise which it makes when beneath the ground. A person, the first time he hears it, is much surprised; for it is not easy to tell whence it comes, nor is it possible to guess what kind of creature utters it. The noise consists in a short, but not rough, nasal grunt, which is monotonously repeated about four times in quick succession;[1] the name Tucutuco is given in imitation of the sound. Where this animal is abundant, it may be heard at all times of the day, and sometimes directly beneath one's feet. When kept in a room, the tucutucos move both slowly and clumsily, which appears owing to the outward action of their hind legs; and they are quite incapable, from the socket of the thigh-bone not having a certain ligament, of jumping even the smallest vertical height. They are very stupid in making any attempt to escape; when angry or frightened they utter the tucu-tuco. Of those I kept alive several, even the first day, became quite tame, not attempting to bite or to run away; others were a little wilder.

The man who caught them asserted that very many are invariably found blind. A specimen which I preserved in spirits was in this state; Mr. Reid considers it to be the effect of inflammation in the nictitating membrane. When the animal was alive I placed my finger within half an inch of its head, and not the slightest notice was taken: it made its way, however, about the room nearly as well as the others. Considering the strictly subterranean habits of the tucutuco, the blindness, though so common, cannot be a very serious evil; yet it appears strange that any animal should possess an organ frequently subject to be injured. Lamarck would have been delighted with this fact, had he known it, when speculating[2] (probably with more truth than usual with him) on the gradually-*acquired* blindness of the Aspalax, a Gnawer living under ground, and of the Proteus, a reptile living in dark caverns filled with water; in both of which animals the eye is in an almost rudimentary state, and is covered by a tendinous membrane

1 At the R. Negro, in Northern Patagonia, there is an animal of the same habits, and probably a closely allied species, but which I never saw. Its noise is different from that of the Maldonado kind; it is repeated only twice instead of three or four times, and is more distinct and sonorous: when heard from a distance it so closely resembles the sound made in cutting down a small tree with an axe, that I have sometimes remained in doubt concerning it.

2 Philosoph. Zoolog., tom. i. p. 242.

and skin. In the common mole the eye is extraordinarily small but perfect, though many anatomists doubt whether it is connected with the true optic nerve; its vision must certainly be imperfect, though probably useful to the animal when it leaves its burrow. In the tucutuco, which I believe never comes to the surface of the ground, the eye is rather larger, but often rendered blind and useless, though without apparently causing any inconvenience to the animal: no doubt Lamarck would have said that the tucutuco is now passing into the state of the Aspalax and Proteus.

Birds of many kinds are extremely abundant on the undulating grassy plains around Maldonado. There are several species of a family allied in structure and manners to our Starling: one of these (Molothrus niger) is remarkable from its habits. Several may often be seen standing together on the back of a cow or horse; and while perched on a hedge, pluming themselves in the sun, they sometimes attempt to sing, or rather to hiss; the noise being very peculiar, resembling that of bubbles of air passing rapidly from a small orifice under water, so as to produce an acute sound. According to Azara, this bird, like the cuckoo, deposits its eggs in other birds' nests. I was several times told by the country people, that there certainly is some bird having this habit; and my assistant in collecting, who is a very accurate person, found a nest of the sparrow of this country (Zonotrichia matutina), with one egg in it larger than the others, and of a different colour and shape. In North America there is another species of Molothrus (M. pecoris), which has a similar cuckoo-like habit, and which is most closely allied in every respect to the species from the Plata, even in such trifling peculiarities as standing on the backs of cattle; it differs only in being a little smaller, and in its plumage and eggs being of a slightly different shade of colour. This close agreement in structure and habits, in representative species coming from opposite quarters of a great continent, always strikes one as interesting, though of common occurrence.

Mr. Swainson has well remarked,[1] that with the exception of the Molothrus pecoris, to which must be added the M. niger, the cuckoos are the only birds which can be called truly parasitical; namely, such as

1 Magazine of Zoology and Botany, vol. i. p. 217.

'fasten themselves, as it were, on another living animal, whose animal heat brings their young into life, whose food they live upon, and whose death would cause theirs during the period of infancy.' It is remarkable that some of the species, but not all, both of the Cuckoo and Molothrus, should agree in this one strange habit of their parasitical propagation, whilst opposed to each other in almost every other habit: the molothrus, like our starling, is eminently sociable, and lives on the open plains without art or disguise: the cuckoo, as every one knows, is a singularly shy bird; it frequents the most retired thickets, and feeds on fruit and caterpillars. In structure also these two genera are widely removed from each other. Many theories, even phrenological theories, have been advanced to explain the origin of the cuckoo laying its eggs in other birds' nests. M. Prévost alone, I think, has thrown light by his observations[1] on this puzzle: he finds that the female cuckoo, which, according to most observers, lays at least from four to six eggs, must pair with the male each time after laying only one or two eggs. Now, if the cuckoo was obliged to sit on her own eggs, she would either have to sit on all together, and therefore leave those first laid so long, that they probably would become addled; or she would have to hatch separately each egg or two eggs, as soon as laid: but as the cuckoo stays a shorter time in this country than any other migratory bird, she certainly would not have time enough for the successive hatchings. Hence we can perceive in the fact of the cuckoo pairing several times, and laying her eggs at intervals, the cause of her depositing her eggs in other birds' nests, and leaving them to the care of foster-parents. I am strongly inclined to believe that this view is correct, from having been independently led (as we shall hereafter see) to an analogous conclusion with regard to the South American ostrich, the females of which are parasitical, if I may so express it, on each other; each female laying several eggs in the nests of several other females, and the male ostrich undertaking all the cares of incubation, like the strange foster-parents with the cuckoo.

I will mention only two other birds, which are very common, and render themselves prominent from their habits. The Saurophagus

---

1 Read before the Academy of Science in Paris. L'Institut, 1834, p. 418.

sulphuratus is typical of the great American tribe of tyrant-flycatchers. In its structure it closely approaches the true shrikes, but in its habits may be compared to many birds. I have frequently observed it, hunting a field, hovering over one spot like a hawk, and then proceeding on to another. When seen thus suspended in the air, it might very readily at a short distance be mistaken for one of the Rapacious order; its stoop, however, is very inferior in force and rapidity to that of a hawk. At other times the Saurophagus haunts the neighbourhood of water, and there, like a kingfisher, remaining stationary, it catches any small fish which may come near the margin. These birds are not unfrequently kept either in cages or in courtyards, with their wings cut. They soon become tame, and are very amusing from their cunning odd manners, which were described to me as being similar to those of the common magpie. Their flight is undulatory, for the weight of the head and bill appear too great for the body. In the evening the Saurophagus takes its stand on a bush, often by the road-side, and continually repeats without change a shrill and rather agreeable cry, which somewhat resembles articulate words: the Spaniards say it is like the words 'Bien te veo' (I see you well), and accordingly have given it this name.

A mocking-bird (Mimus orpheus), called by the inhabitants Calandria, is remarkable, from possessing a song far superior to that of any other bird in the country: indeed, it is nearly the only bird in South America which I have observed to take its stand for the purpose of singing. The song may be compared to that of the Sedge warbler, but is more powerful; some harsh notes and some very high ones, being mingled with a pleasant warbling. It is heard only during the spring. At other times its cry is harsh and far from harmonious. Near Maldonado these birds were tame and bold; they constantly attended the country houses in numbers, to pick the meat which was hung up on the posts or walls: if any other small bird joined the feast, the Calandria soon chased it away. On the wide uninhabited plains of Patagonia another closely allied species, O. Patagonica of d'Orbigny, which frequents the valleys clothed with spiny bushes, is a wilder bird, and has a slightly different tone of voice. It appears to me a curious circumstance, as showing the fine shades of difference in habits, that judging from this latter respect alone, when I first saw this second species, I thought it was different from the Maldonado kind. Having afterwards procured a specimen, and comparing the two without particular care, they appeared so very similar, that I changed my opinion; but now Mr. Gould says that they are certainly distinct; a conclusion in conformity with the trifling difference of habit, of which, however, he was not aware.

The number, tameness, and disgusting habits of the carrion-feeding hawks of South America make them pre-eminently striking to any one accustomed only to the birds of Northern Europe. In this list may be included four species of the Caracara or Polyborus, the Turkey buzzard, the Gallinazo, and the Condor. The Caracaras are, from their structure, placed among the eagles: we shall soon see how ill they become so high a rank. In their habits they well supply the place of our carrion-crows, magpies, and ravens; a tribe of birds widely distributed over the rest of the world, but entirely absent in South America. To begin with the Polyborus Brasiliensis: this is a common bird, and has a wide geographical range; it is most numerous on the grassy savannahs of La Plata (where it goes by the name of Carrancha), and is far from unfrequent throughout the sterile plains of Patagonia. In the desert between the rivers Negro and Colorado, numbers constantly attend the line of road to devour the carcasses of the exhausted animals which chance to perish from fatigue and thirst. Although thus common in these dry and open countries, and likewise on the arid shores of the Pacific, it is nevertheless found inhabiting the damp impervious forests of West Patagonia and Tierra del Fuego. The Carranchas, together with the Chimango, constantly attend in numbers the estancias and slaughtering-houses. If an animal dies on the plain the Gallinazo commences the feast, and then the two species of Polyborus pick the bones clean. These birds, although thus commonly feeding together, are far from being friends. When the Carrancha is quietly seated on the branch of a tree or on the ground, the Chimango often continues for a long time flying backwards and forwards, up and down, in a semicircle, trying each time at the bottom of the curve to strike its larger relative. The Carrancha takes little notice, except by bobbing its head. Although the Carranchas frequently assemble in numbers, they are not gregarious; for in desert places they may be seen solitary, or more commonly by pairs.

The Carranchas are said to be very crafty, and to steal great numbers of eggs. They attempt, also, together with the Chimango, to pick off the scabs from the sore backs of horses and mules. The poor animal, on the one hand, with its ears down and its back arched; and, on the other, the hovering bird, eyeing at the distance of a yard, the disgusting morsel, form a picture, which has been described by Captain Head with his own peculiar spirit and accuracy. These false eagles most rarely kill any living bird or animal; and their vulture-like, necrophagous habits are very evident to any one, who has fallen asleep on the desolate plains of Patagonia, for when he wakes, he will see, on each surrounding hillock, one of these birds patiently watching him with an evil eye: it is a feature in the landscape of these countries, which will be recognized by every one who has wandered over them. If a party of men go out hunting with dogs and horses, they will be accompanied, during the day, by several of these attendants. After feeding, the uncovered craw protrudes; at such times, and indeed generally, the Carrancha is an inactive, tame, and cowardly bird. Its flight is heavy and slow, like that of an English rook. It seldom soars; but I have twice seen one at a great height gliding through the air with much ease. It runs (in contradistinction to hopping), but not quite so quickly as some of its congeners. At times the Carrancha is noisy, but is not generally so: its cry is loud, very harsh and peculiar, and may be likened to the sound of the Spanish guttural *g*, followed by a rough double *r r*; when uttering this cry it elevates its head higher and higher, till at last, with its beak wide open, the crown almost touches the lower part of the back. This fact, which has been doubted, is quite true; I have seen them several times with their heads backwards in a completely inverted position. To these observations I may add, on the high authority of Azara, that the Carrancha feeds on worms, shells, slugs, grasshoppers, and frogs; that it destroys young lambs by tearing the umbilical cord; and that it pursues the Gallinazo, till that bird is compelled to vomit up the carrion it may have recently gorged. Lastly, Azara states that several Carranchas, five or six together, will unite in chace of large birds, even such as herons. All these facts show that it is a bird of very versatile habits and considerable ingenuity.

The Polyborus Chimango is considerably smaller than the last species. It is truly omnivorous, and will eat even bread; and I was assured that it materially injures the potato-crops in Chiloe, by stocking up the roots when first planted. Of all the carrion-feeders it is generally the last which leaves the skeleton of a dead animal; and may often be seen within the ribs of a cow or horse, like a bird in a cage. Another species is the Polyborus Novæ Zelandiæ, which is exceedingly common in the Falkland Islands. These birds in many respects resemble in their habits the Carranchas. They live on the flesh of dead animals and on marine productions; and on the Ramirez rocks their whole sustenance must depend on the sea. They are extraordinarily tame and fearless, and haunt the neighbourhood of houses for offal. If a hunting party kills an animal, a number soon collect and patiently await, standing on the ground on all sides. After eating, their uncovered craws are largely protruded, giving them a disgusting appearance. They readily attack wounded birds: a cormorant in this state having taken to the shore, was immediately seized on by several, and its death hastened by their blows. The *Beagle* was at the Falklands only during the summer, but the officers of the *Adventure*, who were there in the winter, mention many extraordinary instances of the boldness and rapacity of these birds. They actually pounced on a dog that was lying fast asleep close by one of the party; and the sportsmen had difficulty in preventing the wounded geese from being seized before their eyes. It is said that several together (in this respect resembling the Carranchas) wait at the mouth of a rabbit-hole, and together seize on the animal when it comes out. They were constantly flying on board the vessel when in the harbour; and it was necessary to keep a good look-out to prevent the leather being torn from the rigging, and the meat or game from the stern. These birds are very mischievous and inquisitive; they will pick up almost anything from the ground; a large black glazed hat was carried nearly a mile, as was a pair of the heavy balls used in catching cattle. Mr. Usborne experienced during the survey a more severe loss, in their stealing a small Kater's compass in a red morocco leather case, which was never recovered. These birds are, moreover, quarrelsome and very passionate; tearing up the grass with their bills from rage. They are not truly gregarious; they do not soar, and their flight is heavy and clumsy; on the ground they run extremely fast, very much like pheasants. They are noisy, uttering several harsh cries; one of which is like that of the English rook; hence the sealers always call them rooks. It is a curious circumstance that, when crying out, they throw their heads upwards and backwards, after the same manner as the Carrancha. They build in the rocky cliffs of the sea-coast, but only on the small adjoining islets, and not on the two main islands: this is a singular precaution in so tame and fearless a bird. The sealers say that the flesh of these birds, when cooked, is quite white, and very good eating; but bold must the man be who attempts such a meal.

We have now only to mention the turkey-buzzard (Vultur aura), and the Gallinazo. The former is found wherever the country is moderately damp, from Cape Horn to North America. Differently from the Polyborus Brasiliensis and Chimango, it has found its way to the Falkland Islands. The turkey-buzzard is a solitary bird, or at most goes in pairs. It may at once be recognized from a long distance, by its lofty, soaring and most elegant flight. It is well known to be a true carrion-feeder. On the west coast of Patagonia, among the thickly-wooded islets and broken land, it lives exclusively on what the sea throws up, and on the carcasses of dead seals. Wherever these animals are congregated on the rocks, there the vultures may be seen. The Gallinazo (Cathartes atratus) has a different range from the last species, as it never occurs southward of lat. 41°. Azara states that there exists a tradition that these birds, at the time of the conquest, were not found near Monte Video, but that they subsequently followed the inhabitants from more northern districts. At the present day they are numerous in the valley of the Colorado, which is three hundred miles due south of Monte Video. It seems probable that this additional migration has happened since the time of Azara. The Gallinazo generally prefers a humid climate, or rather the neighbourhood of fresh water; hence it is extremely abundant in Brazil and La Plata, while it is never found on the desert and arid plains of Northern Patagonia, excepting near some stream. These birds frequent the whole Pampas to the foot of the Cordillera, but I never saw or heard of one in Chile: in Peru they are preserved as scavengers. These vultures certainly may be called gregarious, for they seem to have pleasure in society, and are not solely brought together by the attraction of a common prey. On a fine day a flock may often be observed at a great height, each bird wheeling round and round without closing its wings, in the most graceful evolutions. This is clearly performed for the mere pleasure of the exercise, or perhaps is connected with their matrimonial alliances.

I have now mentioned all the carrion-feeders, excepting the condor, an account of which will be more appropriately introduced when we visit a country more congenial to its habits than the plains of La Plata.

In a broad band of sand-hillocks which separate the Laguna del Potrero from the shores of the Plata, at the distance of a few miles from Maldonado, I found a group of those vitrified, siliceous tubes, which are formed by lightning entering loose sand. These tubes resemble in every particular those from Drigg in Cumberland, described in the Geological Transactions.[1] The sand-hillocks of Maldonado, not being protected by vegetation, are constantly changing their position. From this cause the tubes projected above the surface; and numerous fragments lying near, showed that they had formerly been buried to a greater depth. Four sets entered the sand perpendicularly: by working with my hands I traced one of them two feet deep; and some fragments which evidently had belonged to the same tube, when added to the other part, measured five feet three inches. The diameter of the whole tube was nearly equal, and therefore we must suppose that originally it extended to a much greater depth. These dimensions are however small, compared to those of the tubes from Drigg, one of which was traced to a depth of not less than thirty feet.

The internal surface is completely vitrified, glossy, and smooth. A small fragment examined under the microscope appeared, from the number of minute entangled air or perhaps steam bubbles, like an assay fused before the blowpipe. The sand is entirely, or in greater part, siliceous; but some points are of a black colour, and from their glossy surface possess a metallic lustre. The thickness of the wall of the tube varies from the thirtieth to a twentieth of an inch, and

1 Geolog. Transact., vol. ii. p. 528. In the Philosoph. Transact. (1790, p. 204) Dr. Priestley has described some imperfect siliceous tubes and a melted pebble of quartz, found in digging into the ground, under a tree, where a man had been killed by lightning.

occasionally even equals a tenth. On the outside the grains of sand are rounded, and have a slightly glazed appearance: I could not distinguish any signs of crystallization. In a similar manner to that described in the Geological Transactions, the tubes are generally compressed, and have deep longitudinal furrows, so as closely to resemble a shrivelled vegetable stalk, or the bark of the elm or cork tree. Their circumference is about two inches, but in some fragments, which are cylindrical and without any furrows, it is as much as four inches. The compression from the surrounding loose sand, acting while the tube was still softened from the effects of the intense heat, has evidently caused the creases or furrows. Judging from the uncompressed fragments, the measure or bore of the lightning (if such a term may be used), must have been about one inch and a quarter. At Paris, M. Hachette and M. Beudant[1] succeeded in making tubes, in most respects similar to these fulgurites, by passing very strong shocks of galvanism through finely-powdered glass: when salt was added, so as to increase its fusibility, the tubes were larger in every dimension. They failed both with powdered felspar and quartz. One tube, formed with pounded glass, was very nearly an inch long, namely .982, and had an internal diameter of .019 of an inch. When we hear that the strongest battery in Paris was used, and that its power on a substance of such easy fusibility as glass was to form tubes so diminutive, we must feel greatly astonished at the force of a shock of lightning which, striking the sand in several places, has formed cylinders, in one instance of at least thirty feet long, and having an internal bore, where not compressed, of full an inch and a half; and this in a material so extraordinarily refractory as quartz!

The tubes, as I have already remarked, enter the sand nearly in a vertical direction. One, however, which was less regular than the others, deviated from a right line, at the most considerable bend, to the amount of thirty-three degrees. From this same tube, two small branches, about a foot apart, were sent off; one pointed downwards, and the other upwards. This latter case is remarkable, as the electric fluid must have turned back at the acute angle of 26°, to the line of its

1 Annales de Chimie et de Physique. tom. xxxvii. p. 319.

main course. Besides the four tubes which I found vertical, and traced beneath the surface, there were several other groups of fragments, the original sites of which without doubt were near. All occurred in a level area of shifting sand, sixty yards by twenty, situated among some high sand-hillocks, and at the distance of about half a mile from a chain of hills four or five hundred feet in height. The most remarkable circumstance, as it appears to me, in this case as well as in that of Drigg, and in one described by M. Ribbentrop in Germany, is the number of tubes found within such limited spaces. At Drigg, within an area of fifteen yards, three were observed, and the same number occurred in Germany. In the case which I have described, certainly more than four existed within the space of the sixty by twenty yards. As it does not appear probable that the tubes are produced by successive distinct shocks, we must believe that the lightning, shortly before entering the ground, divides itself into separate branches.

The neighbourhood of the Rio Plata seems peculiarly subject to electric phenomena. In the year 1793,[1] one of the most destructive thunderstorms perhaps on record happened at Buenos Ayres: thirty-seven places within the city were struck by lightning, and nineteen people killed. From facts stated in several books of travels, I am inclined to suspect that thunderstorms are very common near the mouths of great rivers. Is it not possible that the mixture of large bodies of fresh and salt water may disturb the electrical equilibrium? Even during our occasional visits to this part of South America, we heard of a ship, two churches, and a house, having been struck. Both the church and the house I saw shortly afterwards: the house belonged to Mr. Hood, the consul-general at Monte Video. Some of the effects were curious: the paper, for nearly a foot on each side of the line where the bell-wires had run, was blackened. The metal had been fused, and although the room was about fifteen feet high, the globules, dropping on the chairs and furniture, had drilled in them a chain of minute holes. A part of the wall was shattered as if by gunpowder, and the fragments had been blown off with force sufficient to dent the wall on the opposite side of the room. The frame of a

1　Azara's Voyage, vol. i. p. 36.

looking-glass was blackened, and the gilding must have been volatilized, for a smelling-bottle, which stood on the chimney-piece, was coated with bright metallic particles, which adhered as firmly as if they had been enamelled.

CHAPTER IV

RIO NEGRO TO
BAHIA BLANCA

Rio Negro – Estancias attacked by the Indians – Salt Lakes
– Flamingoes – R. Negro to R. Colorado – Sacred Tree –
Patagonian Hare – Indian Families – General Rosas –
Proceed to Bahia Blanca – Sand Dunes – Negro Lieutenant
– Bahia Blanca – Saline Incrustations – Punta Alta – Zorillo.

JULY 24th, 1833. – The *Beagle* sailed from Maldonado, and on August
the 3rd she arrived off the mouth of the Rio Negro. This is the
principal river on the whole line of coast between the Strait of
Magellan and the Plata. It enters the sea about three hundred miles
south of the estuary of the Plata. About fifty years ago, under the old
Spanish government, a small colony was established here; and it is still
the most southern position (lat. 41°) on this eastern coast of America,
inhabited by civilized man.

The country near the mouth of the river is wretched in the
extreme: on the south side a long line of perpendicular cliffs commences, which exposes a section of the geological nature of
the country. The strata are of sandstone, and one layer was remarkable
from being composed of a firmly-cemented conglomerate of pumice
pebbles, which must have travelled more than four hundred miles,
from the Andes. The surface is everywhere covered up by a thick bed
of gravel, which extends far and wide over the open plain. Water is
extremely scarce, and, where found, is almost invariably brackish.
The vegetation is scanty; and although there are bushes of many
kinds, all are armed with formidable thorns, which seem to warn
the stranger not to enter on these inhospitable regions.

The settlement is situated eighteen miles up the river. The road
follows the foot of the sloping cliff, which forms the northern
boundary of the great valley, in which the Rio Negro flows.
On the way we passed the ruins of some fine 'estancias,' which a
few years since had been destroyed by the Indians. They withstood

several attacks. A man present at one gave me a very lively description of what took place. The inhabitants had sufficient notice to drive all the cattle and horses into the 'corral'[1] which surrounded the house, and likewise to mount some small cannon. The Indians were Araucanians from the south of Chile; several hundreds in number, and highly disciplined. They first appeared in two bodies on a neighbouring hill; having there dismounted, and taken off their fur mantles, they advanced naked to the charge. The only weapon of an Indian is a very long bamboo or chuzo, ornamented with ostrich feathers, and pointed by a sharp spear-head. My informer seemed to remember with the greatest horror the quivering of these chuzos as they approached near. When close, the cacique Pincheira hailed the besieged to give up their arms, or he would cut all their throats. As this would probably have been the result of their entrance under any circumstances, the answer was given by a volley of musketry. The Indians, with great steadiness, came to the very fence of the corral: but to their surprise they found the posts fastened together by iron nails instead of leather thongs, and, of course, in vain attempted to cut them with their knives. This saved the lives of the Christians: many of the wounded Indians were carried away by their companions; and at last one of the under caciques being wounded, the bugle sounded a retreat. They retired to their horses, and seemed to hold a council of war. This was an awful pause for the Spaniards, as all their ammunition, with the exception of a few cartridges, was expended. In an instant the Indians mounted their horses, and galloped out of sight. Another attack was still more quickly repulsed. A cool Frenchman managed the gun; he stopped till the Indians approached close, and then raked their line with grape-shot: he thus laid thirty-nine of them on the ground; and, of course, such a blow immediately routed the whole party.

The town is indifferently called El Carmen or Patagones. It is built on the face of a cliff which fronts the river, and many of the houses are excavated even in the sandstone. The river is about two or three hundred yards wide, and is deep and rapid. The many islands, with

1 The corral is an enclosure made of tall and strong stakes. Every estancia, or farming estate, has one attached to it.

their willow-trees, and the flat headlands, seen one behind the other on the northern boundary of the broad green valley, forms, by the aid of the bright sun, a view almost picturesque. The number of the inhabitants does not exceed a few hundreds. These Spanish colonies do not, like our British ones, carry within themselves the elements of growth. Many Indians of pure blood reside here: the tribe of the Cacique Lucanee constantly have their Toldos[1] on the outskirts of the town. The local government party supplies them with provisions, by giving them all the old worn-out horses, and they earn a little by making horse-rugs and other articles of riding-gear. These Indians are considered civilized; but what their character may have gained by a lesser degree of ferocity, is almost counter-balanced by their entire immorality. Some of the younger men are, however, improving; they are willing to labour, and a short time since a party went on a sealing-voyage, and behaved very well. They were now enjoying the fruits of their labour, by being dressed in very gay, clean clothes, and by being very idle. The taste they showed in their dress was admirable; if you could have turned one of these young Indians into a statue of bronze, his drapery would have been perfectly graceful.

One day I rode to a large salt-lake, or Salina, which is distant fifteen miles from the town. During the winter it consists of a shallow lake of brine, which in summer is converted into a field of snow-white salt. The layer near the margin is from four to five inches thick, but towards the centre its thickness increases. This lake was two and a half miles long, and one broad. Others occur in the neighbourhood many times larger, and with a floor of salt, two and three feet in thickness, even when under water during the winter. One of these brilliantly-white and level expanses, in the midst of the brown and desolate plain, offers an extraordinary spectacle. A large quantity of salt is annually drawn from the salina; and great piles, some hundred tons in weight, were lying ready for exportation. The season for working the salinas forms the harvest of Patagones; for on it, the prosperity of the place depends. Nearly the whole population encamps on the bank of the river, and the people are employed in drawing out the salt in bullock-waggons. This salt is crystallized

1 The hovels of the Indians are thus called.

in great cubes, and is remarkably pure: Mr. Trenham Reeks has kindly analyzed some for me, and he finds in it only 0.26 of gypsum and 0.22 of earthy matter. It is a singular fact, that it does not serve so well for preserving meat as sea-salt from the Cape de Verd islands; and a merchant at Buenos Ayres told me that he considered it as fifty per cent. less valuable. Hence the Cape de Verd salt is constantly imported, and is mixed with that from these salinas. The purity of the Patagonian salt, or absence from it of those other saline bodies found in all sea-water, is the only assignable cause for this inferiority: a conclusion which no one, I think, would have suspected, but which is supported by the fact lately ascertained,[1] that those salts answer best for preserving cheese which contain most of the deliquescent chlorides.

The border of the lake is formed of mud: and in this numerous large crystals of gypsum, some of which are three inches long, lie embedded; whilst on the surface others of sulphate of soda lie scattered about. The Gauchos call the former the 'Padre del sal,' and the latter the 'Madre;' they state that these progenitive salts always occur on the borders of the salinas, when the water begins to evaporate. The mud is black, and has a fetid odour. I could not at first imagine the cause of this, but I afterwards perceived that the froth which the wind drifted on shore was coloured green, as if by confervæ: I attempted to carry home some of this green matter, but from an accident failed. Parts of the lake seen from a short distance appeared of a reddish colour, and this perhaps was owing to some infusorial animalcula. The mud in many places was thrown up by numbers of some kind of worm, or annelidous animal. How surprising it is that any creatures should be able to exist in brine, and that they should be crawling among crystals of sulphate of soda and lime! And what becomes of these worms when, during the long summer, the surface is hardened into a solid layer of salt? Flamingoes in considerable numbers inhabit this lake, and breed here; throughout Patagonia, in Northern Chile, and at the Galapagos Islands, I met with these birds wherever there were lakes of brine. I saw them here wading about in search of food – probably for the worms which burrow in the mud;

1 Report of the Agricult. Chem. Assoc. in the Agricult. Gazette, 1845, p. 93.

and these latter probably feed on infusoria or confervæ. Thus we have a little living world within itself, adapted to these inland lakes of brine. A minute crustaceous animal (Cancer salinus) is said[1] to live in countless numbers in the brine-pans at Lymington; but only in those in which the fluid has attained, from evaporation, considerable strength – namely, about a quarter of a pound of salt to a pint of water. Well may we affirm, that every part of the world is habitable! Whether lakes of brine, or those subterranean ones hidden beneath volcanic mountains – warm mineral springs – the wide expanse and depths of the ocean – the upper regions of the atmosphere, and even the surface of perpetual snow – all support organic beings.

To the northward of the Rio Negro, between it and the inhabited country near Buenos Ayres, the Spaniards have only one small settle-ment, recently established at Bahia Blanca. The distance in a straight line to Buenos Ayres is very nearly five hundred British miles. The wandering tribes of horse Indians, which have always occupied the greater part of this country, having of late much harassed the outlying estancias, the government at Buenos Ayres equipped some time since an army under the command of General Rosas for the purpose of exterminating them. The troops were now encamped on the banks of the Colorado; a river lying about eighty miles northward of the Rio Negro. When General Rosas left Buenos Ayres he struck in a direct line across the unexplored plains: and as the country was thus pretty well cleared of Indians, he left behind him, at wide intervals, a small party of soldiers with a troop of horses (*a posta*), so as to be enabled to keep up a communication with the

1 Linnæan Trans., vol. xi. p. 205. It is remarkable how all the circumstances connected with the salt-lakes in Siberia and Patagonia are similar. Siberia, like Pata-gonia, appears to have been recently elevated above the waters of the sea. In both countries the salt-lakes occupy shallow depressions in the plains; in both the mud on the borders is black and fetid; beneath the crust of common salt, sulphate of soda or of magnesia occurs, imperfectly crystallized; and in both, the muddy sand is mixed with lentils of gypsum. The Siberian salt-lakes are inhabited by small crustaceous animals; and flamingoes (Edin. New Philos. Jour., Jan. 1830) likewise frequent them. As these circumstances, apparently so trifling, occur in two distant continents, we may feel sure that they are the necessary results of common causes. See *Pallas's Travels*, 1793 to 1794, pp. 129–134.

capital. As the *Beagle* intended to call at Bahia Blanca, I determined to proceed there by land; and ultimately I extended my plan to travel the whole way by the postas to Buenos Ayres.

August 11th. – Mr. Harris, an Englishman residing at Patagones, a guide, and five Gauchos, who were proceeding to the army on business, were my companions on the journey. The Colorado, as I have already said, is nearly eighty miles distant: and as we travelled slowly, we were two days and a half on the road. The whole line of country deserves scarcely a better name than that of a desert. Water is found only in two small wells; it is called fresh; but even at this time of the year, during the rainy season, it was quite brackish. In the summer this must be a distressing passage; for now it was sufficiently desolate. The valley of the Rio Negro, broad as it is, has merely been excavated out of the sandstone plain; for immediately above the bank on which the town stands, a level country commences, which is interrupted only by a few trifling valleys and depressions. Everywhere the landscape wears the same sterile aspect; a dry gravelly soil supports tufts of brown withered grass, and low scattered bushes, armed with thorns.

Shortly after passing the first spring we came in sight of a famous tree, which the Indians reverence as the altar of Walleechu. It is situated on a high part of the plain, and hence is a landmark visible at a great distance. As soon as a tribe of Indians come in sight of it, they offer their adorations by loud shouts. The tree itself is low, much branched, and thorny: just above the root it has a diameter of about three feet. It stands by itself without any neighbour, and was indeed the first tree we saw; afterwards we met with a few others of the same kind, but they were far from common. Being winter the tree had no leaves, but in their place numberless threads, by which the various offerings, such as cigars, bread, meat, pieces of cloth, &c. had been suspended. Poor Indians, not having anything better, only pull a thread out of their ponchos, and fasten it to the tree. Richer Indians are accustomed to pour spirits and maté into a certain hole, and likewise to smoke upwards, thinking thus to afford all possible gratification to Walleechu. To complete the scene, the tree was surrounded by the bleached bones of horses which had been slaughtered as sacrifices. All Indians of every age and sex make their offerings; they then think that their horses will not tire, and that they themselves shall be prosperous. The Gaucho who told me this, said that in the time of peace he had witnessed this scene, and that he and others used to wait till the Indians had passed by, for the sake of stealing from Walleechu the offerings.

The Gauchos think that the Indians consider the tree as the god itself; but it seems far more probable, that they regard it as the altar. The only cause which I can imagine for this choice, is its being a landmark in a dangerous passage. The Sierra de la Ventana is visible at an immense distance; and a Gaucho told me that he was once riding with an Indian a few miles to the north of the Rio Colorado, when the Indian commenced making the same loud noise, which is usual at the first sight of the distant tree; putting his hand to his head, and then pointing in the direction of the Sierra. Upon being asked the reason of this, the Indian said in broken Spanish, 'First see the Sierra.' About two leagues beyond this curious tree we halted for the night: at this instant an unfortunate cow was spied by the lynx-eyed Gauchos, who set off in full chace, and in a few minutes dragged her in with their lazos, and slaughtered her. We here had the four necessaries of life 'en el campo,' – pasture for the horses, water (only a muddy puddle), meat and fire-wood. The Gauchos were in high spirits at finding all these luxuries; and we soon set to work at the poor cow. This was the first night which I passed under the open sky, with the gear of the recado for my bed. There is high enjoyment in the independence of the Gaucho life – to be able at any moment to pull up your horse, and say, 'Here we will pass the night.' The death-like stillness of the plain, the dogs keeping watch, the gipsy-group of Gauchos making their beds round the fire, have left in my mind a strongly-marked picture of this first night, which will never be forgotten.

The next day the country continued similar to that above described. It is inhabited by few birds or animals of any kind. Occasionally a deer, or a Guanaco (wild Llama) may be seen; but the Agouti (Cavia Patagonica) is the commonest quadruped. This animal here represents our hares. It differs, however, from that genus in many essential respects; for instance, it has only three toes behind. It is also nearly twice the size, weighing from twenty to twenty-five pounds. The Agouti is a true friend of the desert; it is a common feature in the landscape to see two or three hopping quickly one after the other in a straight line across these wild plains. They are found as far north as the Sierra Tapalguen (lat. 37° 30′), where the plain rather suddenly becomes greener and more humid; and their southern limit is between Port Desire and St. Julian, where there is no change in the nature of the country. It is a singular fact, that although the Agouti is not now found as far south as Port St. Julian, yet that Captain Wood, in his voyage in 1670, talks of them as being numerous there. What cause can have altered, in a wide, uninhabited, and rarely-visited country, the range of an animal like this? It appears also from the number shot by Captain Wood in one day at Port Desire, that they must have been considerably more abundant there formerly than at present. Where the Bizcacha lives and makes its burrows, the Agouti uses them; but where, as at Bahia Blanca, the Bizcacha is not found, the Agouti burrows for itself. The same thing occurs with the little owl of the Pampas (Athene cunicularia), which has so often been described as standing like a sentinel at the mouth of the burrows; for in Banda Oriental, owing to the absence of the Bizcacha, it is obliged to hollow out its own habitation.

The next morning, as we approached the Rio Colorado, the appearance of the country changed; we soon came on a plain covered with turf, which, from its flowers, tall clover, and little owls, resembled the Pampas. We passed also a muddy swamp of considerable extent, which in summer dries, and becomes incrusted with various salts; and hence is called a salitral. It was covered by low succulent plants, of the same kind with those growing on the sea-shore. The Colorado, at the pass where we crossed it, is only about sixty yards wide; generally it must be nearly double that width. Its course is very tortuous, being marked by willow-trees and beds of reeds: in a direct line the distance to the mouth of the river is said to be nine leagues, but by water twenty-five. We were delayed crossing in the canoe by some immense troops of mares, which were swimming the river in order to follow a division of troops into the interior. A more ludicrous spectacle I never beheld than the hundreds and hundreds of heads, all directed one way, with pointed ears and distended snorting nostrils, appearing just above the water like a great shoal of some amphibious animal. Mare's flesh is the only food which the soldiers have when on an expedition. This gives them a great facility of movement; for the distance to which horses can be driven over these plains is quite surprising: I have been assured that an unloaded horse can travel a hundred miles a day for many days successively.

The encampment of General Rosas was close to the river. It consisted of a square formed by waggons, artillery, straw huts, &c. The soldiers were nearly all cavalry; and I should think such a villanous, banditti-like army was never before collected together. The greater number of men were of a mixed breed, between Negro, Indian, and Spaniard. I know not the reason, but men of such origin seldom have a good expression of countenance. I called on the Secretary to show my passport. He began to cross-question me in the most dignified and mysterious manner. By good luck I had a letter of recommendation from the government of Buenos Ayres[1] to the Commandant of Patagones. This was taken to General Rosas, who sent me a very obliging message; and the Secretary returned all smiles and graciousness. We took up our residence in the *rancho*, or hovel, of a curious old Spaniard, who had served with Napoleon in the expedition against Russia.

We stayed two days at the Colorado; I had little to do, for the surrounding country was a swamp, which in summer (December), when the snow melts on the Cordillera, is overflowed by the river. My chief amusement was watching the Indian families as they came to buy little articles at the rancho where we stayed. It was supposed that General Rosas had about six hundred Indian allies. The men were a tall, fine race, yet it was afterwards easy to see in the Fuegian savage the same countenance rendered hideous by cold, want of food, and less civilization. Some authors, in defining the primary races of mankind, have separated these Indians into two classes; but this is certainly incorrect. Among the young women or chinas, some deserve to be called even beautiful. Their hair was coarse, but bright and black; and they wore it in two plaits hanging down to the waist. They had a high colour, and eyes that glistened with brilliancy; their legs, feet, and arms were small and elegantly formed; their ankles, and

1 I am bound to express, in the strongest terms, my obligation to the Government of Buenos Ayres for the obliging manner in which passports to all parts of the country were given me, as naturalist of the *Beagle*.

sometimes their waists, were ornamented by broad bracelets of blue beads. Nothing could be more interesting than some of the family groups. A mother with one or two daughters would often come to our rancho, mounted on the same horse. They ride like men, but with their knees tucked up much higher. This habit, perhaps, arises from their being accustomed, when travelling, to ride the loaded horses. The duty of the women is to load and unload the horses; to make the tents for the night; in short to be, like the wives of all savages, useful slaves. The men fight, hunt, take care of the horses, and make the riding gear. One of their chief indoor occupations is to knock two stones together till they become round, in order to make the bolas. With this important weapon the Indian catches his game, and also his horse, which roams free over the plain. In fighting, his first attempt is to throw down the horse of his adversary with the bolas, and when entangled by the fall to kill him with the chuzo. If the balls only catch the neck or body of an animal, they are often carried away and lost. As the making the stones round is the labour of two days, the manufacture of the balls is a very common employment. Several of the men and women had their faces painted red, but I never saw the horizontal bands which are so common among the Fuegians. Their chief pride consists in having everything made of silver; I have seen a cacique with his spurs, stirrups, handle of his knife, and bridle made of this metal: the head-stall and reins being of wire, were not thicker than whipcord; and to see a fiery steed wheeling about under the command of so light a chain, gave to the horsemanship a remarkable character of elegance.

General Rosas intimated a wish to see me; a circumstance which I was afterwards very glad of. He is a man of an extraordinary character, and has a most predominant influence in the country, which it seems probable he will use to its prosperity and advancement.[1] He is said to be the owner of seventy-four square leagues of land, and to have about three hundred thousand head of cattle. His estates are admirably managed, and are far more productive of corn than those of others. He first gained his celebrity by his laws for

1 This prophecy has turned out entirely and miserably wrong. 1845.

his own estancias, and by disciplining several hundred men, so as
to resist with success the attacks of the Indians. There are many stories
current about the rigid manner in which his laws were enforced. One
of these was, that no man, on penalty of being put into the stocks,
should carry his knife on a Sunday: this being the principal day for
gambling and drinking, many quarrels arose, which from the general
manner of fighting with the knife often proved fatal. One Sunday
the Governor came in great form to pay the estancia a visit, and
General Rosas, in his hurry, walked out to receive him with his knife,
as usual, stuck in his belt. The steward touched his arm, and reminded
him of the law; upon which turning to the Governor, he said he
was extremely sorry, but that he must go into the stocks, and that till
let out, he possessed no power even in his own house. After a little
time the steward was persuaded to open the stocks, and to let him
out, but no sooner was this done, than he turned to the steward and
said, 'You now have broken the laws, so you must take my place
in the stocks.' Such actions as these delighted the Gauchos, who all
possess high notions of their own equality and dignity.

General Rosas is also a perfect horseman – an accomplishment of
no small consequence in a country where an assembled army elected
its general by the following trial: A troop of unbroken horses being
driven into a corral, were let out through a gateway, above which was
a cross-bar: it was agreed whoever should drop from the bar on one
of these wild animals, as it rushed out, and should be able, without
saddle or bridle, not only to ride it, but also to bring it back to
the door of the corral, should be their general. The person who
succeeded was accordingly elected; and doubtless made a fit general
for such an army. This extraordinary feat has also been performed
by Rosas.

By these means, and by conforming to the dress and habits of the
Gauchos, he has obtained an unbounded popularity in the country,
and in consequence a despotic power. I was assured by an English
merchant, that a man who had murdered another, when arrested and
questioned concerning his motive, answered, 'He spoke disrespect-
fully of General Rosas, so I killed him.' At the end of a week the
murderer was at liberty. This doubtless was the act of the general's
party, and not of the general himself.

In conversation he is enthusiastic, sensible, and very grave. His gravity is carried to a high pitch: I heard one of his mad buffoons (for he keeps two, like the barons of old) relate the following anecdote: 'I wanted very much to hear a certain piece of music, so I went to the general two or three times to ask him; he said to me, "Go about your business, for I am engaged." I went a second time; he said, "If you come again I will punish you." A third time I asked, and he laughed. I rushed out of the tent, but it was too late; he ordered two soldiers to catch and stake me. I begged by all the Saints in heaven he would let me off; but it would not do; – when the general laughs he spares neither mad man nor sound.' The poor flighty gentleman looked quite dolorous, at the very recollection of the staking. This is a very severe punishment; four posts are driven into the ground, and the man is extended by his arms and legs horizontally, and there left to stretch for several hours. The idea is evidently taken from the usual method of drying hides. My interview passed away without a smile, and I obtained a passport and order for the government post-horses, and this he gave me in the most obliging and ready manner.

In the morning we started for Bahia Blanca, which we reached in two days. Leaving the regular encampment, we passed by the toldos of the Indians. These are round like ovens, and covered with hides; by the mouth of each, a tapering chuzo was stuck in the ground. The toldos were divided into separate groups, which belonged to the different caciques' tribes, and the groups were again divided into smaller ones, according to the relationship of the owners. For several miles we travelled along the valley of the Colorado. The alluvial plains on the side appeared fertile, and it is supposed that they are well adapted to the growth of corn. Turning northward from the river, we soon entered on a country, differing from the plains south of the river. The land still continued dry and sterile; but it supported many different kinds of plants, and the grass, though brown and withered, was more abundant, as the thorny bushes were less so. These latter in a short space entirely disappeared, and the plains were left without a thicket to cover their nakedness. This change in the vegetation marks the commencement of the grand calcareo-argillaceous deposit, which forms the wide extent of the Pampas, and covers the granitic rocks of Banda Oriental. From the Strait of Magellan to the Colorado, a distance of about eight hundred miles, the face of the country is everywhere composed of shingle: the pebbles are chiefly of porphyry, and probably owe their origin to the rocks of the Cordillera. North of the Colorado this bed thins out, and the pebbles become exceedingly small, and here the characteristic vegetation of Patagonia ceases.

Having ridden about twenty-five miles, we came to a broad belt of sand-dunes, which stretches, as far as the eye can reach, to the east and west. The sand-hillocks resting on the clay, allow small pools of water to collect, and thus afford in this dry country an invaluable supply of fresh water. The great advantage arising from depressions and elevations of the soil, is not often brought home to the mind. The two miserable springs in the long passage between the Rio Negro and Colorado were caused by trifling inequalities in the plain; without them not a drop of water would have been found. The belt of sand-dunes is about eight miles wide; at some former period, it probably formed the margin of a grand estuary, where the Colorado now flows. In this district, where absolute proofs of the recent elevation of the land occur, such speculations can hardly be neglected by any one, although merely considering the physical geography of the country. Having crossed the sandy tract, we arrived in the evening at one of the post-houses; and, as the fresh horses were grazing at a distance, we determined to pass the night there.

The house was situated at the base of a ridge, between one and two hundred feet high – a most remarkable feature in this country. This posta was commanded by a negro lieutenant, born in Africa: to his credit be it said, there was not a rancho between the Colorado and Buenos Ayres in nearly such neat order as his. He had a little room for strangers, and a small corral for the horses, all made of sticks and reeds; he had also dug a ditch round his house, as a defence in case of being attacked. This would, however, have been of little avail, if the Indians had come; but his chief comfort seemed to rest in the thought of selling his life dearly. A short time before, a body of Indians had travelled past in the night; if they had been aware of the posta, our black friend and his four soldiers would assuredly have been slaughtered. I did not any where meet a more civil and obliging man than this negro; it was therefore the more painful to see that he would not sit down and eat with us.

In the morning we sent for the horses very early, and started for another exhilarating gallop. We passed the Cabeza del Buey, an old name given to the head of a large marsh, which extends from Bahia Blanca. Here we changed horses, and passed through some leagues of swamps and saline marshes. Changing horses for the last time, we again began wading through the mud. My animal fell, and I was well soused in black mire – a very disagreeable accident, when one does not possess a change of clothes. Some miles from the fort we met a man, who told us that a great gun had been fired, which is a signal that Indians are near. We immediately left the road, and followed the edge of a marsh, which when chased offers the best mode of escape. We were glad to arrive within the walls, when we found all the alarm was about nothing, for the Indians turned out to be friendly ones, who wished to join General Rosas.

Bahia Blanca scarcely deserves the name of a village. A few houses and the barracks for the troops are enclosed by a deep ditch and fortified wall. The settlement is only of recent standing (since 1828); and its growth has been one of trouble. The government of Buenos Ayres unjustly occupied it by force, instead of following the wise example of the Spanish Viceroys, who purchased the land near the older settlement of the Rio Negro, from the Indians. Hence the need of the fortifications; hence the few houses and little cultivated land without the limits of the walls: even the cattle are not safe from the attacks of the Indians beyond the boundaries of the plain, on which the fortress stands.

The part of the harbour where the *Beagle* intended to anchor being distant twenty-five miles, I obtained from the Commandant a guide and horses, to take me to see whether she had arrived. Leaving the plain of green turf, which extended along the course of a little brook, we soon entered on a wide level waste consisting either of sand, saline marshes, or bare mud. Some parts were clothed by low thickets, and others with those succulent plants, which luxuriate only where salt abounds. Bad as the country was, ostriches, deer, agoutis, and arma-dilloes, were abundant. My guide told me, that two months before he had a most narrow escape of his life: he was out hunting with two other men, at no great distance from this part of the country, when they were suddenly met by a party of Indians, who giving chace, soon overtook and killed his two friends. His own horse's legs were also caught by the bolas; but he jumped off, and with his knife cut them free: while doing this he was obliged to dodge round his horse, and received two severe wounds from their chuzos. Springing on the saddle, he managed, by a most wonderful exertion, just to keep ahead of the long spears of his pursuers, who followed him to within sight of the fort. From that time there was an order that no one should stray far from the settlement. I did not know of this when I started, and was surprised to observe how earnestly my guide watched a deer, which appeared to have been frightened from a distant quarter.

We found the *Beagle* had not arrived, and consequently set out on our return, but the horses soon tiring, we were obliged to bivouac on the plain. In the morning we had caught an armadillo, which, although a most excellent dish when roasted in its shell, did not make a very substantial breakfast and dinner for two hungry men. The ground at the place where we stopped for the night, was incrusted with a layer of sulphate of soda, and hence, of course, was without water. Yet many of the smaller rodents managed to exist even here, and the tucutuco was making its odd little grunt beneath my head, during half the night. Our horses were very poor ones, and in the morning they were soon exhausted from not having had any thing to drink, so that we were obliged to walk. About noon the dogs killed a kid, which we roasted. I ate some of it, but it made me intolerably thirsty. This was the more distressing as the road, from some recent rain, was full of little puddles of clear water, yet not a drop was drinkable. I had scarcely been twenty hours without water, and only part of the time under a hot sun, yet the thirst rendered me very weak. How people survive two or three days under such circumstances, I cannot imagine: at the same time, I must confess that my guide did not suffer at all, and was astonished that one day's deprivation should be so troublesome to me.

I have several times alluded to the surface of the ground being incrusted with salt. This phenomenon is quite different from that of the salinas, and more extraordinary. In many parts of South America, wherever the climate is moderately dry, these incrustations occur; but I have nowhere seen them so abundant as near Bahia Blanca. The salt here, and in other parts of Patagonia, consists chiefly of sulphate of soda with some common salt. As long as the ground remains moist in these salitrales (as the Spaniards improperly call them, mistaking this substance for saltpetre), nothing is to be seen but an extensive plain composed of a black, muddy soil, supporting scattered tufts of succulent plants. On returning through one of these tracts, after a week's hot weather, one is surprised to see square miles of the plain white, as if from a slight fall of snow, here and there heaped up by the wind into little drifts. This latter appearance is chiefly caused by the salts being drawn up, during the slow evaporation of the moisture, round blades of dead grass, stumps of wood, and pieces of broken earth, instead of being crystallized at the bottoms of the puddles of water. The salitrales occur either on level tracts elevated only a few feet above the level of the sea, or on alluvial land bordering rivers. M. Parchappe[1] found that the saline incrustation on the plain, at the distance of some miles from the sea, consisted chiefly of sulphate of soda, with only seven per cent. of common salt; whilst nearer to the coast, the common salt increased to 37 parts in a hundred. This circumstance would tempt one to believe that the sulphate of soda is generated in the soil, from the muriate, left on the surface during the slow and recent elevation of this dry country. The whole phenomenon is well worthy the attention of naturalists. Have the succulent, salt-loving plants, which are well known to contain much soda, the power of decomposing the muriate? Does the black fetid mud, abounding with organic matter, yield the sulphur and ultimately the sulphuric acid?

Two days afterwards I again rode to the harbour: when not far from our destination, my companion, the same man as before, spied three people hunting on horseback. He immediately dismounted, and watching them intently, said, 'They don't ride like Christians, and nobody can leave the fort.' The three hunters joined company, and likewise dismounted from their horses. At last one mounted again and rode over the hill out of sight. My companion said, 'We must now get on our horses: load your pistol;' and he looked to his own sword. I asked, 'Are they Indians?' – 'Quien sabe? (who knows?) if there are no more than three, it does

1 Voyage dans l'Amérique Mérid. par M. A. d'Orbigny. Part. Hist. tom. i. p. 664.

not signify.' It then struck me, that the one man had gone over the hill to fetch the rest of his tribe. I suggested this; but all the answer I could extort was, 'Quien sabe?' His head and eye never for a minute ceased scanning slowly the distant horizon. I thought his uncommon coolness too good a joke, and asked him why he did not return home. I was startled when he answered, 'We are returning, but in a line so as to pass near a swamp, into which we can gallop the horses as far as they can go, and then trust to our own legs; so that there is no danger.' I did not feel quite so confident of this, and wanted to increase our pace. He said, 'No, not until they do.' When any little inequality concealed us, we galloped; but when in sight, continued walking. At last we reached a valley, and turning to the left, galloped quickly to the foot of a hill; he gave me his horse to hold, made the dogs lie down, and then crawled on his hands and knees to reconnoitre. He remained in this position for some time, and at last, bursting out in laughter, exclaimed, 'Mugeres!' (women!) He knew them to be the wife and sister-in-law of the major's son, hunting for ostrich's eggs. I have described this man's conduct, because he acted under the full impression that they were Indians. As soon, however, as the absurd mistake was found out, he gave me a hundred reasons why they could not have been Indians; but all these were forgotten at the time. We then rode on in peace and quietness to a low point called Punta Alta, whence we could see nearly the whole of the great harbour of Bahia Blanca.

The wide expanse of water is choked up by numerous great mud-banks, which the inhabitants call Cangrejales, or *crabberies*, from the number of small crabs. The mud is so soft that it is impossible to walk over them, even for the shortest distance. Many of the banks have their surfaces covered with long rushes, the tops of which alone are visible at high water. On one occasion, when in a boat, we were so entangled by these shallows that we could hardly find our way. Nothing was visible but the flat beds of mud: the day was not very clear, and there was much refraction, or as the sailors expressed it, 'things loomed high.' The only object within our view which was not level was the horizon; rushes looked like bushes unsupported in the air, and water like mud-banks, and mud-banks like water.

We passed the night in Punta Alta, and I employed myself in searching for fossil bones; this point being a perfect catacomb for monsters of extinct races. The evening was perfectly calm and clear; the extreme monotony of the view gave it an interest even in the midst of mud-banks and gulls, sand-hillocks and solitary vultures. In riding back in the morning we came across a very fresh track of a Puma, but did not succeed in finding it. We saw also a couple of Zorillos, or skunks, – odious animals, which are far from uncommon. In general appearance the Zorillo resembles a polecat, but it is rather larger, and much thicker in proportion. Conscious of its power, it roams by day about the open plain, and fears neither dog nor man. If a dog is urged to the attack, its courage is instantly checked by a few drops of the fetid oil, which brings on violent sickness and running at the nose. Whatever is once polluted by it, is for ever useless. Azara says the smell can be perceived at a league distant; more than once, when entering the harbour of Monte Video, the wind being off shore, we have perceived the odour on board the *Beagle*. Certain it is, that every animal most willingly makes room for the Zorillo.

CHAPTER V

BAHIA BLANCA

Bahia Blanca – Geology – Numerous gigantic extinct
Quadrupeds – Recent Extinction – Longevity of Species
– Large Animals do not require a luxuriant Vegetation –
Southern Africa – Siberian Fossils – Two Species of Ostrich
– Habits of Oven-bird – Armadilloes – Venomous Snake,
Toad, Lizard – Hybernation of Animals – Habits of Sea-
Pen – Indian Wars and Massacres – Arrow-head, antiquar-
ian Relic.

THE *Beagle* arrived here on the 24th of August, and a week afterwards sailed for the Plata. With Captain Fitz Roy's consent I was left
behind, to travel by land to Buenos Ayres. I will here add some
observations, which were made during this visit and on a previous
occasion, when the *Beagle* was employed in surveying the harbour.

The plain, at the distance of a few miles from the coast, belongs
to the great Pampean formation, which consists in part of a reddish
clay, and in part of a highly calcareous marly rock. Nearer the coast
there are some plains formed from the wreck of the upper plain, and
from mud, gravel, and sand thrown up by the sea during the slow
elevation of the land, of which elevation we have evidence in upraised beds of recent shells, and in rounded pebbles of pumice scattered over the country. At Punta Alta we have a section of one of
these later-formed little plains, which is highly interesting from the
number and extraordinary character of the remains of gigantic land-
animals embedded in it. These have been fully described by Professor
Owen, in the Zoology of the voyage of the *Beagle*, and are deposited
in the College of Surgeons. I will here give only a brief outline of
their nature.

First, parts of three heads and other bones of the Megatherium, the
huge dimensions of which are expressed by its name. Secondly, the
Megalonyx, a great allied animal. Thirdly, the Scelidotherium, also an
allied animal, of which I obtained a nearly perfect skeleton. It must

have been as large as a rhinoceros: in the structure of its head it comes, according to Mr. Owen, nearest to the Cape Ant-eater, but in some other respects it approaches to the armadilloes. Fourthly, the Mylodon Darwinii, a closely related genus of little inferior size. Fifthly, another gigantic edental quadruped. Sixthly, a large animal, with an osseous coat in compartments, very like that of an armadillo. Seventhly, an extinct kind of horse, to which I shall have again to refer. Eighthly, a tooth of a Pachydermatous animal, probably the same with the Macrauchenia, a huge beast with a long neck like a camel, which I shall also refer to again. Lastly, the Toxodon, perhaps one of the strangest animals ever discovered: in size it equalled an elephant or megatherium, but the structure of its teeth, as Mr. Owen states, proves indisputably that it was intimately related to the Gnawers, the order which, at the present day, includes most of the smallest quadrupeds: in many details it is allied to the Pachydermata: judging from the position of its eyes, ears, and nostrils, it was probably aquatic, like the Dugong and Manatee, to which it is also allied. How wonderfully are the different Orders, at the present time so well separated, blended together in different points of the structure of the Toxodon!

The remains of these nine great quadrupeds, and many detached bones were found embedded on the beach, within the space of about 200 yards square. It is a remarkable circumstance that so many different species should be found together; and it proves how numerous in kind the ancient inhabitants of this country must have been. At the distance of about thirty miles from P. Alta, in a cliff of red earth, I found several fragments of bones, some of large size. Among them were the teeth of a gnawer, equalling in size and closely resembling those of the Capybara, whose habits have been described; and therefore, probably, an aquatic animal. There was also part of the head of a Ctenomys; the species being different from the Tucutuco, but with a close general resemblance. The red earth, like that of the Pampas, in which these remains were embedded, contains, according to Professor Ehrenberg, eight fresh-water and one salt-water infusorial animalcule; therefore, probably, it was an estuary deposit.

The remains at Punta Alta were embedded in stratified gravel and reddish mud, just such as the sea might now wash up on a shallow bank. They were associated with twenty-three species of shells, of which thirteen are recent and four others very closely related to recent forms; whether the remaining ones are extinct or simply unknown, must be doubtful, as few collections of shells have been made on this coast. As, however, the recent species were embedded in nearly the same proportional numbers with those now living in the bay, I think there can be little doubt, that this accumulation belongs to a very late tertiary period. From the bones of the Scelidotherium, including even the knee-cap, being intombed in their proper relative positions, and from the osseous armour of the great armadillo-like animal being so well preserved, together with the bones of one of its legs, we may feel assured that these remains were fresh and united by their ligaments, when deposited in the gravel together with the shells. Hence we have good evidence that the above enumerated gigantic quadrupeds, more different from those of the present day than the oldest of the tertiary quadrupeds of Europe, lived whilst the sea was peopled with most of its present inhabitants; and we have confirmed that remarkable law so often insisted on by Mr. Lyell, namely, that the 'longevity of the species in the mammalia is upon the whole inferior to that of the testacea.'[1]

The great size of the bones of the Megatheroid animals, including the Megatherium, Megalonyx, Scelidotherium, and Mylodon, is truly wonderful. The habits of life of these animals were a complete puzzle to naturalists, until Professor Owen[2] lately solved the problem with remarkable ingenuity. The teeth indicate, by their simple structure, that these Megatheroid animals lived on vegetable food, and probably on the leaves and small twigs of trees; their ponderous forms and great strong curved claws seem so little adapted for locomotion, that some eminent naturalists have actually believed, that, like the sloths, to which they are intimately related, they subsisted by climbing back downwards on trees, and feeding on the leaves. It was a bold, not to say preposterous, idea to conceive even antediluvian trees, with branches strong enough to bear animals as large as elephants. Professor Owen, with far more probability, believes that,

1 Principles of Geology, vol. iv. p. 40.

2 This theory was first developed in the Zoology of the Voyage of the *Beagle*, and subsequently in Professor Owen's Memoir on Mylodon robustus.

instead of climbing on the trees, they pulled the branches down to them, and tore up the smaller ones by the roots, and so fed on the leaves. The colossal breadth and weight of their hinder quarters, which can hardly be imagined without having been seen, become, on this view, of obvious service, instead of being an incumbrance: their apparent clumsiness disappears. With their great tails and their huge heels firmly fixed like a tripod on the ground, they could freely exert the full force of their most powerful arms and great claws. Strongly rooted, indeed, must that tree have been, which could have resisted such force! The Mylodon, moreover, was furnished with a long extensile tongue like that of the giraffe, which, by one of those beautiful provisions of nature, thus reaches with the aid of its long neck its leafy food. I may remark, that in Abyssinia the elephant, according to Bruce, when it cannot reach with its proboscis the branches, deeply scores with its tusks the trunk of the tree, up and down and all round, till it is sufficiently weakened to be broken down.

The beds including the above fossil remains, stand only from fifteen to twenty feet above the level of high-water; and hence the elevation of the land has been small (without there has been an inter-calated period of subsidence, of which we have no evidence) since the great quadrupeds wandered over the surrounding plains; and the external features of the country must then have been very nearly the same as now. What, it may naturally be asked, was the character of the vegetation at that period; was the country as wretchedly sterile as it now is? As so many of the co-embedded shells are the same with those now living in the bay, I was at first inclined to think that the former vegetation was probably similar to the existing one; but this would have been an erroneous inference, for some of these same shells live on the luxuriant coast of Brazil; and generally, the character of the inhabitants of the sea are useless as guides to judge of those on the land. Nevertheless, from the following considerations, I do not believe that the simple fact of many gigantic quadrupeds having lived on the plains round Bahia Blanca, is any sure guide that they formerly were clothed with a luxuriant vegetation: I have no doubt that the sterile country a little southward, near the Rio Negro, with its scattered thorny trees, would support many and large quadrupeds.

★ ★ ★

That large animals require a luxuriant vegetation, has been a general assumption which has passed from one work to another; but I do not hesitate to say that it is completely false, and that it has vitiated the reasoning of geologists on some points of great interest in the ancient history of the world. The prejudice has probably been derived from India, and the Indian islands, where troops of elephants, noble forests, and impenetrable jungles, are associated together in every one's mind. If, however, we refer to any work of travels through the southern parts of Africa, we shall find allusions in almost every page either to the desert character of the country, or to the numbers of large animals inhabiting it. The same thing is rendered evident by the many engravings which have been published of various parts of the interior. When the *Beagle* was at Cape Town, I made an excursion of some days' length into the country, which at least was sufficient to render that which I had read more fully intelligible.

Dr. Andrew Smith, who, at the head of his adventurous party, has lately succeeded in passing the Tropic of Capricorn, informs me that, taking into consideration the whole of the southern part of Africa, there can be no doubt of its being a sterile country. On the southern and south-eastern coasts there are some fine forests, but with these exceptions, the traveller may pass for days together through open plains, covered by a poor and scanty vegetation. It is difficult to convey any accurate idea of degrees of comparative fertility; but it may be safely said that the amount of vegetation supported at any one time[1] by Great Britain, exceeds, perhaps even tenfold, the quantity on an equal area, in the interior parts of Southern Africa. The fact that bullock-waggons can travel in any direction, excepting near the coast, without more than occasionally half an hour's delay in cutting down bushes, gives, perhaps, a more definite notion of the scantiness of the vegetation. Now, if we look to the animals inhabiting these wide plains, we shall find their numbers extraordinarily great, and their bulk immense. We must enumerate the elephant, three species of rhinoceros, and probably, according to Dr. Smith, two others, the hippopotamus, the giraffe, the bos caffer, – as large as a full-grown

---

1 I mean by this to exclude the total amount, which may have been successively produced and consumed during a given period.

bull, and the elan – but little less, two zebras, and the quaccha, two gnus, and several antelopes even larger than these latter animals. It may be supposed that although the species are numerous, the individuals of each kind are few. By the kindness of Dr. Smith, I am enabled to show that the case is very different. He informs me, that in lat. 24°, in one day's march with the bullock-waggons, he saw, without wandering to any great distance on either side, between one hundred and one hundred and fifty rhinoceroses, which belonged to three species: the same day he saw several herds of giraffes, amounting together to nearly a hundred; and that, although no elephant was observed, yet they are found in this district. At the distance of a little more than one hour's march from their place of encampment on the previous night, his party actually killed at one spot eight hippopotamuses, and saw many more. In this same river there were likewise crocodiles. Of course it was a case quite extraordinary, to see so many great animals crowded together, but it evidently proves that they must exist in great numbers. Dr. Smith describes the country passed through that day, as 'being thinly covered with grass, and bushes about four feet high, and still more thinly with mimosa-trees.' The waggons were not prevented travelling in a nearly straight line.

Besides these large animals, every one the least acquainted with the natural history of the Cape, has read of the herds of antelopes, which can be compared only with the flocks of migratory birds. The numbers indeed of the lion, panther, and hyæna, and the multitude of birds of prey, plainly speak of the abundance of the smaller quadrupeds: one evening seven lions were counted at the same time prowling round Dr. Smith's encampment. As this able naturalist remarked to me, the carnage each day in Southern Africa must indeed be terrific! I confess it is truly surprising how such a number of animals can find support in a country producing so little food. The larger quadrupeds no doubt roam over wide tracts in search of it; and their food chiefly consists of underwood, which probably contains much nutriment in a small bulk. Dr. Smith also informs me that the vegetation has a rapid growth; no sooner is a part consumed, than its place is supplied by a fresh stock. There can be no doubt, however, that our ideas respecting the apparent amount of food necessary for the support of large quadrupeds are much exaggerated: it should have been remembered that the camel, an animal of no mean bulk, has always been considered as the emblem of the desert.

The belief that where large quadrupeds exist, the vegetation must necessarily be luxuriant, is the more remarkable, because the converse is far from true. Mr. Burchell observed to me that when entering Brazil, nothing struck him more forcibly than the splendour of the South American vegetation contrasted with that of South Africa, together with the absence of all large quadrupeds. In his Travels,[1] he has suggested that the comparison of the respective weights (if there were sufficient data) of an equal number of the largest herbivorous quadrupeds of each country would be extremely curious. If we take on the one side, the elephant,[2] hippopotamus, giraffe, bos caffer, elan, certainly three, and probably five species of rhinoceros; and on the American side, two tapirs, the guanaco, three deer, the vicuna, peccari, capybara (after which we must choose from the monkeys to complete the number), and then place these two groups alongside each other, it is not easy to conceive ranks more disproportionate in size. After the above facts, we are compelled to conclude, against anterior probability,[3] that among the mammalia

1 Travels in the Interior of South Africa, vol. ii. p. 207.

2 The elephant which was killed at Exeter Change was estimated (being partly weighed) at five tons and a half. The elephant actress, as I was informed, weighed one ton less; so that we may take five as the average of a full-grown elephant. I was told at the Surrey Gardens, that a hippopotamus which was sent to England cut up into pieces was estimated at three tons and a half; we will call it three. From these premises we may give three tons and a half to each of the five rhinoceroses; perhaps a ton to the giraffe, and half to the bos caffer as well as to the elan (a large ox weighs from 1200 to 1500 pounds). This will give an average (from the above estimates) of 2.7 of a ton for the ten largest herbivorous animals of Southern Africa. In South America, allowing 1200 pounds for the two tapirs together, 550 for the guanaco and vicuna, 500 for three deer, 300 for the capybara, peccari, and a monkey, we shall have an average of 250 pounds, which I believe is overstating the result. The ratio will therefore be as 6048 to 250, or 24 to 1, for the ten largest animals from the two continents.

3 If we suppose the case of the discovery of a skeleton of a Greenland whale in a fossil state, not a single cetaceous animal being known to exist, what naturalist would have ventured conjecture on the possibility of a carcass so gigantic being supported on the minute crustacea and mollusca living in the frozen seas of the extreme North?

there exists no close relation between the *bulk* of the species, and the *quantity* of the vegetation, in the countries which they inhabit.

With regard to the number of large quadrupeds, there certainly exists no quarter of the globe which will bear comparison with Southern Africa. After the different statements which have been given, the extremely desert character of that region will not be disputed. In the European division of the world, we must look back to the tertiary epochs, to find a condition of things among the mammalia, resembling that now existing at the Cape of Good Hope. Those tertiary epochs, which we are apt to consider as abounding to an astonishing degree with large animals, because we find the remains of many ages accumulated at certain spots, could hardly boast of more large quadrupeds than Southern Africa does at present. If we speculate on the condition of the vegetation during those epochs, we are at least bound so far to consider existing analogies, as not to urge as absolutely necessary a luxuriant vegetation, when we see a state of things so totally different at the Cape of Good Hope.

We know[1] that the extreme regions of North America, many degrees beyond the limit where the ground at the depth of a few feet remains perpetually congealed, are covered by forests of large and tall trees. In a like manner, in Siberia, we have woods of birch, fir, aspen, and larch, growing in a latitude[2] ($64°$), where the mean temperature of the air falls below the freezing point, and where the earth is so completely frozen, that the carcass of an animal embedded in it is perfectly preserved. With these facts we must grant, as far as *quantity alone* of vegetation is concerned, that the great quadrupeds of the later tertiary epochs might, in most parts of Northern Europe and Asia, have lived on the spots where their remains are now found. I do not

1 See Zoological Remarks to Capt. Back's Expedition, by Dr. Richardson. He says, 'The subsoil north of latitude $56°$ is perpetually frozen, the thaw on the coast not penetrating above three feet, and at Bear Lake, in latitude $64°$, not more than twenty inches. The frozen substratum does not of itself destroy vegetation, for forests flourish on the surface, at a distance from the coast.'

2 See Humboldt, Fragmens Asiatiques, p. 386: Barton's Geography of Plants: and Malte Brun. In the latter work it is said that the limit of the growth of trees in Siberia may be drawn under the parallel of $70°$.

here speak of the *kind* of vegetation necessary for their support; because, as there is evidence of physical changes, and as the animals have become extinct, so may we suppose that the species of plants have likewise been changed.

These remarks, I may be permitted to add, directly bear on the case of the Siberian animals preserved in ice. The firm conviction of the necessity of a vegetation possessing a character of tropical luxuriance, to support such large animals, and the impossibility of reconciling this with the proximity of perpetual congelation, was one chief cause of the several theories of sudden revolutions of climate, and of overwhelming catastrophes, which were invented to account for their entombment. I am far from supposing that the climate has not changed since the period when those animals lived, which now lie buried in the ice. At present I only wish to show, that as far as *quantity* of food *alone* is concerned, the ancient rhinoceroses might have roamed over the *steppes* of central Siberia (the northern parts probably being under water) even in their present condition, as well as the living rhinoceroses and elephants over the *Karros* of Southern Africa.

I will now give an account of the habits of some of the more interesting birds which are common on the wild plains of Northern Patagonia; and first for the largest, or South American ostrich. The ordinary habits of the ostrich are familiar to every one. They live on vegetable matter, such as roots and grass; but at Bahia Blanca I have repeatedly seen three or four come down at low water to the extensive mud-banks which are then dry, for the sake, as the Gauchos say, of feeding on small fish. Although the ostrich in its habits is so shy, wary, and solitary, and although so fleet in its pace, it is caught without much difficulty by the Indian or Gaucho armed with the bolas. When several horsemen appear in a semicircle, it becomes confounded, and does not know which way to escape. They generally prefer running against the wind; yet at the first start they expand their wings, and like a vessel make all sail. On one fine hot day I saw several ostriches enter a bed of tall rushes, where they squatted concealed, till quite closely approached. It is not generally known that ostriches readily take to the water. Mr. King informs me that at the Bay of San Blas, and at Port Valdes in Patagonia, he saw these birds swimming several times from island to island. They ran into the water both when driven down to a point, and likewise of their own accord when not frightened: the distance crossed was about two hundred yards. When swimming, very little of their bodies appear above water; their necks are extended a little forward, and their progress is slow. On two occasions I saw some ostriches swimming across the Santa Cruz river, where its course was about four hundred yards wide, and the stream rapid. Captain Sturt,[1] when descending the Murrumbidgee, in Australia, saw two emus in the act of swimming.

The inhabitants of the country readily distinguish, even at a distance, the cock bird from the hen. The former is larger and darker-coloured,[2] and has a bigger head. The ostrich, I believe the cock, emits a singular, deep-toned, hissing note: when first I heard it, standing in the midst of some sand-hillocks, I thought it was made by some wild beast, for it is a sound that one cannot tell whence it comes, or from how far distant. When we were at Bahia Blanca in the months of September and October, the eggs, in extraordinary numbers, were found all over the country. They lie either scattered and single, in which case they are never hatched, and are called by the Spaniards huachos; or they are collected together into a shallow excavation, which forms the nest. Out of the four nests which I saw, three contained twenty-two eggs each, and the fourth twenty-seven. In one day's hunting on horseback sixty-four eggs were found; forty-four of these were in two nests, and the remaining twenty, scattered huachos. The Gauchos unanimously affirm, and there is no reason to doubt their statement, that the male bird alone hatches the eggs, and for some time afterwards accompanies the young. The cock when on the nest lies very close; I have myself almost ridden over one. It is asserted that at such times they are occasionally fierce, and even dangerous, and that they have been known to attack a man on horseback, trying to kick and leap on him. My informer pointed out to me an old man, whom he had seen much terrified by one chasing him. I observe in Burchell's travels in South Africa, that he remarks, 'Having killed a male ostrich, and the feathers being dirty, it was said by the Hottentots to be a nest bird.' I understand that the

1 Sturt's Travels vol. ii. p. 74.

2 A Gaucho assured me that he had once seen a snow-white or Albino variety, and that it was a most beautiful bird.

male emu in the Zoological Gardens takes charge of the nest; this habit, therefore, is common to the family.

The Gauchos unanimously affirm that several females lay in one nest. I have been positively told that four or five hen birds have been watched to go in the middle of the day, one after the other, to the same nest. I may add, also, that it is believed in Africa, that two or more females lay in one nest.[1] Although the habit at first appears very strange, I think the cause may be explained in a simple manner. The number of eggs in the nest varies from twenty to forty, and even to fifty; and according to Azara, sometimes to seventy or eighty. Now although it is most probable, from the number of eggs found in one district being so extraordinarily great in proportion to the parent birds, and likewise from the state of the ovarium of the hen, that she may in the course of the season lay a large number, yet the time required must be very long. Azara states,[2] that a female in a state of domestication laid seventeen eggs, each at the interval of three days one from another. If the hen was obliged to hatch her own eggs, before the last was laid the first probably would be addled; but if each laid a few eggs at successive periods, in different nests, and several hens, as is stated to be the case, combined together, then the eggs in one collection would be nearly of the same age. If the number of eggs in one of these nests is, as I believe, not greater on an average than the number laid by one female in the season, then there must be as many nests as females, and each cock bird will have its fair share of the labour of incubation; and that during a period when the females probably could not sit, from not having finished laying.[3] I have before mentioned the great numbers of huachos, or deserted eggs; so that in one day's hunting twenty were found in this state. It appears odd that so many should be wasted. Does it not arise from the difficulty of several females associating together, and finding a male ready to undertake the office of incubation? It is evident that there must at

1 Burchell's Travels, vol. i. p. 280.

2 Azara, vol. iv. p. 173.

3 Lichtenstein, however, asserts (Travels, vol. ii. p. 25) that the hens begin sitting when they have laid ten or twelve eggs; and that they continue laying, I presume, in another nest. This appears to me very improbable. He asserts that four or five hens associate for incubation with one cock, who sits only at night.

first be some degree of association between at least two females; otherwise the eggs would remain scattered over the wide plains, at distances far too great to allow of the male collecting them into one nest: some authors have believed that the scattered eggs were deposited for the young birds to feed on. This can hardly be the case in America, because the huachos, although often found addled and putrid, are generally whole.

When at the Rio Negro in Northern Patagonia, I repeatedly heard the Gauchos talking of a very rare bird which they called Avestruz Petise. They described it as being less than the common ostrich (which is there abundant), but with a very close general resemblance. They said its colour was dark and mottled, and that its legs were shorter, and feathered lower down than those of the common ostrich. It is more easily caught by the bolas than the other species. The few inhabitants who had seen both kinds, affirmed they could distinguish them apart from a long distance. The eggs of the small species appeared, however, more generally known; and it was remarked, with surprise, that they were very little less than those of the Rhea, but of a slightly different form, and with a tinge of pale blue. This species occurs most rarely on the plains bordering the Rio Negro; but about a degree and a half further south they are tolerably abundant. When at Port Desire, in Patagonia (lat. 48°), Mr. Martens shot an ostrich; and I looked at it, forgetting at the moment, in the most unaccountable manner, the whole subject of the Petises, and thought it was a not full-grown bird of the common sort. It was cooked and eaten before my memory returned. Fortunately the head, neck, legs, wings, many of the larger feathers, and a large part of the skin, had been preserved; and from these a very nearly perfect specimen has been put together, and is now exhibited in the museum of the Zoological Society. Mr. Gould, in describing this new species, has done me the honour of calling it after my name.

Among the Patagonian Indians in the Strait of Magellan, we found a half Indian, who had lived some years with the tribe, but had been born in the northern provinces. I asked him if he had ever heard of the Avestruz Petise? He answered by saying, 'Why there are none others in these southern countries.' He informed me that the number of eggs in the nest of the petise is considerably less than in that of the other kind, namely, not more than fifteen on an average; but he asserted that more than one female deposited them. At Santa Cruz we saw several of these birds. They were excessively wary: I think they could see a person approaching when too far off to be distinguished themselves. In ascending the river few were seen; but in our quiet and rapid descent, many, in pairs and by fours or fives, were observed. It was remarked that this bird did not expand its wings, when first starting at full speed, after the manner of the northern kind. In conclusion I may observe, that the Struthio rhea inhabits the country of La Plata as far as a little south of the Rio Negro in lat. 41°, and that the Struthio Darwinii takes its place in Southern Patagonia; the part about the Rio Negro being neutral territory. M. A. d'Orbigny,[1] when at the Rio Negro, made great exertions to procure this bird, but never had the good fortune to succeed. Dobrizhoffer[2] long ago was aware of there being two kinds of ostriches; he says, 'You must know, moreover, that Emus differ in size and habits in different tracts of land; for those that inhabit the plains of Buenos Ayres and Tucuman are larger, and have black, white, and grey feathers; those near to the Strait of Magellan are smaller and more beautiful, for their white feathers are tipped with black at the extremity, and their black ones in like manner terminate in white.'

A very singular little bird, Tinochorus rumicivorus, is here common: in its habits and general appearance, it nearly equally partakes of the characters, different as they are, of the quail and snipe. The Tinochorus is found in the whole of southern South America, wherever there are sterile plains, or open dry pasture land. It frequents in pairs or small flocks the most desolate places, where scarcely another living creature can exist. Upon being approached they squat close, and then are very difficult to be distinguished from

1 When at the Rio Negro, we heard much of the indefatigable labours of this naturalist. M. Alcide d'Orbigny, during the years 1825 to 1833, traversed several large portions of South America, and has made a collection, and is now publishing the results on a scale of magnificence, which at once places himself in the list of American travellers second only to Humboldt.

2 Account of the Abipones, A.D. 1749, vol. i. (English translation), p. 314.

the ground. When feeding they walk rather slowly, with their legs wide apart. They dust themselves in roads and sandy places, and frequent particular spots, where they may be found day after day: like partridges, they take wing in a flock. In all these respects, in the muscular gizzard adapted for vegetable food, in the arched beak and fleshy nostrils, short legs and form of foot, the Tinochorus has a close affinity with quails. But as soon as the bird is seen flying, its whole appearance changes; the long pointed wings, so different from those in the gallinaceous order, the irregular manner of flight, and plaintive cry uttered at the moment of rising, recall the idea of a snipe. The sportsmen of the *Beagle* unanimously called it the short-billed snipe. To this genus, or rather to the family of the Waders, its skeleton shows that it is really related.

The Tinochorus is closely related to some other South American birds. Two species of the genus Attagis are in almost every respect ptarmigans in their habits; one lives in Tierra del Fuego, above the limits of the forest land; and the other just beneath the snow-line on the Cordillera of Central Chile. A bird of another closely allied genus, Chionis alba, is an inhabitant of the antarctic regions; it feeds on sea-weed and shells on the tidal rocks. Although not web-footed, from some unaccountable habit, it is frequently met with far out at sea. This small family of birds is one of those which, from its varied relations to other families, although at present offering only difficulties to the systematic naturalist, ultimately may assist in revealing the grand scheme, common to the present and past ages, on which organized beings have been created.

The genus Furnarius contains several species, all small birds, living on the ground, and inhabiting open dry countries. In structure they cannot be compared to any European form. Ornithologists have generally included them among the creepers, although opposed to that family in every habit. The best known species is the common oven-bird of La Plata, the Casara or housemaker of the Spaniards. The nest, whence it takes its name, is placed in the most exposed situations, as on the top of a post, a bare rock, or on a cactus. It is composed of mud and bits of straw, and has strong thick walls: in shape it precisely resembles an oven, or depressed beehive. The opening is large and arched, and directly in front, within the nest, there is a partition, which reaches nearly to the roof, thus forming a passage or antechamber to the true nest.

Another and smaller species of Furnarius (F. cunicularius), resembles the oven-bird in the general reddish tint of its plumage, in a peculiar shrill reiterated cry, and in an odd manner of running by starts. From its affinity, the Spaniards call it Casarita (or little house-builder), although its nidification is quite different. The Casarita builds its nest at the bottom of a narrow cylindrical hole, which is said to extend horizontally to nearly six feet under ground. Several of the country people told me, that when boys, they had attempted to dig out the nest, but had scarcely ever succeeded in getting to the end of the passage. The bird chooses any low bank of firm sandy soil by the side of a road or stream. Here (at Bahia Blanca) the walls round the houses are built of hardened mud; and I noticed that one, which enclosed a courtyard where I lodged, was bored through by round holes in a score of places. On asking the owner the cause of this, he bitterly complained of the little casarita, several of which I afterwards observed at work. It is rather curious to find how incapable these birds must be of acquiring any notion of thickness, for although they were constantly flitting over the low wall, they continued vainly to bore through it, thinking it an excellent bank for their nests. I do not doubt that each bird, as often as it came to daylight on the opposite side, was greatly surprised at the marvellous fact.

I have already mentioned nearly all the mammalia common in this country. Of armadilloes three species occur, namely, the Dasypus minutus or *pichy*, the D. villosus or *peludo*, and the *apar*. The first extends ten degrees further south than any other kind: a fourth species, the *Mulita*, does not come as far south as Bahia Blanca. The four species have nearly similar habits; the *peludo*, however, is nocturnal, while the others wander by day over the open plains, feeding on beetles, larvæ, roots, and even small snakes. The *apar*, commonly called *mataco*, is remarkable by having only three moveable bands; the rest of its tesselated covering being nearly inflexible. It has the power of rolling itself into a perfect sphere, like one kind of English wood-louse. In this state it is safe from the attack of dogs; for the dog not being able to take the whole in its mouth, tries to bite one side, and the ball slips away. The smooth hard covering of the *mataco* offers a better defence than the sharp spines of the hedgehog. The *pichy* prefers a very dry soil; and the sand-dunes near the coast, where for many months it can never taste water, is its favourite resort: it often tries to escape notice, by squatting close to the ground. In the course of a day's ride, near Bahia Blanca, several were generally met with. The instant one was perceived, it was necessary, in order to catch it, almost to tumble off one's horse; for in soft soil the animal burrowed so quickly, that its hinder quarters would almost disappear before one could alight. It seems almost a pity to kill such nice little animals, for as a Gaucho said, while sharpening his knife on the back of one, 'Son tan mansos' (they are so quiet).

Of reptiles there are many kinds: one snake (a Trigonocephalus, or Cophias), from the size of the poison channel in its fangs, must be very deadly. Cuvier, in opposition to some other naturalists, makes this a sub-genus of the rattlesnake, and intermediate between it and the viper. In confirmation of this opinion, I observed a fact, which appears to me very curious and instructive, as showing how every character, even though it may be in some degree independent of structure, has a tendency to vary by slow degrees. The extremity of the tail of this snake is terminated by a point, which is very slightly enlarged; and as the animal glides along, it constantly vibrates the last inch; and this part striking against the dry grass and brushwood, produces a rattling noise, which can be distinctly heard at the distance of six feet. As often as the animal was irritated or surprised, its tail was shaken; and the vibrations were extremely rapid. Even as long as the body retained its irritability, a tendency to this habitual movement was evident. This Trigonocephalus has, therefore, in some respects the structure of a viper, with the habits of a rattlesnake: the noise, however, being produced by a simpler device. The expression of this snake's face was hideous and fierce; the pupil consisted of a vertical slit in a mottled and coppery iris; the jaws were broad at the base, and the nose terminated in a triangular projection. I do not think I ever saw any thing more ugly, excepting, perhaps, some of the vampire bats. I imagine this repulsive aspect originates from the features being placed in positions, with respect to each other, somewhat proportional to those of the human face; and thus we obtain a scale of hideousness.

Amongst the Batrachian reptiles, I found only one little toad (Phryniscus nigricans), which was most singular from its colour. If we imagine, first, that it had been steeped in the blackest ink, and then, when dry, allowed to crawl over a board, freshly painted with the brightest vermilion, so as to colour the soles of its feet and parts of its stomach, a good idea of its appearance will be gained. If it had been an unnamed species, surely it ought to have been called *Diabolicus*, for it is a fit toad to preach in the ear of Eve. Instead of being nocturnal in its habits, as other toads are, and living in damp obscure recesses, it crawls during the heat of the day about the dry sand-hillocks and arid plains, where not a single drop of water can be found. It must necessarily depend on the dew for its moisture; and this probably is absorbed by the skin, for it is known, that these reptiles possess great powers of cutaneous absorption. At Maldonado, I found one in a situation nearly as dry as at Bahia Blanca, and thinking to give it a great treat, carried it to a pool of water; not only was the little animal unable to swim, but, I think without help it would soon have been drowned.

Of lizards there were many kinds, but only one (Proctotretus multimaculatus) remarkable from its habits. It lives on the bare sand near the sea coast, and from its mottled colour, the brownish scales being speckled with white, yellowish red, and dirty blue, can hardly be distinguished from the surrounding surface. When frightened, it attempts to avoid discovery by feigning death, with outstretched legs, depressed body, and closed eyes: if further molested, it buries itself with great quickness in the loose sand. This lizard, from its flattened body and short legs, cannot run quickly.

I will here add a few remarks on the hybernation of animals in this part of South America. When we first arrived at Bahia Blanca, September 7th, 1832, we thought nature had granted scarcely a living creature to this sandy and dry country. By digging, however, in the ground, several insects, large spiders, and lizards were found in a half torpid state. On the 15th, a few animals began to appear, and by the 18th (three days from the equinox), every thing announced the commencement of spring. The plains were ornamented by the flowers of a pink wood-sorrel, wild peas, œnotheræ, and geraniums; and the birds began to lay their eggs. Numerous Lamellicorn and

Heteromerous insects, the latter remarkable for their deeply sculptured bodies, were slowly crawling about; while the lizard tribe, the constant inhabitants of a sandy soil, darted about in every direction. During the first eleven days, whilst nature was dormant, the mean temperature taken from observations made every two hours on board the *Beagle*, was 51°; and in the middle of the day the thermometer seldom ranged above 55°. On the eleven succeeding days, in which all living things became so animated, the mean was 58°, and the range in the middle of the day between sixty and seventy. Here then an increase of seven degrees in mean temperature, but a greater one of extreme heat, was sufficient to awake the functions of life. At Monte Video, from which we had just before sailed, in the twenty-three days included between the 26th of July and the 19th of August, the mean temperature from 276 observations was 58°.4; the mean hottest day being 65°.5, and the coldest 46°. The lowest point to which the thermometer fell was 41°.5, and occasionally in the middle of the day it rose to 69° or 70°. Yet with this high temperature, almost every beetle, several genera of spiders, snails, and land-shells, toads and lizards were all lying torpid beneath stones. But we have seen that at Bahia Blanca, which is four degrees southward, and therefore with a climate only a very little colder, this same temperature with a rather less extreme heat, was sufficient to awake all orders of animated beings. This shows how nicely the stimulus required to arouse hybernating animals is governed by the usual climate of the district, and not by the absolute heat. It is well known that within the tropics, the hybernation, or more properly æstivation, of animals is determined not by the temperature, but by the times of drought. Near Rio de Janeiro, I was at first surprised to observe, that, a few days after some little depressions had been filled with water, they were peopled by numerous full-grown shells and beetles, which must have been lying dormant. Humboldt has related the strange accident of a hovel having been erected over a spot where a young crocodile lay buried in the hardened mud. He adds, 'The Indians often find enormous boas, which they call Uji, or water serpents, in the same lethargic state. To reanimate them, they must be irritated or wetted with water.'

I will only mention one other animal, a zoophyte (I believe Virgularia Patagonica), a kind of sea-pen. It consists of a thin, straight,

fleshy stem, with alternate rows of polypi on each side, and surrounding an elastic stony axis, varying in length from eight inches to two
feet. The stem at one extremity is truncate, but at the other is
terminated by a vermiform fleshy appendage. The stony axis which
gives strength to the stem may be traced at this extremity into a mere
vessel filled with granular matter. At low water hundreds of these
zoophytes might be seen, projecting like stubble, with the truncate
end upwards, a few inches above the surface of the muddy sand.
When touched or pulled they suddenly drew themselves in with
force, so as nearly or quite to disappear. By this action, the highly
elastic axis must be bent at the lower extremity, where it is naturally
slightly curved; and I imagine it is by this elasticity alone that the
zoophyte is enabled to rise again through the mud. Each polypus,
though closely united to its brethren, has a distinct mouth, body, and
tentacula. Of these polypi, in a large specimen, there must be many
thousands; yet we see that they act by one movement: they have also
one central axis connected with a system of obscure circulation, and
the ova are produced in an organ distinct from the separate individuals.[1] Well may one be allowed to ask, what is an individual? It is
always interesting to discover the foundation of the strange tales of the
old voyagers; and I have no doubt but that the habits of this Virgularia
explain one such case. Captain Lancaster, in his voyage[2] in 1601,
narrates that on the sea-sands of the Island of Sombrero, in the East
Indies, he 'found a small twig growing up like a young tree, and on

---

1 The cavities leading from the fleshy compartments of the extremity, were filled
with a yellow pulpy matter, which, examined under a microscope, presented an
extraordinary appearance. The mass consisted of rounded, semi-transparent, irregular
grains, aggregated together into particles of various sizes. All such particles, and the
separate grains, possessed the power of rapid movement; generally revolving around
different axes, but sometimes progressive. The movement was visible with a very weak
power, but even with the highest its cause could not be perceived. It was very different
from the circulation of the fluid in the elastic bag, containing the thin extremity of the
axis. On other occasions, when dissecting small marine animals beneath the microscope, I have seen particles of pulpy matter, some of large size, as soon as they were
disengaged, commence revolving. I have imagined, I know not with how much truth,
that this granulo-pulpy matter was in process of being converted into ova. Certainly in
this zoophyte such appeared to be the case.

2 Kerr's Collection of Voyages, vol. viii. p. 119.

offering to pluck it up it shrinks down to the ground, and sinks, unless held very hard. On being plucked up, a great worm is found to be its root, and as the tree groweth in greatness, so doth the worm diminish; and as soon as the worm is entirely turned into a tree it rooteth in the earth, and so becomes great. This transformation is one of the strangest wonders that I saw in all my travels: for if this tree is plucked up, while young, and the leaves and bark stripped off, it becomes a hard stone when dry, much like white coral: thus is this worm twice transformed into different natures. Of these we gathered and brought home many.'

During my stay at Bahia Blanca, while waiting for the *Beagle*, the place was in a constant state of excitement, from rumours of wars and victories, between the troops of Rosas and the wild Indians. One day an account came that a small party forming one of the postas on the line to Buenos Ayres, had been found all murdered. The next day three hundred men arrived from the Colorado, under the command of Commandant Miranda. A large portion of these men were Indians (*mansos*, or tame), belonging to the tribe of the Cacique Bernantio. They passed the night here; and it was impossible to conceive any thing more wild and savage than the scene of their bivouac. Some drank till they were intoxicated; others swallowed the steaming blood of the cattle slaughtered for their suppers, and then, being sick from drunkenness, they cast it up again, and were besmeared with filth and gore.

> Nam simul expletus dapibus, vinoque sepultus
> Cervicem inflexam posuit, jacuitque per antrum
> Immensus, saniem eructans, ac frusta cruenta
> Per somnum commixta mero.

In the morning they started for the scene of the murder, with orders to follow the 'rastro,' or track, even if it led them to Chile. We subsequently heard that the wild Indians had escaped into the great Pampas, and from some cause the track had been missed. One glance at the rastro tells these people a whole history. Supposing they examine the track of a thousand horses, they will soon guess the number of mounted ones by seeing how many have cantered; by the depth of the other impressions, whether any horses were loaded with cargoes; by the irregularity of the footsteps, how far tired; by the manner in which the food has been cooked, whether the pursued travelled in haste; by the general appearance, how long it has been since they passed. They consider a rastro of ten days or a fortnight, quite recent enough to be hunted out. We also heard that Miranda struck from the west end of the Sierra Ventana, in a direct line to the island of Cholechel, situated seventy leagues up the Rio Negro. This is a distance of between two or three hundred miles, through a country completely unknown. What other troops in the world are so independent? With the sun for their guide, mares' flesh for food, their saddle-cloths for beds, – as long as there is a little water, these men would penetrate to the end of the world.

A few days afterwards I saw another troop of these banditti-like soldiers start on an expedition against a tribe of Indians at the small Salinas, who had been betrayed by a prisoner cacique. The Spaniard who brought the orders for this expedition was a very intelligent man. He gave me an account of the last engagement at which he was present. Some Indians, who had been taken prisoners, gave information of a tribe living north of the Colorado. Two hundred soldiers were sent; and they first discovered the Indians by a cloud of dust from their horses' feet, as they chanced to be travelling. The country was mountainous and wild, and it must have been far in the interior, for the Cordillera were in sight. The Indians, men, women, and children, were about one hundred and ten in number, and they were nearly all taken or killed, for the soldiers sabre every man. The Indians are now so terrified that they offer no resistance in a body, but each flies, neglecting even his wife and children; but when overtaken, like wild animals, they fight against any number to the last moment. One dying Indian seized with his teeth the thumb of his adversary, and allowed his own eye to be forced out sooner than relinquish his hold. Another, who was wounded, feigned death, keeping a knife ready to strike one more fatal blow. My informer said, when he was pursuing an Indian, the man cried out for mercy, at the same time that he was covertly loosing the bolas from his waist, meaning to whirl it round his head and so strike his pursuer. 'I however struck him with my sabre to the ground, and then got off my horse, and cut his throat with my knife.'

This is a dark picture; but how much more shocking is the unquestionable fact, that all the women who appear above twenty years old are massacred in cold blood! When I exclaimed that this appeared rather inhuman, he answered, 'Why, what can be done? they breed so!'

Every one here is fully convinced that this is the most just war, because it is against barbarians. Who would believe in this age that such atrocities could be committed in a Christian civilized country? The children of the Indians are saved, to be sold or given away as servants, or rather slaves for as long a time as the owners can make them believe themselves slaves; but I believe in their treatment there is little to complain of.

In the battle four men ran away together. They were pursued, one was killed, and the other three were taken alive. They turned out to be messengers or ambassadors from a large body of Indians, united in the common cause of defence, near the Cordillera. The tribe to which they had been sent was on the point of holding a grand council; the feast of mare's flesh was ready, and the dance prepared: in the morning the ambassadors were to have returned to the Cordillera. They were remarkably fine men, very fair, above six feet high, and all under thirty years of age. The three survivors of course possessed very valuable information; and to extort this they were placed in a line. The two first being questioned, answered, 'No sé' (I do not know), and were one after the other shot. The third also said 'No sé;' adding, 'Fire, I am a man and can die!' Not one syllable would they breathe to injure the united cause of their country! The conduct of the above-mentioned cacique was very different: he saved his life by betraying the intended plan of warfare, and the point of union in the Andes. It was believed that there were already six or seven hundred Indians together, and that in summer their numbers would be doubled. Ambassadors were to have been sent to the Indians at the small Salinas, near Bahia Blanca, whom I have mentioned that this same cacique had betrayed. The communication, therefore, between the Indians, extends from the Cordillera to the coast of the Atlantic.

General Rosas's plan is to kill all stragglers, and having driven the remainder to a common point, to attack them in a body, in the summer, with the assistance of the Chilenos. This operation is to be repeated for three successive years. I imagine the summer is chosen as the time for the main attack, because the plains are then without water, and the Indians can only travel in particular directions. The escape of the Indians to the south of the Rio Negro, where in such a vast unknown country they would be safe, is prevented by a treaty with the Tehuelches to this effect; – that Rosas pays them so much to slaughter every Indian who passes to the south of the river, but if they fail in so doing, they themselves are to be exterminated. The war is waged chiefly against the Indians near the Cordillera; for many of the tribes on this eastern side are fighting with Rosas. The general, however, like Lord Chesterfield, thinking that his friends may in a future day become his enemies, always places them in the front ranks, so that their numbers may be thinned. Since leaving South America we have heard that this war of extermination completely failed.

Among the captive girls taken in the same engagement, there were two very pretty Spanish ones, who had been carried away by the Indians when young, and could now only speak the Indian tongue. From their account they must have come from Salta, a distance in a straight line of nearly one thousand miles. This gives one a grand idea of the immense territory over which the Indians roam: yet, great as it is, I think there will not, in another half century, be a wild Indian northward of the Rio Negro. The warfare is too bloody to last; the Christians killing every Indian, and the Indians doing the same by the Christians. It is melancholy to trace how the Indians have given way before the Spanish invaders. Schirdel[1] says that in 1535, when Buenos Ayres was founded, there were villages containing two and three thousand inhabitants. Even in Falconer's time (1750) the Indians made inroads as far as Luxan, Areco, and Arrecife, but now they are driven beyond the Salado. Not only have whole tribes been exterminated, but the remaining Indians have become more barbarous: instead of living in large villages, and being employed in the arts of fishing, as well as of the chace, they now wander about the open plains, without home or fixed occupation.

1 Purchas's Collection of Voyages. I believe the date was really 1537.

I heard also some account of an engagement which took place, a few weeks previously to the one mentioned, at Cholechel. This is a very important station on account of being a pass for horses; and it was, in consequence, for some time the headquarters of a division of the army. When the troops first arrived there they found a tribe of Indians, of whom they killed twenty or thirty. The cacique escaped in a manner which astonished every one. The chief Indians always have one or two picked horses, which they keep ready for any urgent occasion. On one of these, an old white horse, the cacique sprung, taking with him his little son. The horse had neither saddle nor bridle. To avoid the shots, the Indian rode in the peculiar method of his nation; namely, with an arm round the horse's neck, and one leg only on its back. Thus hanging on one side he was seen patting the horse's head, and talking to him. The pursuers urged every effort in the chace; the Commandant three times changed his horse, but all in vain. The old Indian father and his son escaped, and were free. What a fine picture one can form in one's mind, – the naked, bronze-like figure of the old man with his little boy, riding like a Mazeppa on the white horse, thus leaving far behind him the host of his pursuers!

I saw one day a soldier striking fire with a piece of flint, which I immediately recognized as having been a part of the head of an arrow. He told me it was found near the island of Cholechel, and that they are frequently picked up there. It was between two and three inches long, and therefore twice as large as those now used in the Tierra del Fuego: it was made of opake cream-coloured flint, but the point and barbs had been intentionally broken off. It is well known that no Pampas Indians now use bows and arrows. I believe a small tribe in Banda Oriental must be excepted; but they are widely separated from the Pampas Indians, and border close on those tribes that inhabit the forest, and live on foot. It appears, therefore, that these arrowheads are antiquarian[1] relics of the Indians, before the great change in habits consequent on the introduction of the horse into South America.

---

1 Azara has even doubted whether the Pampas Indians ever used bows.

CHAPTER VI

BAHIA BLANCA TO
BUENOS AYRES

Set out for Buenos Ayres – Rio Sauce – Sierra Ventana – Third Posta – Driving Horses – Bolas – Partridges and Foxes – Features of the Country – Long-legged Plover – Teru-tero – Hail-storm – Natural Enclosures in the Sierra Tapalguen – Flesh of Puma – Meat Diet – Guardia del Monte – Effects of Cattle on the Vegetation – Cardoon – Buenos Ayres – Corral where Cattle are slaughtered.

SEPTEMBER 8*th*. – I hired a Gaucho to accompany me on my ride to Buenos Ayres, though with some difficulty, as the father of one man was afraid to let him go, and another, who seemed willing, was described to me as so fearful, that I was afraid to take him, for I was told that even if he saw an ostrich at a distance, he would mistake it for an Indian, and would fly like the wind away. The distance to Buenos Ayres is about four hundred miles, and nearly the whole way through an uninhabited country. We started early in the morning; ascending a few hundred feet from the basin of green turf on which Bahia Blanca stands, we entered on a wide desolate plain. It consists of a crumbling argillaceo-calcareous rock, which, from the dry nature of the climate, supports only scattered tufts of withered grass, without a single bush or tree to break the monotonous uniformity. The weather was fine, but the atmosphere remarkably hazy; I thought the appearance foreboded a gale, but the Gauchos said it was owing to the plain, at some great distance in the interior, being on fire. After a long gallop, having changed horses twice, we reached the Rio Sauce: it is a deep, rapid, little stream, not above twenty-five feet wide. The second posta on the road to Buenos Ayres stands on its banks; a little above there is a ford for horses, where the water does not reach to the horse's belly; but from that point, in its course to the sea, it is quite impassable, and hence makes a most useful barrier against the Indians.

Insignificant as this stream is, the Jesuit Falconer, whose information is generally so very correct, figures it as a considerable river, rising at the foot of the Cordillera. With respect to its source, I do not doubt that this is the case; for the Gauchos assured me, that in the middle of the dry summer, this stream, at the same time with the Colorado, has periodical floods; which can only originate in the snow melting on the Andes. It is extremely improbable that a stream so small as the Sauce then was, should traverse the entire width of the continent; and indeed, if it were the residue of a large river, its waters, as in other ascertained cases, would be saline. During the winter we must look to the springs round the Sierra Ventana as the source of its pure and limpid stream. I suspect the plains of Patagonia, like those of Australia, are traversed by many water-courses, which only perform their proper parts at certain periods. Probably this is the case with the water which flows into the head of Port Desire, and likewise with the Rio Chupat, on the banks of which masses of highly cellular scoriæ were found by the officers employed in the survey.

As it was early in the afternoon when we arrived, we took fresh horses, and a soldier for a guide, and started for the Sierra de la Ventana. This mountain is visible from the anchorage at Bahia Blanca; and Capt. Fitz Roy calculates its height to be 3,340 feet – an altitude very remarkable on this eastern side of the continent. I am not aware that any foreigner, previous to my visit, had ascended this mountain; and indeed very few of the soldiers at Bahia Blanca knew anything about it. Hence we heard of beds of coal, of gold and silver, of caves, and of forests, all of which inflamed my curiosity, only to disappoint it. The distance from the posta was about six leagues, over a level plain of the same character as before. The ride was, however, interesting, as the mountain began to show its true form. When we reached the foot of the main ridge, we had much difficulty in finding any water, and we thought we should have been obliged to have passed the night without any. At last we discovered some by looking close to the mountain, for at the distance even of a few hundred yards, the streamlets were buried and entirely lost in the friable calcareous stone and loose detritus. I do not think Nature ever made a more solitary, desolate pile of rock; – it well deserves its name of *Hurtado*, or separated. The mountain is steep, extremely rugged, and broken, and so entirely destitute of trees, and even bushes, that we actually could not make a skewer to stretch out our meat over the fire of thistle-stalks.[1] The strange aspect of this mountain is contrasted by the sea-like plain, which not only abuts against its steep sides, but likewise separates the parallel ranges. The uniformity of the colouring gives an extreme quietness to the view; – the whitish grey of the quartz rock, and the light brown of the withered grass of the plain, being unrelieved by any brighter tint. From custom, one expects to see in the neighbourhood of a lofty and bold mountain, a broken country strewed over with huge fragments. Here nature shows that the last movement before the bed of the sea is changed into dry land may sometimes be one of tranquillity. Under these circumstances I was curious to observe how far from the parent rock any pebbles could be found. On the shores of Bahia Blanca, and near the settlement, there were some of quartz, which certainly must have come from this source: the distance is forty-five miles.

The dew, which in the early part of the night wetted the saddle-cloths under which we slept, was in the morning frozen. The plain, though appearing horizontal, had insensibly sloped up to a height of between 800 and 900 feet above the sea. In the morning (9th of September) the guide told me to ascend the nearest ridge, which he thought would lead me to the four peaks that crown the summit. The climbing up such rough rocks was very fatiguing; the sides were so indented, that what was gained in one five minutes was often lost in the next. At last, when I reached the ridge, my disappointment was extreme in finding a precipitous valley as deep as the plain, which cut the chain transversely in two, and separated me from the four points. This valley is very narrow, but flat-bottomed, and it forms a fine horse-pass for the Indians, as it connects the plains on the northern and southern sides of the range. Having descended, and while crossing it, I saw two horses grazing: I immediately hid myself in the long grass, and began to reconnoitre; but as I could see no signs of Indians I proceeded cautiously on my second ascent. It was late in the day, and this part of the mountain, like the other, was steep and rugged. I was

1 I call these thistle-stalks for the want of a more correct name. I believe it is a species of Eryngium.

on the top of the second peak by two o'clock, but got there with extreme difficulty; every twenty yards I had the cramp in the upper part of both thighs, so that I was afraid I should not have been able to have got down again. It was also necessary to return by another road, as it was out of the question to pass over the saddle-back. I was therefore obliged to give up the two higher peaks. Their altitude was but little greater, and every purpose of geology had been answered; so that the attempt was not worth the hazard of any further exertion. I presume the cause of the cramp was the great change in the kind of muscular action, from that of hard riding to that of still harder climbing. It is a lesson worth remembering, as in some cases it might cause much difficulty.

I have already said the mountain is composed of white quartz rock, and with it a little glossy clay-slate is associated. At the height of a few hundred feet above the plain, patches of conglomerate adhered in several places to the solid rock. They resembled in hardness, and in the nature of the cement, the masses which may be seen daily forming on some coasts. I do not doubt these pebbles were in a similar manner aggregated, at a period when the great calcareous formation was depositing beneath the surrounding sea. We may believe that the jagged and battered forms of the hard quartz yet show the effects of the waves of an open ocean.

I was, on the whole, disappointed with this ascent. Even the view was insignificant; – a plain like the sea, but without its beautiful colour and defined outline. The scene, however, was novel, and a little danger, like salt to meat, gave it a relish. That the danger was very little was certain, for my two companions made a good fire – a thing which is never done when it is suspected that Indians are near. I reached the place of our bivouac by sunset, and drinking much maté, and smoking several cigaritos, soon made up my bed for the night. The wind was very strong and cold, but I never slept more comfortably.

September 10*th*. – In the morning, having fairly scudded before the gale, we arrived by the middle of the day at the Sauce posta. On the road we saw great numbers of deer, and near the mountain a guanaco. The plain, which abuts against the Sierra, is traversed by some curious gulleys, of which one was about twenty feet wide, and at least thirty deep; we were obliged in consequence to make a considerable circuit before we could find a pass. We stayed the night at the posta, the conversation, as was generally the case, being about the Indians. The Sierra Ventana was formerly a great place of resort; and three or four years ago there was much fighting there. My guide had been present when many Indians were killed: the women escaped to the top of the ridge, and fought most desperately with great stones; many thus saving themselves.

September 11*th.* – Proceeded to the third posta in company with the lieutenant who commanded it. The distance is called fifteen leagues; but it is only guess-work, and is generally over-stated. The road was uninteresting, over a dry grassy plain; and on our left hand at a greater or less distance there were some low hills; a continuation of which we crossed close to the posta. Before our arrival we met a large herd of cattle and horses, guarded by fifteen soldiers; but we were told many had been lost. It is very difficult to drive animals across the plains; for if in the night a puma, or even a fox, approaches, nothing can prevent the horses dispersing in every direction; and a storm will have the same effect. A short time since, an officer left Buenos Ayres with five hundred horses, and when he arrived at the army he had under twenty.

Soon afterwards we perceived by the cloud of dust, that a party of horsemen were coming towards us; when far distant my companions knew them to be Indians, by their long hair streaming behind their backs. The Indians generally have a fillet round their heads, but never any covering; and their black hair blowing across their swarthy faces, heightens to an uncommon degree the wildness of their appearance. They turned out to be a party of Bernantio's friendly tribe, going to a salina for salt. The Indians eat much salt, their children sucking it like sugar. This habit is very different from that of the Spanish Gauchos, who, leading the same kind of life, eat scarcely any: according to Mungo Park,[1] it is people who live on vegetable food who have an unconquerable desire for salt. The Indians gave us good-humoured nods as they passed at full gallop, driving before them a troop of horses, and followed by a train of lanky dogs.

1 Travels in Africa, p. 233.

September 12th and 13th. – I staid at this posta two days, waiting for a troop of soldiers which, General Rosas had the kindness to send to inform me, would shortly travel to Buenos Ayres; and he advised me to take the opportunity of the escort. In the morning we rode to some neighbouring hills to view the country, and to examine the geology. After dinner the soldiers divided themselves into two parties for a trial of skill with the bolas. Two spears were stuck in the ground thirty-five yards apart, but they were struck and entangled only once in four or five times. The balls can be thrown fifty or sixty yards, but with little certainty. This, however, does not apply to a man on horseback; for when the speed of the horse is added to the force of the arm, it is said that they can be whirled with effect to the distance of eighty yards. As a proof of their force, I may mention, that at the Falkland Islands, when the Spaniards murdered some of their own countrymen and all the Englishmen, a young friendly Spaniard was running away, when a great tall man, by name Luciano, came at full gallop after him, shouting to him to stop, and saying that he only wanted to speak to him. Just as the Spaniard was on the point of reaching the boat, Luciano threw the balls: they struck him on the legs with such a jerk, as to throw him down and to render him for some time insensible. The man, after Luciano had had his talk, was allowed to escape. He told us that his legs were marked by great weals, where the thong had wound round, as if he had been flogged with a whip. In the middle of the day two men arrived, who brought a parcel from the next posta to be forwarded to the general: so that besides these two, our party consisted this evening of my guide and self, the lieutenant, and his four soldiers. The latter were strange beings; the first a fine young negro; the second half Indian and negro; and the two others nondescripts; namely, an old Chilian miner, the colour of mahogany, and another partly a mulatto; but two such mongrels, with such detestable expressions, I never saw before. At night, when they were sitting round the fire, and playing at cards, I retired to view such a Salvator Rosa scene. They were seated under a low cliff, so that I could look down upon them; around the party were lying dogs, arms, remnants of deer and ostriches; and their long spears were stuck in the turf. Further in the dark background, their horses were tied up, ready for any sudden danger. If the stillness of the desolate plain was broken by one of the dogs barking, a soldier, leaving the fire, would place his head close to the ground, and thus slowly scan the horizon. Even if the noisy teru-tero uttered its scream, there would be a pause in the conversation, and every head, for a moment, a little inclined.

What a life of misery these men appear to us to lead! They were at least ten leagues from the Sauce posta, and since the murder committed by the Indians, twenty from another. The Indians are supposed to have made their attack in the middle of the night; for very early in the morning after the murder, they were luckily seen approaching this posta. The whole party here, however, escaped, together with the troop of horses; each one taking a line for himself, and driving with him as many animals as he was able to manage.

The little hovel, built of thistle-stalks, in which they slept, neither kept out the wind or rain; indeed in the latter case the only effect the roof had, was to condense it into larger drops. They had nothing to eat excepting what they could catch, such as ostriches, deer, armadilloes, &c., and their only fuel was the dry stalks of a small plant, somewhat resembling an aloe. The sole luxury which these men enjoyed was smoking the little paper cigars, and sucking maté. I used to think that the carrion vultures, man's constant attendants on these dreary plains, while seated on the little neighbouring cliffs, seemed by their very patience to say, 'Ah! when the Indians come we shall have a feast.'

In the morning we all sallied forth to hunt, and although we had not much success, there were some animated chaces. Soon after starting the party separated, and so arranged their plans, that at a certain time of the day (in guessing which they show much skill) they should all meet from different points of the compass on a plain piece of ground, and thus drive together the wild animals. One day I went out hunting at Bahia Blanca, but the men there merely rode in a crescent, each being about a quarter of a mile apart from the other. A fine male ostrich being turned by the headmost riders, tried to escape on one side. The Gauchos pursued at a reckless pace, twisting their horses about with the most admirable command, and each man whirling the balls round his head. At length the foremost threw them, revolving through the air: in an instant the ostrich rolled over and over, its legs fairly lashed together by the thong.

The plains abound with three kinds of partridge,[1] two of which are as large as hen pheasants. Their destroyer, a small and pretty fox, was also singularly numerous; in the course of the day we could not have seen less than forty or fifty. They were generally near their earths, but the dogs killed one. When we returned to the posta, we found two of the party returned who had been hunting by themselves. They had killed a puma, and had found an ostrich's nest with twenty-seven eggs in it. Each of these is said to equal in weight eleven hens' eggs; so that we obtained from this one nest as much food as 297 hens' eggs would have given.

September 14th. – As the soldiers belonging to the next posta meant to return, and we should together make a party of five, and all armed, I determined not to wait for the expected troops. My host, the lieutenant, pressed me much to stop. As he had been very obliging – not only providing me with food, but lending me his private horses – I wanted to make him some remuneration. I asked my guide whether I might do so, but he told me certainly not; that the only answer I should receive, probably would be, 'We have meat for the dogs in our country, and therefore do not grudge it to a Christian.' It must not be supposed that the rank of lieutenant in such an army would at all prevent the acceptance of payment: it was only the high sense of hospitality, which every traveller is bound to acknowledge as nearly universal throughout these provinces. After galloping some leagues, we came to a low swampy country, which extends for nearly eighty miles northward, as far as the Sierra Tapalguen. In some parts there were fine damp plains, covered with grass, while others had a soft, black, and peaty soil. There were also many extensive but shallow lakes, and large beds of reeds. The country on the whole resembled the better parts of the Cambridgeshire fens. At night we had some difficulty in finding, amidst the swamps, a dry place for our bivouac.

September 15th. – Rose very early in the morning, and shortly after passed the posta where the Indians had murdered the five soldiers. The officer had eighteen chuzo wounds in his body. By the middle of the day, after a hard gallop, we reached the fifth posta: on account

---

1 Two species of Tinamus, and *Eudromia elegans* of A. d'Orbigny, which can only be called a partridge with regard to its habits.

of some difficulty in procuring horses we stayed there the night. As this point was the most exposed on the whole line, twenty-one soldiers were stationed here; at sunset they returned from hunting, bringing with them seven deer, three ostriches, and many armadilloes and partridges. When riding through the country, it is a common practice to set fire to the plain; and hence at night, as on this occasion, the horizon was illuminated in several places by brilliant conflagrations. This is done partly for the sake of puzzling any stray Indians, but chiefly for improving the pasture. In grassy plains unoccupied by the larger ruminating quadrupeds, it seems necessary to remove the superfluous vegetation by fire, so as to render the new year's growth serviceable.

The rancho at this place did not boast even of a roof, but merely consisted of a ring of thistle-stalks, to break the force of the wind. It was situated on the borders of an extensive but shallow lake, swarming with wild fowl, among which the black-necked swan was conspicuous.

The kind of plover, which appears as if mounted on stilts, (Himantopus nigricollis) is here common in flocks of considerable size. It has been wrongfully accused of inelegance; when wading about in shallow water, which is its favourite resort, its gait is far from awkward. These birds in a flock utter a noise, that singularly resembles the cry of a pack of small dogs in full chace: waking in the night, I have more than once been for a moment startled at the distant sound. The teru-tero (Vanellus cayanus) is another bird which often disturbs the stillness of the night. In appearance and habits it resembles in many respects our peewits; its wings, however, are armed with sharp spurs, like those on the legs of the common cock. As our peewit takes its name from the sound of its voice, so does the teru-tero. While riding over the grassy plains, one is constantly pursued by these birds, which appear to hate mankind, and I am sure deserve to be hated for their never-ceasing, unvaried, harsh screams. To the sportsman they are most annoying, by telling every other bird and animal of his approach: to the traveller in the country, they may possibly, as Molina says, do good, by warning him of the midnight robber. During the breeding season, they attempt, like our peewits, by feigning to be wounded, to draw away from their nests dogs and other enemies. The eggs of this bird are esteemed a great delicacy.

September 16th. – To the seventh posta at the foot of the Sierra Tapalguen. The country was quite level, with a coarse herbage and a soft peaty soil. The hovel was here remarkably neat, the posts and rafters being made of about a dozen dry thistle-stalks bound together with thongs of hide; and by the support of these Ionic-like columns, the roof and sides were thatched with reeds. We were here told a fact, which I would not have credited, if I had not had partly ocular proof of it; namely, that, during the previous night, hail as large as small apples, and extremely hard, had fallen with such violence, as to kill the greater number of the wild animals. One of the men had already found thirteen deer (Cervus campestris) lying dead, and I saw their *fresh* hides; another of the party, a few minutes after my arrival, brought in seven more. Now I well know, that one man without dogs could hardly have killed seven deer in a week. The men believed they had seen about fifteen dead ostriches (part of one of which we had for dinner); and they said that several were running about evidently blind in one eye. Numbers of smaller birds, as ducks, hawks, and partridges, were killed. I saw one of the latter with a black mark on its back, as if it had been struck with a paving-stone. A fence of thistle-stalks round the hovel was nearly broken down, and my informer, putting his head out to see what was the matter, received a severe cut, and now wore a bandage. The storm was said to have been of limited extent: we certainly saw from our last night's bivouac a dense cloud and lightning in this direction. It is marvellous how such strong animals as deer could thus have been killed; but I have no doubt, from the evidence I have given, that the story is not in the least exaggerated. I am glad, however, to have its credibility supported by the Jesuit Drobrizhoffer,[1] who, speaking of a country much to the northward, says, hail fell of an enormous size and killed vast numbers of cattle: the Indians hence called the place *Lalegraicavalca*, meaning 'the little white things.' Dr. Malcolmson, also, informs me that he witnessed in 1831 in India, a hail-storm, which killed numbers of large birds and much injured the cattle. These hail-stones were flat, and one was ten inches in

1 History of the Abipones, vol. ii. p. 6.

circumference, and another weighed two ounces. They ploughed up a gravel-walk like musket-balls, and passed through glass-windows, making round holes, but not cracking them.

Having finished our dinner of hail-stricken meat, we crossed the Sierra Tapalguen; a low range of hills, a few hundred feet in height, which commences at Cape Corrientes. The rock in this part is pure quartz; further eastward I understand it is granitic. The hills are of a remarkable form; they consist of flat patches of table-land, surrounded by low perpendicular cliffs, like the outliers of a sedimentary deposit. The hill which I ascended was very small, not above a couple of hundred yards in diameter; but I saw others larger. One which goes by the name of the 'Corral,' is said to be two or three miles in diameter, and encompassed by perpendicular cliffs between thirty and forty feet high, excepting at one spot, where the entrance lies. Falconer[1] gives a curious account of the Indians driving troops of wild horses into it, and then by guarding the entrance, keeping them secure. I have never heard of any other instance of table-land in a formation of quartz, and which, in the hill I examined, had neither cleavage nor stratification. I was told that the rock of the 'Corral' was white, and would strike fire.

We did not reach the posta on the Rio Tapalguen till after it was dark. At supper, from something which was said, I was suddenly struck with horror at thinking that I was eating one of the favourite dishes of the country, namely, a half-formed calf, long before its proper time of birth. It turned out to be Puma; the meat is very white, and remarkably like veal in taste. Dr. Shaw was laughed at for stating that 'the flesh of the lion is in great esteem, having no small affinity with veal, both in colour, taste, and flavour.' Such certainly is the case with the Puma. The Gauchos differ in their opinion, whether the Jaguar is good eating, but are unanimous in saying that cat is excellent.

September 17th. – We followed the course of the Rio Tapalguen, through a very fertile country, to the ninth posta. Tapalguen itself, or the town of Tapalguen, if it may be so called, consists of a perfectly level plain, studded over, as far as the eye can reach, with the toldos, or oven-shaped huts of the Indians. The families of the friendly

1 Falconer's Patagonia, p. 70.

Indians, who were fighting on the side of Rosas, resided here. We met and passed many young Indian women, riding by two or three together on the same horse: they, as well as many of the young men, were strikingly handsome, – their fine ruddy complexions being the picture of health. Besides the toldos, there were three ranchos; one inhabited by the Commandant, and the two others by Spaniards with small shops.

We were here able to buy some biscuit. I had now been several days without tasting any thing besides meat: I did not at all dislike this new regimen; but I felt as if it would only have agreed with me with hard exercise. I have heard that patients in England, when desired to confine themselves exclusively to an animal diet, even with the hope of life before their eyes, have hardly been able to endure it. Yet the Gaucho in the Pampas, for months together, touches nothing but beef. But they eat, I observe, a very large proportion of fat, which is of a less animalized nature; and they particularly dislike dry meat, such as that of the Agouti. Dr. Richardson,[1] also, has remarked, 'that when people have fed for a long time solely upon lean animal food, the desire for fat becomes so insatiable, that they can consume a large quantity of unmixed and even oily fat without nausea:' this appears to me a curious physiological fact. It is, perhaps, from their meat regimen that the Gauchos, like other carnivorous animals, can abstain long from food. I was told that at Tandeel, some troops voluntarily pursued a party of Indians for three days, without eating or drinking.

We saw in the shops many articles, such as horsecloths, belts, and garters, woven by the Indian women. The patterns were very pretty, and the colours brilliant; the workmanship of the garters was so good that an English merchant at Buenos Ayres maintained they must have been manufactured in England, till he found the tassels had been fastened by split sinew.

September 18*th*. – We had a very long ride this day. At the twelfth posta, which is seven leagues south of the Rio Salado, we came to the first estancia with cattle and white women. Afterwards we had to ride for many miles through a country flooded with water above our horses' knees. By crossing the stirrups, and riding Arab-like with

1 Fauna Boreali-Americana, vol. i. p. 35.

our legs bent up, we contrived to keep tolerably dry. It was nearly dark
when we arrived at the Salado; the stream was deep, and about forty
yards wide; in summer, however, its bed becomes almost dry, and the
little remaining water nearly as salt as that of the sea. We slept at one of
the great estancias of General Rosas. It was fortified, and of such an
extent, that arriving in the dark I thought it was a town and fortress. In
the morning we saw immense herds of cattle, the general here having
seventy-four square leagues of land. Formerly nearly three hundred
men were employed about this estate, and they defied all the attacks of
the Indians.

September 19*th.* – Passed the Guardia del Monte. This is a nice
scattered little town, with many gardens, full of peach and quince
trees. The plain here looked like that around Buenos Ayres; the turf
being short and bright green, with beds of clover and thistles, and with
bizcacha holes. I was very much struck with the marked change in the
aspect of the country after having crossed the Salado. From a coarse
herbage we passed on to a carpet of fine green verdure. I at first
attributed this to some change in the nature of the soil, but the
inhabitants assured me that here, as well as in Banda Oriental, where
there is as great a difference between the country around Monte
Video and the thinly-inhabited savannahs of Colonia, the whole
was to be attributed to the manuring and grazing of the cattle. Exactly
the same fact has been observed in the prairies[1] of North America,
where coarse grass, between five and six feet high, when grazed by
cattle, changes into common pasture land. I am not botanist enough
to say whether the change here is owing to the introduction of new
species, to the altered growth of the same, or to a difference in their
proportional numbers. Azara has also observed with astonishment
this change: he is likewise much perplexed by the immediate appearance of plants not occurring in the neighbourhood, on the borders
of any track that leads to a newly-constructed hovel. In another part
he says,[2] 'ces chevaux (sauvages) ont la manie de préférer les chemins,
et le bord des routes pour déposer leurs excrémens, dont on trouve
des monceaux dans ces endroits.' Does this not partly explain

1 See Mr. Atwater's account of the Prairies, in Silliman's N. A. Journal, vol. i. p. 117.
2 Azara's Voyage, vol. i. p. 373.

the circumstance? We thus have lines of richly-manured land serving as channels of communication across wide districts.

Near the Guardia we find the southern limit of two European plants, now become extraordinarily common. The fennel in great profusion covers the ditch-banks in the neighbourhood of Buenos Ayres, Monte Video, and other towns. But the cardoon (Cynara cardunculus)[1] has a far wider range: it occurs in these latitudes on both sides of the Cordillera, across the continent. I saw it in unfre- quented spots in Chile, Entre Rios, and Banda Oriental. In the latter country alone, very many (probably several hundred) square miles are covered by one mass of these prickly plants, and are impenetrable by man or beast. Over the undulating plains, where these great beds occur, nothing else can now live. Before their introduction, how- ever, the surface must have supported, as in other parts, a rank herbage. I doubt whether any case is on record of an invasion on so grand a scale of one plant over the aborigines. As I have already said, I nowhere saw the cardoon south of the Salado; but it is probable that in proportion as that country becomes inhabited, the cardoon will extend its limits. The case is different with the giant thistle (with variegated leaves) of the Pampas, for I met with it in the valley of the Sauce. According to the principles so well laid down by Mr. Lyell, few countries have undergone more remarkable changes, since the year 1535, when the first colonist of La Plata landed with seventy- two horses. The countless herds of horses, cattle, and sheep, not only have altered the whole aspect of the vegetation, but they have almost banished the guanaco, deer, and ostrich. Numberless other changes must likewise have taken place; the wild pig in some parts probably

1 M. A. d'Orbigny (vol. i. p. 474) says that the cardoon and artichoke are both found wild. Dr. Hooker (Botanical Magazine, vol. lv. p. 2862) has described a variety of the Cynara from this part of South America under the name of *inermis*. He states that botanists are now generally agreed that the cardoon and the artichoke are varieties of one plant. I may add, that an intelligent farmer assured me that he had observed in a deserted garden some artichokes changing into the common cardoon. Dr. Hooker believes that Head's vivid description of the thistle of the Pampas applies to the cardoon; but this is a mistake. Captain Head referred to the plant, which I have mentioned a few lines lower down, under the title of giant thistle. Whether it is a true thistle, I do not know; but it is quite different from the cardoon; and more like a thistle properly so called.

replaces the peccari; packs of wild dogs may be heard howling on the wooded banks of the less frequented streams; and the common cat, altered into a large and fierce animal, inhabits rocky hills. As M. d'Orbigny has remarked, the increase in numbers of the carrion-vulture, since the introduction of the domestic animals, must have been infinitely great; and we have given reasons for believing that they have extended their southern range. No doubt many plants, besides the cardoon and fennel, are naturalized; thus the islands near the mouth of the Parana are thickly clothed with peach and orange trees, springing from seeds carried there by the waters of the river.

While changing horses at the Guardia several people questioned us much about the army, – I never saw any thing like the enthusiasm for Rosas, and for the success of the 'most just of all wars, because against barbarians.' This expression, it must be confessed, is very natural, for till lately, neither man, woman, nor horse, was safe from the attacks of the Indians. We had a long day's ride over the same rich green plain, abounding with various flocks, and with here and there a solitary estancia, and its one *ombu* tree. In the evening it rained heavily: on arriving at a post-house we were told by the owner that if we had not a regular passport we must pass on, for there were so many robbers he would trust no one. When he read, however, my passport, which began with 'El Naturalista Don Carlos,' his respect and civility were as unbounded as his suspicions had been before. What a naturalist might be, neither he nor his countrymen, I suspect, had any idea; but probably my title lost nothing of its value from that cause.

September 20th. – We arrived by the middle of the day at Buenos Ayres. The outskirts of the city looked quite pretty, with the agave hedges, and groves of olive, peach, and willow trees, all just throwing out their fresh green leaves. I rode to the house of Mr. Lumb, an English merchant, to whose kindness and hospitality, during my stay in the country, I was greatly indebted.

The city of Buenos Ayres is large;[1] and I should think one of the most regular in the world. Every street is at right angles to the one it

1 It is said to contain 60,000 inhabitants. Monte Video, the second town of import-ance on the banks of the Plata, has 15,000.

crosses, and the parallel ones being equidistant, the houses are collected into solid squares of equal dimensions, which are called quadras. On the other hand, the houses themselves are hollow squares; all the rooms opening into a neat little courtyard. They are generally only one story high, with flat roofs, which are fitted with seats, and are much frequented by the inhabitants in summer. In the centre of the town is the Plaza, where the public offices, fortress, cathedral, &c., stand. Here also, the old viceroys, before the revolution, had their palaces. The general assemblage of buildings possesses considerable architectural beauty, although none individually can boast of any.

The great *corral*, where the animals are kept for slaughter to supply food to this beef-eating population, is one of the spectacles best worth seeing. The strength of the horse as compared to that of the bullock is quite astonishing: a man on horseback having thrown his lazo round the horns of a beast, can drag it any where he chooses. The animal ploughing up the ground with outstretched legs, in vain efforts to resist the force, generally dashes at full speed to one side; but the horse immediately turning to receive the shock, stands so firmly that the bullock is almost thrown down, and it is surprising that their necks are not broken. The struggle is not, however, one of fair strength; the horse's girth being matched against the bullock's extended neck. In a similar manner a man can hold the wildest horse, if caught with the lazo just behind the ears. When the bullock has been dragged to the spot where it is to be slaughtered, the *matador* with great caution cuts the hamstrings. Then is given the death bellow; a noise more expressive of fierce agony than any I know: I have often distinguished it from a long distance, and have always known that the struggle was then drawing to a close. The whole sight is horrible and revolting: the ground is almost made of bones; and the horses and riders are drenched with gore.

CHAPTER VII

BUENOS AYRES TO ST. FÉ

Excursion to St. Fé – Thistle Beds – Habits of the Bizcacha
– Little Owl – Saline Streams – Level Plains – Mastodon –
St. Fé – Change in Landscape – Geology – Tooth of extinct
Horse – Relation of the Fossil and recent Quadrupeds of
North and South America – Effects of a great Drought –
Parana – Habits of the Jaguar – Scissor-beak – Kingfisher,
Parrot, and Scissor-tail – Revolution – Buenos Ayres –
State of Government.

SEPTEMBER 27*th*. – In the evening I set out on an excursion to St.
Fé, which is situated nearly three hundred English miles from Buenos
Ayres, on the banks of the Parana. The roads in the neighbourhood of
the city, after the rainy weather, were extraordinarily bad. I should
never have thought it possible for a bullock waggon to have crawled
along: as it was, they scarcely went at the rate of a mile an hour, and a
man was kept ahead, to survey the best line for making the attempt.
The bullocks were terribly jaded: it is a great mistake to suppose that
with improved roads, and an accelerated rate of travelling, the sufferings of the animals increase in the same proportion. We passed a train
of waggons and a troop of beasts on their road to Mendoza. The
distance is about 580 geographical miles, and the journey is generally
performed in fifty days. These waggons are very long, narrow, and
thatched with reeds; they have only two wheels, the diameter of
which in some cases is as much as ten feet. Each is drawn by six
bullocks, which are urged on by a goad at least twenty feet long: this is
suspended from within the roof; for the wheel bullocks a smaller one
is kept; and for the intermediate pair, a point projects at right angles
from the middle of the long one. The whole apparatus looked like
some implement of war.

September 28*th*. – We passed the small town of Luxan, where there
is a wooden bridge over the river – a most unusual convenience in
this country. We passed also Areco. The plains appeared level, but

were not so in fact; for in various places the horizon was distant. The estancias are here wide apart; for there is little good pasture, owing to the land being covered by beds either of an acrid clover, or of the great thistle. The latter, well known from the animated description given by Sir F. Head, were at this time of the year two-thirds grown; in some parts they were as high as the horse's back, but in others they had not yet sprung up, and the ground was bare and dusty as on a turnpike-road. The clumps were of the most brilliant green, and they made a pleasing miniature-likeness of broken forest land. When the thistles are full grown, the great beds are impenetrable, except by a few tracks, as intricate as those in a labyrinth. These are only known to the robbers, who at this season inhabit them, and sally forth at night to rob and cut throats with impunity. Upon asking at a house whether robbers were numerous, I was answered, 'The thistles are not up yet;' – the meaning of which reply was not at first very obvious. There is little interest in passing over these tracts, for they are inhabited by few animals or birds, excepting the bizcacha and its friend the little owl.

The bizcacha[1] is well known to form a prominent feature in the zoology of the Pampas. It is found as far south as the Rio Negro, in lat. 41°, but not beyond. It cannot, like the agouti, subsist on the gravelly and desert plains of Patagonia, but prefers a clayey or sandy soil, which produces a different and more abundant vegetation. Near Mendoza, at the foot of the Cordillera, it occurs in close neighbourhood with the allied alpine species. It is a very curious circumstance in its geographical distribution, that it has never been seen, fortunately for the inhabitants of Banda Oriental, to the eastward of the river Uruguay: yet in this province there are plains which appear admirably adapted to its habits. The Uruguay has formed an insuperable obstacle to its migration; although the broader barrier of the Parana has been passed, and the bizcacha is common in Entre Rios, the province between these two great rivers. Near Buenos Ayres these animals are exceedingly common. Their most favourite resort appears to be those parts of the plain which during one half of the year are covered

1 The bizcacha (Lagostomus trichodactylus) somewhat resembles a large rabbit, but with bigger gnawing teeth and a long tail: it has, however, only three toes behind, like the agouti. During the last three or four years the skins of these animals have been sent to England for the sake of the fur.

with giant thistles, to the exclusion of other plants. The Gauchos affirm that it lives on roots; which, from the great strength of its gnawing teeth, and the kind of places frequented by it, seems probable. In the evening the bizcachas come out in numbers, and quietly sit at the mouths of their burrows on their haunches. At such times they are very tame, and a man on horseback passing by seems only to present an object for their grave contemplation. They run very awkwardly, and when running out of danger, from their elevated tails and short front legs, much resemble great rats. Their flesh, when cooked, is very white and good, but it is seldom used.

The bizcacha has one very singular habit; namely, dragging every hard object to the mouth of its burrow: around each group of holes many bones of cattle, stones, thistle-stalks, hard lumps of earth, dry dung, &c., are collected into an irregular heap, which frequently amounts to as much as a wheelbarrow would contain. I was credibly informed that a gentleman, when riding on a dark night, dropped his watch; he returned in the morning, and by searching the neighbourhood of every bizcacha hole on the line of road, as he expected, he soon found it. This habit of picking up whatever may be lying on the ground any where near its habitation, must cost much trouble. For what purpose it is done, I am quite unable to form even the most remote conjecture: it cannot be for defence, because the rubbish is chiefly placed above the mouth of the burrow, which enters the ground at a very small inclination. No doubt there must exist some good reason; but the inhabitants of the country are quite ignorant of it. The only fact which I know analogous to it, is the habit of that extraordinary Australian bird, the Calodera maculata, which makes an elegant vaulted passage of twigs for playing in, and which collects near the spot, land and sea-shells, bones, and the feathers of birds, especially brightly coloured ones. Mr. Gould, who has described these facts, informs me, that the natives, when they lose any hard object, search the playing passages, and he has known a tobacco-pipe thus recovered.

The little owl (Athene cunicularia), which has been so often mentioned, on the plains of Buenos Ayres exclusively inhabits the holes of the bizcacha; but in Banda Oriental it is its own workman. During the open day, but more especially in the evening, these birds may be seen in every direction standing frequently by pairs on the hillock near their burrows. If disturbed they either enter the hole, or, uttering a shrill harsh cry, move with a remarkably undulatory flight to a short distance, and then turning round, steadily gaze at their pursuer. Occasionally in the evening they may be heard hooting. I found in the stomachs of two which I opened the remains of mice, and I one day saw a small snake killed and carried away. It is said that snakes are their common prey during the daytime. I may here mention, as showing on what various kinds of food owls subsist, that a species killed among the islets of the Chonos Archipelago, had its stomach full of good-sized crabs. In India[1] there is a fishing genus of owls, which likewise catches crabs.

In the evening we crossed the Rio Arrecife on a simple raft made of barrels lashed together, and slept at the post-house on the other side. I this day paid horse-hire for thirty-one leagues; and although the sun was glaring hot I was but little fatigued. When Captain Head talks of riding fifty leagues a day, I do not imagine the distance is equal to 150 English miles. At all events, the thirty-one leagues was only 76 miles in a straight line, and in an open country I should think four additional miles for turning would be a sufficient allowance.

29th and 30th. – We continued to ride over plains of the same character. At San Nicolas I first saw the noble river of the Parana. At the foot of the cliff on which the town stands, some large vessels were at anchor. Before arriving at Rozario, we crossed the Saladillo, a stream of fine clear running water, but too saline to drink. Rozario is a large town built on a dead level plain, which forms a cliff about sixty feet high over the Parana. The river here is very broad, with many islands, which are low and wooded, as is also the opposite shore. The view would resemble that of a great lake, if it were not for the linear-shaped islets, which alone give the idea of running water. The cliffs are the most picturesque part; sometimes they are absolutely perpendicular, and of a red colour; at other times in large broken masses, covered with cacti and mimosa-trees. The real grandeur, however, of an immense river like this, is derived from reflecting how important a means of communication and commerce it forms

1 Journal of Asiatic Soc., vol. v. p. 363.

between one nation and another; to what a distance it travels; and from how vast a territory it drains the great body of fresh water which flows past your feet.

For many leagues north and south of San Nicolas and Rozario, the country is really level. Scarcely anything which travellers have written about its extreme flatness, can be considered as exaggeration. Yet I could never find a spot where, by slowly turning round, objects were not seen at greater distances in some directions than in others; and this manifestly proves inequality in the plain. At sea, a person's eye being six feet above the surface of the water, his horizon is two miles and four-fifths distant. In like manner, the more level the plain, the more nearly does the horizon approach within these narrow limits; and this, in my opinion, entirely destroys that grandeur which one would have imagined that a vast level plain would have possessed.

October 1st. – We started by moonlight and arrived at the Rio Tercero by sunrise. This river is also called the Saladillo, and it deserves the name, for the water is brackish. I stayed here the greater part of the day, searching for fossil bones. Besides a perfect tooth of the Toxodon, and many scattered bones, I found two immense skeletons near each other, projecting in bold relief from the perpendicular cliff of the Parana. They were, however, so completely decayed, that I could only bring away small fragments of one of the great molar teeth; but these are sufficient to show that the remains belonged to a Mastodon, probably to the same species with that, which formerly must have inhabited the Cordillera in Upper Peru in such great numbers. The men who took me in the canoe, said they had long known of these skeletons, and had often wondered how they had got there: the necessity of a theory being felt, they came to the conclusion that, like the bizcacha, the mastodon was formerly a burrowing animal! In the evening we rode another stage, and crossed the Monge, another brackish stream, bearing the dregs of the washings of the Pampas.

October 2nd. – We passed through Corunda, which, from the luxuriance of its gardens, was one of the prettiest villages I saw. From this point to St. Fé the road is not very safe. The western side of the Parana northward, ceases to be inhabited; and hence the

Indians sometimes come down thus far, and waylay travellers. The nature of the country also favours this, for instead of a grassy plain, there is an open woodland, composed of low prickly mimosas. We passed some houses that had been ransacked and since deserted; we saw also a spectacle, which my guides viewed with high satisfaction; it was the skeleton of an Indian with the dried skin hanging on the bones, suspended to the branch of a tree.

In the morning we arrived at St. Fé. I was surprised to observe how great a change of climate a difference of only three degrees of latitude between this place and Buenos Ayres had caused. This was evident from the dress and complexion of the men – from the increased size of the ombu-trees – the number of new cacti and other plants – and especially from the birds. In the course of an hour I remarked half-a-dozen birds, which I had never seen at Buenos Ayres. Considering that there is no natural boundary between the two places, and that the character of the country is nearly similar, the difference was much greater than I should have expected.

October 3*rd and* 4*th.* – I was confined for these two days to my bed by a headach. A good-natured old woman, who attended me, wished me to try many odd remedies. A common practice is, to bind an orange-leaf or a bit of black plaster to each temple: and a still more general plan is, to split a bean into halves, moisten them, and place one on each temple, where they will easily adhere. It is not thought proper ever to remove the beans or plaster, but to allow them to drop off; and sometimes, if a man, with patches on his head, is asked, what is the matter? he will answer, 'I had a headach the day before yesterday.' Many of the remedies used by the people of the country are ludicrously strange, but too disgusting to be mentioned. One of the least nasty is to kill and cut open two puppies and bind them on each side of a broken limb. Little hairless dogs are in great request to sleep at the feet of invalids.

St. Fé is a quiet little town, and is kept clean and in good order. The governor, Lopez, was a common soldier at the time of the revolution; but has now been seventeen years in power. This stability of government is owing to his tyrannical habits; for tyranny seems as yet better adapted to these countries than republicanism. The governor's favourite occupation is hunting Indians: a short time since he slaughtered forty-eight, and sold the children at the rate of three or four pounds apiece.

October 5th. – We crossed the Parana to St. Fé Bajada, a town on the opposite shore. The passage took some hours, as the river here consisted of a labyrinth of small streams, separated by low wooded islands. I had a letter of introduction to an old Catalonian Spaniard, who treated me with the most uncommon hospitality. The Bajada is the capital of Entre Rios. In 1825 the town contained 6000 inhabitants, and the province 30,000; yet, few as the inhabitants are, no province has suffered more from bloody and desperate revolutions. They boast here of representatives, ministers, a standing army, and governors: so it is no wonder that they have their revolutions. At some future day this must be one of the richest countries of La Plata. The soil is varied and productive; and its almost insular form gives it two grand lines of communication by the rivers Parana and Uruguay.

I was delayed here five days, and employed myself in examining the geology of the surrounding country, which was very interesting. We here see at the bottom of the cliffs, beds containing sharks' teeth and sea-shells of extinct species, passing above into an indurated marl, and from that into the red clayey earth of the Pampas, with its calcareous concretions and the bones of terrestrial quadrupeds. This vertical section clearly tells us of a large bay of pure salt-water, gradually encroached on, and at last converted into the bed of a muddy estuary, into which floating carcasses were swept. At Punta Gorda, in Banda Oriental, I found an alternation of the Pampæan estuary deposit, with a limestone containing some of the same extinct sea-shells; and this shows either a change in the former currents, or more probably an oscillation of level in the bottom of the ancient estuary. Until lately, my reasons for considering the Pampæan formation to be an estuary deposit were, its general appearance, its position at the mouth of the existing great river the Plata, and the presence of so many bones of terrestrial quadrupeds; but now Professor Ehrenberg has had the kindness to examine for me a little of the red earth, taken from low down in the deposit, close to the skeletons of the mastodon, and he finds in it many infusoria, partly salt-water and partly fresh-water forms, with the latter rather preponderating; and therefore, as he remarks, the water must have been brackish. M. A. d'Orbigny found on the banks of the Parana, at the height of a hundred feet, great beds of an estuary shell, now living a hundred miles lower down nearer the sea; and I found similar shells at a less height on the banks of the Uruguay: this shows that just before the Pampas was slowly elevated into dry land, the water covering it was brackish. Below Buenos Ayres there are upraised beds of sea-shells of existing species, which also proves that the period of elevation of the Pampas was within the recent period.

In the Pampæan deposit at the Bajada I found the osseous armour of a gigantic armadillo-like animal, the inside of which, when the earth was removed, was like a great cauldron; I found also teeth of the Toxodon and Mastodon, and one tooth of a Horse, in the same stained and decayed state. This latter tooth greatly interested me,[1] and I took scrupulous care in ascertaining that it had been embedded contemporaneously with the other remains; for I was not then aware that amongst the fossils from Bahia Blanca there was a horse's tooth hidden in the matrix: nor was it then known with certainty that the remains of horses are common in North America. Mr. Lyell has lately brought from the United States a tooth of a horse; and it is an interesting fact, that Professor Owen could find in no species, either fossil or recent, a slight but peculiar curvature characterizing it, until he thought of comparing it with my specimen found here: he has named this American horse Equus curvidens. Certainly it is a marvellous fact in the history of the Mammalia, that in South America a native horse should have lived and disappeared, to be succeeded in after ages by the countless herds descended from the few introduced with the Spanish colonists!

The existence in South America of a fossil horse, of the mastodon, possibly of an elephant,[2] and of a hollow-horned ruminant, discovered by MM. Lund and Clausen in the caves of Brazil, are highly interesting facts with respect to the geographical distribution of animals. At the present time, if we divide America, not by

1 I need hardly state here that there is good evidence against any horse living in America at the time of Columbus.

2 Cuvier, Ossemens Fossiles, tom. i. p. 158.

the Isthmus of Panama, but by the southern part of Mexico[1] in lat. 20°, where the great table-land presents an obstacle to the migration of species, by affecting the climate, and by forming, with the exception of some valleys and of a fringe of low land on the coast, a broad barrier; we shall then have the two zoological provinces of North and South America strongly contrasted with each other. Some few species alone have passed the barrier, and may be considered as wanderers from the south, such as the puma, opossum, kinkajou, and peccari. South America is characterized by possessing many peculiar gnawers, a family of monkeys, the llama, peccari, tapir, opossums, and, especially, several genera of Edentata, the order which includes the sloths, ant-eaters, and armadilloes. North America, on the other hand, is characterized (putting on one side a few wandering species) by numerous peculiar gnawers, and by four genera (the ox, sheep, goat, and antelope) of hollow-horned ruminants, of which great division South America is not known to possess a single species. Formerly, but within the period when most of the now existing shells were living, North America possessed, besides hollow-horned ruminants, the elephant, mastodon, horse, and three genera of Edentata, namely, the Megatherium, Megalonyx, and Mylodon. Within nearly this same period (as proved by the shells at Bahia Blanca) South America possessed, as we have just seen, a mastodon, horse, hollow-horned ruminant, and the same three genera (as well as several others) of the Edentata. Hence it is evident that North and South America, in having within a late geological period these several genera in common, were much more closely related in the character of their terrestrial inhabitants than they now are. The more I reflect on this case, the more interesting it appears: I know of no other instance where we can almost mark the period and manner of the splitting up of one great region into two well-characterized zoological provinces.

1 This is the geographical division followed by Lichtenstein, Swainson, Erichson, and Richardson. The section from Vera Cruz to Acapulco, given by Humboldt in the Polit. Essay on Kingdom of N. Spain, will show how immense a barrier the Mexican table-land forms. Dr. Richardson, in his admirable Report on the Zoology of N. America read before the Brit. Assoc. 1836 (p. 157), talking of the identification of a Mexican animal with the *Synetheres prehensilis*, says, 'We do not know with what propriety, but if correct, it is, if not a solitary instance, at least very nearly so, of a rodent animal being common to North and South America.'

The geologist, who is fully impressed with the vast oscillations of level which have affected the earth's crust within late periods, will not fear to speculate on the recent elevation of the Mexican platform, or, more probably, on the recent submergence of land in the West Indian Archipelago, as the cause of the present zoological separation of North and South America. The South American character of the West Indian mammals[1] seems to indicate that this archipelago was formerly united to the southern continent, and that it has subsequently been an area of subsidence.

When America, and especially North America, possessed its elephants, mastodons, horse, and hollow-horned ruminants, it was much more closely related in its zoological characters to the temperate parts of Europe and Asia than it now is. As the remains of these genera are found on both sides of Behring's Straits[2] and on the plains of Siberia, we are led to look to the north-western side of North America as the former point of communication between the Old and so-called New World. And as so many species, both living and extinct, of these same genera inhabit and have inhabited the Old World, it seems most probable that the North American elephants, mastodons, horse, and hollow-horned ruminants migrated, on land since submerged near Behring's Straits, from Siberia into North America, and thence, on land since submerged in the West Indies, into South America, where for a time they mingled with the forms characteristic of that southern continent, and have since become extinct.

While travelling through the country, I received several vivid descriptions of the effects of a late great drought; and the account of this may throw some light on the cases where vast numbers of animals of all kinds have been embedded together. The period included

1 See Dr. Richardson's Report, p. 157; also L'Institut, 1837, p. 253. Cuvier says the kinkajou is found in the larger Antilles, but this is doubtful. M. Gervais states that the Didelphis crancrivora is found there. It is certain that the West Indies possess some mammifers peculiar to themselves. A tooth of a mastodon has been brought from Bahama: Edin. New Phil. Journ. 1826, p. 395.

2 See the admirable Appendix by Dr. Buckland to Beechey's Voyage; also the writings of Chamisso in Kotzebue's Voyage.

between the years 1827 and 1830 is called the 'gran seco,' or the great drought. During this time so little rain fell, that the vegetation, even to the thistles, failed; the brooks were dried up, and the whole country assumed the appearance of a dusty high road. This was especially the case in the northern part of the province of Buenos Ayres and the southern part of St. Fé. Very great numbers of birds, wild animals, cattle, and horses perished from the want of food and water. A man told me that the deer[1] used to come into his courtyard to the well, which he had been obliged to dig to supply his own family with water; and that the partridges had hardly strength to fly away when pursued. The lowest estimation of the loss of cattle in the province of Buenos Ayres alone, was taken at one million head. A proprietor at San Pedro had previously to these years 20,000 cattle; at the end not one remained. San Pedro is situated in the middle of the finest country; and even now abounds again with animals; yet, during the latter part of the 'gran seco,' live cattle were brought in vessels for the consumption of the inhabitants. The animals roamed from their estancias, and, wandering far southward, were mingled together in such multitudes, that a government commission was sent from Buenos Ayres to settle the disputes of the owners. Sir Woodbine Parish informed me of another and very curious source of dispute; the ground being so long dry, such quantities of dust were blown about, that in this open country the landmarks became obliterated, and people could not tell the limits of their estates.

I was informed by an eye-witness that the cattle in herds of thousands rushed into the Parana, and being exhausted by hunger they were unable to crawl up the muddy banks, and thus were drowned. The arm of the river which runs by San Pedro was so full of putrid carcasses, that the master of a vessel told me that the

1 In Capt. Owen's Surveying Voyage (vol. ii. p. 274) there is a curious account of the effects of a drought on the elephants, at Benguela (west coast of Africa). 'A number of these animals had some time since entered the town, in a body, to possess themselves of the wells, not being able to procure any water in the country. The inhabitants mustered, when a desperate conflict ensued, which terminated in the ultimate discomfiture of the invaders, but not until they had killed one man, and wounded several others.' The town is said to have a population of nearly three thousand! Dr. Malcolmson informs me that during a great drought in India the wild animals entered the tents of some troops at Ellore, and that a hare drank out of a vessel held by the adjutant of the regiment.

smell rendered it quite impassable. Without doubt several hundred thousand animals thus perished in the river: their bodies when putrid were seen floating down the stream; and many in all probability were deposited in the estuary of the Plata. All the small rivers became highly saline, and this caused the death of vast numbers in particular spots; for when an animal drinks of such water it does not recover. Azara describes[1] the fury of the wild horses on a similar occasion, rushing into the marshes, those which arrived first being overwhelmed and crushed by those which followed. He adds that more than once he has seen the carcasses of upwards of a thousand wild horses thus destroyed. I noticed that the smaller streams in the Pampas were paved with a breccia of bones, but this probably is the effect of a gradual increase, rather than of the destruction at any one period. Subsequently to the drought of 1827 to '32, a very rainy season followed, which caused great floods. Hence it is almost certain that some thousands of the skeletons were buried by the deposits of the very next year. What would be the opinion of a geologist, viewing such an enormous collection of bones, of all kinds of animals and of all ages, thus embedded in one thick earthy mass? Would he not attribute it to a flood having swept over the surface of the land, rather than to the common order of things?[2]

October 12th. – I had intended to push my excursion further, but not being quite well, I was compelled to return by a balandra, or one-masted vessel of about a hundred tons' burden, which was bound to Buenos Ayres. As the weather was not fair, we moored early in the day to a branch of a tree on one of the islands. The Parana is full of islands, which undergo a constant round of decay and renovation. In the memory of the master several large ones had disappeared, and others again had been formed and protected by vegetation. They are composed of muddy sand, without even the smallest pebble, and were then about four feet above the level of the river; but during the periodical floods they are inundated. They all present one character;

1 Travels, vol. i. p. 374.
2 These droughts to a certain degree seem to be almost periodical; I was told the dates of several others, and the intervals were about fifteen years.

numerous willows and a few other trees are bound together by a great variety of creeping plants, thus forming a thick jungle. These thickets afford a retreat for capybaras and jaguars. The fear of the latter animal quite destroyed all pleasure in scrambling through the woods. This evening I had not proceeded a hundred yards, before finding indubitable signs of the recent presence of the tiger, I was obliged to come back. On every island there were tracks; and as on the former excursion 'el rastro de los Indios' had been the subject of conversation, so in this was 'el rastro del tigre.'

The wooded banks of the great rivers appear to be the favourite haunts of the jaguar; but south of the Plata, I was told that they frequented the reeds bordering lakes: wherever they are, they seem to require water. Their common prey is the capybara, so that it is generally said, where capybaras are numerous there is little danger from the jaguar. Falconer states that near the southern side of the mouth of the Plata there are many jaguars, and that they chiefly live on fish; this account I have heard repeated. On the Parana they have killed many wood-cutters, and have even entered vessels at night. There is a man now living in the Bajada, who, coming up from below when it was dark, was seized on the deck; he escaped, however, with the loss of the use of one arm. When the floods drive these animals from the islands, they are most dangerous. I was told that a few years since a very large one found its way into a church at St. Fé: two padres entering one after the other were killed, and a third, who came to see what was the matter, escaped with difficulty. The beast was destroyed by being shot from a corner of the building which was unroofed. They commit also at these times great ravages among cattle and horses. It is said that they kill their prey by breaking their necks. If driven from the carcass, they seldom return to it. The Gauchos say that the jaguar, when wandering about at night, is much tormented by the foxes yelping as they follow him. This is a curious coincidence with the fact which is generally affirmed of the jackals accompanying, in a similarly officious manner, the East Indian tiger. The jaguar is a noisy animal, roaring much by night, and especially before bad weather.

One day, when hunting on the banks of the Uruguay, I was shown certain trees, to which these animals constantly recur for the purpose, as it is said, of sharpening their claws. I saw three well-known trees; in front, the bark was worn smooth, as if by the breast of the animal, and on each side there were deep scratches, or rather grooves, extending in an oblique line, nearly a yard in length. The scars were of different ages. A common method of ascertaining whether a jaguar is in the neighbourhood is to examine these trees. I imagine this habit of the jaguar is exactly similar to one which may any day be seen in the common cat, as with outstretched legs and exserted claws it scrapes the leg of a chair; and I have heard of young fruit-trees in an orchard in England having been thus much injured. Some such habit must also be common to the puma, for on the bare hard soil of Patagonia I have frequently seen scores so deep that no other animal could have made them. The object of this practice is, I believe, to tear off the ragged points of their claws, and not, as the Gauchos think, to sharpen them. The jaguar is killed, without much difficulty, by the aid of dogs baying and driving him up a tree, where he is despatched with bullets.

Owing to bad weather we remained two days at our moorings. Our only amusement was catching fish for our dinner: there were several kinds, and all good eating. A fish called the 'armado' (a Silurus) is remarkable from a harsh grating noise which it makes when caught by hook and line, and which can be distinctly heard when the fish is beneath the water. This same fish has the power of firmly catching hold of any object, such as the blade of an oar or the fishing-line, with the strong spine both of its pectoral and dorsal fin. In the evening the weather was quite tropical, the thermometer standing at 79°. Numbers of fireflies were hovering about, and the musquitoes were very troublesome. I exposed my hand for five minutes, and it was soon black with them; I do not suppose there could have been less than fifty, all busy sucking.

October 15*th.* – We got under weigh and passed Punta Gorda, where there is a colony of tame Indians from the province of Missiones. We sailed rapidly down the current, but before sunset, from a silly fear of bad weather, we brought-to in a narrow arm of the river. I took the boat and rowed some distance up this creek. It was very narrow, winding, and deep; on each side a wall thirty or forty feet high, formed by trees intwined with creepers, gave to the canal a singularly gloomy appearance. I here saw a very extraordinary bird, called the Scissor-beak (Rhynchops nigra). It has short legs, web feet, extremely long-pointed wings, and is of about the size of a tern. The beak is flattened laterally, that is, in a plane at right angles to that of a spoonbill or duck. It is as flat and elastic as an ivory paper-cutter, and the lower mandible, differently from every other bird, is an inch and a half longer than the upper. In a lake near Maldonado, from which the water had been nearly drained, and which, in consequence, swarmed with small fry, I saw several of these birds, generally in small flocks, flying rapidly backwards and forwards close to the surface of the lake. They kept their bills wide open, and the lower mandible half buried in the water. Thus skimming the surface, they ploughed it in their course: the water was quite smooth, and it formed a most curious spectacle to behold a flock, each bird leaving its narrow wake on the mirror-like surface. In their flight they frequently twist about with extreme quickness, and dexterously manage with their projecting lower mandible to plough up small fish, which are secured by the upper and shorter half of their scissor-like bills. This fact I repeatedly saw, as, like swallows, they continued to fly backwards and forwards close before me. Occasionally when leaving the surface of the water their flight was wild, irregular, and rapid; they then uttered loud harsh cries. When these birds are fishing, the advantage of the long primary feathers of their wings, in keeping them dry, is very evident. When thus employed, their forms resemble the symbol by which many artists represent marine birds. Their tails are much used in steering their irregular course.

These birds are common far inland along the course of the Rio Parana; it is said that they remain here during the whole year, and breed in the marshes. During the day they rest in flocks on the grassy plains, at some distance from the water. Being at anchor, as I have said, in one of the deep creeks between the islands of the Parana, as the evening drew to a close, one of these scissor-beaks suddenly appeared. The water was quite still, and many little fish were rising. The bird continued for a long time to skim the surface, flying in its wild and irregular manner up and down the narrow canal, now dark with the growing night and the shadows of the overhanging trees. At Monte Video, I observed that some large flocks during the day remained on the mud-banks at the head of the harbour, in the same manner as on the grassy plains near the Parana; and every evening they took flight seaward. From these facts I suspect that the Rhynchops generally fishes by night, at which time many of the lower animals come most abundantly to the surface. M. Lesson states that he has seen these birds opening the shells of the mactræ buried in the sand-banks on the coast of Chile: from their weak bills, with the lower mandible so much projecting, their short legs and long wings, it is very improbable that this can be a general habit.

In our course down the Parana, I observed only three other birds, whose habits are worth mentioning. One is a small kingfisher (Ceryle Americana); it has a longer tail than the European species, and hence does not sit in so stiff and upright a position. Its flight also, instead of being direct and rapid, like the course of an arrow, is weak and undulatory, as among the soft-billed birds. It utters a low note, like the clicking together of two small stones. A small green parrot (Conurus murinus), with a grey breast, appears to prefer the tall trees on the islands to any other situation for its building-place. A number of nests are placed so close together as to form one great mass of sticks. These parrots always live in flocks, and commit great ravages on the corn-fields. I was told that near Colonia 2500 were killed in the course of one year. A bird with a forked tail, terminated by two long feathers (Tyrannus savana), and named by the Spaniards scissor-tail, is very common near Buenos Ayres: it commonly sits on a branch of the *ombu* tree, near a house, and thence takes a short flight in pursuit of insects, and returns to the same spot. When on the wing it presents in its manner of flight and general appearance a caricature-likeness of the common swallow. It has the power of turning very shortly in the air, and in so doing opens and shuts its tail, sometimes in a horizontal or lateral and sometimes in a vertical direction, just like a pair of scissors.

October 16th. – Some leagues below Rozario, the western shore of the Parana is bounded by perpendicular cliffs, which extend in a long line to below San Nicolas; hence it more resembles a sea-coast than that of a fresh-water river. It is a great drawback to the scenery of the Parana, that, from the soft nature of its banks, the water is very muddy. The Uruguay, flowing through a granitic country, is much clearer; and where the two channels unite at the head of the Plata, the waters may for a long distance be distinguished by their black and red colours. In the evening, the wind being not quite fair, as usual we immediately moored, and the next day, as it blew rather freshly, though with a favouring current, the master was much too indolent to think of starting. At Bajada, he was described to me as 'hombre muy aflicto' – a man always miserable to get on; but certainly he bore all delays with admirable resignation. He was an old Spaniard, and had been many years in this country. He professed a great liking to the English, but stoutly maintained that the battle of Trafalgar was merely won by the Spanish captains having been all bought over; and that the only really gallant action on either side was performed by the Spanish admiral. It struck me as rather characteristic, that this man should prefer his countrymen being thought the worst of traitors, rather than unskilful or cowardly.

18th *and* 19th. – We continued slowly to sail down the noble stream: the current helped us but little. We met, during our descent, very few vessels. One of the best gifts of nature, in so grand a channel of communication, seems here wilfully thrown away – a river in which ships might navigate from a temperate country, as surprisingly abundant in certain productions as destitute of others, to another possessing a tropical climate, and a soil which, according to the best of judges, M. Bonpland, is perhaps unequalled in fertility in any part of the world. How different would have been the aspect of this river if English colonists had by good fortune first sailed up the Plata! What noble towns would now have occupied its shores! Till the death of Francia, the Dictator of Paraguay, these two countries must remain distinct, as if placed on opposite sides of the globe. And when the old bloody-minded tyrant is gone to his long account,

Paraguay will be torn by revolutions, violent in proportion to the previous unnatural calm. That country will have to learn, like every other South American state, that a republic cannot succeed till it contains a certain body of men imbued with the principles of justice and honour.

October 20th. – Being arrived at the mouth of the Parana, and as I was very anxious to reach Buenos Ayres, I went on shore at Las Conchas, with the intention of riding there. Upon landing, I found to my great surprise that I was to a certain degree a prisoner. A violent revolution having broken out, all the ports were laid under an embargo. I could not return to my vessel, and as for going by land to the city, it was out of the question. After a long conversation with the commandant, I obtained permission to go the next day to General Rolor, who commanded a division of the rebels on this side the capital. In the morning I rode to the encampment. The general, officers, and soldiers, all appeared, and I believe really were, great villains. The general, the very evening before he left the city, voluntarily went to the Governor, and with his hand to his heart, pledged his word of honour that he at least would remain faithful to the last. The general told me that the city was in a state of close blockade, and that all he could do was to give me a passport to the commander-in-chief of the rebels at Quilmes. We had therefore to take a great sweep round the city, and it was with much difficulty that we procured horses. My reception at the encampment was quite civil, but I was told it was impossible that I could be allowed to enter the city. I was very anxious about this, as I anticipated the *Beagle*'s departure from the Rio Plata earlier than it took place. Having mentioned, however, General Rosas's obliging kindness to me when at the Colorado, magic itself could not have altered circumstances quicker than did this conversation. I was instantly told that though they could not give me a passport, if I chose to leave my guide and horses, I might pass their sentinels. I was too glad to accept of this, and an officer was sent with me to give directions that I should not be stopped at the bridge. The road for the space of a league was quite deserted. I met one party of soldiers, who were satisfied by gravely looking at an old passport: and at length I was not a little pleased to find myself within the city.

This revolution was supported by scarcely any pretext of grievances: but in a state which, in the course of nine months (from
February to October, 1820), underwent fifteen changes in its government – each governor, according to the constitution, being
elected for three years – it would be very unreasonable to ask for
pretexts. In this case, a party of men – who, being attached to Rosas,
were disgusted with the governor Balcarce – to the number of
seventy left the city, and with the cry of Rosas the whole country
took arms. The city was then blockaded, no provisions, cattle or
horses, were allowed to enter; besides this, there was only a little
skirmishing, and a few men daily killed. The outside party well knew
that by stopping the supply of meat they would certainly be victorious. General Rosas could not have known of this rising; but it appears
to be quite consonant with the plans of his party. A year ago he was
elected governor, but he refused it, unless the Sala would also confer
on him extraordinary powers. This was refused, and since then his
party have shown that no other governor can keep his place. The
warfare on both sides was avowedly protracted till it was possible to
hear from Rosas. A note arrived a few days after I left Buenos Ayres,
which stated that the General disapproved of peace having been
broken, but that he thought the outside party had justice on their
side. On the bare reception of this, the Governor, ministers, and part
of the military, to the number of some hundreds, fled from the city.
The rebels entered, elected a new governor, and were paid for their
services to the number of 5500 men. From these proceedings, it was
clear that Rosas ultimately would become the dictator: to the term
king, the people in this, as in other republics, have a particular dislike.
Since leaving South America, we have heard that Rosas has been
elected, with powers and for a time altogether opposed to the constitutional principles of the republic.

BANDA ORIENTAL
AND PATAGONIA

HAVING been delayed for nearly a fortnight in the city, I was glad to
escape on board a packet bound for Monte Video. A town in a state of
blockade must always be a disagreeable place of residence; in this case
moreover there were constant apprehensions from robbers within.
The sentinels were the worst of all; for, from their office and from
having arms in their hands, they robbed with a degree of authority
which other men could not imitate.

Our passage was a very long and tedious one. The Plata looks like a
noble estuary on the map; but is in truth a poor affair. A wide expanse
of muddy water has neither grandeur nor beauty. At one time of the
day, the two shores, both of which are extremely low, could just be
distinguished from the deck. On arriving at Monte Video I found that
the *Beagle* would not sail for some time, so I prepared for a short
excursion in this part of Banda Oriental. Everything which I have said
about the country near Maldonado is applicable to M. Video; but the
land, with the one exception of the Green Mount, 450 feet high,
from which it takes its name, is far more level. Very little of the
undulating grassy plain is enclosed; but near the town there are a few
hedge-banks, covered with agaves, cacti, and fennel.

November 14th. – We left Monte Video in the afternoon. I intended
to proceed to Colonia del Sacramiento, situated on the northern bank

of the Plata and opposite to Buenos Ayres, and thence, following up the Uruguay, to the village of Mercedes on the Rio Negro (one of the many rivers of this name in South America), and from this point to return direct to Monte Video. We slept at the house of my guide at Canelones. In the morning we rose early, in the hopes of being able to ride a good distance; but it was a vain attempt, for all the rivers were flooded. We passed in boats the streams of Canelones, St. Lucia, and San José, and thus lost much time. On a former excursion I crossed the Lucia near its mouth, and I was surprised to observe how easily our horses, although not used to swim, passed over a width of at least six hundred yards. On mentioning this at Monte Video, I was told that a vessel containing some mountebanks and their horses, being wrecked in the Plata, one horse swam seven miles to the shore. In the course of the day I was amused by the dexterity with which a Gaucho forced a restive horse to swim a river. He stripped off his clothes, and jumping on its back, rode into the water till it was out of its depth; then slipping off over the crupper, he caught hold of the tail, and as often as the horse turned round, the man frightened it back by splashing water in its face. As soon as the horse touched the bottom on the other side, the man pulled himself on, and was firmly seated, bridle in hand, before the horse gained the bank. A naked man on a naked horse is a fine spectacle; I had no idea how well the two animals suited each other. The tail of a horse is a very useful appendage; I have passed a river in a boat with four people in it, which was ferried across in the same way as the Gaucho. If a man and horse have to cross a broad river, the best plan is for the man to catch hold of the pommel or mane, and help himself with the other arm.

We slept and stayed the following day at the post of Cufre. In the evening the postman or letter-carrier arrived. He was a day after his time, owing to the Rio Rozario being flooded. It would not, however, be of much consequence; for, although he had passed through some of the principal towns in Banda Oriental, his luggage consisted of two letters! The view from the house was pleasing; an undulating green surface, with distant glimpses of the Plata. I find that I look at this province with very different eyes from what I did upon my first arrival. I recollect I then thought it singularly level; but now, after galloping over the Pampas, my only surprise is, what could have induced me ever to have called it level. The country is a series of undulations, in themselves perhaps not absolutely great, but, as compared to the plains of St. Fé, real mountains. From these inequalities there is an abundance of small rivulets, and the turf is green and luxuriant.

November 17th. – We crossed the Rozario, which was deep and rapid, and passing the village of Colla, arrived at mid-day at Colonia del Sacramiento. The distance is twenty leagues, through a country covered with fine grass, but poorly stocked with cattle or inhabitants. I was invited to sleep at Colonia, and to accompany on the following day a gentleman to his estancia, where there were some limestone rocks. The town is built on a stony promontory something in the same manner as at Monte Video. It is strongly fortified, but both fortifications and town suffered much in the Brazilian war. It is very ancient; and the irregularity of the streets, and the surrounding groves of old orange and peach trees, gave it a pretty appearance. The church is a curious ruin; it was used as a powder-magazine, and was struck by lightning in one of the ten thousand thunder-storms of the Rio Plata. Two-thirds of the building were blown away to the very foundation; and the rest stands a shattered and curious monument of the united powers of lightning and gunpowder. In the evening I wandered about the half-demolished walls of the town. It was the chief seat of the Brazilian war; – a war most injurious to this country, not so much in its immediate effects, as in being the origin of a multitude of generals and all other grades of officers. More generals are numbered (but not paid) in the United Provinces of La Plata than in the United Kingdom of Great Britain. These gentlemen have learned to like power, and do not object to a little skirmishing. Hence there are many always on the watch to create disturbance and to overturn a government which as yet has never rested on any stable foundation. I noticed, however, both here and in other places, a very general interest in the ensuing election for the President; and this appears a good sign for the prosperity of this little country. The inhabitants do not require much education in their representatives; I heard some men discussing the merits of those for Colonia; and it was said that, 'although they were not men of business, they could all sign their names:' with this they seemed to think every reasonable man ought to be satisfied.

18th. – Rode with my host to his estancia, at the Arroyo de San Juan. In the evening we took a ride round the estate: it contained two square leagues and a half, and was situated in what is called a rincon; that is, one side was fronted by the Plata, and the two others guarded by impassable brooks. There was an excellent port for little vessels, and an abundance of small wood, which is valuable as supplying fuel to Buenos Ayres. I was curious to know the value of so complete an estancia. Of cattle there were 3000, and it would well support three or four times that number; of mares 800, together with 150 broken-in horses, and 600 sheep. There was plenty of water and limestone, a rough house, excellent corrals, and a peach orchard. For all this he had been offered 2000l., and he only wanted 500l. additional, and prob- ably would sell it for less. The chief trouble with an estancia is driving the cattle twice a week to a central spot, in order to make them tame, and to count them. This latter operation would be thought difficult, where there are ten or fifteen thousand head together. It is managed on the principle that the cattle invariably divide themselves into little troops of from forty to one hundred. Each troop is recognized by a few peculiarly marked animals, and its number is known: so that, one being lost out of ten thousand, it is perceived by its absence from one of the tropillas. During a stormy night the cattle all mingle together; but the next morning the tropillas separate as before; so that each animal must know its fellow out of ten thousand others.

On two occasions I met with in this province some oxen of a very curious breed, called nata or niata. They appear externally to hold nearly the same relation to other cattle, which bull or pug dogs do to other dogs. Their forehead is very short and broad, with the nasal end turned up, and the upper lip much drawn back; their lower jaws project beyond the upper, and have a corresponding upward curve; hence their teeth are always exposed. Their nostrils are seated high up and are very open; their eyes project outwards. When walking they carry their heads low, on a short neck; and their hinder legs are rather longer compared with the front legs than is usual. Their bare teeth, their short heads, and upturned nostrils give them the most ludicrous self-confident air of defiance imaginable.

Since my return, I have procured a skeleton head, through the kindness of my friend Captain Sulivan, R.N., which is now deposited in the College of Surgeons.[1] Don F. Muniz, of Luxan, has kindly collected for me all the information which he could respecting this breed. From his account it seems that about eighty or ninety years ago, they were rare and kept as curiosities at Buenos Ayres. The breed is universally believed to have originated amongst the Indians southward of the Plata; and that it was with them the commonest kind. Even to this day, those reared in the provinces near the Plata show their less civilized origin, in being fiercer than common cattle, and in the cow easily deserting her first calf, if visited too often or molested. It is a singular fact that an almost similar structure to the abnormal[2] one of the niata breed, characterizes, as I am informed by Dr. Falconer, that great extinct ruminant of India, the Sivatherium. The breed is very *true*; and a niata bull and cow invariably produce niata calves. A niata bull with a common cow, or the reverse cross, produces offspring having an intermediate character, but with the niata characters strongly displayed: according to Señor Muniz, there is the clearest evidence, contrary to the common belief of agriculturists in analogous cases, that the niata cow when crossed with a common bull transmits her peculiarities more strongly than the niata bull when crossed with a common cow. When the pasture is tolerably long, the niata cattle feed with the tongue and palate as well as common cattle; but during the great droughts, when so many animals perish, the niata breed is under a great disadvantage, and would be exterminated if not attended to; for the common cattle, like horses, are able just to keep alive, by browsing with their lips on twigs of trees and reeds; this the niatas cannot so well do, as their lips do not join, and hence they are found to perish before the common cattle. This strikes me as a good illustration of how little we are able to judge from the ordinary habits of life, on what

1 Mr. Waterhouse has drawn up a detailed description of this head, which I hope he will publish in some Journal.

2 A nearly similar abnormal, but I do not know whether hereditary, structure has been observed in the carp, and likewise in the crocodile of the Ganges: Histoire des Anomalies par M. Isid. Geoffroy St. Hilaire, tom. i. p. 244.

circumstances, occurring only at long intervals, the rarity or extinction of a species may be determined.

November 19*th.* – Passing the valley of Las Vacas, we slept at a house
of a North American, who worked a lime-kiln on the Arroyo de las
Vivoras. In the morning we rode to a projecting headland on the
banks of the river, called Punta Gorda. On the way we tried to find a
jaguar. There were plenty of fresh tracks, and we visited the trees, on
which they are said to sharpen their claws; but we did not succeed in
disturbing one. From this point the Rio Uruguay presented to our
view a noble volume of water. From the clearness and rapidity of
the stream, its appearance was far superior to that of its neighbour the
Parana. On the opposite coast, several branches from the latter river
entered the Uruguay. As the sun was shining, the two colours of the
waters could be seen quite distinct.

In the evening we proceeded on our road towards Mercedes on
the Rio Negro. At night we asked permission to sleep at an estancia at
which we happened to arrive. It was a very large estate, being ten
leagues square, and the owner is one of the greatest landowners in the
country. His nephew had charge of it, and with him there was a
captain in the army, who the other day ran away from Buenos Ayres.
Considering their station, their conversation was rather amusing.
They expressed, as was usual, unbounded astonishment at the globe
being round, and could scarcely credit that a hole would, if deep
enough, come out on the other side. They had, however, heard of a
country where there were six months light and six of darkness, and
where the inhabitants were very tall and thin! They were curious
about the price and condition of horses and cattle in England. Upon
finding out we did not catch our animals with the lazo, they cried out,
'Ah, then, you use nothing but the bolas:' the idea of an enclosed
country was quite new to them. The captain at last said, he had one
question to ask me, which he should be very much obliged if I would
answer with all truth. I trembled to think how deeply scientific it
would be: it was, 'Whether the ladies of Buenos Ayres were not the
handsomest in the world.' I replied, like a renegade, 'Charmingly so.'
He added, 'I have one other question: Do ladies in any other part of
the world wear such large combs?' I solemnly assured him that they
did not. They were absolutely delighted. The captain exclaimed,

'Look there! a man who has seen half the world says it is the case; we always thought so, but now we know it.' My excellent judgment in combs and beauty procured me a most hospitable reception; the captain forced me to take his bed, and he would sleep on his recado.

21st. – Started at sunrise, and rode slowly during the whole day. The geological nature of this part of the province was different from the rest, and closely resembled that of the Pampas. In consequence, there were immense beds of the thistle, as well as of the cardoon: the whole country, indeed, may be called one great bed of these plants. The two sorts grow separate, each plant in company with its own kind. The cardoon is as high as a horse's back, but the Pampas thistle is often higher than the crown of the rider's head. To leave the road for a yard is out of the question; and the road itself is partly, and in some cases entirely, closed. Pasture, of course there is none; if cattle or horses once enter the bed, they are for the time completely lost. Hence it is very hazardous to attempt to drive cattle at this season of the year; for when jaded enough to face the thistles, they rush among them, and are seen no more. In these districts there are very few estancias, and these few are situated in the neighbourhood of damp valleys, where fortunately neither of these overwhelming plants can exist. As night came on before we arrived at our journey's end, we slept at a miserable little hovel inhabited by the poorest people. The extreme though rather formal courtesy of our host and hostess, considering their grade of life, was quite delightful.

November 22nd. – Arrived at an estancia on the Berquelo belonging to a very hospitable Englishman, to whom I had a letter of introduction from my friend Mr. Lumb. I stayed here three days. One morning I rode with my host to the Sierra del Pedro Flaco, about twenty miles up the Rio Negro. Nearly the whole country was covered with good though coarse grass, which was as high as a horse's belly; yet there were square leagues without a single head of cattle. The province of Banda Oriental, if well stocked, would support an astonishing number of animals; at present the annual export of hides from Monte Video amounts to three hundred thousand; and the home consumption, from waste, is very considerable. An estanciero told me that he often had to send large herds of cattle a long journey to a salting establishment, and that the tired beasts were frequently obliged to be killed and skinned; but that he could never persuade the Gauchos to eat of them, and every evening a fresh beast was slaughtered for their suppers! The view of the Rio Negro from the Sierra was more picturesque than any other which I saw in this province. The river, broad, deep and rapid, wound at the foot of a rocky precipitous cliff: a belt of wood followed its course, and the horizon terminated in the distant undulations of the turf-plain.

When in this neighbourhood, I several times heard of the Sierra de las Cuentas: a hill distant many miles to the northward. The name signifies hill of beads. I was assured that vast numbers of little round stones, of various colours, each with a small cylindrical hole, are found there. Formerly the Indians used to collect them, for the purpose of making necklaces and bracelets – a taste, I may observe, which is common to all savage nations, as well as to the most polished. I did not know what to understand from this story, but upon mentioning it at the Cape of Good Hope to Dr. Andrew Smith, he told me that he recollected finding on the south-eastern coast of Africa, about one hundred miles to the eastward of St. John's river, some quartz crystals with their edges blunted from attrition, and mixed with gravel on the sea-beach. Each crystal was about five lines in diameter, and from an inch to an inch and a half in length. Many of them had a small canal extending from one extremity to the other, perfectly cylindrical, and of a size that readily admitted a coarse thread or a piece of fine catgut. Their colour was red or dull white. The natives were acquainted with this structure in crystals. I have mentioned these circumstances because, although no crystallized body is at present known to assume this form, it may lead some future traveller to investigate the real nature of such stones.

While staying at this estancia, I was amused with what I saw and heard of the shepherd-dogs of the country.[1] When riding, it is a common thing to meet a large flock of sheep guarded by one or two dogs, at the distance of some miles from any house or man. I often wondered how so firm a friendship had been established. The method of education consists in separating the puppy, while very young, from

1 M. A. d'Orbigny has given nearly a similar account of these dogs, tom. i. p. 175.

the bitch, and in accustoming it to its future companions. An ewe is held three or four times a day for the little thing to suck, and a nest of wool is made for it in the sheep-pen; at no time is it allowed to associate with other dogs, or with the children of the family. The puppy is, moreover, generally castrated; so that, when grown up, it can scarcely have any feelings in common with the rest of its kind. From this education it has no wish to leave the flock, and just as another dog will defend its master, man, so will these the sheep. It is amusing to observe, when approaching a flock, how the dog immediately advances barking, and the sheep all close in his rear, as if round the oldest ram. These dogs are also easily taught to bring home the flock, at a certain hour in the evening. Their most troublesome fault, when young, is their desire of playing with the sheep; for in their sport they sometimes gallop their poor subjects most unmercifully.

The shepherd-dog comes to the house every day for some meat, and as soon as it is given him, he skulks away as if ashamed of himself. On these occasions the house-dogs are very tyrannical, and the least of them will attack and pursue the stranger. The minute, however, the latter has reached the flock, he turns round and begins to bark, and then all the house-dogs take very quickly to their heels. In a similar manner a whole pack of the hungry wild dogs will scarcely ever (and I was told by some never) venture to attack a flock guarded by even one of these faithful shepherds. The whole account appears to me a curious instance of the pliability of the affections in the dog; and yet, whether wild or however educated, he has a feeling of respect or fear for those that are fulfilling their instinct of association. For we can understand on no principle the wild dogs being driven away by the single one with its flock, except that they consider, from some confused notion, that the one thus associated gains power, as if in company with its own kind. F. Cuvier has observed, that all animals that readily enter into domestication, consider man as a member of their own society, and thus fulfil their instinct of association. In the above case the shepherd-dog ranks the sheep as its fellow-brethren, and thus gains confidence; and the wild dogs, though knowing that the individual sheep are not dogs, but are good to eat, yet partly consent to this view when seeing them in a flock with a shepherd-dog at their head.

One evening a 'domidor' (a subduer of horses) came for the purpose of breaking-in some colts. I will describe the preparatory steps, for I believe they have not been mentioned by other travellers. A troop of wild young horses is driven into the corral, or large enclosure of stakes, and the door is shut. We will suppose that one man alone has to catch and mount a horse, which as yet had never felt bridle or saddle. I conceive, except by a Gaucho, such a feat would be utterly impracticable. The Gaucho picks out a full-grown colt; and as the beast rushes round the circus, he throws his lazo so as to catch both the front legs. Instantly the horse rolls over with a heavy shock, and whilst struggling on the ground, the Gaucho, holding the lazo tight, makes a circle, so as to catch one of the hind legs, just beneath the fetlock, and draws it close to the two front legs: he then hitches the lazo, so that the three are bound together. Then sitting on the horse's neck, he fixes a strong bridle, without a bit, to the lower jaw: this he does by passing a narrow thong through the eye-holes at the end of the reins, and several times round both jaw and tongue. The two front legs are now tied closely together with a strong leathern thong, fastened by a slip-knot. The lazo, which bound the three together, being then loosed, the horse rises with difficulty. The Gaucho now holding fast the bridle fixed to the lower jaw, leads the horse outside the corral. If a second man is present (otherwise the trouble is much greater) he holds the animal's head, whilst the first puts on the horse-cloths and saddle, and girths the whole together. During this operation, the horse, from dread and astonishment at thus being bound round the waist, throws himself over and over again on the ground, and, till beaten, is unwilling to rise. At last, when the saddling is finished, the poor animal can hardly breathe from fear, and is white with foam and sweat. The man now prepares to mount by pressing heavily on the stirrup, so that the horse may not lose its balance; and at the moment that he throws his leg over the animal's back, he pulls the slip-knot binding the front legs, and the beast is free. Some 'domidors' pull the knot while the animal is lying on the ground, and, standing over the saddle, allow him to rise beneath them. The horse, wild with dread, gives a few most violent bounds, and then starts off at full gallop: when quite exhausted, the man, by patience, brings him back to the corral, where, reeking hot and scarcely alive, the poor beast is let free. Those animals which will not gallop away, but obstinately throw themselves on the ground, are by far the most troublesome. This process is tremendously severe, but in two or three trials the horse is tamed. It is not, however, for some weeks that the animal is ridden with the iron bit and solid ring, for it must learn to associate the will of its rider with the feel of the rein, before the most powerful bridle can be of any service.

Animals are so abundant in these countries, that humanity and self-interest are not closely united; therefore I fear it is that the former is here scarcely known. One day, riding in the Pampas with a very respectable 'Estanciero,' my horse, being tired, lagged behind. The man often shouted to me to spur him. When I remonstrated that it was a pity, for the horse was quite exhausted, he cried out, 'Why not? – never mind – spur him – it is *my* horse.' I had then some difficulty in making him comprehend that it was for the horse's sake, and not on his account, that I did not choose to use my spurs. He exclaimed, with a look of great surprise, 'Ah, Don Carlos, que cosa!' It was clear that such an idea had never before entered his head.

The Gauchos are well known to be perfect riders. The idea of being thrown, let the horse do what it likes, never enters their head. Their criterion of a good rider is, a man who can manage an untamed colt, or who, if his horse falls, alights on his own feet, or can perform other such exploits. I have heard of a man betting that he would throw his horse down twenty times, and that nineteen times he would not fall himself. I recollect seeing a Gaucho riding a very stubborn horse, which three times successively reared so high as to fall backwards with great violence. The man judged with uncommon coolness the proper moment for slipping off, not an instant before or after the right time; and as soon as the horse got up, the man jumped on his back, and at last they started at a gallop. The Gaucho never appears to exert any muscular force. I was one day watching a good rider, as we were galloping along at a rapid pace, and thought to myself, 'surely if the horse starts, you appear so careless on your seat, you must fall.' At this moment, a male ostrich sprang from its nest right beneath the horse's nose: the young colt bounded on one side like a stag; but as for the man, all that could be said was, that he started and took fright with his horse.

In Chile and Peru more pains are taken with the mouth of the horse than in La Plata, and this is evidently a consequence of the more intricate nature of the country. In Chile a horse is not considered perfectly broken, till he can be brought up standing, in the midst of his full speed, on any particular spot, – for instance, on a cloak thrown on the ground: or, again, he will charge a wall, and rearing, scrape the surface with his hoofs. I have seen an animal bounding with spirit, yet merely reined by a fore-finger and thumb, taken at full gallop across a courtyard, and then made to wheel round the post of a verandah with great speed, but at so equal a distance, that the rider, with outstretched arm, all the while kept one finger rubbing the post. Then making a demi-volte in the air, with the other arm outstretched in a like manner, he wheeled round, with astonishing force, in an opposite direction.

Such a horse is well broken; and although this at first may appear useless, it is far otherwise. It is only carrying that which is daily necessary into perfection. When a bullock is checked and caught by the lazo, it will sometimes gallop round and round in a circle, and the horse being alarmed at the great strain, if not well broken, will not readily turn like the pivot of a wheel. In consequence many men have been killed; for if the lazo once takes a twist round a man's body, it will instantly, from the power of the two opposed animals, almost cut him in twain. On the same principle the races are managed; the course is only two or three hundred yards long, the wish being to have horses that can make a rapid dash. The race-horses are trained not only to stand with their hoofs touching a line, but to draw all four feet together, so as at the first spring to bring into play the full action of the hind-quarters. In Chile I was told an anecdote, which I believe was true; and it offers a good illustration of the use of a well-broken animal. A respectable man riding one day met two others, one of whom was mounted on a horse which he knew to have been stolen from himself. He challenged them; they answered him by drawing their sabres and giving chase. The man, on his good and fleet beast, kept just ahead: as he passed a thick bush he wheeled round it, and brought up his horse to a dead check. The pursuers were obliged to shoot on one side and ahead. Then instantly dashing on, right behind them, he buried his knife in the back of one, wounded the other, recovered his horse from the dying robber, and rode home. For these feats of horsemanship two things are necessary: a most severe bit, like the Mameluke, the power of which, though seldom used, the horse knows full well; and large blunt spurs, that can be applied either as a mere touch, or as an instrument of extreme pain. I conceive that with English spurs, the slightest touch of which pricks the skin, it would be impossible to break in a horse after the South American fashion.

At an estancia near Las Vacas large numbers of mares are weekly slaughtered for the sake of their hides, although worth only five paper dollars, or about half-a-crown apiece. It seems at first strange that it can answer to kill mares for such a trifle; but as it is thought ridiculous in this country ever to break in or ride a mare, they are of no value except for breeding. The only thing for which I ever saw mares used, was to tread out wheat from the ear; for which purpose they were driven round a circular enclosure, where the wheat-sheaves were strewed. The man employed for slaughtering the mares happened to be celebrated for his dexterity with the lazo. Standing at the distance of twelve yards from the mouth of the corral, he has laid a wager that he would catch by the legs every animal, without missing one, as it rushed past him. There was another man who said he would enter the corral on foot, catch a mare, fasten her front legs together, drive her out, throw her down, kill, skin, and stake the hide for drying (which latter is a tedious job); and he engaged that he would perform this whole operation on twenty-two animals in one day. Or he would kill and take the skin off fifty in the same time. This would have been a prodigious task, for it is considered a good day's work to skin and stake the hides of fifteen or sixteen animals.

November 26th. – I set out on my return in a direct line for Monte Video. Having heard of some giant's bones at a neighbouring farm-house on the Sarandis, a small stream entering the Rio Negro, I rode there accompanied by my host, and purchased for the value of eighteen pence the head of the Toxodon.[1] When found it was quite perfect; but the boys knocked out some of the teeth with stones,

1 I must express my obligation to Mr. Keane, at whose house I was staying on the Berquelo and to Mr. Lumb at Buenos Ayres, for without their assistance these valuable remains would never have reached England.

and then set up the head as a mark to throw at. By a most fortunate chance I found a perfect tooth, which exactly fitted one of the sockets in this skull, embedded by itself on the banks of the Rio Tercero, at the distance of about 180 miles from this place. I found remains of this extraordinary animal at two other places, so that it must formerly have been common. I found here, also, some large portions of the armour of a gigantic armadillo-like animal, and part of the great head of a Mylodon. The bones of this head are so fresh, that they contain, according to the analysis by Mr. T. Reeks, seven per cent. of animal matter; and when placed in a spirit-lamp, they burn with a small flame. The number of the remains embedded in the grand estuary deposit which forms the Pampas and covers the granitic rocks of Banda Oriental, must be extraordinarily great. I believe a straight line drawn in any direction through the Pampas would cut through some skeleton or bones. Besides those which I found during my short excursions, I heard of many others, and the origin of such names as 'the stream of the animal,' 'the hill of the giant,' is obvious. At other times I heard of the marvellous property of certain rivers, which had the power of changing small bones into large; or, as some maintained, the bones themselves grew. As far as I am aware, not one of these animals perished, as was formerly supposed, in the marshes or muddy river-beds of the present land, but their bones have been exposed by the streams intersecting the subaqueous deposit in which they were originally embedded. We may conclude that the whole area of the Pampas is one wide sepulchre of these extinct gigantic quadrupeds.

By the middle of the day, on the 28th, we arrived at Monte Video, having been two days and a half on the road. The country for the whole way was of a very uniform character, some parts being rather more rocky and hilly than near the Plata. Not far from Monte Video we passed through the village of Las Pietras, so named from some large rounded masses of syenite. Its appearance was rather pretty. In this country a few fig-trees round a group of houses, and a site elevated a hundred feet above the general level, ought always to be called picturesque.

During the last six months I have had an opportunity of seeing a little of the character of the inhabitants of these provinces. The

Gauchos, or countrymen, are very superior to those who reside in the towns. The Gaucho is invariably most obliging, polite, and hospitable: I did not meet with even one instance of rudeness or inhospitality. He is modest, both respecting himself and country, but at the same time a spirited, bold fellow. On the other hand, many robberies are committed, and there is much bloodshed: the habit of constantly wearing the knife is the chief cause of the latter. It is lamentable to hear how many lives are lost in trifling quarrels. In fighting, each party tries to mark the face of his adversary by slashing his nose or eyes; as is often attested by deep and horrid-looking scars. Robberies are a natural consequence of universal gambling, much drinking, and extreme indolence. At Mercedes I asked two men why they did not work. One gravely said the days were too long; the other that he was too poor. The number of horses and the profusion of food are the destruction of all industry. Moreover, there are so many feast-days; and again, nothing can succeed without it be begun when the moon is on the increase; so that half the month is lost from these two causes.

Police and justice are quite inefficient. If a man who is poor commits murder and is taken, he will be imprisoned, and perhaps even shot; but if he is rich and has friends, he may rely on it no very severe consequence will ensue. It is curious that the most respectable inhabitants of the country invariably assist a murderer to escape: they seem to think that the individual sins against the government, and not against the people. A traveller has no protection besides his fire-arms; and the constant habit of carrying them is the main check to more frequent robberies.

The character of the higher and more educated classes who reside in the towns, partakes, but perhaps in a lesser degree, of the good parts of the Gaucho, but is, I fear, stained by many vices of which he is free. Sensuality, mockery of all religion, and the grossest corruption, are far from uncommon. Nearly every public officer can be bribed. The head man in the post-office sold forged government franks. The governor and prime minister openly combined to plunder the state. Justice, where gold came into play, was hardly expected by any one. I knew an Englishman, who went to the Chief Justice (he told me that, not then understanding the ways of the place, he trembled as he entered the room), and said, 'Sir, I have come to offer you two hundred (paper) dollars (value about five pounds sterling) if you will arrest before a certain time a man who has cheated me. I know it is against the law, but my lawyer (naming him) recommended me to take this step.' The Chief Justice smiled acquiescence, thanked him, and the man before night was safe in prison. With this entire want of principle in many of the leading men, with the country full of ill-paid turbulent officers, the people yet hope that a democratic form of government can succeed!

On first entering society in these countries, two or three features strike one as particularly remarkable. The polite and dignified manners pervading every rank of life, the excellent taste displayed by the women in their dresses, and the equality amongst all ranks. At the Rio Colorado some men who kept the humblest shops used to dine with General Rosas. A son of a major at Bahia Blanca gained his livelihood by making paper cigars, and he wished to accompany me, as guide or servant, to Buenos Ayres, but his father objected on the score of the danger alone. Many officers in the army can neither read nor write, yet all meet in society as equals. In Entre Rios, the Sala consisted of only six representatives. One of them kept a common shop, and evidently was not degraded by the office. All this is what would be expected in a new country; nevertheless the absence of gentlemen by profession appears to an Englishman something strange.

When speaking of these countries, the manner in which they have been brought up by their unnatural parent, Spain, should always be borne in mind. On the whole, perhaps, more credit is due for what has been done, than blame for that which may be deficient. It is impossible to doubt but that the extreme liberalism of these countries must ultimately lead to good results. The very general toleration of foreign religions, the regard paid to the means of education, the freedom of the press, the facilities offered to all foreigners, and especially, as I am bound to add, to every one professing the humblest pretensions to science, should be recollected with gratitude by those who have visited Spanish South America.

December 6th. – The *Beagle* sailed from the Rio Plata, never again to enter its muddy stream. Our course was directed to Port Desire, on the coast of Patagonia. Before proceeding any further, I will here put together a few observations made at sea.

Several times when the ship has been some miles off the mouth of the Plata, and at other times when off the shores of Northern Patagonia, we have been surrounded by insects. One evening, when we were about ten miles from the Bay of San Blas, vast numbers of butterflies, in bands or flocks of countless myriads, extended as far as the eye could range. Even by the aid of a telescope it was not possible to see a space free from butterflies. The seamen cried out 'it was snowing butterflies,' and such in fact was the appearance. More species than one were present, but the main part belonged to a kind very similar to, but not identical with, the common English Colias edusa. Some moths and hymenoptera accompanied the butterflies; and a fine beetle (Calosoma) flew on board. Other instances are known of this beetle having been caught far out at sea; and this is the more remarkable, as the greater number of the Carabidæ seldom or never take wing. The day had been fine and calm, and the one previous to it equally so, with light and variable airs. Hence we cannot suppose that the insects were blown off the land, but we must conclude that they voluntarily took flight. The great bands of the Colias seem at first to afford an instance like those on record of the migrations of another butterfly, Vanessa cardui;[1] but the presence of other insects makes the case distinct, and even less intelligible. Before sunset a strong breeze sprung up from the north, and this must have caused tens of thousands of the butterflies and other insects to have perished.

On another occasion, when seventeen miles off Cape Corrientes, I had a net overboard to catch pelagic animals. Upon drawing it up, to my surprise I found a considerable number of beetles in it, and although in the open sea, they did not appear much injured by the salt water. I lost some of the specimens, but those which I preserved belonged to the genera Colymbetes, Hydroporus, Hydrobius (two species), Notaphus, Cynucus, Adimonia, and Scarabæus. At first I thought that these insects had been blown from the shore; but upon reflecting that out of the eight species four were aquatic, and

1 Lyell's Principles of Geology, vol. iii. p. 63.

two others partly so in their habits, it appeared to me most probable that they were floated into the sea by a small stream which drains a lake near Cape Corrientes. On any supposition it is an interesting circumstance to find live insects swimming in the open ocean seventeen miles from the nearest point of land. There are several accounts of insects having been blown off the Patagonian shore. Captain Cook observed it, as did more lately Captain King in the *Adventure*. The cause probably is due to the want of shelter, both of trees and hills, so that an insect on the wing, with an off-shore breeze, would be very apt to be blown out to sea. The most remarkable instance I have known of an insect being caught far from the land, was that of a large grasshopper (Acrydium), which flew on board, when the *Beagle* was to windward of the Cape de Verd Islands, and when the nearest point of land, not directly opposed to the trade-wind, was Cape Blanco on the coast of Africa, 370 miles distant.[1]

On several occasions, when the *Beagle* had been within the mouth of the Plata, the rigging has been coated with the web of the Gossamer Spider. One day (November 1st, 1832) I paid particular attention to this subject. The weather had been fine and clear, and in the morning the air was full of patches of the flocculent web, as on an autumnal day in England. The ship was sixty miles distant from the land, in the direction of a steady though light breeze. Vast numbers of a small spider, about one-tenth of an inch in length, and of a dusky red colour, were attached to the webs. There must have been, I should suppose, some thousands on the ship. The little spider, when first coming in contact with the rigging, was always seated on a single thread, and not on the flocculent mass. This latter seems merely to be produced by the entanglement of the single threads. The spiders were all of one species, but of both sexes, together with young ones. These latter were distinguished by their smaller size and more dusky colour. I will not give the description of this spider, but merely state that it does not appear to me to be included in any of Latreille's genera. The little aëronaut as soon as it arrived on board was very active, running about, sometimes letting itself fall, and then reascending

1 The flies which frequently accompany a ship for some days on its passage from harbour to harbour, wandering from the vessel, are soon lost, and all disappear.

the same thread; sometimes employing itself in making a small and very irregular mesh in the corners between the ropes. It could run with facility on the surface of water. When disturbed it lifted up its front legs, in the attitude of attention. On its first arrival it appeared very thirsty, and with exserted maxillæ drank eagerly of drops of water; this same circumstance has been observed by Strack: may it not be in consequence of the little insect having passed through a dry and rarefied atmosphere? Its stock of web seemed inexhaustible. While watching some that were suspended by a single thread, I several times observed that the slightest breath of air bore them away out of sight, in a horizontal line. On another occasion (25th) under similar circumstances, I repeatedly observed the same kind of small spider, either when placed or having crawled on some little eminence, elevate its abdomen, send forth a thread, and then sail away horizontally, but with a rapidity which was quite unaccountable. I thought I could perceive that the spider, before performing the above preparatory steps, connected its legs together with the most delicate threads, but I am not sure whether this observation was correct.

One day, at St. Fé, I had a better opportunity of observing some similar facts. A spider which was about three-tenths of an inch in length, and which in its general appearance resembled a Citigrade (therefore quite different from the gossamer), while standing on the summit of a post, darted forth four or five threads from its spinners. These, glittering in the sunshine, might be compared to diverging rays of light; they were not, however, straight, but in undulations like films of silk blown by the wind. They were more than a yard in length, and diverged in an ascending direction from the orifices. The spider then suddenly let go its hold of the post, and was quickly borne out of sight. The day was hot and apparently quite calm; yet under such circumstances, the atmosphere can never be so tranquil as not to affect a vane so delicate as the thread of a spider's web. If during a warm day we look either at the shadow of any object cast on a bank, or over a level plain at a distant landmark, the effect of an ascending current of heated air is almost always evident: such upward currents, it has been remarked, are also shown by the ascent of soap-bubbles, which will not rise in an in-doors room. Hence I think there is not much difficulty in understanding the ascent of the fine lines projected

from a spider's spinners, and afterwards of the spider itself; the divergence of the lines has been attempted to be explained, I believe by
Mr. Murray, by their similar electrical condition. The circumstance
of spiders of the same species, but of different sexes and ages, being
found on several occasions at the distance of many leagues from the
land, attached in vast numbers to the lines, renders it probable that
the habit of sailing through the air is as characteristic of this tribe, as
that of diving is of the Argyroneta. We may then reject Latreille's
supposition, that the gossamer owes its origin indifferently to the
young of several genera of spiders: although, as we have seen,
the young of other spiders do possess the power of performing aërial
voyages.[1]

During our different passages south of the Plata, I often towed
astern a net made of bunting, and thus caught many curious animals.
Of Crustacea there were many strange and undescribed genera. One,
which in some respects is allied to the Notopods (or those crabs which
have their posterior legs placed almost on their backs, for the purpose
of adhering to the under side of rocks), is very remarkable from the
structure of its hind pair of legs. The penultimate joint, instead of
terminating in a simple claw, ends in three bristle-like appendages of
dissimilar lengths – the longest equalling that of the entire leg. These
claws are very thin, and are serrated with the finest teeth, directed
backwards: their curved extremities are flattened, and on this part five
most minute cups are placed which seem to act in the same manner as
the suckers on the arms of the cuttle-fish. As the animal lives in the
open sea, and probably wants a place of rest, I suppose this beautiful
and most anomalous structure is adapted to take hold of floating
marine animals.

In deep water, far from the land, the number of living creatures is
extremely small: south of the latitude 35°, I never succeeded in
catching anything besides some beroe, and a few species of minute
entomostracous crustacea. In shoaler water, at the distance of a few
miles from the coast, very many kinds of crustacea and some other
animals are numerous, but only during the night. Between latitudes

---

[1] Mr. Blackwall, in his Researches in Zoology, has many excellent observations on
the habits of spiders.

56° and 57° south of Cape Horn, the net was put astern several times; it never, however, brought up anything besides a few of two extremely minute species of Entomostraca. Yet whales and seals, petrels and albatross, are exceedingly abundant throughout this part of the ocean. It has always been a mystery to me on what the albatross, which lives far from the shore, can subsist; I presume that, like the condor, it is able to fast long; and that one good feast on the carcass of a putrid whale lasts for a long time. The central and inter-tropical parts of the Atlantic swarm with Pteropoda, Crustacea, and Radiata, and with their devourers the flying-fish, and again with their devourers the bonitos and albicores; I presume that the numerous lower pelagic animals feed on the Infusoria, which are now known, from the researches of Ehrenberg, to abound in the open ocean: but on what, in the clear blue water, do these Infusoria subsist?

While sailing a little south of the Plata on one very dark night, the sea presented a wonderful and most beautiful spectacle. There was a fresh breeze, and every part of the surface, which during the day is seen as foam, now glowed with a pale light. The vessel drove before her bows two billows of liquid phosphorus, and in her wake she was followed by a milky train. As far as the eye reached, the crest of every wave was bright, and the sky above the horizon, from the reflected glare of these livid flames, was not so utterly obscure as over the vault of the heavens.

As we proceed further southward the sea is seldom phosphorescent; and off Cape Horn I do not recollect more than once having seen it so, and then it was far from being brilliant. This circumstance probably has a close connexion with the scarcity of organic beings in that part of the ocean. After the elaborate paper[1] by Ehrenberg, on the phosphorescence of the sea, it is almost superfluous on my part to make any observations on the subject. I may however add, that the same torn and irregular particles of gelatinous matter, described by Ehrenberg, seem in the southern as well as in the northern hemisphere, to be the common cause of this phenomenon. The particles were so minute as easily to pass through fine gauze; yet many were distinctly visible by the naked eye. The water when placed in a

1 An abstract is given in No. IV. of the Magazine of Zoology and Botany.

tumbler and agitated, gave out sparks, but a small portion in a watch-glass scarcely ever was luminous. Ehrenberg states that these particles all retain a certain degree of irritability. My observations, some of which were made directly after taking up the water, gave a different result. I may also mention, that having used the net during one night, I allowed it to become partially dry, and having occasion twelve hours afterwards to employ it again, I found the whole surface sparkled as brightly as when first taken out of the water. It does not appear probable in this case, that the particles could have remained so long alive. On one occasion having kept a jelly-fish of the genus Dianæa till it was dead, the water in which it was placed became luminous. When the waves scintillate with bright green sparks, I believe it is generally owing to minute crustacea. But there can be no doubt that very many other pelagic animals, when alive, are phosphorescent.

On two occasions I have observed the sea luminous at consider-able depths beneath the surface. Near the mouth of the Plata some circular and oval patches, from two to four yards in diameter, and with defined outlines, shone with a steady but pale light; while the surrounding water only gave out a few sparks. The appearance resembled the reflection of the moon, or some luminous body; for the edges were sinuous from the undulations of the surface. The ship, which drew thirteen feet of water, passed over, without disturb-ing these patches. Therefore we must suppose that some animals were congregated together at a greater depth than the bottom of the vessel.

Near Fernando Noronha the sea gave out light in flashes. The appearance was very similar to that which might be expected from a large fish moving rapidly through a luminous fluid. To this cause the sailors attributed it; at the time, however, I entertained some doubts, on account of the frequency and rapidity of the flashes. I have already remarked that the phenomenon is very much more common in warm than in cold countries; and I have sometimes imagined that a disturbed electrical condition of the atmosphere was most favourable to its production. Certainly I think the sea is most luminous after a few days of more calm weather than ordinary, during which time it has swarmed with various animals. Observing that the water charged with gelatinous particles is in an impure state, and that the luminous appearance in all common cases is produced by the agitation of the fluid in contact with the atmosphere, I am inclined to consider that the phosphorescence is the result of the decomposition of the organic particles, by which process (one is tempted almost to call it a kind of respiration) the ocean becomes purified.

December 23*rd*. – We arrived at Port Desire, situated in lat. 47°, on the coast of Patagonia. The creek runs for about twenty miles inland, with an irregular width. The *Beagle* anchored a few miles within the entrance, in front of the ruins of an old Spanish settlement.

The same evening I went on shore. The first landing in any new country is very interesting, and especially when, as in this case, the whole aspect bears the stamp of a marked and individual character. At the height of between two and three hundred feet above some masses of porphyry a wide plain extends, which is truly characteristic of Patagonia. The surface is quite level, and is composed of well-rounded shingle mixed with a whitish earth. Here and there scattered tufts of brown wiry grass are supported, and, still more rarely, some low thorny bushes. The weather is dry and pleasant, and the fine blue sky is but seldom obscured. When standing in the middle of one of these desert plains and looking towards the interior, the view is generally bounded by the escarpment of another plain, rather higher, but equally level and desolate; and in every other direction the horizon is indistinct from the trembling mirage which seems to rise from the heated surface.

In such a country the fate of the Spanish settlement was soon decided; the dryness of the climate during the greater part of the year, and the occasional hostile attacks of the wandering Indians, compelled the colonists to desert their half-finished buildings. The style, however, in which they were commenced shows the strong and liberal hand of Spain in the old time. The result of all the attempts to colonize this side of America south of 41°, have been miserable. Port Famine expresses by its name the lingering and extreme sufferings of several hundred wretched people, of whom one alone survived to relate their misfortunes. At St. Joseph's Bay, on the coast of Patagonia, a small settlement was made; but during one Sunday the Indians made an attack and massacred the whole party, excepting two men, who remained captives during many years. At the Rio Negro I conversed with one of these men, now in extreme old age.

The zoology of Patagonia is as limited as its Flora.[1] On the arid plains a few black beetles (Heteromera) might be seen slowly crawling about, and occasionally a lizard darted from side to side.Of birds we have three carrion-hawks, and in the valleys a few finches and insect-feeders. An ibis (Theristicus melanops – a species said to be found in central Africa) is not uncommon on the most desert parts: in their stomachs I found grasshoppers, cicadæ, small lizards, and even scorpions.[2] At one time of the year these birds go in flocks, at another in pairs; their cry is very loud and singular, like the neighing of the guanaco.

The guanaco, or wild llama, is the characteristic quadruped of the plains of Patagonia; it is the South American representative of the camel of the East. It is an elegant animal in a state of nature, with a long slender neck and fine legs. It is very common over the whole of the temperate parts of the continent, as far south as the islands near Cape Horn. It generally lives in small herds of from half a dozen to thirty in each; but on the banks of the St. Cruz we saw one herd which must have contained at least five hundred.

They are generally wild and extremely wary. Mr. Stokes told me, that he one day saw through a glass a herd of these animals which evidently had been frightened, and were running away at full speed, although their distance was so great that he could not distinguish them with his naked eye. The sportsman frequently receives the first notice of their presence, by hearing from a long distance their peculiar shrill neighing note of alarm. If he then looks attentively, he will probably see the herd standing in a line on the side of some distant hill. On approaching nearer, a few more squeals are given, and off they set

1 I found here a species of cactus, described by Professor Henslow, under the name of *Opuntia Darwinii* (Magazine of Zoology and Botany, vol. i. p. 466), which was remarkable by the irritability of the stamens, when I inserted either a piece of stick or the end of my finger in the flower. The segments of the perianth also closed on the pistil, but more slowly than the stamens. Plants of this family, generally considered as tropical, occur in North America (Lewis and Clarke's Travels, p. 221), in the same high latitude as here, namely, in both cases, in 47°.

2 These insects were not uncommon beneath stones. I found one cannibal scorpion quietly devouring another.

at an apparently slow, but really quick canter, along some narrow beaten track to a neighbouring hill. If, however, by chance he abruptly meets a single animal, or several together, they will generally stand motionless and intently gaze at him; then perhaps move on a few yards, turn round, and look again. What is the cause of this difference in their shyness? Do they mistake a man in the distance for their chief enemy the puma? Or does curiosity overcome their timidity? That they are curious is certain; for if a person lies on the ground, and plays strange antics, such as throwing up his feet in the air, they will almost always approach by degrees to reconnoitre him. It was an artifice that was repeatedly practised by our sportsmen with success, and it had moreover the advantage of allowing several shots to be fired, which were all taken as parts of the performance. On the mountains of Tierra del Fuego, I have more than once seen a guanaco, on being approached, not only neigh and squeal, but prance and leap about in the most ridiculous manner, apparently in defiance as a challenge. These animals are very easily domesticated, and I have seen some thus kept in northern Patagonia near a house, though not under any restraint. They are in this state very bold, and readily attack a man by striking him from behind with both knees. It is asserted that the motive for these attacks is jealousy on account of their females. The wild guanacos, however, have no idea of defence; even a single dog will secure one of these large animals, till the huntsman can come up. In many of their habits they are like sheep in a flock. Thus when they see men approaching in several directions on horseback, they soon become bewildered, and know not which way to run. This greatly facilitates the Indian method of hunting, for they are thus easily driven to a central point, and are encompassed.

The guanacos readily take to the water: several times at Port Valdes they were seen swimming from island to island. Byron, in his voyage, says he saw them drinking salt water. Some of our officers likewise saw a herd apparently drinking the briny fluid from a salina near Cape Blanco. I imagine in several parts of the country, if they do not drink salt water, they drink none at all. In the middle of the day they frequently roll in the dust, in saucer-shaped hollows. The males fight together; two one day passed quite close to me, squealing and trying to bite each other; and several were shot with their hides deeply scored. Herds sometimes appear to set out on exploring parties: at Bahia Blanca, where, within thirty miles of the coast, these animals are extremely unfrequent, I one day saw the tracks of thirty or forty, which had come in a direct line to a muddy salt-water creek. They then must have perceived that they were approaching the sea, for they had wheeled with the regularity of cavalry, and had returned back in as straight a line as they had advanced. The guanacos have one singular habit, which is to me quite inexplicable; namely, that on successive days they drop their dung in the same defined heap. I saw one of these heaps which was eight feet in diameter, and was composed of a large quantity. This habit, according to M. A. d'Orbigny, is common to all the species of the genus; it is very useful to the Peruvian Indians, who use the dung for fuel, and are thus saved the trouble of collecting it.

The guanacos appear to have favourite spots for lying down to die. On the banks of the St. Cruz, in certain circumscribed spaces, which were generally bushy, and all near the river, the ground was actually white with bones. On one such spot I counted between ten and twenty heads. I particularly examined the bones; they did not appear, as some scattered ones which I had seen, gnawed or broken, as if dragged together by beasts of prey. The animals in most cases must have crawled, before dying, beneath and amongst the bushes. Mr. Bynoe informs me that during a former voyage he observed the same circumstance on the banks of the Rio Gallegos. I do not at all understand the reason of this, but I may observe, that the wounded guanacos at the St. Cruz invariably walked towards the river. At St. Jago in the Cape de Verd islands, I remember having seen in a ravine a retired corner covered with bones of the goat; we at the time exclaimed that it was the burial-ground of all the goats in the island. I mention these trifling circumstances, because in certain cases they might explain the occurrence of a number of uninjured bones in a cave, or buried under alluvial accumulations; and likewise the cause why certain animals are more commonly embedded than others in sedimentary deposits.

One day the yawl was sent under the command of Mr. Chaffers with three days' provisions to survey the upper part of the harbour. In the morning we searched for some watering-places mentioned in an old Spanish chart. We found one creek, at the head of which there was a trickling rill (the first we had seen) of brackish water. Here the tide compelled us to wait several hours; and in the interval I walked some miles into the interior. The plain as usual consisted of gravel, mingled with soil resembling chalk in appearance, but very different from it in nature. From the softness of these materials it was worn into many gulleys. There was not a tree, and, excepting the guanaco, which stood on the hill-top a watchful sentinel over its herd, scarcely an animal or a bird. All was stillness and desolation. Yet in passing over these scenes, without one bright object near, an ill-defined but strong sense of pleasure is vividly excited. One asked how many ages the plain had thus lasted, and how many more it was doomed thus to continue.

> None can reply – all seems eternal now.
> The wilderness has a mysterious tongue,
> Which teaches awful doubt.[1]

In the evening we sailed a few miles further up, and then pitched the tents for the night. By the middle of the next day the yawl was aground, and from the shoalness of the water could not proceed any higher. The water being found partly fresh, Mr. Chaffers took the dingey and went up two or three miles further, where she also grounded, but in a fresh-water river. The water was muddy, and though the stream was most insignificant in size, it would be difficult to account for its origin, except from the melting snow on the Cordillera. At the spot where we bivouacked, we were surrounded by bold cliffs and steep pinnacles of porphyry. I do not think I ever saw a spot which appeared more secluded from the rest of the world, than this rocky crevice in the wide plain.

The second day after our return to the anchorage, a party of officers and myself went to ransack an old Indian grave, which I had found on the summit of a neighbouring hill. Two immense stones, each probably weighing at least a couple of tons, had been placed in front of a ledge of rock about six feet high. At the bottom of the grave on the hard rock there was a layer of earth about a foot deep, which

1 Shelley, Lines on M. Blanc.

must have been brought up from the plain below. Above it a pave-
ment of flat stones was placed, on which others were piled, so as to fill
up the space between the ledge and the two great blocks. To complete the grave, the Indians had contrived to detach from the ledge a
huge fragment, and to throw it over the pile so as to rest on the two
blocks. We undermined the grave on both sides, but could not find
any relics, or even bones. The latter probably had decayed long since
(in which case the grave must have been of extreme antiquity), for
I found in another place some smaller heaps, beneath which a very
few crumbling fragments could yet be distinguished as having
belonged to a man. Falconer states, that where an Indian dies he is
buried, but that subsequently his bones are carefully taken up and
carried, let the distance be ever so great, to be deposited near the sea-
coast. This custom, I think, may be accounted for by recollecting,
that before the introduction of horses, these Indians must have led
nearly the same life as the Fuegians now do, and therefore generally
have resided in the neighbourhood of the sea. The common prejudice of lying where one's ancestors have lain, would make the now
roaming Indians bring the less perishable part of their dead to their
ancient burial-ground on the coast.

January 9th, 1834. – Before it was dark the *Beagle* anchored in the
fine spacious harbour of Port St. Julian, situated about one hundred
and ten miles to the south of Port Desire. We remained here eight
days. The country is nearly similar to that of Port Desire, but perhaps
rather more sterile. One day a party accompanied Captain Fitz Roy
on a long walk round the head of the harbour. We were eleven hours
without tasting any water, and some of the party were quite exhausted. From the summit of a hill (since well named Thirsty Hill) a
fine lake was spied, and two of the party proceeded with concerted
signals to show whether it was fresh water. What was our disappointment to find a snow-white expanse of salt, crystallized in great cubes!
We attributed our extreme thirst to the dryness of the atmosphere;
but whatever the cause might be, we were exceedingly glad late in the
evening to get back to the boats. Although we could nowhere find,
during our whole visit, a single drop of fresh water, yet some must
exist; for by an odd chance I found on the surface of the salt water,
near the head of the bay, a Colymbetes not quite dead, which must

have lived in some not far distant pool. Three other insects (a Cincindela, like *hybrida*, a Cymindis, and a Harpalus, which all live on muddy flats occasionally overflowed by the sea), and one other found dead on the plain, complete the list of the beetles. A good-sized fly (Tabanus) was extremely numerous, and tormented us by its painful bite. The common horsefly, which is so troublesome in the shady lanes of England, belongs to this same genus. We here have the puzzle that so frequently occurs in the case of musquitoes – on the blood of what animals do these insects commonly feed? The guanaco is nearly the only warm-blooded quadruped, and it is found in quite inconsiderable numbers compared with the multitude of flies.

The geology of Patagonia is interesting. Differently from Europe, where the tertiary formations appear to have accumulated in bays, here along hundreds of miles of coast we have one great deposit, including many tertiary shells, all apparently extinct. The most common shell is a massive gigantic oyster, sometimes even a foot in diameter. These beds are covered by others of a peculiar soft white stone, including much gypsum, and resembling chalk, but really of a pumiceous nature. It is highly remarkable, from being composed, to at least one-tenth part of its bulk, of Infusoria: Professor Ehrenberg has already ascertained in it thirty oceanic forms. This bed extends for 500 miles along the coast, and probably for a considerably greater distance. At Port St. Julian its thickness is more than 800 feet! These white beds are everywhere capped by a mass of gravel, forming probably one of the largest beds of shingle in the world: it certainly extends from near the Rio Colorado to between 600 and 700 nautical miles southward; at Santa Cruz (a river a little south of St. Julian), it reaches to the foot of the Cordillera; half way up the river, its thickness is more than 200 feet; it probably everywhere extends to this great chain, whence the well-rounded pebbles of porphyry have been derived: we may consider its average breadth as 200 miles, and its average thickness as about 50 feet. If this great bed of pebbles, without including the mud necessarily derived from their attrition, was piled into a mound, it would form a great mountain chain! When we consider that all these pebbles, countless as the grains of sand in the desert, have been derived from the slow-falling of masses of rock on the old coast-lines and banks of rivers; and that these fragments have been dashed into smaller pieces, and that each of them has since been slowly rolled, rounded, and far transported, the mind is stupefied in thinking over the long, absolutely necessary, lapse of years. Yet all this gravel has been transported, and probably rounded, subsequently to the deposition of the white beds, and long subsequently to the underlying beds with the tertiary shells.

Everything in this southern continent has been effected on a grand scale: the land, from the Rio Plata to Tierra del Fuego, a distance of 1200 miles, has been raised in mass (and in Patagonia to a height of between 300 and 400 feet), within the period of the now existing sea-shells. The old and weathered shells left on the surface of the upraised plain still partially retain their colours. The uprising movement has been interrupted by at least eight long periods of rest, during which the sea ate deeply back into the land, forming at successive levels the long lines of cliffs or escarpments, which separate the different plains as they rise like steps one behind the other. The elevatory movement, and the eating-back power of the sea during the periods of rest, have been equable over long lines of coast; for I was astonished to find that the step-like plains stand at nearly corresponding heights at far distant points. The lowest plain is 90 feet high; and the highest, which I ascended near the coast, is 950 feet; and of this, only relics are left in the form of flat gravel-capped hills. The upper plain of S. Cruz slopes up to a height of 3000 feet at the foot of the Cordillera. I have said that within the period of existing sea-shells Patagonia has been upraised 300 to 400 feet: I may add, that within the period when icebergs transported boulders over the upper plain of Santa Cruz, the elevation has been at least 1500 feet. Nor has Patagonia been affected only by upward movements: the extinct tertiary shells from Port St. Julian and Santa Cruz cannot have lived, according to Professor E. Forbes, in a greater depth of water than from 40 to 250 feet; but they are now covered with sea-deposited strata from 800 to 1000 feet in thickness: hence the bed of the sea, on which these shells once lived, must have sunk downwards several hundred feet, to allow of the accumulation of the superincumbent strata. What a history of geological changes does the simply-constructed coast of Patagonia reveal!

At Port St. Julian,[1] in some red mud capping the gravel on the 90-feet plain, I found half the skeleton of the Macrauchenia Patachonica, a remarkable quadruped, full as large as a camel. It belongs to the same division of the Pachydermata with the rhinoceros, tapir, and palæotherium; but in the structure of the bones of its long neck it shows a clear relation to the camel, or rather to the guanaco and llama. From recent sea-shells being found on two of the higher step-formed plains, which must have been modelled and upraised before the mud was deposited in which the Macrauchenia was intombed, it is certain that this curious quadruped lived long after the sea was inhabited by its present shells. I was at first much surprised how a large quadruped could so lately have subsisted, in lat. 49° 15′, on these wretched gravel plains with their stunted vegetation; but the relationship of the Macrauchenia to the guanaco, now an inhabitant of the most sterile parts, partly explains this difficulty.

The relationship, though distant, between the Macrauchenia and the Guanaco, between the Toxodon and the Capybara, – the closer relationship between the many extinct Edentata and the living sloths, ant-eaters, and armadilloes, now so eminently characteristic of South American zoology, – and the still closer relationship between the fossil and living species of Ctenomys and Hydrochærus, are most interesting facts. This relationship is shown wonderfully – as wonderfully as between the fossil and extinct Marsupial animals of Australia – by the great collection lately brought to Europe from the caves of Brazil by MM. Lund and Clausen. In this collection there are extinct species of all the thirty-two genera, excepting four, of the terrestrial quadrupeds now inhabiting the provinces in which the caves occur; and the extinct species are much more numerous than those now living; there are fossil ant-eaters, armadilloes, tapirs, peccaries, guanacos, opossums, and numerous South American gnawers and monkeys, and other animals. This wonderful relationship in the same continent between the dead and the living, will, I do not doubt,

1 I have lately heard that Capt. Sulivan, R.N., has found numerous fossil bones, embedded in regular strata, on the banks of the R. Gallegos, in lat. 51° 4′. Some of the bones are large; others are small, and appear to have belonged to an armadillo. This is a most interesting and important discovery.

hereafter throw more light on the appearance of organic beings on our earth, and their disappearance from it, than any other class of facts.

It is impossible to reflect on the changed state of the American continent without the deepest astonishment. Formerly it must have swarmed with great monsters: now we find mere pigmies, compared with the antecedent, allied races. If Buffon had known of the gigantic sloth and armadillo-like animals, and of the lost Pachydermata, he might have said with a greater semblance of truth that the creative force in America had lost its power, rather than that it had never possessed great vigour. The greater number, if not all, of these extinct quadrupeds lived at a late period, and were the contemporaries of most of the existing sea-shells. Since they lived, no very great change in the form of the land can have taken place. What, then, has exterminated so many species and whole genera? The mind at first is irresistibly hurried into the belief of some great catastrophe; but thus to destroy animals, both large and small, in Southern Patagonia, in Brazil, on the Cordillera of Peru, in North America up to Behring's Straits, we must shake the entire framework of the globe. An examination, moreover, of the geology of La Plata and Patagonia, leads to the belief that all the features of the land result from slow and gradual changes. It appears from the character of the fossils in Europe, Asia, Australia, and in North and South America, that those conditions which favour the life of the *larger* quadrupeds were lately co-extensive with the world: what those conditions were, no one has yet even conjectured. It could hardly have been a change of temperature, which at about the same time destroyed the inhabitants of tropical, temperate, and arctic latitudes on both sides of the globe. In North America we positively know from Mr. Lyell, that the large quadrupeds lived subsequently to that period, when boulders were brought into latitudes at which icebergs now never arrive: from conclusive but indirect reasons we may feel sure, that in the southern hemisphere the Macrauchenia, also, lived long subsequently to the ice-transporting boulder-period. Did man, after his first inroad into South America, destroy, as has been suggested, the unwieldy Megatherium and the other Edentata? We must at least look to some other cause for the destruction of the little tucutuco at Bahia Blanca, and of the many fossil mice and other small quadrupeds in Brazil. No one will imagine that a drought, even far severer than those which cause such losses in the provinces of La Plata, could destroy every individual of every species from Southern Patagonia to Behring's Straits. What shall we say of the extinction of the horse? Did those plains fail of pasture, which have since been overrun by thousands and hundreds of thousands of the descendants of the stock introduced by the Spaniards? Have the subsequently introduced species consumed the food of the great antecedent races? Can we believe that the Capybara has taken the food of the Toxodon, the Guanaco of the Macrauchenia, the existing small Edentata of their numerous gigantic prototypes? Certainly, no fact in the long history of the world is so startling as the wide and repeated exterminations of its inhabitants.

Nevertheless, if we consider the subject under another point of view, it will appear less perplexing. We do not steadily bear in mind, how profoundly ignorant we are of the conditions of existence of every animal; nor do we always remember, that some check is constantly preventing the too rapid increase of every organized being left in a state of nature. The supply of food, on an average, remains constant; yet the tendency in every animal to increase by propagation is geometrical; and its surprising effects have nowhere been more astonishingly shown, than in the case of the European animals run wild during the last few centuries in America. Every animal in a state of nature regularly breeds; yet in a species long established, any *great* increase in numbers is obviously impossible, and must be checked by some means. We are, nevertheless, seldom able with certainty to tell in any given species, at what period of life, or at what period of the year, or whether only at long intervals, the check falls; or, again, what is the precise nature of the check. Hence probably it is, that we feel so little surprise at one, of two species closely allied in habits, being rare and the other abundant in the same district; or, again, that one should be abundant in one district, and another, filling the same place in the economy of nature, should be scanty in a neighbouring district, differing very little in its conditions. If asked how this is, one immediately replies that it is determined by some slight difference in climate, food, or the number of enemies: yet how rarely, if ever, we can point out the precise cause and manner of action of the check! We are, therefore, driven to the conclusion, that causes generally quite inappreciable by us, determine whether a given species shall be abundant or scanty in numbers.

In the cases where we can trace the extinction of a species through man, either wholly or in one limited district, we know that it becomes rarer and rarer, and is then lost: it would be difficult to point out any just distinction[1] between a species destroyed by man or by the increase of its natural enemies. The evidence of rarity preceding extinction, is more striking in the successive tertiary strata, as remarked by several able observers; it has often been found that a shell very common in a tertiary stratum is now most rare, and has even long been thought to be extinct. If then, as appears probable, species first become rare and then extinct – if the too rapid increase of every species, even the most favoured, is steadily checked, as we must admit, though how and when it is hard to say – and if we see, without the smallest surprise, though unable to assign the precise reason, one species abundant and another closely-allied species rare in the same district – why should we feel such great astonishment at the rarity being carried a step further to extinction? An action going on, on every side of us, and yet barely appreciable, might surely be carried a little further, without exciting our observation. Who would feel any great surprise at hearing that the Megalonyx was formerly rare compared with the Megatherium, or that one of the fossil monkeys was few in number compared with one of the now living monkeys? and yet in this comparative rarity, we should have the plainest evidence of less favourable conditions for their existence. To admit that species generally become rare before they become extinct – to feel no surprise at the comparative rarity of one species with another, and yet to call in some extraordinary agent and to marvel greatly when a species ceases to exist, appears to me much the same as to admit that sickness in the individual is the prelude to death – to feel no surprise at sickness – but when the sick man dies, to wonder, and to believe that he died through violence.

1 See the excellent remarks on this subject by Mr. Lyell, in his Principles of Geology.

CHAPTER IX

SANTA CRUZ, PATAGONIA, AND THE FALKLAND ISLANDS

Santa Cruz – Expedition up the River – Indians – Immense Streams of Basaltic Lava – Fragments not transported by the River – Excavation of the Valley – Condor, habits of – Cordillera – Erratic Boulders of great size – Indian Relics – Return to the Ship – Falkland Islands – Wild Horses, Cattle, Rabbits – Wolf-like Fox – Fire made of Bones – Manner of hunting Wild Cattle – Geology – Streams of Stones – Scenes of Violence – Penguin – Geese – Eggs of Doris – Compound Animals.

APRIL 13*th*, 1834. – The *Beagle* anchored within the mouth of the Santa Cruz. This river is situated about sixty miles south of Port St. Julian. During the last voyage Captain Stokes proceeded thirty miles up it, but then, from the want of provisions, was obliged to return. Excepting what was discovered at that time, scarcely anything was known about this large river. Captain Fitz Roy now determined to follow its course as far as time would allow. On the 18th three whale-boats started, carrying three weeks' provisions; and the party consisted of twenty-five souls – a force which would have been sufficient to have defied a host of Indians. With a strong flood-tide and a fine day we made a good run, soon drank some of the fresh water, and were at night nearly above the tidal influence.

The river here assumed a size and appearance which, even at the highest point we ultimately reached, was scarcely diminished. It was generally from three to four hundred yards broad, and in the middle about seventeen feet deep. The rapidity of the current, which in its whole course runs at the rate of from four to six knots an hour, is perhaps its most remarkable feature. The water is of a fine blue colour, but with a slight milky tinge, and not so transparent as at first sight would have been expected. It flows over a bed of pebbles, like those which compose the beach and the surrounding plains. It runs in a winding course through a valley, which extends in a direct line westward. This valley varies from five to ten miles in breadth; it is bounded by step-formed terraces, which rise in most parts, one above the other, to the height of five hundred feet, and have on the opposite sides a remarkable correspondence.

April 19*th*. – Against so strong a current it was, of course, quite impossible to row or sail: consequently the three boats were fastened together head and stern, two hands left in each, and the rest came on shore to track. As the general arrangements made by Captain Fitz Roy were very good for facilitating the work of all, and as all had a share in it, I will describe the system. The party, including every one, was divided into two spells, each of which hauled at the tracking line alternately for an hour and a half. The officers of each boat lived with, ate the same food, and slept in the same tent with their crew, so that each boat was quite independent of the others. After sunset the first level spot where any bushes were growing, was chosen for our night's lodging. Each of the crew took it in turns to be cook. Immediately the boat was hauled up, the cook made his fire; two others pitched the tent; the coxswain handed the things out of the boat; the rest carried them up to the tents and collected firewood. By this order, in half an hour everything was ready for the night. A watch of two men and an officer was always kept, whose duty it was to look after the boats, keep up the fire, and guard against Indians. Each in the party had his one hour every night.

During this day we tracked but a short distance, for there were many islets, covered by thorny bushes, and the channels between them were shallow.

April 20*th*. – We passed the islands and set to work. Our regular day's march, although it was hard enough, carried us on an average only ten miles in a straight line, and perhaps fifteen or twenty altogether. Beyond the place where we slept last night, the country is completely *terra incognita*, for it was there that Captain Stokes turned back. We saw in the distance a great smoke, and found the skeleton of a horse, so we knew that Indians were in the neighbourhood. On the next morning (21st) tracks of a party of horse, and marks left by the trailing of the chuzos, or long spears, were observed on the ground. It was generally thought that the Indians had reconnoitred us during the night. Shortly afterwards we came to a spot where, from the fresh footsteps of men, children, and horses, it was evident that the party had crossed the river.

April 22nd. – The country remained the same, and was extremely uninteresting. The complete similarity of the productions throughout Patagonia is one of its most striking characters. The level plains of arid shingle support the same stunted and dwarf plants; and in the valleys the same thorn-bearing bushes grow. Everywhere we see the same birds and insects. Even the very banks of the river and of the clear streamlets which entered it, were scarcely enlivened by a brighter tint of green. The curse of sterility is on the land, and the water flowing over a bed of pebbles partakes of the same curse. Hence the number of waterfowl is very scanty; for there is nothing to support life in the stream of this barren river.

Patagonia, poor as she is in some respects, can however boast of a greater stock of small rodents[1] than perhaps any other country in the world. Several species of mice are externally characterized by large thin ears and a very fine fur. These little animals swarm amongst the thickets in the valleys, where they cannot for months together taste a drop of water excepting the dew. They all seem to be cannibals; for no sooner was a mouse caught in one of my traps than it was devoured by others. A small and delicately-shaped fox, which is likewise very abundant, probably derives its entire support from these small animals. The guanaco is also in his proper district; herds of fifty or a hundred were common; and, as I have stated, we saw one which must have contained at least five hundred. The puma, with the condor and other carrion-hawks in its train, follows and preys upon these animals. The footsteps of the puma were to be seen almost everywhere on the banks of the river; and the remains of several guanacos, with their necks dislocated and bones broken, showed how they had met their death.

April 24th. – Like the navigators of old when approaching an unknown land, we examined and watched for the most trivial

1 The deserts of Syria are characterized, according to Volney (tom. i. p. 351), by woody bushes, numerous rats, gazelles, and hares. In the landscape of Patagonia, the guanaco replaces the gazelle, and the agouti the hare.

sign of a change. The drifted trunk of a tree, or a boulder of primitive rock, was hailed with joy, as if we had seen a forest growing on the flanks of the Cordillera. The top, however, of a heavy bank of clouds, which remained almost constantly in one position, was the most promising sign, and eventually turned out a true harbinger. At first the clouds were mistaken for the mountains themselves, instead of the masses of vapour condensed by their icy summits.

April 26th. – We this day met with a marked change in the geological structure of the plains. From the first starting I had carefully examined the gravel in the river, and for the two last days had noticed the presence of a few small pebbles of a very cellular basalt. These gradually increased in number and in size, but none were as large as a man's head. This morning, however, pebbles of the same rock, but more compact, suddenly became abundant, and in the course of half an hour we saw, at the distance of five or six miles, the angular edge of a great basaltic platform. When we arrived at its base we found the stream bubbling among the fallen blocks. For the next twenty-eight miles the river-course was encumbered with these basaltic masses. Above that limit immense fragments of primitive rocks, derived from the surrounding boulder-formation, were equally numerous. None of the fragments of any considerable size had been washed more than three or four miles down the river below their parent-source: considering the singular rapidity of the great body of water in the Santa Cruz, and that no still reaches occur in any part, this example is a most striking one, of the inefficiency of rivers in transporting even moderately-sized fragments.

The basalt is only lava, which has flowed beneath the sea; but the eruptions must have been on the grandest scale. At the point where we first met this formation, it was 120 feet in thickness; following up the river course, the surface imperceptibly rose and the mass became thicker, so that at forty miles above the first station it was 320 feet thick. What the thickness may be close to the Cordillera, I have no means of knowing, but the platform there attains a height of about three thousand feet above the level of the sea: we must therefore look to the mountains of that great chain for its source; and worthy of such a source are streams, that have flowed over the gently inclined bed of the sea to a distance of one hundred miles. At the first glance of the basaltic cliffs on the opposite sides of the valley, it was evident that the strata once were united. What power, then, has removed along a whole line of country, a solid mass of very hard rock, which had an average thickness of nearly three hundred feet, and a breadth varying from rather less than two miles to four miles? The river, though it has so little power in transporting even inconsiderable fragments, yet in the lapse of ages might produce by its gradual erosion an effect, of which it is difficult to judge the amount. But in this case, independently of the insignificance of such an agency, good reasons can be assigned for believing that this valley was formerly occupied by an arm of the sea. It is needless in this work to detail the arguments leading to this conclusion, derived from the form and the nature of the step-formed terraces on both sides of the valley, from the manner in which the bottom of the valley near the Andes expands into a great estuary-like plain with sand-hillocks on it, and from the occurrence of a few sea-shells lying in the bed of the river. If I had space I could prove that South America was formerly here cut off by a strait, joining the Atlantic and Pacific oceans, like that of Magellan. But it may yet be asked, how has the solid basalt been removed? Geologists formerly would have brought into play, the violent action of some overwhelming debacle; but in this case such a supposition would have been quite inadmissible; because, the same step-like plains with existing sea-shells lying on their surface, which front the long line of the Patagonian coast, sweep up on each side of the valley of Santa Cruz. No possible action of any flood could thus have modelled the land, either within the valley or along the open coast; and by the formation of such step-like plains or terraces the valley itself has been hollowed out. Although we know that there are tides, which run within the Narrows of the Strait of Magellan at the rate of eight knots an hour, yet we must confess that it makes the head almost giddy to reflect on the number of years, century after century, which the tides, unaided by a heavy surf, must have required to have corroded so vast an area and thickness of solid basaltic lava. Nevertheless, we must believe that the strata undermined by the waters of this ancient strait, were broken up into huge fragments, and these lying scattered on the beach, were reduced first to smaller blocks, then to pebbles, and lastly to the most impalpable mud, which the tides drifted far into the Eastern or Western Ocean.

With the change in the geological structure of the plains the character of the landscape likewise altered. While rambling up some of the narrow and rocky defiles, I could almost have fancied myself transported back again to the barren valleys of the island of St. Jago. Among the basaltic cliffs, I found some plants which I had seen nowhere else, but others I recognized as being wanderers from Tierra del Fuego. These porous rocks serve as a reservoir for the scanty rain-water; and consequently on the line where the igneous and sediment-ary formations unite, some small springs (most rare occurrences in Patagonia) burst forth; and they could be distinguished at a distance by the circumscribed patches of bright green herbage.

April 27th. – The bed of the river became rather narrower, and hence the stream more rapid. It here ran at the rate of six knots an hour. From this cause, and from the many great angular fragments, tracking the boats became both dangerous and laborious.

This day I shot a condor. It measured from tip to tip of the wings, eight and a half feet, and from beak to tail, four feet. This bird is known to have a wide geographical range, being found on the west coast of South America, from the Strait of Magellan along the Cor-dillera as far as eight degrees N. of the equator. The steep cliff near the mouth of the Rio Negro is its northern limit on the Patagonian coast; and they have there wandered about four hundred miles from the great central line of their habitation in the Andes. Further south, among the bold precipices at the head of Port Desire, the condor is not uncommon; yet only a few stragglers occasionally visit the sea-coast. A line of cliff near the mouth of the Santa Cruz is frequented by these birds, and about eighty miles up the river, where the sides of the valley are formed by steep basaltic precipices, the condor re-appears. From these facts, it seems that the condors require perpen-dicular cliffs. In Chile, they haunt, during the greater part of the year, the lower country near the shores of the Pacific, and at night several roost together in one tree; but in the early part of summer, they retire to the most inaccessible parts of the inner Cordillera, there to breed in peace.

With respect to their propagation, I was told by the country people in Chile, that the condor makes no sort of nest, but in the months of November and December lays two large white eggs on a shelf of bare rock. It is said that the young condors cannot fly for an entire year; and long after they are able, they continue to roost by night, and hunt by day with their parents. The old birds generally live in pairs; but among the inland basaltic cliffs of the Santa Cruz, I found a spot, where scores must usually haunt. On coming suddenly to the brow of the precipice, it was a grand spectacle to see between twenty and thirty of these great birds start heavily from their resting-place, and wheel away in majestic circles. From the quantity of dung on the rocks, they must long have frequented this cliff for roosting and breeding. Having gorged themselves with carrion on the plains below, they retire to these favourite ledges to digest their food. From these facts, the condor, like the gallinazo, must to a certain degree be considered as a gregarious bird. In this part of the country they live altogether on the guanacos which have died a natural death, or, as more commonly happens, have been killed by the pumas. I believe, from what I saw in Patagonia, that they do not on ordinary occasions extend their daily excursions to any great distance from their regular sleeping-places.

The condors may oftentimes be seen at a great height, soaring over a certain spot in the most graceful circles. On some occasions I am sure that they do this only for pleasure, but on others, the Chileno countryman tells you that they are watching a dying animal, or the puma devouring its prey. If the condors glide down, and then suddenly all rise together, the Chileno knows that it is the puma which, watching the carcass, has sprung out to drive away the robbers. Besides feeding on carrion, the condors frequently attack young goats and lambs; and the shepherd-dogs are trained, whenever they pass over, to run out, and looking upwards to bark violently. The Chilenos destroy and catch numbers. Two methods are used; one is to place a carcass on a level piece of ground within an enclosure of sticks with an opening, and when the condors are gorged, to gallop up on horseback to the entrance, and thus enclose them: for when this bird has not space to run, it cannot give its body sufficient momentum to rise from the ground. The second method is to mark the trees in which, frequently to the number of five or six together, they roost, and then at night to climb up and noose them. They are such heavy sleepers, as I have myself witnessed, that this is not a difficult task. At Valparaiso, I have seen a living condor sold for sixpence, but the common price is eight or ten shillings. One which I saw brought in, had been tied with rope, and was much injured; yet, the moment the line was cut by which its bill was secured, although surrounded by people, it began ravenously to tear a piece of carrion. In a garden at the same place, between twenty and thirty were kept alive. They were fed only once a week, but they appeared in pretty good health.[1] The Chileno countrymen assert that the condor will live, and retain its vigour, between five and six weeks without eating: I cannot answer for the truth of this, but it is a cruel experiment, which very likely has been tried.

When an animal is killed in the country, it is well known that the condors, like other carrion-vultures, soon gain intelligence of it, and congregate in an inexplicable manner. In most cases it must not be overlooked, that the birds have discovered their prey, and have picked the skeleton clean, before the flesh is in the least degree tainted. Remembering the experiments of M. Audubon, on the little smelling powers of carrion-hawks, I tried in the above-mentioned garden the following experiment: the condors were tied, each by a rope, in a long row at the bottom of a wall; and having folded up a piece of meat in white paper, I walked backwards and forwards, carrying it in my hand at the distance of about three yards from them, but no notice whatever was taken. I then threw it on the ground, within one yard of an old male bird; he looked at it for a moment with attention, but then regarded it no more. With a stick I pushed it closer and closer, until at last he touched it with his beak; the paper was then instantly torn off with fury, and at the same moment, every bird in the long row began struggling and flapping its wings. Under the same circumstances, it would have been quite impossible to have deceived a dog. The evidence in favour of and against the acute smelling powers of carrion-vultures is singularly balanced. Professor Owen has demonstrated that the olfactory nerves of the turkey-buzzard (Cathartes aura)

---

1 I noticed that several hours before any one of the condors died, all the lice, with which it was infested, crawled to the outside feathers. I was assured that this always happened.

are highly developed; and on the evening when Mr. Owen's paper was read at the Zoological Society, it was mentioned by a gentleman that he had seen the carrion-hawks in the West Indies on two occasions collect on the roof of a house, when a corpse had become offensive from not having been buried: in this case the intelligence could hardly have been acquired by sight. On the other hand, besides the experiments of Audubon and that one by myself, Mr. Bachman has tried in the United States many varied plans, showing that neither the turkey-buzzard (the species dissected by Professor Owen) nor the gallinazo find their food by smell. He covered portions of highly offensive offal with a thin canvas cloth, and strewed pieces of meat on it; these the carrion-vultures ate up, and then remained quietly standing, with their beaks within the eighth of an inch of the putrid mass, without discovering it. A small rent was made in the canvas, and the offal was immediately discovered; the canvas was replaced by a fresh piece, and meat again put on it, and was again devoured by the vultures without their discovering the hidden mass on which they were trampling. These facts are attested by the signatures of six gentlemen, besides that of Mr. Bachman.[1]

Often when lying down to rest on the open plains, on looking upwards, I have seen carrion-hawks sailing through the air at a great height. Where the country is level I do not believe a space of the heavens, of more than fifteen degrees above the horizon, is commonly viewed with any attention by a person either walking or on horseback. If such be the case, and the vulture is on the wing at a height of between three and four thousand feet, before it could come within the range of vision, its distance in a straight line from the beholder's eye, would be rather more than two British miles. Might it not thus readily be overlooked? When an animal is killed by the sportsman in a lonely valley, may he not all the while be watched from above by the sharp-sighted bird? And will not the manner of its descent proclaim throughout the district to the whole family of carrion-feeders, that their prey is at hand?

When the condors are wheeling in a flock round and round any spot, their flight is beautiful. Except when rising from the ground, I do

1 Loudon's Magazine of Nat. Hist., vol. vii.

not recollect ever having seen one of these birds flap its wings. Near Lima, I watched several for nearly half an hour, without once taking off my eyes: they moved in large curves, sweeping in circles, descending and ascending without giving a single flap. As they glided close over my head, I intently watched from an oblique position, the outlines of the separate and great terminal feathers of each wing; and these separate feathers, if there had been the least vibratory movement, would have appeared as if blended together; but they were seen distinct against the blue sky. The head and neck were moved frequently, and apparently with force; and the extended wings seemed to form the fulcrum on which the movements of the neck, body, and tail acted. If the bird wished to descend, the wings were for a moment collapsed; and when again expanded with an altered inclination, the momentum gained by the rapid descent seemed to urge the bird upwards with the even and steady movement of a paper kite. In the case of any bird *soaring*, its motion must be sufficiently rapid, so that the action of the inclined surface of its body on the atmosphere may counter-balance its gravity. The force to keep up the momentum of a body moving in a horizontal plane in the air (in which there is so little friction) cannot be great, and this force is all that is wanted. The movement of the neck and body of the condor, we must suppose, is sufficient for this. However this may be, it is truly wonderful and beautiful to see so great a bird, hour after hour, without any apparent exertion, wheeling and gliding over mountain and river.

April 29th. – From some high land we hailed with joy the white summits of the Cordillera, as they were seen occasionally peeping through their dusky envelope of clouds. During the few succeeding days we continued to get on slowly, for we found the river-course very tortuous, and strewed with immense fragments of various ancient slaty rocks, and of granite. The plain bordering the valley had here attained an elevation of about 1100 feet above the river, and its character was much altered. The well-rounded pebbles of porphyry were mingled with many immense angular fragments of basalt and of primary rocks. The first of these erratic boulders which I noticed, was sixty-seven miles distant from the nearest mountain; another which I measured was five yards square, and projected five feet above the gravel. Its edges were so angular, and its size so great, that I at first mistook it for a rock *in situ*, and took out my compass to observe the direction of its cleavage. The plain here was not quite so level as that nearer the coast, but yet it betrayed no signs of any great violence. Under these circumstances it is, I believe, quite impossible to explain the transportal of these gigantic masses of rock so many miles from their parent-source, on any theory except by that of floating icebergs.

During the last two days we met with signs of horses, and with several small articles which had belonged to the Indians – such as parts of a mantle and a bunch of ostrich feathers – but they appeared to have been lying long on the ground. Between the place where the Indians had so lately crossed the river and this neighbourhood, though so many miles apart, the country appears to be quite unfrequented. At first, considering the abundance of the guanacos, I was surprised at this; but it is explained by the stony nature of the plains, which would soon disable an unshod horse from taking part in the chace. Nevertheless, in two places in this very central region, I found small heaps of stones, which I do not think could have been accidentally thrown together. They were placed on points, projecting over the edge of the highest lava cliff, and they resembled, but on a small scale, those near Port Desire.

May 4th. – Captain Fitz Roy determined to take the boats no higher. The river had a winding course, and was very rapid; and the appearance of the country offered no temptation to proceed any further. Everywhere we met with the same productions, and the same dreary landscape. We were now one hundred and forty miles distant from the Atlantic, and about sixty from the nearest arm of the Pacific. The valley in this upper part expanded into a wide basin, bounded on the north and south by the basaltic platforms, and fronted by the long range of the snow-clad Cordillera. But we viewed these grand mountains with regret, for we were obliged to imagine their nature and productions, instead of standing, as we had hoped, on their summits. Besides the useless loss of time which an attempt to ascend the river any higher would have cost us, we had already been for some days on half allowance of bread. This, although really enough for reasonable men, was, after a hard day's march, rather scanty food: a light stomach and an easy digestion are good things to talk about, but very unpleasant in practice.

5th. – Before sunrise we commenced our descent. We shot down the stream with great rapidity, generally at the rate of ten knots an hour. In this one day we effected what had cost us five-and-a-half hard days' labour in ascending. On the 8th, we reached the *Beagle* after our twenty-one days' expedition. Every one, excepting myself, had cause to be dissatisfied; but to me the ascent afforded a most interesting section of the great tertiary formation of Patagonia.

On *March 1st*, 1833, and again on *March 16th*, 1834, the *Beagle* anchored in Berkeley Sound, in East Falkland Island. This archipelago is situated in nearly the same latitude with the mouth of the Strait of Magellan; it covers a space of one hundred and twenty by sixty geographical miles, and is a little more than half the size of Ireland. After the possession of these miserable islands had been contested by France, Spain, and England, they were left uninhabited. The government of Buenos Ayres then sold them to a private individual, but likewise used them, as old Spain had done before, for a penal settlement. England claimed her right and seized them. The Englishman who was left in charge of the flag was consequently murdered. A British officer was next sent, unsupported by any power: and when we arrived, we found him in charge of a population, of which rather more than half were runaway rebels and murderers.

The theatre is worthy of the scenes acted on it. An undulating land, with a desolate and wretched aspect, is everywhere covered by a peaty soil and wiry grass, of one monotonous brown colour. Here and there a peak or ridge of grey quartz rock breaks through the smooth surface. Every one has heard of the climate of these regions; it may be compared to that which is experienced at the height of between one and two thousand feet, on the mountains of North Wales; having however less sunshine and less frost, but more wind and rain.[1]

16th. – I will now describe a short excursion which I made round a part of this island. In the morning I started with six horses and

1 From accounts published since our voyage, and more especially from several interesting letters from Capt. Sulivan, R.N., employed on the survey, it appears that we took an exaggerated view of the badness of the climate of these islands. But when I

two Gauchos: the latter were capital men for the purpose, and well accustomed to living on their own resources. The weather was very boisterous and cold, with heavy hail-storms. We got on, however, pretty well, but, except the geology, nothing could be less interesting than our day's ride. The country is uniformly the same undulating moorland; the surface being covered by light brown withered grass and a few very small shrubs, all springing out of an elastic peaty soil. In the valleys here and there might be seen a small flock of wild geese, and everywhere the ground was so soft that the snipe were able to feed. Besides these two birds there were few others. There is one main range of hills, nearly two thousand feet in height, and composed of quartz rock, the rugged and barren crests of which gave us some trouble to cross. On the south side we came to the best country for wild cattle; we met, however, no great number, for they had been lately much harassed.

In the evening we came across a small herd. One of my companions, St. Jago by name, soon separated a fat cow; he threw the bolas, and it struck her legs, but failed in becoming entangled. Then dropping his hat to mark the spot where the balls were left, while at full gallop, he uncoiled his lazo, and after a most severe chace, again came up to the cow, and caught her round the horns. The other Gaucho had gone on ahead with the spare horses, so that St. Jago had some difficulty in killing the furious beast. He managed to get her on a level piece of ground, by taking advantage of her as often as she rushed at him; and when she would not move, my horse, from having been trained, would canter up, and with his chest give her a violent push. But when on level ground it does not appear an easy job for one man to kill a beast mad with terror. Nor would it be so, if the horse, when left to itself without its rider, did not soon learn, for its own safety, to keep the lazo tight; so that, if the cow or ox moves forward, the horse moves just as quickly forward; otherwise, it stands motionless leaning on one side. This horse, however, was a young one, and would not stand still, but gave in to the cow as she struggled. It was admirable to see with

---

reflect on the almost universal covering of peat, and on the fact of wheat seldom ripening here, I can hardly believe that the climate in summer is so fine and dry as it has lately been represented.

what dexterity St. Jago dodged behind the beast, till at last he contrived to give the fatal touch to the main tendon of the hind leg; after which, without much difficulty, he drove his knife into the head of the spinal marrow, and the cow dropped as if struck by lightning. He cut off pieces of flesh with the skin to it, but without any bones, sufficient for our expedition. We then rode on to our sleeping-place, and had for supper 'carne con cuero,' or meat roasted with the skin on it. This is as superior to common beef as venison is to mutton. A large circular piece taken from the back is roasted on the embers with the hide downwards and in the form of a saucer, so that none of the gravy is lost. If any worthy alderman had supped with us that evening, 'carne con cuero,' without doubt, would soon have been celebrated in London.

During the night it rained, and the next day (17th) was very stormy, with much hail and snow. We rode across the island to the neck of land which joins the Rincon del Toro (the great peninsula at the S.W. extremity) to the rest of the island. From the great number of cows which have been killed, there is a large proportion of bulls. These wander about single, or two and three together, and are very savage. I never saw such magnificent beasts; they equalled in the size of their huge heads and necks the Grecian marble sculptures. Capt. Sulivan informs me that the hide of an average-sized bull weighs forty-seven pounds, whereas a hide of this weight, less thoroughly dried, is considered as a very heavy one at Monte Video. The young bulls generally run away for a short distance; but the old ones do not stir a step, except to rush at man and horse; and many horses have been thus killed. An old bull crossed a boggy stream, and took his stand on the opposite side to us; we in vain tried to drive him away, and failing, were obliged to make a large circuit. The Gauchos in revenge determined to emasculate him and render him for the future harmless. It was very interesting to see how art completely mastered force. One lazo was thrown over his horns as he rushed at the horse, and another round his hind legs: in a minute the monster was stretched powerless on the ground. After the lazo has once been drawn tightly round the horns of a furious animal, it does not at first appear an easy thing to disengage it again without killing the beast; nor, I apprehend, would it be so if the man was by himself. By the aid, however, of a second person throwing his lazo so as to catch both hind legs, it is quickly managed: for the animal, so long as its hind legs are kept outstretched, is quite helpless, and the first man can with his hands loosen his lazo from the horns, and then quietly mount his horse; but the moment the second man, by backing ever so little, relaxes the strain, the lazo slips off the legs of the struggling beast, which then rises free, shakes himself, and vainly rushes at his antagonist.

During our whole ride we saw only one troop of wild horses. These animals, as well as the cattle, were introduced by the French in 1764, since which time both have greatly increased. It is a curious fact, that the horses have never left the eastern end of the island, although there is no natural boundary to prevent them from roaming, and that part of the island is not more tempting than the rest. The Gauchos whom I asked, though asserting this to be the case, were unable to account for it, except from the strong attachment which horses have to any locality to which they are accustomed. Considering that the island does not appear fully stocked, and that there are no beasts of prey, I was particularly curious to know what has checked their originally rapid increase. That in a limited island some check would sooner or later supervene, is inevitable; but why has the increase of the horse been checked sooner than that of the cattle? Capt. Sulivan has taken much pains for me in this inquiry. The Gauchos employed here attribute it chiefly to the stallions constantly roaming from place to place, and compelling the mares to accompany them, whether or not the young foals are able to follow. One Gaucho told Capt. Sulivan that he had watched a stallion for a whole hour, violently kicking and biting a mare till he forced her to leave her foal to its fate. Capt. Sulivan can so far corroborate this curious account, that he has several times found young foals dead, whereas he has never found a dead calf. Moreover, the dead bodies of full-grown horses are more frequently found, as if more subject to disease or accidents, than those of the cattle. From the softness of the ground their hoofs often grow irregularly to a great length, and this causes lameness. The predominant colours are roan and iron-grey. All the horses bred here, both tame and wild, are rather small-sized, though generally in good condition; and they have lost so much strength, that they are unfit to be used in taking wild cattle with the lazo: in consequence, it is necessary to go to the great expense of importing fresh horses from the Plata. At some future period the southern hemisphere probably will have its breed of Falkland ponies, as the northern has its Shetland breed.

The cattle, instead of having degenerated like the horses, seem, as before remarked, to have increased in size; and they are much more numerous than the horses. Capt. Sulivan informs me that they vary much less in the general form of their bodies and in the shape of their horns than English cattle. In colour they differ much; and it is a remarkable circumstance, that in different parts of this one small island, different colours predominate. Round Mount Usborne, at a height of from 1000 to 1500 feet above the sea, about half of some of the herds are mouse or lead-coloured, a tint which is not common in other parts of the island. Near Port Pleasant dark brown prevails, whereas south of Choiseul Sound (which almost divides the island into two parts), white beasts with black heads and feet are the most common: in all parts black, and some spotted animals may be observed. Capt. Sulivan remarks, that the difference in the prevailing colours was so obvious, that in looking for the herds near Port Pleasant, they appeared from a long distance like black spots, whilst south of Choiseul Sound they appeared like white spots on the hillsides. Capt. Sulivan thinks that the herds do not mingle; and it is a singular fact, that the mouse-coloured cattle, though living on the high land, calve about a month earlier in the season than the other coloured beasts on the lower land. It is interesting thus to find the once domesticated cattle breaking into three colours, of which some one colour would in all probability ultimately prevail over the others, if the herds were left undisturbed for the next several centuries.

The rabbit is another animal which has been introduced, and has succeeded very well; so that they abound over large parts of the island. Yet, like the horses, they are confined within certain limits; for they have not crossed the central chain of hills, nor would they have extended even so far as its base, if, as the Gauchos informed me, small colonies had not been carried there. I should not have supposed that these animals, natives of northern Africa, could have existed in a climate so humid as this, and which enjoys so little sunshine that even wheat ripens only occasionally. It is asserted that in Sweden, which any one would have thought a more favourable climate, the rabbit cannot live out of doors. The first few pair, moreover, had here to contend against pre-existing enemies, in the fox and some large hawks. The French naturalists have considered the black variety a distinct species, and called it Lepus Magellanicus.[1] They imagined that Magellan, when talking of an animal under the name of 'conejos' in the Strait of Magellan, referred to this species; but he was alluding to a small cavy, which to this day is thus called by the Spaniards. The Gauchos laughed at the idea of the black kind being different from the grey, and they said that at all events it had not extended its range any further than the grey kind; that the two were never found separate; and that they readily bred together, and produced piebald offspring. Of the latter I now possess a specimen, and it is marked about the head differently from the French specific description. This circumstance shows how cautious naturalists should be in making species; for even Cuvier, on looking at the skull of one of these rabbits, thought it was probably distinct!

The only quadruped native to the island[2] is a large wolf-like fox (Canis antarcticus), which is common to both East and West Falkland. I have no doubt it is a peculiar species, and confined to this archipelago; because many sealers, Gauchos, and Indians, who have visited these islands, all maintain that no such animal is found in any part of South America. Molina, from a similarity in habits, thought that this was the same with his 'culpeu;'[3] but I have seen both, and they are quite distinct. These wolves are well known, from Byron's account of their tameness and curiosity, which the sailors, who ran into the water to avoid them, mistook for fierceness. To this day their

1 Lesson's Zoology of the Voyage of the *Coquille*, tom. i. p. 168. All the early voyagers, and especially Bougainville, distinctly state that the wolf-like fox was the only native animal on the island. The distinction of the rabbit as a species, is taken from peculiarities in the fur, from the shape of the head, and from the shortness of the ears. I may here observe that the difference between the Irish and English hare rests upon nearly similar characters, only more strongly marked.

2 I have reason, however, to suspect that there is a field-mouse. The common European rat and mouse have roamed far from the habitations of the settlers. The common hog has also run wild on one islet: all are of a black colour: the boars are very fierce, and have great tusks.

3 The "culpeu" is the Canis Magellanicus brought home by Captain King from the Strait of Magellan. It is common in Chile.

manners remain the same. They have been observed to enter a tent, and actually pull some meat from beneath the head of a sleeping seaman. The Gauchos also have frequently in the evening killed them, by holding out a piece of meat in one hand, and in the other a knife ready to stick them. As far as I am aware, there is no other instance in any part of the world, of so small a mass of broken land, distant from a continent, possessing so large an aboriginal quadruped peculiar to itself. Their numbers have rapidly decreased; they are already banished from that half of the island which lies to the eastward of the neck of land between St. Salvador Bay and Berkeley Sound. Within a very few years after these islands shall have become regularly settled, in all probability this fox will be classed with the dodo, as an animal which has perished from the face of the earth.

At night (17th) we slept on the neck of land at the head of Choiseul Sound, which forms the south-west peninsula. The valley was pretty well sheltered from the cold wind; but there was very little brushwood for fuel. The Gauchos, however, soon found what, to my great surprise, made nearly as hot a fire as coals; this was the skeleton of a bullock lately killed, from which the flesh had been picked by the carrion-hawks. They told me that in winter they often killed a beast, cleaned the flesh from the bones with their knives, and then with these same bones roasted the meat for their suppers.

18*th*. – It rained during nearly the whole day. At night we managed, however, with our saddle-cloths to keep ourselves pretty well dry and warm; but the ground on which we slept was on each occasion nearly in the state of a bog, and there was not a dry spot to sit down on after our day's ride. I have in another part stated how singular it is that there should be absolutely no trees on these islands, although Tierra del Fuego is covered by one large forest. The largest bush in the island (belonging to the family of Compositæ) is scarcely so tall as our gorse. The best fuel is afforded by a green little bush about the size of common heath, which has the useful property of burning while fresh and green. It was very surprising to see the Gauchos, in the midst of rain and everything soaking wet, with nothing more than a tinder-box and piece of rag, immediately make a fire. They sought beneath the tufts of grass and bushes for a few dry twigs, and these they rubbed into fibres; then surrounding them with coarser twigs, something like a bird's nest, they put the rag with its spark of fire in the middle and covered it up. The nest being then held up to the wind, by degrees it smoked more and more, and at last burst out in flames. I do not think any other method would have had a chance of succeeding with such damp materials.

19th. – Each morning, from not having ridden for some time previously, I was very stiff. I was surprised to hear the Gauchos, who have from infancy almost lived on horseback, say that, under similar circumstances, they always suffer. St. Jago told me, that having been confined for three months by illness, he went out hunting wild cattle, and in consequence, for the next two days, his thighs were so stiff that he was obliged to lie in bed. This shows that the Gauchos, although they do not appear to do so, yet really must exert much muscular effort in riding. The hunting wild cattle, in a country so difficult to pass as this is on account of the swampy ground, must be very hard work. The Gauchos say they often pass at full speed over ground which would be impassable at a slower pace; in the same manner as a man is able to skate over thin ice. When hunting, the party endeavours to get as close as possible to the herd without being discovered. Each man carries four or five pair of the bolas; these he throws one after the other at as many cattle, which, when once entangled, are left for some days, till they become a little exhausted by hunger and struggling. They are then let free and driven towards a small herd of tame animals, which have been brought to the spot on purpose. From their previous treatment, being too much terrified to leave the herd, they are easily driven, if their strength last out, to the settlement.

The weather continued so very bad that we determined to make a push, and try to reach the vessel before night. From the quantity of rain which had fallen, the surface of the whole country was swampy. I suppose my horse fell at least a dozen times, and sometimes the whole six horses were floundering in the mud together. All the little streams are bordered by soft peat, which makes it very difficult for the horses to leap them without falling. To complete our discomforts we were obliged to cross the head of a creek of the sea, in which the water was as high as our horses' backs; and the little waves, owing to the violence of the wind, broke over us, and made us very wet and cold. Even the iron-framed Gauchos professed themselves glad when they reached the settlement, after our little excursion.

The geological structure of these islands is in most respects simple. The lower country consists of clay-slate and sandstone, containing fossils, very closely related to, but not identical with, those found in the Silurian formations of Europe; the hills are formed of white granular quartz rock. The strata of the latter are frequently arched with perfect symmetry, and the appearance of some of the masses is in consequence most singular. Pernety[1] has devoted several pages to the description of a Hill of Ruins, the successive strata of which he has justly compared to the seats of an amphitheatre. The quartz rock must have been quite pasty when it underwent such remarkable flexures without being shattered into fragments. As the quartz insensibly passes into the sandstone, it seems probable that the former owes its origin to the sandstone having been heated to such a degree that it became viscid, and upon cooling crystallized. While in the soft state it must have been pushed up through the overlying beds.

In many parts of the island the bottoms of the valleys are covered in an extraordinary manner by myriads of great loose angular fragments of the quartz rock, forming 'streams of stones.' These have been mentioned with surprise by every voyager since the time of Pernety. The blocks are not waterworn, their angles being only a little blunted; they vary in size from one or two feet in diameter to ten, or even more than twenty times as much. They are not thrown together into irregular piles, but are spread out into level sheets or great streams. It is not possible to ascertain their thickness, but the water of small streamlets can be heard trickling through the stones many feet below the surface. The actual depth is probably great, because the crevices between the lower fragments must long ago have been filled up with sand. The width of these sheets of stones varies from a few hundred feet to a mile; but the peaty soil daily encroaches on the borders, and even forms islets wherever a few fragments happen to lie close together. In a valley south of Berkeley Sound, which some of

1 Pernety, Voyage aux Isles Malouines, p. 526.

our party called the 'great valley of fragments,' it was necessary to cross an uninterrupted band half a mile wide, by jumping from one pointed stone to another. So large were the fragments, that being overtaken by a shower of rain, I readily found shelter beneath one of them.

Their little inclination is the most remarkable circumstance in these 'streams of stones.' On the hill-sides I have seen them sloping at an angle of ten degrees with the horizon; but in some of the level, broad-bottomed valleys, the inclination is only just sufficient to be clearly perceived. On so rugged a surface there was no means of measuring the angle; but to give a common illustration, I may say that the slope would not have checked the speed of an English mail-coach. In some places, a continuous stream of these fragments followed up the course of a valley, and even extended to the very crest of the hill. On these crests huge masses, exceeding in dimensions any small building, seemed to stand arrested in their headlong course: there, also, the curved strata of the archways lay piled on each other, like the ruins of some vast and ancient cathedral. In endeavouring to describe these scenes of violence one is tempted to pass from one simile to another. We may imagine that streams of white lava had flowed from many parts of the mountains into the lower country, and that when solidified they had been rent by some enormous convulsion into myriads of fragments. The expression 'streams of stones,' which immediately occurred to every one, conveys the same idea. These scenes are on the spot rendered more striking by the contrast of the low, rounded forms of the neighbouring hills.

I was interested by finding on the highest peak of one range (about 700 feet above the sea) a great arched fragment, lying on its convex side, or back downwards. Must we believe that it was fairly pitched up in the air, and thus turned? Or, with more probability, that there existed formerly a part of the same range more elevated than the point on which this monument of a great convulsion of nature now lies. As the fragments in the valleys are neither rounded nor the crevices filled up with sand, we must infer that the period of violence was subsequent to the land having been raised above the waters of the sea. In a transverse section within these valleys, the bottom is nearly level, or rises but very little towards either side. Hence the fragments appear to have travelled from the head of the valley; but in reality it seems more probable that they have been hurled down from the nearest slopes; and that since, by a vibratory movement of overwhelming force,[1] the fragments have been levelled into one continuous sheet. If during the earthquake[2] which in 1835 overthrew Concepcion, in Chile, it was thought wonderful that small bodies should have been pitched a few inches from the ground, what must we say to a movement which has caused fragments many tons in weight, to move onwards like so much sand on a vibrating board, and find their level? I have seen, in the Cordillera of the Andes, the evident marks where stupendous mountains have been broken into pieces like so much thin crust, and the strata thrown on their vertical edges; but never did any scene, like these 'streams of stones,' so forcibly convey to my mind the idea of convulsion, of which in historical records we might in vain seek for any counterpart: yet the progress of knowledge will probably some day give a simple explanation of this phenomenon, as it already has of the so long-thought inexplicable transportal of the erratic boulders, which are strewed over the plains of Europe.

I have little to remark on the zoology of these islands. I have before described the carrion-vulture or Polyborus. There are some other hawks, owls, and a few small land-birds. The waterfowl are particularly numerous, and they must formerly, from the accounts of the old navigators, have been much more so. One day I observed a cormorant playing with a fish which it had caught. Eight times successively the bird let its prey go, then dived after it, and although in deep water, brought it each time to the surface. In the Zoological Gardens I have seen the otter treat a fish in the same manner, much as a cat does a mouse: I do not know of any other instance where dame Nature

1 'Nous n'avons pas été moins saisis d'étonnement à la vûe de l'innombrable quantité de pierres de toutes grandeurs, bouleversées les unes sur les autres, et cependant rangées, comme si elles avoient été amoncelées négligemment pour remplir des ravins. On ne se lassoit pas d'admirer les effets prodigieux de la nature.' – *Pernety*, p. 526.

2 An inhabitant of Mendoza, and hence well capable of judging, assured me that, during the several years he had resided on these islands, he had never felt the slightest shock of an earthquake.

appears so wilfully cruel. Another day, having placed myself between a penguin (Aptenodytes demersa) and the water, I was much amused by watching its habits. It was a brave bird; and till reaching the sea, it regularly fought and drove me backwards. Nothing less than heavy blows would have stopped him: every inch he gained he firmly kept, standing close before me erect and determined. When thus opposed he continually rolled his head from side to side, in a very odd manner, as if the power of distinct vision lay only in the anterior and basal part of each eye. This bird is commonly called the jackass penguin, from its habit, while on shore, of throwing its head backwards, and making a loud strange noise, very like the braying of an ass; but while at sea, and undisturbed, its note is very deep and solemn, and is often heard in the night-time. In diving, its little wings are used as fins; but on the land, as front legs. When crawling, it may be said on four legs, through the tussocks or on the side of a grassy cliff, it moves so very quickly that it might easily be mistaken for a quadruped. When at sea and fishing, it comes to the surface for the purpose of breathing with such a spring, and dives again so instantaneously, that I defy any one at first sight to be sure that it was not a fish leaping for sport.

Two kinds of geese frequent the Falklands. The upland species (Anas Magellanica) is common, in pairs and in small flocks, throughout the island. They do not migrate, but build on the small outlying islets. This is supposed to be from fear of the foxes: and it is perhaps from the same cause that these birds, though very tame by day, are shy and wild in the dusk of the evening. They live entirely on vegetable matter. The rock-goose, so called from living exclusively on the sea-beach (Anas antarctica), is common both here and on the west coast of America, as far north as Chile. In the deep and retired channels of Tierra del Fuego, the snow-white gander, invariably accompanied by his darker consort, and standing close by each other on some distant rocky point, is a common feature in the landscape.

In these islands a great loggerheaded duck or goose (Anas brachyptera), which sometimes weighs twenty-two pounds, is very abundant. These birds were in former days called, from their extraordinary manner of paddling and splashing upon the water, race-horses; but now they are named, much more appropriately, steamers. Their wings are too small and weak to allow of flight, but by their aid, partly swimming and partly flapping the surface of the water, they move very quickly. The manner is something like that by which the common house-duck escapes when pursued by a dog; but I am nearly sure that the steamer moves its wings alternately, instead of both together, as in other birds. These clumsy, loggerheaded ducks make such a noise and splashing, that the effect is exceedingly curious.

Thus we find in South America three birds which use their wings for other purposes besides flight; the penguin as fins, the steamer as paddles, and the ostrich as sails: and the Apteryx of New Zealand, as well as its gigantic extinct prototype the Deinornis, possess only rudimentary representatives of wings. The steamer is able to dive only to a very short distance. It feeds entirely on shell-fish from the kelp and tidal rocks; hence the beak and head, for the purpose of breaking them, are surprisingly heavy and strong: the head is so strong that I have scarcely been able to fracture it with my geological hammer; and all our sportsmen soon discovered how tenacious these birds were of life. When in the evening pluming themselves in a flock, they make the same odd mixture of sounds which bull-frogs do within the tropics.

In Tierra del Fuego, as well as at the Falkland Islands, I made many observations on the lower marine animals,[1] but they are of little general interest. I will mention only one class of facts, relating to certain zoophytes in the more highly organized division of that class. Several genera (Flustra, Eschara, Cellaria, Crisia, and others) agree in having singular moveable organs (like those of Flustra avicularia, found in the European seas) attached to their cells. The organ, in

1 I was surprised to find, on counting the eggs of a large white Doris (this sea-slug was three and a half inches long), how extraordinarily numerous they were. From two to five eggs (each three-thousandths of an inch in diameter) were contained in a spherical little case. These were arranged two deep in transverse rows forming a ribbon. The ribbon adhered by its edge to the rock in an oval spire. One which I found, measured nearly twenty inches in length and half in breadth. By counting how many balls were contained in a tenth of an inch in the row, and how many rows in an equal length of the ribbon, on the most moderate computation there were six hundred thousand eggs. Yet this Doris was certainly not very common: although I was often searching under the stones, I saw only seven individuals. *No fallacy is more common with naturalists, than that the numbers of an individual species depend on its powers of propagation.*

the greater number of cases, very closely resembles the head of a vulture; but the lower mandible can be opened much wider than in a real bird's beak. The head itself possesses considerable powers of movement, by means of a short neck. In one zoophyte the head itself was fixed, but the lower jaw free: in another it was replaced by a triangular hood, with a beautifully-fitted trap-door, which evidently answered to the lower mandible. In the greater number of species, each cell was provided with one head, but in others each cell had two.

The young cells at the end of the branches of these corallines contain quite immature polypi, yet the vulture-heads attached to them, though small, are in every respect perfect. When the polypus was removed by a needle from any of the cells, these organs did not appear in the least affected. When one of the vulture-like heads was cut off from a cell, the lower mandible retained its power of opening and closing. Perhaps the most singular part of their structure is, that when there were more than two rows of cells on a branch, the central cells were furnished with these appendages, of only one-fourth the size of the outside ones. Their movements varied according to the species; but in some I never saw the least motion; while others, with the lower mandible generally wide open, oscillated backwards and forwards at the rate of about five seconds each turn; others moved rapidly and by starts. When touched with a needle, the beak generally seized the point so firmly, that the whole branch might be shaken.

These bodies have no relation whatever with the production of the eggs or gemmules, as they are formed before the young polypi appear in the cells at the end of the growing branches; as they move independently of the polypi, and do not appear to be in any way connected with them; and as they differ in size on the outer and inner rows of cells, I have little doubt, that in their functions, they are related rather to the horny axis of the branches than to the polypi in the cells. The fleshy appendage at the lower extremity of the sea-pen (described at Bahia Blanca) also forms part of the zoophyte, as a whole, in the same manner as the roots of a tree form part of the whole tree, and not of the individual leaf or flower-buds.

In another elegant little coralline (Crisia?), each cell was furnished with a long-toothed bristle, which had the power of moving quickly. Each of these bristles and each of the vulture-like heads generally

moved quite independently of the others, but sometimes all on both
sides of a branch, sometimes only those on one side, moved together
coinstantaneously; sometimes each moved in regular order one after
another. In these actions we apparently behold as perfect a transmis-
sion of will in the zoophyte, though composed of thousands of
distinct polypi, as in any single animal. The case, indeed, is not
different from that of the sea-pens, which, when touched, drew
themselves into the sand on the coast of Bahia Blanca. I will state
one other instance of uniform action, though of a very different
nature, in a zoophyte closely allied to Clytia, and therefore very
simply organized. Having kept a large tuft of it in a basin of salt-
water, when it was dark I found that as often as I rubbed any part of a
branch, the whole became strongly phosphorescent with a green
light: I do not think I ever saw any object more beautifully so. But
the remarkable circumstance was, that the flashes of light always
proceeded up the branches, from the base towards the extremities.

The examination of these compound animals was always very
interesting to me. What can be more remarkable than to see a
plant-like body producing an egg, capable of swimming about and
of choosing a proper place to adhere to, which then sprouts into
branches, each crowded with innumerable distinct animals, often of
complicated organizations? The branches, moreover, as we have just
seen, sometimes possess organs capable of movement and independent of the polypi. Surprising as this union of separate individuals in a
common stock must always appear, every tree displays the same fact,
for buds must be considered as individual plants. It is, however,
natural to consider a polypus, furnished with a mouth, intestines,
and other organs, as a distinct individual, whereas the individuality of
a leaf-bud is not easily realized; so that the union of separate individuals in a common body is more striking in a coralline than in a tree.
Our conception of a compound animal, where in some respects the
individuality of each is not completed, may be aided, by reflecting on
the production of two distinct creatures by bisecting a single one with
a knife, or where Nature herself performs the task of bisection. We
may consider the polypi in a zoophyte, or the buds in a tree, as cases
where the division of the individual has not been completely effected.
Certainly in the case of trees, and judging from analogy in that of

corallines, the individuals propagated by buds seem more intimately related to each other, than eggs or seeds are to their parents. It seems now pretty well established that plants propagated by buds all partake of a common duration of life; and it is familiar to every one, what singular and numerous peculiarities are transmitted with certainty, by buds, layers, and grafts, which by seminal propagation never or only casually reappear.

CHAPTER X

TIERRA DEL FUEGO

Tierra del Fuego, first arrival – Good Success Bay – An Account of the Fuegians on board – Interview with the Savages – Scenery of the Forests – Cape Horn – Wigwam Cove – Miserable Condition of the Savages – Famines – Cannibals – Matricide – Religious Feelings – Great Gale – Beagle Channel – Ponsonby Sound – Build Wigwams and settle the Fuegians – Bifurcation of the Beagle Channel – Glaciers – Return to the Ship – Second Visit in the Ship to the Settlement – Equality of Condition amongst the Natives.

DECEMBER 17*th*, 1832. – Having now finished with Patagonia and the Falkland Islands, I will describe our first arrival in Tierra del Fuego. A little after noon we doubled Cape St. Diego, and entered the famous strait of Le Maire. We kept close to the Fuegian shore, but the outline of the rugged, inhospitable Staten-land was visible amidst the clouds. In the afternoon we anchored in the Bay of Good Success. While entering we were saluted in a manner becoming the inhabitants of this savage land. A group of Fuegians partly concealed by the entangled forest, were perched on a wild point overhanging the sea; and as we passed by, they sprang up and waving their tattered cloaks sent forth a loud and sonorous shout. The savages followed the ship, and just before dark we saw their fire, and again heard their wild cry. The harbour consists of a fine piece of water half surrounded by low rounded mountains of clay-slate, which are covered to the water's edge by one dense gloomy forest. A single glance at the landscape was sufficient to show me how widely different it was from any thing I had ever beheld. At night it blew a gale of wind, and heavy squalls from the mountains swept past us. It would have been a bad time out at sea, and we, as well as others, may call this Good Success Bay.

In the morning the Captain sent a party to communicate with the Fuegians. When we came within hail, one of the four natives who were present advanced to receive us, and began to shout most vehemently, wishing to direct us where to land. When we were on shore the party looked rather alarmed, but continued talking and making gestures with great rapidity. It was without exception the most curious and interesting spectacle I ever beheld: I could not have believed how wide was the difference between savage and civilized man: it is greater than between a wild and domesticated animal, inasmuch as in man there is a greater power of improvement. The chief spokesman was old, and appeared to be the head of the family; the three others were powerful young men, about six feet high. The women and children had been sent away. These Fuegians are a very different race from the stunted, miserable wretches farther westward; and they seem closely allied to the famous Patagonians of the Strait of Magellan. Their only garment consists of a mantle made of guanaco skin, with the wool outside; this they wear just thrown over their shoulders, leaving their persons as often exposed as covered. Their skin is of a dirty coppery red colour.

The old man had a fillet of white feathers tied round his head, which partly confined his black, coarse, and entangled hair. His face was crossed by two broad transverse bars; one, painted bright red, reached from ear to ear and included the upper lip; the other, white like chalk, extended above and parallel to the first, so that even his eyelids were thus coloured. The other two men were ornamented by streaks of black powder, made of charcoal. The party altogether closely resembled the devils which come on the stage in plays like Der Freischutz.

Their very attitudes were abject, and the expression of their countenances distrustful, surprised, and startled. After we had presented them with some scarlet cloth, which they immediately tied round their necks, they became good friends. This was shown by the old man patting our breasts, and making a chuckling kind of noise, as people do when feeding chickens. I walked with the old man, and this demonstration of friendship was repeated several times; it was concluded by three hard slaps, which were given me on the breast and back at the same time. He then bared his bosom for me to return the compliment, which being done, he seemed highly pleased. The language of these people, according to our notions, scarcely deserves to be called articulate. Captain Cook has compared it to a man clearing his throat, but certainly no European ever cleared his throat with so many hoarse, guttural, and clicking sounds.

They are excellent mimics: as often as we coughed or yawned, or made any odd motion, they immediately imitated us. Some of our party began to squint and look awry; but one of the young Fuegians (whose whole face was painted black, excepting a white band across his eyes) succeeded in making far more hideous grimaces. They could repeat with perfect correctness each word in any sentence we addressed them, and they remembered such words for some time. Yet we Europeans all know how difficult it is to distinguish apart the sounds in a foreign language. Which of us, for instance, could follow an American Indian through a sentence of more than three words? All savages appear to possess, to an uncommon degree, this power of mimicry. I was told, almost in the same words, of the same ludicrous habit among the Caffres: the Australians, likewise, have long been notorious for being able to imitate and describe the gait of any man, so that he may be recognized. How can this faculty be explained? is it a consequence of the more practised habits of perception and keener senses, common to all men in a savage state, as compared with those long civilized?

When a song was struck up by our party, I thought the Fuegians would have fallen down with astonishment. With equal surprise they viewed our dancing; but one of the young men, when asked, had no objection to a little waltzing. Little accustomed to Europeans as they appeared to be, yet they knew and dreaded our fire-arms; nothing would tempt them to take a gun in their hands. They begged for knives, calling them by the Spanish word 'cuchilla.' They explained also what they wanted, by acting as if they had a piece of blubber in their mouth, and then pretending to cut instead of tear it.

I have not as yet noticed the Fuegians whom we had on board. During the former voyage of the *Adventure* and *Beagle* in 1826 to 1830, Captain Fitz Roy seized on a party of natives, as hostages for the loss of a boat, which had been stolen, to the great jeopardy of a party employed on the survey; and some of these natives, as well as a child whom he bought for a pearl-button, he took with him to England, determining to educate them and instruct them in religion at his own expense. To settle these natives in their own country, was one chief inducement to Captain Fitz Roy to undertake our present voyage; and before the Admiralty had resolved to send out this expedition, Captain Fitz Roy had generously chartered a vessel, and would himself have taken them back. The natives were accompanied by a missionary, R. Matthews; of whom and of the natives, Captain Fitz Roy has published a full and excellent account. Two men, one of whom died in England of the small-pox, a boy and a little girl, were originally taken; and we had now on board, York Minster, Jemmy Button (whose name expresses his purchase-money), and Fuegia Basket. York Minster was a full-grown, short, thick, powerful man: his disposition was reserved, taciturn, morose, and when excited violently passionate; his affections were very strong towards a few friends on board; his intellect good. Jemmy Button was a universal favourite, but likewise passionate; the expression of his face at once showed his nice disposition. He was merry and often laughed, and was remarkably sympathetic with any one in pain: when the water was rough, I was often a little sea-sick, and he used to come to me and say in a plaintive voice, 'Poor, poor fellow!' but the notion, after his aquatic life, of a man being sea-sick, was too ludicrous, and he was generally obliged to turn on one side to hide a smile or laugh, and then he would repeat his 'Poor, poor fellow!' He was of a patriotic disposition; and he liked to praise his own tribe and country, in which he truly said there were 'plenty of trees,' and he abused all the other tribes: he stoutly declared that there was no Devil in his land. Jemmy was short, thick, and fat, but vain of his personal appearance; he used to wear gloves, his hair was neatly cut, and he was distressed if his well-polished shoes were dirtied. He was fond of admiring himself in a looking-glass; and a merry-faced little Indian boy from the Rio Negro, whom we had for some months on board, soon perceived this, and used to mock him: Jemmy, who was always rather jealous of the attention paid to this little boy, did not at all like this, and used to say, with rather a contemptuous twist of his head, 'Too much sky-lark.' It seems yet wonderful to me, when I think over all his many good qualities, that he should have been of the same race, and doubtless partaken of the same character, with the miserable, de-graded savages whom we first met here. Lastly, Fuegia Basket was a

nice, modest, reserved young girl, with a rather pleasing but sometimes sullen expression, and very quick in learning anything, especially languages. This she showed in picking up some Portuguese and
Spanish, when left on shore for only a short time at Rio de Janeiro and
Monte Video, and in her knowledge of English. York Minster was
very jealous of any attention paid to her; for it was clear he determined to marry her as soon as they were settled on shore.

Although all three could both speak and understand a good deal of
English, it was singularly difficult to obtain much information from
them, concerning the habits of their countrymen: this was partly
owing to their apparent difficulty in understanding the simplest
alternative. Every one accustomed to very young children, knows
how seldom one can get an answer even to so simple a question as
whether a thing is black *or* white; the idea of black or white seems
alternately to fill their minds. So it was with these Fuegians, and hence
it was generally impossible to find out, by cross-questioning, whether
one had rightly understood anything which they had asserted. Their
sight was remarkably acute: it is well known that sailors, from long
practice, can make out a distant object much better than a landsman;
but both York and Jemmy were much superior to any sailor on board:
several times they have declared what some distant object has been,
and though doubted by every one, they have proved right, when it
has been examined through a telescope. They were quite conscious
of this power; and Jemmy, when he had any little quarrel with the
officer on watch, would say, 'Me see ship, me no tell.'

It was interesting to watch the conduct of the savages, when we
landed, towards Jemmy Button: they immediately perceived the
difference between him and ourselves, and held much conversation
one with another on the subject. The old man addressed a long
harangue to Jemmy, which it seems was to invite him to stay with
them. But Jemmy understood very little of their language, and was,
moreover, thoroughly ashamed of his countrymen. When York
Minster afterwards came on shore, they noticed him in the same
way, and told him he ought to shave; yet he had not twenty dwarf
hairs on his face, whilst we all wore our untrimmed beards. They
examined the colour of his skin, and compared it with ours. One
of our arms being bared, they expressed the liveliest surprise and

admiration at its whiteness, just in the same way in which I have seen the ourang-outang do at the Zoological Gardens. We thought that they mistook two or three of the officers, who were rather shorter and fairer, though adorned with large beards, for the ladies of our party. The tallest among the Fuegians was evidently much pleased at his height being noticed. When placed back to back with the tallest of the boat's crew, he tried his best to edge on higher ground, and to stand on tiptoe. He opened his mouth to show his teeth, and turned his face for a side view; and all this was done with such alacrity, that I dare say he thought himself the handsomest man in Tierra del Fuego. After our first feeling of grave astonishment was over, nothing could be more ludicrous than the odd mixture of surprise and imitation which these savages every moment exhibited.

The next day I attempted to penetrate some way into the country. Tierra del Fuego may be described as a mountainous land, partly submerged in the sea, so that deep inlets and bays occupy the place where valleys should exist. The mountain sides, except on the exposed western coast, are covered from the water's edge upwards by one great forest. The trees reach to an elevation of between 1000 and 1500 feet, and are succeeded by a band of peat, with minute alpine plants; and this again is succeeded by the line of perpetual snow, which, according to Captain King, in the Strait of Magellan descends to between 3000 and 4000 feet. To find an acre of level land in any part of the country is most rare. I recollect only one little flat piece near Port Famine, and another of rather larger extent near Goeree Road. In both places, and everywhere else, the surface is covered by a thick bed of swampy peat. Even within the forest, the ground is concealed by a mass of slowly putrefying vegetable matter, which, from being soaked with water, yields to the foot.

Finding it nearly hopeless to push my way through the wood, I followed the course of a mountain torrent. At first, from the waterfalls and number of dead trees, I could hardly crawl along; but the bed of the stream soon became a little more open, from the floods having swept the sides. I continued slowly to advance for an hour along the broken and rocky banks, and was amply repaid by the grandeur of the scene. The gloomy depth of the ravine well accorded with the universal signs of violence. On every side were lying irregular masses of rock and torn-up trees; other trees, though still erect, were decayed to the heart and ready to fall. The entangled mass of the thriving and the fallen reminded me of the forests within the tropics – yet there was a difference: for in these still solitudes, Death, instead of Life, seemed the predominant spirit. I followed the watercourse till I came to a spot, where a great slip had cleared a straight space down the mountain side. By this road I ascended to a considerable elevation, and obtained a good view of the surrounding woods. The trees all belong to one kind, the Fagus betuloides; for the number of the other species of Fagus and of the Winter's Bark, is quite inconsiderable. This beech keeps its leaves throughout the year; but its foliage is of a peculiar brownish-green colour, with a tinge of yellow. As the whole landscape is thus coloured, it has a sombre, dull appearance; nor is it often enlivened by the rays of the sun.

December 20th. – One side of the harbour is formed by a hill about 1500 feet high, which Captain Fitz Roy has called after Sir J. Banks, in commemoration of his disastrous excursion, which proved fatal to two men of his party, and nearly so to Dr. Solander. The snow-storm, which was the cause of their misfortune, happened in the middle of January, corresponding to our July, and in the latitude of Durham! I was anxious to reach the summit of this mountain to collect alpine plants; for flowers of any kind in the lower parts are few in number. We followed the same watercourse as on the previous day, till it dwindled away, and we were then compelled to crawl blindly among the trees. These, from the effects of the elevation and of the impetuous winds, were low, thick, and crooked. At length we reached that which from a distance appeared like a carpet of fine green turf, but which, to our vexation, turned out to be a compact mass of little beech-trees about four or five feet high. They were as thick together as box in the border of a garden, and we were obliged to struggle over the flat but treacherous surface. After a little more trouble we gained the peat, and then the bare slate rock.

A ridge connected this hill with another, distant some miles, and more lofty, so that patches of snow were lying on it. As the day was not far advanced, I determined to walk there and collect plants along the road. It would have been very hard work, had it not been for a well-beaten and straight path made by the guanacos; for these animals, like sheep, always follow the same line. When we reached the hill we found it the highest in the immediate neighbourhood, and the waters flowed to the sea in opposite directions. We obtained a wide view over the surrounding country: to the north a swampy moorland extended, but to the south we had a scene of savage magnificence, well becoming Tierra del Fuego. There was a degree of mysterious grandeur in mountain behind mountain, with the deep intervening valleys, all covered by one thick, dusky mass of forest. The atmosphere, likewise, in this climate, where gale succeeds gale, with rain, hail, and sleet, seems blacker than anywhere else. In the Strait of Magellan, looking due southward from Port Famine, the distant channels between the mountains appeared from their gloominess to lead beyond the confines of this world.

December 21st. – The *Beagle* got under weigh: and on the succeeding day, favoured to an uncommon degree by a fine easterly breeze, we closed in with the Barnevelts, and running past Cape Deceit with its stony peaks, about three o'clock doubled the weather-beaten Cape Horn. The evening was calm and bright, and we enjoyed a fine view of the surrounding isles. Cape Horn, however, demanded his tribute, and before night sent us a gale of wind directly in our teeth. We stood out to sea, and on the second day again made the land, when we saw on our weather-bow this notorious promontory in its proper form – veiled in a mist, and its dim outline surrounded by a storm of wind and water. Great black clouds were rolling across the heavens, and squalls of rain, with hail, swept by us with such extreme violence, that the Captain determined to run into Wigwam Cove. This is a snug little harbour, not far from Cape Horn; and here, at Christmas-eve, we anchored in smooth water. The only thing which reminded us of the gale outside, was every now and then a puff from the mountains, which made the ship surge at her anchors.

December 25th. – Close by the cove, a pointed hill, called Kater's Peak, rises to the height of 1700 feet. The surrounding islands all consist of conical masses of greenstone, associated sometimes with less regular hills of baked and altered clay-slate. This part of Tierra del Fuego may be considered as the extremity of the submerged chain of mountains already alluded to. The cove takes its name of 'Wigwam' from some of the Fuegian habitations; but every bay in the neighbourhood might be so called with equal propriety. The inhabitants, living chiefly upon shell-fish, are obliged constantly to change their place of residence; but they return at intervals to the same spots, as is evident from the piles of old shells, which must often amount to many tons in weight. These heaps can be distinguished at a long distance by the bright green colour of certain plants, which invariably grow on them. Among these may be enumerated the wild celery and scurvy grass, two very serviceable plants, the use of which has not been discovered by the natives.

The Fuegian wigwam resembles, in size and dimensions, a haycock. It merely consists of a few broken branches stuck in the ground, and very imperfectly thatched on one side with a few tufts of grass and rushes. The whole cannot be the work of an hour, and it is only used for a few days. At Goeree Roads I saw a place where one of these naked men had slept, which absolutely offered no more cover than the form of a hare. The man was evidently living by himself, and York Minster said he was 'very bad man,' and that probably he had stolen something. On the west coast, however, the wigwams are rather better, for they are covered with seal-skins. We were detained here several days by the bad weather. The climate is certainly wretched: the summer solstice was now passed, yet every day snow fell on the hills, and in the valleys there was rain, accompanied by sleet. The thermometer generally stood about 45°, but in the night fell to 38° or 40°. From the damp and boisterous state of the atmosphere, not cheered by a gleam of sunshine, one fancied the climate even worse than it really was.

While going one day on shore near Wollaston Island, we pulled alongside a canoe with six Fuegians. These were the most abject and miserable creatures I anywhere beheld. On the east coast the natives, as we have seen, have guanaco cloaks, and on the west, they possess seal-skins. Amongst these central tribes the men generally have an otter-skin, or some small scrap about as large as a pocket-handkerchief, which is barely sufficient to cover their backs as low down as their loins. It is laced across the breast by strings, and according as the wind blows, it is shifted from side to side. But these Fuegians in the canoe were quite naked, and even one full-grown woman was absolutely so. It was raining heavily, and the fresh water, together with the spray, trickled down her body. In another harbour not far distant, a woman, who was suckling a recently-born child, came one day alongside the vessel, and remained there out of mere curiosity, whilst the sleet fell and thawed on her naked bosom, and on the skin of her naked baby! These poor wretches were stunted in their growth, their hideous faces bedaubed with white paint, their skins filthy and greasy, their hair entangled, their voices discordant, and their gestures violent. Viewing such men, one can hardly make oneself believe that they are fellow-creatures, and inhabitants of the same world. It is a common subject of conjecture what pleasure in life some of the lower animals can enjoy: how much more reasonably the same question may be asked with respect to these barbarians! At night, five or six human beings, naked and scarcely protected from the wind and rain of this tempestuous climate, sleep on the wet ground coiled up like animals. Whenever it is low water, winter or summer, night or day, they must rise to pick shell-fish from the rocks; and the women either dive to collect sea-eggs, or sit patiently in their canoes, and with a baited hair-line without any hook, jerk out little fish. If a seal is killed, or the floating carcass of a putrid whale discovered, it is a feast; and such miserable food is assisted by a few tasteless berries and fungi.

They often suffer from famine: I heard Mr. Low, a sealing-master intimately acquainted with the natives of this country, give a curious account of the state of a party of one hundred and fifty natives on the west coast, who were very thin and in great distress. A succession of gales prevented the women from getting shell-fish on the rocks, and they could not go out in their canoes to catch seal. A small party of these men one morning set out, and the other Indians explained to him, that they were going a four days' journey for food: on their return, Low went to meet them, and he found them excessively tired, each man carrying a great square piece of putrid whales-blubber with a hole in the middle, through which they put their heads, like the Gauchos do through their ponchos or cloaks. As soon as the blubber was brought into a wigwam, an old man cut off thin slices, and muttering over them, broiled them for a minute, and distributed them to the famished party, who during this time preserved a profound silence. Mr. Low believes that whenever a whale is cast on shore, the natives bury large pieces of it in the sand, as a resource in time of famine; and a native boy, whom he had on board, once found a stock thus buried. The different tribes when at war are cannibals. From the concurrent, but quite independent evidence of the boy taken by Mr. Low, and of Jemmy Button, it is certainly true, that when pressed in winter by hunger, they kill and devour their old women before they kill their dogs: the boy, being asked by Mr. Low why they did this, answered, 'Doggies catch otters, old women no.' This boy described the manner in which they are killed by being held over smoke and thus choked; he imitated their screams as a joke, and described the parts of their bodies which are considered best to eat. Horrid as such a death by the hands of their friends and relatives must be, the fears of the old women, when hunger begins to press, are more painful to think of; we were told that they then often run away into the mountains, but that they are pursued by the men and brought back to the slaughter-house at their own fire-sides!

Captain Fitz Roy could never ascertain that the Fuegians have any distinct belief in a future life. They sometimes bury their dead in caves, and sometimes in the mountain forests; we do not know what ceremonies they perform. Jemmy Button would not eat land-birds, because 'eat dead men:' they are unwilling even to mention their dead friends. We have no reason to believe that they perform any sort of religious worship; though perhaps the muttering of the old man before he distributed the putrid blubber to his famished party, may be of this nature. Each family or tribe has a wizard or conjuring doctor, whose office we could never clearly ascertain. Jemmy believed in dreams, though not, as I have said, in the devil: I do not think that our Fuegians were much more superstitious than some of the sailors; for an old quarter-master firmly believed that the successive heavy gales, which we encountered off Cape Horn, were caused by our having the Fuegians on board. The nearest approach to a religious feeling which I heard of, was shown by York Minster, who, when Mr. Bynoe shot some very young ducklings as specimens, declared in the most solemn manner, 'Oh Mr. Bynoe, much rain, snow, blow much.' This was evidently a retributive punishment for wasting human food. In a wild and excited manner he also related, that his brother, one day whilst returning to pick up some dead birds which he had left on the coast, observed some feathers blown by the wind. His brother said (York imitating his manner), 'What that?' and crawling onwards, he peeped over the cliff, and saw 'wild man' picking his birds; he crawled a little nearer, and then hurled down a great stone and killed him. York declared for a long time afterwards storms raged, and much rain and snow fell. As far as we could make out, he seemed to consider the elements themselves as the avenging agents: it is evident in this case, how naturally, in a race a little more advanced in culture, the elements would become personified. What the 'bad wild men' were, has always appeared to me most mysterious: from what York said, when we found the place like the form of a hare, where a single man had slept the night before, I should have thought that they were thieves who had been driven from their tribes; but other obscure speeches made me doubt this; I have sometimes imagined that the most probable explanation was that they were insane.

The different tribes have no government or chief; yet each is surrounded by other hostile tribes, speaking different dialects, and separated from each other only by a deserted border or neutral territory: the cause of their warfare appears to be the means of subsistence. Their country is a broken mass of wild rocks, lofty hills, and useless forests: and these are viewed through mists and endless storms. The habitable land is reduced to the stones on the beach; in search of food they are compelled unceasingly to wander from spot to spot, and so steep is the coast, that they can only move about in their wretched canoes. They cannot know the feeling of having a home, and still less that of domestic affection; for the husband is to the wife a brutal master to a laborious slave. Was a more horrid deed ever perpetrated, than that witnessed on the west coast by Byron, who saw a wretched mother pick up her bleeding dying infant-boy, whom her husband had mercilessly dashed on the stones for dropping a basket of sea-eggs! How little can the higher powers of the mind be brought into play: what is there for imagination to picture, for reason to compare, for judgment to decide upon? to knock a limpet from the rock does not require even cunning, that lowest power of the mind. Their skill in some respects may be compared to the instinct of animals; for it is not improved by experience: the canoe, their most ingenious work, poor as it is, has remained the same, as we know from Drake, for the last two hundred and fifty years.

Whilst beholding these savages, one asks, whence have they come? What could have tempted, or what change compelled a tribe of men, to leave the fine regions of the north, to travel down the Cordillera or backbone of America, to invent and build canoes, which are not used by the tribes of Chile, Peru, and Brazil, and then to enter on one of the most inhospitable countries within the limits of the globe? Although such reflections must at first seize on the mind, yet we may feel sure that they are partly erroneous. There is no reason to believe that the Fuegians decrease in number; therefore we must suppose that they enjoy a sufficient share of happiness, of whatever kind it may be, to render life worth having. Nature, by making habit omnipotent, and its effects hereditary, has fitted the Fuegian to the climate and the productions of his miserable country.

After having been detained six days in Wigwam Cove by very bad weather, we put to sea on the 30th of December. Captain Fitz Roy wished to get westward to land York and Fuegia in their own country. When at sea we had a constant succession of gales, and the current was against us: we drifted to 57° 23' south. On the 11th of January, 1833, by carrying a press of sail, we fetched within a few miles of the great rugged mountain of York Minister (so called by Captain Cook, and the origin of the name of the elder Fuegian), when a violent squall compelled us to shorten sail and stand out to sea. The surf was breaking fearfully on the coast, and the spray was carried over a cliff estimated at 200 feet in height. On the 12th the gale was very heavy, and we did not know exactly where we were: it was a most unpleasant sound to hear constantly repeated, 'keep a good look-out to leeward.' On the 13th the storm raged with its full fury: our horizon was narrowly limited by the sheets of spray borne by the wind. The sea looked ominous, like a dreary waving plain with patches of drifted snow: whilst the ship laboured heavily, the albatross glided with its expanded wings right up the wind. At noon a great sea broke over us, and filled one of the whale-boats, which was obliged to be instantly cut away. The poor *Beagle* trembled at the shock, and for a few minutes would not obey her helm; but soon, like a good ship that she was, she righted and came up to the wind again. Had another sea followed the first, our fate would have been decided soon, and for ever. We had now been twenty-four days trying in vain to get westward; the men were worn out with fatigue, and they had not for many nights or days a dry thing to put on. Captain Fitz Roy gave up the attempt to get westward by the outside coast. In the evening we ran in behind False Cape Horn, and dropped our anchor in forty-seven fathoms, fire flashing from the windlass as the chain rushed round it. How delightful was that still night, after having been so long involved in the din of the warring elements!

January 15th, 1833. – The Beagle anchored in Goeree Roads. Captain Fitz Roy having resolved to settle the Fuegians, according to their wishes, in Ponsonby Sound, four boats were equipped to carry them there through the Beagle Channel. This channel, which was discovered by Captain Fitz Roy during the last voyage, is a most remarkable feature in the geography of this, or indeed of any other country: it may be compared to the valley of Lochness in Scotland, with its chain of lakes and friths. It is about one hundred and twenty miles long, with an average breadth, not subject to any very great variation, of about two miles; and is throughout the greater part so perfectly straight, that the view, bounded on each side by a line of mountains, gradually becomes indistinct in the long distance. It crosses the southern part of Tierra del Fuego in an east and west line, and in the middle is joined at right angles on the south side by an irregular channel, which has been called Ponsonby Sound. This is the residence of Jemmy Button's tribe and family.

19th. – Three whale-boats and the yawl, with a party of twenty-eight, started under the command of Captain Fitz Roy. In the afternoon we entered the eastern mouth of the channel, and shortly afterwards found a snug little cove concealed by some surrounding islets. Here we pitched our tents and lighted our fires. Nothing could look more comfortable than this scene. The glassy water of the little harbour, with the branches of the trees hanging over the rocky beach, the boats at anchor, the tents supported by the crossed oars, and the smoke curling up the wooded valley, formed a picture of quiet retirement. The next day (20th) we smoothly glided onwards in our little fleet, and came to a more inhabited district. Few if any of these natives could ever have seen a white man; certainly nothing could exceed their astonishment at the apparition of the four boats. Fires were lighted on every point (hence the name of Tierra del Fuego, or the land of fire), both to attract our attention and to spread far and wide the news. Some of the men ran for miles along the shore. I shall never forget how wild and savage one group appeared: suddenly four or five men came to the edge of an overhanging cliff; they were absolutely naked, and their long hair streamed about their faces; they held rugged staffs in their hands, and, springing from the ground, they waved their arms round their heads, and sent forth the most hideous yells.

At dinner-time we landed among a party of Fuegians. At first they were not inclined to be friendly; for until the Captain pulled in ahead of the other boats, they kept their slings in their hands. We soon, however, delighted them by trifling presents, such as tying red tape round their heads. They liked our biscuit: but one of the savages touched with his finger some of the meat preserved in tin cases which I was eating, and feeling it soft and cold, showed as much disgust at it, as I should have done at putrid blubber. Jemmy was thoroughly ashamed of his countrymen, and declared his own tribe were quite different, in which he was wofully mistaken. It was as easy to please as it was difficult to satisfy these savages. Young and old, men and children, never ceased repeating the word 'yammer-schooner,' which means 'give me.' After pointing to almost every object, one after the other, even to the buttons on our coats, and saying their favourite word in as many intonations as possible, they would then use it in a neuter sense, and vacantly repeat 'yammer-schooner.' After yammerschoonering for any article very eagerly, they would by a simple artifice point to their young women or little children, as much as to say, 'If you will not give it me, surely you will to such as these.'

At night we endeavoured in vain to find an uninhabited cove; and at last were obliged to bivouac not far from a party of natives. They were very inoffensive as long as they were few in numbers, but in the morning (21st) being joined by others they showed symptoms of hostility, and we thought that we should have come to a skirmish.

An European labours under great disadvantages when treating with savages like these, who have not the least idea of the power of fire-arms. In the very act of levelling his musket he appears to the savage far inferior to a man armed with a bow and arrow, a spear, or even a sling. Nor is it easy to teach them our superiority except by striking a fatal blow. Like wild beasts, they do not appear to compare numbers; for each individual, if attacked, instead of retiring, will endeavour to dash your brains out with a stone, as certainly as a tiger under similar circumstances would tear you. Captain Fitz Roy on one occasion being very anxious, from good reasons, to frighten away a small party, first flourished a cutlass near them, at which they only laughed; he then twice fired his pistol close to a native. The man both times looked astounded, and carefully but quickly rubbed his head; he then stared awhile, and gabbled to his companions, but he never seemed to think of running away. We can hardly put ourselves in the position of these savages, and understand their actions. In the case of this Fuegian, the possibility of such a sound as the report of a gun close to his ear could never have entered his mind. He perhaps literally did not for a second know whether it was a sound or a blow, and therefore very naturally rubbed his head. In a similar manner, when a savage sees a mark struck by a bullet, it may be some time before he is able at all to understand how it is effected; for the fact of a body being invisible from its velocity would perhaps be to him an idea totally inconceivable. Moreover, the extreme force of a bullet, that penetrates a hard substance without tearing it, may convince the savage that it has no force at all. Certainly I believe that many savages of the lowest grade, such as these of Tierra del Fuego, have seen objects struck, and even small animals killed by the musket, without being in the least aware how deadly an instrument it is.

22nd. – After having passed an unmolested night, in what would appear to be neutral territory between Jemmy's tribe and the people whom we saw yesterday, we sailed pleasantly along. I do not know anything which shows more clearly the hostile state of the different tribes, than these wide border or neutral tracts. Although Jemmy Button well knew the force of our party, he was, at first, unwilling to land amidst the hostile tribe nearest to his own. He often told us

how the savage Oens men 'when the leaf red,' crossed the mountains
from the eastern coast of Tierra del Fuego, and made inroads on
the natives of this part of the country. It was most curious to watch
him when thus talking, and see his eyes gleaming and his whole
face assume a new and wild expression. As we proceeded along the
Beagle Channel, the scenery assumed a peculiar and very magnificent
character; but the effect was much lessened from the lowness of
the point of view in a boat, and from looking along the valley, and
thus losing all the beauty of a succession of ridges. The mountains
were here about three thousand feet high, and terminated in sharp
and jagged points. They rose in one unbroken sweep from the
water's edge, and were covered to the height of fourteen or fifteen
hundred feet by the dusky-coloured forest. It was most curious to
observe, as far as the eye could range, how level and truly horizontal
the line on the mountain side was, at which trees ceased to grow:
it precisely resembled the high-water mark of drift-weed on a
sea-beach.

At night we slept close to the junction of Ponsonby Sound with
the Beagle Channel. A small family of Fuegians, who were living in
the cove, were quiet and inoffensive, and soon joined our party
round a blazing fire. We were well clothed, and though sitting close
to the fire were far from too warm; yet these naked savages, though
further off, were observed, to our great surprise, to be streaming with
perspiration at undergoing such a roasting. They seemed, however,
very well pleased, and all joined in the chorus of the seamen's songs:
but the manner in which they were invariably a little behindhand was
quite ludicrous.

During the night the news had spread, and early in the morning
(23rd) a fresh party arrived, belonging to the Tekenika, or Jemmy's
tribe. Several of them had run so fast that their noses were
bleeding, and their mouths frothed from the rapidity with which
they talked; and with their naked bodies all bedaubed with black,
white,[1] and red, they looked like so many demoniacs who had been

1 This substance, when dry, is tolerably compact, and of little specific gravity:
Professor Ehrenberg has examined it: he states (König Akad. der Wissen: Berlin, Feb.
1845) that it is composed of infusoria, including fourteen polygastrica, and four
phytolitharia. He says that they are all inhabitants of fresh-water; this is a beautiful

fighting. We then proceeded (accompanied by twelve canoes, each holding four or five people) down Ponsonby Sound to the spot where poor Jemmy expected to find his mother and relatives. He had already heard that his father was dead; but as he had had a 'dream in his head' to that effect, he did not seem to care much about it, and repeatedly comforted himself with the very natural reflection – 'Me no help it.' He was not able to learn any particulars regarding his father's death, as his relations would not speak about it.

Jemmy was now in a district well known to him, and guided the boats to a quiet pretty cove named Woollya, surrounded by islets, every one of which and every point had its proper native name. We found here a family of Jemmy's tribe, but not his relations: we made friends with them; and in the evening they sent a canoe to inform Jemmy's mother and brothers. The cove was bordered by some acres of good sloping land, not covered (as elsewhere) either by peat or by forest-trees. Captain Fitz Roy originally intended, as before stated, to have taken York Minster and Fuegia to their own tribe on the west coast; but as they expressed a wish to remain here, and as the spot was singularly favourable, Captain Fitz Roy determined to settle here the whole party, including Matthews, the missionary. Five days were spent in building for them three large wigwams, in landing their goods, in digging two gardens, and sowing seeds.

The next morning after our arrival (the 24th) the Fuegians began to pour in, and Jemmy's mother and brothers arrived. Jemmy recognized the stentorian voice of one of his brothers at a prodigious distance. The meeting was less interesting than that between a horse, turned out into a field, when he joins an old companion. There was no demonstration of affection; they simply stared for a short time at each other; and the mother immediately went to look after her canoe. We heard, however, through York, that the mother had been inconsolable for the loss of Jemmy, and had searched

example of the results obtainable through Professor Ehrenberg's microscopic re-searches; for Jemmy Button told me that it is always collected at the bottoms of mountain-brooks. It is moreover, a striking fact in the geographical distribution of the infusoria, which are well known to have very wide ranges, that all the species in this substance, although brought from the extreme southern point of Tierra del Fuego, are old, known forms.

everywhere for him, thinking that he might have been left after having been taken in the boat. The women took much notice of and were very kind to Fuegia. We had already perceived that Jemmy had almost forgotten his own language. I should think there was scarcely another human being with so small a stock of language, for his English was very imperfect. It was laughable, but almost pitiable, to hear him speak to his wild brother in English, and then ask him in Spanish ('no sabe?') whether he did not understand him.

Everything went on peaceably during the three next days, whilst the gardens were digging and wigwams building. We estimated the number of natives at about one hundred and twenty. The women worked hard, whilst the men lounged about all day long, watching us. They asked for everything they saw, and stole what they could. They were delighted at our dancing and singing, and were particularly interested at seeing us wash in a neighbouring brook; they did not pay much attention to anything else, not even to our boats. Of all the things which York saw, during his absence from his country, nothing seems more to have astonished him than an ostrich, near Maldonado: breathless with astonishment he came running to Mr. Bynoe, with whom he was out walking – 'Oh, Mr. Bynoe, oh, bird all same horse!' Much as our white skins surprised the natives, by Mr. Low's account a negro-cook to a sealing vessel, did so more effectually; and the poor fellow was so mobbed and shouted at that he would never go on shore again. Everything went on so quietly, that some of the officers and myself took long walks in the surrounding hills and woods. Suddenly, however, on the 27th, every woman and child disappeared. We were all uneasy at this, as neither York nor Jemmy could make out the cause. It was thought by some that they had been frightened by our cleaning and firing off our muskets on the previous evening: by others, that it was owing to offence taken by an old savage, who, when told to keep further off, had coolly spit in the sentry's face, and had then, by gestures acted over a sleeping Fuegian, plainly showed, as it was said, that he should like to cut up and eat our man. Captain Fitz Roy, to avoid the chance of an encounter, which would have been fatal to so many of the Fuegians, thought it advisable for us to sleep at a cove a few miles distant. Matthews, with his usual quiet fortitude (remarkable in a man apparently possessing little energy of character), determined to stay with the Fuegians, who evinced no alarm for themselves; and so we left them to pass their first awful night.

On our return in the morning (28th) we were delighted to find all quiet, and the men employed in their canoes spearing fish. Captain Fitz Roy determined to send the yawl and one whale-boat back to the ship; and to proceed with the two other boats, one under his own command (in which he most kindly allowed me to accompany him), and one under Mr. Hammond, to survey the western parts of the Beagle Channel, and afterwards to return and visit the settlement. The day to our astonishment was over-poweringly hot, so that our skins were scorched: with this beautiful weather, the view in the middle of the Beagle Channel was very remarkable. Looking towards either hand, no object intercepted the vanishing points of this long canal between the mountains. The circumstance of its being an arm of the sea was rendered very evident by several huge whales[1] spouting in different directions. On one occasion I saw two of these monsters, probably male and female, slowly swimming one after the other, within less than a stone's throw of the shore, over which the beech-tree extended its branches.

We sailed on till it was dark, and then pitched our tents in a quiet creek. The greatest luxury was to find for our beds a beach of pebbles, for they were dry and yielded to the body. Peaty soil is damp; rock is uneven and hard; sand gets into one's meat, when cooked and eaten boat-fashion; but when lying in our blanket-bags, on a good bed of smooth pebbles, we passed most comfortable nights.

It was my watch till one o'clock. There is something very solemn in these scenes. At no time does the consciousness in what a remote corner of the world you are then standing, come so strongly before the mind. Everything tends to this effect; the stillness of the night is interrupted only by the heavy breathing of the seamen beneath the tents, and sometimes by the cry of a night-bird. The occasional

1 One day, off the East coast of Tierra del Fuego, we saw a grand sight in several spermaceti whales jumping upright quite out of the water, with the exception of their tail-fins. As they fell down sideways, they splashed the water high up, and the sound reverberated like a distant broadside.

barking of a dog, heard in the distance, reminds one that it is the land of the savage.

January 29th. – Early in the morning we arrived at the point where the Beagle Channel divides into two arms; and we entered the northern one. The scenery here becomes even grander than before. The lofty mountains on the north side compose the granitic axis, or backbone of the country, and boldly rise to a height of between three and four thousand feet, with one peak above six thousand feet. They are covered by a wide mantle of perpetual snow, and numerous cascades pour their waters, through the woods, into the narrow channel below. In many parts, magnificent glaciers extend from the mountain side to the water's edge. It is scarcely possible to imagine any thing more beautiful than the beryl-like blue of these glaciers, and especially as contrasted with the dead white of the upper expanse of snow. The fragments which had fallen from the glacier into the water, were floating away, and the channel with its icebergs presented, for the space of a mile, a miniature likeness of the Polar Sea. The boats being hauled on shore at our dinner-hour, we were admiring from the distance of half a mile a perpendicular cliff of ice, and were wishing that some more fragments would fall. At last, down came a mass with a roaring noise, and immediately we saw the smooth outline of a wave travelling towards us. The men ran down as quickly as they could to the boats; for the chance of their being dashed to pieces was evident. One of the seamen just caught hold of the bows, as the curling breaker reached it: he was knocked over and over, but not hurt; and the boats, though thrice lifted on high and let fall again, received no damage. This was most fortunate for us, for we were a hundred miles distant from the ship, and we should have been left without provisions or fire-arms. I had previously observed that some large fragments of rock on the beach had been lately displaced; but until seeing this wave, I did not understand the cause. One side of the creek was formed by a spur of mica-slate; the head by a cliff of ice about forty feet high; and the other side by a promontory fifty feet high, built up of huge rounded fragments of granite and mica-slate, out of which old trees were growing. This promontory was evidently a moraine, heaped up at a period when the glacier had greater dimensions.

When we reached the western mouth of this northern branch of the Beagle Channel we sailed amongst many unknown desolate islands, and the weather was wretchedly bad. We met with no natives. The coast was almost everywhere so steep, that we had several times to pull many miles before we could find space enough to pitch our two tents: one night we slept on large round boulders, with putrefying sea-weed between them; and when the tide rose, we had to get up and move our blanket-bags. The farthest point westward which we reached was Stewart Island, a distance of about one hundred and fifty miles from our ship. We returned into the Beagle Channel by the southern arm, and thence proceeded, with no adventure, back to Ponsonby Sound.

February 6th. – We arrived at Woollya. Matthews gave so bad an account of the conduct of the Fuegians, that Captain Fitz Roy determined to take him back to the *Beagle*; and ultimately he was left at New Zealand, where his brother was a missionary. From the time of our leaving, a regular system of plunder commenced; fresh parties of the natives kept arriving: York and Jemmy lost many things, and Matthews almost every thing which had not been concealed underground. Every article seemed to have been torn up and divided by the natives. Matthews described the watch he was obliged always to keep as most harassing; night and day he was surrounded by the natives, who tried to tire him out by making an incessant noise close to his head. One day an old man, whom Matthews asked to leave his wigwam, immediately returned with a large stone in his hand: another day a whole party came armed with stones and stakes, and some of the younger men and Jemmy's brother were crying: Matthews met them with presents. Another party showed by signs that they wished to strip him naked and pluck all the hairs out of his face and body. I think we arrived just in time to save his life. Jemmy's relatives had been so vain and foolish, that they had showed to strangers their plunder, and their manner of obtaining it. It was quite melancholy leaving the three Fuegians with their savage countrymen; but it was a great comfort that they had no personal fears. York, being a powerful resolute man, was pretty sure to get on well, together with his wife Fuegia. Poor Jemmy looked rather disconsolate, and would then, I have little doubt, have been glad to have returned with us.

His own brother had stolen many things from him; and as he remarked, 'what fashion call that:' he abused his countrymen, 'all bad men, no sabe (know) nothing,' and, though I never heard him swear before, 'damned fools.' Our three Fuegians, though they had been only three years with civilized men, would, I am sure, have been glad to have retained their new habits; but this was obviously impossible. I fear it is more than doubtful, whether their visit will have been of any use to them.

In the evening, with Matthews on board, we made sail back to the ship, not by the Beagle Channel, but by the southern coast. The boats were heavily laden and the sea rough, and we had a dangerous passage. By the evening of the 7th we were on board the *Beagle* after an absence of twenty days, during which time we had gone three hundred miles in the open boats. On the 11th, Captain Fitz Roy paid a visit by himself to the Fuegians and found them going on well; and that they had lost very few more things.

On the last day of February in the succeeding year (1834), the *Beagle* anchored in a beautiful little cove at the eastern entrance of the Beagle Channel. Captain Fitz Roy determined on the bold, and as it proved successful, attempt to beat against the westerly winds by the same route, which we had followed in the boats to the settlement at Woollya. We did not see many natives until we were near Ponsonby Sound, where we were followed by ten or twelve canoes. The natives did not at all understand the reason of our tacking, and, instead of meeting us at each tack, vainly strove to follow us in our zig-zag course. I was amused at finding what a difference the circumstance of being quite superior in force made, in the interest of beholding these savages. While in the boats I got to hate the very sound of their voices, so much trouble did they give us. The first and last word was 'yammerschooner.' When, entering some quiet little cove, we have looked round and thought to pass a quiet night, the odious word 'yammerschooner' has shrilly sounded from some gloomy nook, and then the little signal-smoke has curled up to spread the news far and wide. On leaving some place we have said to each other, 'Thank Heaven, we have at last fairly left these wretches!' when one more faint halloo from an all-powerful voice, heard at a prodigious distance, would reach our ears, and clearly could we distinguish – 'yammerschooner.' But now, the more Fuegians the merrier; and very merry work it was. Both parties laughing, wondering, gaping at each other; we pitying them, for giving us good fish and crabs for rags, &c.; they grasping at the chance of finding people so foolish as to exchange such splendid ornaments for a good supper. It was most amusing to see the undisguised smile of satisfaction with which one young woman with her face painted black, tied several bits of scarlet cloth round her head with rushes. Her husband, who enjoyed the very universal privilege in this country of possessing two wives, evidently became jealous of all the attention paid to his young wife; and, after a consultation with his naked beauties, was paddled away by them.

Some of the Fuegians plainly showed that they had a fair notion of barter. I gave one man a large nail (a most valuable present) without making any signs for a return; but he immediately picked out two fish, and handed them up on the point of his spear. If any present was designed for one canoe, and it fell near another, it was invariably given to the right owner. The Fuegian boy, whom Mr. Low had on board, showed, by going into the most violent passion, that he quite understood the reproach of being called a liar, which in truth he was. We were this time, as on all former occasions, much surprised at the little notice, or rather none whatever, which was taken of many things, the use of which must have been evident to the natives. Simple circumstances – such as the beauty of scarlet cloth or blue beads, the absence of women, our care in washing ourselves, – excited their admiration far more than any grand or complicated object, such as our ship. Bougainville has well remarked concerning these people, that they treat the 'chef-d'œuvres de l'industrie humaine, comme ils traitent les loix de la nature et ses phénomènes.'

On the 5th of March, we anchored in the cove at Woollya, but we saw not a soul there. We were alarmed at this, for the natives in Ponsonby Sound showed by gestures, that there had been fighting; and we afterwards heard that the dreaded Oens men had made a descent. Soon a canoe, with a little flag flying, was seen approaching, with one of the men in it washing the paint off his face. This man was poor Jemmy, – now a thin haggard savage, with long disordered hair, and naked, except a bit of a blanket round his waist. We did not recognize him till he was close to us; for he was ashamed of himself, and turned his back to the ship. We had left him plump, fat, clean, and well dressed; – I never saw so complete and grievous a change. As soon however as he was clothed, and the first flurry was over, things wore a good appearance. He dined with Captain Fitz Roy, and ate his dinner as tidily as formerly. He told us he had 'too much' (meaning enough) to eat, that he was not cold, that his relations were very good people, and that he did not wish to go back to England: in the evening we found out the cause of this great change in Jemmy's feelings, in the arrival of his young and nice-looking wife. With his usual good feeling, he brought two beautiful otter-skins for two of his best friends, and some spear-heads and arrows made with his own hands for the Captain. He said he had built a canoe for himself, and he boasted that he could talk a little of his own language! But it is a most singular fact, that he appears to have taught all his tribe some English: an old man spontaneously announced 'Jemmy Button's wife.' Jemmy had lost all his property. He told us that York Minster had built a large canoe, and with his wife Fuegia,[1] had several months since gone to his own country, and had taken farewell by an act of consummate villainy; he persuaded Jemmy and his mother to come with him, and then on the way deserted them by night, stealing every article of their property.

Jemmy went to sleep on shore, and in the morning returned, and remained on board till the ship got under weigh, which frightened his wife, who continued crying violently till he got into his canoe. He returned loaded with valuable property. Every soul on board was heartily sorry to shake hands with him for the last time. I do not now doubt that he will be as happy as, perhaps happier than, if he had never left his own country. Every one must sincerely hope that Captain Fitz Roy's noble hope may be fulfilled, of being rewarded for the many generous sacrifices which he made for these Fuegians, by some

1 Captain Sulivan, who, since his voyage in the *Beagle*, has been employed on the survey of the Falkland Islands, heard from a sealer in 1842 (?), that when in the western part of the Strait of Magellan, he was astonished by a native woman coming on board, who could talk some English. Without doubt this was Fuegia Basket. She lived (I fear the term probably bears a double interpretation) some days on board.

shipwrecked sailor being protected by the descendants of Jemmy Button and his tribe! When Jemmy reached the shore, he lighted a signal fire, and the smoke curled up, bidding us a last and long farewell, as the ship stood on her course into the open sea.

The perfect equality among the individuals composing the Fuegian tribes, must for a long time retard their civilization. As we see those animals, whose instinct compels them to live in society and obey a chief, are most capable of improvement, so is it with the races of mankind. Whether we look at it as a cause or a conse-quence, the more civilized always have the most artificial govern-ments. For instance, the inhabitants of Otaheite, who, when first discovered, were governed by hereditary kings, had arrived at a far higher grade than another branch of the same people, the New Zealanders, – who, although benefited by being compelled to turn their attention to agriculture, were republicans in the most absolute sense. In Tierra del Fuego, until some chief shall arise with power sufficient to secure any acquired advantage, such as the domesticated animals, it seems scarcely possible that the political state of the country can be improved. At present, even a piece of cloth given to one is torn into shreds and distributed; and no one individual becomes richer than another. On the other hand, it is difficult to understand how a chief can arise till there is property of some sort by which he might manifest his superiority and increase his power.

I believe, in this extreme part of South America, man exists in a lower state of improvement than in any other part of the world. The South Sea Islanders of the two races inhabiting the Pacific, are comparatively civilized. The Esquimaux, in his subterranean hut, enjoys some of the comforts of life, and in his canoe, when fully equipped, manifests much skill. Some of the tribes of Southern Africa, prowling about in search of roots, and living concealed on the wild and arid plains, are sufficiently wretched. The Australian, in the simplicity of the arts of life, comes nearest the Fuegian: he can, however, boast of his boomerang, his spear and throwing-stick, his method of climbing trees, of tracking animals, and of hunting. Although the Australian may be superior in acquirements, it by no means follows that he is likewise superior in mental capacity: indeed, from what I saw of the Fuegians when on board, and from what I have read of the Australians, I should think the case was exactly the reverse.

CHAPTER XI

STRAIT OF MAGELLAN. – CLIMATE OF THE SOUTHERN COASTS

Strait of Magellan – Port Famine – Ascent of Mount Tarn – Forests – Edible Fungus – Zoology – Great Sea-weed – Leave Tierra del Fuego – Climate – Fruit-trees and Productions of the Southern Coasts – Height of Snow-line on the Cordillera – Descent of Glaciers to the Sea – Icebergs formed – Transportal of Boulders – Climate and Productions of the Antarctic Islands – Preservation of Frozen Carcasses – Recapitulation.

IN the end of May, 1834, we entered for the second time the eastern mouth of the Strait of Magellan. The country on both sides of this part of the Strait consists of nearly level plains, like those of Patagonia. Cape Negro, a little within the second Narrows, may be considered as the point where the land begins to assume the marked features of Tierra del Fuego. On the east coast, south of the Strait, broken park-like scenery in a like manner connects these two countries, which are opposed to each other in almost every feature. It is truly surprising to find in a space of twenty miles such a change in the landscape. If we take a rather greater distance, as between Port Famine and Gregory Bay, that is about sixty miles, the difference is still more wonderful. At the former place, we have rounded mountains concealed by impervious forests, which are drenched with the rain, brought by an endless succession of gales; while at Cape Gregory, there is a clear and bright blue sky over the dry and sterile plains. The atmospheric currents,[1] although rapid, turbulent, and unconfined by any apparent limits,

---

1 The south-westerly breezes are generally very dry. January 29th, being at anchor under Cape Gregory: a very hard gale from W. by S., clear sky with few cumuli; temperature 57°, dew-point 36°, – difference, 21°. On January 15th, at Port St. Julian: in the morning light winds with much rain, followed by a very heavy squall with rain, – settled into heavy gale with large cumuli, – cleared up, blowing very strong from S.S.W. Temperature 60°, dew-point 42°, – difference 18°.

yet seem to follow, like a river in its bed, a regularly determined course.

During our previous visit (in January), we had an interview at Cape Gregory with the famous so-called gigantic Patagonians, who gave us a cordial reception. Their height appears greater than it really is, from their large guanaco mantles, their long flowing hair, and general figure: on an average their height is about six feet, with some men taller and only a few shorter; and the women are also tall; altogether they are certainly the tallest race which we anywhere saw. In features they strikingly resemble the more northern Indians whom I saw with Rosas, but they have a wilder and more formidable appearance: their faces were much painted with red and black, and one man was ringed and dotted with white like a Fuegian. Capt. Fitz Roy offered to take any three of them on board, and all seemed determined to be of the three. It was long before we could clear the boat; at last we got on board with our three giants, who dined with the Captain, and behaved quite like gentlemen, helping themselves with knives, forks, and spoons: nothing was so much relished as sugar. This tribe has had so much communication with sealers and whalers, that most of the men can speak a little English and Spanish; and they are half civilized, and proportionally demoralized.

The next morning a large party went on shore, to barter for skins and ostrich-feathers; fire-arms being refused, tobacco was in greatest request, far more so than axes or tools. The whole population of the toldos, men, women, and children, were arranged on a bank. It was an amusing scene, and it was impossible not to like the so-called giants, they were so thoroughly good-humoured and unsuspecting: they asked us to come again. They seem to like to have Europeans to live with them; and old Maria, an important woman in the tribe, once begged Mr. Low to leave any one of his sailors with them. They spend the greater part of the year here; but in summer they hunt along the foot of the Cordillera: sometimes they travel as far as the Rio Negro, 750 miles to the north. They are well stocked with horses, each man having, according to Mr. Low, six or seven, and all the women, and even children, their one own horse. In the time of Sarmiento (1580), these Indians had bows and arrows, now long since disused; they then also possessed some horses. This is a very curious fact, showing the extraordinarily rapid multiplication of horses in South America. The horse was first landed at Buenos Ayres in 1537, and the colony being then for a time deserted, the horse ran wild;[1] in 1580, only forty-three years afterwards, we hear of them at the Strait of Magellan! Mr. Low informs me, that a neighbouring tribe of foot-Indians is now changing into horse-Indians: the tribe at Gregory Bay giving them their worn-out horses, and sending in winter a few of their best skilled men to hunt for them.

June 1st. – We anchored in the fine bay of Port Famine. It was now the beginning of winter, and I never saw a more cheerless prospect; the dusky woods, piebald with snow, could be only seen indistinctly through a drizzling hazy atmosphere. We were, however, lucky in getting two fine days. On one of these, Mount Sarmiento, a distant mountain 6800 feet high, presented a very noble spectacle. I was frequently surprised, in the scenery of Tierra del Fuego, at the little apparent elevation of mountains really lofty. I suspect it is owing to a cause which would not at first be imagined, namely, that the whole mass, from the summit to the water's edge, is generally in full view. I remember having seen a mountain, first from the Beagle Channel, where the whole sweep from the summit to the base was full in view, and then from Ponsonby Sound across several successive ridges; and it was curious to observe in the latter case, as each fresh ridge afforded fresh means of judging of the distance, how the mountain rose in height.

Before reaching Port Famine, two men were seen running along the shore and hailing the ship. A boat was sent for them. They turned out to be two sailors who had run away from a sealing-vessel, and had joined the Patagonians. These Indians had treated them with their usual disinterested hospitality. They had parted company through accident, and were then proceeding to Port Famine in hopes of finding some ship. I dare say they were worthless vagabonds, but I never saw more miserable-looking ones. They had been living for some days on mussel-shells and berries, and their tattered clothes had been burnt by sleeping so near their fires. They had been exposed

1 Rengger, Natur. der Saeugethiere von Paraguay. S. 334.

night and day, without any shelter, to the late incessant gales, with rain, sleet, and snow, and yet they were in good health.

During our stay at Port Famine, the Fuegians twice came and plagued us. As there were many instruments, clothes, and men on shore, it was thought necessary to frighten them away. The first time a few great guns were fired, when they were far distant. It was most ludicrous to watch through a glass the Indians, as often as the shot struck the water, take up stones, and as a bold defiance, throw them towards the ship, though about a mile and a half distant! A boat was then sent with orders to fire a few musket-shots wide of them. The Fuegians hid themselves behind the trees, and for every discharge of the muskets they fired their arrows; all, however, fell short of the boat, and the officer as he pointed at them laughed. This made the Fuegians frantic with passion, and they shook their mantles in vain rage. At last, seeing the balls cut and strike the trees, they ran away, and we were left in peace and quietness. During the former voyage the Fuegians were here very troublesome, and to frighten them a rocket was fired at night over their wigwams: it answered effectually, and one of the officers told me that the clamour first raised, and the barking of the dogs, was quite ludicrous in contrast with the profound silence which in a minute or two afterwards prevailed. The next morning not a single Fuegian was in the neighbourhood.

When the *Beagle* was here in the month of February, I started one morning at four o'clock to ascend Mount Tarn, which is 2600 feet high, and is the most elevated point in this immediate district. We went in a boat to the foot of the mountain (but unluckily not to the best part), and then began our ascent. The forest commences at the line of high-water mark, and during the first two hours I gave over all hopes of reaching the summit. So thick was the wood, that it was necessary to have constant recourse to the compass; for every land-mark, though in a mountainous country, was completely shut out. In the deep ravines, the death-like scene of desolation exceeded all description; outside it was blowing a gale, but in these hollows, not even a breath of wind stirred the leaves of the tallest trees. So gloomy, cold, and wet was every part, that not even the fungi, mosses, or ferns could flourish. In the valleys it was scarcely possible to crawl along, they were so completely barricaded by great mouldering trunks, which had fallen down in every direction. When passing over these natural bridges, one's course was often arrested by sinking knee deep into the rotten wood; at other times, when attempting to lean against a firm tree, one was startled by finding a mass of decayed matter ready to fall at the slightest touch. We at last found ourselves among the stunted trees, and then soon reached the bare ridge, which conducted us to the summit. Here was a view characteristic of Tierra del Fuego; irregular chains of hills, mottled with patches of snow, deep yellowish-green valleys, and arms of the sea intersecting the land in many directions. The strong wind was piercingly cold, and the atmosphere rather hazy, so that we did not stay long on the top of the mountain. Our descent was not quite so laborious as our ascent; for the weight of the body forced a passage, and all the slips and falls were in the right direction.

I have already mentioned the sombre and dull character of the evergreen forests,[1] in which two or three species of trees grow, to the exclusion of all others. Above the forest land, there are many dwarf alpine plants, which all spring from the mass of peat, and help to compose it: these plants are very remarkable from their close alliance with the species growing on the mountains of Europe, though so many thousand miles distant. The central part of Tierra del Fuego, where the clay-slate formation occurs, is most favourable to the growth of trees; on the outer coast the poorer granitic soil, and a situation more exposed to the violent winds, do not allow of their attaining any great size. Near Port Famine I have seen more large trees than anywhere else: I measured a Winter's Bark which was four feet six inches in girth, and several of the beech were as much as thirteen feet. Captain King also mentions a beech which was seven feet in diameter seventeen feet above the roots.

---

1 Captain Fitz Roy informs me that in April (our October), the leaves of those trees which grow near the base of the mountains, change colour, but not those on the more elevated parts. I remember having read some observations, showing that in England the leaves fall earlier in a warm and fine autumn, than in a late and cold one. The change in the colour being here retarded in the more elevated, and therefore colder situations, must be owing to the same general law of vegetation. The trees of Tierra del Fuego during no part of the year entirely shed their leaves.

There is one vegetable production deserving notice from its importance as an article of food to the Fuegians. It is a globular, bright-yellow fungus, which grows in vast numbers on the beech-trees. When young it is elastic and turgid, with a smooth surface; but when mature, it shrinks, becomes tougher, and has its entire surface deeply pitted or honey-combed, as represented in the accompanying wood-cut. This fungus belongs to a new and curious genus;[1] I found a second species on another species of beech in Chile; and Dr. Hooker informs me, that just lately a third species has been discovered on a third species of beech in Van Diemen's Land. How singular is this relationship between parasitical fungi and the trees on which they grow, in distant parts of the world! In Tierra del Fuego the fungus in its tough and mature state is collected in large quantities by the women and children, and is eaten uncooked. It has a mucilaginous, slightly sweet taste, with a faint smell like that of a mushroom. With the exception of a few berries, chiefly of a dwarf arbutus, the natives eat no vegetable food besides this fungus. In New Zealand, before the introduction of the potato, the roots of the fern were largely consumed; at the present time, I believe, Tierra del Fuego is the only country in the world where a cryptogamic plant affords a staple article of food.

The zoology of Tierra del Fuego, as might have been expected from the nature of its climate and vegetation, is very poor. Of mammalia, besides whales and seals, there is one bat, a kind of mouse (Reithrodon chinchilloides), two true mice, a ctenomys allied to or identical with the tucutuco, two foxes (Canis Magellanicus and C. Azaræ), a sea-otter, the guanaco, and a deer. Most of these animals inhabit only the drier eastern parts of the country; and the deer has never been seen south of the Strait of Magellan. Observing the general correspondence of the cliffs of soft sandstone, mud, and shingle, on the opposite sides of the Strait, and on some intervening

1 Described from my specimens, and notes by the Rev. J. M. Berkeley, in the Linnean Transactions (vol. xix. p. 37), under the name of Cyttaria Darwinii: the Chilian species is the C. Berteroii. This genus is allied to Bulgaria.

islands, one is strongly tempted to believe that the land was once joined, and thus allowed animals so delicate and helpless as the tucu-tuco and Reithrodon to pass over. The correspondence of the cliffs is far from proving any junction; because such cliffs generally are formed by the intersection of sloping deposits, which, before the elevation of the land, had been accumulated near the then existing shores. It is, however, a remarkable coincidence, that in the two large islands cut off by the Beagle Channel from the rest of Tierra del Fuego, one has cliffs composed of matter that may be called stratified alluvium, which front similar ones on the opposite side of the channel, – while the other is exclusively bordered by old crystalline rocks: in the former, called Navarin Island, both foxes and guanacos occur; but in the latter, Hoste Island, although similar in every respect, and only separated by a channel a little more than half a mile wide, I have the word of Jemmy Button for saying, that neither of these animals are found.

The gloomy woods are inhabited by few birds: occasionally the plaintive note of a white-tufted tyrant-flycatcher (Myiobius albiceps) may be heard, concealed near the summit of the most lofty trees; and more rarely the loud strange cry of a black woodpecker, with a fine scarlet crest on its head. A little, dusky-coloured wren (Scytalopus Magellanicus) hops in a skulking manner among the entangled mass of the fallen and decaying trunks. But the creeper (Oxyurus tupinieri) is the commonest bird in the country. Throughout the beech forests, high up and low down, in the most gloomy, wet, and impenetrable ravines, it may be met with. This little bird no doubt appears more numerous than it really is, from its habit of following with seeming curiosity any person who enters these silent woods: continually uttering a harsh twitter, it flutters from tree to tree, within a few feet of the intruder's face. It is far from wishing for the modest concealment of the true creeper (Certhia familiaris); nor does it, like that bird, run up the trunks of trees, but industriously, after the manner of a willow-wren, hops about, and searches for insects on every twig and branch. In the more open parts, three or four species of finches, a thrush, a starling (or Icterus), two Opetiorhynchi, and several hawks and owls occur.

The absence of any species whatever in the whole class of Reptiles, is a marked feature in the zoology of this country, as well as in that of the Falkland Islands. I do not ground this statement merely on my own observation, but I heard it from the Spanish inhabitants of the latter place, and from Jemmy Button with regard to Tierra del Fuego. On the banks of the Santa Cruz, in 50° south, I saw a frog; and it is not improbable that these animals, as well as lizards, may be found as far south as the Strait of Magellan, where the country retains the character of Patagonia; but within the damp and cold limit of Tierra del Fuego not one occurs. That the climate would not have suited some of the orders, such as lizards, might have been foreseen; but with respect to frogs, this was not so obvious.

Beetles occur in very small numbers: it was long before I could believe that a country as large as Scotland, covered with vegetable productions, and with a variety of stations, could be so unproductive. The few which I found were alpine species (Harpalidæ and Heteromidæ) living under stones. The vegetable-feeding Chrysomelidæ, so eminently characteristic of the Tropics, are here almost entirely absent;[1] I saw very few flies, butterflies, or bees, and no crickets or Orthoptera. In the pools of water I found but few aquatic beetles, and not any fresh-water shells: Succinea at first appears an exception; but here it must be called a terrestrial shell, for it lives on the damp herbage far from water. Land-shells could be procured only in the same alpine situations with the beetles. I have already contrasted the climate as well as the general appearance of Tierra del Fuego with that of Patagonia; and the difference is strongly exemplified in the entomology. I do not believe they have one species in common; certainly the general character of the insects is widely different.

If we turn from the land to the sea, we shall find the latter as abundantly stocked with living creatures as the former is poorly so. In all parts of the world a rocky and partially protected shore perhaps

---

1 I believe I must except one alpine Haltica, and a single specimen of a Melasoma. Mr. Waterhouse informs me, that of the Harpalidæ there are eight or nine species – the forms of the greater number being very peculiar; of Heteromera, four or five species; of Rhyncophora six or seven; and of the following families one species in each: Staphylinidæ, Elateridæ, Cebrionidæ, Melolonthidæ. The species in the other orders are even fewer. In all the orders, the scarcity of the individuals is even more remarkable than that of the species. Most of the Coleoptera have been carefully described by Mr. Waterhouse in the Annals of Nat. Hist.

supports, in a given space, a greater number of individual animals than any other station. There is one marine production, which from its importance is worthy of a particular history. It is the kelp, or Macrocystis pyrifera. This plant grows on every rock from low-water mark to a great depth, both on the outer coast and within the channels.[1] I believe, during the voyages of the *Adventure* and *Beagle*, not one rock near the surface was discovered which was not buoyed by this floating weed. The good service it thus affords to vessels navigating near this stormy land is evident; and it certainly has saved many a one from being wrecked. I know few things more surprising than to see this plant growing and flourishing amidst those great breakers of the western ocean, which no mass of rock, let it be ever so hard, can long resist. The stem is round, slimy, and smooth, and seldom has a diameter of so much as an inch. A few taken together are sufficiently strong to support the weight of the large loose stones, to which in the inland channels they grow attached; and yet some of these stones were so heavy that when drawn to the surface, they could scarcely be lifted into a boat by one person. Captain Cook, in his second voyage, says, that this plant at Kerguelen Land rises from a greater depth than twenty-four fathoms; 'and as it does not grow in a perpendicular direction, but makes a very acute angle with the bottom, and much of it afterwards spreads many fathoms on the surface of the sea, I am well warranted to say that some of it grows to the length of sixty fathoms and upwards.' I do not suppose the stem of any other plant attains so great a length as three hundred and sixty feet, as stated by Captain Cook. Captain Fitz Roy, moreover, found it growing[2] up from the greater depth of forty-five fathoms. The beds of this sea-weed, even

1 Its geographical range is remarkably wide; it is found from the extreme southern islets near Cape Horn, as far north on the eastern coast (according to information given me by Mr. Stokes) as lat. 43°, – but on the western coast, as Dr. Hooker tells me, it extends to the R. San Francisco in California, and perhaps even to Kamtschatka. We thus have an immense range in latitude; and as Cook, who must have been well acquainted with the species, found it at Kerguelen Land, no less than 140° in longitude.

2 Voyages of the *Adventure* and *Beagle*, vol. i. p. 363. – It appears that sea-weed grows extremely quick. Mr. Stephenson found (Wilson's Voyage round Scotland, vol. ii. p. 228) that a rock uncovered only at spring-tides, which had been chiselled smooth in November, on the following May, that is within six months afterwards, was thickly covered with Fucus digitatus two feet, and F. esculentus six feet, in length.

when of not great breadth, make excellent natural floating break-waters. It is quite curious to see, in an exposed harbour, how soon the waves from the open sea, as they travel through the straggling stems, sink in height, and pass into smooth water.

The number of living creatures of all Orders, whose existence intimately depends on the kelp, is wonderful. A great volume might be written, describing the inhabitants of one of these beds of sea-weed. Almost all the leaves, excepting those that float on the surface, are so thickly incrusted with corallines as to be of a white colour. We find exquisitely delicate structures, some inhabited by simple hydra-like polypi, others by more organized kinds, and beautiful compound Ascidiæ. On the leaves, also, various patelliform shells, Trochi, un-covered molluscs, and some bivalves are attached. Innumerable crust-acea frequent every part of the plant. On shaking the great entangled roots, a pile of small fish, shells, cuttle-fish, crabs of all orders, sea-eggs, star-fish, beautiful Holuthuriæ, Planariæ, and crawling nerei-dous animals of a multitude of forms, all fall out together. Often as I recurred to a branch of the kelp, I never failed to discover animals of new and curious structures. In Chiloe, where the kelp does not thrive very well, the numerous shells, corallines, and crustacea are absent; but there yet remain a few of the Flustraceæ, and some compound Ascidiæ; the latter, however, are of different species from those in Tierra del Fuego: we here see the fucus possessing a wider range than the animals which use it as an abode. I can only compare these great aquatic forests of the southern hemisphere, with the terrestrial ones in the intertropical regions. Yet if in any country a forest was destroyed, I do not believe nearly so many species of animals would perish as would here, from the destruction of the kelp. Amidst the leaves of this plant numerous species of fish live, which nowhere else could find food or shelter; with their destruction the many cormorants and other fishing birds, the otters, seals, and porpoises, would soon perish also; and lastly, the Fuegian savage, the miserable lord of this miserable land, would redouble his cannibal feast, decrease in numbers, and perhaps cease to exist.

June 8th. – We weighed anchor early in the morning and left Port Famine. Captain Fitz Roy determined to leave the Strait of Magellan by the Magdalen Channel, which had not long been discovered. Our course lay due south, down that gloomy passage which I have before alluded to, as appearing to lead to another and worse world. The wind was fair, but the atmosphere was very thick; so that we missed much curious scenery. The dark ragged clouds were rapidly driven over the mountains, from their summits nearly down to their bases. The glimpses which we caught through the dusky mass, were highly interesting; jagged points, cones of snow, blue glaciers, strong out-lines, marked on a lurid sky, were seen at different distances and heights. In the midst of such scenery we anchored at Cape Turn, close to Mount Sarmiento, which was then hidden in the clouds. At the base of the lofty and almost perpendicular sides of our little cove there was one deserted wigwam, and it alone reminded us that man sometimes wandered into these desolate regions. But it would be difficult to imagine a scene where he seemed to have fewer claims or less authority. The inanimate works of nature – rock, ice, snow, wind, and water – all warring with each other, yet combined against man – here reigned in absolute sovereignty.

June 9th. – In the morning we were delighted by seeing the veil of mist gradually rise from Sarmiento, and display it to our view. This mountain, which is one of the highest in Tierra del Fuego, has an altitude of 6800 feet. Its base, for about an eighth of its total height, is clothed by dusky woods, and above this a field of snow extends to the summit. These vast piles of snow, which never melt, and seem destined to last as long as the world holds together, present a noble and even sublime spectacle. The outline of the mountain was admir-ably clear and defined. Owing to the abundance of light reflected from the white and glittering surface, no shadows were cast on any part; and those lines which intersected the sky could alone be distin-guished: hence the mass stood out in the boldest relief. Several glaciers descended in a winding course from the upper great expanse of snow to the sea-coast: they may be likened to great frozen Niagaras; and perhaps these cataracts of blue ice are full as beautiful as the moving ones of water. By night we reached the western part of the channel; but the water was so deep that no anchorage could be found. We were in consequence obliged to stand off and on in this narrow arm of the sea, during a pitch-dark night of fourteen hours long.

June 10th. – In the morning we made the best of our way into the open Pacific. The Western coast generally consists of low, rounded, quite barren hills of granite and greenstone. Sir. J. Narborough called one part South Desolation, because it is 'so desolate a land to behold:' and well indeed might he say so. Outside the main islands, there are numberless scattered rocks on which the long swell of the open ocean incessantly rages. We passed out between the East and West Furies; and a little farther northward there are so many breakers that the sea is called the Milky Way. One sight of such a coast is enough to make a landsman dream for a week about shipwrecks, peril, and death; and with this sight we bade farewell for ever to Tierra del Fuego.

The following discussion on the climate of the southern parts of the continent with relation to its productions, on the snow-line, on the extraordinarily low descent of the glaciers, and on the zone of perpetual congelation in the antarctic islands, may be passed over by any one not interested in these curious subjects, or the final recapitulation alone may be read. I shall, however, here give only an abstract, and must refer for details to the Thirteenth Chapter and the Appendix of the former edition of this work.

On the Climate and Productions of Tierra del Fuego and of the South-west Coast. – The following table gives the mean temperature of Tierra del Fuego, the Falkland Islands, and, for comparison, that of Dublin: –

	Latitude.	Summer Temp.	Winter Temp.	Mean of Summer and Winter.
Tierra del Fuego . . .	53° 38'S.	50°	33°.08	41°.54
Falkland Islands . . .	51 30 S.	51	–	–
Dublin	53 21 N.	59.54	39.2	49.37

Hence we see that the central part of Tierra del Fuego is colder in winter, and no less than 9½° less hot in summer, than Dublin. According to Von Buch the mean temperature of July (not the hottest month in the year) at Saltenfiord in Norway, is as high as 57°.8, and this place is actually 13° nearer the pole than Port Famine![1]

1 With respect to Tierra del Fuego, the results are deduced from the observations by Capt. King (Geographical Journal, 1830), and those taken on board the *Beagle*. For the Falkland Islands, I am indebted to Capt. Sulivan for the mean of the mean temperature

Inhospitable as this climate appears to our feelings, evergreen trees flourish luxuriantly under it. Humming-birds may be seen sucking the flowers, and parrots feeding on the seeds of the Winter's Bark, in lat. 55° S. I have already remarked to what a degree the sea swarms with living creatures; and the shells (such as the Patellæ, Fissurellæ, Chitons, and Barnacles), according to Mr. G. B. Sowerby, are of a much larger size, and of a more vigorous growth, than the analogous species in the northern hemisphere. A large Voluta is abundant in southern Tierra del Fuego and the Falkland Islands. At Bahia Blanca, in lat. 39° S., the most abundant shells were three species of Oliva (one of large size), one or two Volutas, and a Terebra. Now these are amongst the best characterized tropical forms. It is doubtful whether even one small species of Oliva exists on the southern shores of Europe, and there are no species of the two other genera. If a geologist were to find in lat. 39° on the coast of Portugal, a bed containing numerous shells belonging to three species of Oliva, to a Voluta and Terebra, he would probably assert that the climate at the period of their existence must have been tropical; but judging from South America, such an inference might be erroneous.

The equable, humid, and windy climate of Tierra del Fuego extends, with only a small increase of heat, for many degrees along the west coast of the continent. The forests, for 600 miles northward of Cape Horn, have a very similar aspect. As a proof of the equable climate, even for 300 or 400 miles still further northward, I may mention that in Chiloe (corresponding in latitude with the northern parts of Spain) the peach seldom produces fruit, whilst strawberries and apples thrive to perfection. Even the crops of barley and wheat[1] are often brought into the houses to be dried and ripened. At Valdivia (in the same latitude of 40°, with Madrid) grapes and figs ripen, but are not common; olives seldom ripen even partially, and oranges not at all. These fruits, in corresponding latitudes in Europe, are well known to succeed to perfection; and even in this continent, at the

(reduced from careful observation at midnight, 8 A.M., noon, and 8 P.M.) of the three hottest months, viz. December, January, and February. The temperature of Dublin is taken from Barton.

1 Agüeros, Descrip. Hist. de la Prov. de Chiloé, 1791, p. 94.

Rio Negro, under nearly the same parallel with Valdivia, sweet potatoes (convolvulus) are cultivated; and grapes, figs, olives, oranges, water and musk melons, produce abundant fruit. Although the humid and equable climate of Chiloe, and of the coast northward and southward of it, is so unfavourable to our fruits, yet the native forests, from lat. 45° to 38°, almost rival in luxuriance those of the glowing intertropical regions. Stately trees of many kinds, with smooth and highly coloured barks, are loaded by parasitical mono-cotyledonous plants; large and elegant ferns are numerous, and arborescent grasses intwine the trees into one entangled mass to the height of thirty or forty feet above the ground. Palm-trees grow in lat. 37°; an arborescent grass, very like a bamboo, in 40°; and another closely allied kind, of great length, but not erect, flourishes even as far south as 45° S.

An equable climate, evidently due to the large area of sea com-pared with the land, seems to extend over the greater part of the southern hemisphere; and as a consequence, the vegetation partakes of a semi-tropical character. Tree-ferns thrive luxuriantly in Van Diemen's Land (lat. 45°), and I measured one trunk no less than six feet in circumference. An arborescent fern was found by Forster in New Zealand in 46°, where orchideous plants are parasitical on the trees. In the Auckland Islands, ferns, according to Dr. Dieffenbach,[1] have trunks so thick and high that they may be almost called tree-ferns; and in these islands, and even as far south as lat. 55° in the Macquarrie Islands, parrots abound.

On the Height of the Snow-line, and on the Descent of the Glaciers, in South America. – For the detailed authorities for the following table, I must refer to the former edition: –

Latitude.	Height in feet of Snow-line.	Observer.
Equatorial region; mean result	15,748	Humboldt.
Bolivia, lat. 16° to 18° S.	17,000	Pentland.
Central Chile, lat. 33° S.	14,500 to 15,000	Gillies, and the Author.
Chiloe, lat. 41° to 43° S.	6,000	Officers of the *Beagle*, and the Author.
Tierra del Fuego, 54° S.	3,500 to 4,000	King.

1 See the German Translation of this Journal: and for the other facts Mr. Brown's Appendix to Flinders's Voyage.

As the height of the plane of perpetual snow seems chiefly to be determined by the extreme heat of the summer, rather than by the mean temperature of the year, we ought not to be surprised at its descent in the Strait of Magellan, where the summer is so cool, to only 3500 or 4000 feet above the level of the sea; although in Norway, we must travel to between lat. 67° and 70° N., that is, about 14° nearer the pole, to meet with perpetual snow at this low level. The difference in height, namely, about 9000 feet, between the snow-line on the Cordillera behind Chiloe (with its highest points ranging from only 5600 to 7500 feet) and in central Chile[1] (a distance of only 9° of latitude), is truly wonderful. The land from the south-ward of Chiloe to near Conception (lat. 37°), is hidden by one dense forest dripping with moisture. The sky is cloudy, and we have seen how badly the fruits of southern Europe succeed. In central Chile, on the other hand, a little northward of Conception, the sky is generally clear, rain does not fall for the seven summer months, and southern European fruits succeed admirably; and even the sugar-cane has been cultivated.[2] No doubt the plane of perpetual snow undergoes the above remarkable flexure of 9000 feet, unparalleled in other parts of the world, not far from the latitude of Concepcion, where the land ceases to be covered with forest-trees; for trees in South America indicate a rainy climate, and rain a clouded sky and little heat in summer.

The descent of glaciers to the sea must, I conceive, mainly depend (subject, of course, to a proper supply of snow in the upper region) on the lowness of the line of perpetual snow on steep mountains near the coast. As the snow-line is so low in Tierra del Fuego, we might have expected that many of the glaciers would have reached the sea. Nevertheless I was astonished when I first saw a range, only from

1 On the Cordillera of central Chile, I believe the snow-line varies exceedingly in height in different summers. I was assured that during one very dry and long summer, all the snow disappeared from Aconcagua, although it attains the prodigious height of 23,000 feet. It is probable that much of the snow at these great heights is evaporated, rather than thawed.

2 Miers's Chile, vol. i. p. 415. It is said that the sugar-cane grew at Ingenio, lat. 32° to 33°, but not in sufficient quantity to make the manufacture profitable. In the valley of Quillota, south of Ingenio, I saw some large date palm-trees.

3000 to 4000 feet in height, in the latitude of Cumberland, with every valley filled with streams of ice descending to the sea-coast. Almost every arm of the sea, which penetrates to the interior higher chain, not only in Tierra del Fuego, but on the coast for 650 miles northwards, is terminated by 'tremendous and astonishing glaciers,' as described by one of the officers on the survey. Great masses of ice frequently fall from these icy cliffs, and the crash reverberates like the broadside of a man-of-war, through the lonely channels. These falls, as noticed in the last chapter, produce great waves which break on the adjoining coasts. It is known that earthquakes frequently cause masses of earth to fall from sea-cliffs: how terrific, then, would be the effect of a severe shock (and such occur here)[1] on a body like a glacier, already in motion, and traversed by fissures! I can readily believe that the water would be fairly beaten back out of the deepest channel, and then returning with an overwhelming force, would whirl about huge masses of rock like so much chaff. In Eyre's Sound, in the latitude of Paris, there are immense glaciers, and yet the loftiest

1 Bulkeley's and Cummin's Faithful Narrative of the Loss of the *Wager*. The earthquake happened August 25, 1741.

neighbouring mountain is only 6200 feet high. In this Sound, about fifty icebergs were seen at one time floating outwards, and one of them must have been *at least* 168 feet in total height. Some of the icebergs were loaded with blocks of no inconsiderable size, of granite and other rocks, different from the clay-slate of the surrounding mountains. The glacier furthest from the Pole, surveyed during the voyages of the *Adventure* and *Beagle*, is in lat. 46° 50', in the Gulf of Penas. It is 15 miles long, and in one part 7 broad, and descends to the sea-coast. But even a few miles northward of this glacier, in the Laguna de San Rafael, some Spanish missionaries[1] encountered 'many icebergs, some great, some small, and others middle-sized,' in a narrow arm of the sea, on the 22nd of the month corresponding with our June, and in a latitude corresponding with that of the Lake of Geneva!

In Europe, the most southern glacier which comes down to the sea is met with, according to Von Buch, on the coast of Norway, in lat. 67°. Now this is more than 20° of latitude, or 1230 miles, nearer the pole than the Laguna de San Rafael. The position of the glaciers at this place and in the Gulf of Penas, may be put even in a more striking point of view, for they descend to the sea-coast, within $7\frac{1}{2}°$ of latitude, or 450 miles, of a harbour, where three species of Oliva, a Voluta, and a Terebra, are the commonest shells, within less than 9° from where palms grow, within $4\frac{1}{2}°$ of a region where the jaguar and puma range over the plains, less than $2\frac{1}{2}°$ from arborescent grasses, and (looking to the westward in the same hemisphere) less than 2° from orchideous parasites, and within a single degree of tree-ferns!

These facts are of high geological interest with respect to the climate of the northern hemisphere, at the period when boulders were transported. I will not here detail how simply the theory of icebergs being charged with fragments of rock, explains the origin and position of the gigantic boulders of eastern Tierra del Fuego, on the high plain of Santa Cruz, and on the island of Chiloe. In Tierra del Fuego, the greater number of boulders lie on the lines of old sea-channels, now converted into dry valleys by the elevation of the land.

1 Agüeros, Desc. Hist. de Chiloe, p. 227.

They are associated with a great unstratified formation of mud and sand, containing rounded and angular fragments of all sizes, which has originated[1] in the repeated ploughing up of the sea-bottom by the stranding of icebergs, and by the matter transported on them. Few geologists now doubt that those erratic boulders which lie near lofty mountains, have been pushed forward by the glaciers themselves, and that those distant from mountains, and embedded in subaqueous deposits, have been conveyed thither either on icebergs, or frozen in coast-ice. The connexion between the transportal of boulders and the presence of ice in some form, is strikingly shown by their geographical distribution over the earth. In South America they are not found further than $48°$ of latitude, measured from the southern pole; in North America it appears that the limit of their transportal extends to $53\frac{1}{2}°$ from the northern pole; but in Europe to not more than $40°$ of latitude, measured from the same point. On the other hand, in the intertropical parts of America, Asia, and Africa, they have never been observed; nor at the Cape of Good Hope, nor in Australia.[2]

   On the Climate and Productions of the Antarctic Islands. – Considering the rankness of the vegetation in Tierra del Fuego, and on the coast northward of it, the condition of the islands south and south-west of America is truly surprising. Sandwich Land, in the latitude of the north part of Scotland, was found by Cook, during the hottest month of the year, 'covered many fathoms thick with everlasting snow;' and there seems to be scarcely any vegetation. Georgia, an island 96 miles long and 10 broad, in the latitude of Yorkshire, 'in the very height of summer, is in a manner wholly covered with frozen snow.' It can boast only of moss, some tufts of grass, and wild burnet: it has only one land-bird (*Anthus correndera*), yet Iceland, which is $10°$ nearer the pole, has, according to Mackenzie, fifteen land-birds. The South Shetland Islands, in the same latitude as the southern half of Norway, possess only some lichens, moss, and a little grass;

   1 Geological Transactions, vol. vi. p. 415.

   2 I have given details (the first, I believe, published) on this subject in the first edition, and in the Appendix to it. I have there shown that the apparent exceptions to the absence of erratic boulders in certain hot countries, are due to erroneous observations: several statements there given, I have since found confirmed by various authors.

and Lieut. Kendall[1] found the bay, in which he was at anchor, beginning to freeze at a period corresponding with our 8th of September. The soil here consists of ice and volcanic ashes interstratified; and at a little depth beneath the surface it must remain perpetually congealed, for Lieut. Kendall found the body of a foreign sailor which had long been buried, with the flesh and all the features perfectly preserved. It is a singular fact, that on the two great continents in the northern hemisphere (but not in the broken land of Europe between them), we have the zone of perpetually frozen under-soil in a low latitude – namely, in 56° in North America at the depth of three feet,[2] and in 62° in Siberia at the depth of twelve to fifteen feet – as the result of a directly opposite condition of things, to those of the southern hemisphere. On the northern continents, the winter is rendered excessively cold by the radiation from a large area of land into a clear sky, nor is it moderated by the warmth-bringing currents of the sea; the short summer, on the other hand, is hot. In the Southern Ocean the winter is not so excessively cold, but the summer is far less hot, for the clouded sky seldom allows the sun to warm the ocean, itself a bad absorbent of heat; and hence the mean temperature of the year, which regulates the zone of perpetually congealed under-soil, is low. It is evident that a rank vegetation, which does not so much require heat as it does protection from intense cold, would approach much nearer to this zone of perpetual congelation under the equable climate of the southern hemisphere, than under the extreme climate of the northern continents.

The case of the sailor's body perfectly preserved in the icy soil of the South Shetland Islands (lat. 62° to 63° S.), in a rather lower latitude than that (lat. 64° N.) under which Pallas found the frozen rhinoceros in Siberia, is very interesting. Although it is a fallacy, as I have endeavoured to show in a former chapter, to suppose that the larger quadrupeds require a luxuriant vegetation for their support, nevertheless it is important to find in the South Shetland Islands, a frozen under-soil within 360 miles of the forest-clad islands

1 Geographical Journal, 1830, pp. 65, 66.

2 Richardson's Append. to Back's Exped., and Humboldt's Fragm. Asiat., tom. ii. p. 386.

near Cape Horn, where, as far as the *bulk* of vegetation is concerned, any number of great quadrupeds might be supported. The perfect preservation of the carcasses of the Siberian elephants and rhinoceroses is certainly one of the most wonderful facts in geology; but independently of the imagined difficulty of supplying them with food from the adjoining countries, the whole case is not, I think, so perplexing as it has generally been considered. The plains of Siberia, like those of the Pampas, appear to have been formed under the sea, into which rivers brought down the bodies of many animals; of the greater number of these, only the skeletons have been preserved, but of others the perfect carcass. Now it is known, that in the shallow sea on the arctic coast of America the bottom freezes,[1] and does not thaw in spring so soon as the surface of the land; moreover, at greater depths, where the bottom of the sea does not freeze, the mud a few feet beneath the top layer might remain even in summer below 32°, as is the case on the land with the soil at the depth of a few feet. At still greater depths, the temperature of the mud and water would probably not be low enough to preserve the flesh; and hence, carcasses drifted beyond the shallow parts near an arctic coast, would have only their skeletons preserved: now in the extreme northern parts of Siberia bones are infinitely numerous, so that even islets are said to be almost composed of them;[2] and those islets lie no less than ten degrees of latitude north of the place where Pallas found the frozen rhinoceros. On the other hand, a carcass washed by a flood into a shallow part of the Arctic Sea, would be preserved for an indefinite period, if it were soon afterwards covered with mud, sufficiently thick to prevent the heat of the summer-water penetrating to it; and if, when the sea-bottom was upraised into land, the covering was sufficiently thick to prevent the heat of the summer air and sun thawing and corrupting it.

Recapitulation. – I will recapitulate the principal facts with regard to the climate, ice-action, and organic productions of the southern hemisphere, transposing the places in imagination to Europe, with

---

1 Messrs. Dease and Simpson, in Geograph. Journ., vol. viii. pp. 218 and 220.

2 Cuvier (Ossemens Fossiles, tom. i. p. 151), from Billing's Voyage.

which we are so much better acquainted. Then, near Lisbon, the commonest sea-shells, namely, three species of Oliva, a Voluta and Terebra, would have a tropical character. In the southern provinces of France, magnificent forests, intwined by arborescent grasses and with the trees loaded with parasitical plants, would hide the face of the land. The puma and the jaguar would haunt the Pyrenees. In the latitude of Mont Blanc, but on an island as far westward as central North America, tree-ferns and parasitical Orchideæ would thrive amidst the thick woods. Even as far north as central Denmark, humming-birds would be seen fluttering about delicate flowers, and parrots feeding amidst the evergreen woods; and in the sea there, we should have a Voluta, and all the shells of large size and vigorous growth. Nevertheless, on some islands only 360 miles northward of our new Cape Horn in Denmark, a carcass buried in the soil (or if washed into a shallow sea, and covered up with mud) would be preserved perpetually frozen. If some bold navigator attempted to penetrate northward of these islands, he would run a thousand dangers amidst gigantic icebergs, on some of which he would see great blocks of rock borne far away from their original site. Another island of large size in the latitude of southern Scotland, but twice as far to the west, would be 'almost wholly covered with everlasting snow,' and would have each bay terminated by ice-cliffs, whence great masses would be yearly detached: this island would boast only of a little moss, grass, and burner, and a tit-lark would be its only land inhabitant. From our new Cape Horn in Denmark, a chain of mountains, scarcely half the height of the Alps, would run in a straight line due southward; and on its western flank every deep creek of the sea, or fiord, would end in 'bold and astonishing glaciers.' These lonely channels would frequently reverberate with the falls of ice, and so often would great waves rush along their coasts; numerous icebergs, some as tall as cathedrals, and occasionally loaded with 'no inconsiderable blocks of rock,' would be stranded on the outlying islets; at intervals violent earthquakes would shoot prodigious masses of ice into the waters below. Lastly, some Missionaries attempting to penetrate a long arm of the sea, would behold the not lofty surrounding mountains, sending down their many grand icy streams to the sea-coast, and their progress in the boats would be checked by the innumerable floating icebergs, some small and some great; and this would have occurred on our twenty-second of June, and where the Lake of Geneva is now spread out![1]

---

1 In the former edition and Appendix, I have given some facts on the transportal of erratic boulders and icebergs in the Antarctic Ocean. This subject has lately been treated excellently by Mr. Hayes, in the Boston Journal (vol. iv. p. 426). The author does not appear aware of a case published by me (Geographical Journal, vol. ix. p. 528), of a gigantic boulder embedded in an iceberg in the Antarctic Ocean, almost certainly one hundred miles distant from any land, and perhaps much more distant. In the Appendix I have discussed at length, the probability (at that time hardly thought of) of icebergs, when stranded, grooving and polishing rocks, like glaciers. This is now a very commonly received opinion; and I cannot still avoid the suspicion that it is applicable even to such cases as that of the Jura. Dr. Richardson has assured me, that the icebergs off North America push before them pebbles and sand, and leave the submarine rocky flats quite bare: it is hardly possible to doubt that such ledges must be polished and scored in the direction of the set of the prevailing currents. Since writing that Appendix, I have seen in North Wales (London Phil. Mag., vol. xxi. p. 180) the adjoining action of glaciers and of floating icebergs.

CHAPTER XII

CENTRAL CHILE

Valparaiso – Excursion to the Foot of the Andes – Structure of the Land – Ascend the Bell of Quillota – Shattered Masses of Greenstone – Immense Valleys – Mines – State of Miners – Santiago – Hot-baths of Cauquenes – Gold-mines – Grinding-mills – Perforated Stones – Habits of the Puma – El Turco and Tapacolo – Humming-birds.

JULY 23rd. – The *Beagle* anchored late at night in the bay of Valparaiso, the chief seaport of Chile. When morning came, everything appeared delightful. After Tierra del Fuego, the climate felt quite delicious – the atmosphere so dry, and the heavens so clear and blue with the sun shining brightly, that all nature seemed sparkling with life. The view from the anchorage is very pretty. The town is built at the very foot of a range of hills, about 1600 feet high, and rather steep. From its position, it consists of one long, straggling street, which runs parallel to the beach, and wherever a ravine comes down, the houses are piled up on each side of it. The rounded hills, being only partially protected by a very scanty vegetation, are worn into numberless little gullies, which expose a singularly bright red soil. From this cause, and from the low whitewashed houses with tile roofs, the view reminded me of St. Cruz in Teneriffe. In a north-easterly direction there are some fine glimpses of the Andes: but these mountains appear much grander when viewed from the neighbouring hills; the great distance at which they are situated, can then more readily be perceived. The volcano of Aconcagua is particularly magnificent. This huge and irregularly conical mass has an elevation greater than that of Chimborazo; for, from measurements made by the officers in the *Beagle*, its height is no less than 23,000 feet. The Cordillera, however, viewed from this point, owe the greater part of their beauty to the atmosphere through which they are seen. When the sun was setting in the Pacific, it was admirable to watch how clearly their rugged outlines could be distinguished, yet how varied and how delicate were the shades of their colour.

I had the good fortune to find living here Mr. Richard Corfield, an old schoolfellow and friend, to whose hospitality and kindness I was greatly indebted, in having afforded me a most pleasant residence during the *Beagle*'s stay in Chile. The immediate neighbourhood of Valparaiso is not very productive to the naturalist. During the long summer the wind blows steadily from the southward, and a little off shore, so that rain never falls; during the three winter months, however, it is sufficiently abundant. The vegetation in consequence is very scanty: except in some deep valleys, there are no trees, and only a little grass and a few low bushes are scattered over the less steep parts of the hills. When we reflect, that at the distance of 350 miles to the south, this side of the Andes is completely hidden by one impenetrable forest, the contrast is very remarkable. I took several long walks while collecting objects of natural history. The country is pleasant for exercise. There are many very beautiful flowers; and, as in most other dry climates, the plants and shrubs possess strong and peculiar odours – even one's clothes by brushing through them became scented. I did not cease from wonder at finding each succeeding day as fine as the foregoing. What a difference does climate make in the enjoyment of life! How opposite are the sensations when viewing black mountains half-enveloped in clouds, and seeing another range through the light blue haze of a fine day! The one for a time may be very sublime; the other is all gaiety and happy life.

August 14th. – I set out on a riding excursion, for the purpose of geologizing the basal parts of the Andes, which alone at this time of the year are not shut up by the winter snow. Our first day's ride was northward along the sea-coast. After dark we reached the Hacienda of Quintero, the estate which formerly belonged to Lord Cochrane. My object in coming here was to see the great beds of shells, which stand some yards above the level of the sea, and are burnt for lime. The proofs of the elevation of this whole line of coast are unequivocal: at the height of a few hundred feet old-looking shells are numerous, and I found some at 1300 feet. These shells either lie loose on the surface, or are embedded in a reddish-black vegetable mould. I was much surprised to find under the microscope that this vegetable mould is really marine mud, full of minute particles of organic bodies.

15th. – We returned towards the valley of Quillota. The country was exceedingly pleasant; just such as poets would call pastoral: green open lawns, separated by small valleys with rivulets, and the cottages, we may suppose of the shepherds, scattered on the hill-sides. We were obliged to cross the ridge of the Chilicauquen. At its base there were many fine evergreen forest-trees, but these flourished only in the ravines, where there was running water. Any person who had seen only the country near Valparaiso, would never have imagined that there had been such picturesque spots in Chile. As soon as we reached the brow of the Sierra, the valley of Quillota was immediately under our feet. The prospect was one of remarkable artificial luxuriance. The valley is very broad and quite flat, and is thus easily irrigated in all parts. The little square gardens are crowded with orange and olive trees, and every sort of vegetable. On each side huge bare mountains rise, and this from the contrast renders the patchwork valley the more pleasing. Whoever called 'Valparaiso' the 'Valley of Paradise,' must have been thinking of Quillota. We crossed over to the Hacienda de San Isidro, situated at the very foot of the Bell Mountain.

Chile, as may be seen in the maps, is a narrow strip of land between the Cordillera and the Pacific; and this strip is itself traversed by several mountain-lines, which in this part run parallel to the great range. Between these outer lines and the main Cordillera, a succession of level basins, generally opening into each other by narrow passages, extend far to the southward: in these, the principal towns are situated, as San Felipe, Santiago, San Fernando. These basins or plains, together with the transverse flat valleys (like that of Quillota) which connect them with the coast, I have no doubt are the bottoms of ancient inlets and deep bays, such as at the present day intersect every part of Tierra del Fuego and the western coast. Chile must formerly have resembled the latter country in the configuration of its land and water. The resemblance was occasionally shown strikingly when a level fog-bank covered, as with a mantle, all the lower parts of the country: the white vapour curling into the ravines, beautifully represented little coves and bays; and here and there a solitary hillock peeping up, showed that it had formerly stood there as an islet. The contrast of these flat valleys and basins with the irregular mountains, gave the scenery a character which to me was new and very interesting.

From the natural slope to seaward of these plains, they are very easily irrigated, and in consequence singularly fertile. Without this process the land would produce scarcely anything, for during the whole summer the sky is cloudless. The mountains and hills are dotted over with bushes and low trees, and excepting these the vegetation is very scanty. Each landowner in the valley possesses a certain portion of hill-country, where his half-wild cattle, in considerable numbers, manage to find sufficient pasture. Once every year there is a grand 'rodeo,' when all the cattle are driven down, counted, and marked, and a certain number separated to be fattened in the irrigated fields. Wheat is extensively cultivated, and a good deal of Indian corn: a kind of bean is, however, the staple article of food for the common labourers. The orchards produce an overflowing abundance of peaches, figs, and grapes. With all these advantages, the inhabitants of the country ought to be much more prosperous than they are.

16th. – The major-domo of the Hacienda was good enough to give me a guide and fresh horses; and in the morning we set out to ascend the Campana, or Bell Mountain, which is 6400 feet high. The paths were very bad, but both the geology and scenery amply repaid the trouble. We reached, by the evening, a spring called the Agua del Guanaco, which is situated at a great height. This must be an old name, for it is very many years since a guanaco drank its waters. During the ascent I noticed that nothing but bushes grew on the northern slope, whilst on the southern slope there was a bamboo about fifteen feet high. In a few places there were palms, and I was surprised to see one at an elevation of at least 4500 feet. These palms are, for their family, ugly trees. Their stem is very large, and of a curious form, being thicker in the middle than at the base or top. They are excessively numerous in some parts of Chile, and valuable on account of a sort of treacle made from the sap. On one estate near Petorca they tried to count them, but failed, after having numbered several hundred thousand. Every year in the early spring, in August, very many are cut down, and when the trunk is lying on the ground, the crown of leaves is lopped off. The sap then immediately begins to flow from the upper end, and continues so doing for some months: it is, however, necessary that a thin slice should be shaved off from that end every morning, so as to expose a fresh surface. A good tree will give ninety gallons, and all this must have been contained in the vessels of the apparently dry trunk. It is said that the sap flows much more quickly on those days when the sun is powerful; and likewise, that it is absolutely necessary to take care, in cutting down the tree, that it should fall with its head upwards on the side of the hill; for if it falls down the slope, scarcely any sap will flow; although in that case one would have thought that the action would have been aided, instead of checked, by the force of gravity. The sap is concentrated by boiling, and is then called treacle, which it very much resembles in taste.

We unsaddled our horses near the spring, and prepared to pass the night. The evening was fine, and the atmosphere so clear, that the masts of the vessels at anchor in the bay of Valparaiso, although no less than twenty-six geographical miles distant, could be distinguished clearly as little black streaks. A ship doubling the point under sail, appeared as a bright white speck. Anson expresses much surprise, in his voyage, at the distance at which his vessels were discovered from the coast; but he did not sufficiently allow for the height of the land, and the great transparency of the air.

The setting of the sun was glorious; the valleys being black, whilst the snowy peaks of the Andes yet retained a ruby tint. When it was dark, we made a fire beneath a little arbour of bamboos, fried our charqui (or dried slips of beef), took our maté, and were quite comfortable. There is an inexpressible charm in thus living in the open air. The evening was calm and still; – the shrill noise of the mountain bizcacha, and the faint cry of a goatsucker, were occasionally to be heard. Besides these, few birds, or even insects, frequent these dry, parched mountains.

August 17*th.* – In the morning we climbed up the rough mass of greenstone which crowns the summit. This rock, as frequently happens, was much shattered and broken into huge angular fragments. I observed, however, one remarkable circumstance, namely, that many of the surfaces presented every degree of freshness – some appearing as if broken the day before, whilst on others lichens had either just become, or had long grown, attached. I so fully believed that this was owing to the frequent earthquakes, that I felt inclined to hurry from below each loose pile. As one might very easily be deceived in a fact of this kind, I doubted its accuracy, until ascending Mount Wellington, in Van Diemen's Land, where earthquakes do not occur; and there I saw the summit of the mountain similarly composed and similarly shattered, but all the blocks appeared as if they had been hurled into their present position thousands of years ago.

We spent the day on the summit, and I never enjoyed one more thoroughly. Chile, bounded by the Andes and the Pacific, was seen as in a map. The pleasure from the scenery, in itself beautiful, was heightened by the many reflections which arose from the mere view of the Campana range with its lesser parallel ones, and of the broad valley of Quillota directly intersecting them. Who can avoid wondering at the force which has upheaved these mountains, and even more so at the countless ages which it must have required, to have broken through, removed, and levelled whole masses of them? It is well in this case, to call to mind the vast shingle and sedimentary beds of Patagonia, which, if heaped on the Cordillera, would increase its height by so many thousand feet. When in that country, I wondered how any mountain-chain could have supplied such masses, and not have been utterly obliterated. We must not now reverse the wonder, and doubt whether all-powerful time can grind down mountains – even the gigantic Cordillera – into gravel and mud.

The appearance of the Andes was different from that which I had expected. The lower line of the snow was of course horizontal, and to this line the even summits of the range seemed quite parallel. Only at long intervals, a group of points or a single cone, showed where a volcano had existed, or does now exist. Hence the range resembled a great solid wall, surmounted here and there by a tower, and making a most perfect barrier to the country.

Almost every part of the hill had been drilled by attempts to open gold-mines: the rage for mining has left scarcely a spot in Chile unexamined. I spent the evening as before, talking round the fire with my two companions. The Guasos of Chile, who correspond to the Gauchos of the Pampas, are, however, a very different set of beings. Chile is the more civilized of the two countries, and the inhabitants, in consequence, have lost much individual character. Gradations in rank are much more strongly marked: the Guaso does not by any means consider every man his equal; and I was quite surprised to find that my companions did not like to eat at the same time with myself. This feeling of inequality is a necessary consequence of the existence of an aristocracy of wealth. It is said that some few of the greater landowners possess from five to ten thousand pounds sterling per annum: an inequality of riches which I believe is not met with, in any of the cattle-breeding countries eastward of the Andes. A traveller does not here meet that unbounded hospitality which refuses all payment, but yet is so kindly offered that no scruples can be raised in accepting it. Almost every house in Chile will receive you for the night, but a trifle is expected to be given in the morning; even a rich man will accept two or three shillings. The Gaucho, although he may be a cut-throat, is a gentleman; the Guaso is in few respects better, but at the same time a vulgar, ordinary fellow. The two men, although employed much in the same manner, are different in their habits and attire; and the peculiarities of each are universal in their respective countries. The Gaucho seems part of his horse, and scorns to exert himself excepting when on its back; the Guaso may be hired to work as a labourer in the fields. The former lives entirely on animal food; the latter almost wholly on vegetable. We do not here see the white boots, the broad drawers, and scarlet chilipa; the picturesque costume of the Pampas. Here, common trousers are protected by black and green worsted leggings. The poncho, however, is common to both. The chief pride of the Guaso lies in his spurs; which are absurdly large. I measured one which was six inches in the *diameter* of the rowel, and the rowel itself contained upwards of thirty points. The stirrups are on the same scale, each consisting of a square, carved block of wood, hollowed out, yet weighing three or four pounds. The Guaso is perhaps more expert with the lazo than the Gaucho; but, from the nature of the country, he does not know the use of the bolas.

August 18*th*. – We descended the mountain, and passed some beautiful little spots, with rivulets and fine trees. Having slept at the same hacienda as before, we rode during the two succeeding days up the valley, and passed through Quillota, which is more like a collection of nursery-gardens than a town. The orchards were beautiful, presenting one mass of peach-blossoms. I saw, also, in one or two places the date-palm: it is a most stately tree; and I should think a group of them in their native Asiatic or African deserts must be superb. We passed likewise San Felipe, a pretty straggling town like Quillota. The valley in this part expands into one of those great bays or plains, reaching to the foot of the Cordillera, which have been mentioned as forming so curious a part of the scenery of Chile. In the evening we reached the mines of Jajuel, situated in a ravine at the flank of the great chain. I stayed here five days. My host, the superintendent of the mine, was a shrewd but rather ignorant Cornish miner. He had married a Spanish woman, and did not mean to return home; but his admiration for the mines of Cornwall remained unbounded. Amongst many other questions, he asked me, 'Now that George Rex is dead, how many more of the family of Rexes are yet alive?' This Rex certainly must be a relation of the great author Finis, who wrote all books!

These mines are of copper, and the ore is all shipped to Swansea, to be smelted. Hence the mines have an aspect singularly quiet, as compared to those in England: here no smoke, furnaces, or great steam-engines, disturb the solitude of the surrounding mountains.

The Chilian government, or rather the old Spanish law, encourages by every method the searching for mines. The discoverer may work a mine on any ground, by paying five shillings; and before paying this he may try, even in the garden of another man, for twenty days.

It is now well known that the Chilian method of mining is the cheapest. My host says that the two principal improvements introduced by foreigners have been, first, reducing by previous roasting the copper pyrites – which, being the common ore in Cornwall, the English miners were astounded on their arrival to find thrown away as useless: secondly, stamping and washing the scoriæ from the old furnaces – by which process particles of metal are recovered in abundance. I have actually seen mules carrying to the coast, for transportation to England, a cargo of such cinders. But the first case is much the most curious. The Chilian miners were so convinced that copper pyrites contained not a particle of copper, that they laughed at the Englishmen for their ignorance, who laughed in turn, and bought their richest veins for a few dollars. It is very odd that, in a country where mining had been extensively carried on for many years, so simple a process as gently roasting the ore to expel the sulphur previous to smelting it, had never been discovered. A few improvements have likewise been introduced in some of the simple machinery; but even to the present day, water is removed from some mines by men carrying it up the shaft in leathern bags!

The labouring men work very hard. They have little time allowed for their meals, and during summer and winter they begin when it is light, and leave off at dark. They are paid one pound sterling a month, and their food is given them: this for breakfast consists of sixteen figs and two small loaves of bread; for dinner, boiled beans; for supper, broken roasted wheat grain. They scarcely ever taste meat; as, with the twelve pounds per annum, they have to clothe themselves, and support their families. The miners who work in the mine itself have twenty-five shillings per month, and are allowed a little charqui. But these men come down from their bleak habitations only once in every fortnight or three weeks.

During my stay here I thoroughly enjoyed scrambling about these huge mountains. The geology, as might have been expected, was very interesting. The shattered and baked rocks, traversed by innumerable dykes of greenstone, showed what commotions had formerly taken place. The scenery was much the same as that near the Bell of Quillota – dry barren mountains, dotted at intervals by bushes with a scanty foliage. The cactuses, or rather opuntias, were here very numerous. I measured one of a spherical figure, which, including the spines, was six feet and four inches in circumference. The height of the common cylindrical, branching kind, is from twelve to fifteen feet, and the girth (with spines) of the branches between three and four feet.

A heavy fall of snow on the mountains prevented me, during the last two days, from making some interesting excursions. I attempted to reach a lake which the inhabitants, from some unaccountable reason, believe to be an arm of the sea. During a very dry season, it was proposed to attempt cutting a channel from it for the sake of the water, but the padre, after a consultation, declared it was too dangerous, as all Chile would be inundated, if, as generally supposed, the lake was connected with the Pacific. We ascended to a great height, but becoming involved in the snow-drifts failed in reaching this wonderful lake, and had some difficulty in returning. I thought we should have lost our horses; for there was no means of guessing how deep the drifts were, and the animals, when led, could only move by jumping. The black sky showed that a fresh snow-storm was gathering, and we therefore were not a little glad when we escaped. By the time we reached the base the storm commenced, and it was lucky for us that this did not happen three hours earlier in the day.

August 26th. – We left Jajuel and again crossed the basin of S. Felipe. The day was truly Chilian: glaringly bright, and the atmosphere quite clear. The thick and uniform covering of newly-fallen snow rendered the view of the volcano of Aconcagua and the main chain quite glorious. We were now on the road to Santiago, the capital of Chile. We crossed the Cerro del Talguen, and slept at a little rancho. The host, talking about the state of Chile as compared to other countries, was very humble: 'Some see with two eyes and some with one, but for my part I do not think that Chile sees with any.'

August 27th. – After crossing many low hills we descended into the small land-locked plain of Guitron. In the basins, such as this one, which are elevated from one thousand to two thousand feet above the sea, two species of acacia, which are stunted in their forms, and stand wide apart from each other, grow in large numbers. These trees are never found near the sea-coast; and this gives another characteristic feature to the scenery of these basins. We crossed a low ridge which separates Guitron from the great plain on which Santiago stands. The view was here pre-eminently striking: the dead level surface, covered in parts by woods of acacia, and with the city in the distance, abutting horizontally against the base of the Andes, whose snowy peaks were bright with the evening sun. At the first glance of this view, it was quite evident that the plain represented the extent of a former inland sea. As soon as we gained the level road we pushed our horses into a gallop, and reached the city before it was dark.

I stayed a week in Santiago and enjoyed myself very much. In the morning I rode to various places on the plain, and in the evening dined with several of the English merchants, whose hospitality at this place is well known. A never-failing source of pleasure was to ascend the little hillock of rock (St. Lucia) which projects in the middle of the city. The scenery certainly is most striking, and, as I have said, very peculiar. I am informed that this same character is common to the cities on the great Mexican platform. Of the town I have nothing to say in detail: it is not so fine or so large as Buenos Ayres, but is built after the same model. I arrived here by a circuit to the north; so I resolved to return to Valparaiso by a rather longer excursion to the south of the direct road.

September 5th. – By the middle of the day we arrived at one of the suspension bridges made of hide, which crosses the Maypu, a large turbulent river a few leagues southward of Santiago. These bridges are very poor affairs. The road, following the curvature of the suspending ropes, is made of bundles of sticks placed close together. It was full of holes, and oscillated rather fearfully, even with the weight of a man leading his horse. In the evening we reached a comfortable farm-house, where there were several very pretty señoritas. They were much horrified at my having entered one of their churches out of mere curiosity. They asked me, 'Why do you not become a Christian – for our religion is certain?' I assured them I was a sort of Christian; but they would not hear of it – appealing to my own words, 'Do not your padres, your very bishops, marry?' The absurdity of a bishop having a wife particularly struck them: they scarcely knew whether to be most amused or horror-struck at such an enormity.

6th. – We proceeded due south, and slept at Rancagua. The road passed over the level but narrow plain, bounded on one side by lofty hills, and on the other by the Cordillera. The next day we turned up the valley of the Rio Cachapual, in which the hot-baths of Cauquenes, long celebrated for their medicinal properties, are situated. The suspension bridges, in the less frequented parts, are generally taken down during the winter when the rivers are low. Such was the case in this valley, and we were therefore obliged to cross the stream on horseback. This is rather disagreeable, for the foaming water, though not deep, rushes so quickly over the bed of large rounded stones, that one's head becomes quite confused, and it is difficult even to perceive whether the horse is moving onward or standing still. In summer, when the snow melts, the torrents are quite impassable; their strength and fury is then extremely great, as might be plainly seen by the marks which they had left. We reached the baths in the evening, and stayed there five days, being confined the two last by heavy rain. The buildings consist of a square of miserable little hovels, each with a single table and bench. They are situated in a narrow deep valley just without the central Cordillera. It is a quiet, solitary spot, with a good deal of wild beauty.

The mineral springs of Cauquenes burst forth on a line of dislocation, crossing a mass of stratified rock, the whole of which betrays the action of heat. A considerable quantity of gas is continually escaping from the same orifices with the water. Though the springs are only a few yards apart, they have very different temperatures; and this appears to be the result of an unequal mixture of cold water: for those with the lowest temperature have scarcely any mineral taste. After the great earthquake of 1822 the springs ceased, and the water did not return for nearly a year. They were also much affected by the earthquake of 1835; the temperature being suddenly changed from 118° to 92°.[1] It seems probable that mineral waters rising deep from the bowels of the earth, would always be more deranged by subterranean disturbances than those nearer the surface. The man who had charge of the baths, assured me that in summer the water is hotter and more plentiful than in winter. The former circumstance I should have expected, from the less mixture, during the dry season, of cold water; but the latter statement appears very strange and contradictory. The periodical increase during the summer, when rain never falls, can, I think, only be accounted for by the melting of the snow: yet the mountains which are covered by snow during that season, are three or four leagues distant from the springs. I have no reason to doubt the accuracy of my informer, who, having lived on the spot for several years, ought to be well acquainted with the circumstance, – which, if true, certainly is very curious: for, we must

1 Caldcleugh, in Philosoph. Transact. for 1836.

suppose that the snow-water, being conducted through porous strata to the regions of heat, is again thrown up to the surface by the line of dislocated and injected rocks at Cauquenes; and the regularity of the phenomenon would seem to indicate, that in this district heated rock occurred at a depth not very great.

One day I rode up the valley to the farthest inhabited spot. Shortly above that point, the Cachapual divides into two deep tremendous ravines, which penetrate directly into the great range. I scrambled up a peaked mountain, probably more than six thousand feet high. Here, as indeed everywhere else, scenes of the highest interest presented themselves. It was by one of these ravines, that Pincheira entered Chile and ravaged the neighbouring country. This is the same man whose attack on an estancia at the Rio Negro I have described. He was a renegade half-cast Spaniard, who collected a great body of Indians together and established himself by a stream in the Pampas, which place none of the forces sent after him could ever discover. From this point he used to sally forth, and crossing the Cordillera by passes hitherto unattempted, he ravaged the farm-houses and drove the cattle to his secret rendezvous. Pincheira was a capital horseman, and he made all around him equally good, for he invariably shot any one who hesitated to follow him. It was against this man, and other wandering Indian tribes, that Rosas waged the war of extermination.

September 13*th.* – We left the baths of Cauquenes, and rejoining the main road slept at the Rio Claro. From this place we rode to the town of S. Fernando. Before arriving there, the last land-locked basin had expanded into a great plain, which extended so far to the south, that the snowy summits of the more distant Andes were seen as if above the horizon of the sea. S. Fernando is forty leagues from Santiago; and it was my farthest point southward; for we here turned at right angles towards the coast. We slept at the gold-mines of Yaquil, which are worked by Mr. Nixon, an American gentleman, to whose kindness I was much indebted during the four days I stayed at his house. The next morning we rode to the mines, which are situated at the distance of some leagues, near the summit of a lofty hill. On the way we had a glimpse of the lake Tagua-tagua, celebrated for its floating islands, which have been described by

Mr. Gay.[1] They are composed of the stalks of various dead plants intertwined together, and on the surface of which other living ones take root. Their form is generally circular, and their thickness from four to six feet, of which the greater part is immersed in the water. As the wind blows, they pass from one side of the lake to the other, and often carry cattle and horses as passengers.

When we arrived at the mine, I was struck by the pale appearance of many of the men, and inquired from Mr. Nixon respecting their condition. The mine is 450 feet deep, and each man brings up about 200 pounds weight of stone. With this load they have to climb up the alternate notches cut in the trunks of trees, placed in a zigzag line up the shaft. Even beardless young men, eighteen and twenty years old, with little muscular development of their bodies (they are quite naked excepting drawers) ascend with this great load from nearly the same depth. A strong man, who is not accustomed to this labour, perspires most profusely, with merely carrying up his own body. With this very severe labour, they live entirely on boiled beans and bread. They would prefer having bread alone; but their masters, finding that they cannot work so hard upon this, treat them like horses, and make them eat the beans. Their pay is here rather more than at the mines of Jajuel, being from 24 to 28 shillings per month. They leave the mine only once in three weeks; when they stay with their families for two days. One of the rules in this mine sounds very harsh, but answers pretty well for the master. The only method of stealing gold is to secrete pieces of the ore, and take them out as occasion may offer. Whenever the major-domo finds a lump thus hidden, its full value is stopped out of the wages of all the men; who thus, without they all combine, are obliged to keep watch over each other.

When the ore is brought to the mill, it is ground into an impalpable powder; the process of washing removes all the lighter particles, and amalgamation finally secures the gold-dust. The washing, when described, sounds a very simple process; but it is beautiful to see how the exact adaptation of the current of water to the specific gravity of the gold, so easily separates the powdered matrix from the metal.

1 Annales des Sciences Naturelles, March, 1833. M. Gay, a zealous and able naturalist, was then occupied in studying every branch of natural history throughout the kingdom of Chile.

The mud which passes from the mills is collected into pools, where it subsides, and every now and then is cleared out and thrown into a common heap. A great deal of chemical action then commences, salts of various kinds effloresce on the surface, and the mass becomes hard. After having been left for a year or two, and then rewashed, it yields gold; and this process may be repeated even six or seven times; but the gold each time becomes less in quantity, and the intervals required (as the inhabitants say, to generate the metal) are longer. There can be no doubt that the chemical action, already mentioned, each time liberates fresh gold from some combination. The discovery of a method to effect this before the first grinding, would without doubt raise the value of gold-ores many-fold. It is curious to find how the minute particles of gold, being scattered about and not corroding, at last accumulate in some quantity. A short time since a few miners, being out of work, obtained permission to scrape the ground round the house and mill: they washed the earth thus got together, and so procured thirty dollars' worth of gold. This is an exact counterpart of what takes place in nature. Mountains suffer degradation and wear away, and with them the metallic veins which they contain. The hardest rock is worn into impalpable mud, the ordinary metals oxidate, and both are removed; but gold, platina, and a few others are nearly indestructible, and from their weight, sinking to the bottom, are left behind. After whole mountains have passed through this grinding-mill, and have been washed by the hand of nature, the residue becomes metalliferous, and man finds it worth his while to complete the task of separation.

Bad as the above treatment of the miners appears, it is gladly accepted of by them; for the condition of the labouring agriculturists is much worse. Their wages are lower, and they live almost exclusively on beans. This poverty must be chiefly owing to the feudal-like system on which the land is tilled: the landowner gives a small plot of ground to the labourer, for building on and cultivating, and in return has his services (or those of a proxy) for every day of his life, without any wages. Until a father has a grown-up son, who can by his labour pay the rent, there is no one, except on occasional days, to take care of his own patch of ground. Hence extreme poverty is very common among the labouring classes in this country.

There are some old Indian ruins in this neighbourhood, and I was shown one of the perforated stones, which Molina mentions as being found in many places in considerable numbers. They are of a circular flattened form, from five to six inches in diameter, with a hole passing quite through the centre. It has generally been supposed that they were used as heads to clubs, although their form does not appear at all well adapted for that purpose. Burchell[1] states that some of the tribes in Southern Africa dig up roots, by the aid of a stick pointed at one end, the force and weight of which is increased by a round stone with a hole in it, into which the other end is firmly wedged. It appears probable, that the Indians of Chile formerly used some such rude agricultural instrument.

One day, a German collector in natural history, of the name of Renous, called, and nearly at the same time an old Spanish lawyer. I was amused at being told the conversation which took place between them. Renous speaks Spanish so well, that the old lawyer mistook him for a Chilian. Renous, alluding to me, asked him what he thought of the King of England sending out a collector to their country, to pick up lizards and beetles, and to break stones? The old gentleman thought seriously for some time, and then said, 'It is not well, – *hay un gato encerrado aqui* (there is a cat shut up here). No man is so rich as to send out people to pick up such rubbish. I do not like it: if one of us were to go and do such things in England, do not you think the King of England would very soon send us out of his country?' And this old gentleman, from his profession, belongs to the better informed and more intelligent classes! Renous himself, two or three years before, left in a house at S. Fernando some caterpillars, under charge of a girl to feed, that they might turn into butterflies. This was rumoured through the town, and at last the Padres and Governor consulted together, and agreed it must be some heresy. Accordingly, when Renous returned, he was arrested.

September 19*th*. – We left Yaquil, and followed the flat valley, formed like that of Quillota, in which the Rio Tinderidica flows. Even at these few miles south of Santiago the climate is much damper; in consequence there were fine tracts of pasturage, which were not

1 Burchell's Travels, vol. ii. p. 45.

irrigated. (20th.) We followed this valley till it expanded into a great plain, which reaches from the sea to the mountains west of Rancagua. We shortly lost all trees and even bushes; so that the inhabitants are nearly as badly off for firewood as those in the Pampas. Never having heard of these plains, I was much surprised at meeting with such scenery in Chile. The plains belong to more than one series of different elevations, and they are traversed by broad flat-bottomed valleys; both of which circumstances, as in Patagonia, bespeak the action of the sea on gently rising land. In the steep cliffs bordering these valleys, there are some large caves, which no doubt were originally formed by the waves: one of these is celebrated under the name of Cueva del Obispo; having formerly been consecrated. During the day I felt very unwell, and from that time till the end of October did not recover.

September 22nd. – We continued to pass over green plains without a tree. The next day we arrived at a house near Navedad, on the sea-coast, where a rich Haciendero gave us lodgings. I stayed here the two ensuing days, and although very unwell, managed to collect from the tertiary formation some marine shells.

24th. – Our course was now directed towards Valparaiso, which with great difficulty I reached on the 27th, and was there confined to my bed till the end of October. During this time I was an inmate in Mr. Corfield's house, whose kindness to me I do not know how to express.

I will here add a few observations on some of the animals and birds of Chile. The Puma, or South American Lion, is not uncommon. This animal has a wide geographical range; being found from the equatorial forests, throughout the deserts of Patagonia, as far south as the damp and cold latitudes (53° to 54°) of Tierra del Fuego. I have seen its footsteps in the Cordillera of central Chile, at an elevation of at least 10,000 feet. In La Plata the puma preys chiefly on deer, ostriches, bizcacha, and other small quadrupeds; it there seldom attacks cattle or horses, and most rarely man. In Chile, however, it destroys many young horses and cattle, owing probably to the scarcity of other quadrupeds: I heard, likewise, of two men and a woman who had been thus killed. It is asserted that the puma always kills its prey by springing on the shoulders, and then drawing back the head with one of its paws, until the vertebræ break: I have seen in Patagonia, the skeletons of guanacos, with their necks thus dislocated.

The puma, after eating its fill, covers the carcass with many large bushes, and lies down to watch it. This habit is often the cause of its being discovered; for the condors wheeling in the air, every now and then descend to partake of the feast, and being angrily driven away, rise all together on the wing. The Chileno Guaso then knows there is a lion watching his prey – the word is given – and men and dogs hurry to the chace. Sir F. Head says that a Gaucho in the Pampas, upon merely seeing some condors wheeling in the air, cried 'A lion!' I could never myself meet with any one who pretended to such powers of discrimination. It is asserted, that if a puma has once been betrayed by thus watching the carcass, and has then been hunted, it never resumes this habit; but that having gorged itself, it wanders far away. The puma is easily killed. In an open country, it is first entangled with the bolas, then lazoed, and dragged along the ground till rendered insensible. At Tandeel (south of the Plata) I was told that within three months one hundred were thus destroyed. In Chile they are generally driven up bushes or trees, and are then either shot, or baited to death by dogs. The dogs employed in this chace belong to a particular breed, called Leoneros: they are weak, slight animals, like long-legged terriers, but are born with a particular instinct for this sport. The puma is described as being very crafty: when pursued, it often returns on its former track, and then suddenly making a spring on one side, waits there till the dogs have passed by. It is a very silent animal, uttering no cry even when wounded, and only rarely during the breeding season.

Of birds, two species of the genus Pteroptochos (megapodius and albicollis of Kittlitz) are perhaps the most conspicuous. The former, called by the Chilenos 'el Turco,' is as large as a fieldfare, to which bird it has some alliance; but its legs are much longer, tail shorter, and beak stronger: its colour is a reddish brown. The Turco is not uncommon. It lives on the ground, sheltered among the thickets which are scattered over the dry and sterile hills. With its tail erect, and stilt-like legs, it may be seen every now and then popping from one bush to another with uncommon quickness. It really requires little imagination to believe that the bird is ashamed of itself, and is aware of its most ridiculous figure. On first seeing it, one is tempted to exclaim, 'A vilely stuffed specimen has escaped from some museum, and has come to life again!' It cannot be made to take flight without the greatest trouble, nor does it run, but only hops. The various loud cries which it utters when concealed amongst the bushes, are as strange as its appearance. It is said to build its nest in a deep hole beneath the ground. I dissected several specimens: the gizzard, which was very muscular, contained beetles, vegetable fibres, and pebbles. From this character, from the length of its legs, scratching feet, membranous covering to the nostrils, short and arched wings, this bird seems in a certain degree to connect the thrushes with the gallinaceous order.

The second species (or P. albicollis) is allied to the first in its general form. It is called Tapacolo, or 'cover your posterior;' and well does the shameless little bird deserve its name; for it carries its tail more than erect, that is, inclined backwards towards its head. It is very common, and frequents the bottoms of hedge-rows, and the bushes scattered over the barren hills, where scarcely another bird can exist. In its general manner of feeding, of quickly hopping out of the thickets and back again, in its desire of concealment, unwillingness to take flight, and nidification, it bears a close resemblance to the Turco; but its appearance is not quite so ridiculous. The Tapacolo is very crafty: when frightened by any person, it will remain motionless at the bottom of a bush, and will then, after a little while, try with much address to crawl away on the opposite side. It is also an active bird, and continually making a noise: these noises are various and strangely odd; some are like the cooing of doves, others like the bubbling of water, and many defy all similes. The country people say it changes its cry five times in the year – according to some change of season, I suppose.[1]

1 It is a remarkable fact, that Molina, though describing in detail all the birds and animals of Chile, never once mentions this genus, the species of which are so common, and so remarkable in their habits. Was he at a loss how to classify them, and did he consequently think that silence was the more prudent course? It is one more instance of the frequency of omissions by authors, on those very subjects where it might have been least expected.

Two species of humming-birds are common; Trochilus forficatus is found over a space of 2500 miles on the west coast, from the hot dry country of Lima, to the forests of Tierra del Fuego – where it may be seen flitting about in snow-storms. In the wooded island of Chiloe, which has an extremely humid climate, this little bird, skipping from side to side amidst the dripping foliage, is perhaps more abundant than almost any other kind. I opened the stomachs of several specimens, shot in different parts of the continent, and in all, remains of insects were as numerous as in the stomach of a creeper. When this species migrates in the summer southward, it is replaced by the arrival of another species coming from the north. This second kind (Trochilus gigas) is a very large bird for the delicate family to which it belongs: when on the wing its appearance is singular. Like others of the genus, it moves from place to place with a rapidity which may be compared to that of Syrphus amongst flies, and Sphinx among moths; but whilst hovering over a flower, it flaps its wings with a very slow and powerful movement, totally different from that vibratory one common to most of the species, which produces the humming noise. I never saw any other bird, where the force of its wings appeared (as in a butterfly) so powerful in proportion to the weight of its body. When hovering by a flower, its tail is constantly expanded and shut like a fan, the body being kept in a nearly vertical position. This action appears to steady and support the bird, between the slow movements of its wings. Although flying from flower to flower in search of food, its stomach generally contained abundant remains of insects, which I suspect are much more the object of its search than honey. The note of this species, like that of nearly the whole family, is extremely shrill.

CHAPTER XIII

CHILOE AND CHONOS ISLANDS

Chiloe – General Aspect – Boat Excursion – Native Indians – Castro – Tame Fox – Ascend San Pedro – Chonos Archipelago – Peninsula of Tres Montes – Granitic Range – Boat-wrecked Sailors – Low's Harbour – Wild Potato – Formation of Peat – Myopotamus, Otter and Mice – Cheucau and Barking-bird – Opetiorhynchus – Singular Character of Ornithology – Petrels.

NOVEMBER 10*th*. – The *Beagle* sailed from Valparaiso to the south, for the purpose of surveying the southern part of Chile, the island of Chiloe, and the broken land called the Chonos Archipelago, as far south as the Peninsula of Tres Montes. On the 21st we anchored in the bay of S. Carlos, the capital of Chiloe.

This island is about ninety miles long, with a breadth of rather less than thirty. The land is hilly, but not mountainous, and is covered by one great forest, except where a few green patches have been cleared round the thatched cottages. From a distance the view somewhat resembles that of Tierra del Fuego; but the woods, when seen nearer, are incomparably more beautiful. Many kinds of fine ever-green trees, and plants with a tropical character, here take the place of the gloomy beech of the southern shores. In winter the climate is detestable, and in summer it is only a little better. I should think there are few parts of the world, within the temperate regions, where so much rain falls. The winds are very boisterous, and the sky almost always clouded: to have a week of fine weather is something wonderful. It is even difficult to get a single glimpse of the Cordillera: during our first visit, once only the volcano of Osorno stood out in bold relief, and that was before sunrise; it was curious to watch, as the sun rose, the outline gradually fading away in the glare of the eastern sky.

The inhabitants, from their complexion and low stature, appear to have three-fourths of Indian blood in their veins. They are an humble, quiet, industrious set of men. Although the fertile soil, resulting from the decomposition of the volcanic rocks, supports a rank vegetation, yet the climate is not favourable to any production which requires much sunshine to ripen it. There is very little pasture for the larger quadrupeds; and in consequence, the staple articles of food are pigs, potatoes, and fish. The people all dress in strong woollen garments, which each family makes for itself, and dyes with indigo of a dark blue colour. The arts, however, are in the rudest state; – as may be seen in their strange fashion of ploughing, their method of spinning, grinding corn, and in the construction of their boats. The forests are so impenetrable, that the land is nowhere cultivated except near the coast and on the adjoining islets. Even where paths exist, they are scarcely passable from the soft and swampy state of the soil. The inhabitants, like those of Tierra del Fuego, move about chiefly on the beach or in boats. Although with plenty to eat, the people are very poor: there is no demand for labour, and consequently the lower orders cannot scrape together money sufficient to purchase even the smallest luxuries. There is also a great deficiency of a circulating medium. I have seen a man bringing on his back a bag of charcoal, with which to buy some trifle, and another carrying a plank to exchange for a bottle of wine. Hence every tradesman must also be a merchant, and again sell the goods which he takes in exchange.

November 24th. – The yawl and whale-boat were sent under the command of Mr. (now Captain) Sulivan, to survey the eastern or inland coast of Chiloe; and with orders to meet the *Beagle* at the southern extremity of the island; to which point she would proceed by the outside, so as thus to circumnavigate the whole. I accompanied this expedition, but instead of going in the boats the first day, I hired horses to take me to Chacao, at the northern extremity of the island. The road followed the coast; every now and then crossing promontories covered by fine forests. In these shaded paths it is absolutely necessary that the whole road should be made of logs of wood, which are squared and placed by the side of each other. From the rays of the sun never penetrating the evergreen foliage, the ground is so damp and soft, that except by this means neither man nor horse would be able to pass along. I arrived at the village of Chacao, shortly after the tents belonging to the boats were pitched for the night.

The land in this neighbourhood has been extensively cleared, and there were many quiet and most picturesque nooks in the forest. Chacao was formerly the principal port in the island; but many vessels having been lost, owing to the dangerous currents and rocks in the straits, the Spanish government burnt the church, and thus arbitrarily compelled the greater number of inhabitants to migrate to S. Carlos. We had not long bivouacked, before the barefooted son of the governor came down to reconnoitre us. Seeing the English flag hoisted at the yawl's mast-head, he asked, with the utmost indifference, whether it was always to fly at Chacao. In several places, the inhabitants were much astonished at the appearance of men-of-war's boats, and hoped and believed it was the forerunner of a Spanish fleet, coming to recover the island from the patriot government of Chile. All the men in power, however, had been informed of our intended visit, and were exceedingly civil. While we were eating our supper, the governor paid us a visit. He had been a lieutenant-colonel in the Spanish service, but now was miserably poor. He gave us two sheep, and accepted in return two cotton handkerchiefs, some brass trinkets, and a little tobacco.

25th. – Torrents of rain: we managed, however, to run down the coast as far as Huapi-lenou. The whole of this eastern side of Chiloe has one aspect: it is a plain, broken by valleys and divided into little islands, and the whole thickly covered with one impervious blackish-green forest. On the margins there are some cleared spaces, surrounding the high-roofed cottages.

26th. – The day rose splendidly clear. The volcano of Osorno was spouting out volumes of smoke. This most beautiful mountain, formed like a perfect cone, and white with snow, stands out in front of the Cordillera. Another great volcano, with a saddle-shaped summit, also emitted from its immense crater little jets of steam. Subsequently we saw the lofty-peaked Corcovado – well deserving the name of 'el famoso Corcovado.' Thus we beheld, from one point of view, three great active volcanoes, each about seven thousand feet high. In addition to this, far to the south, there were other lofty cones covered with snow, which, although not known to be active, must be in their origin volcanic. The line of the Andes is not, in this neighbourhood, nearly so elevated as in Chile; neither does it appear to form so perfect a barrier between the regions of the earth. This great range, although running in a straight north and south line, owing to an optical deception, always appeared more or less curved; for the lines drawn from each peak to the beholder's eye, necessarily converged like the radii of a semi-circle, and as it was not possible (owing to the clearness of the atmosphere and the absence of all intermediate objects) to judge how far distant the farthest peaks were off, they appeared to stand in a flattish semicircle.

Landing at midday, we saw a family of pure Indian extraction. The father was singularly like York Minster; and some of the younger boys, with their ruddy complexions, might have been mistaken for Pampas Indians. Everything I have seen, convinces me of the close connexion of the different American tribes, who nevertheless speak distinct languages. This party could muster but little Spanish, and talked to each other in their own tongue. It is a pleasant thing to see the aborigines advanced to the same degree of civilization, however low that may be, which their white conquerors have attained. More to the south we saw many pure Indians: indeed, all the inhabitants of some of the islets retain their Indian surnames. In the census of 1832, there were in Chiloe and its dependencies forty-two thousand souls: the greater number of these appear to be of mixed blood. Eleven thousand retain their Indian surnames, but it is probable that not nearly all of these are of a pure breed. Their manner of life is the same with that of the other poor inhabitants, and they are all Christians; but it is said that they yet retain some strange superstitious ceremonies, and that they pretend to hold communication with the devil in certain caves. Formerly, every one convicted of this offence was sent to the Inquisition at Lima. Many of the inhabitants who are not included in the eleven thousand with Indian surnames, cannot be distinguished by their appearance from Indians. Gomez, the governor of Lemuy, is descended from noblemen of Spain on both sides; but by constant intermarriages with the natives the present man is an Indian. On the other hand, the governor of Quinchao boasts much of his purely kept Spanish blood.

We reached at night a beautiful little cove, north of the island of Caucahue. The people here complained of want of land. This is partly owing to their own negligence in not clearing the woods, and partly to restrictions by the government, which makes it necessary before buying ever so small a piece, to pay two shillings to the surveyor, for measuring each quadra (150 yards square), together with whatever price he fixes for the value of the land. After his valuation, the land must be put up three times to auction, and if no one bids more, the purchaser can have it at that rate. All these exactions must be a serious check to clearing the ground, where the inhabitants are so extremely poor. In most countries, forests are removed without much difficulty by the aid of fire; but in Chiloe, from the damp nature of the climate, and the sort of trees, it is necessary first to cut them down. This is a heavy drawback to the prosperity of Chiloe. In the time of the Spaniards the Indians could not hold land; and a family, after having cleared a piece of ground, might be driven away, and the property seized by the government. The Chilian authorities are now performing an act of justice by making retribution to these poor Indians, giving to each man, according to his grade of life, a certain portion of land. The value of uncleared ground is very little. The government gave Mr. Douglas (the present surveyor, who informed me of these circumstances) eight and a half square miles of forest near San Carlos, in lieu of a debt; and this he sold for 350 dollars, or about 70*l.* sterling.

The two succeeding days were fine, and at night we reached the island of Quinchao. This neighbourhood is the most cultivated part of the Archipelago; for a broad strip of land on the coast of the main island, as well as on many of the smaller adjoining ones, is almost completely cleared. Some of the farm-houses seemed very comfortable. I was curious to ascertain how rich any of these people might be, but Mr. Douglas says that no one can be considered as possessing a regular income. One of the richest landowners might possibly accumulate, in a long industrious life, as much as 1000*l.* sterling; but should this happen, it would all be stowed away in some secret corner, for it is the custom of almost every family to have a jar or treasure-chest buried in the ground.

November 30*th*. – Early on Sunday morning we reached Castro, the ancient capital of Chiloe, but now a most forlorn and deserted place. The usual quadrangular arrangement of Spanish towns could be traced, but the streets and plaza were coated with fine green turf, on which sheep were browsing. The church, which stands in the middle, is entirely built of plank, and has a picturesque and venerable appearance. The poverty of the place may be conceived from the fact, that although containing some hundreds of inhabitants, one of our party was unable anywhere to purchase either a pound of sugar or an ordinary knife. No individual possessed either a watch or a clock; and an old man, who was supposed to have a good idea of time, was employed to strike the church bell by guess. The arrival of our boats was a rare event in this quiet retired corner of the world; and nearly all the inhabitants came down to the beach to see us pitch our tents. They were very civil, and offered us a house; and one man even sent us a cask of cider as a present. In the afternoon we paid our respects to the governor – a quiet old man, who, in his appearance and manner of life, was scarcely superior to an English cottager. At night heavy rain set in, which was hardly sufficient to drive away from our tents the large circle of lookers on. An Indian family, who had come to trade in a canoe from Caylen, bivouacked near us. They had no shelter during the rain. In the morning I asked a young Indian, who was wet to the skin, how he had passed the night. He seemed perfectly content, and answered, 'Muy bien, señor.'

December 1*st*. – We steered for the island of Lemuy. I was anxious to examine a reported coal-mine, which turned out to be lignite of little value, in the sandstone (probably of an ancient tertiary epoch) of which these islands are composed. When we reached Lemuy we had much difficulty in finding any place to pitch our tents, for it was spring-tide, and the land was wooded down to the water's edge. In a short time we were surrounded by a large group of the nearly pure Indian inhabitants. They were much surprised at our arrival, and said one to the other, 'This is the reason we have seen so many parrots lately; the cheucau (an odd red-breasted little bird, which inhabits the thick forest, and utters very peculiar noises) has not cried "beware" for nothing.' They were soon anxious for barter. Money was scarcely worth anything, but their eagerness for tobacco was something quite extraordinary. After tobacco, indigo came next in value; then capsicum, old clothes, and gunpowder. The latter article was required for a very innocent purpose: each parish has a public musket, and the gunpowder was wanted for making a noise on their saint or feast days.

The people here live chiefly on shell-fish and potatoes. At certain seasons they catch also, in 'corrales,' or hedges under water, many fish which are left on the mud-banks as the tide falls. They occasionally possess fowls, sheep, goats, pigs, horses, and cattle; the order in which they are here mentioned, expressing their respective numbers. I never saw anything more obliging and humble than the manners of these people. They generally began with stating, that they were poor natives of the place, and not Spaniards, and that they were in sad want of tobacco and other comforts. At Caylen, the most southern island, the sailors bought with a stick of tobacco, of the value of three-halfpence, two fowls, one of which, the Indian stated, had skin between its toes, and turned out to be a fine duck; and with some cotton handkerchiefs, worth three shillings, three sheep and a large bunch of onions were procured. The yawl at this place was anchored some way from the shore, and we had fears for her safety from robbers during the night. Our pilot, Mr. Douglas, accordingly told the constable of the district that we always placed sentinels with loaded arms, and not understanding Spanish, if we saw any person in the dark, we should assuredly shoot him. The constable, with much humility, agreed to the perfect propriety of this arrangement, and promised us that no one should stir out of his house during that night.

During the four succeeding days we continued sailing southward. The general features of the country remained the same, but it was much less thickly inhabited. On the large island of Tanqui there was scarcely one cleared spot, the trees on every side extending their branches over the sea-beach. I one day noticed, growing on the sandstone cliffs, some very fine plants of the panke (Gunnera scabra), which somewhat resembles the rhubarb on a gigantic scale. The inhabitants eat the stalks, which are subacid, and tan leather with the roots, and prepare a black dye from them. The leaf is nearly circular, but deeply indented on its margin. I measured one which was nearly eight feet in diameter, and therefore no less than twenty-four in circumference! The stalk is rather more than a yard high, and each plant sends out four or five of these enormous leaves, presenting together a very noble appearance.

December 6th. – We reached Caylen, called 'el fin del Cristiandad.' In the morning we stopped for a few minutes at a house on the northern end of Laylec, which was the extreme point of South American Christendom, and a miserable hovel it was. The latitude is 43° 10′, which is two degrees farther south than the Rio Negro on the Atlantic coast. These extreme Christians were very poor, and, under the plea of their situation, begged for some tobacco. As a proof of the poverty of these Indians, I may mention that shortly before this, we had met a man, who had travelled three days and a half on foot, and had as many to return, for the sake of recovering the value of a small axe and a few fish. How very difficult it must be to buy the smallest article, when such trouble is taken to recover so small a debt!

In the evening we reached the island of San Pedro, where we found the *Beagle* at anchor. In doubling the point, two of the officers landed to take a round of angles with the theodolite. A fox (Canis fulvipes), of a kind said to be peculiar to the island, and very rare in it, and which is a new species, was sitting on the rocks. He was so intently absorbed in watching the work of the officers, that I was able, by quietly walking up behind, to knock him on the head with my geological hammer. This fox, more curious or more scientific, but less wise, than the generality of his brethren, is now mounted in the museum of the Zoological Society.

We stayed three days in this harbour, on one of which Captain Fitz Roy, with a party, attempted to ascend to the summit of San Pedro. The woods here had rather a different appearance from those on the northern part of the island. The rock, also, being micaceous slate, there was no beach, but the steep sides dipped directly beneath the water. The general aspect in consequence was more like that of Tierra del Fuego than of Chiloe. In vain we tried to gain the summit: the forest was so impenetrable, that no one who has not beheld it, can imagine so entangled a mass of dying and dead trunks. I am sure that often, for more than ten minutes together, our feet never touched the ground, and we were frequently ten or fifteen feet above it, so that the seamen as a joke called out the soundings. At other times we crept one after another on our hands and knees, under the rotten trunks.

In the lower part of the mountain, noble trees of the Winter's Bark, and a laurel like the sassafras with fragrant leaves, and others, the names of which I do not know, were matted together by a trailing bamboo or cane. Here we were more like fishes struggling in a net than any other animal. On the higher parts, brushwood takes the place of larger trees, with here and there a red cedar or an alerce pine. I was also pleased to see, at an elevation of a little less than 1000 feet, our old friend the southern beech. They were, however, poor stunted trees; and I should think that this must be nearly their northern limit. We ultimately gave up the attempt in despair.

December 10*th*. – The yawl and whale-boat, with Mr. Sulivan, proceeded on their survey, but I remained on board the *Beagle*, which the next day left San Pedro for the southward. On the 13th we ran into an opening in the southern part of Guayatecas, or the Chonos Archipelago; and it was fortunate we did so, for on the following day a storm, worthy of Tierra del Fuego, raged with great fury. White massive clouds were piled up against a dark blue sky, and across them black ragged sheets of vapour were rapidly driven. The successive mountain ranges appeared like dim shadows; and the setting sun cast on the woodland a yellow gleam, much like that produced by the flame of spirits of wine. The water was white with the flying spray, and the wind lulled and roared again through the rigging: it was an ominous, sublime scene. During a few minutes there was a bright rainbow, and it was curious to observe the effect of the spray, which, being carried along the surface of the water, changed the ordinary semi-circle into a circle – a band of prismatic colours being con-tinued, from both feet of the common arch across the bay, close to the vessel's side: thus forming a distorted, but very nearly entire ring.

We stayed here three days. The weather continued bad; but this did not much signify, for the surface of the land in all these islands is all but impassable. The coast is so very rugged that to attempt to walk in that direction requires continued scrambling up and down over the sharp rocks of mica-slate; and as for the woods, our faces, hands, and shin-bones all bore witness to the maltreatment we received, in merely attempting to penetrate their forbidden recesses.

December 18*th*. – We stood out to sea. On the 20th we bade farewell to the south, and with a fair wind turned the ship's head northward. From Cape Tres Montes we sailed pleasantly along the lofty weather-beaten coast, which is remarkable for the bold outline of its hills, and the thick covering of forest even on the almost precipitous flanks. The next day a harbour was discovered, which on this dangerous coast might be of great service to a distressed vessel. It can easily be recognized by a hill 1600 feet high, which is even more perfectly conical than the famous sugar-loaf at Rio de Janeiro. The next day, after anchoring, I succeeded in reaching the summit of this hill. It was a laborious undertaking, for the sides were so steep that in some parts it was necessary to use the trees as ladders. There were also several extensive brakes of the Fuchsia, covered with its beautiful drooping flowers, but very difficult to crawl through. In these wild countries it gives much delight to gain the summit of any mountain. There is an indefinite expectation of seeing something very strange, which, however often it may be balked, never failed with me to recur on each successive attempt. Every one must know the feeling of triumph and pride which a grand view from a height communicates to the mind. In these little frequented countries there is also joined to it some vanity, that you perhaps are the first man who ever stood on this pinnacle or admired this view.

A strong desire is always felt to ascertain whether any human being has previously visited an unfrequented spot. A bit of wood with a nail in it, is picked up and studied as if it were covered with hieroglyphics. Possessed with this feeling, I was much interested by finding, on a wild part of the coast, a bed made of grass beneath a ledge of rock. Close by it there had been a fire, and the man had used an axe. The fire, bed, and situation showed the dexterity of an Indian; but he could scarcely have been an Indian, for the race is in this part extinct, owing to the Catholic desire of making at one blow Christians and Slaves. I had at the time some misgivings that the solitary man who had made his bed on this wild spot, must have been some poor shipwrecked sailor, who, in trying to travel up the coast, had here laid himself down for his dreary night.

December 28th. – The weather continued very bad, but it at last permitted us to proceed with the survey. The time hung heavy on our hands, as it always did when we were delayed from day to day by successive gales of wind. In the evening another harbour was discovered, where we anchored. Directly afterwards a man was seen waving his shirt, and a boat was sent which brought back two seamen. A party of six had run away from an American whaling vessel, and had landed a little to the southward in a boat, which was shortly afterwards knocked to pieces by the surf. They had now been wandering up and down the coast for fifteen months, without knowing which way to go, or where they were. What a singular piece of good fortune it was that this harbour was now discovered! Had it not been for this one chance, they might have wandered till they had grown old men, and at last have perished on this wild coast. Their sufferings had been very great, and one of their party had lost his life by falling from the cliffs. They were sometimes obliged to separate in search of food, and this explained the bed of the solitary man. Considering what they had undergone, I think they had kept a very good reckoning of time, for they had lost only four days.

December 30th. – We anchored in a snug little cove at the foot of some high hills, near the northern extremity of Tres Montes. After breakfast the next morning, a party ascended one of these mountains, which was 2400 feet high. The scenery was remarkable. The chief part of the range was composed of grand, solid, abrupt masses of granite, which appeared as if they had been coeval with the beginning of the world. The granite was capped with mica-slate, and this in the lapse of ages had been worn into strange finger-shaped points. These two formations, thus differing in their outlines, agree in being almost destitute of vegetation. This barrenness had to our eyes a strange appearance, from having been so long accustomed to the sight of an almost universal forest of dark-green trees. I took much delight in examining the structure of these mountains. The complicated and lofty ranges bore a noble aspect of durability – equally profitless, however, to man and to all other animals. Granite to the geologist is classic ground: from its wide-spread limits, and its beautiful and compact texture, few rocks have been more anciently recognized. Granite has given rise, perhaps, to more discussion concerning its origin than any other formation. We generally see it constituting the fundamental rock, and, however formed, we know it is the deepest layer in the crust of this globe to which man has penetrated. The limit of man's knowledge in any subject possesses a high interest, which is perhaps increased by its close neighbourhood to the realms of imagination.

January 1st, 1835. – The new year is ushered in with the ceremonies proper to it in these regions. She lays out no false hopes: a heavy north-western gale, with steady rain, bespeaks the rising year. Thank God, we are not destined here to see the end of it, but hope then to be in the Pacific Ocean, where a blue sky tells one there is a heaven, – a something beyond the clouds above our heads.

The north-west winds prevailing for the next four days, we only managed to cross a great bay, and then anchored in another secure harbour. I accompanied the Captain in a boat to the head of a deep creek. On the way the number of seals which we saw was quite astonishing: every bit of flat rock, and parts of the beach, were covered with them. They appeared to be of a loving disposition, and lay huddled together, fast asleep, like so many pigs; but even pigs would have been ashamed of their dirt, and of the foul smell which came from them. Each herd was watched by the patient but inauspicious eyes of the turkey-buzzard. This disgusting bird, with its bald scarlet head, formed to wallow in putridity, is very common on the west coast, and their attendance on the seals shows on what they rely for their food. We found the water (probably only that of the surface) nearly fresh: this was caused by the number of torrents which, in the form of cascades, came tumbling over the bold granite mountains into the sea. The fresh water attracts the fish, and these bring many terns, gulls, and two kinds of cormorant. We saw also a pair of the beautiful black-necked swans, and several small sea-otters, the fur of which is held in such high estimation. In returning, we were again amused by the impetuous manner in which the heap of seals, old and young, tumbled into the water as the boat passed. They did not remain long under water, but rising, followed us with outstretched necks, expressing great wonder and curiosity.

7th. – Having run up the coast, we anchored near the northern end of the Chonos Archipelago, in Low's Harbour, where we remained a week. The islands were here, as in Chiloe, composed of a stratified, soft, littoral deposit; and the vegetation in consequence was beautifully luxuriant. The woods came down to the sea-beach, just in the manner of an evergreen shrubbery over a gravel walk. We also enjoyed from the anchorage a splendid view of four great snowy cones of the Cordillera, including 'el famoso Corcovado:' the range itself had in this latitude so little height, that few parts of it appeared above the tops of the neighbouring islets. We found here a party of five men from Caylen, 'el fin del Cristiandad,' who had most adventurously crossed in their miserable boat-canoe, for the purpose of fishing, the open space of sea which separates Chonos from Chiloe. These islands will, in all probability, in a short time become peopled like those adjoining the coast of Chiloe.

The wild potato grows on these islands in great abundance, on the sandy, shelly soil near the sea-beach. The tallest plant was four feet in height. The tubers were generally small, but I found one, of an oval shape, two inches in diameter: they resembled in every respect, and had the same smell as English potatoes; but when boiled they shrunk much, and were watery and insipid, without any bitter taste. They are undoubtedly here indigenous: they grow as far south, according to Mr. Low, as lat. 50°, and are called Aquinas by the wild Indians of that part: the Chilotan Indians have a different name for them. Professor Henslow, who has examined the dried specimens which I brought home, says that they are the same with those described by Mr. Sabine[1] from Valparaiso, but that they form a variety which by some botanists has been considered as specifically distinct. It is remarkable that the same plant should be found on the sterile mountains of central Chile, where a drop of rain does not fall for more than six months, and within the damp forests of these southern islands.

In the central parts of the Chonos Archipelago (lat. 45°), the forest has very much the same character with that along the whole west coast, for 600 miles southward to Cape Horn. The arborescent grass of Chiloe is not found here; while the beech of Tierra del Fuego

1 Horticultural Transact., vol. v. p. 249. Mr. Caldcleugh sent home two tubers, which, being well manured, even the first season produced numerous potatoes and an abundance of leaves. See Humboldt's interesting discussion on this plant, which it appears was unknown in Mexico, – in Polit. Essay on New Spain, book iv. chap. ix.

grows to a good size, and forms a considerable proportion of the wood; not, however, in the same exclusive manner as it does farther southward. Cryptogamic plants here find a most congenial climate. In the Strait of Magellan, as I have before remarked, the country appears too cold and wet to allow of their arriving at perfection; but in these islands, within the forest, the number of species and great abundance of mosses, lichens, and small ferns, is quite extraordinary.[1] In Tierra del Fuego trees grow only on the hill-sides; every level piece of land being invariably covered by a thick bed of peat; but in Chiloe flat land supports the most luxuriant forests. Here, within the Chonos Archipelago, the nature of the climate more closely approaches that of Tierra del Fuego than that of northern Chiloe; for every patch of level ground is covered by two species of plants (Astelia pumila and Donatia magellanica), which by their joint decay compose a thick bed of elastic peat.

In Tierra del Fuego, above the region of woodland, the former of these eminently sociable plants is the chief agent in the production of peat. Fresh leaves are always succeeding one to the other round the central tap-root; the lower ones soon decay, and in tracing a root downwards in the peat, the leaves, yet holding their place, can be observed passing through every stage of decomposition, till the whole becomes blended in one confused mass. The Astelia is assisted by a few other plants, – here and there a small creeping Myrtus (M. nummularia), with a woody stem like our cranberry and with a sweet berry, – an Empetrum (E. rubrum), like our heath, – a rush (Juncus grandiflorus), are nearly the only ones that grow on the swampy surface. These plants, though possessing a very close general resemblance to the English species of the same genera, are different. In the more level parts of the country, the surface of the peat is broken up into little pools of water, which stand at different heights, and appear as if artificially excavated. Small streams of water, flowing

---

1 By sweeping with my insect-net, I procured from these situations a considerable number of minute insects, of the family of Staphylinidæ, and others allied to Pselaphus, and minute Hymenoptera. But the most characteristic family in number, both of individuals and species, throughout the more open parts of Chiloe and Chonos, is that of the Telephoridæ.

underground, complete the disorganization of the vegetable matter, and consolidate the whole.

The climate of the southern part of America appears particularly favourable to the production of peat. In the Falkland Islands almost every kind of plant, even the coarse grass which covers the whole surface of the land, becomes converted into this substance: scarcely any situation checks its growth; some of the beds are as much as twelve feet thick, and the lower part becomes so solid when dry, that it will hardly burn. Although every plant lends its aid, yet in most parts the Astelia is the most efficient. It is rather a singular circumstance, as being so very different from what occurs in Europe, that I nowhere saw moss forming by its decay any portion of the peat in South America. With respect to the northern limit, at which the climate allows of that peculiar kind of slow decomposition which is necessary for its production, I believe that in Chiloe (lat. 41° to 42°), although there is much swampy ground, no well characterized peat occurs: but in the Chonos Islands, three degrees farther southward, we have seen that it is abundant. On the eastern coast in La Plata (lat. 35°) I was told by a Spanish resident, who had visited Ireland, that he had often sought for this substance, but had never been able to find any. He showed me, as the nearest approach to it which he had discovered, a black peaty soil, so penetrated with roots as to allow of an extremely slow and imperfect combustion.

The zoology of these broken islets of the Chonos Archipelago is, as might have been expected, very poor. Of quadrupeds two aquatic kinds are common. The Myopotamus Coypus (like a beaver, but with a round tail) is well known from its fine fur, which is an object of trade throughout the tributaries of La Plata. It here, however, exclusively frequents salt water; which same circumstance has been mentioned as sometimes occurring with the great rodent, the Capybara. A small sea-otter is very numerous; this animal does not feed exclusively on fish, but, like the seals, draws a large supply from a small red crab, which swims in shoals near the surface of the water. Mr. Bynoe saw one in Tierra del Fuego eating a cuttle-fish; and at Low's Harbour, another was killed in the act of carrying to its hole a large volute shell. At one place I caught in a trap a singular little mouse (M. brachiotis); it appeared common on several of the islets, but the Chilotans at Low's Harbour said that it was not found in all. What a succession of chances,[1] or what changes of level must have been brought into play, thus to spread these small animals throughout this broken archipelago!

In all parts of Chiloe and Chonos, two very strange birds occur, which are allied to, and replace, the Turco and Tapacolo of central Chile. One is called by the inhabitants 'Cheucau' (Pteroptochos rubecula): it frequents the most gloomy and retired spots within the damp forests. Sometimes, although its cry may be heard close at hand, let a person watch ever so attentively he will not see the cheucau; at other times, let him stand motionless and the red-breasted little bird will approach within a few feet in the most familiar manner. It then busily hops about the entangled mass of rotting canes and branches, with its little tail cocked upwards. The cheucau is held in superstitious fear by the Chilotans, on account of its strange and varied cries. There are three very distinct cries: one is called 'chiduco,' and is an omen of good; another, 'huitreu,' which is extremely unfavourable; and a third, which I have forgotten. These words are given in imitation of the noises; and the natives are in some things absolutely governed by them. The Chilotans assuredly have chosen a most comical little creature for their prophet. An allied species, but rather larger, is called by the natives 'Guid-guid' (Pteroptochos Tarnii), and by the English the barking-bird. This latter name is well given; for I defy any one at first to feel certain that a small dog is not yelping somewhere in the forest. Just as with the cheucau, a person will sometimes hear the bark close by, but in vain may endeavour by watching, and with still less chance by beating the bushes, to see the bird; yet at other times the guid-guid fearlessly comes near. Its manner of feeding and its general habits are very similar to those of the cheucau.

---

1 It is said that some rapacious birds bring their prey alive to their nests. If so, in the course of centuries, every now and then, one might escape from the young birds. Some such agency is necessary, to account for the distribution of the smaller gnawing animals on islands not very near each other.

On the coast,[1] a small dusky-coloured bird (Opetiorhynchus Patagonicus) is very common. It is remarkable from its quiet habits; it lives entirely on the sea-beach, like a sandpiper. Besides these birds only few others inhabit this broken land. In my rough notes I describe the strange noises, which, although frequently heard within these gloomy forests, yet scarcely disturb the general silence. The yelping of the guid-guid, and the sudden whew-whew of the cheucau, sometimes come from afar off, and sometimes from close at hand; the little black wren of Tierra del Fuego occasionally adds its cry; the creeper (Oxyurus) follows the intruder screaming and twittering; the humming-bird may be seen every now and then darting from side to side, and emitting, like an insect, its shrill chirp; lastly, from the top of some lofty tree the indistinct but plaintive note of the white-tufted tyrant-flycatcher (Myiobius) may be noticed. From the great preponderance in most countries of certain common genera of birds, such as the finches, one feels at first surprised at meeting with the peculiar forms above enumerated, as the commonest birds in any district. In central Chile two of them, namely, the Oxyurus and Scytalopus, occur, although most rarely. When finding, as in this case, animals which seem to play so insignificant a part in the great scheme of nature, one is apt to wonder why they were created. But it should always be recollected, that in some other country perhaps they are essential members of society, or at some former period may have been so. If America south of 37° were sunk beneath the waters of the ocean, these two birds might continue to exist in central Chile for a long period, but it is very improbable that their numbers would increase. We should then see a case which must inevitably have happened with very many animals.

These southern seas are frequented by several species of Petrels: the largest kind, Procellaria gigantea, or nelly (quebrantahuesos, or break-bones, of the Spaniards), is a common bird, both in the inland channels and on the open sea. In its habits and manner of flight, there

1 I may mention, as a proof of how great a difference there is between the seasons of the wooded and the open parts of this coast, that on September 20th, in lat. 34°, these birds had young ones in the nest, while among the Chonos Islands, three months later in the summer, they were only laying; the difference in latitude between these two places being about 700 miles.

is a very close resemblance with the albatross; and as with the albatross, a person may watch it for hours together without seeing on what it feeds. The 'break-bones' is, however, a rapacious bird, for it was observed by some of the officers of Port St. Antonio chasing a diver, which tried to escape by diving and flying, but was continually struck down, and at last killed by a blow on its head. At Port St. Julian these great petrels were seen killing and devouring young gulls. A second species (Puffinus cinereus), which is common to Europe, Cape Horn, and the coast of Peru, is of a much smaller size than the P. gigantea, but, like it, of a dirty black colour. It generally frequents the inland sounds in very large flocks: I do not think I ever saw so many birds of any other sort together, as I once saw of these behind the island of Chiloe. Hundreds of thousands flew in an irregular line for several hours in one direction. When part of the flock settled on the water the surface was blackened, and a noise proceeded from them as of human beings talking in the distance.

There are several other species of petrels, but I will only mention one other kind, the Pelacanoides Berardi, which offers an example of those extraordinary cases, of a bird evidently belonging to one well-marked family, yet both in its habits and structure allied to a very distinct tribe. This bird never leaves the quiet inland sounds. When disturbed it dives to a distance, and on coming to the surface, with the same movement takes flight. After flying by the rapid movement of its short wings for a space in a straight line, it drops, as if struck dead, and dives again. The form of its beak and nostrils, length of foot, and even the colouring of its plumage, show that this bird is a petrel: on the other hand, its short wings and consequent little power of flight, its form of body and shape of tail, the absence of a hind toe to its foot, its habit of diving, and its choice of situation, make it at first doubtful whether its relationship is not equally close with the auks. It would undoubtedly be mistaken for an auk, when seen from a distance, either on the wing, or when diving and quietly swimming about the retired channels of Tierra del Fuego.

CHAPTER XIV

CHILOE AND
CONCEPCION: GREAT
EARTHQUAKE

San Carlos, Chiloe – Osorno in eruption, contemporaneously with Aconcagua and Coseguina – Ride to
Cucao – Impenetrable Forests – Valdivia – Indians –
Earthquake – Concepcion – Great Earthquake – Rocks
fissured – Appearance of the former Towns – The Sea black
and boiling – Direction of the Vibrations – Stones twisted
round – Great Wave – Permanent elevation of the Land –
Area of Volcanic Phenomena – The connexion between
the elevatory and eruptive forces – Cause of Earthquakes –
Slow elevation of Mountain-chains.

ON January the 15th we sailed from Low's Harbour, and three days
afterwards anchored a second time in the bay of S. Carlos in Chiloe.
On the night of the 19th the volcano of Osorno was in action. At
midnight the sentry observed something like a large star, which
gradually increased in size till about three o'clock, when it presented
a very magnificent spectacle. By the aid of a glass, dark objects, in
constant succession, were seen, in the midst of a great glare of red
light, to be thrown up and to fall down. The light was sufficient to
cast on the water a long bright reflection. Large masses of molten
matter seem very commonly to be cast out of the craters in this part of
the Cordillera. I was assured that when the Corcovado is in eruption,
great masses are projected upwards and are seen to burst in the air,
assuming many fantastical forms, such as trees: their size must be
immense, for they can be distinguished from the high land behind
S. Carlos, which is no less than ninety-three miles from the Corcovado. In the morning the volcano became tranquil.

I was surprised at hearing afterwards that Aconcagua in Chile, 480
miles northwards, was in action on this same night; and still more
surprised to hear, that the great eruption of Coseguina (2700 miles

north of Aconcagua), accompanied by an earthquake felt over 1000 miles, also occurred within six hours of this same time. This coincidence is the more remarkable, as Coseguina had been dormant for twenty-six years: and Aconcagua most rarely shows any signs of action. It is difficult even to conjecture, whether this coincidence was accidental, or shows some subterranean connexion. If Vesuvius, Etna, and Hecla in Iceland (all three relatively nearer each other, than the corresponding points in South America) suddenly burst forth in eruption on the same night, the coincidence would be thought remarkable; but it is far more remarkable in this case, where the three vents fall on the same great mountain-chain, and where the vast plains along the entire eastern coast, and the upraised recent shells along more than 2000 miles on the western coast, show in how equable and connected a manner the elevatory forces have acted.

Captain Fitz Roy being anxious that some bearings should be taken on the outer coast of Chiloe, it was planned that Mr. King and myself should ride to Castro, and thence across the island to the Capella de Cucao, situated on the west coast. Having hired horses and a guide, we set out on the morning of the 22nd. We had not proceeded far, before we were joined by a woman and two boys, who were bent on the same journey. Every one on this road acts on a 'hail fellow well met fashion;' and one may here enjoy the privilege, so rare in South America, of travelling without fire-arms. At first, the country consisted of a succession of hills and valleys: nearer to Castro it became very level. The road itself is a curious affair: it consists in its whole length, with the exception of very few parts, of great logs of wood, which are either broad and laid longitudinally, or narrow and placed transversely. In summer the road is not very bad: but in winter, when the wood is rendered slippery from rain, travelling is exceedingly difficult. At that time of the year, the ground on each side becomes a morass, and is often overflowed: hence it is necessary that the longitudinal logs should be fastened down by transverse poles, which are pegged on each side into the earth. These pegs render a fall from a horse dangerous; as the chance of alighting on one of them is not small. It is remarkable, however, how active custom has made the Chilotan horses. In crossing bad parts, where the logs had been displaced, they skipped from one to the other, almost with the quickness and certainty of a dog. On both hands the road is bordered by the lofty forest-trees, with their bases matted together by canes. When occasionally a long reach of this avenue could be beheld, it presented a curious scene of uniformity: the white line of logs, narrowing in perspective, became hidden by the gloomy forest, or terminated in a zigzag which ascended some steep hill.

Although the distance from S. Carlos to Castro is only twelve leagues in a straight line, the formation of the road must have been a great labour. I was told that several people had formerly lost their lives in attempting to cross the forest. The first who succeeded was an Indian, who cut his way through the canes in eight days, and reached S. Carlos: he was rewarded by the Spanish government with a grant of land. During the summer, many of the Indians wander about the forests (but chiefly in the higher parts, where the woods are not quite so thick), in search of the half-wild cattle which live on the leaves of the cane and certain trees. It was one of these huntsmen who by chance discovered, a few years since, an English vessel, which had been wrecked on the outer coast. The crew were beginning to fail in provisions, and it is not probable that, without the aid of this man, they would ever have extricated themselves from these scarcely penetrable woods. As it was, one seaman died on the march, from fatigue. The Indians in these excursions steer by the sun: so that if there is a continuance of cloudy weather, they cannot travel.

The day was beautiful, and the number of trees which were in full flower perfumed the air; yet even this could hardly dissipate the effect of the gloomy dampness of the forest. Moreover, the many dead trunks that stand like skeletons, never fail to give to these primeval woods a character of solemnity, absent in those of countries long civilized. Shortly after sunset we bivouacked for the night. Our female companion, who was rather good-looking, belonged to one of the most respectable families in Castro: she rode, however, astride, and without shoes or stockings. I was surprised at the total want of pride shown by her and her brother. They brought food with them, but at all our meals sat watching Mr. King and myself whilst eating, till we were fairly shamed into feeding the whole party. The night was cloudless; and while lying in our beds, we enjoyed the sight (and it is a

high enjoyment) of the multitude of stars which illumined the darkness of the forest.

January 23rd. – We rose early in the morning, and reached the
pretty quiet town of Castro by two o'clock. The old governor had
died since our last visit, and a Chileno was acting in his place. We had
a letter of introduction to Don Pedro, whom we found exceedingly
hospitable and kind, and more disinterested than is usual on this side
of the continent. The next day Don Pedro procured us fresh horses,
and offered to accompany us himself. We proceeded to the south –
generally following the coast, and passing through several hamlets,
each with its large barn-like chapel built of wood. At Vilipilli, Don
Pedro asked the commandant to give us a guide to Cucao. The old
gentleman offered to come himself; but for a long time nothing
would persuade him, that two Englishmen really wished to go to
such an out of the way place as Cucao. We were thus accompanied by
the two greatest aristocrats in the country, as was plainly to be seen in
the manner of all the poorer Indians towards them. At Chonchi we
struck across the island, following intricate winding paths, sometimes
passing through magnificent forests, and sometimes through pretty
cleared spots, abounding with corn and potato crops. This undulating
woody country, partially cultivated, reminded me of the wilder parts
of England, and therefore had to my eye a most fascinating aspect. At
Vilinco, which is situated on the borders of the lake of Cucao, only a
few fields were cleared; and all the inhabitants appeared to be Indians.
This lake is twelve miles long, and runs in an east and west direction.
From local circumstances, the sea-breeze blows very regularly during
the day, and during the night it falls calm: this has given rise to strange
exaggerations, for the phenomenon, as described to us at San Carlos,
was quite a prodigy.

The road to Cucao was so very bad that we determined to embark
in a *periagua*. The commandant, in the most authoritative manner,
ordered six Indians to get ready to pull us over, without deigning to
tell them whether they would be paid. The periagua is a strange rough
boat, but the crew were still stranger: I doubt if six uglier little men
ever got into a boat together. They pulled, however, very well and
cheerfully. The stroke-oarsman gabbled Indian, and uttered strange
cries, much after the fashion of a pig-driver driving his pigs.

We started with a light breeze against us, but yet reached the Capella de Cucao before it was late. The country on each side of the lake was one unbroken forest. In the same periagua with us, a cow was embarked. To get so large an animal into a small boat appears at first a difficulty, but the Indians managed it in a minute. They brought the cow alongside the boat, which was heeled towards her; then placing two oars under her belly, with their ends resting on the gunwale, by the aid of these levers they fairly tumbled the poor beast, heels over head, into the bottom of the boat, and then lashed her down with ropes. At Cucao we found an uninhabited hovel (which is the residence of the padre when he pays this Capella a visit), where, lighting a fire, we cooked our supper, and were very comfortable.

The district of Cucao is the only inhabited part on the whole west coast of Chiloe. It contains about thirty or forty Indian families, who are scattered along four or five miles of the shore. They are very much secluded from the rest of Chiloe, and have scarcely any sort of commerce, except sometimes in a little oil, which they get from seal-blubber. They are tolerably dressed in clothes of their own manufacture, and they have plenty to eat. They seemed, however, discontented, yet humble to a degree which it was quite painful to witness. These feelings are, I think, chiefly to be attributed to the harsh and authoritative manner in which they are treated by their rulers. Our companions, although so very civil to us, behaved to the poor Indians as if they had been slaves, rather than free men. They ordered provisions and the use of their horses, without ever condescending to say how much, or indeed whether the owners should be paid at all. In the morning, being left alone with these poor people, we soon ingratiated ourselves by presents of cigars and maté. A lump of white sugar was divided between all present, and tasted with the greatest curiosity. The Indians ended all their complaints by saying, 'And it is only because we are poor Indians, and know nothing; but it was not so when we had a King.'

The next day after breakfast, we rode a few miles northward to Punta Huantamó. The road lay along a very broad beach, on which, even after so many fine days, a terrible surf was breaking. I was assured that after a heavy gale, the roar can be heard at night even at Castro, a distance of no less than twenty-one sea-miles across a hilly and wooded country. We had some difficulty in reaching the point, owing to the intolerably bad paths; for everywhere in the shade the ground soon becomes a perfect quagmire. The point itself is a bold rocky hill. It is covered by a plant allied, I believe, to Bromelia, and called by the inhabitants Chepones. In scrambling through the beds, our hands were very much scratched. I was amused by observing the precaution our Indian guide took, in turning up his trowsers, thinking that they were more delicate than his own hard skin. This plant bears a fruit, in shape like an artichoke, in which a number of seed-vessels are packed: these contain a pleasant sweet pulp, here much esteemed. I saw at Low's Harbour the Chilotans making chichi, or cider, with this fruit: so true is it, as Humboldt remarks, that almost everywhere man finds means of preparing some kind of beverage from the vegetable kingdom. The savages, however, of Tierra del Fuego, and I believe of Australia, have not advanced thus far in the arts.

The coast to the north of Punta Huantamó is exceedingly rugged and broken, and is fronted by many breakers, on which the sea is eternally roaring. Mr. King and myself were anxious to return, if it had been possible, on foot along this coast; but even the Indians said it was quite impracticable. We were told that men have crossed by striking directly through the woods from Cucao to S. Carlos, but never by the coast. On these expeditions, the Indians carry with them only roasted corn, and of this they eat sparingly twice a day.

26th. – Re-embarking in the periagua, we returned across the lake, and then mounted our horses. The whole of Chiloe took advantage of this week of unusually fine weather, to clear the ground by burning. In every direction volumes of smoke were curling upwards. Although the inhabitants were so assiduous in setting fire to every part of the wood, yet I did not see a single fire which they had succeeded in making extensive. We dined with our friend the commandant, and did not reach Castro till after dark. The next morning we started very early. After having ridden for some time, we obtained from the brow of a steep hill an extensive view (and it is a rare thing on this road) of the great forest. Over the horizon of trees, the volcano of Corcovado, and the great flat-topped one to the north, stood out in proud pre-eminence: scarcely another peak in the long range showed its snow summit. I hope it will be long before I forget this farewell view of the magnificent Cordillera fronting Chiloe. At night we bivouacked under a cloudless sky, and the next morning reached S. Carlos. We arrived on the right day, for before evening heavy rain commenced.

February 4th. – Sailed from Chiloe. During the last week I made several short excursions. One was to examine a great bed of now-existing shells, elevated 350 feet above the level of the sea: from among these shells, large forest-trees were growing. Another ride was to P. Huechucucuy. I had with me a guide who knew the country far too well; for he would pertinaciously tell me endless Indian names for every little point, rivulet, and creek. In the same manner as in Tierra del Fuego, the Indian language appears singularly well adapted for attaching names to the most trivial features of the land. I believe every one was glad to say farewell to Chiloe; yet if we could forget the gloom and ceaseless rain of winter, Chiloe might pass for a charming island. There is also something very attractive in the simplicity and humble politeness of the poor inhabitants.

We steered northward along shore, but owing to thick weather did not reach Valdivia till the night of the 8th. The next morning the boat proceeded to the town, which is distant about ten miles. We followed the course of the river, occasionally passing a few hovels, and patches of ground cleared out of the otherwise unbroken forest; and sometimes meeting a canoe with an Indian family. The town is situated on the low banks of the stream, and is so completely buried in a wood of apple-trees that the streets are merely paths in an orchard. I have never seen any country, where apple-trees appeared to thrive so well as in this damp part of South America: on the borders of the roads there were many young trees evidently self-sown. In Chiloe the inhabitants possess a marvellously short method of making an orch-ard. At the lower part of almost every branch, small, conical, brown, wrinkled points project: these are always ready to change into roots, as may sometimes be seen, where any mud has been accidentally splashed against the tree. A branch as thick as a man's thigh is chosen in the early spring, and is cut off just beneath a group of these points; all the smaller branches are lopped off, and it is then placed about two feet deep in the ground. During the ensuing summer the stump throws out long shoots, and sometimes even bears fruit: I was shown one which had produced as many as twenty-three apples, but this was thought very unusual. In the third season the stump is changed (as I have myself seen) into a well-wooded tree, loaded with fruit. An old man near Valdivia illustrated his motto, 'Necesidad es la madre del invencion,' by giving an account of the several useful things he manufactured from his apples. After making cider, and likewise wine, he extracted from the refuse a white and finely flavoured spirit; by another process he procured a sweet treacle, or, as he called it, honey. His children and pigs seemed almost to live, during this season of the year, in his orchard.

February 11*th*. – I set out with a guide on a short ride, in which, however, I managed to see singularly little, either of the geology of the country or of its inhabitants. There is not much cleared land near Valdivia: after crossing a river at the distance of a few miles, we entered the forest, and then passed only one miserable hovel, before reaching our sleeping-place for the night. The short difference in latitude, of 150 miles, has given a new aspect to the forest, compared with that of Chiloe. This is owing to a slightly different proportion in the kinds of trees. The evergreens do not appear to be quite so numerous; and the forest in consequence has a brighter tint. As in Chiloe, the lower parts are matted together by canes: here also another kind (resembling the bamboo of Brazil and about twenty feet in height) grows in clusters, and ornaments the banks of some of the streams in a very pretty manner. It is with this plant that the Indians make their chuzos, or long tapering spears. Our resting-house was so dirty that I preferred sleeping outside: on these journeys the first night is generally very uncomfortable, because one is not accustomed to the tickling and biting of the fleas. I am sure, in the morning, there was not a space on my legs of the size of a shilling, which had not its little red mark where the flea had feasted.

12*th*. – We continued to ride through the uncleared forest; only occasionally meeting an Indian on horseback, or a troop of fine mules bringing alerce-planks and corn from the southern plains. In the afternoon one of the horses knocked up: we were then on a brow of a hill, which commanded a fine view of the Llanos. The view of these open plains was very refreshing, after being hemmed in and buried in the wilderness of trees. The uniformity of a forest soon becomes very wearisome. This west coast makes me remember with pleasure the free, unbounded plains of Patagonia; yet, with the true spirit of contradiction, I cannot forget how sublime is the silence of the forest. The Llanos are the most fertile and thickly peopled parts of the country; as they possess the immense advantage of being nearly free from trees. Before leaving the forest we crossed some flat little lawns, around which single trees stood, as in an English park: I have often noticed with surprise, in wooded undulatory districts, that the quite level parts have been destitute of trees. On account of the tired horse, I determined to stop at the Mission of Cudico, to the friar of which I had a letter of introduction. Cudico is an intermediate district between the forest and the Llanos. There are a good many cottages, with patches of corn and potatoes, nearly all belonging to Indians. The tribes dependent on Valdivia are 'reducidos y cristianos.' The Indians farther northward, about Arauco and Imperial, are still very wild, and not converted; but they have all much intercourse with the Spaniards. The padre said that the Christian Indians did not much like coming to mass, but that otherwise they showed respect for religion. The greatest difficulty is in making them observe the ceremonies of marriage. The wild Indians take as many wives as they can support, and a cacique will sometimes have more than ten: on entering his house, the number may be told by that of the separate fires. Each wife lives a week in turn with the cacique; but all are employed in weaving ponchos, &c. for his profit. To be the wife of a cacique, is an honour much sought after by the Indian women.

The men of all these tribes wear a coarse woollen poncho: those south of Valdivia wear short trowsers, and those north of it a petticoat, like the chilipa of the Gauchos. All have their long hair bound by a scarlet fillet, but with no other covering on their heads. These Indians are good-sized men; their cheek-bones are prominent, and in general appearance they resemble the great American family to which they belong; but their physiognomy seemed to me to be slightly different from that of any other tribe which I had before seen. Their expression is generally grave, and even austere, and possesses much character: this may pass either for honest bluntness or fierce determination. The long black hair, the grave and much-lined features, and the dark complexion, called to my mind old portraits of James I. On the road we met with none of that humble politeness so universal in Chiloe. Some gave their 'mari-mari' (good morning) with promptness, but the greater number did not seem inclined to offer any salute. This independence of manners is probably a consequence of their long wars, and the repeated victories which they alone, of all the tribes in America, have gained over the Spaniards.

I spent the evening very pleasantly, talking with the padre. He was exceedingly kind and hospitable; and coming from Santiago, had contrived to surround himself with some few comforts. Being a man of some little education, he bitterly complained of the total want of society. With no particular zeal for religion, no business or pursuit, how completely must this man's life be wasted! The next day, on our return, we met seven very wild-looking Indians, of whom some were caciques that had just received from the Chilian government, their yearly small stipend for having long remained faithful. They were fine-looking men, and they rode one after the other, with most gloomy faces. An old cacique, who headed them, had been, I suppose, more excessively drunk than the rest, for he seemed both extremely grave and very crabbed. Shortly before this, two Indians joined us, who were travelling from a distant mission to Valdivia concerning some law-suit. One was a good-humoured old man, but from his wrinkled beardless face looked more like an old woman than a man. I frequently presented both of them with cigars; and though ready to receive them, and I dare say grateful, they would hardly condescend to thank me. A Chilotan Indian would have taken off his hat, and given his 'Dios le page!' The travelling was very tedious, both from the badness of the roads, and from the number of great fallen trees, which it was necessary either to leap over or to avoid by making long circuits. We slept on the road, and next morning reached Valdivia, whence I proceeded on board.

A few days afterwards I crossed the bay with a party of officers, and landed near the fort called Niebla. The buildings were in a most ruinous state, and the gun-carriages quite rotten. Mr. Wickham remarked to the commanding officer, that with one discharge they would certainly all fall to pieces. The poor man, trying to put a good face upon it, gravely replied, 'No, I am sure, sir, they would stand two!' The Spaniards must have intended to have made this place impregnable. There is now lying in the middle of the courtyard a little mountain of mortar, which rivals in hardness the rock on which it is placed. It was brought from Chile, and cost 7000 dollars. The revolution having broken out, prevented its being applied to any purpose, and now it remains a monument of the fallen greatness of Spain.

I wanted to go to a house about a mile and a half distant, but my guide said it was quite impossible to penetrate the wood in a straight line. He offered, however, to lead me, by following obscure cattle-tracks, the shortest way: the walk, nevertheless, took no less than three hours! This man is employed in hunting strayed cattle; yet, well as he must know the woods, he was not long since lost for two whole days, and had nothing to eat. These facts convey a good idea of the impracticability of the forests of these countries. A question often occurred to me – how long does any vestige of a fallen tree remain? This man showed me one which a party of fugitive Royalists had cut down fourteen years ago; and taking this as a criterion, I should think a bole a foot and a half in diameter would in thirty years be changed into a heap of mould.

February 20th. – This day has been memorable in the annals of Valdivia, for the most severe earthquake experienced by the oldest inhabitant. I happened to be on shore, and was lying down in the wood to rest myself. It came on suddenly, and lasted two minutes, but the time appeared much longer. The rocking of the ground was very sensible. The undulations appeared to my companion and myself to come from due east, whilst others thought they proceeded from south-west: this shows how difficult it sometimes is to perceive the direction of the vibrations. There was no difficulty in standing upright, but the motion made me almost giddy: it was something like the movement of a vessel in a little cross-ripple, or still more like that felt by a person skating over thin ice, which bends under the weight of his body.

A bad earthquake at once destroys our oldest associations: the earth, the very emblem of solidity, has moved beneath our feet like a thin crust over a fluid; – one second of time has created in the mind a strange idea of insecurity, which hours of reflection would not have produced. In the forest, as a breeze moved the trees, I felt only the earth tremble, but saw no other effect. Captain Fitz Roy and some officers were at the town during the shock, and there the scene was more striking; for although the houses, from being built of wood, did not fall, they were violently shaken, and the boards creaked and rattled together. The people rushed out of doors in the greatest alarm. It is these accompaniments that create that perfect horror of earthquakes, experienced by all who have thus seen, as well as felt, their effects. Within the forest it was a deeply interesting, but by no means an awe-inspiring phenomenon. The tides were very curiously affected. The great shock took place at the time of low water; and an old woman who was on the beach told me, that the water flowed very quickly, but not in great waves, to high-water mark, and then as quickly returned to its proper level; this was also evident by the line of wet sand. This same kind of quick but quiet movement in the tide, happened a few years since at Chiloe, during a slight earthquake, and created much causeless alarm. In the course of the evening there were many weaker shocks, which seemed to produce in the harbour the most complicated currents, and some of great strength.

March 4th. – We entered the harbour of Concepcion. While the ship was beating up to the anchorage, I landed on the island of Quiriquina. The major-domo of the estate quickly rode down to tell me the terrible news of the great earthquake of the 20th: – 'That not a house in Concepcion or Talcahuano (the port) was standing; that seventy villages were destroyed; and that a great wave had almost washed away the ruins of Talcahuano.' Of this latter statement I soon saw abundant proofs – the whole coast being strewed over with timber and furniture as if a thousand ships had been wrecked. Besides chairs, tables, book-shelves, &c., in great numbers, there were several roofs of cottages, which had been transported almost whole. The store-houses at Talcahuano had been burst open, and great bags of cotton, yerba, and other valuable merchandise were scattered on the shore. During my walk round the island, I observed that numerous fragments of rock, which, from the marine productions adhering to them, must recently have been lying in deep water, had been cast up high on the beach; one of these was six feet long, three broad, and two thick.

The island itself as plainly showed the overwhelming power of the earthquake, as the beach did that of the consequent great wave. The ground in many parts was fissured in north and south lines, perhaps caused by the yielding of the parallel and steep sides of this narrow island. Some of the fissures near the cliffs were a yard wide. Many enormous masses had already fallen on the beach; and the inhabitants thought that when the rains commenced far greater slips would happen. The effect of the vibration on the hard primary slate, which composes the foundation of the island, was still more curious: the superficial parts of some narrow ridges were as completely shivered as if they had been blasted by gunpowder. This effect, which was rendered conspicuous by the fresh fractures and displaced soil, must be confined to near the surface, for otherwise there would not exist a block of solid rock throughout Chile; nor is this improbable, as it is known that the surface of a vibrating body is affected differently from the central part. It is, perhaps, owing to this same reason, that earthquakes do not cause quite such terrific havoc within deep mines as would be expected. I believe this convulsion has been more effectual in lessening the size of the island of Quiriquina, than the ordinary wear-and-tear of the sea and weather during the course of a whole century.

The next day I landed at Talcahuano, and afterwards rode to Concepcion. Both towns presented the most awful yet interesting spectacle I ever beheld. To a person who had formerly known them, it possibly might have been still more impressive; for the ruins were so mingled together, and the whole scene possessed so little the air of a habitable place, that it was scarcely possible to imagine its former condition. The earthquake commenced at half-past eleven o'clock in the forenoon. If it had happened in the middle of the night, the greater number of the inhabitants (which in this one province amount to many thousands) must have perished, instead of less than a hundred: as it was, the invariable practice of running out of doors at the first trembling of the ground, alone saved them. In Concepcion each house, or row of houses, stood by itself, a heap or line of ruins; but in Talcahuano, owing to the great wave, little more than one layer of bricks, tiles, and timber, with here and there part of a wall left standing, could be distinguished. From this circumstance

Concepcion, although not so completely desolated, was a more terrible, and, if I may so call it, picturesque sight. The first shock was very
sudden. The major-domo at Quiriquina told me, that the first notice
he received of it, was finding both the horse he rode and himself,
rolling together on the ground. Rising up, he was again thrown down.
He also told me that some cows which were standing on the steep side
of the island were rolled into the sea. The great wave caused the
destruction of many cattle; on one low island, near the head of the
bay, seventy animals were washed off and drowned. It is generally
thought that this has been the worst earthquake ever recorded in
Chile; but as the very severe ones occur only after long intervals, this
cannot easily be known; nor indeed would a much worse shock have
made any great difference, for the ruin was now complete. Innumerable small tremblings followed the great earthquake, and within the
first twelve days no less than three hundred were counted.

After viewing Concepcion, I cannot understand how the greater
number of inhabitants escaped unhurt. The houses in many parts fell
outwards; thus forming in the middle of the streets little hillocks of
brickwork and rubbish. Mr. Rouse, the English consul, told us that
he was at breakfast when the first movement warned him to run out.
He had scarcely reached the middle of the courtyard, when one side
of his house came thundering down. He retained presence of mind to
remember, that if he once got on the top of that part which had
already fallen, he would be safe. Not being able from the motion of
the ground to stand, he crawled up on his hands and knees; and no
sooner had he ascended this little eminence, than the other side of the
house fell in, the great beams sweeping close in front of his head.
With his eyes blinded, and his mouth choked with the cloud of dust
which darkened the sky, at last he gained the street. As shock succeeded shock, at the interval of a few minutes, no one dared approach
the shattered ruins; and no one knew whether his dearest friends and
relations were not perishing from the want of help. Those who had
saved any property were obliged to keep a constant watch, for thieves
prowled about, and at each little trembling of the ground, with one
hand they beat their breasts and cried 'misericordia!' and then with
the other filched what they could from the ruins. The thatched roofs
fell over the fires, and flames burst forth in all parts. Hundreds knew

themselves ruined, and few had the means of providing food for the day.

Earthquakes alone are sufficient to destroy the prosperity of any country. If beneath England the now inert subterranean forces should exert those powers, which most assuredly in former geological ages they have exerted, how completely would the entire condition of the country be changed! What would become of the lofty houses, thickly packed cities, great manufactories, the beautiful public and private edifices? If the new period of disturbance were first to commence by some great earthquake in the dead of the night, how terrific would be the carnage! England would at once be bankrupt; all papers, records, and accounts would from that moment be lost. Government being unable to collect the taxes, and failing to maintain its authority, the hand of violence and rapine would remain uncontrolled. In every large town famine would go forth, pestilence and death following in its train.

Shortly after the shock, a great wave was seen from the distance of three or four miles, approaching in the middle of the bay with a smooth outline; but along the shore it tore up cottages and trees, as it swept onwards with irresistible force. At the head of the bay it broke in a fearful line of white breakers, which rushed up to a height of 23 vertical feet above the highest spring-tides. Their force must have been prodigious; for at the Fort a cannon with its carriage, estimated at four tons in weight, was moved 15 feet inwards. A schooner was left in the midst of the ruins, 200 yards from the beach. The first wave was followed by two others, which in their retreat carried away a vast wreck of floating objects. In one part of the bay, a ship was pitched high and dry on shore, was carried off, again driven on shore, and again carried off. In another part, two large vessels anchored near together were whirled about, and their cables were thrice wound round each other: though anchored at a depth of 36 feet, they were for some minutes aground. The great wave must have travelled slowly, for the inhabitants of Talcahuano had time to run up the hills behind the town; and some sailors pulled out seaward, trusting successfully to their boat riding securely over the swell, if they could reach it before it broke. One old woman with a little boy, four or five years old, ran into a boat, but there was nobody to row it out: the boat was consequently dashed against an anchor and cut in twain; the old woman was drowned, but the child was picked up some hours afterwards clinging to the wreck. Pools of salt-water were still standing amidst the ruins of the houses, and children, making boats with old tables and chairs, appeared as happy as their parents were miserable. It was, however, exceedingly interesting to observe, how much more active and cheerful all appeared than could have been expected. It was remarked with much truth, that from the destruction being universal, no one individual was humbled more than another, or could suspect his friends of coldness – that most grievous result of the loss of wealth. Mr. Rouse, and a large party whom he kindly took under his protection, lived for the first week in a garden beneath some apple-trees. At first they were as merry as if it had been a picnic; but soon afterwards heavy rain caused much discomfort, for they were absolutely without shelter.

In Captain Fitz Roy's excellent account of the earthquake, it is said that two explosions, one like a column of smoke and another like the blowing of a great whale, were seen in the bay. The water also appeared every where to be boiling; and it 'became black, and exhaled a most disagreeable sulphureous smell.' These latter circumstances were observed in the Bay of Valparaiso during the earthquake of 1822; they may, I think, be accounted for, by the disturbance of the mud at the bottom of the sea containing organic matter in decay. In the Bay of Callao, during a calm day, I noticed, that as the ship dragged her cable over the bottom, its course was marked by a line of bubbles. The lower orders in Talcahuano thought that the earthquake was caused by some old Indian women, who two years ago being offended stopped the volcano of Antuco. This silly belief is curious, because it shows that experience has taught them to observe, that there exists a relation between the suppressed action of the volcanoes, and the trembling of the ground. It was necessary to apply the witchcraft to the point where their perception of cause and effect failed; and this was the closing of the volcanic vent. This belief is the more singular in this particular instance, because, according to Captain Fitz Roy, there is reason to believe that Antuco was noways affected.

The town of Concepcion was built in the usual Spanish fashion, with all the streets running at right angles to each other; one set ranging S.W. by W., and the other set N.W. by N. The walls in the former direction certainly stood better than those in the latter: the greater number of the masses of brickwork were thrown down towards the N.E. Both these circumstances perfectly agree with the general idea, of the undulations having come from the S.W.; in which quarter subterranean noises were also heard: for it is evident that the walls running S.W. and N.E. which presented their ends to the point whence the undulations came, would be much less likely to fall than those walls which, running N.W. and S.E., must in their whole lengths have been at the same instant thrown out of the perpendicular; for the undulations, coming from the S.W., must have extended in N.W. and S.E. waves, as they passed under the foundations. This may be illustrated by placing books edgeways on a carpet, and then, after the manner suggested by Michell, imitating the undulations of an earthquake: it will be found that they fall with more or less readiness, according as their direction more or less nearly coincides with the line of the waves. The fissures in the ground generally, though not uniformly, extended in a S.E. and N.W. direction; and therefore corresponded to the lines of undulation or of principal flexure. Bearing in mind all these circumstances, which so clearly point to the S.W. as the chief focus of disturbance, it is a very interesting fact that the island of S. Maria, situated in that quarter, was, during the general uplifting of the land, raised to nearly three times the height of any other part of the coast.

The different resistance offered by the walls, according to their direction, was well exemplified in the case of the Cathedral. The side which fronted the N.E. presented a grand pile of ruins, in the midst of which door-cases and masses of timber stood up, as if floating in a stream. Some of the angular blocks of brickwork were of great dimensions; and they were rolled to a distance on the level plaza, like fragments of rock at the base of some high mountain. The side walls (running S.W. and N.E.), though exceedingly fractured, yet remained standing; but the vast buttresses (at right angles to them, and therefore parallel to the walls that fell) were in many cases cut clean off, as if by a chisel, and hurled to the ground. Some square ornaments on the coping of these same walls, were moved by the earthquake into a diagonal position. A similar circumstance was observed after an

earthquake at Valparaiso, Calabria, and other places, including some
of the ancient Greek temples.[1] This twisting displacement, at first
appears to indicate a vorticose movement beneath each point thus
affected; but this is highly improbable. May it not be caused by a
tendency in each stone to arrange itself in some particular position,
with respect to the lines of vibration, – in a manner somewhat similar
to pins on a sheet of paper when shaken? Generally speaking, arched
doorways or windows stood much better than any other part of the
buildings. Nevertheless, a poor lame old man, who had been in the
habit, during trifling shocks, of crawling to a certain doorway, was
this time crushed to pieces.

I have not attempted to give any detailed description of the
appearance of Concepcion, for I feel that it is quite impossible to
convey the mingled feelings which I experienced. Several of the
officers visited it before me, but their strongest language failed to
give a just idea of the scene of desolation. It is a bitter and humiliating
thing to see works, which have cost man so much time and labour,
overthrown in one minute; yet compassion for the inhabitants was
almost instantly banished, by the surprise in seeing a state of things
produced in a moment of time, which one was accustomed to
attribute to a succession of ages. In my opinion, we have scarcely
beheld, since leaving England, any sight so deeply interesting.

In almost every severe earthquake, the neighbouring waters of the
sea are said to have been greatly agitated. The disturbance seems
generally, as in the case of Concepcion, to have been of two kinds:
first, at the instant of the shock, the water swells high up on the beach
with a gentle motion, and then as quietly retreats; secondly, some
time afterwards, the whole body of the sea retires from the coast, and
then returns in waves of overwhelming force. The first movement
seems to be an immediate consequence of the earthquake affecting
differently a fluid and a solid, so that their respective levels are slightly
deranged: but the second case is a far more important phenomenon.
During most earthquakes, and especially during those on the west
coast of America, it is certain that the first great movement of the

---

1 M. Arago in L'Institut, 1839, p. 337. See also Miers's Chile, vol. i. p. 392; also
Lyell's Principles of Geology, chap. xv. book ii.

waters has been a retirement. Some authors have attempted to explain this, by supposing that the water retains its level, whilst the land oscillates upwards; but surely the water close to the land, even on a rather steep coast, would partake of the motion of the bottom: moreover, as urged by Mr. Lyell, similar movements of the sea have occurred at islands far distant from the chief line of disturbance, as was the case with Juan Fernandez during this earthquake, and with Madeira during the famous Lisbon shock. I suspect (but the subject is a very obscure one) that a wave, however produced, first draws the water from the shore, on which it is advancing to break: I have observed that this happens with the little waves from the paddles of a steam-boat. It is remarkable that whilst Talcahuano and Callao (near Lima), both situated at the head of large shallow bays, have suffered during every severe earthquake from great waves, Valparaiso, seated close to the edge of profoundly deep water, has never been over-whelmed, though so often shaken by the severest shocks. From the great wave not immediately following the earthquake, but sometimes after the interval of even half an hour, and from distant islands being affected similarly with the coasts near the focus of the disturbance, it appears that the wave first rises in the offing; and as this is of general occurrence, the cause must be general: I suspect we must look to the line, where the less disturbed waters of the deep ocean join the water nearer the coast, which has partaken of the movements of the land, as the place where the great wave is first generated; it would also appear that the wave is larger or smaller, according to the extent of shoal water which has been agitated together with the bottom on which it rested.

The most remarkable effect of this earthquake was the permanent elevation of the land; it would probably be far more correct to speak of it as the cause. There can be no doubt that the land round the Bay of Concepcion was upraised two or three feet; but it deserves notice, that owing to the wave having obliterated the old lines of tidal action on the sloping sandy shores, I could discover no evidence of this fact, except in the united testimony of the inhabitants, that one little rocky shoal, now exposed, was formerly covered with water. At the island of S. Maria (about thirty miles distant) the elevation was greater; on one part, Captain Fitz Roy found beds of putrid mussel-shells *still adhering to the rocks*, ten feet above high-water mark: the inhabitants had formerly dived at low-water spring-tides for these shells. The elevation of this province is particularly interesting, from its having been the theatre of several other violent earthquakes, and from the vast numbers of sea-shells scattered over the land, up to a height of certainly 600, and I believe, of 1000 feet. At Valparaiso, as I have remarked, similar shells are found at the height of 1300 feet: it is hardly possible to doubt that this great elevation has been effected by successive small uprisings, such as that which accompanied or caused the earthquake of this year, and likewise by an insensibly slow rise, which is certainly in progress on some parts of this coast.

The island of Juan Fernandez, 360 miles to the N.E., was, at the time of the great shock of the 20th, violently shaken, so that the trees beat against each other, and a volcano burst forth under water close to the shore: these facts are remarkable because this island, during the earthquake of 1751, was then also affected more violently than other places at an equal distance from Concepcion, and this seems to show some subterranean connexion between these two points. Chiloe, about 340 miles southward of Concepcion, appears to have been shaken more strongly than the intermediate district of Valdivia, where the volcano of Villarica was noways affected, whilst in the Cordillera in front of Chiloe, two of the volcanoes burst forth at the same instant in violent action. These two volcanoes, and some neighbouring ones, continued for a long time in eruption, and ten months afterwards were again influenced by an earthquake at Concepcion. Some men, cutting wood near the base of one of these volcanoes, did not perceive the shock of the 20th, although the whole surrounding Province was then trembling; here we have an eruption relieving and taking the place of an earthquake, as would have happened at Concepcion, according to the belief of the lower orders, if the volcano of Antuco had not been closed by witchcraft. Two years and three quarters afterwards, Valdivia and Chiloe were again shaken, more violently than on the 20th, and an island in the Chonos Archipelago was permanently elevated more than eight feet. It will give a better idea of the scale of these phenomena, if (as in the case of the glaciers) we suppose them to have taken place at corresponding distances in Europe: – then would the land from the North Sea to the Mediterranean have been violently shaken, and at the same instant of time a large tract of the eastern coast of England would have been permanently elevated, together with some outlying islands – a train of volcanoes on the coast of Holland would have burst forth in action, and an eruption taken place at the bottom of the sea, near the northern extremity of Ireland – and lastly, the ancient vents of Auvergne, Cantal, and Mont d'Or would each have sent up to the sky a dark column of smoke, and have long remained in fierce action. Two years and three quarters afterwards, France, from its centre to the English Channel, would have been again desolated by an earthquake, and an island permanently upraised in the Mediterranean.

The space, from under which volcanic matter on the 20th was actually erupted, is 720 miles in one line, and 400 miles in another line at right angles to the first: hence, in all probability, a subterranean lake of lava is here stretched out, of nearly double the area of the Black Sea. From the intimate and complicated manner in which the elevatory and eruptive forces were shown to be connected during this train of phenomena, we may confidently come to the conclusion, that the forces which slowly and by little starts uplift continents, and those which at successive periods pour forth volcanic matter from open orifices, are identical. From many reasons, I believe that the frequent quakings of the earth on this line of coast, are caused by the rending of the strata, necessarily consequent on the tension of the land when upraised, and their injection by fluidified rock. This rending and injection would, if repeated often enough (and we know that earthquakes repeatedly affect the same areas in the same manner), form a chain of hills; – and the linear island of St. Mary, which was upraised thrice the height of the neighbouring country, seems to be undergoing this process. I believe that the solid axis of a mountain differs in its manner of formation from a volcanic hill, only in the molten stone having been repeatedly injected, instead of having been repeatedly ejected. Moreover, I believe that it is impossible to explain the structure of great mountain-chains, such as that of the Cordillera, where the strata, capping the injected axis of plutonic rock, have been thrown on their edges along several parallel and neighbouring lines of elevation, except on this view of the rock of the axis having been repeatedly injected, after intervals sufficiently long to allow the upper parts or wedges to cool and become solid; – for if the strata had been thrown into their present highly-inclined, vertical, and even inverted positions, by a single blow, the very bowels of the earth would have gushed out; and instead of beholding abrupt mountain-axes of rock solidified under great pressure, deluges of lava would have flowed out at innumerable points on every line of elevation.[1]

1 For a full account of the volcanic phenomena which accompanied the earthquake of the 20th, and for the conclusions deducible from them, I must refer to Volume V. of the Geological Transactions.

CHAPTER XV

PASSAGE OF THE
CORDILLERA

Valparaiso – Portillo Pass – Sagacity of Mules – Mountain-
torrents – Mines, how discovered – Proofs of the gradual
Elevation of the Cordillera – Effect of Snow on Rocks –
Geological Structure of the two Main Ranges, their dis-
tinct Origin and Upheaval – Great Subsidence – Red Snow
– Winds – Pinnacles of Snow – Dry and clear Atmosphere
– Electricity – Pampas – Zoology of the opposite sides of
the Andes – Locusts – Great Bugs – Mendoza – Uspallata
Pass – Silicified Trees buried as they grew – Incas Bridge –
Badness of the Passes exaggerated – Cumbre – Casuchas –
Valparaiso.

MARCH 7th, 1835. – We stayed three days at Concepcion, and then
sailed for Valparaiso. The wind being northerly, we only reached
the mouth of the harbour of Concepcion before it was dark. Being
very near the land, and a fog coming on, the anchor was dropped.
Presently a large American whaler appeared close alongside of us; and
we heard the Yankee swearing at his men to keep quiet, whilst
he listened for the breakers. Captain Fitz Roy hailed him, in a loud
clear voice, to anchor where he then was. The poor man must have
thought the voice came from the shore: such a Babel of cries issued
at once from the ship – every one hallooing out, 'Let go the anchor!
veer cable! shorten sail!' It was the most laughable thing I ever heard.
If the ship's crew had been all captains, and no men, there could
not have been a greater uproar of orders. We afterwards found that
the mate stuttered: I suppose all hands were assisting him in giving
his orders.

On the 11th we anchored at Valparaiso, and two days afterwards
I set out to cross the Cordillera. I proceeded to Santiago, where
Mr. Caldcleugh most kindly assisted me in every possible way in
making the little preparations which were necessary. In this part of

Chile there are two passes across the Andes to Mendoza: the one most commonly used – namely, that of Aconcagua or Uspallata – is situated some way to the north; the other, called the Portillo, is to the south, and nearer, but more lofty and dangerous.

March 18th. – We set out for the Portillo pass. Leaving Santiago we crossed the wide burnt-up plain on which that city stands, and in the afternoon arrived at the Maypu, one of the principal rivers in Chile. The valley, at the point where it enters the first Cordillera, is bounded on each side by lofty barren mountains; and although not broad, it is very fertile. Numerous cottages were surrounded by vines, and by orchards of apple, nectarine, and peach trees – their boughs breaking with the weight of the beautiful ripe fruit. In the evening we passed the custom-house, where our luggage was examined. The frontier of Chile is better guarded by the Cordillera, than by the waters of the sea. There are very few valleys which lead to the central ranges, and the mountains are quite impassable in other parts by beasts of burden. The custom-house officers were very civil, which was perhaps partly owing to the passport which the President of the Republic had given me; but I must express my admiration at the natural politeness of almost every Chileno. In this instance, the contrast with the same class of men in most other countries was strongly marked. I may mention an anecdote with which I was at the time much pleased: we met near Mendoza a little and very fat negress, riding astride on a mule. She had a *goître* so enormous that it was scarcely possible to avoid gazing at her for a moment; but my two companions almost instantly, by way of apology, made the common salute of the country by taking off their hats. Where would one of the lower or higher classes in Europe, have shown such feeling politeness to a poor and miserable object of a degraded race?

At night we slept at a cottage. Our manner of travelling was delightfully independent. In the inhabited parts we bought a little firewood, hired pasture for the animals, and bivouacked in the corner of the same field with them. Carrying an iron pot, we cooked and ate our supper under a cloudless sky, and knew no trouble. My companions were Mariano Gonzales, who had formerly accompanied me in Chile, and an 'arriero,' with his ten mules and a 'madrina.' The madrina (or godmother) is a most important personage: she is an old steady mare, with a little bell round her neck; and wherever she goes, the mules, like good children, follow her. The affection of these animals for their madrinas saves infinite trouble. If several large troops are turned into one field to graze, in the morning the muleteers have only to lead the madrinas a little apart, and tinkle their bells; and although there may be two or three hundred together, each mule immediately knows the bell of its own madrina, and comes to her. It is nearly impossible to lose an old mule; for if detained for several hours by force, she will, by the power of smell, like a dog, track out her companions, or rather the madrina, for, according to the muleteer, she is the chief object of affection. The feeling, however, is not of an individual nature; for I believe I am right in saying that any animal with a bell will serve as a madrina. In a troop each animal carries on a level road, a cargo weighing 416 pounds (more than 29 stone), but in a mountainous country 100 pounds less; yet with what delicate slim limbs, without any proportional bulk of muscle, these animals support so great a burden! The mule always appears to me a most surprising animal. That a hybrid should possess more reason, memory, obstinacy, social affection, powers of muscular endurance, and length of life, than either of its parents, seems to indicate that art has here outdone nature. Of our ten animals, six were intended for riding, and four for carrying cargoes, each taking turn about. We carried a good deal of food, in case we should be snowed up, as the season was rather late for passing the Portillo.

March 19*th*. – We rode during this day to the last, and therefore most elevated house in the valley. The number of inhabitants became scanty; but wherever water could be brought on the land, it was very fertile. All the main valleys in the Cordillera are characterized by having, on both sides, a fringe or terrace of shingle and sand, rudely stratified, and generally of considerable thickness. These fringes evidently once extended across the valleys, and were united; and the bottoms of the valleys in northern Chile, where there are no streams, are thus smoothly filled up. On these fringes the roads are generally carried, for their surfaces are even, and they rise with a very gentle slope up the valleys: hence, also, they are easily cultivated by irrigation. They may be traced up to a height of between 7000 and 9000 feet, where they become hidden by the irregular piles of debris. At the lower end or mouths of the valleys, they are continuously united to those land-locked plains (also formed of shingle) at the foot of the main Cordillera, which I have described in a former chapter as characteristic of the scenery of Chile, and which were undoubtedly deposited when the sea penetrated Chile, as it now does the more southern coasts. No one fact in the geology of South America interested me more than these terraces of rudely-stratified shingle. They precisely resemble in composition, the matter which the tor-rents in each valley would deposit, if they were checked in their course by any cause, such as entering a lake or arm of the sea; but the torrents, instead of depositing matter, are now steadily at work wearing away both the solid rock and these alluvial deposits, along the whole line of every main valley and side valley. It is impossible here to give the reasons, but I am convinced that the shingle terraces were accumulated, during the gradual elevation of the Cordillera, by the torrents delivering, at successive levels, their detritus on the beach-heads of long narrow arms of the sea, first high up the valleys, then lower and lower down as the land slowly rose. If this be so, and I cannot doubt it, the grand and broken chain of the Cordillera, instead of having been suddenly thrown up, as was till lately the universal, and still is the common opinion of geologists, has been slowly upheaved in mass, in the same gradual manner as the coasts of the Atlantic and Pacific have risen within the recent period. A multitude of facts in the structure of the Cordillera, on this view receive a simple explanation.

The rivers which flow in these valleys ought rather to be called mountain-torrents. Their inclination is very great, and their water the colour of mud. The roar which the Maypu made, as it rushed over the great rounded fragments, was like that of the sea. Amidst the din of rushing waters, the noise from the stones, as they rattled one over another, was most distinctly audible even from a distance. This rattling noise, night and day, may be heard along the whole course of the torrent. The sound spoke eloquently to the geologist; the thousands and thousands of stones, which, striking against each other, made the one dull uniform sound, were all hurrying in one direction. It was like thinking on time, where the minute that now glides past is irrecoverable. So was it with these stones; the ocean is their eternity, and each note of that wild music told of one more step towards their destiny.

It is not possible for the mind to comprehend, except by a slow process, any effect which is produced by a cause repeated so often, that the multiplier itself conveys an idea, not more definite than the savage implies when he points to the hairs of his head. As often as I have seen beds of mud, sand, and shingle, accumulated to the thickness of many thousand feet, I have felt inclined to exclaim that causes, such as the present rivers and the present beaches, could never have ground down and produced such masses. But, on the other hand, when listening to the rattling noise of these torrents, and calling to mind that whole races of animals have passed away from the face of the earth, and that during this whole period, night and day, these stones have gone rattling onwards in their course, I have thought to myself, can any mountains, any continent, withstand such waste?

In this part of the valley, the mountains on each side were from 3000 to 6000 or 8000 feet high, with rounded outlines and steep bare flanks. The general colour of the rock was dullish purple, and the stratification very distinct. If the scenery was not beautiful, it was remarkable and grand. We met during the day several herds of cattle, which men were driving down from the higher valleys in the Cordillera. This sign of the approaching winter hurried our steps, more than was convenient for geologizing. The house where we slept was situated at the foot of a mountain, on the summit of which are the mines of S. Pedro de Nolasko. Sir F. Head marvels how mines have been discovered in such extraordinary situations, as the bleak summit of the mountain of S. Pedro de Nolasko. In the first place, metallic veins in this country are generally harder than the surrounding strata: hence, during the gradual wear of the hills, they project above the surface of the ground. Secondly, almost every labourer, especially in the northern parts of Chile, understands something about the appearance of ores. In the great mining provinces of Coquimbo and Copiapó, firewood is very scarce, and men search for it over every hill and dale; and by this means nearly all the richest mines have there been discovered. Chanuncillo, from which silver to the value of many hundred thousand pounds has been raised in the course of a few years, was discovered by a man who threw a stone at his loaded donkey, and thinking that it was very heavy, he picked it up, and found it full of pure silver: the vein occurred at no great distance, standing up like a wedge of metal. The miners, also, taking a crowbar with them, often wander on Sundays over the mountains. In this south part of Chile, the men who drive cattle into the Cordillera, and who frequent every ravine where there is a little pasture, are the usual discoverers.

20th. – As we ascended the valley, the vegetation, with the exception of a few pretty alpine flowers, became exceedingly scanty; and of quadrupeds, birds, or insects, scarcely one could be seen. The lofty mountains, their summits marked with a few patches of snow, stood well separated from each other; the valleys being filled up with an immense thickness of stratified alluvium. The features in the scenery of the Andes which struck me most, as contrasted with the other mountain chains with which I am acquainted, were, – the flat fringes sometimes expanding into narrow plains on each side of the valleys, – the bright colours, chiefly red and purple, of the utterly bare and precipitous hills of porphyry, – the grand and continuous wall-like dikes, – the plainly-divided strata which, where nearly vertical, formed the picturesque and wild central pinnacles, but where less inclined, composed the great massive mountains on the outskirts of the range, – and lastly, the smooth conical piles of fine and brightly-coloured detritus, which sloped up at a high angle from the base of the mountains, sometimes to a height of more than 2000 feet.

I frequently observed, both in Tierra del Fuego and within the Andes, that where the rock was covered during the greater part of the year with snow, it was shivered in a very extraordinary manner into small angular fragments. Scoresby[1] has observed the same fact in Spitzbergen. The case appears to me rather obscure: for that part of the mountain which is protected by a mantle of snow, must be less subject to repeated and great changes of temperature than any other part. I have sometimes thought, that the earth and fragments of stone on the surface, were perhaps less effectually removed by slowly

1 Scoresby's Arctic Regions, vol. i. p. 122.

percolating snow water[1] than by rain, and therefore that the appearance of a quicker disintegration of the solid rock under the snow, was deceptive. Whatever the cause may be, the quantity of crumbling stone on the Cordillera is very great. Occasionally in the spring, great masses of this detritus slide down the mountains, and cover the snowdrifts in the valleys, thus forming natural ice-houses. We rode over one, the height of which was far below the limit of perpetual snow.

As the evening drew to a close, we reached a singular basin-like plain, called the Valle del Yeso. It was covered by a little dry pasture, and we had the pleasant sight of a herd of cattle amidst the surrounding rocky deserts. The valley takes its name of Yeso from a great bed, I should think at least 2000 feet thick, of white, and in some parts quite pure, gypsum. We slept with a party of men, who were employed in loading mules with this substance, which is used in the manufacture of wine. We set out early in the morning (21st), and continued to follow the course of the river, which had become very small, till we arrived at the foot of the ridge, that separates the waters flowing into the Pacific and Atlantic Oceans. The road, which as yet had been good with a steady but very gradual ascent, now changed into a steep zigzag track up the great range, dividing the republics of Chile and Mendoza.

I will here give a very brief sketch of the geology of the several parallel lines forming the Cordillera. Of these lines, there are two considerably higher than the others; namely, on the Chilian side, the Peuquenes ridge, which, where the road crosses it, is 13,210 feet above the sea; and the Portillo ridge, on the Mendoza side, which is 14,305 feet. The lower beds of the Peuquenes ridge, and of the several great lines to the westward of it, are composed of a vast pile, many thousand feet in thickness, of porphyries which have flowed as submarine lavas, alternating with angular and rounded fragments of the same rocks, thrown out of the submarine craters. These alternating masses are covered in the central parts, by a great

---

1 I have heard it remarked in Shropshire, that the water, when the Severn is flooded from long-continued rain, is much more turbid than when it proceeds from the snow melting on the Welsh mountains. D'Orbigny (tom. i. p. 184), in explaining the cause of the various colours of the rivers in South America, remarks that those with blue or clear water have their source in the Cordillera, where the snow melts.

thickness of red sandstone, conglomerate, and calcareous clay-slate, associated with, and passing into, prodigious beds of gypsum. In these upper beds shells are tolerably frequent; and they belong to about the period of the lower chalk of Europe. It is an old story, but not the less wonderful, to hear of shells which were once crawling on the bottom of the sea, now standing nearly 14,000 feet above its level. The lower beds in this great pile of strata, have been dislocated, baked, crystallized and almost blended together, through the agency of mountain masses of a peculiar white soda-granitic rock.

The other main line, namely, that of the Portillo, is of a totally different formation: it consists chiefly of grand bare pinnacles of a red potash-granite, which low down on the western flank are covered by a sandstone, converted by the former heat into a quartz-rock. On the quartz, there rest beds of a conglomerate several thousand feet in thickness, which have been upheaved by the red granite, and dip at an angle of 45° towards the Peuquenes line. I was astonished to find that this conglomerate was partly composed of pebbles, derived from the rocks, with their fossil shells, of the Peuquenes range; and partly of red potash-granite, like that of the Portillo. Hence we must conclude, that both the Peuquenes and Portillo ranges were partially upheaved and exposed to wear and tear, when the conglomerate was forming; but as the beds of the conglomerate have been thrown off at an angle of 45° by the red Portillo granite (with the underlying sandstone baked by it), we may feel sure, that the greater part of the injection and upheaval of the already partially formed Portillo line, took place after the accumulation of the conglomerate, and long after the elevation of the Peuquenes ridge. So that the Portillo, the loftiest line in this part of the Cordillera, is not so old as the less lofty line of the Peuquenes. Evidence derived from an inclined stream of lava at the eastern base of the Portillo, might be adduced to show, that it owes part of its great height to elevations of a still later date. Looking to its earliest origin, the red granite seems to have been injected on an ancient pre-existing line of white granite and mica-slate. In most parts, perhaps in all parts, of the Cordillera, it may be concluded that each line has been formed by repeated upheavals and injections; and that the several parallel lines are of different ages. Only thus can we gain time, at all sufficient to explain the truly astonishing amount of denudation, which these great, though comparatively with most other ranges recent, mountains have suffered.

Finally, the shells in the Peuquenes or oldest ridge, prove, as before remarked, that it has been upraised 14,000 feet since a Secondary period, which in Europe we are accustomed to consider as far from ancient; but since these shells lived in a moderately deep sea, it can be shown that the area now occupied by the Cordillera, must have subsided several thousand feet – in northern Chile as much as 6000 feet – so as to have allowed that amount of submarine strata to have been heaped on the bed on which the shells lived. The proof is the same with that by which it was shown, that at a much later period since the tertiary shells of Patagonia lived, there must have been there a subsidence of several hundred feet, as well as an ensuing elevation. Daily it is forced home on the mind of the geologist, that nothing, not even the wind that blows, is so unstable as the level of the crust of this earth.

I will make only one other geological remark: although the Portillo chain is here higher than the Peuquenes, the waters, draining the intermediate valleys, have burst through it. The same fact, on a grander scale, has been remarked in the eastern and loftiest line of the Bolivian Cordillera, through which the rivers pass: analogous facts have also been observed in other quarters of the world. On the supposition of the subsequent and gradual elevation of the Portillo line, this can be understood; for a chain of islets would at first appear, and, as these were lifted up, the tides would be always wearing deeper and broader channels between them. At the present day, even in the most retired Sounds on the coast of Tierra del Fuego, the currents in the transverse breaks which connect the longitudinal channels, are very strong, so that in one transverse channel even a small vessel under sail was whirled round and round.

About noon we began the tedious ascent of the Peuquenes ridge, and then for the first time experienced some little difficulty in our respiration. The mules would halt every fifty yards, and after resting for a few seconds the poor willing animals started of their own accord again. The short breathing from the rarefied atmosphere is called by the Chilenos 'puna;' and they have most ridiculous notions concerning its origin. Some say 'all the waters here have puna;' others that 'where there is snow there is puna;' – and this no doubt is true. The only sensation I experienced was a slight tightness across the head and chest, like that felt on leaving a warm room and running quickly in frosty weather. There was some imagination even in this; for upon finding fossil shells on the highest ridge, I entirely forgot the puna in my delight. Certainly the exertion of walking was extremely great, and the respiration became deep and laborious: I am told that in Potosi (about 13,000 feet above the sea) strangers do not become thoroughly accustomed to the atmosphere for an entire year. The inhabitants all recommend onions for the puna; as this vegetable has sometimes been given in Europe for pectoral complaints, it may possibly be of real service: – for my part I found nothing so good as the fossil shells!

When about half-way up we met a large party with seventy loaded mules. It was interesting to hear the wild cries of the muleteers, and to watch the long descending string of the animals; they appeared so diminutive, there being nothing but the bleak mountains with which they could be compared. When near the summit, the wind, as generally happens, was impetuous and extremely cold. On each side of the ridge we had to pass over broad bands of perpetual snow, which were now soon to be covered by a fresh layer. When we reached the crest and looked backwards, a glorious view was presented. The atmosphere resplendently clear; the sky an intense blue; the profound valleys; the wild broken forms; the heaps of ruins, piled up during the lapse of ages; the bright-coloured rocks, contrasted with the quiet mountains of snow; all these together produced a scene no one could have imagined. Neither plant nor bird, excepting a few condors wheeling around the higher pinnacles, distracted my attention from the inanimate mass. I felt glad that I was alone: it was like watching a thunderstorm, or hearing in full orchestra a chorus of the Messiah.

On several patches of the snow I found the Protococcus nivalis, or red snow, so well known from the accounts of Arctic navigators. My attention was called to it, by observing the footsteps of the mules stained a pale red, as if their hoofs had been slightly bloody. I at first thought that it was owing to dust blown from the surrounding mountains of red porphyry; for from the magnifying power of the crystals of snow, the groups of these microscopical plants appeared like coarse particles. The snow was coloured only where it had thawed very rapidly, or had been accidentally crushed. A little rubbed on paper gave it a faint rose tinge mingled with a little brick-red. I afterwards scraped some off the paper, and found that it consisted of groups of little spheres in colourless cases, each the thousandth part of an inch in diameter.

The wind on the crest of the Peuquenes, as just remarked, is generally impetuous and very cold: it is said[1] to blow steadily from the westward or Pacific side. As the observations have been chiefly made in summer, this wind must be an upper and return current. The Peak of Teneriffe, with a less elevation, and situated in lat. 28°, in like manner falls within an upper return stream. At first it appears rather surprising, that the trade-wind along the northern parts of Chile and on the coast of Peru, should blow in so very southerly a direction as it does; but when we reflect that the Cordillera, running in a north and south line, intercepts, like a great wall, the entire depth of the lower atmospheric current, we can easily see that the trade-wind must be drawn northward, following the line of mountains, towards the equatorial regions, and thus lose part of that easterly movement which it otherwise would have gained from the earth's rotation. At Mendoza, on the eastern foot of the Andes, the climate is said to be subject to long calms, and to frequent though false appearances of gathering rain-storms: we may imagine that the wind, which coming from the eastward is thus banked up by the line of mountains, would become stagnant and irregular in its movements.

Having crossed the Peuquenes, we descended into a mountainous country, intermediate between the two main ranges, and then took up our quarters for the night. We were now in the republic of Mendoza. The elevation was probably not under 11,000 feet, and the vegetation in consequence exceedingly scanty. The root of a small scrubby plant served as fuel, but it made a miserable fire, and

1 Dr. Gillies in Journ. of Nat. and Geograph. Science, Aug. 1830. This author gives the heights of the Passes.

the wind was piercingly cold. Being quite tired with my day's work, I made up my bed as quickly as I could, and went to sleep. About midnight I observed the sky became suddenly clouded: I awakened the arriero to know if there was any danger of bad weather; but he said that without thunder and lightning there was no risk of a heavy snow-storm. The peril is imminent, and the difficulty of subsequent escape great, to any one overtaken by bad weather between the two ranges. A certain cave offers the only place of refuge: Mr. Caldcleugh, who crossed on this same day of the month, was detained there for some time by a heavy fall of snow. Casuchas, or houses of refuge, have not been built in this pass as in that of Uspallata, and therefore, during the autumn, the Portillo is little frequented. I may here remark that within the main Cordillera rain never falls, for during the summer the sky is cloudless, and in winter snow-storms alone occur.

At the place where we slept water necessarily boiled, from the diminished pressure of the atmosphere, at a lower temperature than it does in a less lofty country; the case being the converse of that of a Papin's digester. Hence the potatoes, after remaining for some hours in the boiling water, were nearly as hard as ever. The pot was left on the fire all night, and next morning it was boiled again, but yet the potatoes were not cooked. I found out this, by overhearing my two companions discussing the cause; they had come to the simple conclusion, 'that the cursed pot (which was a new one) did not choose to boil potatoes.'

March 22nd. – After eating our potato-less breakfast, we travelled across the intermediate tract to the foot of the Portillo range. In the middle of summer cattle are brought up here to graze; but they had now all been removed: even the greater number of the guanacos had decamped, knowing well that if overtaken here by a snow-storm, they would be caught in a trap. We had a fine view of a mass of mountains called Tupungato, the whole clothed with unbroken snow, in the midst of which there was a blue patch, no doubt a glacier; – a circumstance of rare occurrence in these mountains. Now commenced a heavy and long climb, similar to that up the Peuquenes. Bold conical hills of red granite rose on each hand; in the valleys there were several broad fields of perpetual snow.

These frozen masses, during the process of thawing, had in some parts been converted into pinnacles or columns,[1] which, as they were high and close together, made it difficult for the cargo mules to pass. On one of these columns of ice, a frozen horse was sticking as on a pedestal, but with its hind legs straight up in the air. The animal, I suppose, must have fallen with its head downward into a hole, when the snow was continuous, and afterwards the surrounding parts must have been removed by the thaw.

When nearly on the crest of the Portillo, we were enveloped in a falling cloud of minute frozen spicula. This was very unfortunate, as it continued the whole day, and quite intercepted our view. The pass takes its name of Portillo, from a narrow cleft or doorway on the highest ridge, through which the road passes. From this point, on a clear day, those vast plains which uninterruptedly extend to the Atlantic Ocean, can be seen. We descended to the upper limit of vegetation, and found good quarters for the night under the shelter of some large fragments of rock. We met here some passengers, who made anxious inquiries about the state of the road. Shortly after it was dark the clouds suddenly cleared away, and the effect was quite magical. The great mountains, bright with the full moon, seemed impending over us on all sides, as over a deep crevice: one morning, very early, I witnessed the same striking effect. As soon as the clouds were dispersed it froze severely; but as there was no wind, we slept very comfortably.

The increased brilliancy of the moon and stars at this elevation, owing to the perfect transparency of the atmosphere, was very remarkable. Travellers having observed the difficulty of judging heights and distances amidst lofty mountains, have generally attributed it to the absence of objects of comparison. It appears to me, that it is fully as much owing to the transparency of the air confounding

1 This structure in frozen snow was long since observed by Scoresby in the icebergs near Spitzbergen, and lately, with more care, by Colonel Jackson (Journ. of Geograph. Soc., vol. v. p. 12) on the Neva. Mr. Lyell (Principles, vol. iv. p. 360) has compared the fissures, by which the columnar structure seems to be determined, to the joints that traverse nearly all rocks, but which are best seen in the non-stratified masses. I may observe, that in the case of the frozen snow, the columnar structure must be owing to a 'metamorphic' action, and not to a process during *deposition*.

objects at different distances, and likewise partly to the novelty of an unusual degree of fatigue arising from a little exertion, – habit being thus opposed to the evidence of the senses. I am sure that this extreme clearness of the air gives a peculiar character to the landscape, all objects appearing to be brought nearly into one plane, as in a drawing or panorama. The transparency is, I presume, owing to the equable and high state of atmospheric dryness. This dryness was shown by the manner in which woodwork shrank (as I soon found by the trouble my geological hammer gave me); by articles of food, such as bread and sugar, becoming extremely hard; and by the preservation of the skin and parts of the flesh of the beasts, which had perished on the road. To the same cause we must attribute the singular facility with which electricity is excited. My flannel-waistcoat, when rubbed in the dark, appeared as if it had been washed with phosphorus; – every hair on a dog's back crackled; – even the linen sheets, and leathern straps of the saddle, when handled, emitted sparks.

March 23rd. – The descent on the eastern side of the Cordillera is much shorter or steeper than on the Pacific side; in other words, the mountains rise more abruptly from the plains than from the alpine country of Chile. A level and brilliantly white sea of clouds was stretched out beneath our feet, shutting out the view of the equally level Pampas. We soon entered the band of clouds, and did not again emerge from it that day. About noon, finding pasture for the animals and bushes for firewood at Los Arenales, we stopped for the night. This was near the uppermost limit of bushes, and the elevation, I suppose, was between seven and eight thousand feet.

I was much struck with the marked difference between the vegetation of these eastern valleys and those on the Chilian side: yet the climate, as well as the kind of soil, is nearly the same, and the difference of longitude very trifling. The same remark holds good with the quadrupeds, and in a lesser degree with the birds and insects. I may instance the mice, of which I obtained thirteen species on the shores of the Atlantic, and five on the Pacific, and not one of them is identical. We must except all those species, which habitually or occasionally frequent elevated mountains; and certain birds, which range as far south as the Strait of Magellan. This fact is in perfect accordance with the geological history of the Andes; for these mountains have existed as a great barrier, since the present races of animals have appeared; and therefore, unless we suppose the same species to have been created in two different places, we ought not to expect any closer similarity between the organic beings on the opposite sides of the Andes, than on the opposite shores of the ocean. In both cases, we must leave out of the question those kinds which have been able to cross the barrier, whether of solid rock or salt-water.[1]

A great number of the plants and animals were absolutely the same as, or most closely allied to those of Patagonia. We here have the agouti, bizcacha, three species of armadillo, the ostrich, certain kinds of partridges and other birds, none of which are ever seen in Chile, but are the characteristic animals of the desert plains of Patagonia. We have likewise many of the same (to the eyes of the person who is not a botanist) thorny stunted bushes, withered grass, and dwarf plants. Even the black slowly-crawling beetles are closely similar, and some, I believe, on rigorous examination, absolutely identical. It had always been to me a subject of regret, that we were unavoidably compelled to give up the ascent of the S. Cruz river, before reaching the mountains: I always had a latent hope of meeting with some great change in the features of the country; but I now feel sure, that it would only have been following the plains of Patagonia up a mountainous ascent.

March 24th. – Early in the morning I climbed up a mountain on one side of the valley, and enjoyed a far extended view over the Pampas. This was a spectacle to which I had always looked forward with interest, but I was disappointed: at the first glance it much resembled a distant view of the ocean, but in the northern parts many irregularities were soon distinguishable. The most striking feature consisted in the rivers, which, facing the rising sun, glittered like silver threads, till lost in the immensity of the distance. At midday we descended the valley, and reached a hovel, where an officer and three soldiers were posted to examine passports. One of these men

1 This is merely an illustration of the admirable laws, first laid down by Mr. Lyell, on the geographical distribution of animals, as influenced by geological changes. The whole reasoning, of course, is founded on the assumption of the immutability of species; otherwise the difference in the species in the two regions, might be considered as superinduced during a length of time.

was a thoroughbred Pampas Indian: he was kept much for the same purpose as a bloodhound, to track out any person who might pass by secretly, either on foot or horseback. Some years ago, a passenger endeavoured to escape detection, by making a long circuit over a neighbouring mountain; but this Indian, having by chance crossed his track, followed it for the whole day over dry and very stony hills, till at last he came on his prey hidden in a gully. We here heard that the silvery clouds, which we had admired from the bright region above, had poured down torrents of rain. The valley from this point gradually opened, and the hills became mere water-worn hillocks compared to the giants behind: it then expanded into a gently-sloping plain of shingle, covered with low trees and bushes. This talus, although appearing narrow, must be nearly ten miles wide before it blends into the apparently dead level Pampas. We passed the only house in this neighbourhood, the Estancia of Chaquaio; and at sunset we pulled up in the first snug corner, and there bivouacked.

March 25th. – I was reminded of the Pampas of Buenos Ayres, by seeing the disk of the rising sun, intersected by an horizon, level as that of the ocean. During the night a heavy dew fell, a circumstance which we did not experience within the Cordillera. The road proceeded for some distance due east across a low swamp; then meeting the dry plain, it turned to the north towards Mendoza. The distance is two very long days' journey. Our first day's journey was called fourteen leagues to Estacado, and the second seventeen to Luxan, near Mendoza. The whole distance is over a level desert plain, with not more than two or three houses. The sun was exceedingly powerful, and the ride devoid of all interest. There is very little water in this 'traversia,' and in our second day's journey we found only one little pool. Little water flows from the mountains, and it soon becomes absorbed by the dry and porous soil; so that, although we travelled at the distance of only ten or fifteen miles from the outer range of the Cordillera, we did not cross a single stream. In many parts the ground was incrusted with a saline efflorescence; hence we had the same salt-loving plants, which are common near Bahia Blanca. The landscape has a uniform character from the Strait of Magellan, along the whole eastern coast of Patagonia, to the Rio Colorado; and it appears that the same kind of country extends inland from this river, in a sweeping line as far as San Luis, and perhaps even further north. To the eastward of this curved line, lies the basin of the comparatively damp and green plains of Buenos Ayres. The sterile plains of Mendoza and Patagonia consist of a bed of shingle, worn smooth and accumulated by the waves of the sea; while the Pampas, covered by thistles, clover, and grass, have been formed by the ancient estuary mud of the Plata.

After our two days' tedious journey, it was refreshing to see in the distance the rows of poplars and willows growing round the village and river of Luxan. Shortly before we arrived at this place, we observed to the south a ragged cloud of a dark reddish-brown colour. At first we thought that it was smoke from some great fire on the plains; but we soon found that it was a swarm of locusts. They were flying northward; and with the aid of a light breeze, they overtook us at a rate of ten or fifteen miles an hour. The main body filled the air from a height of twenty feet, to that, as it appeared, of two or three thousand above the ground; 'and the sound of their wings was as the sound of chariots of many horses running to battle:' or rather, I should say, like a strong breeze passing through the rigging of a ship. The sky, seen through the advanced guard, appeared like a mezzotinto engraving, but the main body was impervious to sight; they were not, however, so thick together, but that they could escape a stick waved backwards and forwards. When they alighted, they were more numerous than the leaves in the field, and the surface became reddish instead of being green: the swarm having once alighted, the individuals flew from side to side in all directions. Locusts are not an uncommon pest in this country: already during this season, several smaller swarms had come up from the south, where, as apparently in all other parts of the world, they are bred in the deserts. The poor cottagers in vain attempted by lighting fires, by shouts, and by waving branches to avert the attack. This species of locust closely resembles, and perhaps is identical with the famous Gryllus migratorius of the East.

We crossed the Luxan, which is a river of considerable size, though its course towards the sea-coast is very imperfectly known: it is even doubtful whether, in passing over the plains, it is not evaporated and lost. We slept in the village of Luxan, which is a small place surrounded by gardens, and forms the most southern

cultivated district in the Province of Mendoza; it is five leagues south
of the capital. At night I experienced an attack (for it deserves no less a
name) of the *Benchuca*, a species of Reduvius, the great black bug of
the Pampas. It is most disgusting to feel soft wingless insects, about an
inch long, crawling over one's body. Before sucking they are quite
thin, but afterwards they become round and bloated with blood, and
in this state are easily crushed. One which I caught at Iquique, (for
they are found in Chile and Peru,) was very empty. When placed on a
table, and though surrounded by people, if a finger was presented, the
bold insect would immediately protrude its sucker, make a charge,
and if allowed, draw blood. No pain was caused by the wound. It was
curious to watch its body during the act of sucking, as in less than ten
minutes it changed from being as flat as a wafer to a globular form.
This one feast, for which the benchuca was indebted to one of the
officers, kept it fat during four whole months; but, after the first
fortnight, it was quite ready to have another suck.

March 27th. – We rode on to Mendoza. The country was beautifully cultivated, and resembled Chile. This neighbourhood is celebrated for its fruit; and certainly nothing could appear more
flourishing than the vineyards and the orchards of figs, peaches, and
olives. We bought water-melons nearly twice as large as a man's head,
most deliciously cool and well-flavoured, for a halfpenny apiece; and
for the value of threepence, half a wheelbarrowful of peaches. The
cultivated and enclosed part of this province is very small; there is
little more than that which we passed through between Luxan and
the Capital. The land, as in Chile, owes its fertility entirely to artificial
irrigation; and it is really wonderful to observe how extraordinarily
productive a barren traversia is thus rendered.

We stayed the ensuing day in Mendoza. The prosperity of
the place has much declined of late years. The inhabitants say 'it is
good to live in, but very bad to grow rich in.' The lower orders have
the lounging, reckless manners of the Gauchos of the Pampas; and
their dress, riding-gear, and habits of life, are nearly the same. To my
mind the town had a stupid, forlorn aspect. Neither the boasted
alameda, nor the scenery, is at all comparable with that of Santiago;
but to those who, coming from Buenos Ayres, have just crossed the
unvaried Pampas, the gardens and orchards must appear delightful.

Sir F. Head, speaking of the inhabitants, says, 'They eat their dinners, and it is so very hot, they go to sleep – and could they do better?' I quite agree with Sir F. Head: the happy doom of the Mendozinos is to eat, sleep, and be idle.

March 29th. – We set out on our return to Chile, by the Uspallata pass situated north of Mendoza. We had to cross a long and most sterile traversia of fifteen leagues. The soil in parts was absolutely bare, in others covered by numberless dwarf cacti, armed with formidable spines, and called by the inhabitants 'little lions.' There were, also, a few low bushes. Although the plain is nearly three thousand feet above the sea, the sun was very powerful; and the heat, as well as the clouds of impalpable dust, rendered the travelling extremely irksome. Our course during the day lay nearly parallel to the Cordillera, but gradually approaching them. Before sunset we entered one of the wide valleys, or rather bays, which open on the plain: this soon narrowed into a ravine, where a little higher up the house of Villa Vicencio is situated. As we had ridden all day without a drop of water, both our mules and selves were very thirsty, and we looked out anxiously for the stream which flows down this valley. It was curious to observe how gradually the water made its appearance: on the plain the course was quite dry; by degrees it became a little damper; then puddles of water appeared; these soon became connected; and at Villa Vicencio there was a nice little rivulet.

30th. – The solitary hovel which bears the imposing name of Villa Vicencio, has been mentioned by every traveller who has crossed the Andes. I stayed here and at some neighbouring mines during the two succeeding days. The geology of the surrounding country is very curious. The Uspallata range is separated from the main Cordillera by a long narrow plain or basin, like those so often mentioned in Chile, but higher, being six thousand feet above the sea. This range has nearly the same geographical position with respect to the Cordillera, which the gigantic Portillo line has, but it is of a totally different origin: it consists of various kinds of submarine lava, alternating with volcanic sandstones and other remarkable sediment-ary deposits; the whole having a very close resemblance to some of the tertiary beds on the shores of the Pacific. From this resemblance

I expected to find silicified wood, which is generally characteristic of those formations. I was gratified in a very extraordinary manner. In the central part of the range, at an elevation of about seven thousand feet, I observed on a bare slope some snow-white projecting columns. These were petrified trees, eleven being silicified, and from thirty to forty converted into coarsely-crystallized white calcareous spar. They were abruptly broken off, the upright stumps projecting a few feet above the ground. The trunks measured from three to five feet each in circumference. They stood a little way apart from each other, but the whole formed one group. Mr. Robert Brown has been kind enough to examine the wood: he says it belongs to the fir tribe, partaking of the character of the Araucarian family, but with some curious points of affinity with the yew. The volcanic sandstone in which the trees were embedded, and from the lower part of which they must have sprung, had accumulated in successive thin layers around their trunks; and the stone yet retained the impression of the bark.

It required little geological practice to interpret the marvellous story which this scene at once unfolded; though I confess I was at first so much astonished, that I could scarcely believe the plainest evidence. I saw the spot where a cluster of fine trees once waved their branches on the shores of the Atlantic, when that ocean (now driven back 700 miles) came to the foot of the Andes. I saw that they had sprung from a volcanic soil which had been raised above the level of the sea, and that subsequently this dry land, with its upright trees, had been let down into the depths of the ocean. In these depths, the formerly dry land was covered by sedimentary beds, and these again by enormous streams of submarine lava – one such mass attaining the thickness of a thousand feet; and these deluges of molten stone and aqueous deposits five times alternately had been spread out. The ocean which received such thick masses, must have been profoundly deep; but again the subterranean forces exerted themselves, and I now beheld the bed of that ocean, forming a chain of mountains more than seven thousand feet in height. Nor had those antagonist forces been dormant, which are always at work wearing down the surface of the land: the great piles of strata had been intersected by many wide valleys, and the trees, now changed into silex, were exposed projecting from the volcanic soil, now changed into rock, whence formerly, in a green and budding state, they had raised their lofty heads. Now, all is utterly irreclaimable and desert; even the lichen cannot adhere to the stony casts of former trees. Vast, and scarcely comprehensible as such changes must ever appear, yet they have all occurred within a period, recent when compared with the history of the Cordillera; and the Cordillera itself is absolutely modern as compared with many of the fossiliferous strata of Europe and America.

April 1st. – We crossed the Uspallata range, and at night slept at the custom-house – the only inhabited spot on the plain. Shortly before leaving the mountains, there was a very extraordinary view; red, purple, green, and quite white sedimentary rocks, alternating with black lavas, were broken up and thrown into all kinds of disorder by masses of porphyry of every shade of colour, from dark brown to the brightest lilac. It was the first view I ever saw, which really resembled those pretty sections which geologists make of the inside of the earth.

The next day we crossed the plain, and followed the course of the same great mountain stream which flows by Luxan. Here it was a furious torrent, quite impassable, and appeared larger than in the low country, as was the case with the rivulet of Villa Vicencio. On the evening of the succeeding day, we reached the Rio de las Vacas, which is considered the worst stream in the Cordillera to cross. As all these rivers have a rapid and short course, and are formed by the melting of the snow, the hour of the day makes a considerable difference in their volume. In the evening the stream is muddy and full, but about day-break it becomes clearer and much less impetuous. This we found to be the case with the Rio Vacas, and in the morning we crossed it with little difficulty.

The scenery thus far was very uninteresting, compared with that of the Portillo pass. Little can be seen beyond the bare walls of the one grand, flat-bottomed valley, which the road follows up to the highest crest. The valley and the huge rocky mountains are extremely barren: during the two previous nights the poor mules had absolutely nothing to eat, for excepting a few low resinous bushes, scarcely a plant can be seen. In the course of this day we crossed some of the worst passes in the Cordillera, but their danger has been much exaggerated. I was told that if I attempted to pass on foot, my head would turn giddy, and that there was no room to dismount; but I did not see a place where any one might not have walked over backwards, or got off his mule on either side. One of the bad passes, called *las Animas* (the Souls), I had crossed, and did not find out till a day afterwards, that it was one of the awful dangers. No doubt there are many parts in which, if the mule should stumble, the rider would be hurled down a great precipice; but of this there is little chance. I dare say, in the spring, the 'laderas,' or roads, which each year are formed anew across the piles of fallen detritus, are very bad; but from what I saw, I suspect the real danger is nothing. With cargo-mules the case is rather different, for the loads project so far, that the animals, occasionally running against each other, or against a point of rock, lose their balance, and are thrown down the precipices. In crossing the rivers I can well believe that the difficulty may be very great: at this season there was little trouble, but in the summer they must be very hazardous. I can quite imagine, as Sir F. Head describes, the different expressions of those who *have* passed the gulf, and those who *are* passing. I never heard of any man being drowned, but with loaded mules it frequently happens. The arriero tells you to show your mule the best line, and allow her to cross as she likes: the cargo-mule takes a bad line, and is often lost.

April 4*th*. – From the Rio de las Vacas to the Puente del Incas, half a day's journey. As there was pasture for the mules, and geology for me, we bivouacked here for the night. When one hears of a natural Bridge, one pictures to oneself some deep and narrow ravine, across which a bold mass of rock has fallen; or a great arch hollowed out like the vault of a cavern. Instead of this, the Incas Bridge consists of a crust of stratified shingle, cemented together by the deposits of the neighbouring hot springs. It appears, as if the stream had scooped out a channel on one side, leaving an overhanging ledge, which was met by earth and stones falling down from the opposite cliff. Certainly an oblique junction, as would happen in such a case, was very distinct on one side. The Bridge of the Incas is by no means worthy of the great monarchs whose name it bears.

5*th*. – We had a long day's ride across the central ridge, from the Incas Bridge to the Ojos del Agua, which are situated near the lowest *casucha* on the Chilian side. These casuchas are round little towers, with steps outside to reach the floor, which is raised some feet above the ground on account of the snow-drifts. They are eight in number, and under the Spanish government were kept during the winter well stored with food and charcoal, and each courier had a master-key. Now they only answer the purpose of caves, or rather dungeons. Seated on some little eminence, they are not, however, ill suited to the surrounding scene of desolation. The zigzag ascent of the Cumbre, or the partition of the waters, was very steep and tedious; its height, according to Mr. Pentland, is 12,454 feet. The road did not pass over any perpetual snow, although there were patches of it on both hands. The wind on the summit was exceedingly cold, but it was impossible not to stop for a few minutes to admire, again and again, the colour of the heavens, and the brilliant transparency of the atmosphere. The scenery was grand: to the westward there was a fine chaos of mountains, divided by profound ravines. Some snow generally falls before this period of the season, and it has even happened that the Cordillera have been finally closed by this time. But we were most fortunate. The sky, by night and by day, was cloudless, excepting a few round little masses of vapour, that floated over the highest pinnacles. I have often seen these islets in the sky, marking the position of the Cordillera, when the far-distant mountains have been hidden beneath the horizon.

April 6th. – In the morning we found some thief had stolen one of our mules, and the bell of the madrina. We therefore rode only two or three miles down the valley, and staid there the ensuing day in hopes of recovering the mule, which the arriero thought had been hidden in some ravine. The scenery in this part had assumed a Chilian character: the lower sides of the mountains, dotted over with the pale evergreen Quillay tree, and with the great chandelier-like cactus, are certainly more to be admired than the bare eastern valleys; but I cannot quite agree with the admiration expressed by some travellers. The extreme pleasure, I suspect, is chiefly owing to the prospect of a good fire and of a good supper, after escaping from the cold regions above: and I am sure I most heartily participated in these feelings.

8th. – We left the valley of the Aconcagua, by which we had descended, and reached in the evening a cottage near the Villa de St. Rosa. The fertility of the plain was delightful: the autumn being advanced, the leaves of many of the fruit-trees were falling; and of the labourers, – some were busy in drying figs and peaches on the roofs of their cottages, while others were gathering the grapes from the vineyards. It was a pretty scene; but I missed that pensive stillness which makes the autumn in England indeed the evening of the year. On the 10th we reached Santiago, where I received a very kind and hospitable reception from Mr. Caldcleugh. My excursion only cost me twenty-four days, and never did I more deeply enjoy an equal space of time. A few days afterwards I returned to Mr. Corfield's house at Valparaiso.

CHAPTER XVI

NORTHERN CHILE
AND PERU

Coast-road to Coquimbo – Great Loads carried by the Miners – Coquimbo – Earthquake – Step-formed Terraces – Absence of recent Deposits – Contemporaneousness of the Tertiary Formations – Excursion up the Valley – Road to Guasco – Deserts – Valley of Copiapó – Rain and Earthquakes – Hydrophobia – The Despoblado – Indian Ruins – Probable Change of Climate – River-bed arched by an Earthquake – Cold Gales of Wind – Noises from a Hill – Iquique – Salt Alluvium – Nitrate of Soda – Lima – Unhealthy Country – Ruins of Callao, overthrown by an Earthquake – Recent Subsidence – Elevated Shells on San Lorenzo, their decomposition – Plain with embedded Shells and fragments of Pottery – Antiquity of the Indian Race.

APRIL 27th. – I set out on a journey to Coquimbo, and thence through Guasco to Copiapó, where Captain Fitz Roy kindly offered to pick me up in the *Beagle*. The distance in a straight line along the shore northward is only 420 miles; but my mode of travelling made it a very long journey. I bought four horses and two mules, the latter carrying the luggage on alternate days. The six animals together only cost the value of twenty-five pounds sterling, and at Copiapó I sold them again for twenty-three. We travelled in the same independent manner as before, cooking our own meals, and sleeping in the open air. As we rode towards the Viño del Mar, I took a farewell view of Valparaiso, and admired its picturesque appearance. For geological purposes I made a detour from the high road to the foot of the Bell of Quillota. We passed through an alluvial district rich in gold, to the neighbourhood of Limache, where we slept. Washing for gold supports the inhabitants of numerous hovels, scattered along the sides of each little rivulet; but, like all those whose gains are uncertain, they are unthrifty in their habits, and consequently poor.

28*th.* – In the afternoon we arrived at a cottage at the foot of the Bell mountain. The inhabitants were freeholders, which is not very usual in Chile. They supported themselves on the produce of a garden and a little field, but were very poor. Capital is here so deficient, that the people are obliged to sell their green corn while standing in the field, in order to buy necessaries for the ensuing year. Wheat in consequence was dearer in the very district of its production than at Valparaiso, where the contractors live. The next day we joined the main road to Coquimbo. At night there was a very light shower of rain: this was the first drop that had fallen since the heavy rain of September 11th and 12th, which detained me a prisoner at the Baths of Cauquenes. The interval was seven and a half months; but the rain this year in Chile was rather later than usual. The distant Andes were now covered by a thick mass of snow; and were a glorious sight.

May 2nd. – The road continued to follow the coast, at no great distance from the sea. The few trees and bushes which are common in central Chile decreased rapidly in numbers, and were replaced by a tall plant, something like a yucca in appearance. The surface of the country, on a small scale, was singularly broken and irregular; abrupt little peaks of rock rising out of small plains or basins. The indented coast and the bottom of the neighbouring sea, studded with breakers, would, if converted into dry land, present similar forms; and such a conversion without doubt has taken place in the part over which we rode.

3rd. – Quilimari to Conchalee. The country became more and more barren. In the valleys there was scarcely sufficient water for any irrigation; and the intermediate land was quite bare, not supporting even goats. In the spring, after the winter showers, a thin pasture rapidly springs up, and cattle are then driven down from the Cordillera to graze for a short time. It is curious to observe how the seeds of the grass and other plants seem to accommodate themselves, as if by an acquired habit, to the quantity of rain which falls on different parts of this coast. One shower far northward at Copiapó produces as great an effect on the vegetation, as two at Guasco, and as three or four in this district. At Valparaiso a winter so dry as greatly to injure the pasture, would at Guasco produce the most unusual abundance.

Proceeding northward, the quantity of rain does not appear to decrease in strict proportion to the latitude. At Conchalee, which is only 67 miles north of Valparaiso, rain is not expected till the end of May; whereas, at Valparaiso some generally falls early in April: the annual quantity is likewise small in proportion to the lateness of the season at which it commences.

4th. – Finding the coast-road devoid of interest of any kind, we turned inland towards the mining district and valley of Illapel. This valley, like every other in Chile, is level, broad, and very fertile: it is bordered on each side, either by cliffs of stratified shingle, or by bare rocky mountains. Above the straight line of the uppermost irrigating ditch, all is brown as on a high road; while all below is of as bright a green as verdigris, from the beds of alfalfa, a kind of clover. We proceeded to Los Hornos, another mining district, where the principal hill was drilled with holes, like a great ants'-nest. The Chilian miners are a peculiar race of men in their habits. Living for weeks together in the most desolate spots, when they descend to the villages on feast-days, there is no excess or extravagance into which they do not run. They sometimes gain a considerable sum, and then, like sailors with prize-money, they try how soon they can contrive to squander it. They drink excessively, buy quantities of clothes, and in a few days return penniless to their miserable abodes, there to work harder than beasts of burden. This thoughtlessness, as with sailors, is evidently the result of a similar manner of life. Their daily food is found them, and they acquire no habits of carefulness; moreover, temptation and the means of yielding to it are placed in their power at the same time. On the other hand, in Cornwall, and some other parts of England, where the system of selling part of the vein is followed, the miners, from being obliged to act and think for themselves, are a singularly intelligent and well-conducted set of men.

The dress of the Chilian miner is peculiar and rather picturesque. He wears a very long shirt of some dark-coloured baize, with a leathern apron; the whole being fastened round his waist by a bright-coloured sash. His trowsers are very broad, and his small cap of scarlet cloth is made to fit the head closely. We met a party of these miners in full costume, carrying the body of one of their companions to be buried. They marched at a very quick trot, four men supporting the corpse. One set having run as hard as they could for about two hundred yards, were relieved by four others, who had previously dashed on ahead on horseback. Thus they proceeded, encouraging each other by wild cries: altogether the scene formed a most strange funeral.

We continued travelling northward, in a zigzag line; sometimes stopping a day to geologize. The country was so thinly inhabited, and the track so obscure, that we often had difficulty in finding our way. On the 12th I stayed at some mines. The ore in this case was not considered particularly good, but from being abundant it was supposed the mine would sell for about thirty or forty thousand dollars (that is, 6000 or 8000 pounds sterling); yet it had been bought by one of the English Associations for an ounce of gold (3*l.* 8*s.*). The ore is yellow pyrites, which, as I have already remarked, before the arrival of the English, was not supposed to contain a particle of copper. On a scale of profits nearly as great as in the above instance, piles of cinders, abounding with minute globules of metallic copper, were purchased; yet with these advantages, the mining associations, as is well known, contrived to lose immense sums of money. The folly of the greater number of the commissioners and shareholders amounted to infatuation; − a thousand pounds per annum given in some cases to entertain the Chilian authorities; libraries of well-bound geological books; miners brought out for particular metals, as tin, which are not found in Chile; contracts to supply the miners with milk, in parts where there are no cows; machinery, where it could not possibly be used; and a hundred similar arrangements, bore witness to our absurdity, and to this day afford amusement to the natives. Yet there can be no doubt, that the same capital well employed in these mines would have yielded an immense return: a confidential man of business, a practical miner and assayer, would have been all that was required.

Captain Head has described the wonderful load which the 'Apires,' truly beasts of burden, carry up from the deepest mines. I confess I thought the account exaggerated; so that I was glad to take an opportunity of weighing one of the loads, which I picked out by hazard. It required considerable exertion on my part, when standing directly over it, to lift it from the ground. The load was considered under weight when found to be 197 pounds. The apire had carried this up eighty perpendicular yards, – part of the way by a steep passage, but the greater part up notched poles, placed in a zigzag line up the shaft. According to the general regulation, the apire is not allowed to halt for breath, except the mine is six hundred feet deep. The average load is considered as rather more than 200 pounds, and I have been assured that one of 300 pounds (twenty-two stone and a half) by way of a trial has been brought up from the deepest mine! At this time the apires were bringing up the usual load twelve times in the day; that is, 2400 pounds from eighty yards deep; and they were employed in the intervals in breaking and picking ore.

These men, excepting from accidents, are healthy, and appear cheerful. Their bodies are not very muscular. They rarely eat meat once a week, and never oftener, and then only the hard dry charqui. Although with a knowledge that the labour was voluntary, it was nevertheless quite revolting to see the state in which they reached the mouth of the mine; their bodies bent forward, leaning with their arms on the steps, their legs bowed, their muscles quivering, the perspiration streaming from their faces over their breasts, their nostrils distended, the corners of their mouth forcibly drawn back, and the expulsion of their breath most laborious. Each time they draw their breath, they utter an articulate cry of 'ay-ay,' which ends in a sound rising from deep in the chest, but shrill like the note of a fife. After staggering to the pile of ore, they emptied the 'carpacho;' in two or three seconds recovering their breath, they wiped the sweat from their brows, and apparently quite fresh descended the mine again at a quick pace. This appears to me a wonderful instance of the amount of labour which habit, for it can be nothing else, will enable a man to endure.

In the evening, talking with the major-domo of these mines, about the number of foreigners now scattered over the whole country, he told me that, though quite a young man, he remembers when he was a boy at school at Coquimbo, a holiday being given to see the captain of an English ship, who was brought to the city to speak to the governor. He believes that nothing would have induced any boy in the school, himself included, to have gone close to the Englishman; so deeply had they been impressed with an idea of the heresy, contamination, and evil to be derived from contact with such a person. To this day they relate the atrocious actions of the bucaniers; and especially of one man, who took away the figure of the Virgin Mary, and returned the year after for that of St. Joseph, saying it was a pity the lady should not have a husband. I heard also of an old lady who, at a dinner in Coquimbo, remarked how wonderfully strange it was that she should have lived to dine in the same room with an Englishman; for she remembered as a girl, that twice, at the mere cry of 'Los Ingleses,' every soul, carrying what valuables they could, had taken to the mountains.

14th. – We reached Coquimbo, where we stayed a few days. The town is remarkable for nothing but its extreme quietness. It is said to contain from 6000 to 8000 inhabitants. On the morning of the 17th it rained lightly, the first time this year, for about five hours. The farmers, who plant corn near the sea-coast where the atmosphere is more humid, taking advantage of this shower, would break up the ground; after a second they would put the seed in; and if a third shower should fall, they would reap a good harvest in the spring. It was interesting to watch the effect of this trifling amount of moisture. Twelve hours afterwards the ground appeared as dry as ever; yet after an interval of ten days, all the hills were faintly tinged with green patches; the grass being sparingly scattered in hair-like fibres a full inch in length. Before this shower every part of the surface was bare as on a high road.

In the evening, Captain Fitz Roy and myself were dining with Mr. Edwards, an English resident well known for his hospitality by all who have visited Coquimbo, when a sharp earthquake happened. I heard the forecoming rumble, but from the screams of the ladies, the running of the servants, and the rush of several of the gentlemen to the doorway, I could not distinguish the motion. Some of the women afterwards were crying with terror, and one gentleman said he should not be able to sleep all night, or if he did, it would only be to dream of falling houses. The father of this person had lately lost all his property at Talcahuano, and he himself had only just escaped a falling roof at Valparaiso, in 1822. He mentioned a curious coincidence which then happened: he was playing at cards, when a German, one of the party, got up, and said he would never sit in a room in these countries with the door shut, as, owing to his having done so, he had nearly lost his life at Copiapó. Accordingly he opened the door; and no sooner had he done this, than he cried out, 'Here it comes again!' and the famous shock commenced. The whole party escaped. The danger in an earthquake is not from the time lost in opening a door, but from the chance of its becoming jammed by the movement of the walls.

It is impossible to be much surprised at the fear which natives and old residents, though some of them known to be men of great command of mind, so generally experience during earthquakes. I think, however, this excess of panic may be partly attributed to a want of habit in governing their fear, as it is not a feeling they are ashamed of. Indeed, the natives do not like to see a person indifferent. I heard of two Englishmen who, sleeping in the open air during a smart shock, knowing that there was no danger, did not rise. The natives cried out indignantly, 'Look at those heretics, they will not even get out of their beds!'

I spent some days in examining the step-formed terraces of shingle, first noticed by Captain B. Hall, and believed by Mr. Lyell to have been formed by the sea, during the gradual rising of the land. This certainly is the true explanation, for I found numerous shells of existing species on these terraces. Five narrow, gently sloping, fringe-like terraces rise one behind the other, and where best developed are formed of shingle: they front the bay, and sweep up both sides of the valley. At Guasco, north of Coquimbo, the phenomenon is displayed on a much grander scale, so as to strike with surprise even some of the inhabitants. The terraces are there much broader, and may be called plains; in some parts there are six of them, but generally only five; they run up the valley for thirty-seven miles from the coast. These step-formed terraces or fringes closely resemble those in the valley of S. Cruz, and except in being on a smaller scale, those great ones along the whole coast-line of Patagonia. They have undoubtedly been formed by the denuding power of the sea, during long periods of rest in the gradual elevation of the continent.

Shells of many existing species not only lie on the surface of the terraces at Coquimbo (to a height of 250 feet), but are embedded in a friable calcareous rock, which in some places is as much as between twenty and thirty feet in thickness, but is of little extent. These modern beds rest on an ancient tertiary formation containing shells, apparently all extinct. Although I examined so many hundred miles of coast on the Pacific, as well as Atlantic side of the continent, I found no regular strata containing sea-shells of recent species, excepting at this place, and at a few points northward on the road to Guasco. This fact appears to me highly remarkable; for the explanation generally given by geologists, of the absence in any district of stratified fossiliferous deposits of a given period, namely, that the surface then existed as dry land, is not here applicable; for we know from the shells strewed on the surface and embedded in loose sand or mould, that the land for thousands of miles along both coasts has lately been submerged. The explanation, no doubt, must be sought in the fact, that the whole southern part of the continent has been for a long time slowly rising; and therefore that all matter deposited along shore in shallow water, must have been soon brought up and slowly exposed to the wearing action of the sea-beach; and it is only in comparatively shallow water that the greater number of marine organic beings can flourish, and in such water it is obviously impossible that strata of any great thickness can accumulate. To show the vast power of the wearing action of sea-beaches, we need only appeal to the great cliffs along the present coast of Patagonia, and to the escarpments or ancient sea-cliffs at different levels, one above another, on that same line of coast.

The old underlying tertiary formation at Coquimbo, appears to be of about the same age with several deposits on the coast of Chile (of which that of Navedad is the principal one), and with the great formation of Patagonia. Both at Navedad and in Patagonia there is evidence, that since the shells (a list of which has been seen by Professor E. Forbes) there intombed were living, there has been a subsidence of several hundred feet, as well as an ensuing elevation. It may naturally be asked, how it comes that, although no extensive fossiliferous deposits of the recent period, nor of any period intermediate between it and the ancient tertiary epoch, have been preserved on either side of the continent, yet that at this ancient tertiary epoch, sedimentary matter containing fossil remains, should have been deposited and preserved at different points in north and south lines, over a space of 1100 miles on the shores of the Pacific, and of at least 1350 miles on the shores of the Atlantic, and in an east and west line of 700 miles across the widest part of the continent? I believe the explanation is not difficult, and that it is perhaps applicable to nearly analogous facts observed in other quarters of the world. Considering the enormous power of denudation which the sea possesses, as shown by numberless facts, it is not probable that a sedimentary deposit, when being upraised, could pass through the ordeal of the beach, so as to be preserved in sufficient masses to last to a distant period, without it were originally of wide extent and of considerable thickness: now it is impossible on a moderately shallow bottom, which alone is favourable to most living creatures, that a thick and widely extended covering of sediment could be spread out, without the bottom sank down to receive the successive layers. This seems to have actually taken place at about the same period in southern Patagonia and Chile, though these places are a thousand miles apart. Hence, if prolonged movements of approximately contemporaneous subsidence are generally widely extensive, as I am strongly inclined to believe from my examination of the Coral Reefs of the great oceans – or if, confining our view to South America, the subsiding movements have been coextensive with those of elevation, by which, within the same period of existing shells, the shores of Peru, Chile, Tierra del Fuego, Patagonia, and La Plata have been upraised – then we can see that at the same time, at far distant points, circumstances would have been favourable to the formation of fossiliferous deposits, of wide extent and of considerable thickness; and such deposits, consequently, would have a good chance of resisting the wear and tear of successive beach-lines, and of lasting to a future epoch.

May 21st. – I set out in company with Don Jose Edwards to the silver-mine of Arqueros, and thence up the valley of Coquimbo. Passing through a mountainous country, we reached by nightfall the mines belonging to Mr. Edwards. I enjoyed my night's rest here from a reason which will not be fully appreciated in England, namely, the absence of fleas! The rooms in Coquimbo swarm with them; but they will not live here at the height of only three or four thousand feet: it can scarcely be the trifling diminution of temperature, but some other cause which destroys these troublesome insects at this place. The mines are now in a bad state, though they formerly yielded about 2000 pounds in weight of silver a year. It has been said that 'a person with a copper-mine will gain; with silver he may gain; but with gold he is sure to lose.' This is not true: all the large Chilian fortunes have been made by mines of the more precious metals. A short time since an English physician returned to England from Copiapó, taking with him the profits of one share in a silver-mine, which amounted to about 24,000 pounds sterling. No doubt a copper-mine with care is a sure game, whereas the other is gambling, or rather taking a ticket in a lottery. The owners lose great quantities of rich ores; for no precautions can prevent robberies. I heard of a gentleman laying a bet with another, that one of his men should rob him before his face. The ore when brought out of the mine is broken into pieces, and the useless stone thrown on one side. A couple of the miners who were thus employed, pitched, as if by accident, two fragments away at the same moment, and then cried out for a joke, 'Let us see which rolls furthest.' The owner, who was standing by, bet a cigar with his friend on the race. The miner by this means watched the very point amongst the rubbish where the stone lay. In the evening he picked it up and carried it to his master, showing him a rich mass of silver-ore, and saying, 'This was the stone on which you won a cigar by its rolling so far.'

May 23rd. – We descended into the fertile valley of Coquimbo, and followed it till we reached an Hacienda belonging to a relation of Don Jose, where we stayed the next day. I then rode one day's journey further, to see what were declared to be some petrified shells and beans, which latter turned out to be small quartz pebbles. We passed through several small villages; and the valley was beautifully cultivated, and the whole scenery very grand. We were here near the main Cordillera, and the surrounding hills were lofty. In all parts of northern Chile, fruit-trees produce much more abundantly at a considerable height near the Andes than in the lower country. The figs and grapes of this district are famous for their excellence, and are cultivated to a great extent. This valley is, perhaps, the most productive one north of Quillota: I believe it contains, including Coquimbo,

25,000 inhabitants. The next day I returned to the Hacienda, and thence, together with Don Jose, to Coquimbo.

June 2nd. – We set out for the valley of Guasco, following the coast-road, which was considered rather less desert than the other. Our first day's ride was to a solitary house, called Yerba Buena, where there was pasture for our horses. The shower mentioned as having fallen a fortnight ago, only reached about half-way to Guasco; we had, therefore, in the first part of our journey a most faint tinge of green, which soon faded quite away. Even where brightest, it was scarcely sufficient to remind one of the fresh turf and budding flowers of the spring of other countries. While travelling through these deserts one feels like a prisoner shut up in a gloomy court, who longs to see something green and to smell a moist atmosphere.

June 3rd. – Yerba Buena to Carizal. During the first part of the day we crossed a mountainous rocky desert, and afterwards a long deep sandy plain, strewed with broken sea-shells. There was very little water, and that little saline: the whole country, from the coast to the Cordillera, is an uninhabited desert. I saw traces only of one living animal in abundance, namely, the shells of a Bulimus, which were collected together in extraordinary numbers on the driest spots. In the spring one humble little plant sends out a few leaves, and on these the snails feed. As they are seen only very early in the morning, when the ground is slightly damp with dew, the Guasos believe that they are bred from it. I have observed in other places that extremely dry and sterile districts, where the soil is calcareous, are extraordinarily favourable to land-shells. At Carizal there were a few cottages, some brackish water, and a trace of cultivation: but it was with difficulty that we purchased a little corn and straw for our horses.

4th. – Carizal to Sauce. We continued to ride over desert plains, tenanted by large herds of guanaco. We crossed also the valley of Chañeral; which, although the most fertile one between Guasco and Coquimbo, is very narrow, and produces so little pasture, that we could not purchase any for our horses. At Sauce we found a very civil old gentleman, superintending a copper-smelting furnace. As an especial favour, he allowed me to purchase at a high price an armful of dirty straw, which was all the poor horses had for supper after their long day's journey. Few smelting-furnaces are now at work in any part of Chile; it is found more profitable, on account of the extreme scarcity of firewood, and from the Chilian method of reduction being so unskilful, to ship the ore for Swansea. The next day we crossed some mountains to Freyrina, in the valley of Guasco. During each day's ride further northward, the vegetation became more and more scanty; even the great chandelier-like cactus was here replaced by a different and much smaller species. During the winter months, both in northern Chile and in Peru, a uniform bank of clouds hangs, at no great height, over the Pacific. From the mountains we had a very striking view of this white and brilliant aërial-field, which sent arms up the valleys, leaving islands and promontories in the same manner, as the sea does in the Chonos archipelago and in Tierra del Fuego.

We stayed two days at Freyrina. In the valley of Guasco there are four small towns. At the mouth there is the port, a spot entirely desert, and without any water in the immediate neighbourhood. Five leagues higher up stands Freyrina, a long straggling village, with decent whitewashed houses. Again, ten leagues further up Ballenar is situated; and above this Guasco Alto, a horticultural village, famous for its dried fruit. On a clear day the view up the valley is very fine; the straight opening terminates in the far-distant snowy Cordillera; on each side an infinity of crossing lines are blended together in a beautiful haze. The foreground is singular from the number of parallel and step-formed terraces; and the included strip of green valley, with its willow-bushes, is contrasted on both hands with the naked hills. That the surrounding country was most barren will be readily believed, when it is known that a shower of rain had not fallen during the last thirteen months. The inhabitants heard with the greatest envy of the rain at Coquimbo; from the appearance of the sky they had hopes of equally good fortune, which, a fortnight afterwards, were realized. I was at Copiapó at the time; and there the people, with equal envy, talked of the abundant rain at Guasco. After two or three very dry years, perhaps with not more than one shower during the whole time, a rainy year generally follows; and this does more harm than even the drought. The rivers swell, and cover with gravel and sand the narrow strips of ground, which alone are fit for cultivation. The floods also injure the irrigating ditches. Great devastation had thus been caused three years ago.

June 8th. − We rode on to Ballenar, which takes its name from Ballenagh in Ireland, the birthplace of the family of O'Higgins, who, under the Spanish government, were presidents and generals in Chile. As the rocky mountains on each hand were concealed by clouds, the terrace-like plains gave to the valley an appearance like that of Santa Cruz in Patagonia. After spending one day at Ballenar I set out, on the 10th, for the upper part of the valley of Copiapó. We rode all day over an uninteresting country. I am tired of repeating the epithets barren and sterile. These words, however, as commonly used, are comparative; I have always applied them to the plains of Patagonia, which can boast of spiny bushes and some tufts of grass; and this is absolute fertility, as compared with northern Chile. Here again, there are not many spaces of two hundred yards square, where some little bush, cactus or lichen, may not be discovered by careful examination; and in the soil seeds lie dormant ready to spring up during the first rainy winter. In Peru real deserts occur over wide tracts of country. In the evening we arrived at a valley, in which the bed of the streamlet was damp: following it up, we came to tolerably good water. During the night, the stream, from not being evaporated and absorbed so quickly, flows a league lower down than during the day. Sticks were plentiful for firewood, so that it was a good place of bivouac for us; but for the poor animals there was not a mouthful to eat.

June 11th. − We rode without stopping for twelve hours, till we reached an old smelting-furnace, where there was water and firewood; but our horses again had nothing to eat, being shut up in an old courtyard. The line of road was hilly, and the distant views interesting from the varied colours of the bare mountains. It was almost a pity to see the sun shining constantly over so useless a country; such splendid weather ought to have brightened fields and pretty gardens. The next day we reached the valley of Copiapó. I was heartily glad of it; for the whole journey was a continued source of anxiety; it was most disagreeable to hear, whilst eating our own suppers, our horses gnawing the posts to which they were tied, and to have no means of relieving their hunger. To all appearance, however, the animals were quite fresh; and no one could have told that they had eaten nothing for the last fifty-five hours.

I had a letter of introduction to Mr. Bingley, who received me very kindly at the Hacienda of Potrero Seco. This estate is between twenty and thirty miles long, but very narrow, being generally only two fields wide, one on each side the river. In some parts the estate is of no width, that is to say, the land cannot be irrigated, and therefore is valueless, like the surrounding rocky desert. The small quantity of cultivated land in the whole line of valley, does not so much depend on inequalities of level, and consequent unfitness for irrigation, as on the small supply of water. The river this year was remarkably full: here, high up the valley, it reached to the horse's belly, and was about fifteen yards wide, and rapid; lower down it becomes smaller and smaller, and is generally quite lost, as happened during one period of thirty years, so that not a drop entered the sea. The inhabitants watch a storm over the Cordillera with great interest; as one good fall of snow provides them with water for the ensuing year. This is of infinitely more consequence than rain in the lower country. Rain, as often as it falls, which is about once in every two or three years, is a great advantage, because the cattle and mules can for some time afterwards find a little pasture on the mountains. But without snow on the Andes, desolation extends throughout the valley. It is on record that three times nearly all the inhabitants have been obliged to emigrate to the south. This year there was plenty of water, and every man irrigated his ground as much as he chose; but it has frequently been necessary to post soldiers at the sluices, to see that each estate took only its proper allowance during so many hours in the week. The valley is said to contain 12,000 souls, but its produce is sufficient only for three months in the year; the rest of the supply being drawn from Valparaiso and the south. Before the discovery of the famous silver-mines of Chanuncillo, Copiapó was in a rapid state of decay; but now it is in a very thriving condition; and the town, which was completely overthrown by an earthquake, has been rebuilt.

The valley of Copiapó, forming a mere ribbon of green in a desert, runs in a very southerly direction; so that it is of considerable length to its source in the Cordillera. The valleys of Guasco and Copiapó may both be considered as long narrow islands, separated from the rest of Chile by deserts of rock instead of by salt water.

Northward of these, there is one other very miserable valley, called Paposo, which contains about two hundred souls; and then there extends the real desert of Atacama – a barrier far worse than the most turbulent ocean. After staying a few days at Potrero Seco, I proceeded up the valley to the house of Don Benito Cruz, to whom I had a letter of introduction. I found him most hospitable; indeed it is impossible to bear too strong testimony to the kindness, with which travellers are received in almost every part of South America. The next day I hired some mules to take me by the ravine of Jolquera into the central Cordillera. On the second night the weather seemed to foretell a storm of snow or rain, and whilst lying in our beds we felt a trifling shock of an earthquake.

The connexion between earthquakes and the weather has been often disputed: it appears to me to be a point of great interest, which is little understood. Humboldt has remarked in one part of the Personal Narrative,[1] that it would be difficult for any person who had long resided in New Andalusia, or in Lower Peru, to deny that there exists some connexion between these phenomena: in another part, however, he seems to think the connexion fanciful. At Guayaquil, it is said that a heavy shower in the dry season is invariably followed by an earthquake. In Northern Chile, from the extreme infrequency of rain, or even of weather foreboding rain, the probability of accidental coincidences becomes very small; yet the inhabitants are here most firmly convinced of some connexion between the state of the atmosphere and of the trembling of the ground: I was much struck by this, when mentioning to some people at Copiapó that there had been a sharp shock at Coquimbo: they immediately cried out, 'How fortunate! there will be plenty of pasture there this year.' To their minds an earthquake foretold rain, as surely as rain foretold abundant pasture. Certainly it did so happen that on the very day of the earthquake, that shower of rain fell, which I have described as in ten days' time producing a thin sprinkling of grass. At other

1 Vol. iv. p. 11, and vol. ii. p. 217. For the remarks on Guayaquil see Silliman's Journ. vol. xxiv. p. 384. For those on Tacna by Mr. Hamilton, see Trans. of British Association, 1840. For those on Coseguina see Mr. Caldcleugh in Phil. Trans., 1835. In the former edition, I collected several references on the coincidences between sudden falls in the barometer and earthquakes; and between earthquakes and meteors.

times, rain has followed earthquakes, at a period of the year when it is a far greater prodigy than the earthquake itself: this happened after the shock of November, 1822, and again in 1829, at Valparaiso; also after that of September, 1833, at Tacna. A person must be somewhat habituated to the climate of these countries, to perceive the extreme improbability of rain falling at such seasons, except as a consequence of some law quite unconnected with the ordinary course of the weather. In the cases of great volcanic eruptions, as that of Coseguina, where torrents of rain fell at a time of the year most unusual for it, and 'almost unprecedented in Central America,' it is not difficult to understand that the volumes of vapour and clouds of ashes might have disturbed the atmospheric equilibrium. Humboldt extends this view to the case of earthquakes unaccompanied by eruptions; but I can hardly conceive it possible, that the small quantity of aëriform fluids which then escape from the fissured ground, can produce such remarkable effects. There appears much probability in the view first proposed by Mr. P. Scrope, that when the barometer is low, and when rain might naturally be expected to fall, the diminished pressure of the atmosphere over a wide extent of country, might well determine the precise day on which the earth, already stretched to the utmost by the subterranean forces, should yield, crack, and consequently tremble. It is, however, doubtful how far this idea will explain the circumstance of torrents of rain falling in the dry season during several days, after an earthquake unaccompanied by an eruption; such cases seem to bespeak some more intimate connexion between the atmospheric and subterranean regions.

Finding little of interest in this part of the ravine, we retraced our steps to the house of Don Benito, where I stayed two days collecting fossil shells and wood. Great prostrate silicified trunks of trees, embedded in a conglomerate, were extraordinarily numerous. I measured one, which was fifteen feet in circumference: how surprising it is that every atom of the woody matter in this great cylinder should have been removed and replaced by silex so perfectly, that each vessel and pore is preserved! These trees flourished at about the period of our lower chalk; they all belonged to the fir-tribe. It was amusing to hear the inhabitants discussing the nature of the fossil shells which

I collected, almost in the same terms as were used a century ago in Europe, – namely, whether or not they had been thus 'born by nature.' My geological examination of the country generally created a good deal of surprise amongst the Chilenos: it was long before they could be convinced that I was not hunting for mines. This was sometimes troublesome: I found the most ready way of explaining my employment, was to ask them how it was that they themselves were not curious concerning earthquakes and volcanoes? – why some springs were hot and others cold? – Why there were mountains in Chile, and not a hill in La Plata? These bare questions at once satisfied and silenced the greater number; some, however (like a few in England who are a century behindhand), thought that all such inquiries were useless and impious; and that it was quite sufficient that God had thus made the mountains.

An order had recently been issued that all stray dogs should be killed, and we saw many lying dead on the road. A great number had lately gone mad, and several men had been bitten and had died in consequence. On several occasions hydrophobia has prevailed in this valley. It is remarkable thus to find so strange and dreadful a disease, appearing time after time in the same isolated spot. It has been remarked that certain villages in England are in like manner much more subject to this visitation than others. Dr. Unanùe states that hydrophobia was first known in South America in 1803: this statement is corroborated by Azara and Ulloa having never heard of it in their time. Dr. Unanùe says that it broke out in Central America, and slowly travelled southward. It reached Arequipa in 1807; and it is said that some men there, who had not been bitten, were affected, as were some negroes, who had eaten a bullock which had died of hydrophobia. At Ica forty-two people thus miserably perished. The disease came on between twelve and ninety days after the bite; and in those cases where it did come on, death ensued invariably within five days. After 1808, a long interval ensued without any cases. On inquiry, I did not hear of hydrophobia in Van Dieman's Land, or in Australia; and Burchell says, that during the five years he was at the Cape of Good Hope, he never heard of an instance of it. Webster asserts that at the Azores hydrophobia has never occurred; and the same assertion has been made with respect to Mauritius and St. Helena.[1] In so strange a disease, some information might possibly be gained by considering the circumstances under which it originates in distant climates; for it is improbable that a dog already bitten, should have been brought to these distant countries.

At night, a stranger arrived at the house of Don Benito, and asked permission to sleep there. He said he had been wandering about the mountains for seventeen days, having lost his way. He started from Guasco, and being accustomed to travelling in the Cordillera, did not expect any difficulty in following the track to Copiapó; but he soon became involved in a labyrinth of mountains, whence he could not escape. Some of his mules had fallen over precipices, and he had been in great distress. His chief difficulty arose from not knowing where to find water in the lower country, so that he was obliged to keep bordering the central ranges.

We returned down the valley, and on the 22nd reached the town of Copiapó. The lower part of the valley is broad, forming a fine plain like that of Quillota. The town covers a considerable space of ground, each house possessing a garden: but it is an uncomfortable place, and the dwellings are poorly furnished. Every one seems bent on the one object of making money, and then migrating as quickly as possible. All the inhabitants are more or less directly concerned with mines; and mines and ores are the sole subjects of conversation. Necessaries of all sorts are extremely dear; as the distance from the town to the port is eighteen leagues, and the land carriage very expensive. A fowl costs five or six shillings; meat is nearly as dear as in England; firewood, or rather sticks, are brought on donkeys from a distance of two and three days' journey within the Cordillera; and pasturage for animals is a shilling a day: all this for South America is wonderfully exorbitant.

June 26th. – I hired a guide and eight mules to take me into the Cordillera by a different line from my last excursion. As the country

1 Observa. sobre el clima de Lima, p. 67. – Azara's Travels, vol. i. p. 381. – Ulloa's Voyage, vol. ii. p. 28. – Burchell's Travels, vol. ii. p. 524. – Webster's Description of the Azores, p. 124. – Voyage à l'Isle de France par un Officier du Roi, tome i. p. 248. – Description of St. Helena, p. 123.

was utterly desert, we took a cargo and a half of barley mixed with chopped straw. About two leagues above the town, a broad valley called the 'Despoblado,' or uninhabited, branches off from that one by which we had arrived. Although a valley of the grandest dimensions, and leading to a pass across the Cordillera, yet it is completely dry, excepting perhaps for a few days during some very rainy winter. The sides of the crumbling mountains were furrowed by scarcely any ravines; and the bottom of the main valley, filled with shingle, was smooth and nearly level. No considerable torrent could ever have flowed down this bed of shingle; for if it had, a great cliff-bounded channel, as in all the southern valleys, would assuredly have been formed. I feel little doubt that this valley, as well as those mentioned by travellers in Peru, were left in the state we now see them by the waves of the sea, as the land slowly rose. I observed in one place, where the Despoblado was joined by a ravine (which in almost any other chain would have been called a grand valley), that its bed, though composed merely of sand and gravel, was higher than that of its tributary. A mere rivulet of water, in the course of an hour, would have cut a channel for itself; but it was evident that ages had passed away, and no such rivulet had drained this great tributary. It was curious to behold the machinery, if such a term may be used, for the drainage, all, with the last trifling exception, perfect, yet without any signs of action. Every one must have remarked how mud-banks, left by the retiring tide, imitate in miniature a country with hill and dale; and here we have the original model in rock, formed as the continent rose during the secular retirement of the ocean, instead of during the ebbing and flowing of the tides. If a shower of rain falls on the mud-bank, when left dry, it deepens the already-formed shallow lines of excavation; and so is it with the rain of successive centuries on the bank of rock and soil, which we call a continent.

We rode on after it was dark, till we reached a side ravine with a small well, called 'Agua amarga.' The water deserved its name, for besides being saline it was most offensively putrid and bitter; so that we could not force ourselves to drink either tea or maté. I suppose the distance from the river of Copiapó to this spot was at least twenty-five or thirty English miles; in the whole space there was not a single drop of water, the country deserving the name of desert in the strictest sense. Yet about half-way we passed some old Indian ruins near Punta Gorda: I noticed also in front of some of the valleys, which branch off from the Despoblado, two piles of stones placed a little way apart, and directed so as to point up the mouths of these small valleys. My companions knew nothing about them, and only answered my queries by their imperturbable 'quien sabe?'

I observed Indian ruins in several parts of the Cordillera: the most perfect, which I saw, were the Ruinas de Tambillos, in the Uspallata Pass. Small square rooms were there huddled together in separate groups: some of the doorways were yet standing; they were formed by a cross slab of stone only about three feet high. Ulloa has remarked on the lowness of the doors in the ancient Peruvian dwellings. These houses, when perfect, must have been capable of containing a considerable number of persons. Tradition says, that they were used as halting places for the Incas, when they crossed the mountains. Traces of Indian habitations have been discovered in many other parts, where it does not appear probable that they were used as mere resting-places, but yet where the land is as utterly unfit for any kind of cultivation as it is near the Tambillos or at the Incas Bridge, or in the Portillo Pass, at all which places I saw ruins. In the ravine of Jajuel, near Aconcagua, where there is no pass, I heard of remains of houses situated at a great height, where it is extremely cold and sterile. At first I imagined that these buildings had been places of refuge, built by the Indians on the first arrival of the Spaniards; but I have since been inclined to speculate on the probability of a small change of climate.

In this northern part of Chile, within the Cordillera, old Indian houses are said to be especially numerous: by digging amongst the ruins, bits of woollen articles, instruments of precious metals, and heads of Indian corn, are not unfrequently discovered: an arrowhead made of agate, and of precisely the same form with those now used in Tierra del Fuego, was given me. I am aware that the Peruvian Indians now frequently inhabit most lofty and bleak situations; but at Copiapó I was assured by men who had spent their lives in travelling through the Andes, that there were very many (*muchisimas*) buildings at heights so great as almost to border on the perpetual snow, and in parts where there exist no passes, and where the land produces absolutely nothing, and what is still more extraordinary, where there is no water. Nevertheless it is the opinion of the people of the country (although they are much puzzled by the circumstance), that, from the appearance of the houses, the Indians must have used them as places of residence. In this valley, at Punta Gorda, the remains consisted of seven or eight square little rooms, which were of a similar form with those at Tambillos, but built chiefly of mud, which the present inhabitants cannot, either here or, according to Ulloa, in Peru, imitate in durability. They were situated in the most conspicuous and defenceless position, at the bottom of the flat broad valley. There was no water nearer than three or four leagues, and that only in very small quantity, and bad: the soil was absolutely sterile; I looked in vain even for a lichen adhering to the rocks. At the present day, with the advantage of beasts of burden, a mine, unless it were very rich, could scarcely be worked here with profit. Yet the Indians formerly chose it as a place of residence! If at the present time two or three showers of rain were to fall annually, instead of one, as now is the case, during as many years, a small rill of water would probably be formed in this great valley; and then, by irrigation (which was formerly so well understood by the Indians), the soil would easily be rendered sufficiently productive to support a few families.

I have convincing proofs that this part of the continent of South America has been elevated near the coast at least from 400 to 500, and in some parts from 1000 to 1300 feet, since the epoch of existing shells; and further inland the rise possibly may have been greater. As the peculiarly arid character of the climate is evidently a consequence of the height of the Cordillera, we may feel almost sure that before the later elevations, the atmosphere could not have been so completely drained of its moisture as it now is; and as the rise has been gradual, so would have been the change in climate. On this notion of a change of climate since the buildings were inhabited, the ruins must be of extreme antiquity, but I do not think their preservation under the Chilian climate any great difficulty. We must also admit on this notion, (and this perhaps is a greater difficulty) that man has inhabited South America for an immensely long period, inasmuch as any change of climate effected by the elevation of the land must have been extremely gradual. At Valparaiso, within the last 220 years, the rise has been somewhat less than 19 feet: at Lima a sea-beach has certainly been upheaved from 80 to 90 feet, within the Indio-human period: but such small elevations could have had little power in deflecting the moisture-bringing atmospheric currents. Dr. Lund, however, found human skeletons in the caves of Brazil, the appearance of which induced him to believe that the Indian race has existed during a vast lapse of time in South America.

When at Lima, I conversed on these subjects[1] with Mr. Gill, a civil engineer, who had seen much of the interior country. He told me that a conjecture of a change of climate had sometimes crossed his mind; but that he thought that the greater portion of land, now incapable of cultivation, but covered with Indian ruins, had been reduced to this state by the water-conduits, which the Indians formerly constructed on so wonderful a scale, having been injured by neglect and by subterranean movements. I may here mention, that the Peruvians actually carried their irrigating streams in tunnels through hills of solid rock. Mr. Gill told me, he had been employed professionally to examine one; he found the passage low, narrow, crooked, and not of uniform breadth, but of very considerable length. Is it not most wonderful that men should have attempted such operations, without the use of iron or gunpowder? Mr. Gill also mentioned to me a most interesting, and, as far as I am aware, quite unparalleled case, of a subterranean disturbance having changed the drainage of a country. Travelling from Casma to Huaraz (not very far distant from Lima), he found a plain covered with ruins and marks of ancient cultivation, but now quite barren. Near it was the dry course of a considerable river, whence the water for irrigation had formerly been conducted. There was nothing in the appearance of the water-course, to indicate that the river had not flowed there a few years previously; in some parts, beds of sand and gravel were spread out; in others, the solid rock had been worn into a broad channel, which in one spot was about 40 yards in breadth and 8 feet deep. It is self-evident that a person

1 Temple, in his travels through Upper Peru, or Bolivia, in going from Potosi to Oruro, says, 'I saw many Indian villages or dwellings in ruins, up even to the very tops of the mountains, attesting a former population where now all is desolate.' He makes similar remarks in another place; but I cannot tell whether this desolation has been caused by a want of population, or by an altered condition of the land.

following up the course of a stream, will always ascend at a greater or less inclination: Mr. Gill, therefore, was much astonished, when walking up the bed of this ancient river, to find himself suddenly going down hill. He imagined that the downward slope had a fall of about 40 or 50 feet perpendicular. We here have unequivocal evidence that a ridge had been uplifted right across the old bed of a stream. From the moment the river-course was thus arched, the water must necessarily have been thrown back, and a new channel formed. From that moment, also, the neighbouring plain must have lost its fertilizing stream, and become a desert.

June 27th. – We set out early in the morning, and by mid-day reached the ravine of Paypote, where there is a tiny rill of water, with a little vegetation, and even a few algarroba trees, a kind of mimosa. From having firewood, a smelting-furnace had formerly been built here: we found a solitary man in charge of it, whose sole employment was hunting guanacos. At night it froze sharply; but having plenty of wood for our fire, we kept ourselves warm.

28th. – We continued gradually ascending, and the valley now changed into a ravine. During the day we saw several guanacos, and the track of the closely-allied species, the Vicuña: this latter animal is pre-eminently alpine in its habits; it seldom descends much below the limit of perpetual snow, and therefore haunts even a more lofty and sterile situation than the guanaco. The only other animal which we saw in any number was a small fox: I suppose this animal preys on the mice and other small rodents, which, as long as there is the least vegetation, subsist in considerable numbers in very desert places. In Patagonia, even on the borders of the salinas, where a drop of fresh water can never be found, excepting dew, these little animals swarm. Next to lizards, mice appear to be able to support existence on the smallest and driest portions of the earth, – even on islets in the midst of great oceans.

The scene on all sides showed desolation, brightened and made palpable by a clear, unclouded sky. For a time such scenery is sublime, but this feeling cannot last, and then it becomes uninteresting. We bivouacked at the foot of the 'primera linea,' or the first line of the partition of the waters. The streams, however, on the east side do not flow to the Atlantic, but into an elevated district, in the middle of which there is a large salina, or salt lake; – thus forming a little Caspian Sea at the height, perhaps, of ten thousand feet. Where we slept, there were some considerable patches of snow, but they do not remain throughout the year. The winds in these lofty regions obey very regular laws: every day a fresh breeze blows up the valley, and at night, an hour or two after sunset, the air from the cold regions above descends as through a funnel. This night it blew a gale of wind, and the temperature must have been considerably below the freezing-point, for water in a vessel soon became a block of ice. No clothes seemed to oppose any obstacle to the air; I suffered very much from the cold, so that I could not sleep, and in the morning rose with my body quite dull and benumbed.

In the Cordillera further southward, people lose their lives from snow-storms; here, it sometimes happens from another cause. My guide, when a boy of fourteen years old, was passing the Cordillera with a party in the month of May; and while in the central parts, a furious gale of wind arose, so that the men could hardly cling on their mules, and stones were flying along the ground. The day was cloud-less, and not a speck of snow fell, but the temperature was low. It is probable that the thermometer would not have stood very many degrees below the freezing-point, but the effect on their bodies, ill protected by clothing, must have been in proportion to the rapidity of the current of cold air. The gale lasted for more than a day; the men began to lose their strength, and the mules would not move onwards. My guide's brother tried to return, but he perished, and his body was found two years afterwards, lying by the side of his mule near the road, with the bridle still in his hand. Two other men in the party lost their fingers and toes; and out of two hundred mules and thirty cows, only fourteen mules escaped alive. Many years ago the whole of a large party are supposed to have perished from a similar cause, but their bodies to this day have never been discovered. The union of a cloudless sky, low temperature, and a furious gale of wind, must be, I should think, in all parts of the world, an unusual occurrence.

June 29th. – We gladly travelled down the valley to our former night's lodging, and thence to near the Agua amarga. On July 1st we reached the valley of Copiapó. The smell of the fresh clover was quite delightful, after the scentless air of the dry sterile Despoblado. Whilst staying in the town I heard an account from several of the inhabitants, of a hill in the neighbourhood which they called 'El Bramador,' – the roarer or bellower. I did not at the time pay sufficient attention to the account; but, as far as I understood, the hill was covered by sand, and the noise was produced only when people, by ascending it, put the sand in motion. The same circumstances are described in detail on the authority of Seetzen and Ehrenberg,[1] as the cause of the sounds which have been heard by many travellers on Mount Sinai near the Red Sea. One person with whom I conversed, had himself heard the noise; he described it as very surprising; and he distinctly stated that, although he could not understand how it was caused, yet it was necessary to set the sand rolling down the acclivity. A horse walking over dry and coarse sand, causes a peculiar chirping noise from the friction of the particles; a circumstance which I several times noticed on the coast of Brazil.

Three days afterwards I heard of the *Beagle*'s arrival at the Port, distant eighteen leagues from the town. There is very little land cultivated down the valley; its wide expanse supports a wretched wiry grass, which even the donkeys can hardly eat. This poorness of the vegetation is owing to the quantity of saline matter with which the soil is impregnated. The Port consists of an assemblage of miserable little hovels, situated at the foot of a sterile plain. At present, as the river contains water enough to reach the sea, the inhabitants enjoy the advantage of having fresh water within a mile and a half. On the beach there were large piles of merchandise, and the little place had an air of activity. In the evening I gave my adios, with a hearty good-will, to my companion Mariano Gonzales, with whom I had ridden so many leagues in Chile. The next morning the *Beagle* sailed for Iquique.

July 12th. – We anchored in the port of Iquique, in lat. 20° 12′, on the coast of Peru. The town contains about a thousand inhabitants, and stands on a little plain of sand at the foot of a great wall of rock,

1 Edinburgh Phil. Journ., Jan. 1830, p. 74; and April, 1830, p. 258. – Also Daubeny on Volcanoes, p. 438; and Bengal Journ., vol. vii. p. 324.

2000 feet in height, here forming the coast. The whole is utterly desert. A light shower of rain falls only once in very many years; and the ravines consequently are filled with detritus, and the mountain-sides covered by piles of fine white sand, even to a height of a thousand feet. During this season of the year a heavy bank of clouds, stretched over the ocean, seldom rises above the wall of rocks on the coast. The aspect of the place was most gloomy; the little port, with its few vessels, and small group of wretched houses, seemed over-whelmed and out of all proportion with the rest of the scene.

The inhabitants live like persons on board a ship: every necessary comes from a distance: water is brought in boats from Pisagua, about forty miles northward, and is sold at the rate of nine reals (4s. 6d.) an eighteen-gallon cask: I bought a wine-bottle full for threepence. In like manner firewood, and of course every article of food, is imported. Very few animals can be maintained in such a place: on the ensuing morning I hired with difficulty, at the price of four pounds sterling, two mules and a guide to take me to the nitrate of soda works. These are at present the support of Iquique. This salt was first exported in 1830: in one year an amount in value of one hundred thousand pounds sterling, was sent to France and England. It is principally used as a manure and in the manufacture of nitric acid: owing to its deliquescent property it will not serve for gunpowder. Formerly there were two exceedingly rich silver-mines in this neigh-bourhood, but their produce is now very small.

Our arrival in the offing caused some little apprehension. Peru was in a state of anarchy; and each party having demanded a contribution, the poor town of Iquique was in tribulation, thinking the evil hour was come. The people had also their domestic troubles; a short time before, three French carpenters had broken open, during the same night, the two churches, and stolen all the plate: one of the robbers, however, subsequently confessed, and the plate was recovered. The convicts were sent to Arequipa, which, though the capital of this province, is two hundred leagues distant; the government there thought it a pity to punish such useful workmen, who could make all sorts of furniture; and accordingly liberated them. Things being in this state, the churches were again broken open, but this time the plate was not recovered. The inhabitants became dreadfully enraged, and declaring that none but heretics would thus 'eat God Almighty,' proceeded to torture some Englishmen, with the intention of afterwards shooting them. At last the authorities interfered, and peace was established.

13*th*. – In the morning I started for the saltpetre-works, a distance of fourteen leagues. Having ascended the steep coast-mountains by a zigzag sandy track, we soon came in view of the mines of Guantajaya and St. Rosa. These two small villages are placed at the very mouths of the mines; and being perched up on hills, they had a still more unnatural and desolate appearance than the town of Iquique. We did not reach the saltpetre-works till after sunset, having ridden all day across an undulating country, a complete and utter desert. The road was strewed with the bones and dried skins of the many beasts of burden which had perished on it from fatigue. Excepting the Vultur aura, which preys on the carcasses, I saw neither bird, quadruped, reptile, nor insect. On the coast-mountains, at the height of about 2000 feet, where during this season the clouds generally hang, a very few cacti were growing in the clefts of rock; and the loose sand was strewed over with a lichen, which lies on the surface quite unattached. This plant belongs to the genus Cladonia, and somewhat resembles the reindeer lichen. In some parts it was in sufficient quantity to tinge the sand, as seen from a distance, of a pale yellowish colour. Further inland, during the whole ride of fourteen leagues, I saw only one other vegetable production, and that was a most minute yellow lichen, growing on the bones of the dead mules. This was the first true desert which I had seen: the effect on me was not impressive; but I believe this was owing to my having become gradually accustomed to such scenes, as I rode northward from Valparaiso, through Coquimbo, to Copiapó. The appearance of the country was remarkable, from being covered by a thick crust of common salt, and of a stratified saliferous alluvium, which seems to have been deposited as the land slowly rose above the level of the sea. The salt is white, very hard, and compact: it occurs in water-worn nodules projecting from the agglutinated sand, and is associated with much gypsum. The appearance of this superficial mass very closely resembled that of a country after snow, before the last dirty patches are thawed. The existence of this crust of a soluble substance over the whole face of the country, shows how extraordinarily dry the climate must have been for a long period.

At night I slept at the house of the owner of one of the saltpetre mines. The country is here as unproductive as near the coast; but water, having rather a bitter and brackish taste, can be procured by digging wells. The well at this house was thirty-six yards deep: as scarcely any rain falls, it is evident the water is not thus derived; indeed if it were, it could not fail to be as salt as brine, for the whole surrounding country is incrusted with various saline substances. We must therefore conclude that it percolates under ground from the Cordillera, though distant many leagues. In that direction there are a few small villages, where the inhabitants, having more water, are enabled to irrigate a little land, and raise hay, on which the mules and asses, employed in carrying the saltpetre, are fed. The nitrate of soda was now selling at the ship's side at fourteen shillings per hundred pounds: the chief expense is its transport to the sea-coast. The mine consists of a hard stratum, between two and three feet thick, of the nitrate mingled with a little of the sulphate of soda and a good deal of common salt. It lies close beneath the surface, and follows for a length of one hundred and fifty miles the margin of a grand basin or plain; this, from its outline, manifestly must once have been a lake, or more probably an inland arm of the sea, as may be inferred from the presence of iodic salts in the saline stratum. The surface of the plain is 3300 feet above the Pacific.

19*th*. – We anchored in the Bay of Callao, the seaport of Lima, the capital of Peru. We stayed here six weeks, but from the troubled state of public affairs, I saw very little of the country. During our whole visit the climate was far from being so delightful, as it is generally represented. A dull heavy bank of clouds constantly hung over the land, so that during the first sixteen days I had only one view of the Cordillera behind Lima. These mountains, seen in stages, one above the other, through openings in the clouds, had a very grand appearance. It is almost become a proverb, that rain never falls in the lower part of Peru. Yet this can hardly be considered correct; for during almost every day of our visit there was a thick drizzling mist, which was sufficient to make the streets muddy and one's clothes damp: this the people are pleased to call Peruvian dew. That much rain does not fall is very certain, for the houses are covered only with flat roofs made of hardened mud; and on the mole ship-loads of wheat were piled up, being thus left for weeks together without any shelter.

I cannot say I liked the very little I saw of Peru: in summer, however, it is said that the climate is much pleasanter. In all seasons, both inhabitants and foreigners suffer from severe attacks of ague. This disease is common on the whole coast of Peru, but is unknown in the interior. The attacks of illness which arise from miasma never fail to appear most mysterious. So difficult is it to judge from the aspect of a country, whether or not it is healthy, that if a person had been told to choose within the tropics a situation appearing favourable for health, very probably he would have named this coast. The plain round the outskirts of Callao is sparingly covered with a coarse grass, and in some parts there are a few stagnant, though very small, pools of water. The miasma, in all probability, arises from these: for the town of Arica was similarly circumstanced, and its healthiness was much improved by the drainage of some little pools. Miasma is not always produced by a luxuriant vegetation with an ardent climate; for many parts of Brazil, even where there are marshes and a rank vegetation, are much more healthy than this sterile coast of Peru. The densest forests in a temperate climate, as in Chiloe, do not seem in the slightest degree to affect the healthy condition of the atmosphere.

The island of St. Jago, at the Cape de Verds, offers another strongly-marked instance of a country, which any one would have expected to find most healthy, being very much the contrary. I have described the bare and open plains as supporting, during a few weeks after the rainy season, a thin vegetation, which directly withers away and dries up: at this period the air appears to become quite poisonous; both natives and foreigners often being affected with violent fevers. On the other hand, the Galapagos Archipelago, in the Pacific, with a similar soil, and periodically subject to the same process of vegetation, is perfectly healthy. Humboldt has observed, that, 'under the torrid zone, the smallest marshes are the most dangerous, being surrounded, as at Vera Cruz and Carthagena, with an arid and sandy soil, which raises the temperature of the ambient air.'[1] On the coast of Peru, however, the temperature is not hot to any excessive degree; and perhaps in consequence, the intermittent fevers are not of the most malignant order. In all unhealthy countries the greatest risk is run by sleeping on shore. Is this owing to the state of the body during sleep, or to a greater abundance of miasma at such times? It appears certain that those who stay on board a vessel, though anchored at only a short distance from the coast, generally suffer less than those actually on shore. On the other hand, I have heard of one remarkable case where a fever broke out among the crew of a man-of-war some hundred miles off the coast of Africa, and at the very same time that one of those fearful periods[2] of death commenced at Sierra Leone.

No State in South America, since the declaration of independence, has suffered more from anarchy than Peru. At the time of our visit, there were four chiefs in arms contending for supremacy in the government: if one succeeded in becoming for a time very powerful, the others coalesced against him; but no sooner were they victorious, than they were again hostile to each other. The other day, at the Anniversary of the Independence, high mass was performed, the President partaking of the sacrament: during the *Te Deum laudamus*, instead of each regiment displaying the Peruvian flag, a black one with death's head was unfurled. Imagine a government under which such a scene could be ordered, on such an occasion, to be typical of their determination of fighting to death! This state of affairs happened at a time very unfortunately for me, as I was precluded from taking any excursions much beyond the limits of the town. The barren island of S. Lorenzo, which forms the harbour, was nearly the only place where one could walk securely. The upper part, which is upwards of 1000 feet in height, during this season of the year (winter), comes within the lower limit of the clouds; and in consequence, an abundant cryptogamic vegetation, and a few flowers, cover

1 Political Essay on the Kingdom of New Spain, vol. iv. p. 199.

2 A similar interesting case is recorded in the Madras Medical Quart. Journ., 1839, p. 340. Dr. Ferguson, in his admirable Paper (see 9th vol. of Edinburgh Royal Trans.), shows clearly that the poison is generated in the drying process; and hence that dry hot countries are often the most unhealthy.

the summit. On the hills near Lima, at a height but little greater, the ground is carpeted with moss, and beds of beautiful yellow lilies, called Amancaes. This indicates a very much greater degree of humidity, than at a corresponding height at Iquique. Proceeding northward of Lima, the climate becomes damper, till on the banks of the Guyaquil, nearly under the equator, we find the most luxuriant forests. The change, however, from the sterile coast of Peru to that fertile land is described as taking place rather abruptly in the latitude of Cape Blanco, two degrees south of Guyaquil.

Callao is a filthy, ill-built, small seaport. The inhabitants, both here and at Lima, present every imaginable shade of mixture, between European, Negro, and Indian blood. They appear a depraved, drunken set of people. The atmosphere is loaded with foul smells, and that peculiar one, which may be perceived in almost every town within the tropics, was here very strong. The fortress, which withstood Lord Cochrane's long siege, has an imposing appearance. But the President, during our stay, sold the brass guns, and proceeded to dismantle parts of it. The reason assigned was, that he had not an officer to whom he could trust so important a charge. He himself had good reasons for thinking so, as he had obtained the presidentship by rebelling while in charge of this same fortress. After we left South America, he paid the penalty in the usual manner, by being conquered, taken prisoner, and shot.

Lima stands on a plain in a valley, formed during the gradual retreat of the sea. It is seven miles from Callao, and is elevated 500 feet above it; but from the slope being very gradual, the road appears absolutely level; so that when at Lima it is difficult to believe one has ascended even one hundred feet: Humboldt has remarked on this singularly deceptive case. Steep, barren hills rise like islands from the plain, which is divided, by straight mud-walls, into large green fields. In these scarcely a tree grows excepting a few willows, and an occasional clump of bananas and of oranges. The city of Lima is now in a wretched state of decay: the streets are nearly unpaved; and heaps of filth are piled up in all directions, where the black gallinazos, tame as poultry, pick up bits of carrion. The houses have generally an upper story, built, on account of the earthquakes, of plastered wood work; but some of the old ones, which are now used by several families, are immensely large, and would rival in suites of apartments the most magnificent in any palace. Lima, the City of the Kings, must formerly have been a splendid town. The extraordinary number of churches gives it, even at the present day, a peculiar and striking character, especially when viewed from a short distance.

One day I went out with some merchants to hunt in the immediate vicinity of the city. Our sport was very poor; but I had an opportunity of seeing the ruins of one of the ancient Indian villages, with its mound like a natural hill in the centre. The remains of houses, enclosures, irrigating streams, and burial mounds, scattered over this plain, cannot fail to give one a high idea of the condition and number of the ancient population. When their earthenware, woollen clothes, utensils of elegant forms cut out of the hardest rocks, tools of copper, ornaments of precious stones, palaces, and hydraulic works, are considered, it is impossible not to respect the considerable advance made by them in the arts of civilization. The burial mounds, called Huacas, are really stupendous; although in some places they appear to be natural hills incased and modelled.

There is also another and very different class of ruins, which possesses some interest, namely, those of old Callao, overwhelmed by the great earthquake of 1746, and its accompanying wave. The destruction must have been more complete even than at Talcahuano. Quantities of shingle almost conceal the foundations of the walls, and vast masses of brickwork appear to have been whirled about like pebbles by the retiring waves. It has been stated that the land subsided during this memorable shock: I could not discover any proof of this; yet it seems far from improbable, for the form of the coast must certainly have undergone some change since the foundation of the old town; as no people in their senses would willingly have chosen for their building place, the narrow spit of shingle on which the ruins now stand. Since our voyage, M. Tschudi has come to the conclusion, by the comparison of old and modern maps, that the coast both north and south of Lima has certainly subsided.

On the island of San Lorenzo, there are very satisfactory proofs of elevation within the recent period; this of course is not opposed to the belief, of a small sinking of the ground having subsequently taken place. The side of this island fronting the Bay of Callao, is worn into three obscure terraces, the lower one of which is covered by a bed a mile in length, almost wholly composed of shells of eighteen species, now living in the adjoining sea. The height of this bed is eighty-five feet. Many of the shells are deeply corroded, and have a much older and more decayed appearance than those at the height of 500 or 600 feet on the coast of Chile. These shells are associated with much common salt, a little sulphate of lime (both probably left by the evaporation of the spray, as the land slowly rose), together with sulphate of soda and muriate of lime. They rest on fragments of the underlying sandstone, and are covered by a few inches thick of detritus. The shells, higher up on this terrace, could be traced scaling off in flakes, and falling into an impalpable powder; and on an upper terrace, at the height of 170 feet, and likewise at some considerably higher points, I found a layer of saline powder of exactly similar appearance, and lying in the same relative position. I have no doubt that this upper layer originally existed as a bed of shells, like that on the eighty-five-feet ledge; but it does not now contain even a trace of organic structure. The powder has been analyzed for me by Mr. T. Reeks; it consists of sulphates and muriates both of lime and soda, with very little carbonate of lime. It is known that common salt and carbonate of lime left in a mass for some time together, partly decompose each other; though this does not happen with small quantities in solution. As the half-decomposed shells in the lower parts are associated with much common salt, together with some of the saline substances composing the upper saline layer, and as these shells are corroded and decayed in a remarkable manner, I strongly suspect that this double decomposition has here taken place. The resultant salts, however, ought to be carbonate of soda and muriate of lime; the latter is present, but not the carbonate of soda. Hence I am led to imagine that by some unexplained means, the carbonate of soda becomes changed into the sulphate. It is obvious that the saline layer could not have been preserved in any country in which abundant rain occasionally fell: on the other hand, this very circumstance, which at first sight appears so highly favourable to the long preservation of exposed shells, has probably been the indirect means, through the common salt not having been washed away, of their decomposition and early decay.

I was much interested by finding on the terrace, at the height of eighty-five feet, *embedded* amidst the shells and much sea-drifted rubbish, some bits of cotton thread, plaited rush, and the head of a stalk of Indian corn: I compared these relics with similar ones taken out of the Huacas, or old Peruvian tombs, and found them identical in appearance. On the mainland in front of San Lorenzo, near Bellavista, there is an extensive and level plain about a hundred feet high, of which the lower part is formed of alternating layers of sand and impure clay, together with some gravel, and the surface, to the depth of from three to six feet, of a reddish loam, containing a few scattered sea-shells and numerous small fragments of coarse red earthenware, more abundant at certain spots than at others. At first I was inclined to believe that this superficial bed, from its wide extent and smoothness, must have been deposited beneath the sea; but I afterwards found in one spot, that it lay on an artificial floor of round stones. It seems, therefore, most probable that at a period when the land stood at a lower level, there was a plain very similar to that now surrounding Callao, which being protected by a shingle beach, is raised but very little above the level of the sea. On this plain, with its underlying red-clay beds, I imagine that the Indians manufactured their earthen vessels; and that, during some violent earthquake, the sea broke over the beach, and converted the plain into a temporary lake, as happened round Callao in 1713 and 1746. The water would then have deposited mud, containing fragments of pottery from the kilns, more abundant at some spots than at others, and shells from the sea. This bed with fossil earthenware, stands at about the same height with the shells on the lower terrace of San Lorenzo, in which the cotton-thread and other relics were embedded. Hence we may safely conclude, that within the Indo-human period there has been an elevation, as before alluded to, of more than eighty-five feet; for some little elevation must have been lost by the coast having subsided since the old maps were engraved. At Valparaiso, although in the 220 years before our visit, the elevation cannot have exceeded nineteen feet, yet subsequently to 1817 there has been a rise, partly insensible and partly by a start during the shock of 1822, of ten or eleven feet. The antiquity of the Indo-human race here, judging by the eighty-five feet rise of the land since the relics were embedded, is the more remarkable, as on the coast of Patagonia, when the land stood about the same number of feet lower, the Macrauchenia was a living beast; but as the Patagonian coast is some way distant from the Cordillera, the rising there may have been slower than here. At Bahia Blanca, the elevation has been only a few feet since the numerous gigantic quadrupeds were there entombed; and, according to the generally received opinion, when these extinct animals were living, man did not exist. But the rising of that part of the coast of Patagonia, is perhaps noways connected with the Cordillera, but rather with a line of old volcanic rocks in Banda Oriental, so that it may have been infinitely slower than on the shores of Peru. All these speculations, however, must be vague; for who will pretend to say, that there may not have been several periods of subsidence, intercalated between the movements of elevation; for we know that along the whole coast of Patagonia, there have certainly been many and long pauses in the upward action of the elevatory forces.

CHAPTER XVII

GALAPAGOS ARCHIPELAGO

The whole Group Volcanic – Number of Craters –
Leafless Bushes – Colony at Charles Island – James
Island – Salt-lake in Crater – Natural History of the
Group – Ornithology, curious Finches – Reptiles –
Great Tortoises, habits of – Marine Lizard, feeds on
Sea-weed – Terrestrial Lizard, burrowing habits, herbivor-
ous – Importance of Reptiles in the Archipelago –
Fish, Shells, Insects – Botany – American type of organiza-
tion – Differences in the Species or Races on different
Islands – Tameness of the Birds – Fear of Man, an acquired
instinct.

SEPTEMBER 15*th*. – This archipelago consists of ten principal
islands, of which five exceed the others in size. They are situated
under the Equator, and between five and six hundred miles westward
of the coast of America. They are all formed of volcanic rocks; a
few fragments of granite curiously glazed and altered by the heat, can
hardly be considered as an exception. Some of the craters, surmount-
ing the larger islands, are of immense size, and they rise to a height
of between three and four thousand feet. Their flanks are studded
by innumerable smaller orifices. I scarcely hesitate to affirm, that
there must be in the whole archipelago at least two thousand craters.
These consist either of lava and scoriæ, or of finely-stratified, sandstone-like tuff. Most of the latter are beautifully symmetrical; they
owe their origin to eruptions of volcanic mud without any lava: it
is a remarkable circumstance that every one of the twenty-eight
tuff-craters which were examined, had their southern sides either
much lower than the other sides, or quite broken down and removed.
As all these craters have apparently been formed when standing
in the sea, and as the waves from the trade-wind and the swell
from the open Pacific here unite their forces on the southern coasts
of all the islands, this singular uniformity in the broken state

384

Culpepper I.

60 Miles

Wenman I.

Abingdon I.

Tower I.

Bindloes I.

Narborough I.

James I.

Indefatigable I.

Albemarle I.

Chatham I.

Barrington I.

Charles I.

Hood's I.

of the craters, composed of the soft and yielding tuff, is easily explained.

Considering that these islands are placed directly under the equator, the climate is far from being excessively hot; this seems chiefly caused by the singularly low temperature of the surrounding water, brought here by the great southern Polar current. Excepting during one short season, very little rain falls, and even then it is irregular; but the clouds generally hang low. Hence, whilst the lower parts of the island are very sterile, the upper parts, at a height of a thousand feet and upwards, possess a damp climate and a tolerably luxuriant vegetation. This is especially the case on the windward sides of the islands, which first receive and condense the moisture from the atmosphere.

In the morning (17th) we landed on Chatham Island, which, like the others, rises with a tame and rounded outline, broken here and there by scattered hillocks, the remains of former craters. Nothing could be less inviting than the first appearance. A broken field of black basaltic lava, thrown into the most rugged waves, and crossed by great fissures, is every where covered by stunted, sun-burnt brushwood, which shows little signs of life. The dry and parched surface, being heated by the noonday sun, gave to the air a close and sultry feeling, like that from a stove: we fancied even that the bushes smelt unpleasantly. Although I diligently tried to collect as many plants as possible, I succeeded in getting very few; and such wretched-looking little weeds would have better become an arctic than an equatorial Flora. The brushwood appears, from a short distance, as leafless as our trees during winter; and it was some time before I discovered that not only almost every plant was now in full leaf, but that the greater number were in flower. The commonest bush is one of the Euphorbiaceæ: an acacia and a great odd-looking cactus are the only trees which afford any shade. After the season of heavy rains, the islands are said to appear for a short time partially green. The volcanic island of Fernando Noronha, placed in many respects under nearly similar conditions, is the only other country where I have seen a vegetation at all like this of the Galapagos islands.

The *Beagle* sailed round Chatham Island, and anchored in several bays. One night I slept on shore on a part of the island, where black truncated cones were extraordinarily numerous: from one small eminence I counted sixty of them, all surmounted by craters more or less perfect. The greater number consisted merely of a ring of red scoriæ or slags, cemented together: and their height above the plain of lava was not more than from fifty to a hundred feet: none had been very lately active. The entire surface of this part of the island seems to have been permeated, like a sieve, by the subterranean vapours: here and there the lava, whilst soft, has been blown into great bubbles; and in other parts, the tops of caverns similarly formed have fallen in, leaving circular pits with steep sides. From the regular form of the many craters, they gave to the country an artificial appearance, which vividly reminded me of those parts of Staffordshire, where the great iron-foundries are most numerous. The day was glowing hot, and the scrambling over the rough surface and through the intricate thickets, was very fatiguing; but I was well repaid by the strange Cyclopean scene. As I was walking along I met two large tortoises, each of which must have weighed at least two hundred pounds: one was eating a piece of cactus, and as I approached, it stared at me and slowly stalked away; the other gave a deep hiss, and drew in its head. These huge reptiles, surrounded by the black lava, the leafless shrubs, and large cacti, seemed to my fancy like some antediluvian animals. The few dull-coloured birds cared no more for me, than they did for the great tortoises.

23rd. – The *Beagle* proceeded to Charles Island. This archipelago has long been frequented, first by the Bucaniers, and latterly by whalers, but it is only within the last six years, that a small colony has been established here. The inhabitants are between two and three hundred in number: they are nearly all people of colour, who have been banished for political crimes from the Republic of the Equator, of which Quito is the capital. The settlement is placed about four and a half miles inland, and at a height probably of a thousand feet. In the first part of the road we passed through leafless thickets, as in Chatham Island. Higher up, the woods gradually became greener: and as soon as we crossed the ridge of the island, we were cooled by a fine southerly breeze, and our sight refreshed by a green and thriving vegetation. In this upper region coarse grasses and ferns abound; but there are no tree-ferns: I saw nowhere any member of the Palm family, which is the more singular, as 360 miles northward, Cocos Island takes its name from the number of cocoa-nuts. The houses are irregularly scattered over a flat space of ground, which is cultivated with sweet potatoes and bananas. It will not easily be imagined how pleasant the sight of black mud was to us, after having been so long accustomed to the parched soil of Peru and northern Chile. The inhabitants, although complaining of poverty, obtain, without much trouble, the means of subsistence. In the woods there are many wild pigs and goats; but the staple article of animal food is supplied by the tortoises. Their numbers have of course been greatly reduced in this island, but the people yet count on two days' hunting giving them food for the rest of the week. It is said that formerly single vessels have taken away as many as seven hundred, and that the ship's company of a frigate some years since brought down in one day two hundred tortoises to the beach.

September 29th. – We doubled the south-west extremity of Albemarle Island, and the next day were nearly becalmed between it and Narborough Island. Both are covered with immense deluges of black naked lava, which have flowed either over the rims of the great caldrons, like pitch over the rim of a pot in which it has been boiled, or have burst forth from smaller orifices on the flanks; in their descent they have spread over miles of the sea-coast. On both of these islands, eruptions are known to have taken place; and in Albemarle, we saw a small jet of smoke curling from the summit of one of the great craters. In the evening we anchored in Bank's Cove, in Albemarle Island. The next morning I went out walking. To the south of the broken tuff-crater, in which the *Beagle* was anchored, there was another beautifully symmetrical one of an elliptic form; its longer axis was a little less than a mile, and its depth about 500 feet. At its bottom there was a shallow lake, in the middle of which a tiny crater formed an islet. The day was overpoweringly hot, and the lake looked clear and blue: I hurried down the cindery slope, and choked with dust eagerly tasted the water – but, to my sorrow, I found it salt as brine.

The rocks on the coast abounded with great black lizards, between three and four feet long; and on the hills, an ugly yellowish-brown species was equally common. We saw many of this latter kind, some clumsily running out of our way, and others shuffling into their burrows. I shall presently describe in more detail the habits of both these reptiles. The whole of this northern part of Albemarle Island is miserably sterile.

October 8th. – We arrived at James Island; this island, as well as Charles Island, were long since thus named after our kings of the Stuart line. Mr. Bynoe, myself, and our servants were left here for a week, with provisions and a tent, whilst the *Beagle* went for water. We found here a party of Spaniards, who had been sent from Charles Island to dry fish, and to salt tortoise-meat. About six miles inland, and at the height of nearly 2000 feet, a hovel had been built in which two men lived, who were employed in catching tortoises, whilst the others were fishing on the coast. I paid this party two visits, and slept there one night. As in the other islands, the lower region was covered by nearly leafless bushes, but the trees were here of a larger growth than elsewhere, several being two feet and some even two feet nine inches in diameter. The upper region being kept damp by the clouds, supports a green and flourishing vegetation. So damp was the ground, that there were large beds of a coarse cyperus, in which great numbers of a very small water-rail lived and bred. While staying in this upper region, we lived entirely upon tortoise-meat: the breast-plate roasted (as the Gauchos do *carne con cuero*), with the flesh on it, is very good; and the young tortoises make excellent soup; but otherwise the meat to my taste is indifferent.

One day we accompanied a party of the Spaniards in their whale-boat to a salina, or lake from which salt is procured. After landing, we had a very rough walk over a rugged field of recent lava, which has almost surrounded a tuff-crater, at the bottom of which the salt-lake lies. The water is only three or four inches deep, and rests on a layer of beautifully crystallized, white salt. The lake is quite circular, and is fringed with a border of bright green succulent plants; the almost precipitous walls of the crater are clothed with wood, so that the scene was altogether both picturesque and curious. A few years since, the sailors belonging to a sealing-vessel murdered their captain in this quiet spot; and we saw his skull lying among the bushes.

During the greater part of our stay of a week, the sky was cloudless, and if the trade-wind failed for an hour, the heat became very oppressive. On two days, the thermometer within the tent stood for some hours at 93°; but in the open air, in the wind and sun, at only 85°. The sand was extremely hot; the thermometer placed in some of a brown colour immediately rose to 137°, and how much above that it would have risen, I do not know, for it was not graduated any higher. The black sand felt much hotter, so that even in thick boots it was quite disagreeable to walk over it.

The natural history of these islands is eminently curious, and well deserves attention. Most of the organic productions are aboriginal creations, found nowhere else; there is even a difference between the inhabitants of the different islands; yet all show a marked relationship with those of America, though separated from that continent by an open space of ocean, between 500 and 600 miles in width. The archipelago is a little world within itself, or rather a satellite attached to America, whence it has derived a few stray colonists, and has received the general character of its indigenous productions. Considering the small size of these islands, we feel the more astonished at the number of their aboriginal beings, and at their confined range. Seeing every height crowned with its crater, and the boundaries of most of the lava-streams still distinct, we are led to believe that within a period, geologically recent, the unbroken ocean was here spread out. Hence, both in space and time, we seem to be brought somewhat near to that great fact – that mystery of mysteries – the first appearance of new beings on this earth.

Of terrestrial mammals, there is only one which must be considered as indigenous, namely, a mouse (Mus Galapagoensis), and this is confined, as far as I could ascertain, to Chatham island, the most easterly island of the group. It belongs, as I am informed by Mr. Waterhouse, to a division of the family of mice characteristic of America. At James island, there is a rat sufficiently distinct from the common kind to have been named and described by Mr. Waterhouse; but as it belongs to the old-world division of the family, and as this island has been frequented by ships for the last hundred and fifty years, I can hardly doubt that this rat is merely a variety, produced by the new and peculiar climate, food, and soil, to which it has been subjected. Although no one has a right to speculate without distinct facts, yet even with respect to the Chatham island mouse, it should be borne in mind, that it may possibly be an American species imported here; for I have seen, in a most unfrequented part of the Pampas, a native mouse living in the roof of a newly-built hovel, and therefore its transportation in a vessel is not improbable: analogous facts have been observed by Dr. Richardson in North America.

Of land-birds I obtained twenty-six kinds, all peculiar to the group and found nowhere else, with the exception of one lark-like finch from North America (Dolichonyx oryzivorus), which ranges on that continent as far north as $54°$, and generally frequents marshes. The other twenty-five birds consist, firstly, of a hawk, curiously intermediate in structure between a Buzzard and the American group of carrion-feeding Polybori; and with these latter birds it agrees most closely in every habit and even tone of voice. Secondly, there are two owls, representing the short-eared and white barn-owls of Europe. Thirdly, a wren, three tyrant-flycatchers (two of them species of Pyrocephalus, one or both of which would be ranked by some ornithologists as only varieties), and a dove – all analogous to, but distinct from, American species. Fourthly, a swallow, which though differing from the Progne purpurea of both Americas, only in being rather duller coloured, smaller, and slenderer, is considered by Mr. Gould as specifically distinct. Fifthly, there are three species of mocking-thrush – a form highly characteristic of America. The remaining land-birds form a most singular group of finches, related to each other in the structure of their beaks, short tails, form of body, and plumage: there are thirteen species, which Mr. Gould has divided into four sub-groups. All these species are peculiar to this archipelago; and so is the whole group, with the exception of one species of the sub-group Cactornis, lately brought from Bow island, in the Low Archipelago. Of Cactornis, the two species may be often seen climbing about the flowers of the great cactus-trees; but all the other species of this group of finches, mingled together in flocks, feed on the dry and sterile ground of the lower districts. The males of all, or certainly of the greater number, are jet black; and the females (with perhaps one or two exceptions) are brown. The most curious fact is the perfect gradation in the size of the beaks in the different species of Geospiza, from one as large as that of a hawfinch to that of a chaffinch, and (if Mr. Gould is right in including his sub-group, Certhidea, in the main group), even to that of a warbler. The largest beak in the genus Geospiza is shown in Fig. 1, and the smallest in Fig. 3; but instead of there being only one intermediate species, with a beak of the size shown in Fig. 2, there are no less than six species with insensibly graduated beaks. The beak of the sub-group Certhidea, is shown in Fig. 4. The beak of Cactornis is somewhat like that of a starling; and that of the fourth sub-group, Camarhynchus, is slightly parrot-shaped. Seeing this gradation and diversity of structure in one small, intimately related group of birds, one might really fancy that from an original paucity of birds in this archipelago, one species had been taken and modified for different ends. In a like manner it might be fancied that a bird originally a buzzard, had been induced here to undertake the office of the carrion-feeding Polybori of the American continent.

Of waders and water-birds I was able to get only eleven kinds, and of these only three (including a rail confined to the damp summits of the islands) are new species. Considering the wandering habits of the gulls, I was surprised to find that the species inhabiting these islands is peculiar, but allied to one from the southern parts of South America. The far greater peculiarity of the land-birds, namely, twenty-five out

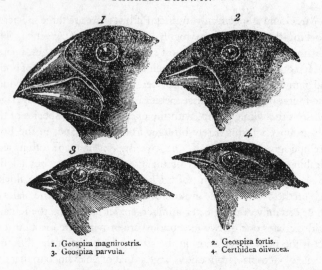

1. Geospiza magnirostris.   2. Geospiza fortis.
3. Geospiza parvula.   4. Certhidea olivacea.

of twenty-six being new species or at least new races, compared with
the waders and web-footed birds, is in accordance with the greater
range which these latter orders have in all parts of the world. We shall
hereafter see this law of aquatic forms, whether marine or fresh-
water, being less peculiar at any given point of the earth's surface
than the terrestrial forms of the same classes, strikingly illustrated in
the shells, and in a lesser degree in the insects of this archipelago.

Two of the waders are rather smaller than the same species
brought from other places: the swallow is also smaller, though it is
doubtful whether or not it is distinct from its analogue. The two owls,
the two tyrant-flycatchers (Pyrocephalus) and the dove, are also
smaller than the analogous but distinct species, to which they are
most nearly related; on the other hand, the gull is rather larger.
The two owls, the swallow, all three species of mocking-thrush, the
dove in its separate colours though not in its whole plumage,
the Totanus, and the gull, are likewise duskier coloured than their
analogous species; and in the case of the mocking-thrush and Totanus, than any other species of the two genera. With the exception of
a wren with a fine yellow breast, and of a tyrant-flycatcher with a
scarlet tuft and breast, none of the birds are brilliantly coloured, as

might have been expected in an equatorial district. Hence it would appear probable, that the same causes which here make the immigrants of some species smaller, make most of the peculiar Galapageian species also smaller, as well as very generally more dusky coloured. All the plants have a wretched, weedy appearance, and I did not see one beautiful flower. The insects, again, are small sized and dull coloured, and, as Mr. Waterhouse informs me, there is nothing in their general appearance which would have led him to imagine that they had come from under the equator. The birds, plants, and insects have a desert character, and are not more brilliantly coloured than those from southern Patagonia; we may, therefore, conclude that the usual gaudy colouring of the intertropical productions, is not related either to the heat or light of those zones, but to some other cause, perhaps to the conditions of existence being generally favourable to life.

We will now turn to the order of reptiles, which gives the most striking character to the zoology of these islands. The species are not numerous, but the numbers of individuals of each species are extraordinarily great. There is one small lizard belonging to a South American genus, and two species (and probably more) of the Amblyrhynchus – a genus confined to the Galapagos islands. There is one snake which is numerous; it is identical, as I am informed by M. Bibron, with the Psammophis Temminckii from Chile. Of seaturtle I believe there is more than one species; and of tortoises there are, as we shall presently show, two or three species or races. Of toads and frogs there are none: I was surprised at this, considering how well suited for them the temperate and damp upper woods appeared to be. It recalled to my mind the remark made by Bory St. Vincent,[1] namely, that none of this family are found on any of the volcanic islands in the great oceans. As far as I can ascertain from various works, this seems to hold good throughout the Pacific, and even in the large

1 Voyage aux Quatre Iles d'Afrique. With respect to the Sandwich Islands, see Tyerman and Bennett's Journal, vol. i. p. 434. For Mauritius, see Voyage par un Officier, &c., Part i. p. 170. There are no frogs in the Canary Islands (Webb et Berthelot, Hist. Nat. des Iles Canaries). I saw none at St. Jago in the Cape de Verds. There are none at St. Helena.

islands of the Sandwich archipelago. Mauritius offers an apparent exception, where I saw the Rana Mascariensis in abundance: this frog is said now to inhabit the Seychelles, Madagascar, and Bourbon; but on the other hand, Du Bois, in his voyage of 1669, states that there were no reptiles in Bourbon except tortoises; and the Officier du Roi asserts that before 1768 it had been attempted, without success, to introduce frogs into Mauritius – I presume, for the purpose of eating: hence it may be well doubted whether this frog is an aboriginal of these islands. The absence of the frog family in the oceanic islands is the more remarkable, when contrasted with the case of lizards, which swarm on most of the smallest islands. May this difference not be caused, by the greater facility with which the eggs of lizards, protected by calcareous shells, might be transported through salt-water, than could the slimy spawn of frogs?

I will first describe the habits of the tortoise (Testudo nigra, formerly called Indica), which has been so frequently alluded to. These animals are found, I believe, on all the islands of the Archipelago; certainly on the greater number. They frequent in preference the high damp parts, but they likewise live in the lower and arid districts. I have already shown, from the numbers which have been caught in a single day, how very numerous they must be. Some grow to an immense size: Mr. Lawson, an Englishman, and vice-governor of the colony, told us that he had seen several so large, that it required six or eight men to lift them from the ground; and that some had afforded as much as two hundred pounds of meat. The old males are the largest, the females rarely growing to so great a size: the male can readily be distinguished from the female by the greater length of its tail. The tortoises which live on those islands where there is no water, or in the lower and arid parts of the others, feed chiefly on the succulent cactus. Those which frequent the higher and damp regions, eat the leaves of various trees, a kind of berry (called guayavita) which is acid and austere, and likewise a pale green filamentous lichen (Usnera plicata), that hangs in tresses from the boughs of the trees.

The tortoise is very fond of water, drinking large quantities, and wallowing in the mud. The larger islands alone possess springs, and these are always situated towards the central parts, and at a considerable height. The tortoises, therefore, which frequent the lower districts, when thirsty, are obliged to travel from a long distance. Hence broad and well-beaten paths branch off in every direction from the wells down to the sea-coast; and the Spaniards by following them up, first discovered the watering-places. When I landed at Chatham Island, I could not imagine what animal travelled so methodically along well-chosen tracks. Near the springs it was a curious spectacle to behold many of these huge creatures, one set eagerly travelling onwards with outstretched necks, and another set returning, after having drunk their fill. When the tortoise arrives at the spring, quite regardless of any spectator, he buries his head in the water above his eyes, and greedily swallows great mouthfuls, at the rate of about ten in a minute. The inhabitants say each animal stays three or four days in the neighbourhood of the water, and then returns to the lower country; but they differed respecting the frequency of these visits. The animal probably regulates them according to the nature of the food on which it has lived. It is, however, certain, that tortoises can subsist even on those islands, where there is no other water than what falls during a few rainy days in the year.

I believe it is well ascertained, that the bladder of the frog acts as a reservoir for the moisture necessary to its existence: such seems to be the case with the tortoise. For some time after a visit to the springs, their urinary bladders are distended with fluid, which is said gradually to decrease in volume, and to become less pure. The inhabitants, when walking in the lower district, and overcome with thirst, often take advantage of this circumstance, and drink the contents of the bladder if full: in one I saw killed, the fluid was quite limpid, and had only a very slightly bitter taste. The inhabitants, however, always first drink the water in the pericardium, which is described as being best.

The tortoises, when purposely moving towards any point, travel by night and day, and arrive at their journey's end much sooner than would be expected. The inhabitants, from observing marked individuals, consider that they travel a distance of about eight miles in two or three days. One large tortoise, which I watched, walked at the rate of sixty yards in ten minutes, that is 360 yards in the hour, or four miles a day, – allowing a little time for it to eat on the road. During the breeding season, when the male and female are together, the male utters a hoarse roar or bellowing, which, it is said, can be heard at the distance of more than a hundred yards. The female never uses her voice, and the male only at these times; so that when the people hear this noise, they know that the two are together. They were at this time (October) laying their eggs. The female, where the soil is sandy, deposits them together, and covers them up with sand; but where the ground is rocky she drops them indiscriminately in any hole: Mr. Bynoe found seven placed in a fissure. The egg is white and spherical; one which I measured was seven inches and three-eighths in circumference, and therefore larger than a hen's egg. The young tortoises, as soon as they are hatched, fall a prey in great numbers to the carrion-feeding buzzard. The old ones seem generally to die from accidents, as from falling down precipices: at least, several of the inhabitants told me, that they had never found one dead without some evident cause.

The inhabitants believe that these animals are absolutely deaf; certainly they do not overhear a person walking close behind them. I was always amused when overtaking one of these great monsters, as it was quietly pacing along, to see how suddenly, the instant I passed, it would draw in its head and legs, and uttering a deep hiss fall to the ground with a heavy sound, as if struck dead. I frequently got on their backs, and then giving a few raps on the hinder part of their shells, they would rise up and walk away; – but I found it very difficult to keep my balance. The flesh of this animal is largely employed, both fresh and salted; and a beautifully clear oil is prepared from the fat. When a tortoise is caught, the man makes a slit in the skin near its tail, so as to see inside its body, whether the fat under the dorsal plate is thick. If it is not, the animal is liberated; and it is said to recover soon from this strange operation. In order to secure the tortoises, it is not sufficient to turn them like turtle, for they are often able to get on their legs again.

There can be little doubt that this tortoise is an aboriginal inhabitant of the Galapagos; for it is found on all, or nearly all, the islands, even on some of the smaller ones where there is no water; had it been an imported species, this would hardly have been the case in a group which has been so little frequented. Moreover, the old Bucaniers found this tortoise in greater numbers even than at present: Wood and Rogers also, in 1708, say that it is the opinion of the

Spaniards, that it is found nowhere else in this quarter of the world. It is now widely distributed; but it may be questioned whether it is in any other place an aboriginal. The bones of a tortoise at Mauritius, associated with those of the extinct Dodo, have generally been considered as belonging to this tortoise: if this had been so, undoubtedly it must have been there indigenous; but M. Bibron informs me that he believes that it was distinct, as the species now living there certainly is.

The Amblyrhynchus, a remarkable genus of lizards, is confined to this archipelago: there are two species, resembling each other in general form, one being terrestrial and the other aquatic. This latter species (A. cristatus) was first characterized by Mr. Bell, who well foresaw, from its short, broad head, and strong claws of equal length, that its habits of life would turn out very peculiar, and different from those of its nearest ally, the Iguana. It is extremely common on all the islands throughout the group, and lives exclusively on the rocky sea-beaches, being never found, at least I never saw one, even ten yards in-shore. It is a hideous-looking creature, of a dirty black colour, stupid, and sluggish in its movements. The usual length of a full-grown one is about a yard, but there are some even four feet long; a large one weighed twenty pounds: on the island of Albemarle they seem to grow to a greater size than elsewhere. Their tails are flattened sideways, and all four feet partially webbed. They are occasionally seen some hundred yards from the shore, swimming about; and Captain Collnett, in his Voyage, says, 'They go to sea in herds

Amblyrhynchus cristatus. *a*, Tooth of natural size, and likewise magnified.

a-fishing, and sun themselves on the rocks; and may be called alligators in miniature.' It must not, however, be supposed that they live on
fish. When in the water this lizard swims with perfect ease and
quickness, by a serpentine movement of its body and flattened tail –
the legs being motionless and closely collapsed on its sides. A seaman
on board sank one, with a heavy weight attached to it, thinking thus
to kill it directly; but when, an hour afterwards, he drew up the line, it
was quite active. Their limbs and strong claws are admirably adapted
for crawling over the rugged and fissured masses of lava, which
everywhere form the coast. In such situations, a group of six or
seven of these hideous reptiles may oftentimes be seen on the black
rocks, a few feet above the surf, basking in the sun with outstretched
legs.

I opened the stomachs of several, and found them largely distended with minced sea-weed (Ulvæ), which grows in thin foli-
aceous expansions of a bright green or a dull red colour. I do not
recollect having observed this sea-weed in any quantity on the tidal
rocks; and I have reason to believe it grows at the bottom of the sea, at
some little distance from the coast. If such be the case, the object of
these animals occasionally going out to sea is explained. The stomach
contained nothing but the sea-weed. Mr. Bynoe, however, found a
piece of a crab in one; but this might have got in accidentally, in the
same manner as I have seen a caterpillar, in the midst of some lichen,
in the paunch of a tortoise. The intestines were large, as in other
herbivorous animals. The nature of this lizard's food, as well as the
structure of its tail and feet, and the fact of its having been seen
voluntarily swimming out at sea, absolutely prove its aquatic habits;
yet there is in this respect one strange anomaly, namely, that when
frightened it will not enter the water. Hence it is easy to drive these
lizards down to any little point overhanging the sea, where they will
sooner allow a person to catch hold of their tails than jump into the
water. They do not seem to have any notion of biting; but when
much frightened they squirt a drop of fluid from each nostril. I threw
one several times as far as I could, into a deep pool left by the retiring
tide; but it invariably returned in a direct line to the spot where
I stood. It swam near the bottom, with a very graceful and rapid
movement, and occasionally aided itself over the uneven ground

with its feet. As soon as it arrived near the edge, but still being under water, it tried to conceal itself in the tufts of sea-weed, or it entered some crevice. As soon as it thought the danger was past, it crawled out on the dry rocks, and shuffled away as quickly as it could. I several times caught this same lizard, by driving it down to a point, and though possessed of such perfect powers of diving and swimming, nothing would induce it to enter the water; and as often as I threw it in, it returned in the manner above described. Perhaps this singular piece of apparent stupidity may be accounted for by the circumstance, that this reptile has no enemy whatever on shore, whereas at sea it must often fall a prey to the numerous sharks. Hence, probably, urged by a fixed and hereditary instinct that the shore is its place of safety, whatever the emergency may be, it there takes refuge.

During our visit (in October), I saw extremely few small individuals of this species, and none I should think under a year old. From this circumstance it seems probable that the breeding season had not then commenced. I asked several of the inhabitants if they knew where it laid its eggs: they said that they knew nothing of its propagation, although well acquainted with the eggs of the land kind – a fact, considering how very common this lizard is, not a little extraordinary.

We will now turn to the terrestrial species (A. Demarlii), with a round tail, and toes without webs. This lizard, instead of being found like the other on all the islands, is confined to the central part of the archipelago, namely to Albemarle, James, Barrington, and Indefatigable islands. To the southward, in Charles, Hood, and Chatham islands, and to the northward, in Towers, Bindloes, and Abingdon, I neither saw nor heard of any. It would appear as if it had been created in the centre of the archipelago, and thence had been dispersed only to a certain distance. Some of these lizards inhabit the high and damp parts of the islands, but they are much more numerous in the lower and sterile districts near the coast. I cannot give a more forcible proof of their numbers, than by stating that when we were left at James Island, we could not for some time find a spot free from their burrows on which to pitch our single tent. Like their brothers the sea-kind, they are ugly animals, of a yellowish orange beneath, and of a brownish red colour above: from their low facial angle they have a singularly stupid appearance. They are, perhaps, of a rather less size than the marine species; but several of them weighed between ten and fifteen pounds. In their movements they are lazy and half torpid. When not frightened, they slowly crawl along with their tails and bellies dragging on the ground. They often stop, and doze for a minute or two, with closed eyes and hind legs spread out on the parched soil.

They inhabit burrows, which they sometimes make between fragments of lava, but more generally on level patches of the soft sandstone-like tuff. The holes do not appear to be very deep, and they enter the ground at a small angle; so that when walking over these lizard-warrens, the soil is constantly giving way, much to the annoyance of the tired walker. This animal, when making its burrow, works alternately the opposite sides of its body. One front leg for a short time scratches up the soil, and throws it towards the hind foot, which is well placed so as to heave it beyond the mouth of the hole. That side of the body being tired, the other takes up the task, and so on alternately. I watched one for a long time, till half its body was buried; I then walked up and pulled it by the tail; at this it was greatly astonished, and soon shuffled up to see what was the matter; and then stared me in the face, as much as to say, 'What made you pull my tail?'

They feed by day, and do not wander far from their burrows; if frightened, they rush to them with a most awkward gait. Except when running down hill, they cannot move very fast, apparently from the lateral position of their legs. They are not at all timorous: when attentively watching any one, they curl their tails, and, raising themselves on their front legs, nod their heads vertically, with a quick movement, and try to look very fierce: but in reality they are not at all so; if one just stamps on the ground, down go their tails, and off they shuffle as quickly as they can. I have frequently observed small fly-eating lizards, when watching anything, nod their heads in precisely the same manner; but I do not at all know for what purpose. If this Amblyrhynchus is held and plagued with a stick, it will bite it very severely; but I caught many by the tail, and they never tried to bite me. If two are placed on the ground and held together, they will fight, and bite each other till blood is drawn.

The individuals, and they are the greater number, which inhabit the lower country, can scarcely taste a drop of water throughout the year; but they consume much of the succulent cactus, the branches of which are occasionally broken off by the wind. I several times threw a piece to two or three of them when together; and it was amusing enough to see them trying to seize and carry it away in their mouths, like so many hungry dogs with a bone. They eat very deliberately, but do not chew their food. The little birds are aware how harmless these creatures are: I have seen one of the thick-billed finches picking at one end of a piece of cactus (which is much relished by all the animals of the lower region), whilst a lizard was eating at the other end; and afterwards the little bird with the utmost indifference hopped on the back of the reptile.

I opened the stomachs of several, and found them full of vegetable fibres and leaves of different trees, especially of an acacia. In the upper region they live chiefly on the acid and astringent berries of the guayavita, under which trees I have seen these lizards and huge tortoises feeding together. To obtain the acacia-leaves they crawl up the low stunted trees; and it is not uncommon to see a pair quietly browsing, whilst seated on a branch several feet above the ground. These lizards, when cooked, yield a white meat, which is liked by those whose stomachs soar above all prejudices. Humboldt has remarked that in intertropical South America, all lizards which inhabit dry regions are esteemed delicacies for the table. The inhabitants state that those which inhabit the upper damp parts drink water, but that the others do not, like the tortoises, travel up for it from the lower sterile country. At the time of our visit, the females had within their bodies numerous large, elongated eggs, which they lay in their burrows: the inhabitants seek them for food.

These two species of Amblyrhynchus agree, as I have already stated, in their general structure, and in many of their habits. Neither have that rapid movement, so characteristic of the genera Lacerta and Iguana. They are both herbivorous, although the kind of vegetation on which they feed is so very different. Mr. Bell has given the name to the genus from the shortness of the snout; indeed, the form of the mouth may almost be compared to that of the tortoise: one is led to suppose that this is an adaptation to their herbivorous appetites. It is very interesting thus to find a well-characterized genus, having its marine and terrestrial species, belonging to so confined a portion of the world. The aquatic species is by far the most remarkable, because it is the only existing lizard which lives on marine vegetable productions. As I at first observed, these islands are not so remarkable for the number of the species of reptiles, as for that of the individuals; when we remember the well-beaten paths made by the thousands of huge tortoises – the many turtles – the great warrens of the terrestrial Amblyrhynchus – and the groups of the marine species basking on the coast-rocks of every island – we must admit that there is no other quarter of the world where this Order replaces the herbivorous mammalia in so extraordinary a manner. The geologist on hearing this will probably refer back in his mind to the Secondary epochs, when lizards, some herbivorous, some carnivorous, and of dimensions comparable only with our existing whales, swarmed on the land and in the sea. It is, therefore, worthy of his observation, that this archipelago, instead of possessing a humid climate and rank vegetation, cannot be considered otherwise than extremely arid, and, for an equatorial region, remarkably temperate.

To finish with the zoology: the fifteen kinds of sea-fish which I procured here are all new species; they belong to twelve genera, all widely distributed, with the exception of Prionotus, of which the four previously known species live on the eastern side of America. Of land-shells I collected sixteen kinds (and two marked varieties), of which, with the exception of one Helix found at Tahiti, all are peculiar to this archipelago: a single fresh-water shell (Paludina) is common to Tahiti and Van Diemen's Land. Mr. Cuming, before our voyage, procured here ninety species of sea-shells, and this does not include several species not yet specifically examined, of Trochus, Turbo, Monodonta, and Nassa. He has been kind enough to give me the following interesting results: of the ninety shells, no less than forty-seven are unknown elsewhere – a wonderful fact, considering how widely distributed sea-shells generally are. Of the forty-three shells found in other parts of the world, twenty-five inhabit the western coast of America, and of these eight are distinguishable as varieties; the remaining eighteen (including one variety) were found by Mr. Cuming in the Low archipelago, and some of them also at the

Philippines. This fact of shells from islands in the central part of the Pacific occurring here, deserves notice, for not one single sea-shell is known to be common to the islands of that ocean and to the west coast of America. The space of open sea running north and south off the west coast, separates two quite distinct conchological provinces; but at the Galapagos Archipelago we have a halting-place, where many new forms have been created, and whither these two great conchological provinces have each sent several colonists. The American province has also sent here representative species; for there is a Galapageian species of Monoceros, a genus only found on the west coast of America; and there are Galapageian species of Fissurella and Cancellaria, genera common on the west coast, but not found (as I am informed by Mr. Cuming) in the central islands of the Pacific. On the other hand, there are Galapageian species of Oniscia and Stylifer, genera common to the West Indies and to the Chinese and Indian seas, but not found either on the west coast of America or in the central Pacific. I may here add, that after the comparison by Messrs. Cuming and Hinds of about 2000 shells from the eastern and western coasts of America, only one single shell was found in common, namely, the Purpura patula, which inhabits the West Indies, the coast of Panama, and the Galapagos. We have, therefore, in this quarter of the world, three great conchological sea-provinces, quite distinct, though surprisingly near each other, being separated by long north and south spaces either of land or of open sea.

I took great pains in collecting the insects, but, excepting Tierra del Fuego, I never saw in this respect so poor a country. Even in the upper and damp region I procured very few, excepting some minute Diptera and Hymenoptera, mostly of common mundane forms. As before remarked, the insects, for a tropical region, are of very small size and dull colours. Of beetles I collected twenty-five species (excluding a Dermestes and Corynetes imported, wherever a ship touches); of these, two belong to the Harpalidæ, two to the Hydrophilidæ, nine to three families of the Heteromera, and the remaining twelve to as many different families. This circumstance of insects (and I may add plants), where few in number, belonging to many different families, is, I believe, very general. Mr. Waterhouse, who has published[1] an account of the insects of this archipelago, and to whom I am indebted for the above details, informs me that there are several new genera; and that of the genera not new, one or two are American, and the rest of mundane distribution. With the exception of a wood-feeding Apate, and of one or probably two water-beetles from the American continent, all the species appear to be new.

The botany of this group is fully as interesting as the zoology. Dr. J. Hooker will soon publish in the 'Linnean Transactions' a full account of the Flora, and I am much indebted to him for the following details. Of flowering plants there are, as far as at present is known, 185 species, and 40 cryptogamic species, making together 225; of this number I was fortunate enough to bring home 193. Of the flowering plants, 100 are new species, and are probably confined to this archipelago. Dr. Hooker conceives that, of the plants not so confined, at least 10 species found near the cultivated ground at Charles Island, have been imported. It is, I think, surprising that more American species have not been introduced naturally, considering that the distance is only between 500 and 600 miles from the continent; and that (according to Collnett, p. 58) drift-wood, bamboos, canes, and the nuts of a palm, are often washed on the south-eastern shores. The proportion of 100 flowering plants out of 185 (or 175 excluding the imported weeds) being new, is sufficient, I conceive, to make the Galapagos Archipelago a distinct botanical province; but this Flora is not nearly so peculiar as that of St. Helena, nor, as I am informed by Dr. Hooker, of Juan Fernandez. The peculiarity of the Galapageian Flora is best shown in certain families; – thus there are 21 species of Compositæ, of which 20 are peculiar to this archipelago; these belong to twelve genera, and of these genera no less than ten are confined to the archipelago! Dr. Hooker informs me that the Flora has an undoubted Western American character; nor can he detect in it any affinity with that of the Pacific. If, therefore, we except the eighteen marine, the one fresh-water, and one land-shell, which have apparently come here as colonists from the central islands of the Pacific, and likewise the one

1 Ann. and Mag. of Nat. Hist., vol. xvi. p. 19.

distinct Pacific species of the Galapageian group of finches, we see that this archipelago, though standing in the Pacific Ocean, is zoologically part of America.

If this character were owing merely to immigrants from America, there would be little remarkable in it; but we see that a vast majority of all the land animals, and that more than half of the flowering plants, are aboriginal productions. It was most striking to be surrounded by new birds, new reptiles, new shells, new insects, new plants, and yet by innumerable trifling details of structure, and even by the tones of voice and plumage of the birds, to have the temperate plains of Patagonia, or the hot dry deserts of Northern Chile, vividly brought before my eyes. Why, on these small points of land, which within a late geological period must have been covered by the ocean, which are formed of basaltic lava, and therefore differ in geological character from the American continent, and which are placed under a peculiar climate, – why were their aboriginal inhabitants, associated, I may add, in different proportions both in kind and number from those on the continent, and therefore acting on each other in a different manner – why were they created on American types of organization? It is probable that the islands of the Cape de Verd group resemble, in all their physical conditions, far more closely the Galapagos Islands than these latter physically resemble the coast of America; yet the aboriginal inhabitants of the two groups are totally unlike; those of the Cape de Verd Islands bearing the impress of Africa, as the inhabitants of the Galapagos Archipelago are stamped with that of America.

I have not as yet noticed by far the most remarkable feature in the natural history of this archipelago; it is, that the different islands to a considerable extent are inhabited by a different set of beings. My attention was first called to this fact by the Vice-Governor, Mr. Lawson, declaring that the tortoises differed from the different islands, and that he could with certainty tell from which island any one was brought. I did not for some time pay sufficient attention to this statement, and I had already partially mingled together the collections from two of the islands. I never dreamed that islands, about fifty or sixty miles apart, and most of them in sight of each other, formed of precisely the same rocks, placed under a quite similar climate, rising to a nearly equal height, would have been differently tenanted; but we shall soon see that this is the case. It is the fate of most voyagers, no sooner to discover what is most interesting in any locality, than they are hurried from it; but I ought, perhaps, to be thankful that I obtained sufficient material to establish this most remarkable fact in the distribution of organic beings.

The inhabitants, as I have said, state that they can distinguish the tortoises from the different islands; and that they differ not only in size, but in other characters. Captain Porter has described[1] those from Charles and from the nearest island to it, namely, Hood Island, as having their shells in front thick and turned up like a Spanish saddle, whilst the tortoises from James Island are rounder, blacker, and have a better taste when cooked. M. Bibron, moreover, informs me that he has seen what he considers two distinct species of tortoise from the Galapagos, but he does not know from which islands. The specimens that I brought from three islands were young ones; and probably owing to this cause, neither Mr. Gray nor myself could find in them any specific differences. I have remarked that the marine Amblyrhynchus was larger at Albemarle Island than elsewhere; and M. Bibron informs me that he has seen two distinct aquatic species of this genus; so that the different islands probably have their representative species or races of the Amblyrhynchus, as well as of the tortoise. My attention was first thoroughly aroused, by comparing together the numerous specimens, shot by myself and several other parties on board, of the mocking-thrushes, when, to my astonishment, I discovered that all those from Charles Island belonged to one species (Mimus trifasciatus); all from Albemarle Island to M. parvulus; and all from James and Chatham Islands (between which two other islands are situated, as connecting links) belonged to M. melanotis. These two latter species are closely allied, and would by some ornithologists be considered as only well-marked races or varieties; but the Mimus trifasciatus is very distinct. Unfortunately most of the specimens of the finch tribe were mingled together; but I have strong reasons to suspect that some of the species of the sub-group Geospiza are confined to separate islands. If the different islands have their

1 Voyage in the U.S. ship *Essex*, vol. i. p. 215.

representatives of Geospiza, it may help to explain the singularly large number of the species of this sub-group in this one small archipelago, and as a probable consequence of their numbers, the perfectly gradu-ated series in the size of their beaks. Two species of the sub-group Cactornis, and two of Camarhynchus, were procured in the archi-pelago; and of the numerous specimens of these two sub-groups shot by four collectors at James Island, all were found to belong to one species of each; whereas the numerous specimens shot either on Chatham or Charles Island (for the two sets were mingled together) all belonged to the two other species: hence we may feel almost sure that these islands possess their representative species of these two sub-groups. In land-shells this law of distribution does not appear to hold good. In my very small collection of insects, Mr. Waterhouse remarks, that of those which were ticketed with their locality, not one was common to any two of the islands.

If we now turn to the Flora, we shall find the aboriginal plants of the different islands wonderfully different. I give all the following results on the high authority of my friend Dr. J. Hooker. I may premise that I indiscriminately collected everything in flower on the different islands, and fortunately kept my collections separate. Too much confidence, however, must not be placed in the propor-tional results, as the small collections brought home by some other naturalists, though in some respects confirming the results, plainly show that much remains to be done in the botany of this group: the Leguminosæ, moreover, have as yet been only approximately worked out (see table overleaf).

Hence we have the truly wonderful fact, that in James Island, of the thirty-eight Galapageian plants, or those found in no other part of the world, thirty are exclusively confined to this one island; and in Albemarle Island, of the twenty-six aboriginal Galapageian plants, twenty-two are confined to this one island, that is, only four are at present known to grow in the other islands of the archipelago; and so on, as shown in the table, with the plants from Chatham and Charles Islands. This fact will, perhaps, be rendered even more striking, by giving a few illustrations: – thus, Scalesia, a remarkable arborescent genus of the Compositæ, is confined to the archipelago: it has six species; one from Chatham, one from Albemarle, one from

Name of Island.	Total No. of Species.	No. of Species found in other parts of the world.	No. of Species confined to the Galapagos Archipelago.	No. confined to the one Island.	No. of Species confined to the Galapagos Archipelago, but found on more than the one Island.
James Island	71	33	38	30	8
Albemarle Island	46	18	26	22	4
Chatham Island	32	16	16	12	4
Charles Island	63	39 (or 29, if the probably imported plants be subtracted)	29	21	8

Charles Island, two from James Island, and the sixth from one of the three latter islands, but it is not known from which: not one of these six species grows on any two islands. Again, Euphorbia, a mundane or widely distributed genus, has here eight species, of which seven are confined to the archipelago, and not one found on any two islands: Acalypha and Borreria, both mundane genera, have respectively six and seven species, none of which have the same species on two islands, with the exception of one Borreria, which does occur on two islands. The species of the Compositæ are particularly local; and Dr. Hooker has furnished me with several other most striking illustrations of the difference of the species on the different islands. He remarks that this law of distribution holds good both with those genera confined to the archipelago, and those distributed in other quarters of the world: in like manner we have seen that the different islands have their proper species of the mundane genus of tortoise, and of the widely distributed American genus of the mocking-thrush, as well as of two of the Galapageian sub-groups of finches, and almost certainly of the Galapageian genus Amblyrhynchus.

The distribution of the tenants of this archipelago would not be nearly so wonderful, if, for instance, one island had a mocking-thrush, and a second island some other quite distinct genus; – if one island had its genus of lizard, and a second island another distinct genus, or none whatever; – or if the different islands were inhabited, not by representative species of the same genera of plants, but by totally different genera, as does to a certain extent hold good; for, to give one instance, a large berry-bearing tree at James Island has no representative species in Charles Island. But it is the circumstance, that several of the islands possess their own species of the tortoise, mocking-thrush, finches, and numerous plants, these species having the same general habits, occupying analogous situations, and obviously filling the same place in the natural economy of this archipelago, that strikes me with wonder. It may be suspected that some of these representative species, at least in the case of the tortoise and of some of the birds, may hereafter prove to be only well-marked races; but this would be of equally great interest to the philosophical naturalist. I have said that most of the islands are in sight of each other: I may specify that Charles Island is fifty miles from the nearest part of Chatham Island, and thirty-three miles from the nearest part of Albemarle Island. Chatham Island is sixty miles from the nearest part of James Island, but there are two intermediate islands between them which were not visited by me. James Island is only ten miles from the nearest part of Albemarle Island, but the two points where the collections were made are thirty-two miles apart. I must repeat, that neither the nature of the soil, nor height of the land, nor the climate, nor the general character of the associated beings, and therefore their action one on another, can differ much in the different islands. If there be any sensible difference in their climates, it must be between the windward group (namely Charles and Chatham Islands), and that to leeward; but there seems to be no corresponding difference in the productions of these two halves of the archipelago.

The only light which I can throw on this remarkable difference in the inhabitants of the different islands is, that very strong currents of the sea running in a westerly and W.N.W. direction must separate, as far as transportal by the sea is concerned, the southern islands from the northern ones; and between these northern islands a strong N.W. current was observed, which must effectually separate James and Albemarle Islands. As the archipelago is free to a most remarkable degree from gales of wind, neither the birds, insects, nor lighter seeds, would be blown from island to island. And lastly, the profound depth of the ocean between the islands, and their apparently recent (in a geological sense) volcanic origin, render it highly unlikely that they were ever united; and this, probably, is a far more important consideration than any other, with respect to the geographical distribution of their inhabitants. Reviewing the facts here given, one is astonished at the amount of creative force, if such an expression may be used, displayed on these small, barren, and rocky islands; and still more so, at its diverse yet analogous action on points so near each other. I have said that the Galapagos Archipelago might be called a satellite attached to America, but it should rather be called a group of satellites, physically similar, organically distinct, yet intimately related to each other, and all related in a marked, though much lesser degree, to the great American continent.

I will conclude my description of the natural history of these islands, by giving an account of the extreme tameness of the birds.

This disposition is common to all the terrestrial species; namely, to the mocking-thrushes, the finches, wrens, tyrant-flycatchers, the dove, and carrion-buzzard. All of them often approached sufficiently near to be killed with a switch, and sometimes, as I myself tried, with a cap or hat. A gun is here almost superfluous; for with the muzzle I pushed a hawk off the branch of a tree. One day, whilst lying down, a mocking-thrush alighted on the edge of a pitcher, made of the shell of a tortoise, which I held in my hand, and began very quietly to sip the water; it allowed me to lift it from the ground whilst seated on the vessel: I often tried, and very nearly succeeded, in catching these birds by their legs. Formerly the birds appear to have been even tamer than at present. Cowley (in the year 1684) says that the 'Turtle-doves were so tame, that they would often alight upon our hats and arms, so as that we could take them alive: they not fearing man, until such time as some of our company did fire at them, whereby they were rendered more shy.' Dampier also, in the same year, says that a man in a morning's walk might kill six or seven dozen of these doves. At present, although certainly very tame, they do not alight on people's arms, nor do they suffer themselves to be killed in such large numbers.

It is surprising that they have not become wilder; for these islands during the last hundred and fifty years have been frequently visited by bucaniers and whalers; and the sailors, wandering through the woods in search of tortoises, always take cruel delight in knocking down the little birds.

These birds, although now still more persecuted, do not readily become wild: in Charles Island, which had then been colonized about six years, I saw a boy sitting by a well with a switch in his hand, with which he killed the doves and finches as they came to drink. He had already procured a little heap of them for his dinner; and he said that he had constantly been in the habit of waiting by this well for the same purpose. It would appear that the birds of this archipelago, not having as yet learnt that man is a more dangerous animal than the tortoise or the Amblyrhynchus, disregard him, in the same manner as in England shy birds, such as magpies, disregard the cows and horses grazing in our fields.

The Falkland Islands offer a second instance of birds with a similar disposition. The extraordinary tameness of the little Opetiorhynchus has been remarked by Pernety, Lesson, and other voyagers. It is not, however, peculiar to that bird: the Polyborus, snipe, upland and lowland goose, thrush, bunting, and even some true hawks, are all more or less tame. As the birds are so tame there, where foxes, hawks, and owls occur, we may infer that the absence of all rapacious animals at the Galapagos, is not the cause of their tameness here. The upland geese at the Falklands show, by the precaution they take in building on the islets, that they are aware of their danger from the foxes; but they are not by this rendered wild towards man. This tameness of the birds, especially of the water-fowl, is strongly contrasted with the habits of the same species in Tierra del Fuego, where for ages past they have been persecuted by the wild inhabitants. In the Falklands, the sportsman may sometimes kill more of the upland geese in one day than he can carry home; whereas in Tierra del Fuego, it is nearly as difficult to kill one, as it is in England to shoot the common wild goose.

In the time of Pernety (1763), all the birds there appear to have been much tamer than at present; he states that the Opetiorhynchus would almost perch on his finger; and that with a wand he killed ten in half an hour. At that period the birds must have been about as tame, as they now are at the Galapagos. They appear to have learnt caution more slowly at these latter islands than at the Falklands, where they have had proportionate means of experience; for besides frequent visits from vessels, those islands have been at intervals colonized during the entire period. Even formerly, when all the birds were so tame, it was impossible by Pernety's account to kill the black-necked swan – a bird of passage, which probably brought with it the wisdom learnt in foreign countries.

I may add that, according to Du Bois, all the birds at Bourbon in 1571–72, with the exception of the flamingoes and geese, were so extremely tame, that they could be caught by the hand, or killed in any number with a stick. Again, at Tristan d'Acunha in the Atlantic, Carmichael[1] states that the only two land-birds, a thrush and a bunting, were 'so tame as to suffer themselves to be caught with a hand-net.' From these several facts we may, I think, conclude, first, that the wildness of birds with regard to man, is a particular instinct directed against *him*, and not dependent on any general degree of caution arising from other sources of danger; secondly, that it is not acquired by individual birds in a short time, even when much persecuted; but that in the course of successive generations it becomes hereditary. With domesticated animals we are accustomed to see new mental habits or instincts acquired and rendered hereditary; but with animals in a state of nature, it must always be most difficult to discover instances of acquired hereditary knowledge. In regard to the wildness of birds towards man, there is no way of accounting for it, except as an inherited habit: comparatively few young birds, in any one year, have been injured by man in England, yet almost all, even nestlings, are afraid of him; many individuals, on

1 Linn. Trans., vol. xii. p. 496. The most anomalous fact on this subject which I have met with, is the wildness of the small birds in the Arctic parts of North America (as described by Richardson, Fauna Bor., vol. ii. p. 332), where they are said never to be persecuted. This case is the more strange, because it is asserted that some of the same species in their winter-quarters in the United States are tame. There is much, as Dr. Richardson well remarks, utterly inexplicable connected with the different degrees of shyness and care with which birds conceal their nests. How strange it is that the English wood-pigeon, generally so wild a bird, should very frequently rear its young in shrubberies close to houses!

the other hand, both at the Galapagos and at the Falklands, have been pursued and injured by man, but yet have not learned a salutary dread of him. We may infer from these facts, what havoc the introduction of any new beast of prey must cause in a country, before the instincts of the indigenous inhabitants have become adapted to the stranger's craft or power.

CHAPTER XVIII

TAHITI AND NEW ZEALAND

Pass through the Low Archipelago – Tahiti – Aspect – Vegetation on the Mountains – View of Eimeo – Excursion into the Interior – Profound Ravines – Succession of Waterfalls – Number of wild useful Plants – Temperance of the Inhabitants – Their moral state – Parliament convened – New Zealand – Bay of Islands – Hippahs – Excursion to Waimate – Missionary Establishment – English Weeds now run wild – Waiomio – Funeral of a New Zealand Woman – Sail for Australia.

OCTOBER 20*th*. – The survey of the Galapagos Archipelago being concluded, we steered towards Tahiti and commenced our long passage of 3200 miles. In the course of a few days we sailed out of the gloomy and clouded ocean-district which extends during the winter far from the coast of South America. We then enjoyed bright and clear weather, while running pleasantly along at the rate of 150 or 160 miles a day before the steady trade-wind. The temperature in this more central part of the Pacific is higher than near the American shore. The thermometer in the poop cabin, by night and day, ranged between 80° and 83°, which feels very pleasant; but with one degree or two higher, the heat becomes oppressive. We passed through the Low or Dangerous Archipelago, and saw several of those most curious rings of coral land, just rising above the water's edge, which have been called Lagoon Islands. A long and brilliantly-white beach is capped by a margin of green vegetation; and the strip, looking either way, rapidly narrows away in the distance, and sinks beneath the horizon. From the mast-head a wide expanse of smooth water can be seen within the ring. These low hollow coral islands bear no proportion to the vast ocean out of which they abruptly rise; and it seems wonderful, that such weak invaders are not overwhelmed, by the all-powerful and never-tiring waves of that great sea, miscalled the Pacific.

November 15th. – At daylight, Tahiti, an island which must for ever remain classical to the voyager in the South Sea, was in view. At a distance the appearance was not attractive. The luxuriant vegetation of the lower part could not yet be seen, and as the clouds rolled past, the wildest and most precipitous peaks showed themselves towards the centre of the island. As soon as we anchored in Matavai Bay, we were surrounded by canoes. This was our Sunday, but the Monday of Tahiti: if the case had been reversed, we should not have received a single visit; for the injunction not to launch a canoe on the sabbath is rigidly obeyed. After dinner we landed to enjoy all the delights produced by the first impressions of a new country, and that country the charming Tahiti. A crowd of men, women, and children, was collected on the memorable Point Venus, ready to receive us with laughing, merry faces. They marshalled us towards the house of Mr. Wilson, the missionary of the district, who met us on the road, and gave us a very friendly reception. After sitting a short time in his house, we separated to walk about, but returned there in the evening.

The land capable of cultivation, is scarcely in any part more than a fringe of low alluvial soil, accumulated round the base of the mountains, and protected from the waves of the sea by a coral reef, which encircles the entire line of coast. Within the reef there is an expanse of smooth water, like that of a lake, where the canoes of the natives can ply with safety and where ships anchor. The low land which comes down to the beach of coral-sand, is covered by the most beautiful productions of the intertropical regions. In the midst of bananas, orange, cocoa-nut, and bread-fruit trees, spots are cleared where yams, sweet potatoes, the sugar-cane, and pine-apples, are cultivated. Even the brushwood is an imported fruit-tree, namely, the guava, which from its abundance has become as noxious as a weed. In Brazil I have often admired the varied beauty of the bananas, palms, and orange-trees contrasted together; and here we also have the bread-fruit, conspicuous from its large, glossy, and deeply digitated leaf. It is admirable to behold groves of a tree, sending forth its branches with the vigour of an English oak, loaded with large and most nutritious fruit. However seldom the usefulness of an object can account for the pleasure of beholding it, in the case of these beautiful woods, the knowledge of their high productiveness no doubt enters largely into the feeling of admiration. The little winding paths, cool from the surrounding shade, led to the scattered houses; the owners of which every where gave us a cheerful and most hospitable reception.

I was pleased with nothing so much as with the inhabitants. There is a mildness in the expression of their countenances which at once banishes the idea of a savage; and an intelligence which shows that they are advancing in civilization. The common people, when working, keep the upper part of their bodies quite naked; and it is then that the Tahitians are seen to advantage. They are very tall, broad-shouldered, athletic, and well-proportioned. It has been remarked, that it requires little habit to make a dark skin more pleasing and natural to the eye of an European than his own colour. A white man bathing by the side of a Tahitian, was like a plant bleached by the gardener's art compared with a fine dark green one growing vigorously in the open fields. Most of the men are tattooed, and the ornaments follow the curvature of the body so gracefully, that they have a very elegant effect. One common pattern, varying in its details, is somewhat like the crown of a palm-tree. It springs from the central line of the back, and gracefully curls round both sides. The simile may be a fanciful one, but I thought the body of a man thus ornamented was like the trunk of a noble tree embraced by a delicate creeper.

Many of the elder people had their feet covered with small figures, so placed as to resemble a sock. This fashion, however, is partly gone by, and has been succeeded by others. Here, although fashion is far from immutable, every one must abide by that prevailing in his youth. An old man has thus his age for ever stamped on his body, and he cannot assume the airs of a young dandy. The women are tattooed in the same manner as the men, and very commonly on their fingers. One unbecoming fashion is now almost universal: namely, shaving the hair from the upper part of the head, in a circular form, so as to leave only an outer ring. The missionaries have tried to persuade the people to change this habit; but it is the fashion, and that is a sufficient answer at Tahiti, as well as at Paris. I was much disappointed in the personal appearance of the women: they are far inferior in every respect to the men. The custom of wearing a white or scarlet flower in the back of the head, or through a small hole in each ear, is pretty. A crown of woven cocoa-nut leaves is also worn as a shade for the eyes. The women appear to be in greater want of some becoming costume even than the men.

Nearly all the natives understand a little English – that is, they know the names of common things; and by the aid of this, together with signs, a lame sort of conversation could be carried on. In returning in the evening to the boat, we stopped to witness a very pretty scene. Numbers of children were playing on the beach, and had lighted bonfires which illumined the placid sea and surrounding trees; others, in circles, were singing Tahitian verses. We seated ourselves on the sand, and joined their party. The songs were im-promptu, and I believe related to our arrival: one little girl sang a line, which the rest took up in parts, forming a very pretty chorus. The whole scene made us unequivocally aware that we were seated on the shores of an island in the far-famed South Sea.

17*th*. – This day is reckoned in the log-book as Tuesday the 17th, instead of Monday the 16th, owing to our, so far, successful chase of the sun. Before breakfast the ship was hemmed in by a flotilla of canoes; and when the natives were allowed to come on board, I suppose there could not have been less than two hundred. It was the opinion of every one that it would have been difficult to have picked out an equal number from any other nation, who would have given so little trouble. Everybody brought something for sale: shells were the main article of trade. The Tahitians now fully understand the value of money, and prefer it to old clothes or other articles. The various coins, however, of English and Spanish denomination puzzle them, and they never seemed to think the small silver quite secure until changed into dollars. Some of the chiefs have accumulated consider-able sums of money. One chief, not long since, offered 800 dollars (about 160*l*. sterling) for a small vessel; and frequently they purchase whale-boats and horses at the rate of from 50 to 100 dollars.

After breakfast I went on shore, and ascended the nearest slope to a height of between two and three thousand feet. The outer mountains are smooth and conical, but steep; and the old volcanic rocks, of which they are formed, have been cut through by many profound ravines, diverging from the central broken parts of the island to the coast. Having crossed the narrow low girt of inhabited and fertile land, I followed a smooth steep ridge between two of the deep ravines. The vegetation was singular, consisting almost exclusively of small dwarf ferns, mingled, higher up, with coarse grass; it was not very dissimilar from that on some of the Welsh hills, and this so close above the orchard of tropical plants on the coast was very surprising. At the highest point, which I reached, trees again appeared. Of the three zones of comparative luxuriance, the lower one owes its moisture, and therefore fertility, to its flatness; for, being scarcely raised above the level of the sea, the water from the higher land drains away slowly. The intermediate zone does not, like the upper one, reach into a damp and cloudy atmosphere, and therefore remains sterile. The woods in the upper zone are very pretty, tree-ferns replacing the cocoa-nuts on the coast. It must not, however, be supposed that these woods at all equal in splendour the forests of Brazil. The vast number of productions, which characterize a continent, cannot be expected to occur in an island.

From the highest point which I attained, there was a good view of the distant island of Eimeo, dependent on the same sovereign with Tahiti. On the lofty and broken pinnacles, white massive clouds were piled up, which formed an island in the blue sky, as Eimeo itself did in the blue ocean. The island, with the exception of one small gateway, is completely encircled by a reef. At this distance, a narrow but well-defined brilliantly white line was alone visible, where the waves first encountered the wall of coral. The mountains rose abruptly out of the glassy expanse of the lagoon, included within this narrow white line, outside which the heaving waters of the ocean were dark-coloured. The view was striking: it may aptly be compared to a framed engraving, where the frame represents the breakers, the marginal paper the smooth lagoon, and the drawing the island itself. When in the evening I descended from the mountain, a man, whom I had pleased with a trifling gift, met me, bringing with him hot roasted bananas, a pine-apple, and cocoa-nuts. After walking under a burning sun, I do not know anything more delicious than the milk of a young cocoa-nut. Pine-apples are here so abundant that the people eat them in the same wasteful manner as we might turnips. They are of an excellent flavour – perhaps even better than those cultivated in England; and this I believe is the highest compliment which can be paid to any fruit. Before going on board, Mr. Wilson interpreted for me to the Tahitian who had paid me so adroit an attention, that I wanted him and another man to accompany me on a short excursion into the mountains.

18th. – In the morning I came on shore early, bringing with me some provisions in a bag, and two blankets for myself and servant. These were lashed to each end of a long pole, which was alternately carried by my Tahitian companions on their shoulders. These men are accustomed thus to carry, for a whole day, as much as fifty pounds at each end of their poles. I told my guides to provide themselves with food and clothing; but they said that there was plenty of food in the mountains, and for clothing, that their skins were sufficient. Our line of march was the valley of Tia-auru, down which a river flows into the sea by Point Venus. This is one of the principal streams in the island, and its source lies at the base of the loftiest central pinnacles, which rise to a height of about 7000 feet. The whole island is so mountainous that the only way to penetrate into the interior is to follow up the valleys. Our road, at first, lay through woods which bordered each side of the river; and the glimpses of the lofty central peaks, seen as through an avenue, with here and there a waving cocoa-nut tree on one side, were extremely picturesque. The valley soon began to narrow, and the sides to grow lofty and more precipitous. After having walked between three and four hours, we found the width of the ravine scarcely exceeded that of the bed of the stream. On each hand the walls were nearly vertical; yet from the soft nature of the volcanic strata, trees and a rank vegetation sprung from every projecting ledge. These precipices must have been some thousand feet high; and the whole formed a mountain gorge far more magnificent than anything which I had ever before beheld. Until the mid-day sun stood vertically over the ravine, the air felt cool and damp, but now it became very sultry. Shaded by a ledge of rock, beneath a façade of columnar lava, we ate our dinner. My guides had already procured a dish of small fish and fresh-water prawns. They carried with them a small net stretched on a hoop; and where the water was deep and in eddies, they dived, and like otters, with their eyes open followed the fish into holes and corners, and thus caught them.

The Tahitians have the dexterity of amphibious animals in the water. An anecdote mentioned by Ellis shows how much they feel at home in this element. When a horse was landing for Pomarre in 1817, the slings broke, and it fell into the water: immediately the natives jumped overboard, and by their cries and vain efforts at assistance almost drowned it. As soon, however, as it reached the shore, the whole population took to flight, and tried to hide themselves from the man-carrying pig, as they christened the horse.

A little higher up, the river divided itself into three little streams. The two northern ones were impracticable, owing to a succession of waterfalls which descended from the jagged summit of the highest mountain; the other to all appearance was equally inaccessible, but we managed to ascend it by a most extraordinary road. The sides of the valley were here nearly precipitous; but, as frequently happens with stratified rocks, small ledges projected, which were thickly covered by wild bananas, liliaceous plants, and other luxuriant productions of the tropics. The Tahitians, by climbing amongst these ledges, search-ing for fruit, had discovered a track by which the whole precipice could be scaled. The first ascent from the valley was very dangerous; for it was necessary to pass a steeply-inclined face of naked rock, by the aid of ropes which we brought with us. How any person dis-covered that this formidable spot was the only point where the side of the mountain was practicable, I cannot imagine. We then cautiously walked along one of the ledges till we came to one of the three streams. This ledge formed a flat spot, above which a beautiful cascade, some hundred feet in height, poured down its waters, and beneath, another high cascade fell into the main stream in the valley below. From this cool and shady recess we made a circuit to avoid the overhanging waterfall. As before, we followed little projecting ledges, the danger being partly concealed by the thickness of the vegetation. In passing from one of the ledges to another, there was a vertical wall of rock. One of the Tahitians, a fine active man, placed the trunk of a tree against this, climbed up it, and then by the aid of crevices reached the summit. He fixed the ropes to a projecting point, and lowered them for our dog and luggage, and then we clambered up ourselves. Beneath the ledge on which the dead tree was placed, the precipice must have been five or six hundred feet deep; and if the abyss had not been partly concealed by the overhanging ferns and lilies, my head would have turned giddy, and nothing should have induced me to have attempted it. We continued to ascend, sometimes along ledges, and sometimes along knife-edged ridges, having on each hand profound ravines. In the Cordillera I have seen mountains on a far grander scale, but for abruptness, nothing at all comparable with this. In the evening we reached a flat little spot on the banks of the same stream, which we had continued to follow, and which descends in a chain of waterfalls: here we bivouacked for the night. On each side of the ravine there were great beds of the mountain-banana, covered with ripe fruit. Many of these plants were from twenty to twenty-five feet high, and from three to four in circumference. By the aid of strips of bark for rope, the stems of bamboos for rafters, and the large leaf of the banana for a thatch, the Tahitians in a few minutes built us an excellent house; and with withered leaves made a soft bed.

They then proceeded to make a fire, and cook our evening meal. A light was procured, by rubbing a blunt-pointed stick in a groove made in another, as if with intention of deepening it, until by the friction the dust became ignited. A peculiarly white and very light wood (the Hibiscus tiliaceus) is alone used for this purpose: it is the same which serves for poles to carry any burden, and for the floating outriggers to their canoes. The fire was produced in a few seconds: but to a person who does not understand the art, it requires, as I found, the greatest exertion; but at last, to my great pride, I succeeded in igniting the dust. The Gaucho in the Pampas uses a different method: taking an elastic stick about eighteen inches long, he presses one end on his breast, and the other pointed end into a hole in a piece of wood, and then rapidly turns the curved part, like a carpenter's centre-bit. The Tahitians having made a small fire of sticks, placed a score of stones, of about the size of cricket-balls, on the burning wood. In about ten minutes the sticks were consumed, and the stones hot. They had previously folded up in small parcels of leaves, pieces of beef, fish, ripe and unripe bananas, and the tops of the wild arum. These green parcels were laid in a layer between two layers of the hot stones, and the whole then covered up with earth, so that no smoke or steam could escape. In about a quarter of an hour, the whole was most deliciously cooked. The choice green parcels were now laid on a cloth of banana leaves, and with a cocoa-nut shell we drank the cool water of the running stream; and thus we enjoyed our rustic meal.

I could not look on the surrounding plants without admiration. On every side were forests of banana; the fruit of which, though serving for food in various ways, lay in heaps decaying on the ground. In front of us there was an extensive brake of wild sugar-cane; and the stream was shaded by the dark green knotted stem of the Ava, – so famous in former days for its powerful intoxicating effects. I chewed a piece, and found that it had an acrid and unpleasant taste, which would have induced any one at once to have pronounced it poisonous. Thanks to the missionaries, this plant now thrives only in these deep ravines, innocuous to every one. Close by I saw the wild arum, the roots of which, when well baked, are good to eat, and the young leaves better than spinach. There was the wild yam, and a liliaceous plant called Ti, which grows in abundance, and has a soft brown root, in shape and size like a huge log of wood: this served us for dessert, for it is as sweet as treacle, and with a pleasant taste. There were, moreover, several other wild fruits, and useful vegetables. The little stream, besides its cool water, produced eels and cray-fish. I did indeed admire this scene, when I compared it with an uncultivated one in the temperate zones. I felt the force of the remark, that man, at least savage man, with his reasoning powers only partly developed, is the child of the tropics.

As the evening drew to a close, I strolled beneath the gloomy shade of the bananas up the course of the stream. My walk was soon brought to a close, by coming to a waterfall between two and three hundred feet high; and again above this there was another. I mention all these waterfalls in this one brook, to give a general idea of the inclination of the land. In the little recess where the water fell, it did not appear that a breath of wind had ever blown. The thin edges of the great leaves of the banana, damp with spray, were unbroken, instead of being, as is so generally the case, split into a thousand shreds. From our position, almost suspended on the mountain-side, there were glimpses into the depths of the neighbouring valleys; and the lofty points of the central mountains, towering up within sixty degrees of the zenith, hid half the evening sky. Thus seated, it was a sublime spectacle to watch the shades of night gradually obscuring the last and highest pinnacles.

Before we laid ourselves down to sleep, the elder Tahitian fell on his knees, and with closed eyes repeated a long prayer in his native tongue. He prayed as a Christian should do, with fitting reverence, and without the fear of ridicule or any ostentation of piety. At our meals neither of the men would taste food, without saying before-hand a short grace. Those travellers who think that a Tahitian prays only when the eyes of the missionary are fixed on him, should have slept with us that night on the mountain-side. Before morning it rained very heavily; but the good thatch of banana-leaves kept us dry.

November 19*th*. – At daylight my friends, after their morning prayer, prepared an excellent breakfast in the same manner as in the evening. They themselves certainly partook of it largely; indeed I never saw any men eat near so much. I suppose such enormously capacious stomachs must be the effect of a large part of their diet consisting of fruit and vegetables, which contain, in a given bulk, a comparatively small portion of nutriment. Unwittingly, I was the means of my companions breaking, as I afterwards learned, one of their own laws and resolutions: I took with me a flask of spirits, which they could not refuse to partake of; but as often as they drank a little, they put their fingers before their mouths, and uttered the word 'Missionary.' About two years ago, although the use of the ava was prevented, drunkenness from the introduction of spirits became very prevalent. The missionaries prevailed on a few good men, who saw that their country was rapidly going to ruin, to join with them in a Temperance Society. From good sense or shame, all the chiefs and the queen were at last persuaded to join. Immediately a law was passed, that no spirits should be allowed to be introduced into the island, and that he who sold and he who bought the forbidden article should be punished by a fine. With remarkable justice, a certain period was allowed for stock in hand to be sold, before the law came into effect. But when it did, a general search was made, in which even the houses of the missionaries were not exempted, and all the ava (as the natives call all ardent spirits) was poured on the ground. When one reflects on the effect of intemperance on the aborigines of the two Americas, I think it will be acknowledged that every well-wisher of Tahiti owes no common debt of gratitude to the missionaries. As long as the little island of St. Helena remained under the government of the East India Company, spirits, owing to the great injury they had produced, were not allowed to be imported; but wine was supplied from the Cape of Good Hope. It is rather a striking, and not very gratifying fact, that in the same year that spirits were allowed to be sold in St. Helena, their use was banished from Tahiti by the free will of the people.

After breakfast we proceeded on our journey. As my object was merely to see a little of the interior scenery, we returned by another track, which descended into the main valley lower down. For some distance we wound, by a most intricate path, along the side of the mountain which formed the valley. In the less precipitous parts we passed through extensive groves of the wild banana. The Tahitians, with their naked, tattooed bodies, their heads ornamented with flowers, and seen in the dark shade of these groves, would have formed a fine picture of man inhabiting some primeval land. In our descent we followed the line of ridges; these were exceedingly narrow, and for considerable lengths steep as a ladder; but all clothed with vegetation. The extreme care necessary in poising each step rendered the walk fatiguing. I did not cease to wonder at these ravines and precipices: when viewing the country from one of the knife-edged ridges, the point of support was so small, that the effect was nearly the same as it must be from a balloon. In this descent we had occasion to use the ropes only once, at the point where we entered the main valley. We slept under the same ledge of rock where we had dined the day before: the night was fine, but from the depth and narrowness of the gorge, profoundly dark.

Before actually seeing this country, I found it difficult to understand two facts mentioned by Ellis; namely, that after the murderous battles of former times, the survivors on the conquered side retired into the mountains, where a handful of men could resist a multitude. Certainly half-a-dozen men, at the spot where the Tahitian reared the old tree, could easily have repulsed thousands. Secondly, that after the introduction of Christianity, there were wild men who lived in the mountains, and whose retreats were unknown to the more civilized inhabitants.

November 20th. – In the morning we started early, and reached Matavai at noon. On the road we met a large party of noble athletic men, going for wild bananas. I found that the ship, on account of the difficulty in watering, had moved to the harbour of Papawa, to which place I immediately walked. This is a very pretty spot. The cove is surrounded by reefs, and the water as smooth as in a lake. The cultivated ground, with its beautiful productions, interspersed with cottages, comes close down to the water's edge.

From the varying accounts which I had read before reaching these islands, I was very anxious to form, from my own observation, a judgment of their moral state, – although such judgment would necessarily be very imperfect. First impressions at all times very much depend on one's previously-acquired ideas. My notions were drawn from Ellis's 'Polynesian Researches' – an admirable and most interesting work, but naturally looking at every thing under a favourable point of view; from Beechey's Voyage; and from that of Kotzebue, which is strongly adverse to the whole missionary system. He who compares these three accounts will, I think, form a tolerably accurate conception of the present state of Tahiti. One of my impressions, which I took from the two last authorities, was decidedly incorrect; viz., that the Tahitians had become a gloomy race, and lived in fear of the missionaries. Of the latter feeling I saw no trace, unless, indeed, fear and respect be confounded under one name. Instead of discontent being a common feeling, it would be difficult in Europe to pick out of a crowd half so many merry and happy faces. The prohibition of the flute and dancing is inveighed against as wrong and foolish; – the more than presbyterian manner of keeping the sabbath is looked at in a similar light. On these points I will not pretend to offer any opinion, in opposition to men who have resided as many years as I was days on the island.

On the whole, it appears to me that the morality and religion of the inhabitants are highly creditable. There are many who attack, even more acrimoniously than Kotzebue, both the missionaries, their system, and the effects produced by it. Such reasoners never compare the present state with that of the island only twenty years ago; nor even with that of Europe at this day; but they compare it with the high standard of Gospel perfection. They expect the missionaries to effect that which the Apostles themselves failed to do. In as much as the condition of the people falls short of this high standard, blame is attached to the missionary, instead of credit for that which he has effected. They forget, or will not remember, that human sacrifices, and the power of an idolatrous priesthood – a system of profligacy unparalleled in any other part of the world – infanticide a consequence of that system – bloody wars, where the conquerors spared neither women nor children – that all these have been abolished; and that dishonety, intemperance, and licentiousness have been greatly reduced by the introduction of Christianity. In a voyager to forget these things is base ingratitude; for should he chance to be at the point of shipwreck on some unknown coast, he will most devoutly pray that the lesson of the missionary may have extended thus far.

In point of morality, the virtue of the women, it has been often said, is most open to exception. But before they are blamed too severely, it will be well distinctly to call to mind the scenes described by Captain Cook and Mr. Banks, in which the grandmothers and mothers of the present race played a part. Those who are most severe, should consider how much of the morality of the women in Europe, is owing to the system early impressed by mothers on their daughters, and how much in each individual case to the precepts of religion. But it is useless to argue against such reasoners; – I believe that, disappointed in not finding the field of licentiousness quite so open as formerly, they will not give credit to a morality which they do not wish to practise, or to a religion which they undervalue, if not despise.

Sunday, 22nd. – The harbour of Papiéte, where the queen resides, may be considered as the capital of the island: it is also the seat of government, and the chief resort of shipping. Captain Fitz Roy took a party there this day to hear divine service, first in the Tahitian language, and afterwards in our own. Mr. Pritchard, the leading missionary in the island, performed the service. The chapel consisted of a large airy framework of wood; and it was filled to excess by tidy, clean people, of all ages and both sexes. I was rather disappointed in the apparent degree of attention; but I believe my expectations were raised too high. At all events the appearance was quite equal to that in a country church in England. The singing of the hymns was decidedly very pleasing; but the language from the pulpit, although fluently delivered, did not sound well: a constant repetition of words, like '*tata ta, mata mai*,' rendered it monotonous. After English service, a party returned on foot to Matavai. It was a pleasant walk, sometimes along the sea-beach and sometimes under the shade of the many beautiful trees.

About two years ago, a small vessel under English colours was plundered by some of the inhabitants of the Low Islands, which were then under the dominion of the Queen of Tahiti. It was believed that the perpetrators were instigated to this act by some indiscreet laws issued by her majesty. The British government demanded compensation; which was acceded to, and a sum of nearly three thousand dollars was agreed to be paid on the first of last September. The Commodore at Lima ordered Captain Fitz Roy to inquire concerning this debt, and to demand satisfaction if it were not paid. Captain Fitz Roy accordingly requested an interview with the Queen Pomarre, since famous from the ill-treatment she has received from the French; and a parliament was held to consider the question, at which all the principal chiefs of the island, and the queen, were assembled. I will not attempt to describe what took place, after the interesting account given by Captain Fitz Roy. The money, it appeared, had not been paid; perhaps the alleged reasons were rather equivocal; but otherwise I cannot sufficiently express our general surprise at the extreme good sense, the reasoning powers, moderation, candour, and prompt resolution, which were displayed on all sides. I believe we all left the meeting with a very different opinion of the Tahitians, from what we entertained when we entered. The chiefs and people resolved to subscribe and complete the sum which was wanting; Captain Fitz Roy urged that it was hard that their private property should be sacrificed for the crimes of distant islanders. They replied, that they were grateful for his consideration, but that Pomarre was their Queen, and that they were determined to help her in this her difficulty. This resolution and its prompt execution, for a book was opened early the next morning, made a perfect conclusion to this very remarkable scene of loyalty and good feeling.

After the main discussion was ended, several of the chiefs took the opportunity of asking Captain Fitz Roy many intelligent questions on international customs and laws, relating to the treatment of ships and foreigners. On some points, as soon as the decision was made, the law was issued verbally on the spot. This Tahitian parliament lasted for several hours; and when it was over Captain Fitz Roy invited Queen Pomarre to pay the *Beagle* a visit.

November 25*th.* – In the evening four boats were sent for her majesty; the ship was dressed with flags, and the yards manned on her coming on board. She was accompanied by most of the chiefs. The behaviour of all was very proper: they begged for nothing, and seemed much pleased with Captain Fitz Roy's presents. The Queen is a large awkward woman, without any beauty, grace, or dignity. She has only one royal attribute: a perfect immoveability of expression under all circumstances, and that rather a sullen one. The rockets were most admired; and a deep 'Oh!' could be heard from the shore, all round the dark bay, after each explosion. The sailors' songs were also much admired; and the queen said she thought that one of the most boisterous ones certainly could not be a hymn! The royal party did not return on shore till past midnight.

26*th.* – In the evening, with a gentle land-breeze, a course was steered for New Zealand; and as the sun set, we had a farewell view of the mountains of Tahiti – the island to which every voyager has offered up his tribute of admiration.

December 19*th.* – In the evening we saw in the distance New Zealand. We may now consider that we have nearly crossed the Pacific. It is necessary to sail over this great ocean to comprehend its immensity. Moving quickly onwards for weeks together, we meet with nothing but the same blue, profoundly deep, ocean. Even within the archipelagoes, the islands are mere specks, and far distant one from the other. Accustomed to look at maps drawn on a small scale, where dots, shading, and names are crowded together, we do not rightly judge how infinitely small the proportion of dry land is to the water of this vast expanse. The meridian of the Antipodes has likewise been passed; and now every league, it made us happy to think, was one league nearer to England. These Antipodes call to one's mind old recollections of childish doubt and wonder. Only the other day I looked forward to this airy barrier as a definite point in our voyage homewards; but now I find it, and all such resting-places for the imagination, are like shadows, which a man moving onwards cannot catch. A gale of wind lasting for some days, has lately given us full leisure to measure the future stages in our long homeward voyage, and to wish most earnestly for its termination.

December 21st. – Early in the morning we entered the Bay of Islands, and being becalmed for some hours near the mouth, we did not reach the anchorage till the middle of the day. The country is hilly, with a smooth outline, and is deeply intersected by numerous arms of the sea extending from the bay. The surface appears from a distance as if clothed with coarse pasture, but this in truth is nothing but fern. On the more distant hills, as well as in parts of the valleys, there is a good deal of woodland. The general tint of the landscape is not a bright green; and it resembles the country a short distance to the south of Concepcion in Chile. In several parts of the bay, little villages of square tidy-looking houses are scattered close down to the water's edge. Three whaling-ships were lying at anchor, and a canoe every now and then crossed from shore to shore; with these exceptions, an air of extreme quietness reigned over the whole district. Only a single canoe came alongside. This, and the aspect of the whole scene, afforded a remarkable, and not very pleasing contrast, with our joyful and boisterous welcome at Tahiti.

In the afternoon we went on shore to one of the larger groups of houses, which yet hardly deserves the title of a village. Its name is Pahia: it is the residence of the missionaries; and there are no native residents except servants and labourers. In the vicinity of the Bay of Islands, the number of Englishmen, including their families, amounts to between two and three hundred. All the cottages, many of which are white-washed and look very neat, are the property of the English. The hovels of the natives are so diminutive and paltry, that they can scarcely be perceived from a distance. At Pahia, it was quite pleasing to behold the English flowers in the gardens before the houses; there were roses of several kinds, honeysuckle, jasmine, stocks, and whole hedges of sweetbriar.

December 22nd. – In the morning I went out walking; but I soon found that the country was very impracticable. All the hills are thickly covered with tall fern, together with a low bush which grows like a cypress; and very little ground has been cleared or cultivated. I then tried the sea-beach; but proceeding towards either hand, my walk was soon stopped by salt-water creeks and deep brooks. The communication between the inhabitants of the different parts of the bay, is (as in Chiloe) almost entirely kept up by boats. I was surprised to find that almost every hill which I ascended, had been at some former time more or less fortified. The summits were cut into steps or successive terraces, and frequently they had been protected by deep trenches. I afterwards observed that the principal hills inland in like manner showed an artificial outline. These are the Pas, so frequently mentioned by Captain Cook under the name of 'hippah;' the difference of sound being owing to the prefixed article.

That the Pas had formerly been much used, was evident from the piles of shells, and the pits in which, as I was informed, sweet potatoes used to be kept as a reserve. As there was no water on these hills, the defenders could never have anticipated a long siege, but only a hurried attack for plunder, against which the successive terraces would have afforded good protection. The general introduction of fire-arms has changed the whole system of warfare; and an exposed situation on the top of a hill is now worse than useless. The Pas in consequence are, at the present day, always built on a level piece of ground. They consist of a double stockade of thick and tall posts, placed in a zigzag line, so that every part can be flanked. Within the stockade a mound of earth is thrown up, behind which the defenders can rest in safety, or use their fire-arms over it. On the level of the ground little archways sometimes pass through this breastwork, by which means the defenders can crawl out to the stockade to reconnoitre their enemies. The Rev. W. Williams, who gave me this account, added, that in one Pas he had noticed spurs or buttresses projecting on the inner and protected side of the mound of earth. On asking the chief the use of them, he replied, that if two or three of his men were shot, their neighbours would not see the bodies, and so be discouraged.

These Pas are considered by the New Zealanders as very perfect means of defence: for the attacking force is never so well disciplined as to rush in a body to the stockade, cut it down, and effect their entry.

When a tribe goes to war, the chief cannot order one party to go here and another there; but every man fights in the manner which best pleases himself; and to each separate individual to approach a stockade defended by fire-arms must appear certain death. I should think a more warlike race of inhabitants could not be found in any part of the world than the New Zealanders. Their conduct on first seeing a ship, as described by Captain Cook, strongly illustrates this: the act of throwing volleys of stones at so great and novel an object, and their defiance of 'Come on shore and we will kill and eat you all,' shows uncommon boldness. This warlike spirit is evident in many of their customs, and even in their smallest actions. If a New Zealander is struck, although but in joke, the blow must be returned; and of this I saw an instance with one of our officers.

At the present day, from the progress of civilization, there is much less warfare, except among some of the southern tribes. I heard a characteristic anecdote of what took place some time ago in the south. A missionary found a chief and his tribe in preparation for war; – their muskets clean and bright, and their ammunition ready. He reasoned long on the inutility of the war, and the little provocation which had been given for it. The chief was much shaken in his resolution, and seemed in doubt: but at length it occurred to him that a barrel of his gunpowder was in a bad state, and that it would not keep much longer. This was brought forward as an unanswerable argument for the necessity of immediately declaring war: the idea of allowing so much good gunpowder to spoil was not to be thought of; and this settled the point. I was told by the missionaries that in the life of Shongi, the chief who visited England, the love of war was the one and lasting spring of every action. The tribe in which he was a principal chief, had at one time been much oppressed by another tribe, from the Thames River. A solemn oath was taken by the men, that when their boys should grow up, and they should be powerful enough, they would never forget or forgive these injuries. To fulfil this oath appears to have been Shongi's chief motive for going to England; and when there it was his sole object. Presents were valued only as they could be converted into arms; of the arts, those alone interested him which were connected with the manufacture of arms. When at Sydney, Shongi, by a strange coincidence, met the hostile chief of the Thames River at the house of Mr. Marsden: their conduct was civil to each other; but Shongi told him that when again in New Zealand he would never cease to carry war into his country. The challenge was accepted; and Shongi on his return fulfilled the threat to the utmost letter. The tribe on the Thames River was utterly overthrown, and the chief to whom the challenge had been given was himself killed. Shongi, although harbouring such deep feelings of hatred and revenge, is described as having been a goodnatured person.

In the evening I went with Captain Fitz Roy and Mr. Baker, one of the missionaries, to pay a visit to Kororadika: we wandered about the village, and saw and conversed with many of the people, both men, women, and children. Looking at the New Zealander, one naturally compares him with the Tahitian; both belonging to the same family of mankind. The comparison, however, tells heavily against the New Zealander. He may, perhaps, be superior in energy, but in every other respect his character is of a much lower order. One glance at their respective expressions, brings conviction to the mind that one is a savage, the other a civilized man. It would be vain to seek in the whole of New Zealand a person with the face and mien of the old Tahitian chief Utamme. No doubt the extraordinary manner in which tattooing is here practised, gives a disagreeable expression to their countenances. The complicated but symmetrical figures covering the whole face, puzzle and mislead an unaccustomed eye: it is moreover probable, that the deep incisions, by destroying the play of the superficial muscles, give an air of rigid inflexibility. But, besides this, there is a twinkling in the eye, which cannot indicate any thing but cunning and ferocity. Their figures are tall and bulky; but not comparable in elegance with those of the working-classes in Tahiti.

Both their persons and houses are filthily dirty and offensive: the idea of washing either their bodies or their clothes never seems to enter their heads. I saw a chief, who was wearing a shirt black and matted with filth, and when asked how it came to be so dirty, he replied, with surprise, 'Do not you see it is an old one?' Some of the men have shirts; but the common dress is one or two large blankets, generally black with dirt, which are thrown over their shoulders in a very inconvenient and awkward fashion. A few of the principal chiefs have decent suits of English clothes; but these are only worn on great occasions.

December 23rd. – At a place called Waimate, about fifteen miles from the Bay of Islands, and midway between the eastern and western coasts, the missionaries have purchased some land for agricultural purposes. I had been introduced to the Rev. W. Williams, who, upon my expressing a wish, invited me to pay him a visit there. Mr. Bushby, the British resident, offered to take me in his boat by a creek, where I should see a pretty waterfall, and by which means my walk would be shortened. He likewise procured for me a guide. Upon asking a neighbouring chief to recommend a man, the chief himself offered to go; but his ignorance of the value of money was so complete, that at first he asked how many pounds I would give him, but afterwards was well contented with two dollars. When I showed the chief a very small bundle, which I wanted carried, it became absolutely necessary for him to take a slave. These feelings of pride are beginning to wear away; but formerly a leading man would sooner have died, than undergone the indignity of carrying the smallest burden. My companion was a light active man, dressed in a dirty blanket, and with his face completely tattooed. He had formerly been a great warrior. He appeared to be on very cordial terms with Mr. Bushby; but at various times they had quarrelled violently. Mr. Bushby remarked that a little quiet irony would frequently silence any one of these natives in their most blustering moments. This chief has come and harangued Mr. Bushby in a hectoring manner, saying, 'A great chief, a great man, a friend of mine, has come to pay me a visit – you must give him something good to eat, some fine presents, &c.' Mr. Bushby has allowed him to finish his discourse, and then has quietly replied by some such answer as, 'What else shall your slave do for you?' The man would then instantly, with a very comical expression, cease his braggadocio.

Some time ago, Mr. Bushby suffered a far more serious attack. A chief and a party of men tried to break into his house in the middle of the night, and not finding this so easy, commenced a brisk firing with their muskets. Mr. Bushby was slightly wounded; but the party was at length driven away. Shortly afterwards it was discovered who was the aggressor; and a general meeting of the chiefs was convened to consider the case. It was considered by the New Zealanders as very atrocious, inasmuch as it was a night attack, and that Mrs. Bushby was lying ill in the house: this latter circumstance, much to their honour, being considered in all cases as a protection. The chiefs agreed to confiscate the land of the aggressor to the King of England. The whole proceeding, however, in thus trying and punishing a chief was entirely without precedent. The aggressor, moreover, lost caste in the estimation of his equals; and this was considered by the British as of more consequence than the confiscation of his land.

As the boat was shoving off, a second chief stepped into her, who only wanted the amusement of the passage up and down the creek. I never saw a more horrid and ferocious expression than this man had. It immediately struck me I had somewhere seen his likeness: it will be found in Retzsch's outlines to Schiller's ballad of Fridolin, where two men are pushing Robert into the burning iron furnace. It is the man who has his arm on Robert's breast. Physiognomy here spoke the truth; this chief had been a notorious murderer, and was an arrant coward to boot. At the point where the boat landed, Mr. Bushby accompanied me a few hundred yards on the road: I could not help admiring the cool impudence of the hoary old villain, whom we left lying in the boat, when he shouted to Mr. Bushby, 'Do not you stay long, I shall be tired of waiting here.'

We now commenced our walk. The road lay along a well-beaten path, bordered on each side by the tall fern, which covers the whole country. After travelling some miles, we came to a little country village, where a few hovels were collected together, and some patches of ground cultivated with potatoes. The introduction of the potato has been the most essential benefit to the island; it is now much more used than any native vegetable. New Zealand is favoured by one great natural advantage; namely, that the inhabitants can never perish from famine. The whole country abounds with fern; and the roots of this plant, if not very palatable, yet contain much nutriment. A native can always subsist on these, and on the shell-fish, which are abundant on all parts of the sea-coast. The villages are chiefly conspicuous by the platforms which are raised on four posts ten or twelve feet above the ground, and on which the produce of the fields is kept secure from all accidents.

On coming near one of the huts I was much amused by seeing in due form the ceremony of rubbing, or, as it ought to be called, pressing noses. The women, on our first approach, began uttering something in a most dolorous voice, they then squatted themselves down and held up their faces; my companion standing over them, one after another, placed the bridge of his nose at right angles to theirs, and commenced pressing. This lasted rather longer than a cordial shake of the hand with us; and as we vary the force of the grasp of the hand in shaking, so do they in pressing. During the process they uttered comfortable little grunts, very much in the same manner as two pigs do, when rubbing against each other. I noticed that the slave would press noses with any one he met, indifferently either before or after his master the chief. Although among these savages, the chief has absolute power of life and death over his slave, yet there is an entire absence of ceremony between them. Mr. Burchell has remarked the same thing in Southern Africa, with the rude Bachapins. Where civilization has arrived at a certain point, complex formalities soon arise between the different grades of society: thus at Tahiti all were formerly obliged to uncover themselves as low as the waist in presence of the king.

The ceremony of pressing noses having been duly completed with all present, we seated ourselves in a circle in the front of one of the hovels, and rested there half-an-hour. All the hovels have nearly the same form and dimensions, and all agree in being filthily dirty. They resemble a cow-shed with one end open, but having a partition a little way within, with a square hole in it, making a small gloomy chamber. In this the inhabitants keep all their property, and when the weather is cold they sleep there. They eat, however, and pass their time in the open part in front. My guides having finished their pipes, we continued our walk. The path led through the same undulating country, the whole uniformly clothed as before with fern. On our right hand we had a serpentine river, the banks of which were fringed with trees, and here and there on the hill-sides there was a clump of wood. The whole scene, in spite of its green colour, had rather a desolate aspect. The sight of so much fern impresses the mind with an idea of sterility: this, however, is not correct; for wherever the fern grows thick and breast-high, the land by tillage becomes productive. Some of the residents think that all this extensive open country originally was covered with forests, and that it has been cleared by fire. It is said, that by digging in the barest spots, lumps of the kind of resin which flows from the kauri pine are frequently found. The natives had an evident motive in clearing the country; for the fern, formerly a staple article of food, flourishes only in the open cleared tracks. The almost entire absence of associated grasses, which forms so remarkable a feature in the vegetation of this island, may perhaps be accounted for by the land having been aboriginally covered with forest-trees.

The soil is volcanic; in several parts we passed over slaggy lavas, and craters could clearly be distinguished on several of the neighbouring hills. Although the scenery is nowhere beautiful, and only occasionally pretty, I enjoyed my walk. I should have enjoyed it more, if my companion, the chief, had not possessed extraordinary conversational powers. I knew only three words; 'good,' 'bad,' and 'yes:' and with these I answered all his remarks, without of course having understood one word he said. This, however, was quite sufficient: I was a good listener, an agreeable person, and he never ceased talking to me.

At length we reached Waimate. After having passed over so many miles of an uninhabited useless country, the sudden appearance of an English farm-house, and its well-dressed fields, placed there as if by an enchanter's wand, was exceedingly pleasant. Mr. Williams not being at home, I received in Mr. Davies's house a cordial welcome. After drinking tea with his family party, we took a stroll about the farm. At Waimate there are three large houses, where the missionary gentlemen, Messrs. Williams, Davies, and Clarke, reside; and near them are the huts of the native labourers. On an adjoining slope, fine crops of barley and wheat were standing in full ear; and in another part, fields of potatoes and clover. But I cannot attempt to describe all I saw; there were large gardens, with every fruit and vegetable which England produces; and many belonging to a warmer clime. I may instance asparagus, kidney beans, cucumbers, rhubarb, apples, pears, figs, peaches, apricots, grapes, olives, gooseberries, currants, hops, gorse for fences, and English oaks; also many kinds of flowers. Around the farm-yard there were stables, a thrashing-barn with its winnowing machine, a blacksmith's forge, and on the ground ploughshares and other tools: in the middle was that happy mixture of pigs and poultry, lying comfortably together, as in every English farm-yard. At the distance of a few hundred yards, where the water of a little rill had been dammed up into a pool, there was a large and substantial water-mill.

All this is very surprising, when it is considered that five years ago nothing but the fern flourished here. Moreover, native workmanship, taught by the missionaries, has effected this change; – the lesson of the missionary is the enchanter's wand. The house had been built, the windows framed, the fields ploughed, and even the trees grafted, by the New Zealander. At the mill, a New Zealander was seen powdered white with flour, like his brother miller in England. When I looked at this whole scene, I thought it admirable. It was not merely that England was brought vividly before my mind; yet, as the evening drew to a close, the domestic sounds, the fields of corn, the distant undulating country with its trees might well have been mistaken for our father-land: nor was it the triumphant feeling at seeing what Englishmen could effect; but rather the high hopes thus inspired for the future progress of this fine island.

Several young men, redeemed by the missionaries from slavery, were employed on the farm. They were dressed in a shirt, jacket, and trousers, and had a respectable appearance. Judging from one trifling anecdote, I should think they must be honest. When walking in the fields, a young labourer came up to Mr. Davies, and gave him a knife and gimlet, saying that he had found them on the road, and did not know to whom they belonged! These young men and boys appeared very merry and good-humoured. In the evening I saw a party of them at cricket: when I thought of the austerity of which the missionaries have been accused, I was amused by observing one of their own sons taking an active part in the game. A more decided and pleasing change was manifested in the young women, who acted as servants within the houses. Their clean, tidy, and healthy appearance, like that of dairy-maids in England, formed a wonderful contrast with the women of the filthy hovels in Kororadika. The wives of the missionaries tried to persuade them not to be tattooed; but a famous operator having arrived from the south, they said, 'We really must just have a few lines on our lips; else when we grow old, our lips will shrivel, and we shall be so very ugly.' There is not nearly so much tattooing as formerly; but as it is a badge of distinction between the chief and the slave, it will probably long be practised. So soon does any train of ideas become habitual, that the missionaries told me that even in their eyes a plain face looked mean, and not like that of a New Zealand gentleman.

Late in the evening I went to Mr. Williams's house, where I passed the night. I found there a large party of children, collected together for Christmas-day, and all sitting round a table at tea. I never saw a nicer or more merry group; and to think that this was in the centre of the land of cannibalism, murder, and all atrocious crimes! The cordiality and happiness so plainly pictured in the faces of the little circle, appeared equally felt by the older persons of the mission.

December 24th. – In the morning, prayers were read in the native tongue to the whole family. After breakfast I rambled about the gardens and farm. This was a market-day, when the natives of the surrounding hamlets bring their potatoes, Indian corn, or pigs, to exchange for blankets, tobacco, and sometimes, through the persuasions of the missionaries, for soap. Mr. Davies's eldest son, who manages a farm of his own, is the man of business in the market. The children of the missionaries, who came while young to the island, understand the language better than their parents, and can get anything more readily done by the natives.

A little before noon Messrs. Williams and Davies walked with me to part of a neighbouring forest, to show me the famous kauri pine. I measured one of these noble trees, and found it thirty-one feet in circumference above the roots. There was another close by, which I did not see, thirty-three feet; and I heard of one no less than forty feet. These trees are remarkable for their smooth cylindrical boles, which run up to a height of sixty, and even ninety feet, with a nearly equal diameter, and without a single branch. The crown of branches at the summit is out of all proportion small to the trunk; and the leaves are likewise small compared with the branches. The forest was here almost composed of the kauri; and the largest trees, from the parallelism of their sides, stood up like gigantic columns of wood. The timber of the kauri is the most valuable production of the island; moreover, a quantity of resin oozes from the bark, which is sold at a penny a pound to the Americans, but its use was then unknown. Some of the New Zealand forests must be impenetrable to an extraordinary degree. Mr. Matthews informed me that one forest only thirty-four miles in width, and separating two inhabited districts, had only lately, for the first time, been crossed. He and another missionary, each with a party of about fifty men, undertook to open a road; but it cost them more than a fortnight's labour! In the woods I saw very few birds. With regard to animals, it is a most remarkable fact, that so large an island, extending over more than 700 miles in latitude, and in many parts ninety broad, with varied stations, a fine climate, and land of all heights, from 14,000 feet downwards, with the exception of a small rat, did not possess one indigenous animal. The several species of that gigantic genus of birds, the Deinornis, seem here to have replaced mammiferous quadrupeds, in the same manner as the reptiles still do at the Galapagos archipelago. It is said that the common Norway rat, in the short space of two years, annihilated in this northern end of the island, the New Zealand species. In many places I noticed several sorts of weeds, which, like the rats, I was forced to own as countrymen. A leek has overrun whole districts, and will prove very troublesome, but it was imported as a favour by a French vessel. The common dock is also widely disseminated, and will, I fear, for ever remain a proof of the rascality of an Englishman, who sold the seeds for those of the tobacco plant.

On returning from our pleasant walk to the house, I dined with Mr. Williams; and then, a horse being lent me, I returned to the Bay of Islands. I took leave of the missionaries with thankfulness for their kind welcome, and with feelings of high respect for their gentleman-like, useful, and upright characters. I think it would be difficult to find a body of men better adapted for the high office which they fulfil.

Christmas-Day. – In a few more days the fourth year of our absence from England will be completed. Our first Christmas-day was spent at Plymouth; the second at St. Martin's Cove, near Cape Horn; the third at Port Desire, in Patagonia; the fourth at anchor in a wild harbour in the peninsula of Tres Montes; this fifth here; and the next,

I trust in Providence, will be in England. We attended divine service in the chapel of Pahia; part of the service being read in English, and part in the native language. Whilst at New Zealand we did not hear of any recent acts of cannibalism; but Mr. Stokes found burnt human bones strewed round a fire-place on a small island near the anchorage; but these remains of a comfortable banquet might have been lying there for several years. It is probable that the moral state of the people will rapidly improve. Mr. Bushby mentioned one pleasing anecdote as a proof of the sincerity of some, at least, of those who profess Christianity. One of his young men left him, who had been accustomed to read prayers to the rest of the servants. Some weeks afterwards, happening to pass late in the evening by an outhouse, he saw and heard one of his men reading the Bible with difficulty by the light of the fire, to the others. After this the party knelt and prayed: in their prayers they mentioned Mr. Bushby and his family, and the missionaries, each separately in his respective district.

December 26th. – Mr. Bushby offered to take Mr. Sulivan and myself in his boat some miles up the river to Cawa-Cawa; and proposed afterwards to walk on to the village of Waiomio, where there are some curious rocks. Following one of the arms of the bay, we enjoyed a pleasant row, and passed through pretty scenery, until we came to a village, beyond which the boat could not pass. From this place a chief and a party of men volunteered to walk with us to Waiomio, a distance of four miles. The chief was at this time rather notorious from having lately hung one of his wives and a slave for adultery. When one of the missionaries remonstrated with him he seemed surprised, and said he thought he was exactly following the English method. Old Shongi, who happened to be in England during the Queen's trial, expressed great disapprobation at the whole proceeding: he said he had five wives, and he would rather cut off all their heads than be so much troubled about one. Leaving this village, we crossed over to another, seated on a hill-side at a little distance. The daughter of a chief, who was still a heathen, had died there five days before. The hovel in which she had expired had been burnt to the ground: her body being enclosed between two small canoes, was placed upright on the ground, and protected by an enclosure bearing wooden images of their gods, and the whole was painted bright red, so as to be conspicuous from afar. Her gown was fastened to the coffin, and her hair being cut off was cast at its foot. The relatives of the family had torn the flesh of their arms, bodies, and faces, so that they were covered with clotted blood; and the old women looked most filthy, disgusting objects. On the following day some of the officers visited this place, and found the women still howling and cutting themselves.

We continued our walk, and soon reached Waiomio. Here there are some singular masses of limestone, resembling ruined castles. These rocks have long served for burial-places, and in consequence are held too sacred to be approached. One of the young men, however, cried out, 'Let us all be brave,' and ran on ahead; but when within a hundred yards, the whole party thought better of it, and stopped short. With perfect indifference, however, they allowed us to examine the whole place. At this village we rested some hours, during which time there was a long discussion with Mr. Bushby, concerning the right of sale of certain lands. One old man, who appeared a perfect genealogist, illustrated the successive possessors by bits of stick driven into the ground. Before leaving the houses a little basketful of roasted sweet potatoes was given to each of our party; and we all, according to the custom, carried them away to eat on the road. I noticed that among the women employed in cooking, there was a man-slave: it must be a humiliating thing for a man in this warlike country to be employed in doing that which is considered as the lowest woman's work. Slaves are not allowed to go to war; but this perhaps can hardly be considered as a hardship. I heard of one poor wretch who, during hostilities, ran away to the opposite party; being met by two men, he was immediately seized; but as they could not agree to whom he should belong, each stood over him with a stone hatchet, and seemed determined that the other at least should not take him away alive. The poor man, almost dead with fright, was only saved by the address of a chief's wife. We afterwards enjoyed a pleasant walk back to the boat, but did not reach the ship till late in the evening.

December 30*th*. – In the afternoon we stood out of the Bay of Islands, on our course to Sydney. I believe we were all glad to leave New Zealand. It is not a pleasant place. Amongst the natives there is absent that charming simplicity which is found at Tahiti; and the greater part of the English are the very refuse of society. Neither is the country itself attractive. I look back but to one bright spot, and that is Waimate, with its Christian inhabitants.

CHAPTER XIX

AUSTRALIA

Sydney – Excursion to Bathurst – Aspect of the Woods –
Party of Natives – Gradual Extinction of the Aborigines –
Infection generated by associated men in health – Blue
Mountains – View of the grand gulf-like Valleys – Their
Origin and Formation – Bathurst, general civility of the
lower orders – State of Society – Van Diemen's Land –
Hobart Town – Aborigines all banished – Mount Welling-
ton – King George's Sound – Cheerless aspect of the
Country – Bald Head, calcareous casts of Branches of
Trees – Party of Natives – Leave Australia.

JANUARY 12th, 1836. – Early in the morning a light air carried us
towards the entrance of Port Jackson. Instead of beholding a
verdant country, interspersed with fine houses, a straight line of
yellowish cliff brought to our minds the coast of Patagonia. A solitary
lighthouse, built of white stone, alone told us that we were near a
great and populous city. Having entered the harbour, it appears fine
and spacious, with cliff-formed shores of horizontally stratified sandstone. The nearly level country is covered with thin scrubby trees,
bespeaking the curse of sterility. Proceeding further inland, the country improves: beautiful villas and nice cottages are here and there
scattered along the beach. In the distance stone houses, two and three
stories high, and windmills standing on the edge of a bank, pointed
out to us the neighbourhood of the capital of Australia.

At last we anchored within Sydney Cove. We found the little
basin occupied by many large ships, and surrounded by warehouses.
In the evening I walked through the town, and returned full of
admiration at the whole scene. It is a most magnificent testimony to
the power of the British nation. Here, in a less promising country,
scores of years have done many times more than an equal number of
centuries have effected in South America. My first feeling was to
congratulate myself that I was born an Englishman. Upon seeing

more of the town afterwards, perhaps my admiration fell a little; but yet it is a fine town. The streets are regular, broad, clean, and kept in excellent order; the houses are of a good size, and the shops well furnished. It may be faithfully compared to the large suburbs which stretch out from London and a few other great towns in England; but not even near London or Birmingham is there an appearance of such rapid growth. The number of large houses and other buildings just finished was truly surprising; nevertheless, every one complained of the high rents and difficulty in procuring a house. Coming from South America, where in the towns every man of property is known, no one thing surprised me more than not being able to ascertain at once to whom this or that carriage belonged.

I hired a man and two horses to take me to Bathurst, a village about one hundred and twenty miles in the interior, and the centre of a great pastoral district. By this means I hoped to gain a general idea of the appearance of the country. On the morning of the 16th (January) I set out on my excursion. The first stage took us to Paramatta, a small country town, next to Sydney in importance. The roads were excellent, and made upon the MacAdam principle, whinstone having been brought for the purpose from the distance of several miles. In all respects there was a close resemblance to England: perhaps the ale-houses here were more numerous. The iron gangs, or parties of convicts who have committed here some offence, appeared the least like England: they were working in chains, under the charge of sentries with loaded arms. The power which the Government possesses, by means of forced labour, of at once opening good roads throughout the country, has been, I believe, one main cause of the early prosperity of this colony. I slept at night at a very comfortable inn at Emu ferry, thirty-five miles from Sydney, and near the ascent of the Blue Mountains. This line of road is the most frequented, and has been the longest inhabited of any in the colony. The whole land is enclosed with high railings, for the farmers have not succeeded in rearing hedges. There are many substantial houses and good cottages scattered about; but although considerable pieces of land are under cultivation, the greater part yet remains as when first dis-covered.

The extreme uniformity of the vegetation is the most remarkable feature in the landscape of the greater part of New South Wales. Everywhere we have an open woodland, the ground being partially covered with a very thin pasture, with little appearance of verdure. The trees nearly all belong to one family, and mostly have their leaves placed in a vertical, instead of, as in Europe, in a nearly horizontal position: the foliage is scanty, and of a peculiar pale green tint, without any gloss. Hence the woods appear light and shadowless: this, although a loss of comfort to the traveller under the scorching rays of summer, is of importance to the farmer, as it allows grass to grow where it otherwise would not. The leaves are not shed periodically: this character appears common to the entire southern hemisphere, namely, South America, Australia, and the Cape of Good Hope. The inhabitants of this hemisphere, and of the intertropical regions, thus lose perhaps one of the most glorious, though to our eyes common, spectacles in the world – the first bursting into full foliage of the leafless tree. They may, however, say that we pay dearly for this by having the land covered with mere naked skeletons for so many months. This is too true; but our senses thus acquire a keen relish for the exquisite green of the spring, which the eyes of those living within the tropics, sated during the long year with the gorgeous productions of those glowing climates, can never experience. The greater number of the trees, with the exception of some of the Blue-gums, do not attain a large size; but they grow tall and tolerably straight, and stand well apart. The bark of some of the Eucalypti falls annually, or hangs dead in long shreds which swing about with the wind, and give to the woods a desolate and untidy appearance. I cannot imagine a more complete contrast, in every respect, than between the forests of Valdivia or Chiloe, and the woods of Australia.

At sunset, a party of a score of the black aborigines passed by, each carrying, in their accustomed manner, a bundle of spears and other weapons. By giving a leading young man a shilling, they were easily detained, and threw their spears for my amusement. They were all partly clothed, and several could speak a little English: their countenances were good-humoured and pleasant, and they appeared far from being such utterly degraded beings as they have usually been represented. In their own arts they are admirable. A cap being fixed at thirty yards distance, they transfixed it with a spear, delivered by the throwing-stick with the rapidity of an arrow from the bow of a practised archer. In tracking animals or men they show most wonderful sagacity; and I heard of several of their remarks which manifested considerable acuteness. They will not, however, cultivate the ground, or build houses and remain stationary, or even take the trouble of tending a flock of sheep when given to them. On the whole they appear to me to stand some few degrees higher in the scale of civilization than the Fuegians.

It is very curious thus to see in the midst of a civilized people, a set of harmless savages wandering about without knowing where they shall sleep at night, and gaining their livelihood by hunting in the woods. As the white man has travelled onwards, he has spread over the country belonging to several tribes. These, although thus enclosed by one common people, keep up their ancient distinctions, and sometimes go to war with each other. In an engagement which took place lately, the two parties most singularly chose the centre of the village of Bathurst for the field of battle. This was of service to the defeated side, for the runaway warriors took refuge in the barracks.

The number of aborigines is rapidly decreasing. In my whole ride, with the exception of some boys brought up by Englishmen, I saw only one other party. This decrease, no doubt, must be partly owing to the introduction of spirits, to European diseases (even the milder ones of which, such as the measles,[1] prove very destructive), and to the gradual extinction of the wild animals. It is said that numbers of their children invariably perish in very early infancy from the effects of their wandering life; and as the difficulty of procuring food increases, so must their wandering habits increase; and hence the population, without any apparent deaths from famine, is repressed in a manner extremely sudden compared to what happens in civilized

1 It is remarkable how the same disease is modified in different climates. At the little island of St. Helena, the introduction of scarlet-fever is dreaded as a plague. In some countries, foreigners and natives are as differently affected by certain contagious disorders, as if they had been different animals; of which fact some instances have occurred in Chile; and, according to Humboldt, in Mexico. (Polit. Essay, New Spain, vol. iv.)

countries, where the father, though in adding to his labour he may injure himself, does not destroy his offspring.

Besides these several evident causes of destruction, there appears to be some more mysterious agency generally at work. Wherever the European has trod, death seems to pursue the aboriginal. We may look to the wide extent of the Americas, Polynesia, the Cape of Good Hope, and Australia, and we find the same result. Nor is it the white man alone that thus acts the destroyer; the Polynesian of Malay extraction has in parts of the East Indian archipelago, thus driven before him the dark-coloured native. The varieties of man seem to act on each other in the same way as different species of animals – the stronger always extirpating the weaker. It was melancholy at New Zealand to hear the fine energetic natives saying, that they knew the land was doomed to pass from their children. Every one has heard of the inexplicable reduction of the population in the beautiful and healthy island of Tahiti since the date of Captain Cook's voyages: although in that case we might have expected that it would have been increased; for infanticide, which formerly prevailed to so extraordinary a degree, has ceased, profligacy has greatly diminished, and the murderous wars become less frequent.

The Rev. J. Williams, in his interesting work,[1] says, that the first intercourse between natives and Europeans, 'is invariably attended with the introduction of fever, dysentery, or some other disease, which carries off numbers of the people.' Again he affirms, 'It is certainly a fact, which cannot be controverted, that most of the diseases which have raged in the islands during my residence there, have been introduced by ships;[2] and what renders this fact remarkable is, that there might be no appearance of disease among the crew of the

1 Narrative of Missionary Enterprise, p. 282.

2 Captain Beechey (chap. iv. vol. i.) states that the inhabitants of Pitcairn Island are firmly convinced that after the arrival of every ship they suffer cutaneous and other disorders. Captain Beechey attributes this to the change of diet during the time of the visit. Dr. Macculloch (Western Isles, vol. ii. p. 32) says, 'It is asserted, that on the arrival of a stranger (at St. Kilda) all the inhabitants, in the common phraseology, catch a cold.' Dr. Macculloch considers the whole case, although often previously affirmed, as ludicrous. He adds, however, that 'the question was put by us to the inhabitants who unanimously agreed in the story.' In Vancouver's Voyage, there is a somewhat similar statement with respect to Otaheite. Dr. Dieffenbach, in a note to his translation of this

ship which conveyed this destructive importation.' This statement is not quite so extraordinary as it at first appears; for several cases are on record of the most malignant fevers having broken out, although the parties themselves, who were the cause, were not affected. In the early part of the reign of George III., a prisoner who had been confined in a dungeon, was taken in a coach with four constables before a magistrate; and, although the man himself was not ill, the four constables died from a short putrid fever; but the contagion extended to no others. From these facts it would almost appear as if the effluvium of one set of men shut up for some time together was poisonous when inhaled by others; and possibly more so, if the men be of different races. Mysterious as this circumstance appears to be, it is not more surprising than that the body of one's fellow-creature, directly after death, and before putrefaction has commenced, should often be of so deleterious a quality, that the mere puncture from an instrument used in its dissection, should prove fatal.

17th. – Early in the morning we passed the Nepean in a ferry-boat. The river, although at this spot both broad and deep, had a very small body of running water. Having crossed a low piece of land on the opposite side, we reached the slope of the Blue Mountains. The ascent is not steep, the road having been cut with much care on the side of a sandstone cliff. On the summit an almost level plain extends, which, rising imperceptibly to the westward, at last attains a height of more than 3000 feet. From so grand a title as Blue Mountains, and from their absolute altitude, I expected to have seen a bold chain of mountains crossing the country; but instead of this, a sloping plain presents merely an inconsiderable front to the low land near the coast. From this first slope, the view of the extensive woodland to the east

Journal, states that the same fact is universally believed by the inhabitants of the Chatham Islands, and in parts of New Zealand. It is impossible that such a belief should have become universal in the northern hemisphere, at the Antipodes, and in the Pacific, without some good foundation. Humboldt (Polit. Essay on King of New Spain, vol. iv.) says, that the great epidemics at Panama and Callao are 'marked' by the arrival of ships from Chile, because the people from that temperate region, first experience the fatal effects of the torrid zones. I may add, that I have heard it stated in Shropshire, that sheep, which have been imported from vessels, although themselves in a healthy condition, if placed in the same fold with others, frequently produce sickness in the flock.

was striking, and the surrounding trees grew bold and lofty. But when once on the sandstone platform, the scenery becomes exceedingly monotonous; each side of the road is bordered by scrubby trees of the never-failing Eucalyptus family; and with the exception of two or three small inns, there are no houses or cultivated land: the road, moreover, is solitary; the most frequent object being a bullock-waggon, piled up with bales of wool.

In the middle of the day we baited our horses at a little inn, called the Weatherboard. The country here is elevated 2800 feet above the sea. About a mile and a half from this place there is a view exceedingly well worth visiting. Following down a little valley and its tiny rill of water, an immense gulf unexpectedly opens through the trees which border the pathway, at the depth of perhaps 1500 feet. Walking on a few yards, one stands on the brink of a vast precipice, and below one sees a grand bay or gulf, for I know not what other name to give it, thickly covered with forest. The point of view is situated as if at the head of a bay, the line of cliff diverging on each side, and showing headland behind headland, as on a bold sea-coast. These cliffs are composed of horizontal strata of whitish sandstone; and are so absolutely vertical, that in many places a person standing on the edge and throwing down a stone, can see it strike the trees in the abyss below. So unbroken is the line of cliff, that in order to reach the foot of the waterfall, formed by this little stream, it is said to be necessary to go sixteen miles round. About five miles distant in front, another line of cliff extends, which thus appears completely to encircle the valley; and hence the name of bay is justified, as applied to this grand amphitheatrical depression. If we imagine a winding harbour, with its deep water surrounded by bold cliff-like shores, to be laid dry, and a forest to spring up on its sandy bottom, we should then have the appearance and structure here exhibited. This kind of view was to me quite novel, and extremely magnificent.

In the evening we reached the Blackheath. The sandstone plateau has here attained the height of 3400 feet; and is covered, as before, with the same scrubby woods. From the road, there were occasional glimpses into a profound valley, of the same character as the one described; but from the steepness and depth of its sides, the bottom was scarcely ever to be seen. The Blackheath is a very comfortable inn, kept by an old soldier; and it reminded me of the small inns in North Wales.

18*th*. – Very early in the morning, I walked about three miles to see Govett's Leap: a view of a similar character with that near the Weatherboard, but perhaps even more stupendous. So early in the day the gulf was filled with a thin blue haze, which, although destroying the general effect of the view, added to the apparent depth at which the forest was stretched out beneath our feet. These valleys, which so long presented an insuperable barrier to the attempts of the most enterprising of the colonists to reach the interior, are most remarkable. Great arm-like bays, expanding at their upper ends, often branch from the main valleys and penetrate the sandstone platform; on the other hand, the platform often sends promontories into the valleys, and even leaves in them great, almost insulated, masses. To descend into some of these valleys, it is necessary to go round twenty miles; and into others, the surveyors have only lately penetrated, and the colonists have not yet been able to drive in their cattle. But the most remarkable feature in their structure is, that although several miles wide at their heads, they generally contract towards their mouths to such a degree as to become impassable. The Surveyor-General, Sir T. Mitchell,[1] endeavoured in vain, first walking and then by crawling between the great fallen fragments of sandstone, to ascend through the gorge by which the river Grose joins the Nepean; yet the valley of the Grose in its upper part, as I saw, forms a magnificent level basin some miles in width, and is on all sides surrounded by cliffs, the summits of which are believed to be no-where less than 3000 feet above the level of the sea. When cattle are driven into the valley of the Wolgan by a path (which I descended), partly natural and partly made by the owner of the land, they cannot escape; for this valley is in every other part surrounded by perpendicular cliffs, and eight miles lower down, it contracts from an average width of half a mile, to a mere chasm, impassable to man or beast. Sir T. Mitchell states that the great valley of the Cox river with all its branches, contracts, where it unites with the Nepean, into a gorge

1 Travels in Australia, vol. i. p. 154. I must express my obligation to Sir T. Mitchell, for several interesting personal communications, on the subject of these great valleys of New South Wales.

2200 yards in width, and about 1000 feet in depth. Other similar cases might have been added.

The first impression, on seeing the correspondence of the horizontal strata on each side of these valleys and great amphitheatrical depressions, is that they have been hollowed out, like other valleys, by the action of water; but when one reflects on the enormous amount of stone, which on this view must have been removed through mere gorges or chasms, one is led to ask whether these spaces may not have subsided. But considering the form of the irregularly branching valleys, and of the narrow promontories projecting into them from the platforms, we are compelled to abandon this notion. To attribute these hollows to the present alluvial action would be preposterous; nor does the drainage from the summit-level always fall, as I remarked near the Weatherboard, into the head of these valleys, but into one side of their bay-like recesses. Some of the inhabitants remarked to me that they never viewed one of those bay-like recesses, with the headlands receding on both hands, without being struck with their resemblance to a bold sea-coast. This is certainly the case; moreover, on the present coast of New South Wales, the numerous fine, widely-branching harbours, which are generally connected with the sea by a narrow mouth worn through the sandstone coast-cliffs, varying from one mile in width to a quarter of a mile, present a likeness, though on a miniature scale, to the great valleys of the interior. But then immediately occurs the startling difficulty, why has the sea worn out these great, though circumscribed depressions on a wide platform, and left mere gorges at the openings, through which the whole vast amount of triturated matter must have been carried away? The only light I can throw upon this enigma, is by remarking that banks of the most irregular forms appear to be now forming in some seas, as in parts of the West Indies and in the Red Sea, and that their sides are exceedingly steep. Such banks, I have been led to suppose, have been formed by sediment heaped by strong currents on an irregular bottom. That in some cases the sea, instead of spreading out sediment in a uniform sheet, heaps it round submarine rocks and islands, it is hardly possible to doubt, after examining the charts of the West Indies; and that the waves have power to form high and precipitous cliffs, even in land-locked harbours, I have noticed in many parts of South America. To apply these ideas to the sandstone platforms of New South Wales, I imagine that the strata were heaped by the action of strong currents, and of the undulations of an open sea, on an irregular bottom; and that the valley-like spaces thus left unfilled had their steeply sloping flanks worn into cliffs, during a slow elevation of the land; the worn-down sandstone being removed, either at the time when the narrow gorges were cut by the retreating sea, or subsequently by alluvial action.

Soon after leaving the Blackheath, we descended from the sandstone platform by the pass of Mount Victoria. To effect this pass, an enormous quantity of stone has been cut through; the design, and its manner of execution, being worthy of any line of road in England. We now entered upon a country less elevated by nearly a thousand feet, and consisting of granite. With the change of rock, the vegetation improved; the trees were both finer and stood farther apart; and the pasture between them was a little greener and more plentiful. At Hassan's Walls, I left the high road, and made a short détour to a farm called Walerawang; to the superintendent of which I had a letter of introduction from the owner in Sydney. Mr. Browne had the kindness to ask me to stay the ensuing day, which I had much pleasure in doing. This place offers an example of one of the large farming, or rather sheep-grazing, establishments of the colony. Cattle and horses are, however, in this case rather more numerous than usual, owing to some of the valleys being swampy and producing a coarser pasture. Two or three flat pieces of ground near the house were cleared and cultivated with corn, which the harvest-men were now reaping: but no more wheat is sown than sufficient for the annual support of the labourers employed on the establishment. The usual number of assigned convict-servants here is about forty, but at the present time there were rather more. Although the farm was well stocked with every necessary, there was an apparent absence of comfort; and not one single woman resided here. The sunset of a fine day will generally cast an air of happy contentment on any scene; but here, at this retired farm-house, the brightest tints on the surrounding woods could not make me forget that forty hardened, profligate men were ceasing from their daily labours, like the slaves from Africa, yet without their holy claim for compassion.

Early on the next morning, Mr. Archer, the joint superintendent, had the kindness to take me out kangaroo-hunting. We continued riding the greater part of the day, but had very bad sport, not seeing a kangaroo, or even a wild dog. The greyhounds pursued a kangaroo rat into a hollow tree, out of which we dragged it: it is an animal as large as a rabbit, but with the figure of a kangaroo. A few years since this country abounded with wild animals; but now the emu is banished to a long distance, and the kangaroo is become scarce; to both the English greyhound has been highly destructive. It may be long before these animals are altogether exterminated, but their doom is fixed. The aborigines are always anxious to borrow the dogs from the farm-houses: the use of them, the offal when an animal is killed, and some milk from the cows, are the peace-offerings of the settlers, who push farther and farther towards the interior. The thoughtless aboriginal, blinded by these trifling advantages, is delighted at the approach of the white man, who seems predestined to inherit the country of his children.

Although having poor sport, we enjoyed a pleasant ride. The woodland is generally so open that a person on horseback can gallop through it. It is traversed by a few flat-bottomed valleys, which are green and free from trees: in such spots the scenery was pretty like that of a park. In the whole country I scarcely saw a place without the marks of a fire; whether these had been more or less recent – whether the stumps were more or less black, was the greatest change which varied the uniformity, so wearisome to the traveller's eye. In these woods there are not many birds; I saw, however, some large flocks of the white cockatoo feeding in a corn-field, and a few most beautiful parrots; crows like our jackdaws were not uncommon, and another bird something like the magpie. In the dusk of the evening I took a stroll along a chain of ponds, which in this dry country represented the course of a river, and had the good fortune to see several of the famous Ornithorhynchus paradoxus. They were diving and playing about the surface of the water, but showed so little of their bodies, that they might easily have been mistaken for water-rats. Mr. Browne shot one: certainly it is a most extraordinary animal; a stuffed specimen does not at all give a good idea of the appearance of the head and beak when fresh; the latter becoming hard and con-tracted.[1]

20th. – A long day's ride to Bathurst. Before joining the high road we followed a mere path through the forest; and the country, with the exception of a few squatters' huts, was very solitary. We experienced this day the sirocco-like wind of Australia, which comes from the parched deserts of the interior. Clouds of dust were travelling in every direction; and the wind felt as if it had passed over a fire. I afterwards heard that the thermometer out of doors had stood at 119°, and in a closed room at 96°. In the afternoon we came in view of the downs of Bathurst. These undulating but nearly smooth plains are very remark-able in this country, from being absolutely destitute of trees. They support only a thin brown pasture. We rode some miles over this country, and then reached the township of Bathurst, seated in the middle of what may be called either a very broad valley, or narrow plain. I was told at Sydney not to form too bad an opinion of Australia by judging of the country from the road-side, nor too good a one from Bathurst; in this latter respect, I did not feel myself in the least danger of being prejudiced. The season, it must be owned, had been one of great drought, and the country did not wear a favourable aspect; although I understand it was incomparably worse two or three months before. The secret of the rapidly growing prosperity of Bathurst is, that the brown pasture which appears to the stranger's eye so wretched, is excellent for sheep-grazing. The town stands, at the height of 2200 feet above the sea, on the banks of the Macquarie: this is one of the rivers flowing into the vast and scarcely known interior. The line of water-shed, which divides the inland streams from those on the coast, has a height of about 3000 feet, and runs in a

---

1 I was interested by finding here the hollow conical pitfall of the lion-ant, or some other insect; first a fly fell down the treacherous slope and immediately disappeared; then came a large but unwary ant; its struggles to escape being very violent, those curious little jets of sand, described by Kirby and Spence (Entomol., vol. i. p. 425) as being flirted by the insect's tail, were promptly directed against the expected victim. But the ant enjoyed a better fate than the fly, and escaped the fatal jaws which lay concealed at the base of the conical hollow. This Australian pitfall was only about half the size of that made by the European lion-ant.

north and south direction at the distance of from eighty to a hundred miles from the sea-side. The Macquarie figures in the map as a respectable river, and it is the largest of those draining this part of the water-shed; yet to my surprise I found it a mere chain of ponds, separated from each other by spaces almost dry. Generally a small stream is running; and sometimes there are high and impetuous floods. Scanty as the supply of the water is throughout this district, it becomes still scantier further inland.

22nd. – I commenced my return, and followed a new road called Lockyer's Line, along which the country is rather more hilly and picturesque. This was a long day's ride; and the house where I wished to sleep was some way off the road, and not easily found. I met on this occasion, and indeed on all others, a very general and ready civility among the lower orders, which, when one considers what they are, and what they have been, would scarcely have been expected. The farm where I passed the night, was owned by two young men who had only lately come out, and were beginning a settler's life. The total want of almost every comfort was not very attractive; but future and certain prosperity was before their eyes, and that not far distant.

The next day we passed through large tracts of country in flames, volumes of smoke sweeping across the road. Before noon we joined our former road, and ascended Mount Victoria. I slept at the Weatherboard, and before dark took another walk to the amphi-theatre. On the road to Sydney I spent a very pleasant evening with Captain King at Dunheved; and thus ended my little excursion in the colony of New South Wales.

Before arriving here the three things which interested me most were – the state of society amongst the higher classes, the condition of the convicts, and the degree of attraction sufficient to induce persons to emigrate. Of course, after so very short a visit, one's opinion is worth scarcely anything; but it is as difficult not to form some opinion, as it is to form a correct judgment. On the whole, from what I heard, more than from what I saw, I was disappointed in the state of society. The whole community is rancorously divided into parties on almost every subject. Among those who, from their station in life, ought to be the best, many live in such open profligacy that respectable people cannot associate with them. There is much jealousy between the children of the rich emancipist and the free settlers, the former being pleased to consider honest men as interlopers. The whole population, poor and rich, are bent on acquiring wealth: amongst the higher orders, wool and sheep-grazing form the constant subject of conversation. There are many serious drawbacks to the comforts of a family, the chief of which, perhaps, is being surrounded by convict servants. How thoroughly odious to every feeling, to be waited on by a man who the day before, perhaps, was flogged, from your representation, for some trifling misdemeanour. The female servants are of course much worse: hence children learn the vilest expressions, and it is fortunate, if not equally vile ideas.

On the other hand, the capital of a person, without any trouble on his part, produces him treble interest to what it will in England; and with care he is sure to grow rich. The luxuries of life are in abundance, and very little dearer than in England, and most articles of food are cheaper. The climate is splendid, and perfectly healthy; but to my mind its charms are lost by the uninviting aspect of the country. Settlers possess a great advantage in finding their sons of service when very young. At the age of from sixteen to twenty, they frequently take charge of distant farming stations. This, however, must happen at the expense of their boys associating entirely with convict servants. I am not aware that the tone of society has assumed any peculiar character; but with such habits, and without intellectual pursuits, it can hardly fail to deteriorate. My opinion is such, that nothing but rather sharp necessity should compel me to emigrate.

The rapid prosperity and future prospects of this colony are to me, not understanding these subjects, very puzzling. The two main exports are wool and whale-oil, and to both of these productions there is a limit. The country is totally unfit for canals, therefore there is a not very distant point, beyond which the land-carriage of wool will not repay the expense of shearing and tending sheep. Pasture everywhere is so thin that settlers have already pushed far into the interior: moreover, the country further inland becomes extremely poor. Agriculture, on account of the droughts, can never succeed on an extended scale: therefore, so far as I can see, Australia must ultimately depend upon being the centre of commerce for the southern hemisphere, and perhaps on her future manufactories. Possessing coal, she always has the moving power at hand. From the habitable country extending along the coast, and from her English extraction, she is sure to be a maritime nation. I formerly imagined that Australia would rise to be as grand and powerful a country as North America, but now it appears to me that such future grandeur is rather problematical.

With respect to the state of the convicts, I had still fewer opportunities of judging than on the other points. The first question is, whether their condition is at all one of punishment: no one will maintain that it is a very severe one. This, however, I suppose, is of little consequence as long as it continues to be an object of dread to criminals at home. The corporeal wants of the convicts are tolerably well supplied: their prospect of future liberty and comfort is not distant, and after good conduct certain. A 'ticket of leave,' which, as long as a man keeps clear of suspicion as well as of crime, makes him free within a certain district, is given upon good conduct, after years proportional to the length of the sentence; yet with all this, and overlooking the previous imprisonment and wretched passage out, I believe the years of assignment are passed away with discontent and unhappiness. As an intelligent man remarked to me, the convicts know no pleasure beyond sensuality, and in this they are not gratified. The enormous bribe which Government possesses in offering free pardons, together with the deep horror of the secluded penal settlements, destroys confidence between the convicts, and so prevents crime. As to a sense of shame, such a feeling does not appear to be known, and of this I witnessed some very singular proofs. Though it is a curious fact, I was universally told that the character of the convict population is one of arrant cowardice: not unfrequently some become desperate, and quite indifferent as to life, yet a plan requiring cool or continued courage is seldom put into execution. The worst feature in the whole case is, that although there exists what may be called a legal reform, and comparatively little is committed which the law can touch, yet that any moral reform should take place appears to be quite out of the question. I was assured by well-informed people, that a man who should try to improve, could not while living with other assigned servants; – his life would be one of intolerable misery and persecution. Nor must the contamination of the convict-ships and prisons, both here and in England, be forgotten. On the whole, as a place of punishment, the object is scarcely gained; as a real system of reform it has failed, as perhaps would every other plan; but as a means of making men outwardly honest, – of converting vagabonds, most useless in one hemisphere, into active citizens of another, and thus giving birth to a new and splendid country – a grand centre of civilization – it has succeeded to a degree perhaps unparalleled in history.

30*th*. – The *Beagle* sailed for Hobart Town in Van Diemen's Land. On the 5th of February, after a six days' passage, of which the first part was fine, and the latter very cold and squally, we entered the mouth of Storm Bay: the weather justified this awful name. The bay should rather be called an estuary, for it receives at its head the waters of the Derwent. Near the mouth, there are some extensive basaltic platforms; but higher up the land becomes mountainous, and is covered by a light wood. The lower parts of the hills which skirt the bay are cleared; and the bright yellow fields of corn, and dark green ones of potatoes, appeared very luxuriant. Late in the evening we anchored in the snug cove, on the shores of which stands the capital of Tasmania. The first aspect of the place was very inferior to that of Sydney; the latter might be called a city, this only a town. It stands at the base of Mount Wellington, a mountain 3100 feet high, but of little picturesque beauty: from this source, however, it receives a good supply of water. Round the cove there are some fine warehouses, and on one side a small fort. Coming from the Spanish settlements, where such magnificent care has generally been paid to the fortifications, the means of defence in these colonies appeared very contemptible. Comparing the town with Sydney, I was chiefly struck with the comparative fewness of the large houses, either built or building. Hobart Town, from the census of 1835, contained 13,826 inhabitants, and the whole of Tasmania 36,505.

All the aborigines have been removed to an island in Bass's Straits, so that Van Diemen's Land enjoys the great advantage of being free from a native population. This most cruel step seems to have been quite unavoidable, as the only means of stopping a fearful succession of robberies, burnings, and murders, committed by the blacks;

and which sooner or later would have ended in their utter destruction. I fear there is no doubt, that this train of evil and its consequences, originated in the infamous conduct of some of our countrymen. Thirty years is a short period, in which to have banished the last aboriginal from his native island, – and that island nearly as large as Ireland. The correspondence on this subject, which took place between the government at home and that of Van Diemen's Land, is very interesting. Although numbers of natives were shot and taken prisoners in the skirmishing, which was going on at intervals for several years; nothing seems fully to have impressed them with the idea of our overwhelming power, until the whole island, in 1830, was put under martial law, and by proclamation the whole population commanded to assist in one great attempt to secure the entire race. The plan adopted was nearly similar to that of the great hunting-matches in India: a line was formed reaching across the island, with the intention of driving the natives into a *cul-de-sac* on Tasman's peninsula. The attempt failed; the natives, having tied up their dogs, stole during one night through the lines. This is far from surprising, when their practised senses, and usual manner of crawling after wild animals is considered. I have been assured that they can conceal themselves on almost bare ground, in a manner which until witnessed is scarcely credible; their dusky bodies being easily mistaken for the blackened stumps which are scattered all over the country. I was told of a trial between a party of Englishmen and a native, who was to stand in full view on the side of a bare hill; if the Englishmen closed their eyes for less than a minute, he would squat down, and then they were never able to distinguish him from the surrounding stumps. But to return to the hunting-match; the natives understanding this kind of warfare, were terribly alarmed, for they at once perceived the power and numbers of the whites. Shortly afterwards a party of thirteen belonging to two tribes came in; and, conscious of their unprotected condition, delivered themselves up in despair. Subsequently by the intrepid exertions of Mr. Robinson, an active and benevolent man, who fearlessly visited by himself the most hostile of the natives, the whole were induced to act in a similar manner. They were then removed to an island, where food and clothes were provided them.

Count Strzelecki states,[1] that 'at the epoch of their deportation in 1835, the number of natives amounted to 210. In 1842, that is after the interval of seven years, they mustered only fifty-four individuals; and, while each family of the interior of New South Wales, uncontaminated by contact with the whites, swarms with children, those of Flinders' Island had during eight years, an accession of only fourteen in number!'

The *Beagle* staid here ten days, and in this time I made several pleasant little excursions, chiefly with the object of examining the geological structure of the immediate neighbourhood. The main points of interest consist, first in some highly fossiliferous strata, belonging to the Devonian or Carboniferous period; secondly, in proofs of a late small rise of the land; and lastly, in a solitary and superficial patch of yellowish limestone or travertin, which contains numerous impressions of leaves of trees, together with land-shells, not now existing. It is not improbable that this one small quarry, includes the only remaining record of the vegetation of Van Diemen's Land during one former epoch.

The climate here is damper than in New South Wales, and hence the land is more fertile. Agriculture flourishes: the cultivated fields look well, and the gardens abound with thriving vegetables and fruit-trees. Some of the farm-houses, situated in retired spots, had a very attractive appearance. The general aspect of the vegetation is similar to that of Australia; perhaps it is a little more green and cheerful; and the pasture between the trees rather more abundant. One day I took a long walk on the side of the bay opposite to the town: I crossed in a steam-boat, two of which are constantly plying backwards and forwards. The machinery of one of these vessels was entirely manufactured in this colony, which, from its very foundation, then numbered only three and thirty years! Another day I ascended Mount Wellington; I took with me a guide, for I failed in a first attempt, from the thickness of the wood. Our guide, however, was a stupid fellow, and conducted us to the southern and damp side of the mountain, where the vegetation was very luxuriant; and where the labour of the ascent, from the number of rotten trunks, was almost

1 Physical Description of New South Wales and Van Diemen's Land, p. 354.

as great as on a mountain in Tierra del Fuego or in Chiloe. It cost us five and a half hours of hard climbing before we reached the summit. In many parts the Eucalypti grew to a great size, and composed a noble forest. In some of the dampest ravines, tree-ferns flourished in an extraordinary manner; I saw one which must have been at least twenty feet high to the base of the fronds, and was in girth exactly six feet. The fronds forming the most elegant parasols, produced a gloomy shade, like that of the first hour of night. The summit of the mountain is broad and flat, and is composed of huge angular masses of naked greenstone. Its elevation is 3100 feet above the level of the sea. The day was splendidly clear, and we enjoyed a most extensive view; to the north, the country appeared a mass of wooded mountains, of about the same height with that on which we were standing, and with an equally tame outline: to the south the broken land and water, forming many intricate bays, was mapped with clearness before us. After staying some hours on the summit, we found a better way to descend, but did not reach the *Beagle* till eight o'clock, after a severe day's work.

February 7th. – The *Beagle* sailed from Tasmania, and, on the 6th of the ensuing month, reached King George's Sound, situated close to the S.W. corner of Australia. We staid there eight days; and we did not during our voyage pass a more dull and uninteresting time. The country, viewed from an eminence, appears a woody plain, with here and there rounded and partly bare hills of granite protruding. One day I went out with a party, in hopes of seeing a kangaroo hunt, and walked over a good many miles of country. Everywhere we found the soil sandy, and very poor; it supported either a coarse vegetation of thin, low brushwood and wiry grass, or a forest of stunted trees. The scenery resembled that of the high sandstone platform of the Blue Mountains; the Casuarina (a tree somewhat resembling a Scotch fir) is, however, here in greater number, and the Eucalyptus in rather less. In the open parts there were many grass-trees, – a plant which, in appearance, has some affinity with the palm; but, instead of being surmounted by a crown of noble fronds, it can boast merely of a tuft of very coarse grass-like leaves. The general bright green colour of the brushwood and other plants, viewed from a distance, seemed to promise fertility. A single walk, however, was enough to dispel such an illusion; and he who thinks with me will never wish to walk again in so uninviting a country.

One day I accompanied Captain Fitz Roy to Bald Head; the place mentioned by so many navigators, where some imagined that they saw corals, and others that they saw petrified trees, standing in the position in which they had grown. According to our view, the beds have been formed by the wind having heaped up fine sand, composed of minute rounded particles of shells and corals, during which process branches and roots of trees, together with many land-shells, became enclosed. The whole then became consolidated by the percolation of calcareous matter; and the cylindrical cavities left by the decaying of the wood, were thus also filled up with a hard pseudo-stalactitical stone. The weather is now wearing away the softer parts, and in consequence the hard casts of the roots and branches of the trees project above the surface, and, in a singularly deceptive manner, resemble the stumps of a dead thicket.

A large tribe of natives, called the White Cockatoo men, happened to pay the settlement a visit while we were there. These men, as well as those of the tribe belonging to King George's Sound, being tempted by the offer of some tubs of rice and sugar, were persuaded to hold a 'corrobery,' or great dancing-party. As soon as it grew dark, small fires were lighted, and the men commenced their toilet, which consisted in painting themselves white in spots and lines. As soon as all was ready, large fires were kept blazing, round which the women and children were collected as spectators; the Cockatoo and King George's men formed two distinct parties, and generally danced in answer to each other. The dancing consisted in their running either sideways or in Indian file into an open space, and stamping the ground with great force as they marched together. Their heavy footsteps were accompanied by a kind of grunt, by beating their clubs and spears together, and by various other gesticulations, such as extending their arms and wriggling their bodies. It was a most rude, barbarous scene, and, to our ideas, without any sort of meaning; but we observed that the black women and children watched it with the greatest pleasure. Perhaps these dances originally represented actions, such as wars and victories; there was one called the Emu dance, in which each man extended his arm in a bent manner, like the neck of that bird. In another dance, one man imitated the movements of a kangaroo grazing in the woods, whilst a second crawled up, and pretended to spear him. When both tribes mingled in the dance, the ground trembled with the heaviness of their steps, and the air resounded with their wild cries. Every one appeared in high spirits, and the group of nearly naked figures, viewed by the light of the blazing fires, all moving in hideous harmony, formed a perfect display of a festival amongst the lowest barbarians. In Tierra del Fuego, we have beheld many curious scenes in savage life, but never, I think, one where the natives were in such high spirits, and so perfectly at their ease. After the dancing was over, the whole party formed a great circle on the ground, and the boiled rice and sugar was distributed, to the delight of all.

After several tedious delays from clouded weather, on the 14th of March, we gladly stood out of King George's Sound on our course to Keeling Island. Farewell, Australia! you are a rising child, and doubtless some day will reign a great princess in the South: but you are too great and ambitious for affection, yet not great enough for respect. I leave your shores without sorrow or regret.

CHAPTER XX

KEELING ISLAND: – CORAL FORMATIONS

Keeling Island – Singular appearance – Scanty Flora – Transport of Seeds – Birds and Insects – Ebbing and flowing Springs – Fields of dead Coral – Stones transported in the Roots of Trees – Great Crab – Stinging Corals – Coral-eating Fish – Coral Formations – Lagoon Islands, or Atolls – Depth at which reef-building Corals can live – Vast Areas interspersed with low Coral Islands – Subsidence of their Foundations – Barrier Reefs – Fringing Reefs – Conversion of Fringing Reefs into Barrier Reefs, and into Atolls – Evidence of Changes in Level – Breaches in Barrier Reefs – Maldiva Atolls; their peculiar structure – Dead and submerged Reefs – Areas of Subsidence and Elevation – Distribution of Volcanoes – Subsidence slow, and vast in amount.

APRIL 1st. – We arrived in view of the Keeling or Cocos Islands, situated in the Indian Ocean, and about six hundred miles distant from the coast of Sumatra. This is one of the lagoon-islands (or atolls) of coral formation, similar to those in the Low Archipelago which we passed near. When the ship was in the channel at the entrance, Mr. Liesk, an English resident, came off in his boat. The history of the inhabitants of this place, in as few words as possible, is as follows. About nine years ago, Mr. Hare, a worthless character, brought from the East Indian archipelago a number of Malay slaves, which now, including children, amount to more than a hundred. Shortly afterwards, Captain Ross, who had before visited these islands in his merchant-ship, arrived from England, bringing with him his family and goods for settlement: along with him came Mr. Liesk, who had been a mate in his vessel. The Malay slaves soon ran away from the islet on which Mr. Hare was settled, and joined Captain Ross's party. Mr. Hare upon this was ultimately obliged to leave the place.

The Malays are now nominally in a state of freedom, and certainly are so, as far as regards their personal treatment; but in most other points they are considered as slaves. From their discontented state, from the repeated removals from islet to islet, and perhaps also from a little mismanagement, things are not very prosperous. The island has no domestic quadruped, excepting the pig, and the main vegetable production is the cocoa-nut. The whole prosperity of the place depends on this tree: the only exports being oil from the nut, and the nuts themselves, which are taken to Singapore and Mauritius, where they are chiefly used, when grated, in making curries. On the cocoa-nut, also, the pigs, which are loaded with fat, almost entirely subsist, as do the ducks and poultry. Even a huge land-crab is furnished by nature with the means to open and feed on this most useful production.

The ring-formed reef of the lagoon-island is surmounted in the greater part of its length by linear islets. On the northern or leeward side, there is an opening through which vessels can pass to the anchorage within. On entering, the scene was very curious and rather pretty; its beauty, however, entirely depended on the brilliancy of the surrounding colours. The shallow, clear, and still water of the lagoon, resting in its greater part on white sand, is, when illumined by a vertical sun, of the most vivid green. This brilliant expanse, several miles in width, is on all sides divided, either by a line of snow-white breakers from the dark heaving waters of the ocean, or from the blue vault of heaven by the strips of land, crowned by the level tops of the cocoa-nut trees. As a white cloud here and there affords a pleasing contrast with the azure sky, so in the lagoon, bands of living coral darken the emerald green water.

The next morning after anchoring, I went on shore on Direction Island. The strip of dry land is only a few hundred yards in width; on the lagoon side there is a white calcareous beach, the radiation from which under this sultry climate was very oppressive; and on the outer coast, a solid broad flat of coral-rock served to break the violence of the open sea. Excepting near the lagoon, where there is some sand, the land is entirely composed of rounded fragments of coral. In such a loose, dry, stony soil, the climate of the intertropical regions alone could produce a vigorous vegetation. On some of the smaller islets, nothing could be more elegant than the manner in which the young and full-grown cocoa-nut trees, without destroying each other's symmetry, were mingled into one wood. A beach of glittering white sand formed a border to these fairy spots.

I will now give a sketch of the natural history of these islands, which, from its very paucity, possesses a peculiar interest. The cocoa-nut tree, at the first glance, seems to compose the whole wood; there are, however, five or six other trees. One of these grows to a very large size, but, from the extreme softness of its wood, is useless; another sort affords excellent timber for ship-building. Besides the trees, the number of plants is exceedingly limited, and consists of insignificant weeds. In my collection, which includes, I believe, nearly the perfect Flora, there are twenty species, without reckoning a moss, lichen, and fungus. To this number two trees must be added; one of which was not in flower, and the other I only heard of. The latter is a solitary tree of its kind, and grows near the beach, where, without doubt, the one seed was thrown up by the waves. A Guilandina also grows on only one of the islets. I do not include in the above list the sugar-cane, banana, some other vegetables, fruit-trees, and imported grasses. As the islands consist entirely of coral, and at one time must have existed as mere water-washed reefs, all their terrestrial productions must have been transported here by the waves of the sea. In accordance with this, the Florula has quite the character of a refuge for the destitute: Professor Henslow informs me that of the twenty species nineteen belong to different genera, and these again to no less than sixteen families![1]

In Holman's[2] Travels an account is given, on the authority of Mr. A.S. Keating, who resided twelve months on these islands, of the various seeds and other bodies which have been known to have been washed on shore. 'Seeds and plants from Sumatra and Java have been driven up by the surf on the windward side of the islands. Among them have been found the Kimiri, native of Sumatra and the peninsula of Malacca; the cocoa-nut of Balci, known by its shape and size; the Dadass, which is planted by the Malays with the pepper-vine,

---

1 These plants are described in the Annals of Nat. Hist., vol. i. 1838, p. 337.
2 Holman's Travels, vol. iv. p. 378.

the latter intwining round its trunk, and supporting itself by the prickles on its stem; the soap-tree; the castor-oil plant; trunks of the sago palm; and various kinds of seeds unknown to the Malays settled on the islands. These are all supposed to have been driven by the N.W. monsoon to the coast of New Holland, and thence to these islands by the S.E. trade-wind. Large masses of Java teak and Yellow wood have also been found, besides immense trees of red and white cedar, and the blue gum-wood of New Holland, in a perfectly sound condition. All the hardy seeds, such as creepers, retain their germinating power, but the softer kinds, among which is the mangostin, are destroyed in the passage. Fishing-canoes, apparently from Java, have at times been washed on shore.' It is interesting thus to discover how numerous the seeds are, which, coming from several countries, are drifted over the wide ocean. Professor Henslow tells me, he believes that nearly all the plants which I brought from these islands, are common littoral species in the East Indian archipelago. From the direction, however, of the winds and currents, it seems scarcely possible that they could have come here in a direct line. If, as suggested with much probability by Mr. Keating, they were first carried towards the coast of New Holland, and thence drifted back together with the productions of that country, the seeds, before germinating, must have travelled between 1800 and 2400 miles.

Chamisso,[1] when describing the Radack Archipelago, situated in the western part of the Pacific, states that 'the sea brings to these islands the seeds and fruits of many trees, most of which have not yet grown here. The greater part of these seeds appear to have not yet lost the capability of growing.' It is also said that palms and bamboos from somewhere in the torrid zone, and trunks of northern firs, are washed on shore: these firs must have come from an immense distance. These facts are highly interesting. It cannot be doubted that if there were land-birds to pick up the seeds when first cast on shore, and a soil better adapted for their growth than the loose blocks of coral, that the most isolated of the lagoon-islands would in time possess a far more abundant Flora than they now have.

1 Kotzebue's First Voyage, vol. iii. p. 155.

The list of land animals is even poorer than that of the plants. Some of the islets are inhabited by rats, which were brought in a ship from the Mauritius, wrecked here. These rats are considered by Mr. Waterhouse as identical with the English kind, but they are smaller, and more brightly coloured. There are no true land-birds; for a snipe and a rail (Rallus Phillippensis), though living entirely in the dry herbage, belong to the order of Waders. Birds of this order are said to occur on several of the small low islands in the Pacific. At Ascension, where there is no land-bird, a rail (Por-phyrio simplex) was shot near the summit of the mountain, and it was evidently a solitary straggler. At Tristan d'Acunha, where, according to Carmichael, there are only two land-birds, there is a coot. From these facts I believe that the waders, after the innumerable web-footed species, are generally the first colonists of small isolated islands. I may add, that whenever I noticed birds, not of oceanic species, very far out at sea, they always belonged to this order; and hence they would naturally become the earliest colonists of any remote point of land.

Of reptiles I saw only one small lizard. Of insects I took pains to collect every kind. Exclusive of spiders, which were numerous, there were thirteen species.[1] Of these, one only was a beetle. A small ant swarmed by thousands under the loose dry blocks of coral, and was the only true insect which was abundant. Although the productions of the land are thus scanty, if we look to the waters of the surrounding sea, the number of organic beings is indeed infinite. Chamisso has described[2] the natural history of a lagoon-island in the Radack Archipelago; and it is remarkable how closely its inhabitants, in number and kind, resemble those of Keeling Island. There is one lizard and two waders, namely, a snipe and curlew. Of plants there are nineteen species, including a fern; and some of these are the same with those growing here, though on a spot so immensely remote, and in a different ocean.

1 The thirteen species belong to the following orders: – In the *Coleoptera*, a minute Elater; *Orthoptera*, a Gryllus and a Blatta; *Hemiptera*, one species; *Homoptera*, two; *Neuroptera*, a Chrysopa; *Hymenoptera*, two ants; *Lepidoptera nocturna*, a Diopæa, and a Pterophorus (?); *Diptera*, two species.
2 Kotzebue's First Voyage, vol. iii. p. 222.

The long strips of land, forming the linear islets, have been raised only to that height to which the surf can throw fragments of coral, and the wind heap up calcareous sand. The solid flat of coral rock on the outside, by its breadth, breaks the first violence of the waves, which otherwise, in a day, would sweep away these islets and all their productions. The ocean and the land seem here struggling for mastery: although terra firma has obtained a footing, the denizens of the water think their claim at least equally good. In every part one meets hermit crabs of more than one species,[1] carrying on their backs the shells which they have stolen from the neighbouring beach. Overhead, numerous gannets, frigate-birds, and terns, rest on the trees; and the wood, from the many nests and from the smell of the atmosphere, might be called a sea-rookery. The gannets, sitting on their rude nests, gaze at one with a stupid yet angry air. The noddies, as their name expresses, are silly little creatures. But there is one charming bird: it is a small snow-white tern, which smoothly hovers at the distance of a few feet above one's head, its large black eye scanning, with quiet curiosity, your expression. Little imagination is required to fancy that so light and delicate a body must be tenanted by some wandering fairy spirit.

Sunday, April 3rd. – After service I accompanied Captain Fitz Roy to the settlement, situated at the distance of some miles, on the point of an islet thickly covered with tall cocoa-nut trees. Captain Ross and Mr. Liesk live in a large barn-like house open at both ends, and lined with mats made of woven bark. The houses of the Malays are arranged along the shore of the lagoon. The whole place had rather a desolate aspect, for there were no gardens to show the signs of care and cultivation. The natives belong to different islands in the East Indian archipelago, but all speak the same language: we saw the inhabitants of Borneo, Celebes, Java, and Sumatra. In colour they resemble the Tahitians, from whom they do not widely differ in features. Some of the women, however, show a good deal of the

---

1 The large claws or pincers of some of these crabs are most beautifully adapted, when drawn back, to form an operculum to the shell, nearly as perfect as the proper one originally belonging to the molluscous animal. I was assured, and as far as my observation went I found it so, that certain species of the hermit crabs always use certain species of shells.

Chinese character. I liked both their general expressions and the sound of their voices. They appeared poor, and their houses were destitute of furniture; but it was evident, from the plumpness of the little children, that cocoa-nuts and turtle afford no bad sustenance.

On this island the wells are situated, from which ships obtain water. At first sight it appears not a little remarkable that the fresh water should regularly ebb and flow with the tides; and it has even been imagined that sand has the power of filtering the salt from the sea-water. These ebbing wells are common on some of the low islands in the West Indies. The compressed sand, or porous coral rock, is permeated like a sponge with the salt water; but the rain which falls on the surface must sink to the level of the surrounding sea, and must accumulate there, displacing an equal bulk of the salt water. As the water in the lower part of the great sponge-like coral mass rises and falls with the tides, so will the water near the surface; and this will keep fresh, if the mass be sufficiently compact to prevent much mechanical admixture; but where the land consists of great loose blocks of coral with open interstices, if a well be dug, the water, as I have seen, is brackish.

After dinner we stayed to see a curious half superstitious scene acted by the Malay women. A large wooden spoon dressed in garments, and which had been carried to the grave of a dead man, they pretend becomes inspired at the full of the moon, and will dance and jump about. After the proper preparations, the spoon, held by two women, became convulsed, and danced in good time to the song of the surrounding children and women. It was a most foolish spectacle; but Mr. Liesk maintained that many of the Malays believed in its spiritual movements. The dance did not commence till the moon had risen, and it was well worth remaining to behold her bright orb so quietly shining through the long arms of the cocoa-nut trees as they waved in the evening breeze. These scenes of the tropics are in themselves so delicious, that they almost equal those dearer ones at home, to which we are bound by each best feeling of the mind.

The next day I employed myself in examining the very interesting, yet simple structure and origin of these islands. The water being unusually smooth, I waded over the outer flat of dead rock as far as the living mounds of coral, on which the swell of the open sea breaks. In some of the gullies and hollows there were beautiful green and other coloured fishes, and the forms and tints of many of the zoophytes were admirable. It is excusable to grow enthusiastic over the infinite numbers of organic beings with which the sea of the tropics, so prodigal of life, teems; yet I must confess I think those naturalists who have described, in well-known words, the submarine grottoes decked with a thousand beauties, have indulged in rather exuberant language.

April 6th. – I accompanied Captain Fitz Roy to an island at the head of the lagoon: the channel was exceedingly intricate, winding through fields of delicately branched corals. We saw several turtle, and two boats were then employed in catching them. The water was so clear and shallow, that although at first a turtle quickly dives out of sight, yet in a canoe or boat under sail, the pursuers after no very long chase come up to it. A man standing ready in the bow, at this moment dashes through the water upon the turtle's back; then clinging with both hands by the shell of its neck, he is carried away till the animal becomes exhausted and is secured. It was quite an interesting chase to see the two boats thus doubling about, and the men dashing head foremost into the water trying to seize their prey. Captain Moresby informs me that in the Chagos archipelago in this same ocean, the natives, by a horrible process, take the shell from the back of the living turtle. 'It is covered with burning charcoal, which causes the outer shell to curl upwards; it is then forced off with a knife, and before it becomes cold flattened between boards. After this barbarous process the animal is suffered to regain its native element, where, after a certain time, a new shell is formed; it is, however, too thin to be of any service, and the animal always appears languishing and sickly.'

When we arrived at the head of the lagoon, we crossed a narrow islet, and found a great surf breaking on the windward coast. I can hardly explain the reason, but there is to my mind much grandeur in the view of the outer shores of these lagoon-islands. There is a simplicity in the barrier-like beach, the margin of green bushes and tall cocoa-nuts, the solid flat of dead coral-rock, strewed here and there with great loose fragments, and the line of furious breakers, all rounding away towards either hand. The ocean throwing its waters over the broad reef appears an invincible, all-powerful enemy; yet we see it resisted, and even conquered, by means which at first seem most weak and inefficient. It is not that the ocean spares the rock of coral; the great fragments scattered over the reef, and heaped on the beach, whence the tall cocoa-nut springs, plainly bespeak the unrelenting power of the waves. Nor are any periods of repose granted. The long swell caused by the gentle but steady action of the trade-wind, always blowing in one direction over a wide area, causes breakers, almost equalling in force those during a gale of wind in the temperate regions, and which never cease to rage. It is impossible to behold these waves without feeling a conviction that an island, though built of the hardest rock, let it be porphyry, granite, or quartz, would ultimately yield and be demolished by such an irresistible power. Yet these low, insignificant coral-islets stand and are victorious: for here another power, as an antagonist, takes part in the contest. The organic forces separate the atoms of carbonate of lime, one by one, from the foaming breakers, and unite them into a symmetrical structure. Let the hurricane tear up its thousand huge fragments; yet what will that tell against the accumulated labour of myriads of architects at work night and day, month after month? Thus do we see the soft and gelatinous body of a polypus, through the agency of the vital laws, conquering the great mechanical power of the waves of an ocean which neither the art of man nor the inanimate works of nature could successfully resist.

We did not return on board till late in the evening, for we staid a long time in the lagoon, examining the fields of coral and the gigantic shells of the chama, into which, if a man were to put his hand, he would not, as long as the animal lived, be able to withdraw it. Near the head of the lagoon, I was much surprised to find a wide area, considerably more than a mile square, covered with a forest of delicately branching corals, which, though standing upright, were all dead and rotten. At first I was quite at a loss to understand the cause; afterwards it occurred to me that it was owing to the following rather curious combination of circumstances. It should, however, first be stated, that corals are not able to survive even a short exposure in the air to the sun's rays, so that their upward limit of growth is determined by that of lowest water at spring tides. It appears, from some old charts, that the long island to windward was formerly separated by wide channels into several islets; this fact is likewise indicated by the trees being younger on these portions. Under the former condition of the reef, a strong breeze, by throwing more water over the barrier, would tend to raise the level of the lagoon. Now it acts in a directly contrary manner; for the water within the lagoon not only is not increased by currents from the outside, but is itself blown outwards by the force of the wind. Hence it is observed, that the tide near the head of the lagoon does not rise so high during a strong breeze as it does when it is calm. This difference of level, although no doubt very small, has, I believe, caused the death of those coral-groves, which under the former and more open condition of the outer reef had attained the utmost possible limit of upward growth.

A few miles north of Keeling there is another small atoll, the lagoon of which is nearly filled up with coral-mud. Captain Ross found embedded in the conglomerate on the outer coast, a well-rounded fragment of greenstone, rather larger than a man's head: he and the men with him were so much surprised at this, that they brought it away and preserved it as a curiosity. The occurrence of this one stone, where every other particle of matter is calcareous, certainly is very puzzling. The island has scarcely ever been visited, nor is it probable that a ship had been wrecked there. From the absence of any better explanation, I came to the conclusion that it must have come entangled in the roots of some large tree: when, however, I considered the great distance from the nearest land, the combination of chances against a stone thus being entangled, the tree washed into the sea, floated so far, then landed safely, and the stone finally so embedded as to allow of its discovery, I was almost afraid of imagining a means of transport apparently so improbable. It was therefore with great interest that I found Chamisso, the justly distinguished naturalist who accompanied Kotzebue, stating that the inhabitants of the Radack archipelago, a group of lagoon-islands in the midst of the Pacific, obtained stones for sharpening their instruments by searching the roots of trees which are cast upon the beach. It will be evident that this must have happened several times, since laws have been established that such stones belong to the chief, and a punishment is inflicted on any one who attempts to steal them. When the isolated position of these small islands in the midst of a vast ocean – their great distance from any land excepting that of coral formation, attested by the value which the inhabitants, who are such bold navigators, attach to a stone of any kind,[1] – and the slowness of the currents of the open sea, are all considered, the occurrence of pebbles thus transported does appear wonderful. Stones may often be thus carried; and if the island on which they are stranded is constructed of any other substance besides coral, they would scarcely attract attention, and their origin at least would never be guessed.

Moreover, this agency may long escape discovery from the probability of trees, especially those loaded with stones, floating beneath the surface. In the channels of Tierra del Fuego large quantities of drift timber are cast upon the beach, yet it is extremely rare to meet a tree swimming on the water. These facts may possibly throw light on single stones, whether angular or rounded, occasionally found embedded in fine sedimentary masses.

During another day I visited West Islet, on which the vegetation was perhaps more luxuriant than on any other. The cocoa-nut trees generally grow separate, but here the young ones flourished beneath their tall parents, and formed with their long and curved fronds the most shady arbours. Those alone who have tried it, know how delicious it is to be seated in such shade, and drink the cool pleasant fluid of the cocoa-nut. In this island there is a large bay-like space, composed of the finest white sand: it is quite level, and is only covered by the tide at high water; from this large bay smaller creeks penetrate the surrounding woods. To see a field of glittering white sand, representing water, with the cocoa-nut trees extending their tall and waving trunks round the margin, formed a singular and very pretty view.

I have before alluded to a crab which lives on the cocoa-nuts: it is very common on all parts of the dry land, and grows to a monstrous size: it is closely allied or identical with the Birgos latro. The front pair of legs terminate in very strong and heavy pincers, and the last pair are fitted with others weaker and much narrower. It would at first be

1 Some natives carried by Kotzebue to Kamtschatka collected stones to take back to their country.

thought quite impossible for a crab to open a strong cocoa-nut covered with the husk; but Mr. Liesk assures me that he has repeatedly seen this effected. The crab begins by tearing the husk, fibre by fibre, and always from that end under which the three eye-holes are situated; when this is completed, the crab commences hammering with its heavy claws on one of the eye-holes till an opening is made. Then turning round its body, by the aid of its posterior and narrow pair of pincers, it extracts the white albuminous substance. I think this is as curious a case of instinct as ever I heard of, and likewise of adaptation in structure between two objects apparently so remote from each other in the scheme of nature, as a crab and a cocoa-nut tree. The Birgos is diurnal in its habits; but every night it is said to pay a visit to the sea, no doubt for the purpose of moistening its branchiæ. The young are likewise hatched, and live for some time, on the coast. These crabs inhabit deep burrows, which they hollow out beneath the roots of trees; and where they accumulate surprising quantities of the picked fibres of the cocoa-nut husk, on which they rest as on a bed. The Malays sometimes take advantage of this, and collect the fibrous mass to use as junk. These crabs are very good to eat; moreover, under the tail of the larger ones there is a great mass of fat, which, when melted, sometimes yields as much as a quart bottle full of limpid oil. It has been stated by some authors that the Birgos crawls up the cocoa-nut trees for the purpose of stealing the nuts: I very much doubt the possibility of this; but with the Pandanus[1] the task would be very much easier. I was told by Mr. Liesk that on these islands the Birgos lives only on the nuts which have fallen to the ground.

Captain Moresby informs me that this crab inhabits the Chagos and Seychelle groups, but not the neighbouring Maldiva archipelago. It formerly abounded at Mauritius, but only a few small ones are now found there. In the Pacific, this species, or one with closely allied habits, is said[2] to inhabit a single coral island, north of the Society group. To show the wonderful strength of the front pair of pincers, I may mention, that Captain Moresby confined one in a strong

1 See Proceedings of Zoological Society, 1832, p. 17.
2 Tyerman and Bennett. Voyage, &c., vol. ii. p. 33.

tin-box, which had held biscuits, the lid being secured with wire; but the crab turned down the edges and escaped. In turning down the edges, it actually punched many small holes quite through the tin!

I was a good deal surprised by finding two species of coral of the genus Millepora (M. complanata and alcicornis), possessed of the power of stinging. The stony branches or plates, when taken fresh from the water, have a harsh feel and are not slimy, although possessing a strong and disagreeable smell. The stinging property seems to vary in different specimens: when a piece was pressed or rubbed on the tender skin of the face or arm, a pricking sensation was usually caused, which came on after the interval of a second, and lasted only for a few minutes. One day, however, by merely touching my face with one of the branches, pain was instantaneously caused; it increased as usual after a few seconds, and remaining sharp for some minutes, was perceptible for half an hour afterwards. The sensation was as bad as that from a nettle, but more like that caused by the Physalia or Portuguese man-of-war. Little red spots were produced on the tender skin of the arm, which appeared as if they would have formed watery pustules, but did not. M. Quoy mentions this case of the Millepora; and I have heard of stinging corals in the West Indies. Many marine animals seem to have this power of stinging: besides the Portuguese man-of-war, many jelly-fish, and the Aplysia or sea-slug of the Cape de Verd Islands, it is stated in the voyage of the *Astrolabe*, that an Actinia or sea-anemone, as well as a flexible coralline allied to Sertularia, both possess this means of offence or defence. In the East Indian sea, a stinging sea-weed is said to be found.

Two species of fish, of the genus Scarus, which are common here, exclusively feed on coral: both are coloured of a splendid bluish-green, one living invariably in the lagoon, and the other amongst the outer breakers. Mr. Liesk assured us, that he had repeatedly seen whole shoals grazing with their strong bony jaws on the tops of the coral branches: I opened the intestines of several, and found them distended with yellowish calcareous sandy mud. The slimy disgusting Holuthuriæ (allied to our star-fish), which the Chinese gourmands are so fond of, also feed largely, as I am informed by Dr. Allan, on corals; and the bony apparatus within their bodies seems well adapted for this end. These holuthuriæ, the fish, the numerous burrowing shells, and nereidous worms, which perforate every block of dead coral, must be very efficient agents in producing the fine white mud which lies at the bottom and on the shores of the lagoon. A portion, however, of this mud, which when wet strikingly resembled pounded chalk, was found by Professor Ehrenberg to be partly composed of siliceous-shielded infusoria.

April 12*th.* – In the morning we stood out of the lagoon on our passage to the Isle of France. I am glad we have visited these islands: such formations surely rank high amongst the wonderful objects of this world. Captain Fitz Roy found no bottom with a line 7200 feet in length, at the distance of only 2200 yards from the shore; hence this island forms a lofty submarine mountain, with sides steeper even than those of the most abrupt volcanic cone. The saucer-shaped summit is nearly ten miles across; and every single atom,[1] from the least particle to the largest fragment of rock, in this great pile, which however is small compared with very many other lagoon-islands, bears the stamp of having been subjected to organic arrangement. We feel surprise when travellers tell us of the vast dimensions of the Pyramids and other great ruins, but how utterly insignificant are the greatest of these, when compared to these mountains of stone accumulated by the agency of various minute and tender animals! This is a wonder which does not at first strike the eye of the body, but, after reflection, the eye of reason.

I will now give a very brief account of the three great classes of coral-reefs; namely, Atolls, Barrier, and Fringing-reefs, and will explain my views[2] on their formation. Almost every voyager who has crossed the Pacific has expressed his unbounded astonishment at the lagoon-islands, or as I shall for the future call them by their Indian name of atolls, and has attempted some explanation. Even as long ago

1 I exclude, of course, some soil which has been imported here in vessels from Malacca and Java, and likewise some small fragments of pumice, drifted here by the waves. The one block of greenstone, moreover, on the northern island must be excepted.

2 These were first read before the Geological Society in May, 1837, and have since been developed in a separate volume on the 'Structure and Distribution of Coral Reefs.'

as the year 1605, Pyrard de Laval well exclaimed, 'C'est une meruille de voir chacun de ces atollons, enuironné d'un grand banc de pierre tout autour, n'y ayant point d'artifice humain.' The accompanying sketch of Whitsunday Island in the Pacific, copied from Capt. Beechey's admirable Voyage, gives but a faint idea of the singular aspect of an atoll: it is one of the smallest size, and has its narrow islets united together in a ring. The immensity of the ocean, the fury of the breakers, contrasted with the lowness of the land and the smoothness of the bright green water within the lagoon, can hardly be imagined without having been seen.

The earlier voyagers fancied that the coral-building animals instinctively built up their great circles to afford themselves protection in the inner parts; but so far is this from the truth, that those massive kinds, to whose growth on the exposed outer shores the very existence of the reef depends, cannot live within the lagoon, where other delicately-branching kinds flourish. Moreover, on this view, many species of distinct genera and families are supposed to combine for one end; and of such a combination, not a single instance can be found in the whole of nature. The theory that has been most generally received is, that atolls are based on submarine craters; but when we consider the form and size of some, the number, proximity, and relative positions of others, this idea loses its plausible character: thus, Suadiva atoll is 44 geographical miles in diameter in one line, by 34 miles in another line; Rimsky is 54 by 20 miles across, and it has a strangely sinuous margin; Bow atoll is 30 miles long, and on an average only 6 in width; Menchicoff atoll consists of three atolls united or tied together. This theory, moreover, is totally inapplicable to the northern Maldiva atolls in the Indian Ocean (one of which is 88 miles in length, and between 10 and 20 in breadth), for they are not bounded like ordinary atolls by narrow reefs, but by a vast number of separate little atolls; other little atolls rising out of the great central lagoon-like spaces. A third and better theory was advanced by Chamisso, who thought that from the corals growing more vigorously where exposed to the open sea, as undoubtedly is the case, the outer edges would grow up from the general foundation before any other part, and that this would account for the ring or cup-shaped structure. But we shall immediately see, that in this, as well as in the crater-theory, a most important consideration has been overlooked, namely, on what have the reef-building corals, which cannot live at a great depth, based their massive structures?

Numerous soundings were carefully taken by Captain Fitz Roy on the steep outside of Keeling atoll, and it was found that within ten fathoms, the prepared tallow at the bottom of the lead, invariably came up marked with the impressions of living corals, but as perfectly clean as if it had been dropped on a carpet of turf; as the depth increased, the impressions became less numerous, but the adhering particles of sand more and more numerous, until at last it was evident that the bottom consisted of a smooth sandy layer: to carry on the analogy of the turf, the blades of grass grew thinner and thinner, till at last the soil was so sterile, that nothing sprang from it. From these observations, confirmed by many others, it may be safely inferred that the utmost depth at which corals can construct reefs is between 20 and 30 fathoms. Now there are enormous areas in the Pacific and Indian Oceans, in which every single island is of coral formation, and is raised only to that height to which the waves can throw up fragments, and the winds pile up sand. Thus the Radack group of atolls is an irregular square, 520 miles long and 240 broad; the Low archipelago is elliptic-formed, 840 miles in its longer, and 420 in its shorter axis: there are other small groups and single low islands between these two archipelagoes, making a linear space of ocean actually more than 4000 miles in length, in which not one single island rises above the specified height. Again, in the Indian Ocean there is a space of ocean 1500 miles in length, including three archipelagoes, in which every island is low and of coral formation. From the fact of the reef-building corals not living at great depths, it is absolutely certain that throughout these vast areas, wherever there is now an atoll, a foundation must have originally existed within a depth of from 20 to 30 fathoms from the surface. It is improbable in the highest degree that broad, lofty, isolated, steep-sided banks of sediment, arranged in groups and lines hundreds of leagues in length, could have been deposited in the central and profoundest parts of the Pacific and Indian Oceans, at an immense distance from any continent, and where the water is perfectly limpid. It is equally improbable that the elevatory forces should have uplifted throughout the above vast areas, innumerable great rocky banks within 20 to 30 fathoms, or 120 to 180 feet, of the surface of the sea, and not one single point above that level; for where on the whole face of the globe can we find a single chain of mountains, even a few hundred miles in length, with their many summits rising within a few feet of a given level, and not one pinnacle above it? If then the foundations, whence the atoll-building corals sprang, were not formed of sediment, and if they were not lifted up to the required level, they must of necessity have subsided into it; and this at once solves the difficulty. For as mountain after mountain, and island after island, slowly sank beneath the water, fresh bases would be successively afforded for the growth of the corals. It is impossible here to enter into all the necessary details, but I venture to defy[1] any one to explain in any other manner, how it is possible that numerous islands should be distributed throughout vast areas – all the islands being low – all being built of corals, absolutely requiring a foundation within a limited depth from the surface.

Before explaining how atoll-formed reefs acquire their peculiar structure, we must turn to the second great class, namely, Barrier-reefs. These either extend in straight lines in front of the shores of a continent or of a large island, or they encircle smaller islands; in both cases, being separated from the land by a broad and rather deep channel of water, analogous to the lagoon within an atoll. It is

1 It is remarkable that Mr. Lyell, even in the first Edition of his 'Principles of Geology,' inferred that the amount of subsidence in the Pacific must have exceeded that of elevation, from the area of land being very small relatively to the agents there tending to form it, namely, the growth of coral and volcanic action.

remarkable how little attention has been paid to encircling barrier-reefs; yet they are truly wonderful structures. The above sketch represents part of the barrier encircling the island of Bolabola in the Pacific, as seen from one of the central peaks. In this instance the whole line of reef has been converted into land; but usually a snow-white line of great breakers, with only here and there a single low islet crowned with cocoa-nut trees, divides the dark heaving waters of the ocean from the light-green expanse of the lagoon-channel. And the quiet waters of this channel generally bathe a fringe of low alluvial soil, loaded with the most beautiful productions of the tropics, and lying at the foot of the wild, abrupt, central mountains.

Encircling barrier-reefs are of all sizes, from three miles to no less than forty-four miles in diameter; and that which fronts one side, and encircles both ends, of New Caledonia, is 400 miles long. Each reef includes one, two, or several rocky islands of various heights; and in one instance, even as many as twelve separate islands. The reef runs at a greater or less distance from the included land; in the Society archipelago generally from one to three or four miles; but at Hogoleu the reef is 20 miles on the southern side, and 14 miles on the opposite or northern side, from the included islands. The depth within the lagoon-channel also varies much; from 10 to 30 fathoms may be taken as an average; but at Vanikoro there are spaces no less than 56 fathoms or 336 feet deep. Internally the reef either slopes gently into the lagoon-channel, or ends in a perpendicular wall sometimes between two and three hundred feet under water in height: externally the reef rises, like an atoll, with extreme abruptness out of the profound depths of the ocean. What can be more singular than these structures? We see an island, which may be compared to a castle situated on the summit of a lofty submarine mountain, protected by a great wall of coral-rock, always steep externally and sometimes internally, with a broad level summit, here and there breached by narrow gateways, through which the largest ships can enter the wide and deep encircling moat.

As far as the actual reef of coral is concerned, there is not the smallest difference, in general size, outline, grouping, and even in quite trifling details of structure, between a barrier and an atoll. The geographer Balbi has well remarked, that an encircled island is an atoll with high land rising out of its lagoon; remove the land from within, and a perfect atoll is left.

But what has caused these reefs to spring up at such great distances from the shores of the included islands? It cannot be that the corals will not grow close to the land; for the shores within the lagoon-channel, when not surrounded by alluvial soil, are often fringed by living reefs; and we shall presently see that there is a whole class, which I have called Fringing-reefs from their close attachment to the shores both of continents and of islands. Again, on what have the reef-building corals, which cannot live at great depths, based their encircling structures? This is a great apparent difficulty, analogous to that in the case of atolls, which has generally been overlooked. It will be perceived more clearly by inspecting the following sections, which are real ones, taken in north and south lines, through the islands with their barrier-reefs, of Vanikoro, Gambier, and Maurua; and they are laid down, both vertically and horizontally, on the same scale of a quarter of an inch to a mile.

It should be observed that the sections might have been taken in any direction through these islands, or through many other encircled islands, and the general features would have been the same. Now bearing in mind that reef-building coral cannot live at a greater depth than from 20 to 30 fathoms, and that the scale is so small that the plummets on the right hand show a depth of 200 fathoms, on what are these barrier-reefs based? Are we to suppose that each island is surrounded by a collar-like submarine ledge of rock, or by a great bank of sediment, ending abruptly where the reef ends? If the sea had

1. Vanikoro.   2. Gambier Islands.   3. Maurua.

The horizontal shading shows the barrier-reefs and lagoon-channels. The inclined shading above the level of the sea (AA), shows the actual form of the land; the inclined shading below this line, shows its probable prolongation under water.

formerly eaten deeply into the islands, before they were protected by the reefs, thus having left a shallow ledge round them under water, the present shores would have been invariably bounded by great precipices; but this is most rarely the case. Moreover, on this notion, it is not possible to explain why the corals should have sprung up, like a wall, from the extreme outer margin of the ledge, often leaving a broad space of water within, too deep for the growth of corals. The accumulation of a wide bank of sediment all round these islands, and generally widest where the included islands are smallest, is highly improbable, considering their exposed positions in the central and deepest parts of the ocean. In the case of the barrier-reef of New Caledonia, which extends for 150 miles beyond the northern point of the island, in the same straight line with which it fronts the west coast, it is hardly possible to believe, that a bank of sediment could thus have been straightly deposited in front of a lofty island, and so far beyond its termination in the open sea. Finally, if we look to other oceanic islands of about the same height and of similar geological constitution, but not encircled by coral-reefs, we may in vain search for so trifling a circumambient depth as 30 fathoms, except quite near to their shores; for usually land that rises abruptly out of water, as do most of the encircled and non-encircled oceanic islands, plunges abruptly under it. On what then, I repeat, are these barrier-reefs based? Why, with their wide and deep moat-like channels, do they stand so far from the included land? We shall soon see how easily these difficulties disappear.

We come now to our third class of Fringing-reefs, which will require a very short notice. Where the land slopes abruptly under water, these reefs are only a few yards in width, forming a mere ribbon or fringe round the shores: where the land slopes gently under the water the reef extends further, sometimes even as much as a mile from the land; but in such cases the soundings outside the reef always show that the submarine prolongation of the land is gently inclined. In fact the reefs extend only to that distance from the shore, at which a foundation within the requisite depth from 20 to 30 fathoms is found. As far as the actual reef is concerned, there is no essential difference between it and that forming a barrier or an atoll: it is, however, generally of less width, and consequently few islets have been formed on it. From the corals growing more vigorously on the outside, and from the noxious effect of the sediment washed inwards, the outer edge of the reef is the highest part, and between it and the land there is generally a shallow sandy channel a few feet in depth. Where banks of sediment have accumulated near to the surface, as in parts of the West Indies, they sometimes become fringed with corals, and hence in some degree resemble lagoon-islands or atolls; in the same manner as fringing-reefs, surrounding gently-sloping islands, in some degree resemble barrier-reefs.

No theory on the formation of coral-reefs can be considered satisfactory which does not include the three great classes. We have seen that we are driven to believe in the subsidence of these vast areas, interspersed with low islands, of which not one rises above the height to which the wind and waves can throw up matter, and yet are constructed by animals requiring a foundation, and that foundation to lie at no great depth. Let us then take an island surrounded by fringing-reefs, which offer no difficulty in their structure; and let this island with its reef, represented by the unbroken lines in the woodcut overleaf, slowly subside. Now as the island sinks down, either a few feet at a time or quite insensibly, we may safely infer, from what is known of the conditions favourable to the growth of coral, that the living masses, bathed by the surf on the margin of the reef, will soon regain the surface. The water, however, will encroach little by little on the shore, the island becoming lower and smaller, and the space between the inner edge of the reef and the beach proportionally broader. A section of the reef and island in this state, after a subsidence of several hundred feet, is given by the dotted lines. Coral islets are supposed to have been formed on the reef; and a ship is anchored in the lagoon-channel. This channel will be more or less deep, according to the rate of subsidence, to the amount of sediment accumulated in it, and to the growth of the delicately branched corals which can live there. The section in this state resembles in every respect one drawn through an encircled island: in fact, it is a real section (on the scale of .517 of an inch to a mile) through Bolabola in the Pacific. We can now at once see why encircling barrier-reefs stand so far from the shores which they front. We can also perceive, that a line drawn perpendicularly down from the outer edge of the new reef, to the foundation of solid rock beneath the old fringing-reef, will exceed by as many feet as there have been feet of subsidence, that small limit of depth at which the effective corals can live: – the little architects having built up their great wall-like mass, as the whole sank down, upon a basis formed of other corals and their consolidated fragments. Thus the difficulty on this head, which appeared so great, disappears.

AA. Outer edges of the fringing-reef, at the level of the sea. BB. The shores of the fringed island.

A'A'. Outer edges of the reef, after its upward growth during a period of subsidence, now converted into a barrier, with islets on it. B'B'. The shores of the now encircled island. CC. Lagoon-channel.

N.B. In this and the following woodcut, the subsidence of the land could be represented only by an apparent rise in the level of the sea.

If, instead of an island, we had taken the shore of a continent fringed with reefs, and had imagined it to have subsided, a great straight barrier, like that of Australia or New Caledonia, separated from the land by a wide and deep channel, would evidently have been the result.

Let us take our new encircling barrier-reef, of which the section is now represented by unbroken lines, and which, as I have said, is a real section through Bolabola, and let it go on subsiding. As the barrier-reef slowly sinks down, the corals will go on vigorously growing upwards; but as the island sinks, the water will gain inch by inch on the shore – the separate mountains first forming separate islands within one great reef – and finally, the last and highest pinnacle disappearing. The instant this takes place, a perfect atoll is formed: I have said, remove the high land from within an encircling barrier-reef, and an atoll is left, and the land has been removed. We can now perceive how it comes that atolls, having sprung from encircling barrier-reefs, resemble them in general size, form, in the manner in which they are grouped together, and in their arrangement in single or double lines; for they may be called rude outline charts of the sunken islands over which they stand. We can further see how it arises that the atolls in the Pacific and Indian oceans extend in lines parallel to the generally prevailing strike of the high islands and the great coast-lines of those oceans. I venture, therefore, to affirm, that on the

A'A'. Outer edges of the barrier-reef at the level of the sea, with islets on it. B'B'. The shores of the included island. CC. The lagoon-channel.

A''A''. Outer edges of the reef, now converted into an atoll. C'. The lagoon of the new atoll.

N.B. According to the true scale, the depths of the lagoon-channel and lagoon are much exaggerated.

theory of the upward growth of the corals during the sinking of the land,[1] all the leading features in those wonderful structures, the lagoon-islands or atolls, which have so long excited the attention of voyagers, as well as in the no less wonderful barrier-reefs, whether encircling small islands or stretching for hundreds of miles along the shores of a continent, are simply explained.

It may be asked, whether I can offer any direct evidence of the subsidence of barrier-reefs or atolls; but it must be borne in mind how difficult it must ever be to detect a movement, the tendency of which is to hide under water the part affected. Nevertheless, at Keeling atoll I observed on all sides of the lagoon old cocoa-nut trees undermined and falling; and in one place the foundation-posts of a shed, which the inhabitants asserted had stood seven years before just above high-water mark, but now was daily washed by every tide: on inquiry I found that three earthquakes, one of them severe, had been felt here during the last ten years. At Vanikoro, the lagoon-channel is remarkably deep, scarcely any alluvial soil has accumulated at the foot of the lofty included mountains, and remarkably few islets have been formed by the heaping of fragments and sand on the wall-like barrier-reef; these facts, and some analogous ones, led me to believe that this island must lately have subsided and the reef grown upwards: here again earthquakes are frequent and very severe. In the Society archipelago, on the other hand, where the lagoon-channels are almost choked up, where much low alluvial land has accumulated, and where in some cases long islets have been formed on the barrier-reefs – facts all showing that the islands have not very lately subsided – only feeble shocks are most rarely felt. In these coral formations, where the land and water seem struggling for mastery, it must be ever difficult to decide between the effects of a change in the set of the tides and of a slight subsidence: that many of these reefs and atolls

1 It has been highly satisfactory to me to find the following passage in a pamphlet by Mr. Couthouy, one of the naturalists in the great Antarctic Expedition of the United States: – 'Having personally examined a large number of coral-islands, and resided eight months among the volcanic class having shore and partially encircling reefs, I may be permitted to state that my own observations have impressed a conviction of the correctness of the theory of Mr. Darwin.' – The naturalists, however, of this expedition differ with me on some points respecting coral formations.

are subject to changes of some kind is certain; on some atolls the islets appear to have increased greatly within a late period; on others they have been partially or wholly washed away. The inhabitants of parts of the Maldiva archipelago know the date of the first formation of some islets; in other parts, the corals are now flourishing on water-washed reefs, where holes made for graves attest the former existence of inhabited land. It is difficult to believe in frequent changes in the tidal currents of an open ocean; whereas, we have in the earthquakes recorded by the natives on some atolls, and in the great fissures observed on other atolls, plain evidence of changes and disturbances in progress in the subterranean regions.

It is evident, on our theory, that coasts merely fringed by reefs cannot have subsided to any perceptible amount; and therefore they must, since the growth of their corals, either have remained stationary or have been upheaved. Now it is remarkable how generally it can be shown, by the presence of upraised organic remains, that the fringed islands have been elevated: and so far, this is indirect evidence in favour of our theory. I was particularly struck with this fact, when I found to my surprise, that the descriptions given by MM. Quoy and Gaimard were applicable, not to reefs in general as implied by them, but only to those of the fringing-class; my surprise, however, ceased when I afterwards found that, by a strange chance, all the several islands visited by these eminent naturalists, could be shown by their own statements to have been elevated within a recent geological era.

Not only the grand features in the structure of barrier-reefs and of atolls, and of their likeness to each other in form, size, and other characters, are explained on the theory of subsidence – which theory we are independently forced to admit in the very areas in question, from the necessity of finding bases for the corals within the requisite depth – but many details in structure and exceptional cases can thus also be simply explained. I will give only a few instances. In barrier-reefs it has long been remarked with surprise, that the passages through the reef exactly face valleys in the included land, even in cases where the reef is separated from the land by a lagoon-channel so wide and so much deeper than the actual passage itself, that it seems hardly possible that the very small quantity of water or sediment brought down could injure the corals on the reef. Now, every reef of the fringing-class is breached by a narrow gateway in front of the smallest rivulet, even if dry during the greater part of the year, for the mud, sand, or gravel, occasionally washed down, kills the corals on which it is deposited. Consequently, when an island thus fringed subsides, though most of the narrow gateways will probably become closed by the outward and upward growth of the corals, yet any that are not closed (and some must always be kept open by the sediment and impure water flowing out of the lagoon-channel) will still continue to front exactly the upper parts of those valleys, at the mouths of which the original basal fringing-reef was breached.

We can easily see how an island fronted only on one side, or on one side with one end or both ends encircled by barrier-reefs, might after long-continued subsidence be converted either into a single wall-like reef, or into an atoll with a great straight spur projecting from it, or into two or three atolls tied together by straight reefs – all of which exceptional cases actually occur. As the reef-building corals require food, are preyed upon by other animals, are killed by sediment, cannot adhere to a loose bottom, and may be easily carried down to a depth whence they cannot spring up again, we need feel no surprise at the reefs both of atolls and barriers becoming in parts imperfect. The great barrier of New Caledonia is thus imperfect and broken in many parts; hence, after long subsidence, this great reef would not produce one great atoll 400 miles in length, but a chain or archipelago of atolls, of very nearly the same dimensions with those in the Maldiva archipelago. Moreover, in an atoll once breached on opposite sides, from the likelihood of the oceanic and tidal currents passing straight through the breaches, it is extremely improbable that the corals, especially during continued subsidence, would ever be able again to unite the rim; if they did not, as the whole sank downwards, one atoll would be divided into two or more. In the Maldiva archipelago there are distinct atolls so related to each other in position, and separated by channels either unfathomable or very deep (the channel between Ross and Ari atolls is 150 fathoms, and that between the north and south Nillandoo atolls is 200 fathoms in depth), that it is impossible to look at a map of them without believing that they were once more intimately related. And in this same archipelago, Mahlos-Mahdoo atoll is divided by a bifurcating channel from 100 to 132 fathoms in depth, in such a manner, that it is scarcely possible to say whether it ought strictly to be called three separate atolls, or one great atoll not yet finally divided.

I will not enter on many more details; but I must remark that the curious structure of the northern Maldiva atolls receives (taking into consideration the free entrance of the sea through their broken margins) a simple explanation in the upward and outward growth of the corals, originally based both on small detached reefs in their lagoons, such as occur in common atolls, and on broken portions of the linear marginal reef, such as bounds every atoll of the ordinary form. I cannot refrain from once again remarking on the singularity of these complex structures – a great sandy and generally concave disk rises abruptly from the unfathomable ocean, with its central expanse studded, and its edge symmetrically bordered with oval basins of coral-rock just lipping the surface of the sea, sometimes clothed with vegetation, and each containing a lake of clear water!

One more point in detail: as in two neighbouring archipelagoes corals flourish in one and not in the other, and as so many conditions before enumerated must affect their existence, it would be an inexplicable fact if, during the changes to which earth, air, and water are subjected, the reef-building corals were to keep alive for perpetuity on any one spot or area. And as by our theory the areas including atolls and barrier-reefs are subsiding, we ought occasionally to find reefs both dead and submerged. In all reefs, owing to the sediment being washed out of the lagoon or lagoon-channel to leeward, that side is least favourable to the long-continued vigorous growth of the corals; hence dead portions of reef not unfrequently occur on the leeward side; and these, though still retaining their proper wall-like form, are now in several instances sunk several fathoms beneath the surface. The Chagos group appears from some cause, possibly from the subsidence having been too rapid, at present to be much less favourably circumstanced for the growth of reefs than formerly: one atoll has a portion of its marginal reef, nine miles in length, dead and submerged; a second has only a few quite small living points which rise to the surface; a third and fourth are entirely dead and submerged; a fifth is a mere wreck, with its structure almost obliterated. It is remarkable that in all these cases, the dead reefs and portions of reef lie at nearly the same depth, namely, from six to eight fathoms beneath the surface, as if they had been carried down by one uniform movement. One of these 'half-drowned atolls,' so called by Capt. Moresby (to whom I am indebted for much invaluable information), is of vast size, namely, ninety nautical miles across in one direction, and seventy miles in another line; and is in many respects eminently curious. As by our theory it follows that new atolls will generally be formed in each new area of subsidence, two weighty objections might have been raised, namely, that atolls must be increasing indefinitely in number; and secondly, that in old areas of subsidence each separate atoll must be increasing indefinitely in thickness, if proofs of their occasional destruction could not have been adduced. Thus have we traced the history of these great rings of coral-rock, from their first origin through their normal changes, and through the occasional accidents of their existence, to their death and final obliteration.

In my volume on 'Coral Formations' I have published a map, in which I have coloured all the atolls dark-blue, the barrier-reefs pale-blue, and the fringing-reefs red. These latter reefs have been formed whilst the land has been stationary, or, as appears from the frequent presence of upraised organic remains, whilst it has been slowly rising: atolls and barrier-reefs, on the other hand, have grown up during the directly opposite movement of subsidence, which movement must have been very gradual, and in the case of atolls so vast in amount as to have buried every mountain-summit over wide ocean-spaces. Now in this map we see that the reefs tinted pale- and dark-blue, which have been produced by the same order of movement, as a general rule manifestly stand near each other. Again we see, that the areas with the two blue tints are of wide extent; and that they lie separate from extensive lines of coast coloured red, both of which circumstances might naturally have been inferred, on the theory of the nature of the reefs having been governed by the nature of the earth's movement. It deserves notice, that in more than one instance where single red and blue circles approach near each other, I can show that there have been oscillations of level; for in such cases the red or fringed circles consist of atolls, originally by our theory formed during subsidence, but subsequently upheaved; and on the other hand, some of the pale-blue or encircled islands are composed of coral-rock, which must have been uplifted to its present height before that subsidence took place, during which the existing barrier-reefs grew upwards.

Authors have noticed with surprise, that although atolls are the commonest coral-structures throughout some enormous oceanic tracts, they are entirely absent in other seas, as in the West Indies: we can now at once perceive the cause, for where there has not been subsidence, atolls cannot have been formed; and in the case of the West Indies and parts of the East Indies, these tracts are known to have been rising within the recent period. The larger areas, coloured red and blue, are all elongated; and between the two colours there is a degree of rude alternation, as if the rising of one had balanced the sinking of the other. Taking into consideration the proofs of recent elevation both on the fringed coasts and on some others (for instance, in South America) where there are no reefs, we are led to conclude that the great continents are for the most part rising areas; and from the nature of the coral-reefs, that the central parts of the great oceans are sinking areas. The East Indian archipelago, the most broken land in the world, is in most parts an area of elevation, but surrounded and penetrated, probably in more lines than one, by narrow areas of subsidence.

I have marked with vermilion spots all the many known active volcanoes within the limits of this same map. Their entire absence from every one of the great subsiding areas, coloured either pale or dark blue, is most striking; and not less so is the coincidence of the chief volcanic chains with the parts coloured red, which we are led to conclude have either long remained stationary, or more generally have been recently upraised. Although a few of the vermilion spots occur within no great distance of single circles tinted blue, yet not one single active volcano is situated within several hundred miles of an archipelago, or even small group of atolls. It is, therefore, a striking fact that in the Friendly Archipelago, which consists of a group of atolls upheaved and since partially worn down, two volcanoes, and perhaps more, are historically known to have been in action. On the other hand, although most of the islands in the Pacific which are encircled by barrier-reefs, are of volcanic origin, often with the remnants of craters still distinguishable, not one of them is known to have ever been in eruption. Hence in these cases it would appear, that volcanoes burst forth into action and become extinguished on the same spots, according as elevatory or subsiding movements prevail there. Numberless facts could be adduced to prove that upraised organic remains are common wherever there are active volcanoes; but until it could be shown that in areas of subsidence, volcanoes were either absent or inactive, the inference, however probable in itself, that their distribution depended on the rising or falling of the earth's surface, would have been hazardous. But now, I think, we may freely admit this important deduction.

Taking a final view of the map, and bearing in mind the statement made with respect to the upraised organic remains, we must feel astonished at the vastness of the areas, which have suffered changes in level either downwards or upwards, within a period not geologically remote. It would appear, also, that the elevatory and subsiding movements follow nearly the same laws. Throughout the spaces interspersed with atolls, where not a single peak of high land has been left above the level of the sea, the sinking must have been immense in amount. The sinking, moreover, whether continuous, or recurrent with intervals sufficiently long for the corals again to bring up their living edifices to the surface, must necessarily have been extremely slow. This conclusion is probably the most important one which can be deduced from the study of coral formations; – and it is one which it is difficult to imagine, how otherwise could ever have been arrived at. Nor can I quite pass over the probability of the former existence of large archipelagoes of lofty islands, where now only rings of coral-rock scarcely break the open expanse of the sea, throwing some light on the distribution of the inhabitants of the other high islands, now left standing so immensely remote from each other in the midst of the great oceans. The reef-constructing corals have indeed reared and preserved wonderful memorials of the subterranean oscillations of level; we see in each barrier-reef a proof that the land has there subsided, and in each atoll a monument over an island now lost. We may thus, like unto a geologist who had lived his ten thousand years and kept a record of the passing changes, gain some insight into the great system by which the surface of this globe has been broken up, and land and water interchanged.

CHAPTER XXI

MAURITIUS TO ENGLAND

Mauritius, beautiful appearance of – Great crateriform
Ring of Mountains – Hindoos – St. Helena – History of
the Changes in the Vegetation – Cause of the Extinction of
Land-shells – Ascension – Variation in the Imported Rats –
Volcanic Bombs – Beds of Infusoria – Bahia – Brazil –
Splendour of Tropical Scenery – Pernambuco – Singular
Reef – Slavery – Return to England – Retrospect on our
Voyage.

APRIL 29th. – In the morning we passed round the northern end of
Mauritius, or the Isle of France. From this point of view the aspect
of the island equalled the expectations raised by the many well-
known descriptions of its beautiful scenery. The sloping plain of
the Pamplemousses, interspersed with houses, and coloured by the
large fields of sugar-cane of a bright green, composed the foreground.
The brilliancy of the green was the more remarkable, because it is a
colour which generally is conspicuous only from a very short distance. Towards the centre of the island groups of wooded mountains
rose out of this highly-cultivated plain; their summits, as so commonly happens with ancient volcanic rocks, being jagged into the
sharpest points. Masses of white clouds were collected around these
pinnacles, as if for the sake of pleasing the stranger's eye. The whole
island, with its sloping border and central mountains, was adorned
with an air of perfect elegance: the scenery, if I may use such an
expression, appeared to the sight harmonious.

I spent the greater part of the next day in walking about the
town, and visiting different people. The town is of considerable
size, and is said to contain 20,000 inhabitants; the streets are very
clean and regular. Although the island has been so many years
under the English government, the general character of the place is
quite French: Englishmen speak to their servants in French, and
the shops are all French; indeed I should think that Calais or

Boulogne was much more Anglified. There is a very pretty little theatre, in which operas are excellently performed. We were also surprised at seeing large booksellers' shops, with well-stored shelves; – music and reading bespeak our approach to the old world of civilization; for in truth both Australia and America are new worlds.

The various races of men walking in the streets afford the most interesting spectacle in Port Louis. Convicts from India are banished here for life; at present there are about 800, and they are employed in various public works. Before seeing these people, I had no idea that the inhabitants of India were such noble-looking figures. Their skin is extremely dark, and many of the older men had large mustaches and beards of a snow-white colour; this, together with the fire of their expression, gave them quite an imposing aspect. The greater number had been banished for murder and the worst crimes; others for causes which can scarcely be considered as moral faults, such as for not obeying, from superstitious motives, the English laws. These men are generally quiet and well conducted; from their outward conduct, their cleanliness, and faithful observance of their strange religious rites, it was impossible to look at them with the same eyes as on our wretched convicts in New South Wales.

May 1st. – Sunday. I took a quiet walk along the sea-coast to the north of the town. The plain in this part is quite uncultivated; it consists of a field of black lava, smoothed over with coarse grass and bushes, the latter being chiefly Mimosas. The scenery may be described as intermediate in character between that of the Galapagos and of Tahiti; but this will convey a definite idea to very few persons. It is a very pleasant country, but it has not the charms of Tahiti, or the grandeur of Brazil. The next day I ascended La Pouce, a mountain so called from a thumb-like projection, which rises close behind the town to a height of 2600 feet. The centre of the island consists of a great platform, surrounded by old broken basaltic mountains, with their strata dipping seawards. The central platform, formed of comparatively recent streams of lava, is of an oval shape, thirteen geographical miles across, in the line of its shorter axis. The exterior bounding mountains come into that class of structures called Craters of Elevation, which are supposed to have been formed not like ordinary craters, but by a great and sudden upheaval. There appear to me to be insuperable objections to this view: on the other hand, I can hardly believe, in this and in some other cases, that these marginal crateriform mountains are merely the basal remnants of immense volcanoes, of which the summits either have been blown off, or swallowed up in subterranean abysses.

From our elevated position we enjoyed an excellent view over the island. The country on this side appears pretty well cultivated, being divided into fields and studded with farm-houses. I was however assured that of the whole land, not more than half is yet in a productive state; if such be the case, considering the present large export of sugar, this island, at some future period when thickly peopled, will be of great value. Since England has taken possession of it, a period of only twenty-five years, the export of sugar is said to have increased seventy-five fold. One great cause of its prosperity is the excellent state of the roads. In the neighbouring Isle of Bourbon, which remains under the French government, the roads are still in the same miserable state as they were here only a few years ago. Although the French residents must have largely profited by the increased prosperity of their island, yet the English government is far from popular.

3rd. – In the evening Captain Lloyd, the Surveyor-general so well known from his examination of the Isthmus of Panama, invited Mr. Stokes and myself to his country-house, which is situated on the edge of Wilhelm Plains, and about six miles from the Port. We staid at this delightful place two days; standing nearly 800 feet above the sea, the air was cool and fresh, and on every side there were delightful walks. Close by, a grand ravine has been worn to a depth of about 500 feet through the slightly inclined streams of lava, which have flowed from the central platform.

5th. – Captain Lloyd took us to the Rivière Noire, which is several miles to the southward, that I might examine some rocks of elevated coral. We passed through pleasant gardens, and fine fields of sugar-cane growing amidst huge blocks of lava. The roads were bordered by hedges of Mimosa, and near many of the houses there were avenues of the mango. Some of the views, where the peaked hills and the cultivated farms were seen together, were exceedingly picturesque; and we were constantly tempted to exclaim, 'How pleasant it would be to pass one's life in such quiet abodes!' Captain Lloyd possessed an elephant, and he sent it halfway with us, that we might enjoy a ride in true Indian fashion. The circumstance which surprised me most was its quite noiseless step. This elephant is the only one at present on the island; but it is said others will be sent for.

May 9*th.* – We sailed from Port Louis, and, calling at the Cape of Good Hope, on the 8th of July we arrived off St. Helena. This island, the forbidding aspect of which has been so often described, rises abruptly like a huge black castle from the ocean. Near the town, as if to complete nature's defence, small forts and guns fill up every gap in the rugged rocks. The town runs up a flat and narrow valley; the houses look respectable, and are interspersed with a very few green trees. When approaching the anchorage there was one striking view: an irregular castle perched on the summit of a lofty hill, and surrounded by a few scattered fir-trees, boldly projected against the sky.

The next day I obtained lodgings within a stone's throw of Napoleon's tomb:[1] it was a capital central situation, whence I could make excursions in every direction. During the four days I staid here, I wandered over the island from morning to night, and examined its geological history. My lodgings were situated at a height of about 2000 feet; here the weather was cold and boisterous, with constant showers of rain; and every now and then the whole scene was veiled in thick clouds.

Near the coast the rough lava is quite bare: in the central and higher parts, feldspathic rocks by their decomposition have produced a clayey soil, which, where not covered by vegetation, is stained in broad bands of many bright colours. At this season, the land moistened by constant showers, produces a singularly bright green pasture, which lower and lower down, gradually fades away and at last

1 After the volumes of eloquence which have poured forth on this subject, it is dangerous even to mention the tomb. A modern traveller, in twelve lines, burdens the poor little island with the following titles, – it is a grave, tomb, pyramid, cemetery, sepulchre, catacomb, sarcophagus, minaret, and mausoleum!

disappears. In latitude 16°, and at the trifling elevation of 1500 feet, it is surprising to behold a vegetation possessing a character decidedly British. The hills are crowned with irregular plantations of Scotch firs; and the sloping banks are thickly scattered over with thickets of gorse, covered with its bright yellow flowers. Weeping-willows are common on the banks of the rivulets, and the hedges are made of the blackberry, producing its well-known fruit. When we consider that the number of plants now found on the island is 746, and that out of these fifty-two alone are indigenous species, the rest having been imported, and most of them from England, we see the reason of the British character of the vegetation. Many of these English plants appear to flourish better than in their native country; some also from the opposite quarter of Australia succeed remarkably well. The many imported species must have destroyed some of the native kinds; and it is only on the highest and steepest ridges, that the indigenous Flora is now predominant.

The English, or rather Welsh character of the scenery, is kept up by the numerous cottages and small white houses; some buried at the bottom of the deepest valleys, and others mounted on the crests of the lofty hills. Some of the views are striking, for instance that from near Sir W. Doveton's house, where the bold peak called Lot is seen over a dark wood of firs, the whole being backed by the red water-worn mountains of the southern coast. On viewing the island from an eminence, the first circumstance which strikes one, is the number of the roads and forts: the labour bestowed on the public works, if one forgets its character as a prison, seems out of all proportion to its extent or value. There is so little level or useful land, that it seems surprising how so many people, about 5000, can subsist here. The lower orders, or the emancipated slaves, are I believe extremely poor: they complain of the want of work. From the reduction in the number of public servants, owing to the island having been given up by the East India Company, and the consequent emigration of many of the richer people, the poverty probably will increase. The chief food of the working class is rice with a little salt meat; as neither of these articles are the products of the island, but must be purchased with money, the low wages tell heavily on the poor people. Now that the people are blessed with freedom, a right which I believe they value fully, it seems probable that their numbers will quickly increase: if so, what is to become of the little state of St. Helena?

My guide was an elderly man, who had been a goatherd when a boy, and knew every step amongst the rocks. He was of a race many times crossed, and although with a dusky skin, he had not the disagreeable expression of a mulatto. He was a very civil, quiet old man, and such appears the character of the greater number of the lower classes. It was strange to my ears to hear a man, nearly white and respectably dressed, talking with indifference of the times when he was a slave. With my companion, who carried our dinners and a horn of water, which is quite necessary, as all the water in the lower valleys is saline, I every day took long walks.

Beneath the upper and central green circle, the wild valleys are quite desolate and untenanted. Here, to the geologist, there were scenes of high interest, showing successive changes and complicated disturbances. According to my views, St. Helena has existed as an island from a very remote epoch: some obscure proofs, however, of the elevation of the land are still extant. I believe that the central and highest peaks form parts of the rim of a great crater, the southern half of which has been entirely removed by the waves of the sea: there is, moreover, an external wall of black basaltic rocks, like the coast-mountains of Mauritius, which are older than the central volcanic streams. On the higher parts of the island, considerable numbers of a shell, long thought a marine species, occur embedded in the soil. It proves to be a Cochlogena, or land-shell of a very peculiar form;[1] with it I found six other kinds; and in another spot an eighth species. It is remarkable that none of them are now found living. Their extinction has probably been caused by the entire destruction of the woods, and the consequent loss of food and shelter, which occurred during the early part of the last century.

The history of the changes, which the elevated plains of Longwood and Deadwood have undergone, as given in General Beatson's account of the island, is extremely curious. Both plains, it is said, in former times were covered with wood, and were therefore called the

---

1 It deserves notice, that all the many specimens of this shell found by me in one spot differ, as a marked variety, from another set of specimens procured from a different spot.

Great Wood. So late as the year 1716 there were many trees, but in 1724 the old trees had mostly fallen; and as goats and hogs had been suffered to range about, all the young trees had been killed. It appears also from the official records, that the trees were unexpectedly, some years afterwards, succeeded by a wire grass, which spread over the whole surface.[1] General Beatson adds that now this plain 'is covered with fine sward, and is become the finest piece of pasture on the island.' The extent of surface, probably covered by wood at a former period, is estimated at no less than two thousand acres; at the present day scarcely a single tree can be found there. It is also said that in 1709 there were quantities of dead trees in Sandy Bay; this place is now so utterly desert, that nothing but so well attested an account could have made me believe that they could ever have grown there. The fact, that the goats and hogs destroyed all the young trees as they sprang up, and that in the course of time the old ones, which were safe from their attacks, perished from age, seems clearly made out. Goats were introduced in the year 1502; eighty-six years afterwards, in the time of Cavendish, it is known that they were exceedingly numerous. More than a century afterwards, in 1731, when the evil was complete and irretrievable, an order was issued that all stray animals should be destroyed. It is very interesting thus to find, that the arrival of animals at St. Helena in 1501, did not change the whole aspect of the island, until a period of two hundred and twenty years had elapsed: for the goats were introduced in 1502, and in 1724 it is said 'the old trees had mostly fallen.' There can be little doubt that this great change in the vegetation affected not only the land-shells, causing eight species to become extinct, but likewise a multitude of insects.

St. Helena, situated so remote from any continent, in the midst of a great ocean, and possessing a unique Flora, excites our curiosity. The eight land-shells, though now extinct, and one living Succinea, are peculiar species found nowhere else. Mr. Cuming, however, informs me that an English Helix is common here, its eggs no doubt having been imported in some of the many introduced plants. Mr. Cuming collected on the coast sixteen species of sea-shells, of which seven, as far as he knows, are confined to this island. Birds and

---

1 Beatson's St. Helena. Introductory chapter, p. 4.

insects,[1] as might have been expected, are very few in number; indeed I believe all the birds have been introduced within late years. Partridges and pheasants are tolerably abundant: the island is much too English not to be subject to strict game-laws. I was told of a more unjust sacrifice to such ordinances than I ever heard of even in England. The poor people formerly used to burn a plant, which grows on the coast-rocks, and export the soda from its ashes; but a peremptory order came out prohibiting this practice, and giving as a reason that the partridges would have nowhere to build!

In my walks I passed more than once over the grassy plain, bounded by deep valleys, on which Longwood stands. Viewed from a short distance, it appears like a respectable gentleman's

1 Among these few insects, I was surprised to find a small Aphodius (*nov. spec.*) and an Oryctes, both extremely numerous under dung. When the island was discovered it certainly possessed no quadruped, excepting *perhaps* a mouse: it becomes, therefore, a difficult point to ascertain, whether these stercovorous insects have since been imported by accident, or if aborigines, on what food they formerly subsisted. On the banks of the Plata, where, from the vast number of cattle and horses, the fine plains of turf are richly manured, it is vain to seek the many kinds of dung-feeding beetles, which occur so abundantly in Europe. I observed only an Oryctes (the insects of this genus in Europe generally feed on decayed vegetable matter) and two species of Phanæus, common in such situations. On the opposite side of the Cordillera in Chiloe, another species of Phanæus is exceedingly abundant, and it buries the dung of the cattle in large earthen balls beneath the ground. There is reason to believe that the genus Phanæus, before the introduction of cattle, acted as scavengers to man. In Europe, beetles, which find support in the matter which has already contributed towards the life of other and larger animals, are so numerous, that there must be considerably more than one hundred different species. Considering this, and observing what a quantity of food of this kind is lost on the plains of La Plata, I imagined I saw an instance where man had disturbed that chain, by which so many animals are linked together in their native country. In Van Diemen's Land, however, I found four species of Onthophagus, two of Aphodius, and one of the third genus, very abundant under the dung of cows; yet these latter animals had been then introduced only thirty-three years. Previously to that time, the Kangaroo and some other small animals were the only quadrupeds; and their dung is of a very different quality from that of their successors introduced by man. In England the greater number of stercovorous beetles are confined in their appetites; that is, they do not depend indifferently on any quadruped for the means of subsistence. The change, therefore, in habits, which must have taken place in Van Diemen's Land, is highly remarkable. I am indebted to the Rev. F. W. Hope, who, I hope, will permit me to call him my master in Entomology, for giving me the names of the foregoing insects.

country-seat. In front there are a few cultivated fields, and beyond them the smooth hill of coloured rocks called the Flagstaff, and the rugged square black mass of the Barn. On the whole the view was rather bleak and uninteresting. The only inconvenience I suffered during my walks was from the impetuous winds. One day I noticed a curious circumstance: standing on the edge of a plain, terminated by a great cliff of about a thousand feet in depth, I saw at the distance of a few yards right to windward, some tern, struggling against a very strong breeze, whilst, where I stood, the air was quite calm. Approaching close to the brink, where the current seemed to be deflected upwards from the face of the cliff, I stretched out my arm, and immediately felt the full force of the wind: an invisible barrier, two yards in width, separated perfectly calm air from a strong blast.

I so much enjoyed my rambles among the rocks and mountains of St. Helena, that I felt almost sorry on the morning of the 14th to descend to the town. Before noon I was on board, and the *Beagle* made sail.

On the 19th of July we reached Ascension. Those who have beheld a volcanic island, situated under an arid climate, will at once be able to picture to themselves the appearance of Ascension. They will imagine smooth conical hills of a bright red colour, with their summits generally truncated, rising separately out of a level surface of black rugged lava. A principal mound in the centre of the island, seems the father of the lesser cones. It is called Green Hill; its name being taken from the faintest tinge of that colour, which at this time of the year is barely perceptible from the anchorage. To complete the desolate scene, the black rocks on the coast are lashed by a wild and turbulent sea.

The settlement is near the beach; it consists of several houses and barracks placed irregularly, but well built of white freestone. The only inhabitants are marines, and some negroes liberated from slave-ships, who are paid and victualled by government. There is not a private person on the island. Many of the marines appeared well contented with their situation; they think it better to serve their one-and-twenty years on shore, let it be what it may, than in a ship; in this choice, if I were a marine, I should most heartily agree.

The next morning I ascended Green Hill, 2840 feet high, and thence walked across the island to the windward point. A good cart-road leads from the coast-settlement to the houses, gardens, and fields, placed near the summit of the central mountain. On the roadside there are milestones, and likewise cisterns, where each thirsty passer-by can drink some good water. Similar care is displayed in each part of the establishment, and especially in the management of the springs, so that a single drop of water may not be lost: indeed the whole island may be compared to a huge ship kept in first-rate order. I could not help, when admiring the active industry which had created such effects out of such means, at the same time regretting that it had been wasted on so poor and trifling an end. M. Lesson has remarked with justice, that the English nation alone would have thought of making the island of Ascension a productive spot; any other people would have held it as a mere fortress in the ocean.

Near this coast nothing grows; further inland, an occasional green castor-oil plant, and a few grasshoppers, true friends of the desert, may be met with. Some grass is scattered over the surface of the central elevated region, and the whole much resembles the worse parts of the Welsh mountains. But scanty as the pasture appears, about six hundred sheep, many goats, a few cows and horses, all thrive well on it. Of native animals, land-crabs and rats swarm in numbers. Whether the rat is really indigenous, may well be doubted; there are two varieties as described by Mr. Waterhouse; one is of a black colour, with fine glossy fur, and lives on the grassy summit; the other is brown-coloured and less glossy, with longer hairs, and lives near the settlement on the coast. Both these varieties are one-third smaller than the common black rat (M. rattus); and they differ from it both in the colour and character of their fur, but in no other essential respect. I can hardly doubt that these rats (like the common mouse, which has also run wild) have been imported, and, as at the Galapagos, have varied from the effect of the new conditions to which they have been exposed: hence the variety on the summit of the island differs from that on the coast. Of native birds there are none; but the guinea-fowl, imported from the Cape de Verd Islands, is abundant, and the common fowl has likewise run wild. Some cats, which were origin-ally turned out to destroy the rats and mice, have increased, so as to become a great plague. The island is entirely without trees, in which, and in every other respect, it is very far inferior to St. Helena.

One of my excursions took me towards the S.W. extremity of the island. The day was clear and hot, and I saw the island, not smiling with beauty, but staring with naked hideousness. The lava streams are covered with hummocks, and are rugged to a degree which, geologically speaking, is not of easy explanation. The intervening spaces are concealed with layers of pumice, ashes, and volcanic tuff. Whilst passing this end of the island at sea, I could not imagine what the white patches were with which the whole plain was mottled; I now found that they were sea-fowl, sleeping in such full confidence, that even in mid-day a man could walk up and seize hold of them. These birds were the only living creatures I saw during the whole day. On the beach a great surf, although the breeze was light, came tumbling over the broken lava rocks.

The geology of this island is in many respects interesting. In several places I noticed volcanic bombs, that is, masses of lava which have been shot through the air whilst fluid, and have consequently assumed a spherical or pear-shape. Not only their external form, but, in several cases, their internal structure shows in a very curious manner that they have revolved in their aërial course. The internal structure of one of these bombs, when broken, is represented very accurately in the woodcut on the previous page. The central part is coarsely cellular, the cells decreasing in size towards the exterior; where there is a shell-like case about the third of an inch in thickness, of compact stone, which again is overlaid by the outside crust of finely cellular lava. I think there can be little doubt, first, that the external crust cooled rapidly in the state in which we now see it; secondly, that the still fluid lava within, was packed by the centrifugal force, generated by the revolving of the bomb, against the external cooled crust, and so produced the solid shell of stone; and lastly, that the centrifugal force, by relieving the pressure in the more central parts of the bomb, allowed the heated vapours to expand their cells, thus forming the coarsely cellular mass of the centre.

A hill, formed of the older series of volcanic rocks, and which has been incorrectly considered as the crater of a volcano, is remarkable from its broad, slightly hollowed, and circular summit having been filled up with many successive layers of ashes and fine scoriæ. These saucer-shaped layers crop out on the margin, forming perfect rings of many different colours, giving to the summit a most fantastic appearance; one of these rings is white and broad, and resembles a course round which horses have been exercised; hence the hill has been called the Devil's Riding School. I brought away specimens of one of the tufaceous layers of a pinkish colour; and it is a most extraordinary fact, that Professor Ehrenberg[1] finds it almost wholly composed of matter which has been organized: he detects in it some siliceous-shielded, fresh-water infusoria, and no less than twenty-five different kinds of the siliceous tissue of plants, chiefly of grasses. From the absence of all carbonaceous matter, Professor Ehrenberg believes that these organic bodies have passed through the volcanic fire, and have been erupted in the state in which we now see them. The appearance of the layers induced me to believe that they had been deposited under water, though from the extreme dryness of the climate I was forced to imagine, that torrents of rain had probably fallen during some great eruption, and that thus a temporary lake had been formed, into which the ashes fell. But it may now be suspected that the lake

1 Monats. der König. Akad. d. Wiss. zu Berlin. Vom April, 1845.

was not a temporary one. Anyhow, we may feel sure, that at some former epoch, the climate and productions of Ascension were very different from what they now are. Where on the face of the earth can we find a spot, on which close investigation will not discover signs of that endless cycle of change, to which this earth has been, is, and will be subjected?

On leaving Ascension we sailed for Bahia, on the coast of Brazil, in order to complete the chronometrical measurement of the world. We arrived there on August 1st, and stayed four days, during which I took several long walks. I was glad to find my enjoyment in tropical scenery had not decreased from the want of novelty, even in the slightest degree. The elements of the scenery are so simple, that they are worth mentioning, as a proof on what trifling circumstances exquisite natural beauty depends.

The country may be described as a level plain of about three hundred feet in elevation, which in all parts has been worn into flat-bottomed valleys. This structure is remarkable in a granitic land, but is nearly universal in all those softer formations of which plains are usually composed. The whole surface is covered by various kinds of stately trees, interspersed with patches of cultivated ground, out of which houses, convents, and chapels arise. It must be remembered that within the tropics, the wild luxuriance of nature is not lost even in the vicinity of large cities; for the natural vegetation of the hedges and hill-sides overpowers in picturesque effect the artificial labour of man. Hence, there are only a few spots where the bright red soil affords a strong contrast with the universal clothing of green. From the edges of the plain there are distant views either of the ocean, or of the great Bay with its low-wooded shores, and on which numerous boats and canoes show their white sails. Excepting from these points, the scene is extremely limited; following the level pathways, on each hand, only glimpses into the wooded valleys below can be obtained. The houses, I may add, and especially the sacred edifices, are built in a peculiar and rather fantastic style of architecture. They are all whitewashed; so that when illumined by the brilliant sun of mid-day, and as seen against the pale-blue sky of the horizon, they stand out more like shadows than real buildings.

Such are the elements of the scenery, but it is a hopeless attempt to paint the general effect. Learned naturalists describe these scenes of the tropics by naming a multitude of objects, and mentioning some characteristic feature of each. To a learned traveller this possibly may communicate some definite ideas: but who else from seeing a plant in an herbarium can imagine its appearance when growing in its native soil? Who from seeing choice plants in a hothouse, can magnify some into the dimensions of forest trees, and crowd others into an entangled jungle? Who when examining in the cabinet of the entomologist the gay exotic butterflies, and singular cicadas, will associate with these lifeless objects, the ceaseless harsh music of the latter, and the lazy flight of the former, – the sure accompaniments of the still, glowing noonday of the tropics? It is when the sun has attained its greatest height, that such scenes should be viewed: then the dense splendid foliage of the mango hides the ground with its darkest shade, whilst the upper branches are rendered from the profusion of light of the most brilliant green. In the temperate zones the case is different – the vegetation there is not so dark or so rich, and hence the rays of the declining sun, tinged of a red, purple, or bright yellow colour, add most to the beauties of those climes.

When quietly walking along the shady pathways, and admiring each successive view, I wished to find language to express my ideas. Epithet after epithet was found too weak to convey to those who have not visited the intertropical regions, the sensation of delight which the mind experiences. I have said that the plants in a hothouse fail to communicate a just idea of the vegetation, yet I must recur to it. The land is one great wild, untidy, luxuriant hothouse, made by Nature for herself, but taken possession of by man, who has studded it with gay houses and formal gardens. How great would be the desire in every admirer of nature to behold, if such were possible, the scenery of another planet! yet to every person in Europe, it may be truly said, that at the distance of only a few degrees from his native soil, the glories of another world are opened to him. In my last walk I stopped again and again to gaze on these beauties, and endeavoured to fix in my mind for ever, an impression which at the time I knew sooner or later must fail. The form of the orange-tree, the cocoa-nut, the palm, the mango, the tree-fern, the banana, will remain clear and separate; but the thousand beauties which unite these into one perfect scene must fade away; yet they will leave, like a tale heard in childhood, a picture full of indistinct, but most beautiful figures.

August 6th. – In the afternoon we stood out to sea, with the intention of making a direct course to the Cape de Verd Islands. Unfavourable winds, however, delayed us, and on the 12th we ran into Pernambuco, – a large city on the coast of Brazil, in latitude 8° south. We anchored outside the reef; but in a short time a pilot came on board and took us into the inner harbour, where we lay close to the town.

Pernambuco is built on some narrow and low sand-banks, which are separated from each other by shoal channels of salt water. The three parts of the town are connected together by two long bridges built on wooden piles. The town is in all parts disgusting, the streets being narrow, ill-paved, and filthy; the houses, tall and gloomy. The season of heavy rains had hardly come to an end, and hence the surrounding country, which is scarcely raised above the level of the sea, was flooded with water; and I failed in all my attempts to take long walks.

The flat swampy land on which Pernambuco stands is surrounded, at the distance of a few miles, by a semicircle of low hills, or rather by the edge of a country elevated perhaps two hundred feet above the sea. The old city of Olinda stands on one extremity of this range. One day I took a canoe, and proceeded up one of the channels to visit it; I found the old town from its situation both sweeter and cleaner than that of Pernambuco. I must here commemorate what happened for the first time during our nearly five years' wandering, namely, having met with a want of politeness: I was refused in a sullen manner at two different houses, and obtained with difficulty from a third, permission to pass through their gardens to an uncultivated hill, for the purpose of viewing the country. I feel glad that this happened in the land of the Brazilians, for I bear them no good will – a land also of slavery, and therefore of moral debasement. A Spaniard would have felt ashamed at the very thought of refusing such a request, or of behaving to a stranger with rudeness. The channel by which we went to and returned from Olinda, was bordered on each side by mangroves, which sprang like a miniature forest out of the greasy mud-banks. The bright green colour of these bushes always reminded me of the rank grass in a churchyard: both are nourished by putrid exhalations; the one speaks of death past, and the other too often of death to come.

The most curious object which I saw in this neighbourhood, was the reef that forms the harbour. I doubt whether in the whole world any other natural structure has so artificial an appearance.[1] It runs for a length of several miles in an absolutely straight line, parallel to, and not far distant from, the shore. It varies in width from thirty to sixty yards, and its surface is level and smooth; it is composed of obscurely-stratified hard sandstone. At high water the waves break over it; at low water its summit is left dry, and it might then be mistaken for a breakwater erected by Cyclopean workmen. On this coast the currents of the sea tend to throw up in front of the land, long spits and bars of loose sand, and on one of these, part of the town of Pernambuco stands. In former times a long spit of this nature seems to have become consolidated by the percolation of calcareous matter, and afterwards to have been gradually upheaved; the outer and loose parts during this process having been worn away by the action of the sea, and the solid nucleus left as we now see it. Although night and day the waves of the open Atlantic, turbid with sediment, are driven against the steep outside edges of this wall of stone, yet the oldest pilots know of no tradition of any change in its appearance. This durability is much the most curious fact in its history: it is due to a tough layer, a few inches thick, of calcareous matter, wholly formed by the successive growth and death of the small shells of Serpulæ, together with some few barnacles and nulliporæ. These nulliporæ, which are hard, very simply-organized sea-plants, play an analogous and important part in protecting the upper surfaces of coral-reefs, behind and within the breakers, where the true corals, during the outward growth of the mass, become killed by exposure to the sun and air. These insignificant organic beings, especially the Serpulæ, have done good service to the people of Pernambuco; for without their protective aid the bar of sandstone would inevitably have been long ago worn away, and without the bar, there would have been no harbour.

On the 19th of August we finally left the shores of Brazil. I thank God, I shall never again visit a slave-country. To this day, if I hear a

1 I have described this Bar in detail, in the Lond. and Edin. Phil. Mag., vol. xix. (1841) p. 257.

distant scream, it recalls with painful vividness my feelings, when passing a house near Pernambuco, I heard the most pitiable moans, and could not but suspect that some poor slave was being tortured, yet knew that I was as powerless as a child even to remonstrate. I suspected that these moans were from a tortured slave, for I was told that this was the case in another instance. Near Rio de Janeiro I lived opposite to an old lady, who kept screws to crush the fingers of her female slaves. I have staid in a house where a young household mulatto, daily and hourly, was reviled, beaten, and persecuted enough to break the spirit of the lowest animal. I have seen a little boy, six or seven years old, struck thrice with a horse-whip (before I could interfere) on his naked head, for having handed me a glass of water not quite clean; I saw his father tremble at a mere glance from his master's eye. These latter cruelties were witnessed by me in a Spanish colony, in which it has always been said, that slaves are better treated than by the Portuguese, English, or other European nations. I have seen at Rio Janeiro a powerful negro afraid to ward off a blow directed, as he thought, at his face. I was present when a kind-hearted man was on the point of separating for ever the men, women, and little children of a large number of families who had long lived together. I will not even allude to the many heart-sickening atrocities which I authentically heard of; – nor would I have mentioned the above revolting details, had I not met with several people, so blinded by the constitutional gaiety of the negro, as to speak of slavery as a tolerable evil. Such people have generally visited at the houses of the upper classes, where the domestic slaves are usually well treated; and they have not, like myself, lived amongst the lower classes. Such enquirers will ask slaves about their condition; they forget that the slave must indeed be dull, who does not calculate on the chance of his answer reaching his master's ears.

It is argued that self-interest will prevent excessive cruelty; as if self-interest protected our domestic animals, which are far less likely than degraded slaves, to stir up the rage of their savage masters. It is an argument long since protested against with noble feeling, and strikingly exemplified, by the ever illustrious Humboldt. It is often attempted to palliate slavery by comparing the state of slaves with our poorer countrymen: if the misery of our poor be caused not by the laws of nature, but by our institutions, great is our sin; but how this bears on slavery, I cannot see; as well might the use of the thumbscrew be defended in one land, by showing that men in another land suffered from some dreadful disease. Those who look tenderly at the slave-owner, and with a cold heart at the slave, never seem to put themselves into the position of the latter; – what a cheerless prospect, with not even a hope of change! picture to yourself the chance, ever hanging over you, of your wife and your little children – those objects which nature urges even the slave to call his own – being torn from you and sold like beasts to the first bidder! And these deeds are done and palliated by men, who profess to love their neighbours as themselves, who believe in God, and pray that his Will be done on earth! It makes one's blood boil, yet heart tremble, to think that we Englishmen and our American descendants, with their boastful cry of liberty, have been and are so guilty: but it is a consolation to reflect, that we at least have made a greater sacrifice, than ever made by any nation, to expiate our sin.

On the last day of August we anchored for the second time at Porto Praya in the Cape de Verd archipelago; thence we proceeded to the Azores, where we staid six days. On the 2nd of October we made the shores of England; and at Falmouth I left the *Beagle*, having lived on board the good little vessel nearly five years.

Our Voyage having come to an end, I will take a short retrospect of the advantages and disadvantages, the pains and pleasures, of our circumnavigation of the world. If a person asked my advice, before undertaking a long voyage, my answer would depend upon his possessing a decided taste for some branch of knowledge, which could by this means be advanced. No doubt it is a high satisfaction to behold various countries and the many races of mankind, but the pleasures gained at the time do not counterbalance the evils. It is necessary to look forward to a harvest, however distant that may be, when some fruit will be reaped, some good effected.

Many of the losses which must be experienced are obvious; such as that of the society of every old friend, and of the sight of those places with which every dearest remembrance is so intimately connected.

These losses, however, are at the time partly relieved by the exhaust-less delight of anticipating the long wished-for day of return. If, as poets say, life is a dream, I am sure in a voyage these are the visions which best serve to pass away the long night. Other losses, although not at first felt, tell heavily after a period: these are the want of room, of seclusion, of rest; the jading feeling of constant hurry; the privation of small luxuries, the loss of domestic society, and even of music and the other pleasures of imagination. When such trifles are mentioned, it is evident that the real grievances, excepting from accidents, of a sea-life are at an end. The short space of sixty years has made an astonishing difference in the facility of distant navigation. Even in the time of Cook, a man who left his fireside for such expeditions underwent severe privations. A yacht now, with every luxury of life, can circumnavigate the globe. Besides the vast improvements in ships and naval resources, the whole western shores of America are thrown open, and Australia has become the capital of a rising continent. How different are the circumstances to a man ship-wrecked at the present day in the Pacific, to what they were in the time of Cook! Since his voyage a hemisphere has been added to the civilized world.

If a person suffer much from sea-sickness, let him weigh it heavily in the balance. I speak from experience: it is no trifling evil, cured in a week. If, on the other hand, he take pleasure in naval tactics, he will assuredly have full scope for his taste. But it must be borne in mind, how large a proportion of the time, during a long voyage, is spent on the water, as compared with the days in harbour. And what are the boasted glories of the illimitable ocean? A tedious waste, a desert of water, as the Arabian calls it. No doubt there are some delightful scenes. A moon-light night, with the clear heavens and the dark glittering sea, and the white sails filled by the soft air of a gently-blowing trade-wind; a dead calm, with the heaving surface polished like a mirror, and all still except the occasional flapping of the canvas. It is well once to behold a squall with its rising arch and coming fury, or the heavy gale of wind and mountainous waves. I confess, how-ever, my imagination had painted something more grand, more terrific in the full-grown storm. It is an incomparably finer spectacle when beheld on shore, where the waving trees, the wild flight of the birds, the dark shadows and bright lights, the rushing of the torrents, all proclaim the strife of the unloosed elements. At sea the albatross and little petrel fly as if the storm were their proper sphere, the water rises and sinks as if fulfilling its usual task, the ship alone and its inhabitants seem the objects of wrath. On a forlorn and weather-beaten coast, the scene is indeed different, but the feelings partake more of horror than of wild delight.

Let us now look at the brighter side of the past time. The pleasure derived from beholding the scenery and the general aspect of the various countries we have visited, has decidedly been the most constant and highest source of enjoyment. It is probable that the picturesque beauty of many parts of Europe exceeds anything which we beheld. But there is a growing pleasure in comparing the character of the scenery in different countries, which to a certain degree is distinct from merely admiring its beauty. It depends chiefly on an acquaintance with the individual parts of each view: I am strongly induced to believe that, as in music, the person who under-stands every note will, if he also possesses a proper taste, more thoroughly enjoy the whole, so he who examines each part of a fine view, may also thoroughly comprehend the full and combined effect. Hence, a traveller should be a botanist, for in all views plants form the chief embellishment. Group masses of naked rock even in the wildest forms, and they may for a time afford a sublime spectacle, but they will soon grow monotonous. Paint them with bright and varied colours, as in Northern Chile, they will become fantastic; clothe them with vegetation, they must form a decent, if not a beautiful picture.

When I say that the scenery of parts of Europe is probably superior to anything which we beheld, I except, as a class by itself, that of the intertropical zones. The two classes cannot be compared together; but I have already often enlarged on the grandeur of those regions. As the force of impressions generally depends on preconceived ideas, I may add, that mine were taken from the vivid descriptions in the Personal Narrative of Humboldt, which far exceed in merit anything else which I have read. Yet with these high-wrought ideas, my feelings were far from partaking of a tinge of disappointment on my first and final landing on the shores of Brazil.

Among the scenes which are deeply impressed on my mind, none exceed in sublimity the primeval forests undefaced by the hand of man; whether those of Brazil, where the powers of Life are predominant, or those of Tierra del Fuego, where Death and Decay prevail. Both are temples filled with the varied productions of the God of Nature: — no one can stand in these solitudes unmoved, and not feel that there is more in man than the mere breath of his body. In calling up images of the past, I find that the plains of Patagonia frequently cross before my eyes; yet these plains are pronounced by all wretched and useless. They can be described only by negative characters; without habitations, without water, without trees, without mountains, they support merely a few dwarf plants. Why then, and the case is not peculiar to myself, have these arid wastes taken so firm a hold on my memory? Why have not the still more level, the greener and more fertile Pampas, which are serviceable to mankind, produced an equal impression? I can scarcely analyze these feelings: but it must be partly owing to the free scope given to the imagination. The plains of Patagonia are boundless, for they are scarcely passable, and hence unknown: they bear the stamp of having lasted, as they are now, for ages, and there appears no limit to their duration through future time. If, as the ancients supposed, the flat earth was surrounded by an impassable breadth of water, or by deserts heated to an intolerable excess, who would not look at these last boundaries to man's knowledge with deep but ill-defined sensations?

Lastly, of natural scenery, the views from lofty mountains, though certainly in one sense not beautiful, are very memorable. When looking down from the highest crest of the Cordillera, the mind, undisturbed by minute details, was filled with the stupendous dimensions of the surrounding masses.

Of individual objects, perhaps nothing is more certain to create astonishment than the first sight in his native haunt of a barbarian, – of man in his lowest and most savage state. One's mind hurries back over past centuries, and then asks, could our progenitors have been men like these? – men, whose very signs and expressions are less intelligible to us than those of the domesticated animals; men, who do not possess the instinct of those animals, nor yet appear to boast of human reason, or at least of arts consequent on that reason. I do not believe it is possible to describe or paint the difference between savage and civilized man. It is the difference between a wild and tame animal: and part of the interest in beholding a savage, is the same which would lead every one to desire to see the lion in his desert, the tiger tearing his prey in the jungle, or the rhinoceros wandering over the wild plains of Africa.

Among the other most remarkable spectacles which we have beheld, may be ranked the Southern Cross, the cloud of Magellan, and the other constellations of the southern hemisphere – the water-spout – the glacier leading its blue stream of ice, overhanging the sea in a bold precipice – a lagoon-island raised by the reef-building corals – an active volcano – and the overwhelming effects of a violent earthquake. These latter phenomena, perhaps, possess for me a peculiar interest, from their intimate connexion with the geological structure of the world. The earthquake, however, must be to every one a most impressive event: the earth, considered from our earliest childhood as the type of solidity, has oscillated like a thin crust beneath our feet; and in seeing the laboured works of man in a moment overthrown, we feel the insignificance of his boasted power.

It has been said, that the love of the chase is an inherent delight in man – a relic of an instinctive passion. If so, I am sure the pleasure of living in the open air, with the sky for a roof and the ground for a table, is part of the same feeling; it is the savage returning to his wild and native habits. I always look back to our boat cruises, and my land journeys, when through unfrequented countries, with an extreme delight, which no scenes of civilization could have created. I do not doubt that every traveller must remember the glowing sense of happiness which he experienced, when he first breathed in a foreign clime, where the civilized man had seldom or never trod.

There are several other sources of enjoyment in a long voyage, which are of a more reasonable nature. The map of the world ceases to be a blank; it becomes a picture full of the most varied and animated figures. Each part assumes its proper dimensions: continents are not looked at in the light of islands, or islands considered as mere specks, which are, in truth, larger than many kingdoms of Europe. Africa, or North and South America, are well-sounding names, and easily pronounced; but it is not until having sailed for weeks along small portions of their shores, that one is thoroughly convinced what vast spaces on our immense world these names imply.

From seeing the present state, it is impossible not to look forward with high expectations to the future progress of nearly an entire hemisphere. The march of improvement, consequent on the introduction of Christianity throughout the South Sea, probably stands by itself in the records of history. It is the more striking when we remember that only sixty years since, Cook, whose excellent judgment none will dispute, could foresee no prospect of a change. Yet these changes have now been effected by the philanthropic spirit of the British nation.

In the same quarter of the globe Australia is rising, or indeed may be said to have risen, into a grand centre of civilization, which, at some not very remote period, will rule as empress over the southern hemisphere. It is impossible for an Englishman to behold these distant colonies, without a high pride and satisfaction. To hoist the British flag, seems to draw with it as a certain consequence, wealth, prosperity, and civilization.

In conclusion, it appears to me that nothing can be more improving to a young naturalist, than a journey in distant countries. It both sharpens, and partly allays that want and craving, which, as Sir J. Herschel remarks, a man experiences although every corporeal sense be fully satisfied. The excitement from the novelty of objects, and the chance of success, stimulate him to increased activity. Moreover, as a number of isolated facts soon become uninteresting, the habit of comparison leads to generalization. On the other hand, as the traveller stays but a short time in each place, his descriptions must generally consist of mere sketches, instead of detailed observations. Hence arises, as I have found to my cost, a constant tendency to fill up the wide gaps of knowledge, by inaccurate and superficial hypotheses.

But I have too deeply enjoyed the voyage, not to recommend any naturalist, although he must not expect to be so fortunate in his companions as I have been, to take all chances, and to start, on travels by land if possible, if otherwise on a long voyage. He may feel assured, he will meet with no difficulties or dangers, excepting in rare cases, nearly so bad as he beforehand anticipates. In a moral point of view, the effect ought to be, to teach him good-humoured patience, freedom from selfishness, the habit of acting for himself, and of making the best of every occurrence. In short, he ought to partake of the characteristic qualities of most sailors. Travelling ought also to teach him distrust; but at the same time he will discover, how many truly kind-hearted people there are, with whom he never before had, or ever again will have any further communication, who yet are ready to offer him the most disinterested assistance.

NOTE. – The snake, described at page 111, with the curious habit of vibrating its tail, is a new species of Trigonocephalus, which M. Bibron proposes to call T. crepitans.

ON THE ORIGIN OF SPECIES

BY MEANS OF NATURAL SELECTION

OR

THE PRESERVATION OF FAVOURED RACES IN THE STRUGGLE FOR LIFE

BUT with regard to the material world, we can at least go so far as this – we can perceive that events are brought about not by insulated interpositions of Divine power, exerted in each particular case, but by the establishment of general laws.

W. WHEWELL: *Bridgewater Treatise*

TO conclude, therefore, let no man out of a weak conceit of sobriety, or an ill-applied moderation, think or maintain, that a man can search too far or be too well studied in the book of God's word, or in the book of God's works; divinity or philosophy; but rather let men endeavour an endless progress or proficience in both.

BACON: *Advancement of Learning*

CONTENTS

CHAPTER I
VARIATION UNDER
DOMESTICATION   543

CHAPTER II
VARIATION UNDER NATURE   572

AN HISTORICAL SKETCH

OF THE PROGRESS OF OPINION ON
THE ORIGIN OF SPECIES

PREVIOUSLY TO THE PUBLICATION OF THE
FIRST EDITION OF THIS WORK

I WILL here give a brief sketch of the progress of opinion on the Origin of Species. Until recently the great majority of naturalists believed that species were immutable productions, and had been separately created. This view has been ably maintained by many authors. Some few naturalists, on the other hand, have believed that species undergo modification, and that the existing forms of life are the descendants by true generation of pre-existing forms. Passing over allusions to the subject in the classical writers,[*] the first author who in modern times has treated it in a scientific spirit was Buffon. But as his opinions fluctuated greatly at different periods, and as he does not enter on the causes or means of the transformation of species, I need not here enter on details.

---

[*]Aristotle, in his 'Physicae Auscultationes' (lib. 2, cap. 8, s. 2), after remarking that rain does not fall in order to make the corn grow, any more than it falls to spoil the farmer's corn when threshed out of doors, applies the same argument to organization: and adds (as translated by Mr. Clair Grece, who first pointed out the passage to me), 'So what hinders the different parts [of the body] from having this merely accidental relation in nature? as the teeth, for example, grow by necessity, the front ones sharp, adapted for dividing, and the grinders flat, and serviceable for masticating the food; since they were not made for the sake of this, but it was the result of accident. And in like manner as to the other parts in which there appears to exist an adaptation to an end. Wheresoever, therefore, all things together (that is all the parts of one whole) happened like as if they were made for the sake of something, these were preserved, having been appropriately constituted by an internal spontaneity; and whatsoever things were not thus constituted, perished, and still perish.' We here see the principle of natural selection shadowed forth, but how little Aristotle fully comprehended the principle, is shown by his remarks on the formation of the teeth.

527

Lamarck was the first man whose conclusions on the subject excited much attention. This justly-celebrated naturalist first published his views in 1801; he much enlarged them in 1809 in his 'Philosophie Zoologique,' and subsequently, in 1815, in the Introduction to his 'Hist. Nat. des Animaux sans Vertèbres.' In these works he upholds the doctrine that species, including man, are descended from other species. He first did the eminent service of arousing attention to the probability of all change in the organic, as well as in the inorganic world, being the result of law, and not of miraculous interposition. Lamarck seems to have been chiefly led to his conclusion on the gradual change of species, by the difficulty of distinguishing species and varieties, by the almost perfect gradation of forms in certain groups, and by the analogy of domestic productions. With respect to the means of modification, he attributed something to the direct action of the physical conditions of life, something to the crossing of already existing forms, and much to use and disuse, that is, to the effects of habit. To this latter agency he seemed to attribute all the beautiful adaptations in nature; – such as the long neck of the giraffe for browsing on the branches of trees. But he likewise believed in a law of progressive development; and as all the forms of life thus tend to progress, in order to account for the existence at the present day of simple productions, he maintains that such forms are now spontaneously generated.[*]

[*] I have taken the date of the first publication of Lamarck from Isid. Geoffroy Saint-Hilaire's ('Hist. Nat. Générale,' tom. ii. p. 405, 1859) excellent history of opinion on this subject. In this work a full account is given of Buffon's conclusions on the same subject. It is curious how largely my grandfather, Dr. Erasmus Darwin, anticipated the views and erroneous grounds of opinion of Lamarck in his 'Zoonomia' (vol. i. pp. 500–510), published in 1794. According to Isid. Geoffroy there is no doubt that Goethe was an extreme partisan of similar views, as shown in the Introduction to a work written in 1794 and 1795, but not published till long afterwards: he has pointedly remarked ('Goethe als Naturforscher,' von Dr. Karl Meding, s. 34) that the future question for naturalists will be how, for instance, cattle got their horns, and not for what they are used. It is rather a singular instance of the manner in which similar views arise at about the same time, that Goethe in Germany, Dr. Darwin in England, and Geoffroy Saint-Hilaire (as we shall immediately see) in France, came to the same conclusion on the origin of species, in the years 1794–5.

Geoffroy Saint-Hilaire, as is stated in his 'Life,' written by his son, suspected, as early as 1795, that what we call species are various degenerations of the same type. It was not until 1828 that he published his conviction that the same forms have not been perpetuated since the origin of all things. Geoffroy seems to have relied chiefly on the conditions of life, or the '*monde ambiant*' as the cause of change. He was cautious in drawing conclusions, and did not believe that existing species are now undergoing modification; and, as his son adds, 'C'est donc un problème à réserver entièrement à l'avenir, supposé même que l'avenir doive avoir prise sur lui.'

In 1813, Dr. W. C. Wells read before the Royal Society 'An Account of a White female, part of whose skin resembled that of a Negro;' but his paper was not published until his famous 'Two Essays upon Dew and Single Vision' appeared in 1818. In this paper he distinctly recognizes the principle of natural selection, and this is the first recognition which has been indicated; but he applies it only to the races of man, and to certain characters alone. After remarking that negroes and mulattoes enjoy an immunity from certain tropical diseases, he observes, firstly, that all animals tend to vary in some degree, and, secondly, that agriculturists improve their domesticated animals by selection; and then, he adds, but what is done in this latter case 'by art, seems to be done with equal efficacy, though more slowly, by nature, in the formation of varieties of mankind, fitted for the country which they inhabit. Of the accidental varieties of man, which would occur among the first few and scattered inhabitants of the middle regions of Africa, some one would be better fitted than the others to bear the diseases of the country. This race would consequently multiply, while the others would decrease; not only from their inability to sustain the attacks of disease, but from their incapacity of contending with their more vigorous neighbours. The colour of this vigorous race I take for granted, from what has been already said, would be dark. But the same disposition to form varieties still existing, a darker and a darker race would in the course of time occur: and as the darkest would be the best fitted for the climate, this would at length become the most prevalent, if not the only race, in the particular country in which it had originated.' He then extends these same views to the white inhabitants of colder climates. I am indebted to Mr. Rowley, of the United States, for having called my attention, through Mr. Brace, to the above passage in Dr. Wells' work.

The Hon. and Rev. W. Herbert, afterwards Dean of Manchester, in the fourth volume of the 'Horticultural Transactions,' 1822, and in his work on the 'Amaryllidaceae' (1837, pp. 19, 339), declares that 'horticultural experiments have established, beyond the possibility of refutation, that botanical species are only a higher and more permanent class of varieties.' He extends the same view to animals. The Dean believes that single species of each genus were created in an originally highly plastic condition, and that these have produced, chiefly by intercrossing, but likewise by variation, all our existing species.

In 1826 Professor Grant, in the concluding paragraph in his well-known paper ('Edinburgh Philosophical Journal,' vol. xiv. p. 283) on the Spongilla, clearly declares his belief that species are descended from other species, and that they become improved in the course of modification. This same view was given in his 55th Lecture, published in the 'Lancet' in 1834.

In 1831 Mr. Patrick Matthew published his work on 'Naval Timber and Arboriculture,' in which he gives precisely the same view on the origin of species as that (presently to be alluded to) propounded by Mr. Wallace and myself in the 'Linnean Journal,' and as that enlarged in the present volume. Unfortunately the view was given by Mr. Matthew very briefly in scattered passages in an Appendix to a work on a different subject, so that it remained unnoticed until Mr. Matthew himself drew attention to it in the 'Gardener's Chronicle,' on April 7th, 1860. The differences of Mr. Matthew's view from mine are not of much importance; he seems to consider that the world was nearly depopulated at successive periods, and then re-stocked; and he gives as an alternative, that new forms may be generated 'without the presence of any mould or germ of former aggregates.' I am not sure that I understand some passages; but it seems that he attributes much influence to the direct action of the conditions of life. He clearly saw, however, the full force of the principle of natural selection.

The celebrated geologist and naturalist, Von Buch, in his excellent 'Description Physique des Isles Canaries' (1836, p. 147), clearly expresses his belief that varieties slowly become changed into permanent species, which are no longer capable of intercrossing.

Rafinesque, in his 'New Flora of North America,' published in 1836, wrote (p. 6) as follows: – 'All species might have been varieties once, and many varieties are gradually becoming species by assuming constant and peculiar characters;' but farther on (p. 18) he adds, 'except the original types or ancestors of the genus.'

In 1843–44 Professor Haldeman ('Boston Journal of Nat. Hist. U. States', vol. iv. p. 468) has ably given the arguments for and against the hypothesis of the development and modification of species: he seems to lean towards the side of change.

The 'Vestiges of Creation' appeared in 1844. In the tenth and much improved edition (1853) the anonymous author says (p. 155): – 'The proposition determined on after much consideration is, that the several series of animated beings, from the simplest and oldest up to the highest and most recent, are, under the providence of God, the results, *first*, of an impulse which has been imparted to the forms of life, advancing them, in definite times, by generation, through grades of organisation terminating in the highest dicotyledons and vertebrata, these grades being few in number, and generally marked by intervals of organic character, which we find to be a practical difficulty in ascertaining affinities; *second*, of another impulse connected with the vital forces, tending, in the course of generations, to modify organic structures in accordance with external circumstances, as food, the nature of the habitat, and the meteoric agencies, these being the "adaptations" of the natural theologian.' The author apparently believes that organization progresses by sudden leaps, but that the effects produced by the conditions of life are gradual. He argues with much force on general grounds that species are not immutable productions. But I cannot see how the two supposed 'impulses' account in a scientific sense for the numerous and beautiful co-adaptations which we see throughout nature; I cannot see that we thus gain any insight how, for instance, a woodpecker has become adapted to its peculiar habits of life. The work, from its powerful and brilliant style, though displaying in the earlier editions little accurate knowledge and a great want of scientific caution, immediately had a very wide circulation. In my opinion it has done excellent service in this country in calling attention to the subject, in removing prejudice, and in thus preparing the ground for the reception of analogous views.

In 1846 the veteran geologist M. J. d'Omalius d'Halloy published in an excellent though short paper ('Bulletins de l'Acad. Roy Bruxelles,' tom. xiii. p. 581) his opinion that it is more probable that new species have been produced by descent with modification than that they have been separately created: the author first promulgated this opinion in 1831.

Professor Owen, in 1849 ('Nature of Limbs,' p. 86), wrote as follows: – 'The archetypal idea was manifested in the flesh under diverse such modifications, upon this planet, long prior to the existence of those animal species that actually exemplify it. To what natural laws or secondary causes the orderly succession and progression of such organic phenomena may have been committed, we, as yet, are ignorant.' In his Address to the British Association, in 1858, he speaks (p. li.) of 'the axiom of the continuous operation of creative power, or of the ordained becoming of living things.' Farther on (p. xc.), after referring to geographical distribution, he adds, 'These phenomena shake our confidence in the conclusion that the Apteryx of New Zealand and the Red Grouse of England were distinct creations in and for those islands respectively. Always, also, it may be well to bear in mind that by the word "creation" the zoologist means "a process he knows not what."' He amplifies this idea by adding that when such cases as that of the Red Grouse are 'enumerated by the zoologists as evidence of distinct creation of the bird in and for such islands, he chiefly expresses that he knows not how the Red Grouse came to be there, and there exclusively; signifying also, by this mode of expressing such ignorance, his belief that both the bird and the islands owed their origin to a great first Creative Cause.' If we interpret these sentences given in the same Address, one by the other, it appears that this eminent philosopher felt in 1858 his confidence shaken that the Apteryx and the Red Grouse first appeared in their respective homes, 'he knew not how,' or by some process 'he knew not what.'

This Address was delivered after the papers by Mr. Wallace and myself on the Origin of Species, presently to be referred to, had been read before the Linnean Society. When the first edition of this work was published, I was so completely deceived, as were many others, by such expressions as 'the continuous operation of creative power,' that I included Professor Owen with other palaeontologists as being firmly convinced of the immutability of species; but it appears ('Anat. of Vertebrates,' vol. iii. p. 796) that this was on my part a preposterous error. In the last edition of this work I inferred, and the inference still seems to me perfectly just, from a passage beginning with the words 'no doubt the type-form,' &c. (Ibid. vol. i. p. xxxv.), that Professor Owen admitted that natural selection may have done something in the formation of a new species; but this it appears (Ibid. vol. iii. p. 798) is inaccurate and without evidence. I also gave some extracts from a correspondence between Professor Owen and the Editor of the 'London Review,' from which it appeared manifest to the Editor as well as to myself, that Professor Owen claimed to have promulgated the theory of natural selection before I had done so; and I expressed my surprise and satisfaction at this announcement; but as far as it is possible to understand certain recently published passages (Ibid. vol. iii. p. 798) I have either partially or wholly again fallen into error. It is consolatory to me that others find Professor Owen's controversial writings as difficult to understand and to reconcile with each other, as I do. As far as the mere enunciation of the principle of natural selection is concerned, it is quite immaterial whether or not Professor Owen preceded me, for both of us, as shown in this historical sketch, were long ago preceded by Dr. Wells and Mr. Matthews.

M. Isidore Geoffroy Saint-Hilaire, in his lectures delivered in 1850 (of which a Résumé appeared in the 'Revue et Mag. de Zoolog.,' Jan. 1851), briefly gives his reason for believing that specific characters 'sont fixés, pour chaque espèce, tant qu'elle se perpétue au milieu des mêmes circonstances: ils se modifient, si les circonstances ambiantes viennent à changer.' 'En résumé, *l'observation* des animaux sauvages démontre déjà la variabilité *limitée* des espèces. Les *expériences* sur les animaux sauvages devenus domestiques, et sur les animaux domestiques redevenus sauvages, la démontrent plus clairement encore. Ces mêmes expériences prouvent, de plus, que les différences produites peuvent etre de *valeur générique.*' In his 'Hist. Nat. Générale' (tom. ii. p. 430, 1859) he amplifies analogous conclusions.

From a circular lately issued it appears that Dr. Freke, in 1851 ('Dublin Medical Press,' p. 322), propounded the doctrine that all organic beings have descended from one primordial form. His grounds of belief and treatment of the subject are wholly different from mine; but as Dr. Freke has now (1861) published his Essay on the 'Origin of Species by means of Organic Affinity,' the difficult attempt to give any idea of his views would be superfluous on my part.

Mr. Herbert Spencer, in an Essay (originally published in the 'Leader,' March, 1852, and republished in his 'Essays,' in 1858), has contrasted the theories of the Creation and the Development of organic beings with remarkable skill and force. He argues from the analogy of domestic productions, from the changes which the embryos of many species undergo, from the difficulty of distinguishing species and varieties, and from the principle of general gradation, that species have been modified; and he attributes the modification to the change of circumstances. The author (1855) has also treated Psychology on the principle of the necessary acquirement of each mental power and capacity by gradation.

In 1852 M. Naudin, a distinguished botanist, expressly stated, in an admirable paper on the Origin of Species ('Revue Horticole,' p. 102; since partly republished in the 'Nouvelles Archives du Muséum,' tom. i. p. 171), his belief that species are formed in an analogous manner as varieties are under cultivation; and the latter process he attributes to man's power of selection. But he does not show how selection acts under nature. He believes, like Dean Herbert, that species, when nascent, were more plastic than at present. He lays weight on what he calls the principle of finality, 'puissance mystérieuse, indéterminée; fatalité pour les uns; pour les autres, volonté providentielle, dont l'action incessante sur les êtres vivants détermine, à toutes les époques de l'existence du monde, la forme, le volume, et la durée de chacun d'eux, en raison de sa destinée dans l'ordre de choses dont il fait partie. C'est cette puissance qui harmonise chaque membre à l'ensemble, en l'appropriant à la fonction qu'il doit remplir dans l'organisme général de la nature, fonction qui est pour lui sa raison d'être.'*

In 1853 a celebrated geologist, Count Keyserling ('Bulletin de la Soc. Géolog.,' 2nd Ser., tom. x. p. 357), suggested that as new diseases, supposed to have been caused by some miasma, have arisen and spread over the world, so at certain periods the germs of existing species may have been chemically affected by circumambient molecules of a particular nature, and thus have given rise to new forms.

In this same year, 1853, Dr. Schaaffhausen published an excellent pamphlet ('Verhand. des Naturhist. Vereins der Preuss. Rheinlands,' &c.), in which he maintains the development of organic forms on the earth. He infers that many species have kept true for long periods, whereas a few have become modified. The distinction of species he explains by the destruction of intermediate graduated forms. 'Thus living plants and animals are not separated from the extinct by new creations, but are to be regarded as their descendants through continued reproduction.'

A well-known French botanist, M. Lecoq, writes in 1854 ('Etudes sur Géograph. Bot.,' tom. i. p. 250), 'On voit que nos recherches sur la fixité ou la variation de l'espèce, nous conduisent directement aux idées émises, par deux hommes justement célèbres, Geoffroy Saint-Hilaire et Goethe.' Some other passages scattered through M. Lecoq's large work, make it a little doubtful how far he extends his views on the modification of species.

The 'Philosophy of Creation' has been treated in a masterly manner by the Rev. Baden Powell, in his 'Essays on the Unity of Worlds,' 1855. Nothing can be more striking than the manner in

---

*From references in Bronn's 'Untersuchungen über die Entwicke-lungs-Gesetze,' it appears that the celebrated botanist and palaeontologist Unger published, in 1852, his belief that species undergo development and modification. Dalton, likewise, in Pander and Dalton's work on Fossil Sloths, expressed, in 1821, a similar belief. Similar views have, as is well known, been maintained by Oken in his mystical 'Natur-Philosophie.' From other references in Godron's work 'Sur l'Espèce,' it seems that Bory St. Vincent, Burdach, Poiret, and Fries, have all admitted that new species are continually being produced.

I may add, that of the thirty-four authors named in this Historical Sketch, who believe in the modification of species, or at least disbelieve in separate acts of creation, twenty-seven have written on special branches of natural history or geology.

which he shows that the introduction of new species is 'a regular, not a casual phenomenon,' or, as Sir John Herschel expresses it, 'a natural in contradistinction to a miraculous process.'

The third volume of the 'Journal of the Linnean Society' contains papers, read July 1st, 1858, by Mr. Wallace and myself, in which, as stated in the introductory remarks to this volume, the theory of Natural Selection is promulgated by Mr. Wallace with admirable force and clearness.

Von Baer, towards whom all zoologists feel so profound a respect, expressed about the year 1859 (see Prof. Rudolph Wagner, a 'Zoologisch-Anthropologische Untersuchungen,' 1861, s. 51) his conviction, chiefly grounded on the laws of geographical distribution, that forms now perfectly distinct have descended from a single parent-form.

In June, 1859, Professor Huxley gave a lecture before the Royal Institution on the 'Persistent Types of Animal Life.' Referring to such cases, he remarks, 'It is difficult to comprehend the meaning of such facts as these, if we suppose that each species of animal and plant, or each great type of organisation, was formed and placed upon the surface of the globe at long intervals by a distinct act of creative power; and it is well to recollect that such an assumption is as unsupported by tradition or revelation as it is opposed to the general analogy of nature. If, on the other hand, we view "Persistent Types" in relation to that hypothesis which supposes the species living at any time to be the result of the gradual modification of pre-existing species, a hypothesis which, though unproven, and sadly damaged by some of its supporters, is yet the only one to which physiology lends any countenance; their existence would seem to show that the amount of modification which living beings have undergone during geological time is but very small in relation to the whole series of changes which they have suffered.'

In December, 1859, Dr. Hooker published his 'Introduction to the Australian Flora.' In the first part of this great work he admits the truth of the descent and modification of species, and supports this doctrine by many original observations.

The first edition of this work was published on November 24th, 1859, and the second edition on January 7th, 1860.

AUTHOR'S INTRODUCTION

WHEN on board H.M.S. *Beagle*, as naturalist, I was much struck with certain facts in the distribution of the inhabitants of South America, and in the geological relations of the present to the past inhabitants of that continent. These facts seemed to me to throw some light on the origin of species – that mystery of mysteries, as it has been called by one of our greatest philosophers. On my return home, it occurred to me, in 1837, that something might perhaps be made out on this question by patiently accumulating and reflecting on all sorts of facts which could possibly have any bearing on it. After five years' work I allowed myself to speculate on the subject, and drew up some short notes; these I enlarged in 1844 into a sketch of the conclusions, which then seemed to me probable: from that period to the present day I have steadily pursued the same object. I hope that I may be excused for entering on these personal details, as I give them to show that I have not been hasty in coming to a decision.

My work is now nearly finished; but as it will take me two or three more years to complete it, and as my health is far from strong, I have been urged to publish this Abstract. I have more especially been induced to do this, as Mr. Wallace, who is now studying the natural history of the Malay archipelago, has arrived at almost exactly the same general conclusions that I have on the origin of species. Last year he sent to me a memoir on this subject, with a request that I would forward it to Sir Charles Lyell, who sent it to the Linnean Society, and it is published in the third volume of the Journal of that Society. Sir C. Lyell and Dr. Hooker, who both knew of my work – the latter having read my sketch of 1844 – honoured me by thinking it advisable to publish, with Mr. Wallace's excellent memoir, some brief extracts from my manuscripts.

This Abstract, which I now publish, must necessarily be imperfect. I cannot here give references and authorities for my several statements; and I must trust to the reader reposing some confidence in my accuracy. No doubt errors will have crept in, though I hope

I have always been cautious in trusting to good authorities alone. I can here give only the general conclusions at which I have arrived, with a few facts in illustration, but which, I hope, in most cases will suffice. No one can feel more sensible than I do of the necessity of hereafter publishing in detail all the facts, with references, on which my conclusions have been grounded; and I hope in a future work to do this. For I am well aware that scarcely a single point is discussed in this volume on which facts cannot be adduced, often apparently leading to conclusions directly opposite to those at which I have arrived. A fair result can be obtained only by fully stating and balancing the facts and arguments on both sides of each question; and this cannot possibly be here done.

I much regret that want of space prevents my having the satisfaction of acknowledging the generous assistance which I have received from very many naturalists, some of them personally unknown to me. I cannot, however, let this opportunity pass without expressing my deep obligations to Dr. Hooker, who for the last fifteen years has aided me in every possible way by his large stores of knowledge and his excellent judgement.

In considering the Origin of Species, it is quite conceivable that a naturalist, reflecting on the mutual affinities of organic beings, on their embryological relations, their geographical distribution, geological succession, and other such facts, might come to the conclusion that each species had not been independently created, but had descended, like varieties, from other species. Nevertheless, such a conclusion, even if well founded, would be unsatisfactory, until it could be shown how the innumerable species inhabiting this world have been modified, so as to acquire that perfection of structure and coadaptation which most justly excites our admiration. Naturalists continually refer to external conditions, such as climate, food, &c., as the only possible cause of variation. In one very limited sense, as we shall hereafter see, this may be true; but it is preposterous to attribute to mere external conditions, the structure, for instance, of the woodpecker, with its feet, tail, beak, and tongue, so admirably adapted to catch insects under the bark of trees. In the case of the misseltoe, which draws its nourishment from certain trees, which has seeds that must be transported by certain birds, and which has flowers with separate sexes absolutely requiring the agency of certain insects to bring pollen from one flower to the other, it is equally preposterous to account for the structure of this parasite, with its relations to several distinct organic beings, by the effects of external conditions, or of habit, or of the volition of the plant itself.

The author of the 'Vestiges of Creation' would, I presume, say that, after a certain unknown number of generations, some bird had given birth to a woodpecker, and some plant to the misseltoe, and that these had been produced perfect as we now see them; but this assumption seems to me to be no explanation, for it leaves the case of the coadaptations of organic beings to each other and to their physical conditions of life, untouched and unexplained.

It is, therefore, of the highest importance to gain a clear insight into the means of modification and coadaptation. At the commencement of my observations it seemed to me probable that a careful study of domesticated animals and of cultivated plants would offer the best chance of making out this obscure problem. Nor have I been disappointed; in this and in all other perplexing cases I have invariably found that our knowledge, imperfect thought it be, of variation under domestication, afforded the best and safest clue. I may venture to express my conviction of the high value of such studies, although they have been very commonly neglected by naturalists.

From these considerations, I shall devote the first chapter of this Abstract to Variation under Domestication. We shall thus see that a large amount of hereditary modification is at least possible; and, what is equally or more important, we shall see how great is the power of man in accumulating by his Selection successive slight variations. I will then pass on to the variability of species in a state of nature; but I shall, unfortunately, be compelled to treat this subject far too briefly, as it can be treated properly only by giving long catalogues of facts. We shall, however, be enabled to discuss what circumstances are most favourable to variation. In the next chapter the Struggle for Existence amongst all organic beings throughout the world, which inevitably follows from their high geometrical powers of increase, will be treated of. This is the doctrine of Malthus, applied to the whole animal and vegetable kingdoms. As many more individuals of each species are born than can possibly survive; and as, consequently, there is a frequently recurring struggle for existence, it follows that any being, if it vary however slightly in any manner profitable to itself, under the complex and sometimes varying conditions of life, will have a better chance of surviving, and thus be *naturally selected*. From the strong principle of inheritance, any selected variety will tend to propagate its new and modified form.

This fundamental subject of Natural Selection will be treated at some length in the fourth chapter; and we shall then see how Natural Selection almost inevitably causes much Extinction of the less improved forms of life, and induces what I have called Divergence of Character. In the next chapter I shall discuss the complex and little known laws of variation and of correlation of growth. In the four succeeding chapters, the most apparent and gravest difficulties on the theory will be given: namely, first, the difficulties of transitions, or in understanding how a simple being or a simple organ can be changed and perfected into a highly developed being or elaborately constructed organ; secondly, the subject of Instinct, or the mental powers of animals; thirdly, Hybridism, or the infertility of species and the fertility of varieties when intercrossed; and fourthly, the imperfection of the Geological Record. In the next chapter I shall consider the geological succession of organic beings throughout time; in the eleventh and twelfth, their geographical distribution throughout space; in the thirteenth, their classification or mutual affinities, both when mature and in an embryonic condition. In the last chapter I shall give a brief recapitulation of the whole work, and a few concluding remarks.

No one ought to feel surprise at much remaining as yet unexplained in regard to the origin of species and varieties, if he makes due allowance for our profound ignorance in regard to the mutual relations of all the beings which live around us. Who can explain why one species ranges widely and is very numerous, and why another allied species has a narrow range and is rare? Yet these relations are of the highest importance, for they determine the present welfare, and, as I believe, the future success and modification of every inhabitant of this world. Still less do we know of the mutual relations of the innumerable inhabitants of the world during the many past geological epochs in its history. Although much remains obscure, and will long remain obscure, I can entertain no doubt, after the most deliberate study and dispassionate judgement of which I am capable, that the view which most naturalists entertain, and which I formerly entertained – namely, that each species has been independently created – is erroneous. I am fully convinced that species are not immutable; but that those belonging to what are called the same genera are lineal descendants of some other and generally extinct species, in the same manner as the acknowledged varieties of any one species are the descendants of that species. Furthermore, I am convinced that Natural Selection has been the main but not exclusive means of modification.

CHAPTER I

VARIATION UNDER DOMESTICATION

Causes of Variability – Effects of Habit – Correlation
of Growth – Inheritance – Character of Domestic Varieties
– Difficulty of distinguishing between Varieties and Species
– Origin of Domestic Varieties from one or more Species –
Domestic Pigeons, their Differences and Origin – Principle
of Selection anciently followed, its Effects – Methodical
and Unconscious Selection – Unknown Origin of our
Domestic Productions – Circumstances favourable to
Man's power of Selection

WHEN we look to the individuals of the same variety or sub-variety
of our older cultivated plants and animals, one of the first points
which strikes us, is, that they generally differ much more from
each other, than do the individuals of any one species or variety in
a state of nature. When we reflect on the vast diversity of the
plants and animals which have been cultivated, and which have
varied during all ages under the most different climates and treatment, I think we are driven to conclude that this greater variability
is simply due to our domestic productions having been raised
under conditions of life not so uniform as, and somewhat different
from, those to which the parent-species have been exposed under
nature. There is, also, I think, some probability in the view propounded by Andrew Knight, that this variability may be partly
connected with excess of food. It seems pretty clear that organic
beings must be exposed during several generations to the new
conditions of life to cause any appreciable amount of variation; and
that when the organization has once begun to vary, it generally
continues to vary for many generations. No case is on record of a
variable being ceasing to be variable under cultivation. Our oldest
cultivated plants, such as wheat, still often yield new varieties: our
oldest domesticated animals are still capable of rapid improvement
or modification.

It has been disputed at what period of life the causes of variability, whatever they may be, generally act; whether during the early or late period of development of the embryo, or at the instant of conception. Geoffroy St. Hilaire's experiments show that unnatural treatment of the embryo causes monstrosities; and monstrosities cannot be separated by any clear line of distinction from mere variations. But I am strongly inclined to suspect that the most frequent cause of variability may be attributed to the male and female reproductive elements having been affected prior to the act of conception. Several reasons make me believe in this; but the chief one is the remarkable effect which confinement or cultivation has on the functions of the reproductive system; this system appearing to be far more susceptible than any other part of the organization, to the action of any change in the conditions of life. Nothing is more easy than to tame an animal, and few things more difficult than to get it to breed freely under confinement, even in the many cases when the male and female unite. How many animals there are which will not breed, though living long under not very close confinement in their native country! This is generally attributed to vitiated instincts; but how many cultivated plants display the utmost vigour, and yet rarely or never seed! In some few such cases it has been found out that very trifling changes, such as a little more or less water at some particular period of growth, will determine whether or not the plant sets a seed. I cannot here enter on the copious details which I have collected on this curious subject; but to show how singular the laws are which determine the reproduction of animals under confinement, I may just mention that carnivorous animals, even from the tropics, breed in this country pretty freely under confinement, with the exception of the plantigrades or bear family; whereas, carnivorous birds, with the rarest exceptions, hardly ever lay fertile eggs. Many exotic plants have pollen utterly worthless, in the same exact condition as in the most sterile hybrids. When, on the one hand, we see domesticated animals and plants, though often weak and sickly, yet breeding quite freely under confinement; and when, on the other hand, we see individuals, though taken young from a state of nature, perfectly tamed, long-lived, and healthy (of which I could give numerous instances), yet having their reproductive system so seriously affected by unperceived causes as to fail in acting, we need not be surprised at this system, when it does act under confinement, acting not quite regularly, and producing offspring not perfectly like their parents or variable.

Sterility has been said to be the bane of horticulture; but on this view we owe variability to the same cause which produces sterility; and variability is the source of all the choicest productions of the garden. I may add, that as some organisms will breed most freely under the most unnatural conditions (for instance, the rabbit and ferret kept in hutches), showing that their reproductive system has not been thus affected; so will some animals and plants withstand domestication or cultivation, and vary very slightly – perhaps hardly more than in a state of nature.

A long list could easily be given of 'sporting plants;' by this term gardeners mean a single bud or offset, which suddenly assumes a new and sometimes very different character from that of the rest of the plant. Such buds can be propagated by grafting, &c., and sometimes by seed. These 'sports' are extremely rare under nature, but far from rare under cultivation; and in this case we see that the treatment of the parent has affected a bud or offset, and not the ovules or pollen. But it is the opinion of most physiologists that there is no essential difference between a bud and an ovule in their earliest stages of formation; so that, in fact, 'sports' support my view, that variability may be largely attributed to the ovules or pollen, or to both, having been affected by the treatment of the parent prior to the act of conception. These cases anyhow show that variation is not necessarily connected, as some authors have supposed, with the act of generation.

Seedlings from the same fruit, and the young of the same litter, sometimes differ considerably from each other, though both the young and the parents, as Müller has remarked, have apparently been exposed to exactly the same conditions of life; and this shows how unimportant the direct effects of the conditions of life are in comparison with the laws of reproduction, and of growth, and of inheritance; for had the action of the conditions been direct, if any of the young had varied, all would probably have varied in the same manner. To judge how much, in the case of any variation, we should attribute to the direct action of heat, moisture, light, food, &c., is most difficult: my impression is, that with animals such agencies have produced very little direct effect, though apparently more in the case of plants. Under this point of view, Mr. Buckman's recent experiments on plants seem extremely valuable. When all or nearly all the individuals exposed to certain conditions are affected in the same way, the change at first appears to be directly due to such conditions; but in some cases it can be shown that quite opposite conditions produce similar changes of structure. Nevertheless some slight amount of change may, I think, be attributed to the direct action of the conditions of life – as, in some cases, increased size from amount of food, colour from particular kinds of food and from light, and perhaps the thickness of fur from climate.

Habit also has a deciding influence, as in the period of flowering with plants when transported from one climate to another. In animals it has a more marked effect; for instance, I find in the domestic duck that the bones of the wing weigh less and the bones of the leg more, in proportion to the whole skeleton, than do the same bones in the wild-duck; and I presume that this change may be safely attributed to the domestic duck flying much less, and walking more, than its wild parent. The great and inherited development of the udders in cows and goats in countries where they are habitually milked, in comparison with the state of these organs in other countries, is another instance of the effect of use. Not a single domestic animal can be named which has not in some country drooping ears; and the view suggested by some authors, that the drooping is due to the disuse of the muscles of the ear, from the animals not being much alarmed by danger, seems probable.

There are many laws regulating variation, some few of which can be dimly seen, and will be hereafter briefly mentioned. I will here only allude to what may be called correlation of growth. Any change in the embryo or larva will almost certainly entail changes in the mature animal. In monstrosities, the correlations between quite distinct parts are very curious; and many instances are given in Isidore Geoffroy St. Hilaire's great work on this subject. Breeders believe that long limbs are almost always accompanied by an elongated head. Some instances of correlation are quite whimsical; thus cats with blue eyes are invariably deaf; colour and constitutional peculiarities go together, of which many remarkable cases could be given amongst animals and plants. From the facts collected by Heusinger, it appears that white sheep and pigs are differently affected from coloured individuals by certain vegetable poisons. Hairless dogs have imperfect teeth; long-haired and coarse-haired animals are apt to have, as is asserted, long or many horns; pigeons with feathered feet have skin between their outer toes; pigeons with short beaks have small feet, and those with long beaks large feet. Hence, if man goes on selecting, and thus augmenting, any peculiarity, he will almost certainly unconsciously modify other parts of the structure, owing to the mysterious laws of the correlation of growth.

The result of the various, quite unknown, or dimly seen laws of variation is infinitely complex and diversified. It is well worth while carefully to study the several treatises published on some of our old cultivated plants, as on the hyacinth, potato, even the dahlia, &c.; and it is really surprising to note the endless points in structure and constitution in which the varieties and sub-varieties differ slightly from each other. The whole organization seems to have become plastic, and tends to depart in some small degree from that of the parental type.

Any variation which is not inherited is unimportant for us. But the number and diversity of inheritable deviations of structure, both those of slight and those of considerable physiological importance, is endless. Dr. Prosper Lucas's treatise, in two large volumes, is the fullest and the best on this subject. No breeder doubts how strong is the tendency to inheritance: like produces like is his fundamental belief: doubts have been thrown on this principle by theoretical writers alone. When a deviation appears not unfrequently, and we see it in the father and child, we cannot tell whether it may not be due to the same original cause acting on both; but when amongst individuals, apparently exposed to the same conditions, any very rare deviation, due to some extraordinary combination of circumstances, appears in the parent – say, once amongst several million individuals – and it reappears in the child, the mere doctrine of chances almost compels us to attribute its reappearance to inheritance. Every one must have heard of cases of albinism, prickly skin, hairy bodies, &c., appearing in several members of the same family. If strange and rare deviations of structure are truly inherited, less strange and commoner deviations may be freely admitted to be inheritable. Perhaps the correct way of viewing the whole subject, would be, to look at the inheritance of every character whatever as the rule, and non-inheritance as the anomaly.

The laws governing inheritance are quite unknown; no one can say why the same peculiarity in different individuals of the same species, and in individuals of different species, is sometimes inherited and sometimes not so; why the child often reverts in certain characters to its grandfather or grandmother or other much more remote ancestor; why a peculiarity is often transmitted from one sex to both sexes, or to one sex alone, more commonly but not exclusively to the like sex. It is a fact of some little importance to us, that peculiarities appearing in the males of our domestic breeds are often transmitted either exclusively, or in a much greater degree, to males alone. A much more important rule, which I think may be trusted, is that, at whatever period of life a peculiarity first appears, it tends to appear in the offspring at a corresponding age, though sometimes earlier. In many cases this could not be otherwise: thus the inherited peculiarities in the horns of cattle could appear only in the offspring when nearly mature; peculiarities in the silkworm are known to appear at the corresponding caterpillar or cocoon stage. But hereditary diseases and some other facts make me believe that the rule has a wider extension, and that when there is no apparent reason why a peculiarity should appear at any particular age, yet that it does tend to appear in the offspring at the same period at which it first appeared in the parent. I believe this rule to be of the highest importance in explaining the laws of embryology. These remarks are of course confined to the first *appearance* of the peculiarity, and not to its primary cause, which may have acted on the ovules or male element; in nearly the same manner as in the crossed offspring from a short-horned cow by a long-horned bull, the greater length of horn, though appearing late in life, is clearly due to the male element.

Having alluded to the subject of reversion, I may here refer to a statement often made by naturalists – namely, that our domestic varieties, when run wild, gradually but certainly revert in character to their aboriginal stocks. Hence it has been argued that no deductions can be drawn from domestic races to species in a state of nature.

I have in vain endeavoured to discover on what decisive facts the above statement has so often and so boldly been made. There would be great difficulty in proving its truth: we may safely conclude that very many of the most strongly-marked domestic varieties could not possibly live in a wild state. In many cases we do not know what the aboriginal stock was, and so could not tell whether or not nearly perfect reversion had ensued. It would be quite necessary, in order to prevent the effects of intercrossing, that only a single variety should be turned loose in its new home. Nevertheless, as our varieties certainly do occasionally revert in some of their characters to ancestral forms, it seems to me not improbable, that if we could succeed in naturalizing, or were to cultivate, during many generations, the several races, for instance, of the cabbage, in very poor soil (in which case, however, some effect would have to be attributed to the direct action of the poor soil), that they would to a large extent, or even wholly, revert to the wild aboriginal stock. Whether or not the experiment would succeed, is not of great importance for our line of argument; for by the experiment itself the conditions of life are changed. If it could be shown that our domestic varieties manifested a strong tendency to reversion, – that is, to lose their acquired characters, whilst kept under unchanged conditions, and whilst kept in a considerable body, so that free intercrossing might check, by blending together, any slight deviations of structure, in such case, I grant that we could deduce nothing from domestic varieties in regard to species. But there is not a shadow of evidence in favour of this view: to assert that we could not breed our cart and race-horses, long and short-horned cattle, and poultry of various breeds, and esculent vegetables, for an almost infinite number of generations, would be opposed to all experience. I may add, that when under nature the conditions of life do change, variations and reversions of character probably do occur; but natural selection, as will hereafter be explained, will determine how far the new characters thus arising shall be preserved.

When we look to the hereditary varieties or races of our domestic animals and plants, and compare them with species closely allied together, we generally perceive in each domestic race, as already remarked, less uniformity of character than in true species. Domestic races of the same species, also, often have a somewhat monstrous character; by which I mean, that, although differing from each other, and from the other species of the same genus, in several trifling respects, they often differ in an extreme degree in some one part, both when compared one with another, and more especially when compared with all the species in nature to which they are nearest allied. With these exceptions (and with that of the perfect fertility of varieties when crossed, – a subject hereafter to be discussed), domestic races of the same species differ from each other in the same manner as, only in most cases in a lesser degree than, do closely-allied species of the same genus in a state of nature. I think this must be admitted, when we find that there are hardly any domestic races, either amongst animals or plants, which have not been ranked by some competent judges as mere varieties, and by other competent judges as the descendants of aboriginally distinct species. If any marked distinction existed between domestic races and species, this source of doubt could not so perpetually recur. It has often been stated that domestic races do not differ from each other in characters of generic value. I think it could be shown that this statement is hardly correct; but naturalists differ most widely in determining what characters are of generic value; all such valuations being at present empirical. More-over, on the view of the origin of genera which I shall presently give, we have no right to expect often to meet with generic differences in our domesticated productions.

When we attempt to estimate the amount of structural difference between the domestic races of the same species, we are soon involved in doubt, from not knowing whether they have descended from one or several parent-species. This point, if it could be cleared up, would be interesting; if, for instance, it could be shown that the greyhound, bloodhound, terrier, spaniel, and bull-dog, which we all know propagate their kind so truly, were the offspring of any single species, then such facts would have great weight in making us doubt about the immutability of the many very closely allied and natural species – for instance, of the many foxes – inhabiting different quarters of the world. I do not believe, as we shall presently see, that all our dogs have descended from any one wild species; but, in the case of some other domestic races, there is presumptive, or even strong, evidence in favour of this view.

It has often been assumed that man has chosen for domestication animals and plants having an extraordinary inherent tendency to vary, and likewise to withstand diverse climates. I do not dispute that these capacities have added largely to the value of most of our domesticated productions; but how could a savage possibly know, when he first tamed an animal, whether it would vary in succeeding generations, and whether it would endure other climates? Has the little variability of the ass or guinea-fowl, or the small power of endurance of warmth by the reindeer, or of cold by the common camel, prevented their domestication? I cannot doubt that if other animals and plants, equal in number to our domesticated productions, and belonging to equally diverse classes and countries, were taken from a state of nature, and could be made to breed for an equal number of generations under domestication, they would vary on an average as largely as the parent species of our existing domesticated productions have varied.

In the case of most of our anciently domesticated animals and plants, I do not think it is possible to come to any definite conclusion, whether they have descended from one or several species. The argument mainly relied on by those who believe in the multiple origin of our domestic animals is, that we find in the most ancient records, more especially on the monuments of Egypt, much diversity in the breeds; and that some of the breeds closely resemble, perhaps are identical with, those still existing. Even if this latter fact were found more strictly and generally true than seems to me to be the case, what does it show, but that some of our breeds originated there, four or five thousand years ago? But Mr. Horner's researches have rendered it in some degree probable that man sufficiently civilized to have manufactured pottery existed in the valley of the Nile thirteen or fourteen thousand years ago; and who will pretend to say how long before these ancient periods, savages, like those of Tierra del Fuego or Australia, who possess a semi-domestic dog, may not have existed in Egypt?

The whole subject must, I think, remain vague; nevertheless, I may, without here entering on any details, state that, from geographical and other considerations, I think it highly probable that our domestic dogs have descended from several wild species. In regard to sheep and goats I can form no opinion. I should think, from facts communicated to me by Mr. Blyth, on the habits, voice, and constitution, &c., of the humped Indian cattle, that these had descended from a different aboriginal stock from our European cattle; and several competent judges believe that these latter have had more than one wild parent. With respect to horses, from reasons which I cannot give here, I am doubtfully inclined to believe, in opposition to several authors, that all the races have descended from one wild stock. Mr. Blyth, whose opinion, from his large and varied stores of knowledge, I should value more than that of almost any one, thinks that all the breeds of poultry have proceeded from the common wild Indian fowl (Gallus bankiva). In regard to ducks and rabbits, the breeds of which differ considerably from each other in structure, I do not doubt that they all have descended from the common wild duck and rabbit.

The doctrine of the origin of our several domestic races from several aboriginal stocks, has been carried to an absurd extreme by some authors. They believe that every race which breeds true, let the distinctive characters be ever so slight, has had its wild prototype. At this rate there must have existed at least a score of species of wild cattle, as many sheep, and several goats in Europe alone, and several even within Great Britain. One author believes that there formerly existed in Great Britain eleven wild species of sheep peculiar to it! When we bear in mind that Britain has now hardly one peculiar mammal, and France but few distinct from those of Germany and conversely, and so with Hungary, Spain, &c., but that each of these kingdoms possesses several peculiar breeds of cattle, sheep, &c., we must admit that many domestic breeds have originated in Europe; for whence could they have been derived, as these several countries do not possess a number of peculiar species as distinct parent-stocks? So it is in India. Even in the case of the domestic dogs of the whole world, which I fully admit have probably descended from several wild species, I cannot doubt that there has been an immense amount of inherited variation. Who can believe that animals closely resembling the Italian greyhound, the bloodhound, the bull-dog, or Blenheim spaniel, &c. – so unlike all wild Canidae – ever existed freely in a state of nature? It has often been loosely said that all our races of dogs have been produced by the crossing of a few aboriginal species; but by crossing we can get only forms in some degree intermediate between their parents; and if we account for our several domestic races by this process, we must admit the former existence of the most extreme forms, as the Italian greyhound, bloodhound, bull-dog, &c., in the wild state. Moreover, the possibility of making distinct races by crossing has been greatly exaggerated. There can be no doubt that a race may be modified by occasional crosses, if aided by the careful selection of those individual mongrels, which present any desired character; but that a race could be obtained nearly intermediate between two extremely different races or species, I can hardly believe. Sir J. Sebright expressly experimentized for this object, and failed. The offspring from the first cross between two pure breeds is tolerably and sometimes (as I have found with pigeons) extremely uniform, and everything seems simple enough; but when these mongrels are crossed one with another for several generations, hardly two of them will be alike, and then the extreme difficulty, or rather utter hopelessness, of the task becomes apparent. Certainly, a breed intermediate between *two very distinct* breeds could not be got without extreme care and long-continued selection; nor can I find a single case on record of a permanent race having been thus formed.

On the Breeds of the Domestic Pigeon. Believing that it is always best to study some special group, I have, after deliberation, taken up domestic pigeons. I have kept every breed which I could purchase or obtain, and have been most kindly favoured with skins from several quarters of the world, more especially by the Hon. W. Elliot from India, and by the Hon. C. Murray from Persia. Many treatises in different languages have been published on pigeons, and some of them are very important, as being of considerable antiquity. I have associated with several eminent fanciers, and have been permitted to join two of the London Pigeon Clubs. The diversity of the breeds is something astonishing. Compare the English carrier and the short-faced tumbler, and see the wonderful difference in their beaks, entailing corresponding differences in their skulls. The carrier, more especially the male bird, is also remarkable from the wonderful development of the carunculated skin about the head, and this is accompanied by greatly elongated eyelids, very large external orifices to the nostrils, and a wide gape of mouth. The short-faced tumbler has a beak in outline almost like that of a finch; and the common tumbler has the singular and strictly inherited habit of flying at a great height in a compact flock, and tumbling in the air head over heels. The runt is a bird of great size, with long, massive beak and large feet; some of the sub-breeds of runts have very long necks, others very long wings and tails, others singularly short tails. The barb is allied to the carrier, but, instead of a very long beak, has a very short and very broad one. The pouter has a much elongated body, wings, and legs; and its enormously developed crop, which it glories in inflating, may well excite astonishment and even laughter. The turbit has a very short and conical beak, with a line of reversed feathers down the breast; and it has the habit of continually expanding slightly the upper part of the oesophagus. The jacobin has the feathers so much reversed along the back of the neck that they form a hood, and it has, proportionally to its size, much elongated wing and tail feathers. The trumpeter and laugher, as their names express, utter a very different coo from the other breeds. The fantail has thirty or even forty tail-feathers, instead of twelve or fourteen, the normal number in all members of the great pigeon family; and these feathers are kept expanded, and are carried so erect that in good birds the head and tail touch; the oil-gland is quite aborted. Several other less distinct breeds might have been specified.

In the skeletons of the several breeds, the development of the bones of the face in length and breadth and curvature differs enormously. The shape, as well as the breadth and length of the ramus of the lower jaw, varies in a highly remarkable manner. The number of the caudal and sacral vertebrae varies; as does the number of the ribs, together with their relative breadth and the presence of processes. The size and shape of the apertures in the sternum are highly variable; so is the degree of divergence and relative size of the two arms of the furcula. The proportional width of the gape of mouth, the proportional length of the eyelids, of the orifice of the nostrils, of the tongue (not always in strict correlation with the length of beak), the size of the crop and of the upper part of the oesophagus; the development and abortion of the oil-gland; the number of the primary wing and caudal feathers; the relative length of wing and tail to each other and to the body; the relative length of leg and of the feet; the number of scutellae on the toes, the development of skin between the toes, are all points of structure which are variable. The period at which the perfect plumage is acquired varies, as does the state of the down with which the nestling birds are clothed when hatched. The shape and size of the eggs vary. The manner of flight differs remarkably; as does in some breeds the voice and disposition. Lastly, in certain breeds, the males and females have come to differ to a slight degree from each other.

Altogether at least a score of pigeons might be chosen, which if shown to an ornithologist, and he were told that they were wild birds, would certainly, I think, be ranked by him as well-defined species. Moreover, I do not believe that any ornithologist would place the English carrier, the short-faced tumbler, the runt, the barb, pouter, and fantail in the same genus; more especially as in each of these breeds several truly-inherited sub-breeds, or species as he might have called them, could be shown him.

Great as the differences are between the breeds of pigeons, I am fully convinced that the common opinion of naturalists is correct, namely, that all have descended from the rock-pigeon (Columba livia), including under this term several geographical races or sub-species, which differ from each other in the most trifling respects. As several of the reasons which have led me to this belief are in some degree applicable in other cases, I will here briefly give them. If the several breeds are not varieties, and have not proceeded from the rock-pigeon, they must have descended from at least seven or eight aboriginal stocks; for it is impossible to make the present domestic breeds by the crossing of any lesser number: how, for instance, could a pouter be produced by crossing two breeds unless one of the parent-stocks possessed the characteristic enormous crop? The supposed aboriginal stocks must all have been rock-pigeons, that is, not breeding or willingly perching on trees. But besides C. livia, with its geographical sub-species, only two or three other species of rock-pigeons are known; and these have not any of the characters of the domestic breeds. Hence the supposed aboriginal stocks must either still exist in the countries where they were originally domesticated, and yet be unknown to ornithologists; and this, considering their size, habits, and remarkable characters, seems very improbable; or they must have become extinct in the wild state. But birds breeding on precipices, and good fliers, are unlikely to be exterminated; and the common rock-pigeon, which has the same habits with the domestic breeds, has not been exterminated even on several of the smaller British islets, or on the shores of the Mediterranean. Hence the supposed extermination of so many species having similar habits with the rock-pigeon seems to me a very rash assumption. Moreover, the several above-named domesticated breeds have been transported to all parts of the world, and, therefore, some of them must have been carried back again into their native country; but not one has ever become wild or feral, though the dovecot-pigeon, which is the rock-pigeon in a very slightly altered state, has become feral in several places. Again, all recent experience shows that it is most difficult to get any wild animal to breed freely under domestication; yet on the hypothesis of the multiple origin of our pigeons, it must be assumed that at least seven or eight species were so thoroughly domesticated in ancient times by half-civilized man, as to be quite prolific under confinement.

An argument, as it seems to me, of great weight, and applicable in several other cases, is, that the above-specified breeds, though agreeing generally in constitution, habits, voice, colouring, and in most parts of their structure, with the wild rock-pigeon, yet are certainly highly abnormal in other parts of their structure: we may look in vain throughout the whole great family of Columbidae for a beak like that of the English carrier, or that of the short-faced tumbler, or barb; for reversed feathers like those of the jacobin; for a crop like that of the pouter; for tail-feathers like those of the fantail. Hence it must be assumed not only that half-civilized man succeeded in thoroughly domesticating several species, but that he intentionally or by chance picked out extraordinarily abnormal species; and further, that these very species have since all become extinct or unknown. So many strange contingencies seem to me improbable in the highest degree.

Some facts in regard to the colouring of pigeons well deserve consideration. The rock-pigeon is of a slaty-blue, and has a white rump (the Indian sub-species, C. intermedia of Strickland, having it bluish); the tail has a terminal dark bar, with the bases of the outer feathers externally edged with white; the wings have two black bars;

some semi-domestic breeds and some apparently truly wild breeds have, besides the two black bars, the wings chequered with black. These several marks do not occur together in any other species of the whole family. Now, in every one of the domestic breeds, taking thoroughly well-bred birds, all the above marks, even to the white edging of the outer tail-feathers, sometimes concur perfectly developed. Moreover, when two birds belonging to two distinct breeds are crossed, neither of which is blue or has any of the above-specified marks, the mongrel offspring are very apt suddenly to acquire these characters; for instance, I crossed some uniformly white fantails with some uniformly black barbs, and they produced mottled brown and black birds; these I again crossed together, and one grandchild of the pure white fantail and pure black barb was of as beautiful a blue colour, with the white rump, double black wing-bar, and barred and white-edged tail-feathers, as any wild rock-pigeon! We can understand these facts, on the well-known principle of reversion to ancestral characters, if all the domestic breeds have descended from the rock-pigeon. But if we deny this, we must make one of the two following highly improbable suppositions. Either, firstly, that all the several imagined aboriginal stocks were coloured and marked like the rock-pigeon, although no other existing species is thus coloured and marked, so that in each separate breed there might be a tendency to revert to the very same colours and markings. Or, secondly, that each breed, even the purest, has within a dozen or, at most, within a score of generations, been crossed by the rock-pigeon: I say within a dozen or twenty generations, for we know of no fact countenancing the belief that the child ever reverts to some one ancestor, removed by a greater number of generations. In a breed which has been crossed only once with some distinct breed, the tendency to reversion to any character derived from such cross will naturally become less and less, as in each succeeding generation there will be less of the foreign blood; but when there has been no cross with a distinct breed, and there is a tendency in both parents to revert to a character, which has been lost during some former generation, this tendency, for all that we can see to the contrary, may be transmitted undiminished for an indefinite number of generations. These two distinct cases are often confounded in treatises on inheritance.

Lastly, the hybrids or mongrels from between all the domestic breeds of pigeons are perfectly fertile. I can state this from my own observations, purposely made on the most distinct breeds. Now, it is difficult, perhaps impossible, to bring forward one case of the hybrid offspring of two animals *clearly distinct* being themselves perfectly fertile. Some authors believe that long-continued domestication eliminates this strong tendency to sterility: from the history of the dog I think there is some probability in this hypothesis, if applied to species closely related together, though it is unsupported by a single experiment. But to extend the hypothesis so far as to suppose that species, aboriginally as distinct as carriers, tumblers, pouters, and fantails now are, should yield offspring perfectly fertile, *inter se*, seems to me rash in the extreme.

From these several reasons, namely, the improbability of man having formerly got seven or eight supposed species of pigeons to breed freely under domestication; these supposed species being quite unknown in a wild state, and their becoming nowhere feral; these species having very abnormal characters in certain respects, as compared with all other Columbidae, though so like in most other respects to the rock-pigeon; the blue colour and various marks occasionally appearing in all the breeds, both when kept pure and when crossed; the mongrel offspring being perfectly fertile; – from these several reasons, taken together, I can feel no doubt that all our domestic breeds have descended from the Columbia livia with its geographical sub-species.

In favour of this view, I may add, firstly, that C. livia, or the rock-pigeon, has been found capable of domestication in Europe and in India; and that it agrees in habits and in a great number of points of structure with all the domestic breeds. Secondly, although an English carrier or short-faced tumbler differs immensely in certain characters from the rock-pigeon, yet by comparing the several sub-breeds of these breeds, more especially those brought from distant countries, we can make an almost perfect series between the extremes of structure. Thirdly, those characters which are mainly distinctive of each breed, for instance the wattle and length of beak of the carrier, the shortness of that of the tumbler, and the number of tail-feathers in the fantail, are in each breed eminently variable; and the explanation of this fact will be obvious when we come to treat of selection. Fourthly, pigeons have been watched, and tended with the utmost care, and loved by many people. They have been domesticated for thousands of years in several quarters of the world; the earliest known record of pigeons is in the fifth Egyptian dynasty, about 3000 B.C., as was pointed out to me by Professor Lepsius; but Mr. Birch informs me that pigeons are given in a bill of fare in the previous dynasty. In the time of the Romans, as we hear from Pliny, immense prices were given for pigeons; 'nay, they are come to this pass, that they can reckon up their pedigree and race.' Pigeons were much valued by Akber Khan in India, about the year 1600; never less than 20,000 pigeons were taken with the court. 'The monarchs of Iran and Turan sent him some very rare birds;' and, continues the courtly historian, 'His Majesty by crossing the breeds, which method was never prac- tised before, has improved them astonishingly.' About this same period the Dutch were as eager about pigeons as were the old Romans. The paramount importance of these considerations in explaining the immense amount of variation which pigeons have undergone, will be obvious when we treat of Selection. We shall then, also, see how it is that the breeds so often have a somewhat monstrous character. It is also a most favourable circumstance for the production of distinct breeds, that male and female pigeons can be easily mated for life; and thus different breeds can be kept together in the same aviary.

I have discussed the probable origin of domestic pigeons at some, yet quite insufficient, length; because when I first kept pigeons and watched the several kinds, knowing well how true they bred, I felt fully as much difficulty in believing that they could ever have des- cended from a common parent, as any naturalist could in coming to a similar conclusion in regard to the many species of finches, or other large groups of birds, in nature. One circumstance has struck me much; namely, that all the breeders of the various domestic animals and the cultivators of plants, with whom I have ever conversed, or whose treatises I have read, are firmly convinced that the several breeds to which each has attended, are descended from so many aboriginally distinct species. Ask, as I have asked, a celebrated raiser of Hereford cattle, whether his cattle might not have descended from long-horns, and he will laugh you to scorn. I have never met a pigeon, or poultry, or duck, or rabbit fancier, who was not fully convinced that each main breed was descended from a distinct species. Van Mons, in his treatise on pears and apples, shows how utterly he disbelieves that the several sorts, for instance a Ribston-pippin or Codlin-apple, could ever have proceeded from the seeds of the same tree. Innumerable other examples could be given. The explanation, I think, is simple: from long-continued study they are strongly impressed with the differences between the several races; and though they well know that each race varies slightly, for they win their prizes by selecting such slight differences, yet they ignore all general arguments, and refuse to sum up in their minds slight differences accumulated during many successive generations. May not those naturalists who, knowing far less of the laws of inheritance than does the breeder, and knowing no more than he does of the intermediate links in the long lines of descent, yet admit that many of our domestic races have descended from the same parents – may they not learn a lesson of caution, when they deride the idea of species in a state of nature being lineal descendants of other species?

Selection. Let us now briefly consider the steps by which domestic races have been produced, either from one or from several allied species. Some little effect may, perhaps, be attributed to the direct action of the external conditions of life, and some little to habit; but he would be a bold man who would account by such agencies for the differences of a dray and race-horse, a greyhound and bloodhound, a carrier and tumbler pigeon. One of the most remarkable features in our domesticated races is that we see in them adaptation, not indeed to the animal's or plant's own good, but to man's use or fancy. Some variations useful to him have probably arisen suddenly, or by one step; many botanists, for instance, believe that the fuller's teazle, with its hooks, which cannot be rivalled by any mechanical contrivance, is only a variety of the wild Dipsacus; and this amount of change may have suddenly arisen in a seedling. So it has probably been with the turnspit dog; and this is known to have been the case with the ancon sheep. But when we compare the dray-horse and race-horse, the dromedary and camel, the various breeds of sheep fitted either for cultivated land or mountain pasture, with the wool of one breed good for one purpose, and that of another breed for another purpose; when we compare the many breeds of dogs, each good for man in very different ways; when we compare the game-cock, so pertinacious in battle, with other breeds so little quarrelsome, with 'everlasting layers' which never desire to sit, and with the bantam so small and elegant; when we compare the host of agricultural, culinary, orchard, and flower-garden races of plants, most useful to man at different seasons and for different purposes, or so beautiful in his eyes, we must, I think, look further than to mere variability. We cannot suppose that all the breeds were suddenly produced as perfect and as useful as we now see them; indeed, in several cases, we know that this has not been their history. The key is man's power of accumulative selection: nature gives successive variations; man adds them up in certain directions useful to him. In this sense he may be said to make for himself useful breeds.

The great power of this principle of selection is not hypothetical. It is certain that several of our eminent breeders have, even within a single lifetime, modified to a large extent some breeds of cattle and sheep. In order fully to realize what they have done, it is almost necessary to read several of the many treatises devoted to this subject, and to inspect the animals. Breeders habitually speak of an animal's organization as something quite plastic, which they can model almost as they please. If I had space I could quote numerous passages to this effect from highly competent authorities. Youatt, who was probably better acquainted with the works of agriculturalists than almost any other individual, and who was himself a very good judge of an animal, speaks of the principle of selection as 'that which enables the agriculturist, not only to modify the character of his flock, but to change it altogether. It is the magician's wand, by means of which he may summon into life whatever form and mould he pleases.' Lord Somerville, speaking of what breeders have done for sheep, says: – 'It would seem as if they had chalked out upon a wall a form perfect in itself, and then had given it existence.' That most skilful breeder, Sir John Sebright, used to say, with respect to pigeons, that 'he would produce any given feather in three years, but it would take him six years to obtain head and beak.' In Saxony the importance of the principle of selection in regard to merino sheep is so fully recognized, that men follow it as a trade: the sheep are placed on a table and are studied, like a picture by a connoisseur; this is done three times at intervals of months, and the sheep are each time marked and classed, so that the very best may ultimately be selected for breeding.

What English breeders have actually effected is proved by the enormous prices given for animals with a good pedigree; and these have now been exported to almost every quarter of the world. The improvement is by no means generally due to crossing different breeds; all the best breeders are strongly opposed to this practice, except sometimes amongst closely allied sub-breeds. And when a cross has been made, the closest selection is far more indispensable even than in ordinary cases. If selection consisted merely in separating some very distinct variety, and breeding from it, the principle would be so obvious as hardly to be worth notice; but its importance consists in the great effect produced by the accumulation in one direction, during successive generations, of differences absolutely inappreciable by an uneducated eye – differences which I for one have vainly attempted to appreciate. Not one man in a thousand has accuracy of eye and judgement sufficient to become an eminent breeder. If gifted with these qualities, and he studies his subject for years, and devotes his lifetime to it with indomitable perseverance, he will succeed, and may make great improvements; if he wants any of these qualities, he will assuredly fail. Few would readily believe in the natural capacity and years of practice requisite to become even a skilful pigeon-fancier.

The same principles are followed by horticulturists; but the variations are here often more abrupt. No one supposes that our choicest productions have been produced by a single variation from the aboriginal stock. We have proofs that this is not so in some cases, in which exact records have been kept; thus, to give a very trifling instance, the steadily-increasing size of the common gooseberry may be quoted. We see an astonishing improvement in many florists' flowers, when the flowers of the present day are compared with drawings made only twenty or thirty years ago. When a race of plants is once pretty well established, the seed-raisers do not pick out the best plants, but merely go over their seed-beds, and pull up the 'rogues,' as they call the plants that deviate from the proper standard.

With animals this kind of selection is, in fact, also followed; for hardly any one is so careless as to allow his worst animals to breed.

In regard to plants, there is another means of observing the accumulated effects of selection – namely, by comparing the diversity of flowers in the different varieties of the same species in the flower-garden; the diversity of leaves, pods, or tubers, or whatever part is valued, in the kitchen-garden, in comparison with the flowers of the same varieties; and the diversity of fruit of the same species in the orchard, in comparison with the leaves and flowers of the same set of varieties. See how different the leaves of the cabbage are, and how extremely alike the flowers; how unlike the flowers of the heartsease are, and how alike the leaves; how much the fruit of the different kinds of gooseberries differ in size, colour, shape, and hairiness, and yet the flowers present very slight differences. It is not that the varieties which differ largely in some one point do not differ at all in other points; this is hardly ever, perhaps never, the case. The laws of correlation of growth, the importance of which should never be overlooked, will ensure some differences; but, as a general rule, I cannot doubt that the continued selection of slight variations, either in the leaves, the flowers, or the fruit, will produce races differing from each other chiefly in these characters.

It may be objected that the principle of selection has been reduced to methodical practice for scarcely more than three-quarters of a century; it has certainly been more attended to of late years, and many treatises have been published on the subject; and the result, I may add, has been, in a corresponding degree, rapid and important. But it is very far from true that the principle is a modern discovery. I could give several references to the full acknowledgement of the importance of the principle in works of high antiquity. In rude and barbarous periods of English history choice animals were often imported, and laws were passed to prevent their exportation: the destruction of horses under a certain size was ordered, and this may be compared to the 'roguing' of plants by nurserymen. The principle of selection I find distinctly given in an ancient Chinese encyclopaedia. Explicit rules are laid down by some of the Roman classical writers. From passages in Genesis, it is clear that the colour of domestic animals was at that early period attended to.

Savages now sometimes cross their dogs with wild canine animals, to improve the breed, and they formerly did so, as is attested by passages in Pliny. The savages in South Africa match their draught cattle by colour, as do some of the Esquimaux their teams of dogs. Livingstone shows how much good domestic breeds are valued by the negroes of the interior of Africa who have not associated with Europeans. Some of these facts do not show actual selection, but they show that the breeding of domestic animals was carefully attended to in ancient times, and is now attended to by the lowest savages. It would, indeed, have been a strange fact, had attention not been paid to breeding, for the inheritance of good and bad qualities is so obvious.

At the present time, eminent breeders try by methodical selection, with a distinct object in view, to make a new strain or sub-breed, superior to anything existing in the country. But, for our purpose, a kind of Selection, which may be called Unconscious, and which results from every one trying to possess and breed from the best individual animals, is more important. Thus, a man who intends keeping pointers naturally tries to get as good dogs as he can, and afterwards breeds from his own best dogs, but he has no wish or expectation of permanently altering the breed. Nevertheless I cannot doubt that this process, continued during centuries, would improve and modify any breed, in the same way as Bakewell, Collins, &c., by this very same process, only carried on more methodically, did greatly modify, even during their own lifetimes, the forms and qualities of their cattle. Slow and insensible changes of this kind could never be recognized unless actual measurements or careful drawings of the breeds in question had been made long ago, which might serve for comparison. In some cases, however, unchanged or but little changed individuals of the same breed may be found in less civilized districts, where the breed has been less improved. There is reason to believe that King Charles's spaniel has been unconsciously modified to a large extent since the time of that monarch. Some highly competent authorities are convinced that the setter is directly derived from the spaniel, and has probably been slowly altered from it. It is known that the English pointer has been greatly changed within the last century, and in this case the change has, it is believed, been chiefly effected by crosses with the fox-hound; but what concerns us is, that the change has been effected unconsciously and gradually, and yet so effectually, that, though the old Spanish pointer certainly came from Spain, Mr. Barrow has not seen, as I am informed by him, any native dog in Spain like our pointer.

By a similar process of selection, and by careful training, the whole body of English race-horses have come to surpass in fleetness and size the parent Arab stock, so that the latter, by the regulations for the Goodwood Races, are favoured in the weights they carry. Lord Spencer and others have shown how the cattle of England have increased in weight and in early maturity, compared with the stock formerly kept in this country. By comparing the accounts given in old pigeon treatises of carriers and tumblers with these breeds as now existing in Britain, India, and Persia, we can, I think, clearly trace the stages through which they have insensibly passed, and come to differ so greatly from the rock-pigeon.

Youatt gives an excellent illustration of the effects of a course of selection, which may be considered as unconsciously followed, in so far that the breeders could never have expected or even have wished to have produced the result which ensued – namely, the production of two distinct strains. The two flocks of Leicester sheep kept by Mr. Buckley and Mr. Burgess, as Mr. Youatt remarks, 'have been purely bred from the original stock of Mr. Bakewell for upwards of fifty years. There is not a suspicion existing in the mind of any one at all acquainted with the subject that the owner of either of them has deviated in any one instance from the pure blood of Mr. Bakewell's flock, and yet the difference between the sheep possessed by these two gentlemen is so great that they have the appearance of being quite different varieties.'

If there exist savages so barbarous as never to think of the inherited character of the offspring of their domestic animals, yet any one animal particularly useful to them, for any special purpose, would be carefully preserved during famines and other accidents, to which savages are so liable, and such choice animals would thus generally leave more offspring than the inferior ones; so that in this case there would be a kind of unconscious selection going on. We see the value set on animals even by the barbarians of Tierra del Fuego, by their killing and devouring their old women, in times of dearth, as of less value than their dogs.

In plants the same gradual process of improvement, through the occasional preservation of the best individuals, whether or not sufficiently distinct to be ranked at their first appearance as distinct varieties, and whether or not two or more species or races have become blended together by crossing, may plainly be recognized in the increased size and beauty which we now see in the varieties of the heartsease, rose, pelargonium, dahlia, and other plants, when compared with the older varieties or with their parent-stocks. No one would ever expect to get a first-rate heartsease or dahlia from the seed of a wild plant. No one would expect to raise a first-rate melting pear from the seed of a wild pear, though he might succeed from a poor seedling growing wild, if it had come from a garden-stock. The pear, though cultivated in classical times, appears, from Pliny's description, to have been a fruit of very inferior quality. I have seen great surprise expressed in horticultural works at the wonderful skill of gardeners, in having produced such splendid results from such poor materials; but the art, I cannot doubt, has been simple, and, as far as the final result is concerned, has been followed almost unconsciously. It has consisted in always cultivating the best known variety, sowing its seeds, and, when a slightly better variety has chanced to appear, selecting it, and so onwards. But the gardeners of the classical period, who cultivated the best pear they could procure, never thought what splendid fruit we should eat; though we owe our excellent fruit, in some small degree, to their having naturally chosen and preserved the best varieties they could anywhere find.

A large amount of change in our cultivated plants, thus slowly and unconsciously accumulated, explains, as I believe, the well-known fact, that in a vast number of cases we cannot recognize, and therefore do not know, the wild parent-stocks of the plants which have been longest cultivated in our flower and kitchen gardens. If it has taken centuries or thousands of years to improve or modify most of our plants up to their present standard of usefulness to man, we can understand how it is that neither Australia, the Cape of Good Hope, nor any other region inhabited by quite uncivilized man, has afforded us a single plant worth culture. It is not that these countries, so rich in species, do not by a strange chance possess the aboriginal stocks of any useful plants, but that the native plants have not been improved by continued selection up to a standard of perfection comparable with that given to the plants in countries anciently civilized.

In regard to the domestic animals kept by uncivilized man, it should not be overlooked that they almost always have to struggle for their own food, at least during certain seasons. And in two countries very differently circumstanced, individuals of the same species, having slightly different constitutions or structure, would often succeed better in the one country than in the other, and thus by a process of 'natural selection,' as will hereafter be more fully explained, two sub-breeds might be formed. This, perhaps, partly explains what has been remarked by some authors, namely, that the varieties kept by savages have more of the character of species than the varieties kept in civilized countries.

On the view here given of the all-important part which selection by man has played, it becomes at once obvious, how it is that our domestic races show adaptation in their structure or in their habits to man's wants or fancies. We can, I think, further understand the frequently abnormal character of our domestic races, and likewise their differences being so great in external characters and relatively so slight in internal parts or organs. Man can hardly select, or only with much difficulty, any deviation of structure excepting such as is externally visible; and indeed he rarely cares for what is internal. He can never act by selection, excepting on variations which are first given to him in some slight degree by nature. No man would ever try to make a fantail, till he saw a pigeon with a tail developed in some slight degree in an unusual manner, or a pouter till he saw a pigeon with a crop of somewhat unusual size; and the more abnormal or unusual any character was when it first appeared, the more likely it would be to catch his attention. But to use such an expression as trying to make a fantail, is, I have no doubt, in most cases, utterly incorrect. The man who first selected a pigeon with a slightly larger tail, never dreamed what the descendants of that pigeon would become through long-continued, partly unconscious and partly methodical selection. Perhaps the parent bird of all fantails had only fourteen tail-feathers somewhat expanded, like the present Java fantail, or like individuals of other and distinct breeds, in which as many as seventeen tail-feathers have been counted. Perhaps the first pouter-pigeon did not inflate its crop much more than the turbit now does the upper part of its oesophagus, – a habit which is disregarded by all fanciers, as it is not one of the points of the breed.

Nor let it be thought that some great deviation of structure would be necessary to catch the fancier's eye: he perceives extremely small differences, and it is in human nature to value any novelty, however slight, in one's own possession. Nor must the value which would formerly be set on any slight differences in the individuals of the same species, be judged of by the value which would now be set on them, after several breeds have once fairly been established. Many slight differences might, and indeed do now, arise amongst pigeons, which are rejected as faults or deviations from the standard of perfection of each breed. The common goose has not given rise to any marked varieties; hence the Thoulouse and the common breed, which differ only in colour, that most fleeting of characters, have lately been exhibited as distinct at our poultry-shows.

I think these views further explain what has sometimes been noticed – namely, that we know nothing about the origin or history of any of our domestic breeds. But, in fact, a breed, like a dialect of a language, can hardly be said to have had a definite origin. A man preserves and breeds from an individual with some slight deviation of structure, or takes more care than usual in matching his best animals and thus improves them, and the improved individuals slowly spread in the immediate neighbourhood. But as yet they will hardly have a distinct name, and from being only slightly valued, their history will be disregarded. When further improved by the same slow and gradual process, they will spread more widely, and will get recognized as something distinct and valuable, and will then probably first receive a provincial name. In semi-civilized countries, with little free communication, the spreading and knowledge of any new sub-breed will be a slow process. As soon as the points of value of the new sub-breed are once fully acknowledged, the principle, as I have called it, of unconscious selection will always tend, – perhaps more at one period than at another, as the breed rises or falls in fashion, – perhaps more in one district than in another, according to the state of civilization of the inhabitants – slowly to add to the characteristic features of the breed, whatever they may be. But the chance will be infinitely small of any record having been preserved of such slow, varying, and insensible changes.

I must now say a few words on the circumstances, favourable, or the reverse, to man's power of selection. A high degree of variability is obviously favourable, as freely giving the materials for selection to work on; not that mere individual differences are not amply suffi-cient, with extreme care, to allow of the accumulation of a large amount of modification in almost any desired direction. But as variations manifestly useful or pleasing to man appear only occasion-ally, the chance of their appearance will be much increased by a large number of individuals being kept; and hence this comes to be of the highest importance to success. On this principle Marshall has remarked, with respect to the sheep of parts of Yorkshire, that 'as they generally belong to poor people, and are mostly *in small lots*, they never can be improved.' On the other hand, nurserymen, from raising large stocks of the same plants, are generally far more successful than amateurs in getting new and valuable varieties. The keeping of a large number of individuals of a species in any country requires that the species should be placed under favourable conditions of life, so as to breed freely in that country. When the individuals of any species are scanty, all the individuals, whatever their quality may be, will generally be allowed to breed, and this will effectually prevent selec-tion. But probably the most important point of all, is, that the animal or plant should be so highly useful to man, or so much valued by him, that the closest attention should be paid to even the slightest deviation in the qualities or structure of each individual. Unless such attention be paid nothing can be effected. I have seen it gravely remarked, that it was most fortunate that the strawberry began to vary just when gardeners began to attend closely to this plant. No doubt the straw-berry had always varied since it was cultivated, but the slight varieties had been neglected. As soon, however, as gardeners picked out individual plants with slightly larger, earlier, or better fruit, and raised seedlings from them, and again picked out the best seedlings and bred from them, then, there appeared (aided by some crossing with distinct species) those many admirable varieties of the strawberry which have been raised during the last thirty or forty years.

In the case of animals with separate sexes, facility in preventing crosses is an important element of success in the formation of new races, – at least, in a country which is already stocked with other races. In this respect enclosure of the land plays a part. Wandering savages or the inhabitants of open plains rarely possess more than one breed of the same species. Pigeons can be mated for life, and this is a great convenience to the fancier, for thus many races may be kept true, though mingled in the same aviary; and this circumstance must have largely favoured the improvement and formation of new breeds. Pigeons, I may add, can be propagated in great numbers and at a very quick rate, and inferior birds may be freely rejected, as when killed they serve for food. On the other hand, cats, from their nocturnal rambling habits, cannot be matched, and, although so much valued by women and children, we hardly ever see a distinct breed kept up; such breeds as we do sometimes see are almost always imported from some other country, often from islands. Although I do not doubt that some domestic animals vary less than others, yet the rarity or absence of distinct breeds of the cat, the donkey, peacock, goose, &c., may be attributed in main part to selection not having been brought into play: in cats, from the difficulty in pairing them; in donkeys, from only a few being kept by poor people, and little attention paid to their breeding; in peacocks, from not being very easily reared and a large stock not kept; in geese, from being valuable only for two purposes, food and feathers, and more especially from no pleasure having been felt in the display of distinct breeds.

To sum up on the origin of our Domestic Races of animals and plants. I believe that the conditions of life, from their action on the reproductive system, are so far of the highest importance as causing variability. I do not believe that variability is an inherent and necessary contingency, under all circumstances, with all organic beings, as some authors have thought. The effects of variability are modified by various degrees of inheritance and of reversion. Variability is governed by many unknown laws, more especially by that of correlation of growth. Something may be attributed to the direct action of the conditions of life. Something must be attributed to use and disuse. The final result is thus rendered infinitely complex. In some cases, I do not doubt that the intercrossing of species, aboriginally distinct, has played an important part in the origin of our domestic productions. When in any country several domestic breeds have once been established, their occasional intercrossing, with the aid of selection, has, no doubt, largely aided in the formation of new sub-breeds; but the importance of the crossing of varieties has, I believe, been greatly exaggerated, both in regard to animals and to those plants which are propagated by seed. In plants which are temporarily propagated by cuttings, buds, &c., the importance of the crossing both of distinct species and of varieties is immense; for the cultivator here quite disregards the extreme variability both of hybrids and mongrels, and the frequent sterility of hybrids; but the cases of plants not propagated by seed are of little importance to us, for their endurance is only temporary. Over all these causes of Change I am convinced that the accumulative action of Selection, whether applied methodically and more quickly, or unconsciously and more slowly, but more efficiently, is by far the predominant Power.

CHAPTER II

VARIATION UNDER NATURE

Variability – Individual differences – Doubtful species –
Wide ranging, much diffused, and common species vary
most – Species of the larger genera in any country
vary more than the species of the smaller genera – Many
of the species of the larger genera resemble varieties in
being very closely, but unequally, related to each other,
and in having restricted ranges

BEFORE applying the principles arrived at in the last chapter
to organic beings in a state of nature, we must briefly discuss
whether these latter are subject to any variation. To treat this subject
at all properly, a long catalogue of dry facts should be given; but
these I shall reserve for my future work. Nor shall I here discuss
the various definitions which have been given of the term species.
No one definition has as yet satisfied all naturalists; yet every naturalist knows vaguely what he means when he speaks of a species.
Generally the term includes the unknown element of a distinct act of
creation. The term 'variety' is almost equally difficult to
define; but here community of descent is almost universally implied,
though it can rarely be proved. We have also what are called monstrosities; but they graduate into varieties. By a monstrosity I presume
is meant some considerable deviation of structure in one part,
either injurious to or not useful to the species, and not generally
propagated. Some authors use the term 'variation' in a technical
sense, as implying a modification directly due to the physical conditions of life; and 'variations' in this sense are supposed not to be
inherited: but who can say that the dwarfed condition of shells in
the brackish waters of the Baltic, or dwarfed plants on Alpine
summits, or the thicker fur of an animal from far northwards,
would not in some cases be inherited for at least some few generations? and in this case I presume that the form would be called
a variety.

Again, we have many slight differences which may be called individual differences, such as are known frequently to appear in the offspring from the same parents, or which may be presumed to have thus arisen, from being frequently observed in the individuals of the same species inhabiting the same confined locality. No one supposes that all the individuals of the same species are cast in the very same mould. These individual differences are highly important for us, as they afford materials for natural selection to accumulate, in the same manner as man can accumulate in any given direction individual differences in his domesticated productions. These individual differences generally affect what naturalists consider unimportant parts; but I could show by a long catalogue of facts, that parts which must be called important, whether viewed under a physiological or classificatory point of view, sometimes vary in the individuals of the same species. I am convinced that the most experienced naturalist would be surprised at the number of the cases of variability, even in important parts of structure, which he could collect on good authority, as I have collected, during a course of years. It should be remembered that systematists are far from pleased at finding variability in important characters, and that there are not many men who will laboriously examine internal and important organs, and compare them in many specimens of the same species. I should never have expected that the branching of the main nerves close to the great central ganglion of an insect would have been variable in the same species; I should have expected that changes of this nature could have been effected only by slow degrees: yet quite recently Mr. Lubbock has shown a degree of variability in these main nerves in Coccus, which may almost be compared to the irregular branching of the stem of a tree. This philosophical naturalist, I may add, has also quite recently shown that the muscles in the larvae of certain insects are very far from uniform. Authors sometimes argue in a circle when they state that important organs never vary; for these same authors practically rank that character as important (as some few naturalists have honestly confessed) which does not vary; and, under this point of view, no instance of any important part varying will ever be found: but under any other point of view many instances assuredly can be given.

There is one point connected with individual differences, which seems to me extremely perplexing: I refer to those genera which have sometimes been called 'protean' or 'polymorphic,' in which the species present an inordinate amount of variation; and hardly two naturalists can agree which forms to rank as species and which as varieties. We may instance Rubus, Rosa, and Hieracium amongst plants, several genera of insects, and several genera of Brachiopod shells. In most polymorphic genera some of the species have fixed and definite characters. Genera which are polymorphic in one country seem to be, with some few exceptions, polymorphic in other countries, and likewise, judging from Brachiopod shells, at former periods of time. These facts seem to be very perplexing, for they seem to show that this kind of variability is independent of the conditions of life. I am inclined to suspect that we see in these polymorphic genera variations in points of structure which are of no service or disservice to the species, and which consequently have not been seized on and rendered definite by natural selection, as hereafter will be explained.

Those forms which possess in some considerable degree the character of species, but which are so closely similar to some other forms, or are so closely linked to them by intermediate gradations, that naturalists do not like to rank them as distinct species, are in several respects the most important for us. We have every reason to believe that many of these doubtful and closely-allied forms have permanently retained their characters in their own country for a long time; for as long, as far as we know, as have good and true species. Practically, when a naturalist can unite two forms together by others having intermediate characters, he treats the one as a variety of the other, ranking the most common, but sometimes the one first described, as the species, and the other as the variety. But cases of great difficulty, which I will not here enumerate, sometimes occur in deciding whether or not to rank one form as a variety of another, even when they are closely connected by intermediate links; nor will the commonly-assumed hybrid nature of the intermediate links always remove the difficulty. In very many cases, however, one form is ranked as a variety of another, not because the intermediate links have actually been found, but because analogy leads the observer

THE ORIGIN OF SPECIES

to suppose either that they do now somewhere exist, or may formerly have existed; and here a wide door for the entry of doubt and conjecture is opened.

Hence, in determining whether a form should be ranked as a species or a variety, the opinion of naturalists having sound judgement and wide experience seems the only guide to follow. We must, however, in many cases, decide by a majority of naturalists, for few well-marked and well-known varieties can be named which have not been ranked as species by at least some competent judges.

That varieties of this doubtful nature are far from uncommon cannot be disputed. Compare the several floras of Great Britain, of France or of the United States, drawn up by different botanists, and see what a surprising number of forms have been ranked by one botanist as good species, and by another as mere varieties. Mr. H. C. Watson, to whom I lie under deep obligation for assistance of all kinds, has marked for me 182 British plants, which are generally considered as varieties, but which have all been ranked by botanists as species; and in making this list he has omitted many trifling varieties, but which nevertheless have been ranked by some botanists as species, and he has entirely omitted several highly polymorphic genera. Under genera, including the most polymorphic forms, Mr. Babington gives 251 species, whereas Mr. Bentham gives only 112, – a difference of 139 doubtful forms! Amongst animals which unite for each birth, and which are highly locomotive, doubtful forms, ranked by one zoologist as a species and by another as a variety, can rarely be found within the same country, but are common in separated areas. How many of those birds and insects in North America and Europe, which differ very slightly from each other, have been ranked by one eminent naturalist as undoubted species, and by another as varieties, or, as they are often called, as geographical races! Many years ago, when comparing, and seeing others compare, the birds from the separate islands of the Galapagos Archipelago, both one with another, and with those from the American mainland, I was much struck how entirely vague and arbitrary is the distinction between species and varieties. On the islets of the little Madeira group there are many insects which are characterized as varieties in

Mr. Wollaston's admirable work, but which it cannot be doubted would be ranked as distinct species by many entomologists. Even Ireland has a few animals, now generally regarded as varieties, but which have been ranked as species by some zoologists. Several most experienced ornithologists consider our British red grouse as only a strongly-marked race of a Norwegian species, whereas the greater number rank it as an undoubted species peculiar to Great Britain. A wide distance between the homes of two doubtful forms leads many naturalists to rank both as distinct species; but what distance, it has been well asked, will suffice? If that between America and Europe is ample, will that between the Continent and the Azores, or Madeira, or the Canaries, or Ireland, be sufficient? It must be admitted that many forms, considered by highly-competent judges as varieties, have so perfectly the character of species that they are ranked by other highly-competent judges as good and true species. But to discuss whether they are rightly called species or varieties, before any definition of these terms has been generally accepted, is vainly to beat the air.

Many of the cases of strongly-marked varieties or doubtful species well deserve consideration; for several interesting lines of argument, from geographical distribution, analogical variation, hybridism, &c., have been brought to bear on the attempt to determine their rank. I will here give only a single instance, – the well-known one of the primrose and cowslip, or Primula veris and elatior. These plants differ considerably in appearance; they have a different flavour and emit a different odour; they flower at slightly different periods; they grow in somewhat different stations; they ascend mountains to different heights; they have different geographical ranges; and lastly, according to very numerous experiments made during several years by that most careful observer Gärtner, they can be crossed only with much difficulty. We could hardly wish for better evidence of the two forms being specifically distinct. On the other hand, they are united by many intermediate links, and it is very doubtful whether these links are hybrids; and there is, as it seems to me, an overwhelming amount of experimental evidence, showing that they descend from common parents, and consequently must be ranked as varieties.

Close investigation, in most cases, will bring naturalists to an agreement how to rank doubtful forms. Yet it must be confessed, that it is in the best-known countries that we find the greatest number of forms of doubtful value. I have been struck with the fact, that if any animal or plant in a state of nature be highly useful to man, or from any cause closely attract his attention, varieties of it will almost universally be found recorded. These varieties, moreover, will be often ranked by some authors as species. Look at the common oak, how closely it has been studied; yet a German author makes more than a dozen species out of forms, which are very generally considered as varieties; and in this country the highest botanical authorities and practical men can be quoted to show that the sessile and pedunculated oaks are either good and distinct species or mere varieties.

When a young naturalist commences the study of a group of organisms quite unknown to him, he is at first much perplexed to determine what differences to consider as specific, and what as varieties; for he knows nothing of the amount and kind of variation to which the group is subject; and this shows, at least, how very generally there is some variation. But if he confine his attention to one class within one country, he will soon make up his mind now to rank most of the doubtful forms. His general tendency will be to make many species, for he will become impressed, just like the pigeon or poultry-fancier before alluded to, with the amount of difference in the forms which he is continually studying; and he has little general knowledge of analogical variation in other groups and in other countries, by which to correct his first impressions. As he extends the range of his observations, he will meet with more cases of diffi-culty; for he will encounter a greater number of closely-allied forms. But if his observations be widely extended, he will in the end generally be enabled to make up his own mind which to call varieties and which species; but he will succeed in this at the expense of admitting much variation, − and the truth of this admission will often be disputed by other naturalists. When, moreover, he comes to study allied forms brought from countries not now continuous, in which case he can hardly hope to find the intermediate links between his doubtful forms, he will have to trust almost entirely to analogy, and his difficulties will rise to a climax.

Certainly no clear line of demarcation has as yet been drawn between species and sub-species – that is, the forms which in the opinion of some naturalists come very near to, but do not quite arrive at the rank of species; or, again, between sub-species and well-marked varieties, or between lesser varieties and individual differences. These differences blend into each other in an insensible series; and a series impresses the mind with the idea of an actual passage.

Hence I look at individual differences, though of small interest to the systematist, as of high importance for us, as being the first step towards such slight varieties as are barely thought worth recording in works on natural history. And I look at varieties which are in any degree more distinct and permanent, as steps leading to more strongly marked and more permanent varieties; and at these latter, as leading to sub-species, and to species. The passage from one stage of difference to another and higher stage may be, in some cases, due merely to the long-continued action of different physical conditions in two different regions; but I have not much faith in this view; and I attribute the passage of a variety, from a state in which it differs very slightly from its parent to one in which it differs more, to the action of natural selection in accumulating (as will hereafter be more fully explained) differences of structure in certain definite directions. Hence I believe a well-marked variety may be justly called an incipient species; but whether this belief be justifiable must be judged of by the general weight of the several facts and views given throughout this work.

It need not be supposed that all varieties or incipient species necessarily attain the rank of species. They may whilst in this incipient state become extinct, or they may endure as varieties for very long periods, as has been shown to be the case by Mr. Wollaston with the varieties of certain fossil land-shells in Madeira. If a variety were to flourish so as to exceed in numbers the parent species, it would then rank as the species, and the species as the variety; or it might come to supplant and exterminate the parent species; or both might co-exist, and both rank as independent species. But we shall hereafter have to return to this subject.

From these remarks it will be seen that I look at the term species, as one arbitrarily given for the sake of convenience to a set of individuals closely resembling each other, and that it does not essentially differ from the term variety, which is given to less distinct and more fluctuating forms. The term variety, again, in comparison with mere individual differences, is also applied arbitrarily, and for mere convenience sake.

Guided by theoretical considerations, I thought that some interesting results might be obtained in regard to the nature and relations of the species which vary most, by tabulating all the varieties in several well-worked floras. At first this seemed a simple task; but Mr. H. C. Watson, to whom I am much indebted for valuable advice and assistance on this subject, soon convinced me that there were many difficulties, as did subsequently Dr. Hooker, even in stronger terms. I shall reserve for my future work the discussion of these difficulties, and the tables themselves of the proportional numbers of the varying species. Dr. Hooker permits me to add, that after having carefully read my manuscript, and examined the tables, he thinks that the following statements are fairly well established. The whole subject, however, treated as it necessarily here is with much brevity, is rather perplexing, and allusions cannot be avoided to the 'struggle for existence,' 'divergence of character,' and other questions, hereafter to be discussed.

Alph. De Candolle and others have shown that plants which have very wide ranges generally present varieties; and this might have been expected, as they become exposed to diverse physical conditions, and as they come into competition (which, as we shall hereafter see, is a far more important circumstance) with different sets of organic beings. But my tables further show that, in any limited country, the species which are most common, that is abound most in individuals, and the species which are most widely diffused within their own country (and this is a different consideration from wide range, and to a certain extent from commonness), often give rise to varieties sufficiently well-marked to have been recorded in botanical works. Hence it is the most flourishing, or, as they may be called, the dominant species, – those which range widely over the world, are the most diffused in their own country, and are the most numerous in individuals, – which oftenest produce well-marked varieties, or, as I consider them, incipient species. And this, perhaps, might have been

anticipated; for, as varieties, in order to become in any degree permanent, necessarily have to struggle with the other inhabitants of the
country, the species which are already dominant will be the most
likely to yield offspring which, though in some slight degree modified, will still inherit those advantages that enabled their parents to
become dominant over their compatriots.

If the plants inhabiting a country and described in any Flora be
divided into two equal masses, all those in the larger genera being
placed on one side, and all those in the smaller genera on the other
side, a somewhat larger number of the very common and much
diffused or dominant species will be found on the side of the larger
genera. This, again, might have been anticipated; for the mere fact
of many species of the same genus inhabiting any country, shows
that there is something in the organic or inorganic conditions of that
country favourable to the genus; and, consequently, we might have
expected to have found in the larger genera, or those including many
species, a large proportional number of dominant species. But so
many causes tend to obscure this result, that I am surprised that my
tables show even a small majority on the side of the larger genera.
I will here allude to only two causes of obscurity. Fresh-water and
salt-loving plants have generally very wide ranges and are much
diffused, but this seems to be connected with the nature of the stations
inhabited by them, and has little or no relation to the size of the
genera to which the species belong. Again, plants low in the scale of
organization are generally much more widely diffused than plants
higher in the scale; and here again there is no close relation to the size
of the genera. The cause of lowly-organized plants ranging widely
will be discussed in our chapter on geographical distribution.

From looking at species as only strongly-marked and well-defined
varieties, I was led to anticipate that the species of the larger genera
in each country would oftener present varieties, than the species of the
smaller genera; for wherever many closely related species (*i.e.* species
of the same genus) have been formed, many varieties or incipient
species ought, as a general rule, to be now forming. Where many
large trees grow, we expect to find saplings. Where many species of
a genus have been formed through variation, circumstances have
been favourable for variation; and hence we might expect that the

circumstances would generally be still favourable to variation. On the other hand, if we look at each species as a special act of creation, there is no apparent reason why more varieties should occur in a group having many species, than in one having few.

To test the truth of this anticipation I have arranged the plants of twelve countries, and the coleopterous insects of two districts, into two nearly equal masses, the species of the larger genera on one side, and those of the smaller genera on the other side, and it has invariably proved to be the case that a larger proportion of the species on the side of the larger genera present varieties, than on the side of the smaller genera. Moreover, the species of the large genera which present any varieties, invariably present a larger average number of varieties than do the species of the small genera. Both these results follow when another division is made, and when all the smallest genera, with from only one to four species, are absolutely excluded from the tables. These facts are of plain signification on the view that species are only strongly marked and permanent varieties; for whenever many species of the same genus have been formed, or where, if we may use the expression, the manufactory of species has been active, we ought generally to find the manufactory still in action, more especially as we have every reason to believe the process of manufacturing new species to be a slow one. And this certainly is the case, if varieties be looked at as incipient species; for my tables clearly show as a general rule that, wherever many species of a genus have been formed, the species of that genus present a number of varieties, that is of incipient species, beyond the average. It is not that all large genera are now varying much, and are thus increasing in the number of their species, or that no small genera are now varying and increasing; for if this had been so, it would have been fatal to my theory; inasmuch as geology plainly tells us that small genera have in the lapse of time often increased greatly in size; and that large genera have often come to their maxima, declined, and disappeared. All that we want to show is, that where many species of a genus have been formed, on an average many are still forming; and this holds good.

There are other relations between the species of large genera and their recorded varieties which deserve notice. We have seen that there is no infallible criterion by which to distinguish species and well-marked varieties; and in those cases in which intermediate links have not been found between doubtful forms, naturalists are compelled to come to a determination by the amount of difference between them, judging by analogy whether or not the amount suffices to raise one or both to the rank of species. Hence the amount of difference is one very important criterion in settling whether two forms should be ranked as species or varieties. Now Fries has remarked in regard to plants, and Westwood in regard to insects, that in large genera the amount of difference between the species is often exceedingly small. I have endeavoured to test this numerically by averages, and, as far as my imperfect results go, they always confirm the view. I have also consulted some sagacious and most experienced observers, and, after deliberation, they concur in this view. In this respect, therefore, the species of the larger genera resemble varieties, more than do the species of the smaller genera. Or the case may be put in another way, and it may be said, that in the larger genera, in which a number of varieties or incipient species greater than the average are now manufacturing, many of the species already manufactured still to a certain extent resemble varieties, for they differ from each other by a less than usual amount of difference.

Moreover, the species of the large genera are related to each other, in the same manner as the varieties of any one species are related to each other. No naturalist pretends that all the species of a genus are equally distinct from each other; they may generally be divided into sub-genera, or sections, or lesser groups. As Fries has well remarked, little groups of species are generally clustered like satellites around certain other species. And what are varieties but groups of forms, unequally related to each other, and clustered round certain forms – that is, round their parent-species? Undoubtedly there is one most important point of difference between varieties and species; namely, that the amount of difference between varieties, when compared with each other or with their parent-species, is much less than that between the species of the same genus. But when we come to discuss the principle, as I call it, of Divergence of Character, we shall see how this may be explained, and how the lesser differences between varieties will tend to increase into the greater differences between species.

There is one other point which seems to me worth notice. Varieties generally have much restricted ranges: this statement is indeed scarcely more than a truism, for if a variety were found to have a wider range than that of its supposed parent-species, their denominations ought to be reversed. But there is also reason to believe, that those species which are very closely allied to other species, and in so far resemble varieties, often have much restricted ranges. For instance, Mr. H. C. Watson has marked for me in the well-sifted London Catalogue of plants (4th edition) 63 plants which are therein ranked as species, but which he considers as so closely allied to other species as to be of doubtful value: these 63 reputed species range on an average over 6.9 of the provinces into which Mr. Watson has divided Great Britain. Now, in this same catalogue, 53 acknowledged varieties are recorded, and these range over 7.7 provinces; whereas, the species to which these varieties belong range over 14.3 provinces. So that the acknowledged varieties have very nearly the same restricted average range, as have those very closely allied forms, marked for me by Mr. Watson as doubtful species, but which are almost universally ranked by British botanists as good and true species.

Finally, then, varieties have the same general characters as species, for they cannot be distinguished from species, – except, firstly, by the discovery of intermediate linking forms, and the occurrence of such links cannot affect the actual characters of the forms which they connect; and except, secondly, by a certain amount of difference, for two forms, if differing very little, are generally ranked as varieties, notwithstanding that intermediate linking forms have not been discovered; but the amount of difference considered necessary to give to two forms the rank of species is quite indefinite. In genera having more than the average number of species in any country, the species of these genera have more than the average number of varieties. In large genera the species are apt to be closely, but unequally, allied together, forming little clusters round certain species. Species very closely allied to other species apparently have restricted ranges. In all these several respects the species of large genera present a strong analogy with varieties. And we can clearly understand these analogies, if species have once existed as varieties, and have thus originated: whereas, these analogies are utterly inexplicable if each species has been independently created.

We have, also, seen that it is the most flourishing and dominant species of the larger genera which on an average vary most; and varieties, as we shall hereafter see, tend to become converted into new and distinct species. The larger genera thus tend to become larger; and throughout nature the forms of life which are now dominant tend to become still more dominant by leaving many modified and dominant descendants. But by steps hereafter to be explained, the larger genera also tend to break up into smaller genera. And thus, the forms of life throughout the universe become divided into groups subordinate to groups.

CHAPTER III

STRUGGLE FOR EXISTENCE

Bears on natural selection – The term used in a wide
sense – Geometrical powers of increase – Rapid increase
of naturalized animals and plants – Nature of the checks
to increase – Competition universal – Effects of climate –
Protection from the number of individuals – Complex
relations of all animals and plants throughout nature –
Struggle for life most severe between individuals and
varieties of the same species; often severe between species
of the same genus – The relation of organism to organism
the most important of all relations

BEFORE entering on the subject of this chapter, I must make a few
preliminary remarks, to show how the struggle for existence bears on
Natural Selection. It has been seen in the last chapter that amongst
organic beings in a state of nature there is some individual variability;
indeed I am not aware that this has ever been disputed. It is immaterial
for us whether a multitude of doubtful forms be called species or sub-
species or varieties; what rank, for instance, the two or three hundred
doubtful forms of British plants are entitled to hold, if the existence of
any well-marked varieties be admitted. But the mere existence
of individual variability and of some few well-marked varieties,
though necessary as the foundation for the work, helps us but little
in understanding how species arise in nature. How have all those
exquisite adaptations of one part of the organization to another part,
and to the conditions of life, and of one distinct organic being to
another being, been perfected? We see these beautiful co-adaptations
most plainly in the woodpecker and misseltoe; and only a little less
plainly in the humblest parasite which clings to the hairs of a quadruped or feathers of a bird; in the structure of the beetle which dives
through the water; in the plumed seed which is wafted by the gentlest
breeze; in short, we see beautiful adaptations everywhere and in every
part of the organic world.

Again, it may be asked, how is it that varieties, which I have called incipient species, become ultimately converted into good and distinct species, which in most cases obviously differ from each other far more than do the varieties of the same species? How do those groups of species, which constitute what are called distinct genera, and which differ from each other more than do the species of the same genus, arise? All these results, as we shall more fully see in the next chapter, follow inevitably from the struggle for life. Owing to this struggle for life, any variation, however slight and from whatever cause proceeding, if it be in any degree profitable to an individual of any species, in its infinitely complex relations to other organic beings and to external nature, will tend to the preservation of that individual, and will generally be inherited by its offspring. The offspring, also, will thus have a better chance of surviving, for, of the many individuals of any species which are periodically born, but a small number can survive. I have called this principle, by which each slight variation, if useful, is preserved, by the term of Natural Selection, in order to mark its relation to man's power of selection. We have seen that man by selection can certainly produce great results, and can adapt organic beings to his own uses, through the accumulation of slight but useful variations, given to him by the hand of Nature. But Natural Selection, as we shall hereafter see, is a power incessantly ready for action, and is as immeasurably superior to man's feeble efforts, as the works of Nature are to those of Art.

We will now discuss in a little more detail the struggle for existence. In my future work this subject shall be treated, as it well deserves, at much greater length. The elder De Candolle and Lyell have largely and philosophically shown that all organic beings are exposed to severe competition. In regard to plants, no one has treated this subject with more spirit and ability than W. Herbert, Dean of Manchester, evidently the result of his great horticultural knowledge. Nothing is easier than to admit in words the truth of the universal struggle for life, or more difficult – at least I have found it so – than constantly to bear this conclusion in mind. Yet unless it be thoroughly engrained in the mind, I am convinced that the whole economy of nature, with every fact on distribution, rarity, abundance, extinction, and variation, will be dimly seen or quite misunderstood. We behold the face of nature bright with gladness, we often see superabundance of food; we do not see, or we forget, that the birds which are idly singing round us mostly live on insects or seeds, and are thus constantly destroying life; or we forget how largely these songsters, or their eggs, or their nestlings, are destroyed by birds and beasts of prey; we do not always bear in mind, that though food may be now superabundant, it is not so at all seasons of each recurring year.

I should premise that I use the term Struggle for Existence in a large and metaphorical sense, including dependence of one being on another, and including (which is more important) not only the life of the individual, but success in leaving progeny. Two canine animals in a time of dearth, may be truly said to struggle with each other which shall get food and live. But a plant on the edge of a desert is said to struggle for life against the drought, though more properly it should be said to be dependent on the moisture. A plant which annually produces a thousand seeds, of which on an average only one comes to maturity, may be more truly said to struggle with the plants of the same and other kinds which already clothe the ground. The misseltoe is dependent on the apple and a few other trees, but can only in a far-fetched sense be said to struggle with these trees, for if too many of these parasites grow on the same tree, it will languish and die. But several seedling misseltoes, growing close together on the same branch, may more truly be said to struggle with each other. As the misseltoe is disseminated by birds, its existence depends on birds; and it may metaphorically be said to struggle with other fruit-bearing plants, in order to tempt birds to devour and thus disseminate its seeds rather than those of other plants. In these several senses, which pass into each other, I use for convenience sake the general term of struggle for existence.

A struggle for existence inevitably follows from the high rate at which all organic beings tend to increase. Every being, which during its natural lifetime produces several eggs or seeds, must suffer destruction during some period of its life, and during some season or occasional year, otherwise, on the principle of geometrical increase, its numbers would quickly become so inordinately great that no country could support the product. Hence, as more individuals are produced than can possibly survive, there must in every case be a struggle for existence, either one individual with another of the same species, or with the individuals of distinct species, or with the physical conditions of life. It is the doctrine of Malthus applied with manifold force to the whole animal and vegetable kingdoms; for in this case there can be no artificial increase of food, and no prudential restraint from marriage. Although some species may be now increasing, more or less rapidly, in numbers, all cannot do so, for the world would not hold them.

There is no exception to the rule that every organic being naturally increases at so high a rate, that if not destroyed, the earth would soon be covered by the progeny of a single pair. Even slow-breeding man has doubled in twenty-five years, and at this rate, in a few thousand years, there would literally not be standing room for his progeny. Linnaeus has calculated that if an annual plant produced only two seeds – and there is no plant so unproductive as this – and their seedlings next year produced two, and so on, then in twenty years there would be a million plants. The elephant is reckoned to be the slowest breeder of all known animals, and I have taken some pains to estimate its probable minimum rate of natural increase: it will be under the mark to assume that it breeds when thirty years old, and goes on breeding till ninety years old, bringing forth three pairs of young in this interval; if this be so, at the end of the fifth century there would be alive fifteen million elephants, descended from the first pair.

But we have better evidence on this subject than mere theoretical calculations, namely, the numerous recorded cases of the astonishingly rapid increase of various animals in a state of nature, when circumstances have been favourable to them during two or three following seasons. Still more striking is the evidence from our domestic animals of many kinds which have run wild in several parts of the world: if the statements of the rate of increase of slow-breeding cattle and horses in South America, and latterly in Australia, had not been well authenticated, they would have been quite incredible. So it is with plants: cases could be given of introduced plants which have become common throughout whole islands in a period of less than ten years. Several of the plants now most numerous over the wide plains of La Plata, clothing square leagues of surface almost to the exclusion of all other plants, have been introduced from

Europe; and there are plants which now range in India, as I hear from Dr. Falconer, from Cape Comorin to the Himalaya, which have been imported from America since its discovery. In such cases, and endless instances could be given, no one supposes that the fertility of these animals or plants has been suddenly and temporarily increased in any sensible degree. The obvious explanation is that the conditions of life have been very favourable, and that there has consequently been less destruction of the old and young, and that nearly all the young have been enabled to breed. In such cases the geometrical ratio of increase, the result of which never fails to be surprising, simply explains the extraordinarily rapid increase and wide diffusion of naturalized productions in their new homes.

In a state of nature almost every plant produces seed, and amongst animals there are very few which do not annually pair. Hence we may confidently assert, that all plants and animals are tending to increase at a geometrical ratio, that all would most rapidly stock every station in which they could any how exist, and that the geometrical tendency to increase must be checked by destruction at some period of life. Our familiarity with the larger domestic animals tends, I think, to mislead us: we see no great destruction falling on them, and we forget that thousands are annually slaughtered for food, and that in a state of nature an equal number would have somehow to be disposed of.

The only difference between organisms which annually produce eggs or seeds by the thousand, and those which produce extremely few, is, that the slow-breeders would require a few more years to people, under favourable conditions, a whole district, let it be ever so large. The condor lays a couple of eggs and the ostrich a score, and yet in the same country the condor may be the more numerous of the two: the Fulmar petrel lays but one egg, yet it is believed to be the most numerous bird in the world. One fly deposits hundreds of eggs, and another, like the hippobosca, a single one; but this difference does not determine how many individuals of the two species can be supported in a district. A large number of eggs is of some importance to those species, which depend on a rapidly fluctuating amount of food, for it allows them rapidly to increase in number. But the real importance of a large number of eggs or seeds is to make up for much destruction at some period of life; and this period in the great majority of cases is an early one. If an animal can in any way protect its own eggs or young, a small number may be produced, and yet the average stock be fully kept up; but if many eggs or young are destroyed, many must be produced, or the species will become extinct. It would suffice to keep up the full number of a tree, which lived on an average for a thousand years, if a single seed were produced once in a thousand years, supposing that this seed were never destroyed, and could be ensured to germinate in a fitting place. So that in all cases, the average number of any animal or plant depends only indirectly on the number of its eggs or seeds.

In looking at Nature, it is most necessary to keep the foregoing considerations always in mind – never to forget that every single organic being around us may be said to be striving to the utmost to increase in numbers; that each lives by a struggle at some period of its life; that heavy destruction inevitably falls either on the young or old, during each generation or at recurrent intervals. Lighten any check, mitigate the destruction ever so little, and the number of the species will almost instantaneously increase to any amount. The face of Nature may be compared to a yielding surface, with ten thousand sharp wedges packed close together and driven inwards by incessant blows, sometimes one wedge being struck, and then another with greater force.

What checks the natural tendency of each species to increase in number is most obscure. Look at the most vigorous species; by as much as it swarms in numbers, by so much will its tendency to increase be still further increased. We know not exactly what the checks are in even one single instance. Nor will this surprise any one who reflects how ignorant we are on this head, even in regard to mankind, so incomparably better known than any other animal. This subject has been ably treated by several authors, and I shall, in my future work, discuss some of the checks at considerable length, more especially in regard to the feral animals of South America. Here I will make only a few remarks, just to recall to the reader's mind some of the chief points. Eggs or very young animals seem generally to suffer most, but this is not invariably the case. With plants there is a vast destruction of seeds, but, from some observations which I have made, I believe that it is the seedlings which suffer most from germinating in ground already thickly stocked with other plants. Seedlings, also, are destroyed in vast numbers by various enemies; for instance, on a piece of ground three feet long and two wide, dug and cleared, and where there could be no choking from other plants, I marked all the seedlings of our native weeds as they came up, and out of the 357 no less than 295 were destroyed, chiefly by slugs and insects. If turf which has long been mown, and the case would be the same with turf closely browsed by quadrupeds, be let to grow, the more vigorous plants gradually kill the less vigorous, though fully grown, plants: thus out of twenty species growing on a little plot of turf (three feet by four) nine species perished from the other species being allowed to grow up freely.

The amount of food for each species of course gives the extreme limit to which each can increase; but very frequently it is not the obtaining food, but the serving as prey to other animals, which determines the average numbers of a species. Thus, there seems to be little doubt that the stock of partridges, grouse, and hares on any large estate depends chiefly on the destruction of vermin. If not one head of game were shot during the next twenty years in England, and, at the same time, if no vermin were destroyed, there would, in all probability, be less game than at present, although hundreds of thousands of game animals are now annually killed. On the other hand, in some cases, as with the elephant and rhinoceros, none are destroyed by beasts of prey: even the tiger in India most rarely dares to attack a young elephant protected by its dam.

Climate plays an important part in determining the average numbers of a species, and periodical seasons of extreme cold or drought, I believe to be the most effective of all checks. I estimated that the winter of 1854–55 destroyed four-fifths of the birds in my own grounds; and this is a tremendous destruction, when we remember that ten per cent. is an extraordinarily severe mortality from epidemics with man. The action of climate seems at first sight to be quite independent of the struggle for existence; but in so far as climate chiefly acts in reducing food, it brings on the most severe struggle between the individuals, whether of the same or of distinct species, which subsist on the same kind of food. Even when climate, for instance extreme cold, acts directly, it will be the least vigorous, or those which have got least food through the advancing winter, which will suffer most. When we travel from south to north, or from a damp region to a dry, we invariably see some species gradually getting rarer and rarer, and finally disappearing; and the change of climate being conspicuous, we are tempted to attribute the whole effect to its direct action. But this is a very false view: we forget that each species, even where it most abounds, is constantly suffering enormous destruction at some period of its life, from enemies or from competitors for the same place and food; and if these enemies or competitors be in the least degree favoured by any slight change of climate, they will increase in numbers, and, as each area is already fully stocked with inhabitants, the other species will decrease. When we travel southward and see a species decreasing in numbers, we may feel sure that the cause lies quite as much in other species being favoured, as in this one being hurt. So it is when we travel northward, but in a somewhat lesser degree, for the number of species of all kinds, and therefore of competitors, decreases north-wards; hence in going northward, or in ascending a mountain, we far oftener meet with stunted forms, due to the *directly* injurious action of climate, than we do in proceeding southwards or in descending a mountain. When we reach the Arctic regions, or snow-capped summits, or absolute deserts, the struggle for life is almost exclusively with the elements.

That climate acts in main part indirectly by favouring other species, we may clearly see in the prodigious number of plants in our gardens which can perfectly well endure our climate, but which never become naturalized, for they cannot compete with our native plants, nor resist destruction by our native animals.

When a species, owing to highly favourable circumstances, in-creases inordinately in numbers in a small tract, epidemics – at least, this seems generally to occur with our game animals – often ensue: and here we have a limiting check independent of the struggle for life. But even some of these so-called epidemics appear to be due to parasitic worms, which have from some cause, possibly in part through facility of diffusion amongst the crowded animals, been disproportionably favoured: and here comes in a sort of struggle between the parasite and its prey.

On the other hand, in many cases, a large stock of individuals of the same species, relatively to the numbers of its enemies, is absolutely necessary for its preservation. Thus we can easily raise plenty of corn and rape-seed, &c., in our fields, because the seeds are in great excess compared with the number of birds which feed on them; nor can the birds, though having a superabundance of food at this one season, increase in number proportionally to the supply of seed, as their numbers are checked during winter: but any one who has tried, knows how troublesome it is to get seed from a few wheat or other such plants in a garden; I have in this case lost every single seed. This view of the necessity of a large stock of the same species for its preservation, explains, I believe, some singular facts in nature, such as that of very rare plants being sometimes extremely abundant in the few spots where they do occur; and that of some social plants being social, that is, abounding in individuals, even on the extreme confines of their range. For in such cases, we may believe, that a plant could exist only where the conditions of its life were so favourable that many could exist together, and thus save each other from utter destruction. I should add that the good effects of frequent intercrossing, and the ill effects of close interbreeding, probably came into play in some of these cases; but on this intricate subject I will not here enlarge.

Many cases are on record showing how complex and unexpected are the checks and relations between organic beings, which have to struggle together in the same country. I will give only a single instance, which, though a simple one, has interested me. In Staffordshire, on the estate of a relation where I had ample means of investigation, there was a large and extremely barren heath, which had never been touched by the hand of man; but several hundred acres of exactly the same nature had been enclosed twenty-five years previously and planted with Scotch fir. The change in the native vegetation of the planted part of the heath was most remarkable, more than is generally seen in passing from one quite different soil to another: not only the proportional numbers of the heath-plants were wholly changed, but twelve species of plants (not counting grasses and carices) flourished in the plantations, which could not be found on the heath. The effect on the insects must have been still greater, for six insectivorous birds were very common in the planta-tions, which were not to be seen on the heath; and the heath was frequented by two or three distinct insectivorous birds. Here we see how potent has been the effect of the introduction of a single tree, nothing whatever else having been done, with the exception that the land had been enclosed, so that cattle could not enter. But how important an element enclosure is, I plainly saw near Farnham, in Surrey. Here there are extensive heaths, with a few clumps of old Scotch firs on the distant hill-tops: within the last ten years large spaces have been enclosed, and self-sown firs are now springing up in multitudes, so close together that all cannot live. When I ascertained that these young trees had not been sown or planted, I was so much surprised at their numbers that I went to several points of view, whence I could examine hundreds of acres of the unenclosed heath, and literally I could not see a single Scotch fir, except the old planted clumps. But on looking closely between the stems of the heath, I found a multitude of seedlings and little trees, which had been perpetually browsed down by the cattle. In one square yard, at a point some hundreds yards distant from one of the old clumps, I counted thirty-two little trees; and one of them, judging from the rings of growth, had during twenty-six years tried to raise its head above the stems of the heath, and had failed. No wonder that, as soon as the land was enclosed, it became thickly clothed with vigorously growing young firs. Yet the heath was so extremely barren and so extensive that no one would ever have imagined that cattle would have so closely and effectually searched it for food.

Here we see that cattle absolutely determine the existence of the Scotch fir; but in several parts of the world insects determine the existence of cattle. Perhaps Paraguay offers the most curious instance of this; for here neither cattle nor horses nor dogs have ever run wild, though they swarm southward and northward in a feral state; and Azara and Rengger have shown that this is caused by the greater number in Paraguay of a certain fly, which lays its eggs in the navels of these animals when first born. The increase of these flies, numerous as they are, must be habitually checked by some means, probably by birds. Hence, if certain insectivorous birds (whose numbers are probably regulated by hawks or beasts of prey) were to increase in

Paraguay, the flies would decrease – then cattle and horses would become feral, and this would certainly greatly alter (as indeed I have observed in parts of South America) the vegetation: this again would largely affect the insects; and this, as we just have seen in Staffordshire, the insectivorous birds, and so onwards in ever-increasing circles of complexity. We began this series by insectivorous birds, and we have ended with them. Not that in nature the relations can ever be as simple as this. Battle within battle must ever be recurring with varying success; and yet in the long-run the forces are so nicely balanced, that the face of nature remains uniform for long periods of time, though assuredly the merest trifle would often give the victory to one organic being over another. Nevertheless so profound is our ignorance, and so high our presumption, that we marvel when we hear of the extinction of an organic being; and as we do not see the cause, we invoke cataclysms to desolate the world, or invent laws on the duration of the forms of life!

I am tempted to give one more instance showing how plants and animals, most remote in the scale of nature, are bound together by a web of complex relations. I shall hereafter have occasion to show that the exotic Lobelia fulgens, in this part of England, is never visited by insects, and consequently, from its peculiar structure, never can set a seed. Many of our orchidaceous plants absolutely require the visits of moths to remove their pollen-masses and thus to fertilize them. I have, also, reason to believe that humble-bees are indispensable to the fertilization of the heartsease (Viola tricolor), for other bees do not visit this flower. From experiments which I have tried, I have found that the visits of bees, if not indispensable, are at least highly beneficial to the fertilization of our clovers; but humble-bees alone visit the common red clover (Trifolium pratense), as other bees cannot reach the nectar. Hence I have very little doubt, that if the whole genus of humble-bees became extinct or very rare in England, the heartsease and red clover would become very rare, or wholly disappear. The number of humble-bees in any district depends in a great degree on the number of field-mice, which destroy their combs and nests; and Mr. H. Newman, who has long attended to the habits of humble-bees, believes that 'more than two thirds of them are thus destroyed all over England.' Now the number of mice is largely dependent, as every one knows, on the number of cats; and Mr. Newman says, 'Near villages and small towns I have found the nests of humble-bees more numerous than elsewhere, which I attribute to the number of cats that destroy the mice.' Hence it is quite credible that the presence of a feline animal in large numbers in a district might determine, through the intervention first of mice and then of bees, the frequency of certain flowers in that district!

In the case of every species, many different checks, acting at different periods of life, and during different seasons or years, probably come into play; some one check or some few being generally the most potent, but all concurring in determining the average number or even the existence of the species. In some cases it can be shown that widely-different checks act on the same species in different districts. When we look at the plants and bushes clothing an entangled bank, we are tempted to attribute their proportional numbers and kinds to what we call chance. But how false a view is this! Every one has heard that when an American forest is cut down, a very different vegetation springs up; but it has been observed that the trees now growing on the ancient Indian mounds, in the Southern United States, display the same beautiful diversity and proportion of kinds as in the surrounding virgin forests. What a struggle between the several kinds of trees must here have gone on during long centuries, each annually scattering its seeds by the thousand; what war between insect and insect – between insects, snails, and other animals with birds and beasts of prey – all striving to increase, and all feeding on each other or on the trees or their seeds and seedlings, or on the other plants which first clothed the ground and thus checked the growth of the trees! Throw up a handful of feathers, and all must fall to the ground according to definite laws; but how simple is this problem compared to the action and reaction of the innumerable plants and animals which have determined, in the course of centuries, the proportional numbers and kinds of trees now growing on the old Indian ruins!

The dependency of one organic being on another, as of a parasite on its prey, lies generally between beings remote in the scale of nature. This is often the case with those which may strictly be said to struggle with each other for existence, as in the case of locusts and grass-feeding quadrupeds. but the struggle almost invariably will be most severe between the individuals of the same species, for they frequent the same districts, require the same food, and are exposed to the same dangers. In the case of varieties of the same species, the struggle will generally be almost equally severe, and we sometimes see the contest soon decided: for instance, if several varieties of wheat be sown together, and the mixed seed be resown, some of the varieties which best suit the soil or climate, or are naturally the most fertile, will beat the others and so yield more seed, and will consequently in a few years quite supplant the other varieties. To keep up a mixed stock of even such extremely close varieties as the variously coloured sweet-peas, they must be each year harvested separately, and the seed then mixed in due proportion, otherwise the weaker kinds will steadily decrease in numbers and disappear. So again with the varieties of sheep: it has been asserted that certain mountain-varieties will starve out other mountain-varieties, so that they cannot be kept together. The same result has followed from keeping together differ- ent varieties of the medicinal leech. It may even be doubted whether the varieties of any one of our domestic plants or animals have so exactly the same strength, habits, and constitution, that the original proportions of a mixed stock could be kept up for half a dozen generations, it they were allowed to struggle together, like beings in a state of nature, and if the seed or young were not annually sorted.

As species of the same genus have usually, though by no means invariably, some similarity in habits and constitution, and always in structure, the struggle will generally be more severe between species of the same genus, when they come into competition with each other, than between species of distinct genera. We see this in the recent extension over parts of the United States of one species of swallow having caused the decrease of another species. The recent increase of the missel-thrush in parts of Scotland has caused the decrease of the song-thrush. How frequently we hear of one species of rat taking the place of another species under the most different climates! In Russia the small Asiatic cockroach has everywhere driven before it its great congener. One species of charlock will supplant another, and so in other cases. We can dimly see why the competition should be most severe between allied forms, which fill nearly the same place in the economy of nature; but probably in no one case could we precisely say why one species has been victorious over another in the great battle of life.

A corollary of the highest importance may be deduced from the foregoing remarks, namely, that the structure of every organic being is related, in the most essential yet often hidden manner, to that of all other organic beings, with which it comes into competition for food or residence, or from which it has to escape, or on which it preys. This is obvious in the structure of the teeth and talons of the tiger; and in that of the legs and claws of the parasite which clings to the hair on the tiger's body. But in the beautifully plumed seed of the dandelion, and in the flattened and fringed legs of the water-beetle, the relation seems at first confined to the elements of air and water. Yet the advantage of plumed seeds no doubt stands in the closest relation to the land being already thickly clothed by other plants; so that the seeds may be widely distributed and fall on unoccupied ground. In the water-beetle, the structure of its legs, so well adapted for diving, allows it to compete with other aquatic insects, to hunt for its own prey, and to escape serving as prey to other animals.

The store of nutriment laid up within the seeds of many plants seems at first sight to have no sort of relation to other plants. But from the strong growth of young plants produced from such seeds (as peas and beans), when sown in the midst of long grass, I suspect that the chief use of the nutriment in the seed is to favour the growth of the young seedling, whilst struggling with other plants growing vigorously all around.

Look at a plant in the midst of its range, why does it not double or quadruple its numbers? We know that it can perfectly well withstand a little more heat or cold, dampness or dryness, for elsewhere it ranges into slightly hotter or colder, damper or drier districts. In this case we can clearly see that if we wished in imagination to give the plant the power of increasing in number, we should have to give it some advantage over its competitors, or over the animals which preyed on it. On the confines of its geographical range, a change of constitution with respect to climate would clearly be an advantage to our plant; but we have reason to believe that only a few plants or animals range so far, that they are destroyed by the rigour of the climate alone. Not until we reach the extreme confines of life, in the arctic regions or on the borders of an utter desert, will competition cease. The land may be extremely cold or dry, yet there will be competition between some few species, or between the individuals of the same species, for the warmest or dampest spots.

Hence, also, we can see that when a plant or animal is placed in a new country amongst new competitors, though the climate may be exactly the same as in its former home, yet the conditions of its life will generally be changed in an essential manner. If we wished to increase its average numbers in its new home, we should have to modify it in a different way to what we should have done in its native country; for we should have to give it some advantage over a different set of competitors or enemies.

It is good thus to try in our imagination to give any form some advantage over another. Probably in no single instance should we know what to do, so as to succeed. It will convince us of our ignorance on the mutual relations of all organic beings; a conviction as necessary, as it seems to be difficult to acquire. All that we can do, is to keep steadily in mind that each organic being is striving to increase at a geometrical ratio; that each at some period of its life, during some season of the year, during each generation or at intervals, has to struggle for life, and to suffer great destruction. When we reflect on this struggle, we may console ourselves with the full belief, that the war of nature is not incessant, that no fear is felt, that death is generally prompt, and that the vigorous, the healthy, and the happy survive and multiply.

CHAPTER IV

NATURAL SELECTION

Natural Selection – its power compared with man's selection – its power on characters of trifling importance – its power at all ages and on both sexes – Sexual Selection – On the generality of intercrosses between individuals of the same species – Circumstances favourable and unfavourable to Natural Selection, namely, intercrossing, isolation, number of individuals – Slow action – Extinction caused by Natural Selection – Divergence of Character, related to the diversity of inhabitants of any small area, and to naturalization – Action of Natural Selection, through Divergence of Character and Extinction, on the descendants from a common parent – Explains the Grouping of all organic beings

HOW will the struggle for existence, discussed too briefly in the last chapter, act in regard to variation? Can the principle of selection, which we have seen is so potent in the hands of man, apply in nature? I think we shall see that it can act most effectually. Let it be borne in mind in what an endless number of strange peculiarities our domestic productions, and, in a lesser degree, those under nature, vary; and how strong the hereditary tendency is. Under domestication, it may be truly said that the whole organization becomes in some degree plastic. Let it be borne in mind how infinitely complex and close-fitting are the mutual relations of all organic beings to each other and to their physical conditions of life. Can it, then, be thought improbable, seeing that variations useful to man have undoubtedly occurred, that other variations useful in some way to each being in the great and complex battle of life, should sometimes occur in the course of thousands of generations? If such do occur, can we doubt (remembering that many more individuals are born than can possibly survive) that individuals having any advantage, however slight, over others, would have the best chance of surviving and of procreating their kind? On the other hand, we may feel sure that any variation in the least degree injurious would be rigidly destroyed. This preservation of favourable variations and the rejection of injurious variations, I call Natural Selection. Variations neither useful nor injurious would not be affected by natural selection, and would be left a fluctuating element, as perhaps we see in the species called polymorphic.

We shall best understand the probable course of natural selection by taking the case of a country undergoing some physical change, for instance, of climate. The proportional numbers of its inhabitants would almost immediately undergo a change, and some species might become extinct. We may conclude, from what we have seen of the intimate and complex manner in which the inhabitants of each country are bound together, that any change in the numerical proportions of some of the inhabitants, independently of the change of climate itself, would most seriously affect many of the others. If the country were open on its borders, new forms would certainly immigrate, and this also would seriously disturb the relations of some of the former inhabitants. Let it be remembered how powerful the influence of a single introduced tree or mammal has been shown to be. But in the case of an island, or of a country partly surrounded by barriers, into which new and better adapted forms could not freely enter, we should then have places in the economy of nature which would assuredly be better filled up, if some of the original inhabitants were in some manner modified; for, had the area been open to immigration, these same places would have been seized on by intruders. In such case, every slight modification, which in the course of ages chanced to arise, and which in any way favoured the individuals of any of the species, by better adapting them to their altered conditions, would tend to be preserved; and natural selection would thus have free scope for the work of improvement.

We have reason to believe, as stated in the first chapter, that a change in the conditions of life, by specially acting on the reproductive system, causes or increases variability; and in the foregoing case the conditions of life are supposed to have undergone a change, and this would manifestly be favourable to natural selection, by giving a better chance of profitable variations occurring; and unless profitable variations do occur, natural selection can do nothing. Not that, as I believe, any extreme amount of variability is necessary; as man can certainly produce great results by adding up in any given direction mere individual differences, so could Nature, but far more easily, from having incomparably longer time at her disposal. Nor do I believe that any great physical change, as of climate, or any unusual degree of isolation to check immigration, is actually necessary to produce new and unoccupied places for natural selection to fill up by modifying and improving some of the varying inhabitants. For as all the inhabitants of each country are struggling together with nicely balanced forces, extremely slight modifications in the structure or habits of one inhabitant would often give it an advantage over others; and still further modifications of the same kind would often still further increase the advantage. No country can be named in which all the native inhabitants are now so perfectly adapted to each other and to the physical conditions under which they live, that none of them could anyhow be improved; for in all countries, the natives have been so far conquered by naturalized productions, that they have allowed foreigners to take firm possession of the land. And as foreigners have thus everywhere beaten some of the natives, we may safely conclude that the natives might have been modified with advantage, so as to have better resisted such intruders.

As man can produce and certainly has produced a great result by his methodical and unconscious means of selection, what may not nature effect? Man can act only on external and visible characters: nature cares nothing for appearances, except in so far as they may be useful to any being. She can act on every internal organ, on every shade of constitutional difference, on the whole machinery of life. Man selects only for his own good; Nature only for that of the being which she tends. Every selected character is fully exercised by her; and the being is placed under well-suited conditions of life. Man keeps the natives of many climates in the same country; he seldom exercises each selected character in some peculiar and fitting manner; he feeds a long and a short beaked pigeon on the same food; he does not exercise a long-backed or long-legged quadruped in any peculiar manner; he exposes sheep with long and short wool to the same climate. He does not allow the most vigorous males to struggle for the females. He does not rigidly destroy all inferior animals, but protects during each varying season, as far as lies in his power, all his productions. He often begins his selection by some half-monstrous form; or at least by some modification prominent enough to catch his eye, or to be plainly useful to him. Under nature, the slightest difference of structure or constitution may well turn the nicely-balanced scale in the struggle for life, and so be preserved. How fleeting are the wishes and efforts of man! how short his time! and consequently how poor will his products be, compared with those accumulated by nature during whole geological periods. Can we wonder, then, that nature's productions should be far 'truer' in character than man's productions; that they should be infinitely better adapted to the most complex conditions of life, and should plainly bear the stamp of far higher workmanship?

It may be said that natural selection is daily and hourly scrutinizing, throughout the world, every variation, even the slightest; rejecting that which is bad, preserving and adding up all that is good; silently and insensibly working, whenever and wherever opportunity offers, at the improvement of each organic being in relation to its organic and inorganic conditions of life. We see nothing of these slow changes in progress, until the hand of time has marked the long lapses of ages, and then so imperfect is our view into long past geological ages, that we only see that the forms of life are now different from what they formerly were.

Although natural selection can act only through and for the good of each being, yet characters and structures, which we are apt to consider as of very trifling importance, may thus be acted on. When we see leaf-eating insects green, and bark-feeders mottled-grey; the alpine ptarmigan white in winter, the red-grouse the colour of heather, and the black-grouse that of peaty earth, we must believe that these tints are of service to these birds and insects in preserving them from danger. Grouse, if not destroyed at some period of their lives, would increase in countless numbers; they are known to suffer largely from birds of prey; and hawks are guided by eyesight to their prey, – so much so, that on parts of the Continent persons are warned not to keep white pigeons, as being the most liable to destruction. Hence I can see no reason to doubt that natural selection might be most effective in giving the proper colour to each kind of grouse, and in keeping that colour, when once acquired, true and constant. Nor ought we to think that the occasional destruction of an animal of any particular colour would produce little effect: we should remember how essential it is in a flock of white sheep to destroy every lamb with the faintest trace of black. In plants the down on the fruit and the colour of the flesh are considered by botanists as characters of the most trifling importance: yet we hear from an excellent horticulturist, Downing, that in the United States smooth-skinned fruits suffer far more from a beetle, a curculio, than those with down; that purple plums suffer far more from a certain disease than yellow plums; whereas another disease attacks yellow-fleshed peaches far more than those with other coloured flesh. If, with all the aids of art, these slight differences make a great difference in cultivating the several varieties, assuredly, in a state of nature, where the trees would have to struggle with other trees and with a host of enemies, such differences would effectually settle which variety, whether a smooth or downy, a yellow or purple-fleshed fruit, should succeed.

In looking at many small points of difference between species, which, as far as our ignorance permits us to judge, seem to be quite unimportant, we must not forget that climate, food, &c., probably produce some slight and direct effect. It is, however, far more necessary to bear in mind that there are many unknown laws of correlation of growth, which, when one part of the organization is modified through variation, and the modifications are accumulated by natural selection for the good of the being, will cause other modifications, often of the most unexpected nature.

As we see that those variations which under domestication appear at any particular period of life, tend to reappear in the offspring at the same period; – for instance, in the seeds of the many varieties of our culinary and agricultural plants; in the caterpillar and cocoon stages of the varieties of the silkworm; in the eggs of poultry, and in the colour of the down of their chickens; in the horns of our sheep and cattle when nearly adult; – so in a state of nature, natural selection will be enabled to act on and modify organic beings at any age, by the accumulation of profitable variations at that age, and by their inheritance at a corresponding age. If it profit a plant to have its seeds more and more widely disseminated by the wind, I can see no greater difficulty in this being effected through natural selection, than in the cotton-planter increasing and improving by selection the down in the pods on his cotton-trees. Natural selection may modify and adapt the larva of an insect to a score of contingencies, wholly different from those which concern the mature insect. These modifications will no doubt affect, through the laws of correlation, the structure of the adult; and probably in the case of those insects which live only for a few hours, and which never feed, a large part of their structure is merely the correlated result of successive changes in the structure of their larvae. So, conversely, modifications in the adult will probably often affect the structure of the larva; but in all cases natural selection will ensure that modifications consequent on other modifications at a different period of life, shall not be in the least degree injurious: for if they became so, they would cause the extinction of the species.

Natural selection will modify the structure of the young in relation to the parent, and of the parent in relation to the young. In social animals it will adapt the structure of each individual for the benefit of the community; if each in consequence profits by the selected change. What natural selection cannot do, is to modify the structure of one species, without giving it any advantage, for the good of another species; and though statements to this effect may be found in works of natural history, I cannot find one case which will bear investigation. A structure used only once in an animal's whole life, if of high importance to it, might be modified to any extent by natural selection; for instance, the great jaws possessed by certain insects, and used exclusively for opening the cocoon – or the hard tip to the beak of nestling birds, used for breaking the egg. It has been asserted, that of the best short-beaked tumbler-pigeons more perish in the egg than are able to get out of it; so that fanciers assist in the act of hatching. Now, if nature had to make the beak of a full-grown pigeon very short for the bird's own advantage, the process of modification would be very slow, and there would be simultaneously the most rigorous selection of the young birds within the egg, which had the most powerful and hardest beaks, for all with weak beaks would inevitably perish: or, more delicate and more easily broken shells might be selected, the thickness of the shell being known to vary like every other structure.

Sexual Selection. Inasmuch as peculiarities often appear under domestication in one sex and become hereditarily attached to that sex, the same fact probably occurs under nature, and if so, natural selection will be able to modify one sex in its functional relations to the other sex, or in relation to wholly different habits of life in the two sexes, as is sometimes the case with insects. And this leads me to say a few words on what I call Sexual Selection. This depends, not on a struggle for existence, but on a struggle between the males for possession of the females; the result is not death to the unsuccessful competitor, but few or no offspring. Sexual selection is, therefore, less rigorous than natural selection. Generally, the most vigorous males, those which are best fitted for their places in nature, will leave most progeny. But in many cases, victory will depend not on general vigour, but on having special weapons, confined to the male sex. A hornless stag or spurless cock would have a poor chance of leaving offspring. Sexual selection by always allowing the victor to breed might surely give indomitable courage, length to the spur, and strength to the wing to strike in the spurred leg, as well as the brutal cock-fighter, who knows well that he can improve his breed by careful selection of the best cocks. How low in the scale of nature this law of battle descends, I know not; male alligators have been described as fighting, bellowing, and whirling round, like Indians in a war-dance, for the possession of the females; male salmons have been seen fighting all day long; male stag-beetles often bear wounds from the huge mandibles of other males. The war is, perhaps, severest between the males of polygamous animals, and these seem oftenest provided with special weapons. The males of carnivorous animals are already well armed; though to them and to others, special means of defence may be given through means of sexual selection, as the mane to the lion, the shoulder-pad to the boar, and the hooked jaw to the male salmon; for the shield may be as important for victory, as the sword or spear.

Amongst birds, the contest is often of a more peaceful character. All those who have attended to the subject, believe that there is the severest rivalry between the males of many species to attract by singing the females. The rock-thrush of Guiana, birds of Paradise, and some others, congregate; and successive males display their gorgeous plumage and perform strange antics before the females, which standing by as spectators, at last choose the most attractive partner. Those who have closely attended to birds in confinement well know that they often take individual preferences and dislikes: thus Sir R. Heron has described how one pied peacock was eminently attractive to all his hen birds. It may appear childish to attribute any effect to such apparently weak means: I cannot here enter on the details necessary to support this view; but if man can in a short time give elegant carriage and beauty to his bantams, according to his standard of beauty, I can see no good reason to doubt that female birds, by selecting, during thousands of generations, the most melodious or beautiful males, according to their standard of beauty, might produce a marked effect. I strongly suspect that some well-known laws with respect to the plumage of male and female birds, in comparison with the plumage of the young, can be explained on the view of plumage having been chiefly modified by sexual selection, acting when the birds have come to the breeding age or during the breeding season; the modifications thus produced being inherited at corresponding ages or seasons, either by the males alone, or by the males and females; but I have not space here to enter on this subject.

Thus it is, as I believe, that when the males and females of any animal have the same general habits of life, but differ in structure, colour, or ornament, such differences have been mainly caused by sexual selection; that is, individual males have had, in successive generations, some slight advantage over other males, in their weapons, means of defence, or charms; and have transmitted these advantages to their male offspring. Yet, I would not wish to attribute all such sexual differences to this agency: for we see peculiarities arising and becoming attached to the male sex in our domestic animals (as the wattle in male carriers, horn-like protuberances in the cocks of certain fowls, &c.), which we cannot believe to be either useful to the males in battle, or attractive to the females. We see analogous cases under nature, for instance, the tuft of hair on the breast of the turkey-cock, which can hardly be either useful or ornamental to this bird; – indeed, had the tuft appeared under domestication, it would have been called a monstrosity.

Illustrations of the action of Natural Selection. In order to make it clear how, as I believe, natural selection acts, I must beg permission to give one or two imaginary illustrations. Let us take the case of a wolf, which preys on various animals, securing some by craft, some by strength, and some by fleetness; and let us suppose that the fleetest prey, a deer for instance, had from any change in the country increased in numbers, or that other prey had decreased in numbers, during that season of the year when the wolf is hardest pressed for food. I can under such circumstances see no reason to doubt that the swiftest and slimmest wolves would have the best chance of surviving, and so be preserved or selected, – provided always that they retained strength to master their prey at this or at some other period of the year, when they might be compelled to prey on other animals. I can see no more reason to doubt this, than that man can improve the fleetness of his greyhounds by careful and methodical selection, or by that unconscious selection which results from each man trying to keep the best dogs without any thought of modifying the breed.

Even without any change in the proportional numbers of the animals on which our wolf preyed, a cub might be born with an innate tendency to pursue certain kinds of prey. Nor can this be thought very improbable; for we often observe great differences in the natural tendencies of our domestic animals; one cat, for instance, taking to catch rats, another mice; one cat, according to Mr. St. John, bringing home winged game, another hares or rabbits, and another hunting on marshy ground and almost nightly catching woodcocks or snipes. The tendency to catch rats rather than mice is known to be inherited. Now, if any slight innate change of habit or of structure benefited an individual wolf, it would have the best chance of surviving and of leaving offspring. Some of its young would probably inherit the same habits or structure, and by the repetition of this process, a new variety might be formed which would either supplant or co-exist with the parent-form of wolf. Or, again, the wolves inhabiting a mountainous district, and those frequenting the lowlands, would naturally be forced to hunt different prey; and from the continued preservation of the individuals best fitted for the two sites, two varieties might slowly be formed. These varieties would cross and blend where they met; but to this subject of intercrossing we shall soon have to return. I may add, that, according to Mr. Pierce, there are two varieties of the wolf inhabiting the Catskill Mountains in the United States, one with a light greyhound-like form, which pursues deer, and the other more bulky, with shorter legs, which more frequently attacks the shepherd's flocks.

Let us now take a more complex case. Certain plants excrete a sweet juice, apparently for the sake of eliminating something injurious from their sap: this is effected by glands at the base of the stipules in some Leguminosae, and at the back of the leaf of the common laurel. This juice, though small in quantity, is greedily sought by insects. Let us now suppose a little sweet juice or nectar to be excreted by the inner bases of the petals of a flower. In this case insects in seeking the nectar would get dusted with pollen, and would certainly often transport the pollen from one flower to the stigma of another flower. The flowers of two distinct individuals of the same species would thus get crossed; and the act of crossing, we have good reason to believe (as will hereafter be more fully alluded to), would produce very vigorous seedlings, which consequently would have the best chance of flourishing and surviving. Some of these seedlings would probably inherit the nectar-excreting power. Those individual flowers which had the largest glands or nectaries, and which excreted most nectar, would be oftenest visited by insects, and would be oftenest crossed; and so in the long-run would gain the upper hand. Those flowers, also, which had their stamens and pistils placed, in relation to the size and habits of the particular insects which visited them, so as to favour in any degree the transportal of their pollen from flower to flower, would likewise be favoured or selected. We might have taken the case of insects visiting flowers for the sake of collecting pollen instead of nectar; and as pollen is formed for the sole object of fertilization, its destruction appears a simple loss to the plant; yet if a little pollen were carried, at first occasionally and then habitually, by the pollen-devouring insects from flower to flower, and a cross thus effected, although nine-tenths of the pollen were destroyed, it might still be a great gain to the plant; and those individuals which produced more and more pollen, and had larger and larger anthers, would be selected.

When our plant, by this process of the continued preservation or natural selection of more and more attractive flowers, had been rendered highly attractive to insects, they would, unintentionally on their part, regularly carry pollen from flower to flower; and that they can most effectually do this, I could easily show by many striking instances. I will give only one – not as a very striking case, but as likewise illustrating one step in the separation of the sexes of plants, presently to be alluded to. Some holly-trees bear only male flowers, which have four stamens producing rather a small quantity of pollen, and a rudimentary pistil; other holly-trees bear only female flowers; these have a full-sized pistil, and four stamens with shrivelled anthers, in which not a grain of pollen can be detected. Having found a female tree exactly sixty yards from a male tree, I put the stigmas of twenty flowers, taken from different branches, under the microscope, and on all, without exception, there were pollen-grains, and on some a profusion of pollen. As the wind had set for several days from the female to the male tree, the pollen could not thus have been carried. The weather had been cold and boisterous, and therefore not favourable to bees, nevertheless every female flower which I examined had been effectually fertilized by the bees, acciden-tally dusted with pollen, having flown from tree to tree in search of nectar. But to return to our imaginary case: as soon as the plant had been rendered so highly attractive to insects that pollen was regularly carried from flower to flower, another process might commence. No naturalist doubts the advantage of what has been called the 'physio-logical division of labour;' hence we may believe that it would be advantageous to a plant to produce stamens alone in one flower or on one whole plant, and pistils alone in another flower or on another plant. In plants under culture and placed under new conditions of life, sometimes the male organs and sometimes the female organs become more or less impotent; now if we suppose this to occur in ever so slight a degree under nature, then as pollen is already carried regularly from flower to flower, and as a more complete separation of the sexes of our plant would be advantageous on the principle of the division of labour, individuals with this tendency more and more increased, would be continually favoured or selected, until at last a complete separation of the sexes would be effected.

Let us now turn to the nectar-feeding insects in our imaginary case: we may suppose the plant of which we have been slowly increasing the nectar by continued selection, to be a common plant; and that certain insects depended in main part on its nectar for food. I could give many facts, showing how anxious bees are to save time; for instance, their habit of cutting holes and sucking the nectar at the bases of certain flowers, which they can, with a very little more trouble, enter by the mouth. Bearing such facts in mind, I can see no reason to doubt that an accidental deviation in the size and form of the body, or in the curvature and length of the proboscis, &c., far too slight to be appreciated by us, might profit a bee or other insect, so that an individual so characterized would be able to obtain its food more quickly, and so have a better chance of living and leaving descendants. Its descendants would probably inherit a tendency to a similar slight deviation of structure. The tubes of the corollas of the common red and incarnate clovers (Trifolium pratense and incarnatum) do not on a hasty glance appear to differ in length; yet the hive-bee can easily suck the nectar out of the incarnate clover, but not out of the common red clover, which is visited by humble-bees alone; so that whole fields of the red clover offer in vain an abundant supply of precious nectar to the hive-bee. Thus it might be a great advantage to the hive-bee to have a slightly longer or differently constructed proboscis. On the other hand, I have found by experiment that the fertility of clover greatly depends on bees visiting and moving parts of the corolla, so as to push the pollen on to the stigmatic surface. Hence, again, if humble-bees were to become rare in any country, it might be a great advantage to the red clover to have a shorter or more deeply divided tube to its corolla, so that the hive-bee could visit its flowers. Thus I can understand how a flower and a bee might slowly become, either simultaneously or one after the other, modified and adapted in the most perfect manner to each other, by the continued preservation of individuals presenting mutual and slightly favourable deviations of structure.

I am well aware that this doctrine of natural selection, exemplified in the above imaginary instances, is open to the same objections which were at first urged against Sir Charles Lyell's noble views on 'the modern changes of the earth, as illustrative of geology;' but we now very seldom hear the action, for instance, of the coast-waves, called a trifling and insignificant cause, when applied to the excavation of gigantic valleys or to the formation of the longest lines of inland cliffs. Natural selection can act only by the preservation and accumulation of infinitesimally small inherited modifications, each profitable to the preserved being; and as modern geology has almost banished such views as the excavation of a great valley by a single diluvial wave, so will natural selection, if it be a true principle, banish the belief of the continued creation of new organic beings, or of any great and sudden modification in their structure.

On the Intercrossing of Individuals. I must here introduce a short digression. In the case of animals and plants with separated sexes, it is of course obvious that two individuals must always unite for each birth; but in the case of hermaphrodites this is far from obvious. Nevertheless I am strongly inclined to believe that with all hermaphrodites two individuals, either occasionally or habitually, concur for the reproduction of their kind. This view, I may add, was first suggested by Andrew Knight. We shall presently see its importance; but I must here treat the subject with extreme brevity, though I have the materials prepared for an ample discussion. All vertebrate animals, all insects, and some other large groups of animals, pair for each birth. Modern research has much diminished the number of supposed hermaphrodites, and of real hermaphrodites a large number pair; that is, two individuals regularly unite for reproduction, which is all that concerns us. But still there are many hermaphrodite animals which certainly do not habitually pair, and a vast majority of plants are hermaphrodites. What reason, it may be asked, is there for supposing in these cases that two individuals ever concur in reproduction? As it is impossible here to enter on details, I must trust to some general considerations alone.

In the first place, I have collected so large a body of facts, showing, in accordance with the almost universal belief of breeders, that with animals and plants a cross between different varieties, or between individuals of the same variety but of another strain, gives vigour and fertility to the offspring; and on the other hand, that *close* interbreeding diminishes vigour and fertility; that these facts alone incline me to believe that it is a general law of nature (utterly ignorant though we be of the meaning of the law) that no organic being self-fertilizes itself for an eternity of generations; but that a cross with another individual is occasionally – perhaps at very long intervals – indispensable.

On the belief that this is a law of nature, we can, I think, understand several large classes of facts, such as the following, which on any other view are inexplicable. Every hybridizer knows how unfavourable exposure to wet is to the fertilization of a flower, yet what a multitude of flowers have their anthers and stigmas fully exposed to the weather! but if an occasional cross be indispensable, the fullest freedom for the entrance of pollen from another individual will explain this state of exposure, more especially as the plant's own anthers and pistil generally stand so close together that self-fertilization seems almost inevitable. Many flowers, on the other hand, have their organs of fructification closely enclosed, as in the great papilionaceous or pea-family; but in several, perhaps in all, such flowers, there is a very curious adaptation between the structure of the flower and the manner in which bees suck the nectar; for, in doing this, they either push the flower's own pollen on the stigma, or bring pollen from another flower. So necessary are the visits of bees to papilionaceous flowers, that I have found, by experiments published elsewhere, that their fertility is greatly diminished if these visits be prevented. Now, it is scarcely possible that bees should fly from flower to flower, and not carry pollen from one to the other, to the great good, as I believe, of the plant. Bees will act like a camel-hair pencil, and it is quite sufficient just to touch the anthers of one flower and then the stigma of another with the same brush to ensure fertilization; but it must not be supposed that bees would thus produce a multitude of hybrids between distinct species; for if you bring on the same brush a plant's own pollen and pollen from another species, the former will have such a prepotent effect, that it will invariably and completely destroy, as has been shown by Gärtner, any influence from the foreign pollen.

When the stamens of a flower suddenly spring towards the pistil, or slowly move one after the other towards it, the contrivance seems adapted solely to ensure self-fertilization; and no doubt it is useful for this end: but, the agency of insects is often required to cause the stamens to spring forward, as Kölreuter has shown to be the case with the barberry; and curiously in this very genus, which seems to have a special contrivance for self-fertilization, it is well known that if very closely-allied forms or varieties are planted near each other, it is hardly possible to raise pure seedlings, so largely do they naturally cross. In many other cases, far from there being any aids for self-fertilization, there are special contrivances, as I could show from the writings of C. C. Sprengel and from my own observations, which effectually prevent the stigma receiving pollen from its own flower: for instance, in Lobelia fulgens, there is a really beautiful and elaborate contrivance by which every one of the infinitely numerous pollen-granules is swept out of the conjoined anthers of each flower, before the stigma of that individual flower is ready to receive them; and as this flower is never visited, at least in my garden, by insects, it never sets a seed, though by placing pollen from one flower on the stigma of another, I raised plenty of seedlings; and whilst another species of Lobelia growing close by, which is visited by bees, seeds freely. In very many other cases, though there be no special mechanical contrivance to prevent the stigma of a flower receiving its own pollen, yet, as C. C. Sprengel has shown, and as I can confirm, either the anthers burst before the stigma is ready for fertilization, or the stigma is ready before the pollen of that flower is ready, so that these plants have in fact separated sexes, and must habitually be crossed. How strange are these facts! How strange that the pollen and stigmatic surface of the same flower, though placed so close together, as if for the very purpose of self-fertilization, should in so many cases be mutually useless to each other! How simply are these facts explained on the view of an occasional cross with a distinct individual being advantageous or indispensable!

If several varieties of the cabbage, radish, onion, and of some other plants, be allowed to seed near each other, a large majority, as I have found, of the seedlings thus raised will turn out mongrels: for instance, I raised 233 seedling cabbages from some plants of different varieties growing near each other, and of these only 78 were true to their kind, and some even of these were not perfectly true. Yet the pistil of each cabbage-flower is surrounded not only by its own six stamens, but by those of the many other flowers on the same plant. How, then, comes it that such a vast number of the seedlings are mongrelized? I suspect that it must arise from the pollen of a distinct *variety* having a prepotent effect over a flower's own pollen; and that this is part of the general law of good being derived from the intercrossing of distinct individuals of the same species. When distinct *species* are crossed the case is directly the reverse, for a plant's own pollen is always prepotent over foreign pollen; but to this subject we shall return in a future chapter.

In the case of a gigantic tree covered with innumerable flowers, it may be objected that pollen could seldom be carried from tree to tree, and at most only from flower to flower on the same tree, and that flowers on the same tree can be considered as distinct individuals only in a limited sense. I believe this objection to be valid, but that nature has largely provided against it by giving to trees a strong tendency to bear flowers with separated sexes. When the sexes are separated, although the male and female flowers may be produced on the same tree, we can see that pollen must be regularly carried from flower to flower; and this will give a better chance of pollen being occasionally carried from tree to tree. That trees belonging to all Orders have their sexes more often separated than other plants, I find to be the case in this country; and at my request Dr. Hooker tabulated the trees of New Zealand, and Dr. Asa Gray those of the United States, and the result was as I anticipated. On the other hand, Dr. Hooker has recently informed me that he finds that the rule does not hold in Australia; and I have made these few remarks on the sexes of trees simply to call attention to the subject.

Turning for a very brief space to animals: on the land there are some hermaphrodites, as land-mollusca and earth-worms; but these all pair. As yet I have not found a single case of a terrestrial animal which fertilizes itself. We can understand this remarkable fact, which offers so strong a contrast with terrestrial plants, on the view of an occasional cross being indispensable, by considering the medium in which terrestrial animals live, and the nature of the fertilizing element; for we know of no means, analogous to the action of insects and of the wind in the case of plants, by which an occasional cross could be effected with terrestrial animals without the concurrence of two individuals. Of aquatic animals, there are many self-fertilizing hermaphrodites; but here currents in the water offer an obvious means for an occasional cross. And, as in the case of flowers, I have as yet failed, after consultation with one of the highest authorities, namely, Professor Huxley, to discover a single case of an hermaphrodite animal with the organs of reproduction so perfectly enclosed within the body, that access from without and the occasional influence of a distinct individual can be shown to be physically impossible. Cirripedes long appeared to me to present a case of very great difficulty under this point of view; but I have been enabled, by a fortunate chance, elsewhere to prove that two individuals, though both are self-fertilizing hermaphrodites, do sometimes cross.

It must have struck most naturalists as a strange anomaly that, in the case of both animals and plants, species of the same family and even of the same genus, though agreeing closely with each other in almost their whole organization, yet are not rarely, some of them hermaphrodites, and some of them unisexual. But if, in fact, all hermaphrodites do occasionally intercross with other individuals, the difference between hermaphrodites and unisexual species, as far as function is concerned, becomes very small.

From these several considerations and from the many special facts which I have collected, but which I am not here able to give, I am strongly inclined to suspect that, both in the vegetable and animal kingdoms, an occasional intercross with a distinct individual is a law of nature. I am well aware that there are, on this view, many cases of difficulty, some of which I am trying to investigate. Finally then, we may conclude that in many organic beings, a cross between two individuals is an obvious necessity for each birth; in many others it occurs perhaps only at long intervals; but in none, as I suspect, can self-fertilization go on for perpetuity.

Circumstances favourable to Natural Selection. This is an extremely intricate subject. A large amount of inheritable and diversified variability is favourable, but I believe mere individual differences suffice for the work. A large number of individuals, by giving a better chance for the appearance within any given period of profitable variations, will compensate for a lesser amount of variability in each individual, and is, I believe, an extremely important element of success. Though nature grants vast periods of time for the work of natural selection, she does not grant an indefinite period; for as all organic beings are striving, it may be said, to seize on each place in the economy of nature, if any one species does not become modified and improved in a corresponding degree with its competitors, it will soon be exterminated.

In man's methodical selection, a breeder selects for some definite object, and free intercrossing will wholly stop his work. But when many men, without intending to alter the breed, have a nearly common standard of perfection, and all try to get and breed from the best animals, much improvement and modification surely but slowly follow from this unconscious process of selection, notwithstanding a large amoung of crossing with inferior animals. Thus it will be in nature; for within a confined area, with some place in its polity not so perfectly occupied as might be, natural selection will always tend to preserve all the individuals varying in the right direction, though in different degrees, so as better to fill up the unoccupied place. But if the area be large, its several districts will almost certainly present different conditions of life; and then if natural selection be modifying and improving a species in the several districts, there will be intercrossing with the other individuals of the same species on the confines of each. And in this case the effects of intercrossing can hardly be counterbalanced by natural selection always tending to modify all the individuals in each district in exactly the same manner to the conditions of each; for in a continuous area, the conditions will generally graduate away insensibly from one district to another. The intercrossing will most affect those animals which unite for each birth, which wander much, and which do not breed at a very quick rate. Hence in animals of this nature, for instance in birds, varieties will generally be confined to separated countries; and this I believe to be the case. In hermaphrodite organisms which cross only occasionally, and likewise in animals which unite for each birth, but which wander little and which can increase at a very rapid rate, a new and improved variety might be quickly formed on any one spot, and might there maintain itself in a body, so that whatever intercrossing took place would be chiefly between the individuals of the same new variety. A local variety when once thus formed might subsequently slowly spread to other districts. On the above principle, nurserymen always prefer getting seed from a large body of plants of the same variety, as the chance of intercrossing with other varieties is thus lessened.

Even in the case of slow-breeding animals, which unite for each birth, we must not overrate the effects of intercrosses in retarding natural selection; for I can bring a considerable catalogue of facts, showing that within the same area, varieties of the same animal can long remain distinct, from haunting different stations, from breeding at slightly different seasons, or from varieties of the same kind preferring to pair together.

Intercrossing plays a very important part in nature in keeping the individuals of the same species, or of the same variety, true and uniform in character. It will obviously thus act far more efficiently with those animals which unite for each birth; but I have already attempted to show that we have reason to believe that occasional intercrosses take place with all animals and with all plants. Even if these take place only at long intervals, I am convinced that the young thus produced will gain so much in vigour and fertility over the offspring from long-continued self-fertilization, that they will have a better chance of surviving and propagating their kind; and thus, in the long run, the influence of intercrosses, even at rare intervals, will be great. If there exist organic beings which never intercross, uniformity of character can be retained amongst them, as long as their conditions of life remain the same, only through the principle of inheritance, and through natural selection destroying any which depart from the proper type; but if their conditions of life change and they undergo modification, uniformity of character can be given to their modified offspring, solely by natural selection preserving the same favourable variations.

Isolation, also, is an important element in the process of natural selection. In a confined or isolated area, if not very large, the organic and inorganic conditions of life will generally be in a great degree uniform; so that natural selection will tend to modify all the individuals of a varying species throughout the area in the same manner in relation to the same conditions. Intercrosses, also, with the individuals of the same species, which otherwise would have inhabited the surrounding and differently circumstanced districts, will be prevented. But isolation probably acts more efficiently in checking the immigration of better adapted organisms, after any physical change, such as of climate or elevation of the land, &c.; and thus new places in the natural economy of the country are left open for the old inhabitants to struggle for, and become adapted to, through modifications in their structure and constitution. Lastly, isolation, by checking immigration and consequently competition, will give time for any new variety to be slowly improved; and this may sometimes be of importance in the production of new species. If, however, an isolated area be very small, either from being surrounded by barriers, or from having very peculiar physical conditions, the total number of the individuals supported on it will necessarily be very small; and fewness of individuals will greatly retard the production of new species through natural selection, by decreasing the chance of the appearance of favourable variations.

If we turn to nature to test the truth of these remarks, and look at any small isolated area, such as an oceanic island, although the total number of the species inhabiting it, will be found to be small, as we shall see in our chapter on geographical distribution; yet of these species a very large proportion are endemic, – that is, have been produced there, and nowhere else. Hence an oceanic island at first sight seems to have been highly favourable for the production of new species. But we may thus greatly deceive ourselves, for to ascertain whether a small isolated area, or a large open area like a continent, has been most favourable for the production of new organic forms, we ought to make the comparison within equal times; and this we are incapable of doing.

Although I do not doubt that isolation is of considerable importance in the production of new species, on the whole I am inclined to believe that largeness of area is of more importance, more especially in the production of species, which will prove capable of enduring for a long period, and of spreading widely. Throughout a great and open area, not only will there be a better chance of favourable variations arising from the large number of individuals of the same species there supported, but the conditions of life are infinitely complex from the large number of already existing species; and if some of these many species become modified and improved, others will have to be improved in a corresponding degree or they will be exterminated. Each new form, also, as soon as it has been much improved, will be able to spread over the open and continuous area, and will thus come into competition with many others. Hence more new places will be formed, and the competition to fill them will be more severe, on a large than on a small and isolated area. Moreover, great areas, though now continuous, owing to oscillations of level, will often have recently existed in a broken condition, so that the good effects of isolation will generally, to a certain extent, have concurred. Finally, I conclude that, although small isolated areas probably have been in some respects highly favourable for the pro-duction of new species, yet that the course of modification will generally have been more rapid on large areas; and what is more important, that the new forms produced on large areas, which already have been victorious over many competitors, will be those that will spread most widely, will give rise to most new varieties and species, and will thus play an important part in the changing history of the organic world.

We can, perhaps, on these views, understand some facts which will be again alluded to in our chapter on geographical distribution; for instance, that the productions of the smaller continent of Australia have formerly yielded, and apparently are now yielding, before those of the larger Europaeo-Asiatic area. Thus, also, it is that continental productions have everywhere become so largely naturalized on islands. On a small island, the race for life will have been less severe, and there will have been less modification and less extermination. Hence, perhaps, it comes that the flora of Madeira, according to Oswald Heer, resembles the extinct tertiary flora of Europe. All fresh-water basins, taken together, make a small area compared with that of the sea or of the land; and, consequently, the competition between fresh-water productions will have been less severe than elsewhere; new forms will have been more slowly formed, and old forms more slowly exterminated. And it is in fresh water that we find seven genera of Ganoid fishes, remnants of a once preponderant order: and in fresh water we find some of the most anomalous forms now known in the world, as the Ornithorhynchus and Lepidosiren, which, like fossils, connect to a certain extent orders now widely separated in the natural scale. These anomalous forms may almost be called living fossils; they have endured to the present day, from having inhabited a confined area, and from having thus been exposed to less severe competition.

To sum up the circumstances favourable and unfavourable to natural selection, as far as the extreme intricacy of the subject permits. I conclude, looking to the future, that for terrestrial productions a large continental area, which will probably undergo many oscillations of level, and which consequently will exist for long periods in a broken condition, will be the most favourable for the production of many new forms of life, likely to endure long and to spread widely. For the area will first have existed as a continent, and the inhabitants, at this period numerous in individuals and kinds, will have been subjected to very severe competition. When converted by subsidence into large separate islands, there will still exist many individuals of the same species on each island: intercrossing on the confines of the range of each species will thus be checked: after physical changes of any kind, immigration will be prevented, so that new places in the polity of each island will have to be filled up by modifications of the old inhabitants; and time will be allowed for the varieties in each to become well modified and perfected. When, by renewed elevation, the islands shall be re-converted into a continental area, there will again be severe competition: the most favoured or improved varieties will be enabled to spread: there will be much extinction of the less improved forms, and the relative proportional numbers of the various inhabitants of the renewed continent will again be changed; and again there will be a fair field for natural selection to improve still further the inhabitants, and thus produce new species.

That natural selection will always act with extreme slowness, I fully admit. Its action depends on there being places in the polity of nature, which can be better occupied by some of the inhabitants of the country undergoing modification of some kind. The existence of such places will often depend on physical changes, which are gener-ally very slow, and on the immigration of better adapted forms having been checked. But the action of natural selection will probably still oftener depend on some of the inhabitants becoming slowly modified; the mutual relations of many of the other inhabitants being thus disturbed. Nothing can be effected, unless favourable variations occur, and variation itself is apparently always a very slow process. The process will often be greatly retarded by free intercrossing. Many will exclaim that these several causes are amply sufficient wholly to stop the action of natural selection. I do not believe so. On the other hand, I do believe that natural selection will always act very slowly, often only at long intervals of time, and generally on only a very few of the inhabitants of the same region at the same time. I further believe, that this very slow, intermittent action of natural selection accords perfectly well with what geology tells us of the rate and manner at which the inhabitants of this world have changed.

Slow though the process of selection may be, if feeble man can do much by his powers of artificial selection, I can see no limit to the amount of change, to the beauty and infinite complexity of the coadaptations between all organic beings, one with another and with their physical conditions of life, which may be effected in the long course of time by nature's power of selection.

Extinction. This subject will be more fully discussed in our chapter on Geology; but it must be here alluded to from being intimately connected with natural selection. Natural selection acts solely through the preservation of variations in some way advantageous, which consequently endure. But as from the high geometrical powers of increase of all organic beings, each area is already fully stocked with inhabitants, it follows that as each selected and favoured form increases in number, so will the less favoured forms decrease and become rare. Rarity, as geology tells us, is the precursor to extinction. We can, also, see that any form represented by few individuals will, during fluctuations in the seasons or in the number of its enemies, run a good chance of utter extinction. But we may go further than this; for as new forms are continually and slowly being produced, unless we believe that the number of specific forms goes on perpetually and almost indefinitely increasing, numbers inevitably must become extinct. That the number of specific forms has not indefinitely increased, geology shows us plainly; and indeed we can see reason why they should not have thus increased, for the number of places in the polity of nature is not indefinitely great, – not that we have any means of knowing that any one region has as yet got its maximum of species. Probably no region is as yet fully stocked, for at the Cape of Good Hope, where more species of plants are crowded together than in any other quarter of the world, some foreign plants have become naturalized, without causing, as far as we know, the extinction of any natives.

Furthermore, the species which are most numerous in individuals will have the best chance of producing within any given period favourable variations. We have evidence of this, in the facts given in the second chapter, showing that it is the common species which afford the greatest number of recorded varieties, or incipient species. Hence, rare species will be less quickly modified or improved within any given period, and they will consequently be beaten in the race for life by the modified descendants of the commoner species.

From these several considerations I think it inevitably follows, that as new species in the course of time are formed through natural selection, others will become rarer and rarer, and finally extinct. The forms which stand in closest competition with those undergoing modification and improvement, will naturally suffer most. And we have seen in the chapter on the Struggle for Existence that it is the most closely-allied forms, – varieties of the same species, and species of the same genus or of related genera, – which, from having nearly the same structure, constitution, and habits, generally come into the severest competition with each other. Consequently, each new variety or species, during the progress of its formation, will generally press hardest on its nearest kindred, and tend to exterminate them. We see the same process of extermination amongst our domesticated productions, through the selection of improved forms by man. Many curious instances could be given showing how quickly new breeds of cattle, sheep, and other animals, and varieties of flowers, take the place of older and inferior kinds. In Yorkshire, it is historically known that the ancient black cattle were displaced by the long-horns, and that these 'were swept away by the short-horns' (I quote the words of an agricultural writer) 'as if by some murderous pestilence.'

Divergence of Character. The principle, which I have designated by this term, is of high importance on my theory, and explains, as I believe, several important facts. In the first place, varieties, even strongly-marked ones, though having somewhat of the character of species – as is shown by the hopeless doubts in many cases how to rank them – yet certainly differ from each other far less than do good and distinct species. Nevertheless, according to my view, varieties are species in the process of formation, or are, as I have called them, incipient species. How, then, does the lesser difference between varieties become augmented into the greater difference between species? That this does habitually happen, we must infer from most of the innumerable species throughout nature presenting well-marked differences; whereas varieties, the supposed prototypes and parents of future well-marked species, present slight and ill-defined differences. Mere chance, as we may call it, might cause one variety to differ in some character from its parents, and the offspring of this variety again to differ from its parent in the very same character and in a greater degree; but this alone would never account for so habitual and large an amount of difference as that between varieties of the same species and species of the same genus.

As has always been my practice, let us seek light on this head from our domestic productions. We shall here find something analogous. A fancier is struck by a pigeon having a slightly shorter beak; another fancier is struck by a pigeon having a rather longer beak; and on the acknowledged principle that 'fanciers do not and will not admire a medium standard, but like extremes,' they both go on (as has actually occurred with tumbler-pigeons) choosing and breeding from birds with longer and longer beaks, or with shorter and shorter beaks. Again, we may suppose that at an early period one man preferred swifter horses; another stronger and more bulky horses. The early differences would be very slight; in the course of time, from the continued selection of swifter horses by some breeders, and of stronger ones by others, the differences would become greater, and would be noted as forming two sub-breeds; finally, after the lapse of centuries, the sub-breeds would become converted into two well-established and distinct breeds. As the differences slowly become greater, the inferior animals with intermediate characters, being neither very swift nor very strong, will have been neglected, and will have tended to disappear. Here, then, we see in man's productions the action of what may be called the principle of divergence, causing differences, at first barely appreciable, steadily to increase, and the breeds to diverge in character both from each other and from their common parent.

But how, it may be asked, can any analogous principle apply in nature? I believe it can and does apply most efficiently, from the simple circumstance that the more diversified the descendants from any one species become in structure, constitution, and habits, by so much will they be better enabled to seize on many and widely diversified places in the polity of nature, and so be enabled to increase in numbers.

We can clearly see this in the case of animals with simple habits. Take the case of a carnivorous quadruped, of which the number that can be supported in any country has long ago arrived at its full average. If its natural powers of increase be allowed to act, it can succeed in increasing (the country not undergoing any change in its conditions) only by its varying descendants seizing on places at present occupied by other animals: some of them, for instance, being enabled to feed on new kinds of prey, either dead or alive; some inhabiting new stations, climbing trees, frequenting water, and some perhaps becoming less carnivorous. The more diversified in habits and structure the descendants of our carnivorous animal became, the more places they would be enabled to occupy. What applies to one animal will apply throughout all time to all animals – that is, if they vary – for otherwise natural selection can do nothing. So it will be with plants. It has been experimentally proved, that if a plot of ground be sown with one species of grass, and a similar plot be sown with several distinct genera of grasses, a greater number of plants and a greater weight of dry herbage can thus be raised. The same has been found to hold good when first one variety and then several mixed varieties of wheat have been sown on equal spaces of ground. Hence, if any one species of grass were to go on varying, and those varieties were continually selected which differed from each other in at all the same manner as distinct species and genera of grasses differ from each other, a greater number of individual plants of this species of grass, including its modified descendants, would succeed in living on the same piece of ground. And we well know that each species and each variety of grass is annually sowing almost countless seeds; and thus, as it may be said, is striving its utmost to increase its numbers. Consequently, I cannot doubt that in the course of many thousands of generations, the most distinct varieties of any one species of grass would always have the best chance of succeeding and of increasing in numbers, and thus of supplanting the less distinct varieties; and varieties, when rendered very distinct from each other, take the rank of species.

The truth of the principle, that the greatest amount of life can be supported by great diversification of structure, is seen under many natural circumstances. In an extremely small area, especially if freely open to immigration, and where the contest between individual and individual must be severe, we always find great diversity in its inhabitants. For instance, I found that a piece of turf, three feet by four in size, which had been exposed for many years to exactly the same conditions, supported twenty species of plants, and these belonged to eighteen genera and to eight orders, which shows how much these plants differed from each other. So it is with the plants and insects on small and uniform islets; and so in small ponds of fresh water. Farmers find that they can raise most food by a rotation of plants belonging to the most different orders: nature follows what may be called a simultaneous rotation. Most of the animals and plants which live close round any small piece of ground, could live on it (supposing it not to be in any way peculiar in its nature), and may be said to be striving to the utmost to live there; but, it is seen, that where they come into the closest competition with each other, the advantages of diversification of structure, with the accompanying differences of habit and constitution, determine that the inhabitants, which thus jostle each other most closely, shall, as a general rule, belong to what we call different genera and orders.

The same principle is seen in the naturalization of plants through man's agency in foreign lands. It might have been expected that the plants which have succeeded in becoming naturalized in any land would generally have been closely allied to the indigenes; for these are commonly looked at as specially created and adapted for their own country. It might, also, perhaps have been expected that naturalized plants would have belonged to a few groups more especially adapted to certain stations in their new homes. But the case is very different; and Alph. De Candolle has well remarked in his great and admirable work, that floras gain by naturalization, proportionally with the number of the native genera and species, far more in new genera than in new species. To give a single instance: in the last edition of Dr. Asa Gray's 'Manual of the Flora of the Northern United States,' 260 naturalized plants are enumerated, and these belong to 162 genera. We thus see that these naturalized plants are of a highly diversified nature. They differ, moreover, to a large extent from the indigenes, for out of the 162 genera, no less than 100 genera are not there indigenous, and thus a large proportional addition is made to the genera of these States.

By considering the nature of the plants or animals which have struggled successfully with the indigenes of any country, and have there become naturalized, we can gain some crude idea in what manner some of the natives would have had to be modified, in order to have gained an advantage over the other natives; and we may, I think, at least safely infer that diversification of structure, amounting to new generic differences, would have been profitable to them.

The advantage of diversification in the inhabitants of the same region is, in fact, the same as that of the physiological division of labour in the organs of the same individual body – a subject so well elucidated by Milne Edwards. No physiologist doubts that a stomach by being adapted to digest vegetable matter alone, or flesh alone, draws most nutriment from these substances. So in the general economy of any land, the more widely and perfectly the animals and plants are diversified for different habits of life, so will a greater number of individuals be capable of there supporting themselves. A set of animals, with their organization but little diversified, could hardly compete with a set more perfectly diversified in structure. It may be doubted, for instance, whether the Australian marsupials, which are divided into groups differing but little from each other, and feebly representing, as Mr. Waterhouse and others have remarked, our carnivorous, ruminant, and rodent mammals, could successfully compete with these well-pronounced orders. In the Australian mammals, we see the process of diversification in an early and incomplete stage of development.

After the foregoing discussion, which ought to have been much amplified, we may, I think, assume that the modified descendants of any one species will succeed by so much the better as they become more diversified in structure, and are thus enabled to encroach on places occupied by other beings. Now let us see how this principle of great benefit being derived from divergence of character, combined with the principles of natural selection and of extinction, will tend to act.

The accompanying diagram will aid us in understanding this rather perplexing subject. Let A to L represent the species of a genus large in its own country; these species are supposed to resemble each other in unequal degrees, as is so generally the case in nature, and as is represented in the diagram by the letters standing at unequal distances. I have said a large genus, because we have seen in the second chapter, that on an average more of the species of large genera vary than of small genera; and the varying species of the large genera present a greater number of varieties. We have, also, seen that the species, which are the commonest and the most widely-diffused, vary more than rare species with restricted ranges. Let (A) be a common, widely-diffused, and varying species, belonging to a genus large in its own country. The little fan of diverging dotted lines of unequal lengths proceeding from (A), may represent its varying offspring. The variations are supposed to be extremely slight, but of the most diversified nature; they are not supposed all to appear simultaneously, but often after long intervals of time; nor are they all supposed to endure for equal periods. Only those variations which are in some way profitable will be preserved or naturally selected. And here the importance of the principle of benefit being derived from divergence of character comes in; for this will generally lead to the most different or divergent variations (represented by the outer dotted lines) being preserved and accumulated by natural selection. When a dotted line reaches one of the horizontal lines, and is there marked by a small numbered letter, a sufficient amount of variation is supposed to have been accumulated to have formed a fairly well-marked variety, such as would be thought worthy of record in a systematic work.

The intervals between the horizontal lines in the diagram, may represent each a thousand generations; but it would have been better if each had represented ten thousand generations. After a thousand generations, species (A) is supposed to have produced two fairly well-marked varieties, namely a^1 and m^1. These two varieties will generally continue to be exposed to the same conditions which made their parents variable, and the tendency to variability is in itself hereditary, consequently they will tend to vary, and generally to vary in nearly the same manner as their parents varied. Moreover, these two varieties, being only slightly modified forms, will tend to inherit those advantages which made their common parent (A) more numerous than most of the other inhabitants of the same country; they will likewise partake of those more general advantages which made the genus to which the parent-species belonged, a large genus in its own country. And these circumstances we know to be favourable to the production of new varieties.

If, then, these two varieties be variable, the most divergent of their variations will generally be preserved during the next thousand generations. And after this interval, variety a^1 is supposed in the diagram to have produced variety a^2, which will, owing to the principle of divergence, differ more from (A) than did variety a^1. Variety m^1 is supposed to have produced two varieties, namely m^2 and s^2, differing from each other, and more considerably from their common parent (A). We may continue the process by similar steps for any length of time; some of the varieties, after each thousand generations, producing only a single variety, but in a more and more modified condition, some producing two or three varieties, and some failing to produce any. Thus the varieties or modified descendants, proceeding from the common parent (A), will generally go on increasing in number and diverging in character. In the diagram the process is represented up to the ten-thousandth generation, and under a condensed and simplified form up to the fourteen-thousandth generation.

But I must here remark that I do not suppose that the process ever goes on so regularly as is represented in the diagram, though in itself

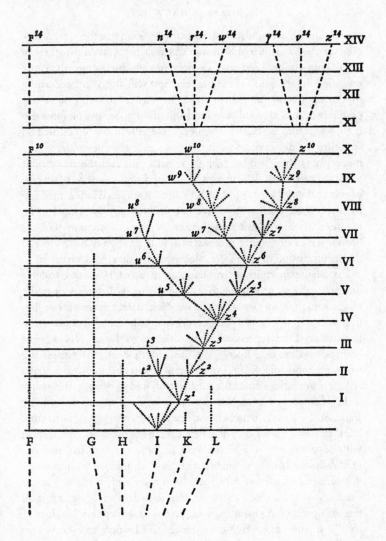

made somewhat irregular. I am far from thinking that the most divergent varieties will invariably prevail and multiply: a medium form may
often long endure, and may or may not produce more than one
modified descendant; for natural selection will always act according
to the nature of the places which are either unoccupied or not
perfectly occupied by other beings; and this will depend on infinitely
complex relations. But as a general rule, the more diversified in
structure the descendants from any one species can be rendered, the
more places they will be enabled to seize on, and the more their
modified progeny will be increased. In our diagram the line of succession is broken at regular intervals by small numbered letters marking
the successive forms which have become sufficiently distinct to be
recorded as varieties. But these breaks are imaginary, and might have
been inserted anywhere, after intervals long enough to have allowed
the accumulation of a considerable amount of divergent variation.

As all the modified descendants from a common and widely-
diffused species, belonging to a large genus, will tend to partake
of the same advantages which made their parent successful in life,
they will generally go on multiplying in number as well as diverging
in character: this is represented in the diagram by the several divergent
branches proceeding from (A). The modified offspring from the
later and more highly improved branches in the lines of descent,
will, it is probable, often take the place of, and so destroy, the earlier
and less improved branches: this is represented in the diagram by
some of the lower branches not reaching to the upper horizontal
lines. In some cases I do not doubt that the process of modification
will be confined to a single line of descent, and the number
of the descendants will not be increased; although the amount of
divergent modification may have been increased in the successive
generations. This case would be represented in the diagram, if all
the lines proceeding from (A) were removed, excepting that from a^1
to a^{10}. In the same way, for instance, the English race-horse and
English pointer have apparently both gone on slowly diverging in
character from their original stocks, without either having given
off any fresh branches or races.

After ten thousand generations, species (A) is supposed to have
produced three forms, a^{10}, f^{10}, and m^{10}, which, from having diverged

in character during the successive generations, will have come to differ largely, but perhaps unequally, from each other and from their common parent. If we suppose the amount of change between each horizontal line in our diagram to be excessively small, these three forms may still be only well-marked varieties; or they may have arrived at the doubtful category of sub-species; but we have only to suppose the steps in the process of modification to be more numerous or greater in amount, to convert these three forms into well-defined species: thus the diagram illustrates the steps by which the small differences distinguishing varieties are increased into the larger differences distinguishing species. By continuing the same process for a greater number of generations (as shown in the diagram in a condensed and simplified manner), we get eight species, marked by the letters between a^{14} and m^{14}, all descended from (A). Thus, as I believe, species are multiplied and genera are formed.

In a large genus it is probable that more than one species would vary. In the diagram I have assumed that a second species (I) has produced, by analogous steps, after ten thousand generations, either two well-marked varieties (w^{10} and z^{10}) or two species, according to the amount of change supposed to be represented between the horizontal lines. After fourteen thousand generations, six new species, marked by the letters n^{14} to z^{14}, are supposed to have been produced. In each genus, the species, which are already extremely different in character, will generally tend to produce the greatest number of modified descendants; for these will have the best chance of filling new and widely different places in the polity of nature: hence in the diagram I have chosen the extreme species (A), and the nearly extreme species (I), as those which have largely varied, and have given rise to new varieties and species. The other nine species (marked by capital letters) of our original genus, may for a long period continue transmitting unaltered descendants; and this is shown in the diagram by the dotted lines not prolonged far upwards from want of space.

But during the process of modification, represented in the diagram, another of our principles, namely that of extinction, will have played an important part. As in each fully stocked country natural selection necessarily acts by the selected form having some advantage in the struggle for life over other forms, there will be a constant tendency in the improved descendants of any one species to supplant and exterminate in each stage of descent their predecessors and their original parent. For it should be remembered that the competition will generally be most severe between those forms which are most nearly related to each other in habits, constitution, and structure. Hence all the intermediate forms between the earlier and later states, that is between the less and more improved state of a species, as well as the original parent-species itself, will generally tend to become extinct. So it probably will be with many whole collateral lines of descent, which will be conquered by later and improved lines of descent. If, however, the modified offspring of a species get into some distinct country, or become quickly adapted to some quite new station, in which child and parent do not come into competition, both may continue to exist.

If then our diagram be assumed to represent a considerable amount of modification, species (A) and all the earlier varieties will have become extinct, having been replaced by eight new species (a^{14} to m^{14}); and (I) will have been replaced by six (n^{14} to z^{14}) new species.

But we may go further than this. The original species of our genus were supposed to resemble each other in unequal degrees, as is so generally the case in nature; species (A) being more nearly related to B, C, and D, than to the other species; and species (I) more to G, H, K, L, than to the others. These two species (A) and (I), were also supposed to be very common and widely diffused species, so that they must originally have had some advantage over most of the other species of the genus. Their modified descendants, fourteen in number at the fourteen-thousandth generation, will probably have inherited some of the same advantages: they have also been modified and improved in a diversified manner at each stage of descent, so as to have become adapted to many related places in the natural economy of their country. It seems, therefore, to me extremely probable that they will have taken the places of, and thus exterminated, not only their parents (A) and (I), but likewise some of the original species which were most nearly related to their parents. Hence very few of the original species will have transmitted offspring to the fourteen-thousandth generation. We may suppose that only one (F), of the two species which were least closely related to the other nine original species, has transmitted descendants to this late stage of descent.

The new species in our diagram descended from the original eleven species, will now be fifteen in number. Owing to the divergent tendency of natural selection, the extreme amount of difference in character between species a^{14} and z^{14} will be much greater than that between the most different of the original eleven species. The new species, moreover, will be allied to each other in a widely different manner. Of the eight descendants from (A) the three marked a^{14}, q^{14}, p^{14}, will be nearly related from having recently branched off from a^{10}; b^{14} and f^{14}, from having diverged at an earlier period from a^5, will be in some degree distinct from the three first-named species; and lastly, o^{14}, e^{14}, and m^{14}, will be nearly related one to the other, but from having diverged at the first commencement of the process of modification, will be widely different from the other five species, and may constitute a sub-genus or even a distinct genus.

The six descendants from (I) will form two sub-genera or even genera. But as the original species (I) differed largely from (A), standing nearly at the extreme points of the original genus, the six descendants from (I) will, owing to inheritance, differ considerably from the eight descendants from (A); the two groups, moreover, are supposed to have gone on diverging in different directions. The intermediate species, also (and this is a very important consideration), which connected the original species (A) and (I), have all become, excepting (F), extinct, and have left no descendants. Hence the six new species descended from (I), and the eight descended from (A), will have to be ranked as very distinct genera, or even as distinct sub-families.

Thus it is, as I believe, that two or more genera are produced by descent, with modification, from two or more species of the same genus. And the two or more parent-species are supposed to have descended from some one species of an earlier genus. In our diagram, this is indicated by the broken lines, beneath the capital letters, converging in sub-branches downwards towards a single point; this point representing a single species, the supposed single parent of our several new sub-genera and genera.

It is worth while to reflect for a moment on the character of the new species f^{14}, which is supposed not to have diverged much in character, but to have retained the form of (F), either unaltered or altered only in a slight degree. In this case, its affinities to the other fourteen new species will be of a curious and circuitous nature. Having descended from a form which stood between the two parent-species (A) and (I), now supposed to be extinct and unknown, it will be in some degree intermediate in character between the two groups descended from these species. But as these two groups have gone on diverging in character from the type of their parents, the new species (F^{14}) will not be directly inter-mediate between them, but rather between types of the two groups; and every naturalist will be able to bring some such case before his mind.

In the diagram, each horizontal line has hitherto been supposed to represent a thousand generations, but each may represent a million or hundred million generations, and likewise a section of the successive strata of the earth's crust including extinct remains. We shall, when we come to our chapter on Geology, have to refer again to this subject, and I think we shall then see that the diagram throws light on the affinities of extinct beings, which, though generally belonging to the same orders, or families, or genera, with those now living, yet are often, in some degree, intermediate in character between existing groups; and we can understand this fact, for the extinct species lived at very ancient epochs when the branching lines of descent had diverged less.

I see no reason to limit the process of modification, as now explained, to the formation of genera alone. If, in our diagram, we suppose the amount of change represented by each successive group of diverging dotted lines to be very great, the forms marked a^{14} to p^{14}, those marked b^{14} and f^{14}, and those marked o^{14} to m^{14}, will form three very distinct genera. We shall also have two very distinct genera descended from (I); and as these latter two genera, both from con-tinued divergence of character and from inheritance from a different parent, will differ widely from the three genera descended from (A), the two little groups of genera will form two distinct families, or even orders, according to the amount of divergent modification supposed to be represented in the diagram. And the two new families, or orders, will have descended from two species of the original genus;

and these two species are supposed to have descended from one species of a still more ancient and unknown genus.

We have seen that in each country it is the species of the larger genera which oftenest present varieties or incipient species. This, indeed, might have been expected; for as natural selection acts through one form having some advantage over other forms in the struggle for existence, it will chiefly act on those which already have some advantage; and the largeness of any group shows that its species have inherited from a common ancestor some advantage in common. Hence, the struggle for the production of new and modified descendants, will mainly lie between the larger groups, which are all trying to increase in number. One large group will slowly conquer another large group, reduce its numbers, and thus lessen its chance of further variation and improvement. Within the same large group, the later and more highly perfected sub-groups, from branching out and seizing on many new places in the polity of Nature, will constantly tend to supplant and destroy the earlier and less improved sub-groups. Small and broken groups and sub-groups will finally tend to disappear. Looking to the future, we can predict that the groups of organic beings which are now large and triumphant, and which are least broken up, that is, which as yet have suffered least extinction, will for a long period continue to increase. But which groups will ultimately prevail, no man can predict; for we well know that many groups, formerly most extensively developed, have now become extinct. Looking still more remotely to the future, we may predict that, owing to the continued and steady increase of the larger groups, a multitude of smaller groups will become utterly extinct, and leave no modified descendants; and consequently that of the species living at any one period, extremely few will transmit descendants to a remote futurity. I shall have to return to this subject in the chapter on Classification, but I may add that on this view of extremely few of the more ancient species having transmitted descendants, and on the view of all the descendants of the same species making a class, we can understand how it is that there exist but very few classes in each main division of the animal and vegetable kingdoms. Although extremely few of the most ancient species may now have living and modified descendants, yet at the most remote geological period, the earth may have been as well peopled with many species of many genera, families, orders, and classes, as at the present day.

Summary of Chapter. If during the long course of ages and under varying conditions of life, organic beings vary at all in the several parts of their organization, and I think this cannot be disputed; if there be, owing to the high geometrical powers of increase of each species, at some age, season, or year, a severe struggle for life, and this certainly cannot be disputed; then, considering the infinite complexity of the relations of all organic beings to each other and to their conditions of existence, causing an infinite diversity in structure, constitution, and habits, to be advantageous to them, I think it would be a most extraordinary fact if no variation ever had occurred useful to each being's own welfare, in the same way as so many variations have occurred useful to man. But if variations useful to any organic being do occur, assuredly individuals thus characterized will have the best chance of being preserved in the struggle for life; and from the strong principle of inheritance they will tend to produce offspring similarly characterized. This principle of preservation, I have called, for the sake of brevity, Natural Selection. Natural selection, on the principle of qualities being inherited at corresponding ages, can modify the egg, seed, or young, as easily as the adult. Amongst many animals, sexual selection will give its aid to ordinary selection, by assuring to the most vigorous and best adapted males the greatest number of offspring. Sexual selection will also give characters useful to the males alone, in their struggles with other males.

Whether natural selection has really thus acted in nature, in modifying and adapting the various forms of life to their several conditions and stations, must be judged of by the general tenour and balance of evidence given in the following chapters. But we already see how it entails extinction; and how largely extinction has acted in the world's history, geology plainly declares. Natural selection, also, leads to divergence of character; for more living beings can be supported on the same area the more they diverge in structure, habits, and constitution, of which we see proof by looking at the inhabitants of any small spot or at naturalized productions. Therefore during the modification of the descendants of any one species, and during the incessant struggle of all species to increase in numbers, the more diversified these descendants become, the better will be their chance of succeeding in the battle of life. Thus the small differences distinguishing varieties of the same species, will steadily tend to increase till they come to equal the greater differences between species of the same genus, or even of distinct genera.

We have seen that it is the common, the widely-diffused, and widely-ranging species, belonging to the larger genera, which vary most; and these will tend to transmit to their modified offspring that superiority which now makes them dominant in their own countries. Natural selection, as has just been remarked, leads to divergence of character and to much extinction of the less improved and intermediate forms of life. On these principles, I believe, the nature of the affinities of all organic beings may be explained. It is a truly wonderful fact – the wonder of which we are apt to overlook from familiarity – that all animals and all plants throughout all time and space should be related to each other in group subordinate to group, in the manner which we everywhere behold – namely, varieties of the same species most closely related together, species of the same genus less closely and unequally related together, forming sections and sub-genera, species of distinct genera much less closely related, and genera related in different degrees, forming sub-families, families, orders, sub-classes, and classes. The several subordinate groups in any class cannot be ranked in a single file, but seem rather to be clustered round points, and these round other points, and so on in almost endless cycles. On the view that each species has been independently created, I can see no explanation of this great fact in the classification of all organic beings; but, to the best of my judgement, it is explained through inheritance and the complex action of natural selection, entailing extinction and divergence of character, as we have seen illustrated in the diagram.

The affinities of all the beings of the same class have sometimes been represented by a great tree. I believe this simile largely speaks the truth. The green and budding twigs may represent existing species; and those produced during each former year may represent the long succession of extinct species. At each period of growth all the growing twigs have tried to branch out on all sides, and to overtop and kill the surrounding twigs and branches, in the same manner as species and groups of species have tried to overmaster other species in the great battle for life. The limbs divided into great branches, and these into lesser and lesser branches, were themselves once, when the tree was small, budding twigs; and this connexion of the former and present buds by ramifying branches may well represent the classification of all extinct and living species in groups subordinate to groups. Of the many twigs which flourished when the tree was a mere bush, only two or three, now grown into great branches, yet survive and bear all the other branches; so with the species which lived during long-past geological periods, very few now have living and modified descendants. From the first growth of the tree, many a limb and branch has decayed and dropped off; and these lost branches of various sizes may represent those whole orders, families, and genera which have now no living representatives, and which are known to us only from having been found in a fossil state. As we here and there see a thin straggling branch springing from a fork low down in a tree, and which by some chance has been favoured and is still alive on its summit, so we occasionally see an animal like the Ornithorhynchus or Lepidosiren, which in some small degree connects by its affinities two large branches of life, and which has apparently been saved from fatal competition by having inhabited a protected station. As buds give rise by growth to fresh buds, and these, if vigorous, branch out and overtop on all sides many a feebler branch, so by generation I believe it has been with the great Tree of Life, which fills with its dead and broken branches the crust of the earth, and covers the surface with its ever branching and beautiful ramifications.

CHAPTER V

LAWS OF VARIATION

Effects of external conditions – Use and disuse, combined with natural selection; organs of flight and of vision – Acclimatization – Correlation of growth – Compensation and economy of growth – False correlations – Multiple, rudimentary, and lowly organized structures variable – Parts developed in an unusual manner are highly variable: specific characters more variable than generic: secondary sexual characters variable – Species of the same genus vary in an analogous manner – Reversions to long-lost characters – Summary

I HAVE hitherto sometimes spoken as if the variations – so common and multiform in organic beings under domestication, and in a lesser degree in those in a state of nature – had been due to chance. This, of course, is a wholly incorrect expression, but it serves to acknowledge plainly our ignorance of the cause of each particular variation. Some authors believe it to be as much the function of the reproductive system to produce individual differences, or very slight deviations of structure, as to make the child like its parents. But the much greater variability, as well as the greater frequency of monstrosities, under domestication or cultivation, than under nature, leads me to believe that deviations of structure are in some way due to the nature of the conditions of life, to which the parents and their more remote ancestors have been exposed during several generations. I have remarked in the first chapter – but a long catalogue of facts which cannot be here given would be necessary to show the truth of the remark – that the reproductive system is eminently susceptible to changes in the conditions of life; and to this system being functionally disturbed in the parents, I chiefly attribute the varying or plastic condition of the offspring. The male and female sexual elements seem to be affected before that union takes place which is to form a new being. In the case of 'sporting' plants, the bud, which in its earliest condition does not apparently differ essentially from an ovule, is alone affected. But why, because the reproductive system is disturbed, this or that part should vary more or less, we are profoundly ignorant. Nevertheless, we can here and there dimly catch a faint ray of light, and we may feel sure that there must be some cause for each deviation of structure, however slight.

How much direct effect difference of climate, food, &c., produces on any being is extremely doubtful. My impression is, that the effect is extremely small in the case of animals, but perhaps rather more in that of plants. We may, at least, safely conclude that such influences cannot have produced the many striking and complex co-adaptations of structure between one organic being and another, which we see everywhere throughout nature. Some little influence may be attributed to climate, food, &c.: thus, E. Forbes speaks confidently that shells at their southern limit, and when living in shallow water, are more brightly coloured than those of the same species further north or from greater depths. Gould believes that birds of the same species are more brightly coloured under a clear atmosphere, than when living on islands or near the coast. So with insects, Wollaston is convinced that residence near the sea affects their colours. Moquin-Tandon gives a list of plants which when growing near the sea-shore have their leaves in some degree fleshy, though not elsewhere fleshy. Several other such cases could be given.

The fact of varieties of one species, when they range into the zone of habitation of other species, often acquiring in a very slight degree some of the characters of such species, accords with our view that species of all kinds are only well-marked and permanent varieties. Thus the species of shells which are confined to tropical and shallow seas are generally brighter-coloured than those confined to cold and deeper seas. The birds which are confined to continents are, according to Mr. Gould, brighter-coloured than those of islands. The insect-species confined to sea-coasts, as every collector knows, are often brassy or lurid. Plants which live exclusively on the sea-side are very apt to have fleshy leaves. He who believes in the creation of each species, will have to say that this shell, for instance, was created with bright colours for a warm sea; but that this other shell became bright-coloured by variation when it ranged into warmer or shallower waters.

When a variation is of the slightest use to a being, we cannot tell how much of it to attribute to the accumulative action of natural selection, and how much to the conditions of life. Thus, it is well known to furriers that animals of the same species have thicker and better fur the more severe the climate is under which they have lived; but who can tell how much of this difference may be due to the warmest-clad individuals having been favoured and preserved during many generations, and how much to the direct action of the severe climate? for it would appear that climate has some direct action on the hair of our domestic quadrupeds.

Instances could be given of the same variety being produced under conditions of life as different as can well be conceived; and, on the other hand, of different varieties being produced from the same species under the same conditions. Such facts show how indirectly the conditions of life must act. Again, innumerable instances are known to every naturalist of species keeping true, or not varying at all, although living under the most opposite climates. Such considerations as these incline me to lay very little weight on the direct action of the conditions of life. Indirectly, as already remarked, they seem to play an important part in affecting the reproductive system, and in thus inducing variability; and natural selection will then accumulate all profitable variations, however slight, until they become plainly developed and appreciable by us.

Effects of Use and Disuse. From the facts alluded to in the first chapter, I think there can be little doubt that use in our domestic animals strengthens and enlarges certain parts, and disuse diminishes them; and that such modifications are inherited. Under free nature, we can have no standard of comparison, by which to judge of the effects of long-continued use or disuse, for we know not the parent-forms; but many animals have structures which can be explained by the effects of disuse. As Professor Owen has remarked, there is no greater anomaly in nature than a bird that cannot fly; yet there are several in this state. The logger-headed duck of South America can only flap along the surface of the water, and has its wings in nearly the same condition as the domestic Aylesbury duck. As the larger ground-feeding birds seldom take flight except to escape danger, I believe that the nearly wingless condition of several birds, which now inhabit or have lately inhabited several oceanic islands, tenanted by no beast of prey, has been caused by disuse. The ostrich indeed inhabits continents and is exposed to danger from which it cannot escape by flight, but by kicking it can defend itself from enemies, as well as any of the smaller quadrupeds. We may imagine that the early progenitor of the ostrich had habits like those of a bustard, and that as natural selection increased in successive generations the size and weight of its body, its legs were used more, and its wings less, until they became incapable of flight.

Kirby has remarked (and I have observed the same fact) that the anterior tarsi, or feet, of many male dung-feeding beetles are very often broken off; he examined seventeen specimens in his own collection, and not one had even a relic left. In the Onites apelles the tarsi are so habitually lost, that the insect has been described as not having them. In some other genera they are present, but in a rudimentary condition. In the Ateuchus or sacred beetle of the Egyptians, they are totally deficient. There is not sufficient evidence to induce us to believe that mutilations are ever inherited; and I should prefer explaining the entire absence of the anterior tarsi in Ateuchus, and their rudimentary condition in some other genera, by the long-continued effects of disuse in their progenitors; for as the tarsi are almost always lost in many dung-feeding beetles, they must be lost early in life, and therefore cannot be much used by these insects.

In some cases we might easily put down to disuse modifications of structure which are wholly, or mainly, due to natural selection. Mr. Wollaston has discovered the remarkable fact that 200 beetles, out of the 550 species inhabiting Madeira, are so far deficient in wings that they cannot fly; and that of the twenty-nine endemic genera, no less than twenty-three genera have all their species in this condition! Several facts, namely, that beetles in many parts of the world are very frequently blown to sea and perish; that the beetles in Madeira, as observed by Mr. Wollaston, lie much concealed, until the wind lulls and the sun shines; that the proportion of wingless beetles is larger on the exposed Dezertas than in Madeira itself; and especially the extraordinary fact, so strongly insisted on by Mr. Wollaston, of the almost entire absence of certain large groups of beetles, elsewhere excessively numerous, and which groups have habits of life almost necessitating frequent flight; – these several considerations have made me believe that the wingless condition of so many Madeira beetles is mainly due to the action of natural selection, but combined probably with disuse. For during thousands of successive generations each individual beetle which flew least, either from its wings having been ever so little less perfectly developed or from indolent habit, will have had the best chance of surviving from not being blown out to sea; and, on the other hand, those beetles which most readily took to flight will oftenest have been blown to sea and thus have been destroyed.

The insects in Madeira which are not ground-feeders, and which, as the flower-feeding coleoptera and lepidoptera, must habitually use their wings to gain their subsistence, have, as Mr. Wollaston suspects, their wings not at all reduced, but even enlarged. This is quite compatible with the action of natural selection. For when a new insect first arrived on the island, the tendency of natural selection to enlarge or to reduce the wings, would depend on whether a greater number of individuals were saved by successfully battling with the winds, or by giving up the attempt and rarely or never flying. As with mariners shipwrecked near a coast, it would have been better for the good swimmers if they had been able to swim still further, whereas it would have been better for the bad swimmers if they had not been able to swim at all and had stuck to the wreck.

The eyes of moles and of some burrowing rodents are rudimentary in size, and in some cases are quite covered up by skin and fur. This state of the eyes is probably due to gradual reduction from disuse, but aided perhaps by natural selection. In South America, a burrowing rodent, the tuco-tuco, or Ctenomys, is even more subterranean in its habits than the mole; and I was assured by a Spaniard, who had often caught them, that they were frequently blind; one which I kept alive was certainly in this condition, the cause, as appeared on dissection, having been inflammation of the nictitating membrane. As frequent inflammation of the eyes must be injurious to any animal, and as eyes are certainly not indispensable to animals with subterranean habits, a reduction in their size with the adhesion of the eyelids and growth of fur over them, might in such case be an advantage; and if so, natural selection would constantly aid the effects of disuse.

It is well known that several animals, belonging to the most different classes, which inhabit the caves of Styria and of Kentucky, are blind. In some of the crabs the foot-stalk for the eye remains, though the eye is gone; the stand for the telescope is there, though the telescope with its glasses has been lost. As it is difficult to imagine that eyes, though useless, could be in any way injurious to animals living in darkness, I attribute their loss wholly to disuse. In one of the blind animals, namely, the cave-rat, the eyes are of immense size; and Professor Silliman thought that it regained, after living some days in the light, some slight power of vision. In the same manner as in Madeira the wings of some of the insects have been enlarged, and the wings of others have been reduced by natural selection aided by use and disuse, so in the case of the cave-rat natural selection seems to have struggled with the loss of light and to have increased the size of the eyes; whereas with all the other inhabitants of the caves, disuse by itself seems to have done its work.

It is difficult to imagine conditions of life more similar than deep limestone caverns under a nearly similar climate; so that on the common view of the blind animals having been separately created for the American and European caverns, close similarity in their organization and affinities might have been expected; but, as Schiödte and others have remarked, this is not the case, and the cave-insects of the two continents are not more closely allied than might have been anticipated from the general resemblance of the other inhabit-ants of North America and Europe. On my view we must suppose that American animals, having ordinary powers of vision, slowly migrated by successive generations from the outer world into the deeper and deeper recesses of the Kentucky caves, as did European animals into the caves of Europe. We have some evidence of this gradation of habit; for, as Schiödte remarks, 'animals not far remote from ordinary forms, prepare the transition from light to darkness. Next follow those that are constructed for twilight; and, last of all, those destined for total darkness.' By the time that an animal had reached, after numberless generations, the deepest recesses, disuse will on this view have more or less perfectly obliterated its eyes, and natural selection will often have effected other changes, such as an increase in the length of the antennae or palpi, as a compensation for blindness. Notwithstanding such modifications, we might expect still to see in the cave-animals of America, affinities to the other inhabitants of that continent, and in those of Europe, to the inhabitants of the European continent. And this is the case with some of the American cave-animals, as I hear from Professor Dana; and some of the European cave-insects are very closely allied to those of the surrounding country. It would be most difficult to give any rational explanation of the affinities of the blind cave-animals to the other inhabitants of the two continents on the ordinary view of their independent creation. That several of the inhabitants of the caves of the Old and New Worlds should be closely related, we might expect from the well-known relationship of most of their other productions. Far from feeling any surprise that some of the cave-animals should be very anomalous, as Agassiz has remarked in regard to the blind fish, the Amblyopsis, and as is the case with the blind Proteus with reference to the reptiles of Europe, I am only surprised that more wrecks of ancient life have not been preserved, owing to the less severe competition to which the inhabitants of these dark abodes will probably have been exposed.

Acclimatization. Habit is hereditary with plants, as in the period of flowering, in the amount of rain requisite for seeds to germinate, in the time of sleep, &c., and this leads me to say a few words on acclimatization. As it is extremely common for species of the same genus to inhabit very hot and very cold countries, and as I believe that all the species of the same genus have descended from a single parent, if this view be correct, acclimatization must be readily effected during long-continued descent. It is notorious that each species is adapted to the climate of its own home: species from an arctic or even from a temperate region cannot endure a tropical climate, or conversely. So again, many succulent plants cannot endure a damp climate. But the degree of adaptation of species to the climates under which they live is often overrated. We may infer this from our frequent inability to predict whether or not an imported plant will endure our climate, and from the number of plants and animals brought from warmer countries which here enjoy good health. We have reason to believe that species in a state of nature are limited in their ranges by the competition of other organic beings quite as much as, or more than, by adaptation to particular climates. But whether or not the adaptation be generally very close, we have evidence, in the case of some few plants, of their becoming, to a certain extent, naturally habituated to different temperatures, or becoming acclimatized: thus the pines and rhododendrons, raised from seed collected by Dr. Hooker from trees growing at different heights on the Himalaya, were found in this country to possess different constitutional powers of resisting cold. Mr. Thwaites informs me that he has observed similar facts in Ceylon, and analogous observations have been made by Mr. H. C. Watson on European species of plants brought from the Azores to England. In regard to animals, several authentic cases could be given of species within historical times having largely extended their range from warmer to cooler latitudes, and conversely; but we do not positively know that these animals were strictly adapted to their native climate, but in all ordinary cases we assume such to be the case; nor do we know that they have subsequently become acclimatized to their new homes.

As I believe that our domestic animals were originally chosen by uncivilized man because they were useful and bred readily under confinement, and not because they were subsequently found capable of far-extended transportation, I think the common and extraordinary capacity in our domestic animals of not only withstanding the most different climates but of being perfectly fertile (a far severer test) under them, may be used as an argument that a large proportion of other animals, now in a state of nature, could easily be brought to bear widely different climates. We must not, however, push the foregoing argument too far, on account of the probable origin of some of our domestic animals from several wild stocks: the blood, for instance, of a tropical and arctic wolf or wild dog may perhaps be mingled in our domestic breeds. The rat and mouse cannot be considered as domestic animals, but they have been transported by man to many parts of the world, and now have a far wider range than any other rodent, living free under the cold climate of Faroe in the north and of the Falklands in the south, and on many islands in the torrid zones. Hence I am inclined to look at adaptation to any special climate as a quality readily grafted on an innate wide flexibility of constitution, which is common to most animals. On this view, the capacity of enduring the most different climates by man himself and by his domestic animals, and such facts as that former species of the elephant and rhinoceros were capable of enduring a glacial climate, whereas the living species are now all tropical or sub-tropical in their habits, ought not to be looked at as anomalies, but merely as examples of a very common flexibility of constitution, brought, under peculiar circumstances, into play.

How much of the acclimatization of species to any peculiar climate is due to mere habit, and how much to the natural selection of varieties having different innate constitutions, and how much to both means combined, is a very obscure question. That habit or custom has some influence I must believe, both from analogy, and from the incessant advice given in agricultural works, even in the ancient Encyclopaedias of China, to be very cautious in transposing animals from one district to another; for it is not likely that man should have succeeded in selecting so many breeds and sub-breeds with constitutions specially fitted for their own districts: the result must, I think, be due to habit. On the other hand, I can see no reason to doubt that natural selection will continually tend to preserve those individuals which are born with constitutions best adapted to their native countries. In treatises on many kinds of cultivated plants, certain varieties are said to withstand certain climates better than others: this is very strikingly shown in works on fruit trees published in the United States, in which certain varieties are habitually recommended for the northern, and others for the southern States; and as most of these varieties are of recent origin, they cannot owe their constitutional differences to habit. The case of the Jerusalem artichoke, which is never propagated by seed, and of which consequently new varieties have not been produced, has even been advanced – for it is now as tender as ever it was – as proving that acclimatization cannot be effected! The case, also, of the kidney-bean has been often cited for a similar purpose, and with much greater weight; but until some one will sow, during a score of generations, his kidney-beans so early that a very large proportion are destroyed by frost, and then collect seed from the few survivors, with care to prevent accidental crosses, and then again get seed from these seedlings, with the same precautions, the experiment cannot be said to have been even tried. Nor let it be supposed that no differences in the constitution of seedling kidney-beans ever appear, for an account has been published how much more hardy some seedlings appeared to be than others.

On the whole, I think we may conclude that habit, use, and disuse, have, in some cases, played a considerable part in the modification of the constitution, and of the structure of various organs; but that the effects of use and disuse have often been largely combined with, and sometimes overmastered by, the natural selection of innate differences.

Correlation of Growth. I mean by this expression that the whole organization is so tied together during its growth and development, that when slight variations in any one part occur, and are accumulated through natural selection, other parts become modified. This is a very important subject, most imperfectly understood. The most obvious case is, that modifications accumulated solely for the good of the young or larva, will, it may safely be concluded, affect the structure of the adult; in the same manner as any malconformation affecting the early embryo, seriously affects the whole organization of the adult. The several parts of the body which are homologous, and which, at an early embryonic period, are alike, seem liable to vary in an allied manner: we see this in the right and left sides of the body varying in the same manner; in the front and hind legs, and even in the jaws and limbs, varying together, for the lower jaw is believed to be homologous with the limbs. These tendencies, I do not doubt, may be mastered more or less completely by natural selection: thus a family of stags once existed with an antler only on one side; and if this had been of any great use to the breed it might probably have been rendered permanent by natural selection.

Homologous parts, as has been remarked by some authors, tend to cohere; this is often seen in monstrous plants: and nothing is more common than the union of homologous parts in normal structures, as the union of the petals of the corolla into a tube. Hard parts seem to affect the form of adjoining soft parts; it is believed by some authors that the diversity in the shape of the pelvis in birds causes the remarkable diversity in the shape of their kidneys. Others believe that the shape of the pelvis in the human mother influences by pressure the shape of the head of the child. In snakes, according to Schlegel, the shape of the body and the manner of swallowing determine the position of several of the most important viscera.

The nature of the bond of correlation is very frequently quite obscure. M. Is. Geoffroy St. Hilaire has forcibly remarked, that certain malconformations very frequently, and that others rarely co-exist, without our being able to assign any reason. What can be more singular than the relation between blue eyes and deafness in cats, and the tortoise-shell colour with the female sex; the feathered feet and skin between the outer toes in pigeons, and the presence of more or less down on the young birds when first hatched, with the future colour of their plumage; or, again, the relation between the hair and teeth in the naked Turkish dog, though here probably homology comes into play? With respect to this latter case of correlation, I think it can hardly be accidental, that if we pick out the two orders of mammalia which are most abnormal in their dermal coverings, viz. Cetacea (whales) and Edentata (armadilloes, scaly ant-eaters, &c.), that these are likewise the most abnormal in their teeth.

I know of no case better adapted to show the importance of the laws of correlation in modifying important structures, independently of utility and, therefore, of natural selection, than that of the difference between the outer and inner flowers in some Compositous and Umbelliferous plants. Every one knows the difference in the ray and central florets of, for instance, the daisy, and this difference is often accompanied with the abortion of parts of the flower. But, in some Compositous plants, the seeds also differ in shape and sculpture: and even the ovary itself, with its accessory parts, differs, as has been described by Cassini. These differences have been attributed by some authors to pressure, and the shape of the seeds in the ray-florets in some Compositae countenances this idea; but, in the case of the corolla of the Umbelliferae, it is by no means, as Dr. Hooker informs me, in species with the densest heads that the inner and outer flowers most frequently differ. It might have been thought that the development of the ray-petals by drawing nourishment from certain other parts of the flower had caused their abortion; but in some Compositae there is a difference in the seeds of the outer and inner florets without any difference in the corolla. Possibly, these several differences may be connected with some difference in the flow of nutriment towards the central and external flowers: we know, at least, that in irregular flowers, those nearest to the axis are oftenest subject to peloria, and become regular. I may add, as an instance of this, and of a striking case of correlation, that I have recently observed in some garden pelargoniums, that the central flower of the truss often loses the patches of darker colour in the two upper petals; and that when this occurs, the adherent nectary is quite aborted; when the colour is absent from only one of the two upper petals, the nectary is only much shortened.

With respect to the difference in the corolla of the central and exterior flowers of a head or umbel, I do not feel at all sure that C. C. Sprengel's idea that the ray-florets serve to attract insects, whose agency is highly advantageous in the fertilization of plants of these two orders, is so far-fetched, as it may at first appear: and if it be advantageous, natural selection may have come into play. But in regard to the differences both in the internal and external structure of the seeds, which are not always correlated with any differences in the flowers, it seems impossible that they can be in any way advantageous to the plant: yet in the Umbelliferae these differences are of such apparent importance – the seeds being in some cases, according to Tausch, orthospermous in the exterior flowers and coelospermous in the central flowers, – that the elder De Candolle founded his main divisions of the order on analogous differences. Hence we see that modifications of structure, viewed by systematists as of high value, may be wholly due to unknown laws of correlated growth, and without being, as far as we can see, of the slightest service to the species.

We may often falsely attribute to correlation of growth structures which are common to whole groups of species, and which in truth are simply due to inheritance; for an ancient progenitor may have acquired through natural selection some one modification in structure, and, after thousands of generations, some other and independent modification; and these two modifications, having been transmitted to a whole group of descendants with diverse habits, would naturally be thought to be correlated in some necessary manner. So, again, I do not doubt that some apparent correlations, occurring throughout whole orders, are entirely due to the manner alone in which natural selection can act. For instance, Alph. De Candolle has remarked that winged seeds are never found in fruits which do not open: I should explain the rule by the fact that seeds could not gradually become winged through natural selection, except in fruits which opened; so that the individual plants producing seeds which were a little better fitted to be wafted further, might get an advantage over those producing seed less fitted for dispersal; and this process could not possibly go on in fruit which did not open.

The elder Geoffroy and Goethe propounded, at about the same period, their law of compensation or balancement of growth; or, as Goethe expressed it, 'in order to spend on one side, nature is forced to economize on the other side.' I think this holds true to a certain extent with our domestic productions: if nourishment flows to one part or organ in excess, it rarely flows, at least in excess, to another part; thus it is difficult to get a cow to give much milk and to fatten readily. The same varieties of the cabbage do not yield abundant and nutritious foliage and a copious supply of oil-bearing seeds. When the seeds in our fruits become atrophied, the fruit itself gains largely in size and quality. In our poultry, a large tuft of feathers on the head is generally accompanied by a diminished comb, and a large beard by diminished wattles. With species in a state of nature it can hardly be maintained that the law is of universal application; but many good observers, more especially botanists, believe in its truth. I will not, however, here give any instances, for I see hardly any way of distinguishing between the effects, on the one hand, of a part being largely developed through natural selection and another and adjoining part being reduced by this same process or by disuse, and, on the other hand, the actual withdrawal of nutriment from one part owing to the excess of growth in another and adjoining part.

I suspect, also, that some of the cases of compensation which have been advanced, and likewise some other facts, may be merged under a more general principle, namely, that natural selection is continually trying to economize in every part of the organization. If under changed conditions of life a structure before useful becomes less useful, any diminution, however slight, in its development, will be seized on by natural selection, for it will profit the individual not to have its nutriment wasted in building up an useless structure. I can thus only understand a fact with which I was much struck when examining cirripedes, and of which many other instances could be given: namely, that when a cirripede is parasitic within another and is thus protected, it loses more or less completely its own shell or carapace. This is the case with the male Ibla, and in a truly extraordinary manner with the Proteolepas: for the carapace in all other cirripedes consists of the three highly-important anterior segments of the head enormously developed, and furnished with great nerves and muscles; but in the parasitic and protected Proteolepas, the whole anterior part of the head is reduced to the merest rudiment attached to the bases of the prehensile antennae. Now the saving of a large and complex structure, when rendered superfluous by the parasitic habits of the Proteolepas, though effected by slow steps, would be a decided advantage to each successive individual of the species; for in the struggle for life to which every animal is exposed, each individual Proteolepas would have a better chance of supporting itself, by less nutriment being wasted in developing a structure now become useless.

Thus, as I believe, natural selection will always succeed in the long run in reducing and saving every part of the organization, as soon as it is rendered superfluous, without by any means causing some other part to be largely developed in a corresponding degree. And, conversely, that natural selection may perfectly well succeed in largely developing any organ, without requiring as a necessary compensation the reduction of some adjoining part.

It seems to be a rule, as remarked by Is. Geoffroy St. Hilaire, both in varieties and in species, that when any part or organ is repeated many times in the structure of the same individual (as the vertebrae in snakes, and the stamens in polyandrous flowers) the number is variable; whereas the number of the same part or organ, when it occurs in lesser numbers, is constant. The same author and some botanists have further remarked that multiple parts are also very liable to variation in structure. Inasmuch as this 'vegetative repetition,' to use Prof. Owen's expression, seems to be a sign of low organization; the foregoing remark seems connected with the very general opinion of naturalists, that beings low in the scale of nature are more variable than those which are higher. I presume that lowness in this case means that the several parts of the organization have been but little specialized for particular funtions; and as long as the same part has to perform diversified work, we can perhaps see why it should remain variable, that is, why natural selection should have preserved or rejected each little deviation of form less carefully than when the part has to serve for one special purpose alone. In the same way that a knife which has to cut all sorts of things may be of almost any shape; whilst a tool for some particular object had better be of some particular shape. Natural selection, it should never be forgotten, can act on each part of each being, solely through and for its advantage.

Rudimentary parts, it has been stated by some authors, and I believe with truth, are apt to be highly variable. We shall have to recur to the general subject of rudimentary and aborted organs; and I will here only add that their variability seems to be owing to their uselessness, and therefore to natural selection having no power to check deviations in their structure. Thus rudimentary parts are left to the free play of the various laws of growth, to the effects of long-continued disuse, and to the tendency to reversion.

A part developed in any species in an extraordinary degree or manner, in comparison with the same part in allied species, tends to be highly variable. Several years ago I was much struck with a remark, nearly to the above effect, published by Mr. Waterhouse. I infer also from an observation made by Professor Owen, with respect to the length of the arms of the ourang-outang, that he has come to a nearly similar conclusion. It is hopeless to attempt to convince any one of the truth of this proposition without giving the long array of facts which I have collected, and which cannot possibly be here introduced. I can only state my conviction that it is a rule of high generality. I am aware of several causes of error, but I hope that I have made due allowance for them. It should be understood that the rule by no means applies to any part, however unusually developed, unless it be unusually developed in comparison with the same part in closely allied species. Thus, the bat's wing is a most abnormal structure in the class mammalia; but the rule would not here apply, because there is a whole group of bats having wings; it would apply only if some one species of bat had its wings developed in some remarkable manner in comparison with the other species of the same genus. The rule applies very strongly in the case of secondary sexual characters, when displayed in any unusual manner. The term, secondary sexual characters, used by Hunter, applies to characters which are attached to one sex, but are not directly connected with the act of reproduction. The rule applies to males and females; but as females more rarely offer remarkable secondary sexual characters, it applies more rarely to them. The rule being so plainly applicable in the case of secondary sexual characters, may be due to the great variability of these characters, whether or not displayed in any unusual manner – of which fact I think there can be little doubt. But that our rule is not confined to secondary sexual characters is clearly shown in the case of hermaphrodite cirripedes; and I may here add, that I particularly attended to Mr. Waterhouse's remark, whilst investigating this Order, and I am fully convinced that the rule almost invariably holds good with cirripedes. I shall, in my future work, give a list of the more remarkable cases; I will here only briefly give one, as it illustrates the rule in its largest application. The opercular valves of sessile cirripedes (rock barnacles) are, in every sense of the word, very important structures, and they differ extremely little even in different genera; but in the several species of one genus, Pyrgoma, these valves present a marvellous amount of diversification: the homologous valves in the different species being sometimes wholly unlike in shape; and the amount of variation in the individuals of several of the species is so great, that it is no exaggeration to state that the varieties differ more from each other in the characters of these important valves than do other species of distinct genera.

As birds within the same country vary in a remarkably small degree, I have particularly attended to them, and the rule seems to me certainly to hold good in this class. I cannot make out that it applies to plants, and this would seriously have shaken my belief in its truth, had not the great variability in plants made it particularly difficult to compare their relative degrees of variability.

When we see any part or organ developed in a remarkable degree or manner in any species, the fair presumption is that it is of high importance to that species; nevertheless the part in this case is eminently liable to variation. Why should this be so? On the view that each species has been independently created, with all its parts as we now see them, I can see no explanation. But on the view that groups of species have descended from other species, and have been modified through natural selection, I think we can obtain some light. In our domestic animals, if any part, or the whole animal, be neglected and no selection be applied, that part (for instance, the comb in the Dorking fowl) or the whole breed will cease to have a nearly uniform character. The breed will then be said to have degenerated. In rudimentary organs, and in those which have been but little specialized for any particular purpose, and perhaps in polymorphic groups, we see a nearly parallel natural case; for in such cases natural selection either has not or cannot come into full play, and thus the organization is left in a fluctuating condition. But what here more especially concerns us is, that in our domestic animals those points, which at the present time are undergoing rapid change by continued selection, are also eminently liable to variation. Look at the breeds of the pigeon; see what a prodigious amount of difference there is in the beak of the different tumblers, in the beak and wattle of the different carriers, in the carriage and tail of our fantails, &c., these being the points now mainly attended to by English fanciers. Even in the sub-breeds, as in the short-faced tumbler, it is notoriously difficult to breed them nearly to perfection, and frequently individuals are born which depart widely from the standard. There may be truly said to be a constant struggle going on between, on the one hand, the tendency to reversion to a less modified state, as well as an innate tendency to further variability of all kinds, and, on the other hand, the power of steady selection to keep the breed true. In the long run selection gains the day, and we do not expect to fail so far as to breed a bird as coarse as a common tumbler from a good short-faced strain. But as long as selection is rapidly going on, there may always be expected to be much variability in the structure undergoing modification. It further deserves notice that these variable characters, produced by man's selection, sometimes become attached, from causes quite unknown to us, more to one sex than to the other, generally to the male sex, as with the wattle of carriers and the enlarged crop of pouters.

Now let us turn to nature. When a part has been developed in an extraordinary manner in any one species, compared with the other species of the same genus, we may conclude that this part has undergone an extraordinary amount of modification, since the period when the species branched off from the common progenitor of the genus. This period will seldom be remote in any extreme degree, as species very rarely endure for more than one geological period. An extraordinary amount of modification implies an unusually large and long-continued amount of variability, which has continually been accumulated by natural selection for the benefit of the species. But as the variability of the extraordinarily-developed part or organ has been so great and long-continued within a period not excessively remote, we might, as a general rule, expect still to find more variability in such parts than in other parts of the organization, which have remained for a much longer period nearly constant. And this, I am convinced, is the case. That the struggle between natural selection on the one hand, and the tendency to reversion and variability on the other hand, will in the course of time cease; and that the most abnormally developed organs may be made constant, I can see no reason to doubt. Hence when an organ, however abnormal it may be, has been transmitted in approximately the same condition to many modified descendants, as in the case of the wing of the bat, it must have existed, according to my theory, for an immense period in nearly the same state; and thus it comes to be no more variable than any other structure. It is only in those cases in which the modification has been comparatively recent and extraordinarily great that we ought to find the *generative variability*, as it may be called, still present in a high degree. For in this case the variability will seldom as yet have been fixed by the continued selection of the individuals varying in the required manner and degree, and by the continued rejection of those tending to revert to a former and less modified condition.

The principle included in these remarks may be extended. It is notorious that specific characters are more variable than generic. To explain by a simple example what is meant. If some species in a large genus of plants had blue flowers and some had red, the colour would be only a specific character, and no one would be surprised at one of the blue species varying into red, or conversely; but if all the species had blue flowers, the colour would become a generic character, and its variation would be a more unusual circumstance. I have chosen this example because an explanation is not in this case applicable, which most naturalists would advance, namely, that specific characters are more variable than generic, because they are taken from parts of less physiological importance than those commonly used for classing genera. I believe this explanation is partly, yet only indirectly, true; I shall, however, have to return to this subject in our chapter on Classification. It would be almost superfluous to adduce evidence in support of the above statement, that specific characters are more variable than generic; but I have repeatedly noticed in works on natural history, that when an author has remarked with surprise that some *important* organ or part, which is generally very constant throughout large groups of species, has *differed* considerably in closely-allied species, that it has, also, been *variable* in the individuals of some of the species. And this fact shows that a character, which is generally of generic value, when it sinks in value and becomes only of specific value, often becomes variable, though its physiological importance may remain the same. Something of the same kind applies to monstrosities: at least Is. Geoffroy St. Hilaire seems to entertain no doubt, that the more an organ normally differs in the different species of the same group, the more subject it is to individual anomalies.

On the ordinary view of each species having been independently created, why should that part of the structure, which differs from the same part in other independently-created species of the same genus, be more variable than those parts which are closely alike in the several species? I do not see that any explanation can be given. But on the view of species being only strongly marked and fixed varieties, we might surely expect to find them still often continuing to vary in those parts of their structure which have varied within a moderately recent period, and which have thus come to differ. Or to state the case in another manner: – the points in which all the species of a genus resemble each other, and in which they differ from the species of some other genus, are called generic characters; and these characters in common I attribute to inheritance from a common progenitor, for it can rarely have happened that natural selection will have modified several species, fitted to more or less widely-different habits, in exactly the same manner: and as these so-called generic characters have been inherited from a remote period, since that period when the species first branched off from their common progenitor, and subsequently have not varied or come to differ in any degree, or only in a slight degree, it is not probable that they should vary at the present day. On the other hand, the points in which species differ from other species of the same genus, are called specific characters; and as these specific characters have varied and come to differ within the period of the branching off of the species from a common progenitor, it is probable that they should still often be in some degree variable, – at least more variable than those parts of the organization which have for a very long period remained constant.

In connexion with the present subject, I will make only two other remarks. I think it will be admitted, without my entering on details, that secondary sexual characters are very variable; I think it also will be admitted that species of the same group differ from each other more widely in their secondary sexual characters, than in other parts of their organization; compare, for instance, the amount of difference between the males of gallinaceous birds, in which secondary sexual characters are strongly displayed, with the amount of difference between their females; and the truth of this proposition will be granted. The cause of the original variability of secondary sexual characters is not manifest; but we can see why these characters should not have been rendered as constant and uniform as other parts of the organization; for secondary sexual characters have been accumulated by sexual selection, which is less rigid in its action than ordinary selection, as it does not entail death, but only gives fewer offspring to the less favoured males. Whatever the cause may be of the variability of secondary sexual characters, as they are highly variable, sexual selection will have had a wide scope for action, and may thus readily have succeeded in giving to the species of the same group a greater amount of difference in their sexual characters, than in other parts of their structure.

It is a remarkable fact, that the secondary sexual differences between the two sexes of the same species are generally displayed in the very same parts of the organization in which the different species of the same genus differ from each other. Of this fact I will give in illustration two instances, the first which happen to stand on my list; and as the differences in these cases are of a very unusual nature, the relation can hardly be accidental. The same number of joints in the tarsi is a character generally common to very large groups of beetles, but in the Engidae, as Westwood has remarked, the number varies greatly; and the number likewise differs in the two sexes of the same species: again in fossorial hymenoptera, the manner of neuration of the wings is a character of the highest importance, because common to large groups; but in certain genera the neuration differs in the different species, and likewise in the two sexes of the same species. This relation has a clear meaning on my view of the subject: I look at all the species of the same genus as having as certainly descended from the same progenitor, as have the two sexes of any one of the species. Consequently, whatever part of the structure of the common progenitor, or of its early descendants, became variable; variations of this part would, it is highly probable, be taken advantage of by natural and sexual selection, in order to fit the several species to their several places in the economy of nature, and likewise to fit the two sexes of the same species to each other, or to fit the males and females to different habits of life, or the males to struggle with other males for the possession of the females.

Finally, then, I conclude that the greater variability of specific characters, or those which distinguish species from species, than of generic characters, or those which the species possess in common; – that the frequent extreme variability of any part which is developed in a species in an extraordinary manner in comparison with the same part in its congeners; and the not great degree of variability in a part, however extraordinarily it may be developed, if it be common to a whole group of species; – that the great variability of secondary sexual characters, and the great amount of difference in these same characters between closely allied species; – that secondary sexual and ordinary specific differences are generally displayed in the same parts of the organization, – are all principles closely connected together. All being mainly due to the species of the same group having descended from a common progenitor, from whom they have inherited much in common, – to parts which have recently and largely varied being more likely still to go on varying than parts which have long been inherited and have not varied, – to natural selection having more or less completely, according to the lapse of time, overmastered the tendency to reversion and to further variability, – to sexual selection being less rigid than ordinary selection, – and to variations in the same parts having been accumulated by natural and sexual selection, and thus adapted for secondary sexual, and for ordinary specific purposes.

Distinct species present analogous variations; and a variety of one species often assumes some of the characters of an allied species, or reverts to some of the characters of an early progenitor. These propositions will be most readily understood by looking to our domestic races. The most distinct breeds of pigeons, in countries most widely apart, present sub-varieties with reversed feathers on the head and feathers on the feet, – characters not possessed by the aboriginal rock-pigeon; these then are analogous variations in two or more distinct races. The frequent presence of fourteen or even sixteen tail-feathers in the pouter, may be considered as a variation representing the normal structure of another race, the fantail. I presume that no one will doubt that all such analogous variations are due to the several races of the pigeon having inherited from a common parent the same constitution and tendency to variation, when acted on by similar unknown influences. In the vegetable kingdom we have a case of analogous variation, in the enlarged stems, or roots as commonly called, of the Swedish turnip and Ruta baga, plants which several botanists rank as varieties produced by cultivation from a common parent: if this be not so, the case will then be one of analogous variation in two so-called distinct species; and to these a third may be added, namely, the common turnip. According to the ordinary view of each species having been independently created, we should have to attribute this similarity in the enlarged stems of these three plants, not to the *vera causa* of community of descent, and a consequent tendency to vary in a like manner, but to three separate yet closely related acts of creation.

With pigeons, however, we have another case, namely, the occasional appearance in all the breeds, of slaty-blue birds with two black bars on the wings, a white rump, a bar at the end of the tail, with the outer feathers externally edged near their bases with white. As all these marks are characteristic of the parent rock-pigeon, I presume that no one will doubt that this is a case of reversion, and not of a new yet analogous variation appearing in the several breeds. We may I think confidently come to this conclusion, because, as we have seen, these coloured marks are eminently liable to appear in the crossed offspring of two distinct and differently coloured breeds; and in this case there is nothing in the external conditions of life to cause the reappearance of the slaty-blue, with the several marks, beyond the influence of the mere act of crossing on the laws of inheritance.

No doubt it is a very surprising fact that characters should reappear after having been lost for many, perhaps for hundreds of generations. But when a breed has been crossed only once by some other breed, the offspring occasionally show a tendency to revert in character to the foreign breed for many generations – some say, for a dozen or even a score of generations. After twelve generations, the proportion of blood, to use a common expression, of any one ancestor, is only 1 in 2048; and yet, as we see, it is generally believed that a tendency to reversion is retained by this very small proportion of foreign blood. In a breed which has not been crossed, but in which *both* parents have lost some character which their progenitor possessed, the tendency, whether strong or weak, to reproduce the lost character might be, as was formerly remarked, for all that we can see to the contrary, transmitted for almost any number of generations. When a character which has been lost in a breed, reappears after a great number of generations, the most probable hypothesis is, not that the offspring suddenly takes after an ancestor some hundred generations distant, but that in each successive generation there has been a tendency to reproduce the character in question, which at last, under unknown favourable conditions, gains an ascendancy. For instance, it is probable that in each generation of the barb-pigeon, which produces most rarely a blue and black-barred bird, there has been a tendency in each generation in the plumage to assume this colour. This view is hypothetical, but could be supported by some facts; and I can see no more abstract improbability in a tendency to produce any character being inherited for an endless number of generations, than in quite useless or rudimentary organs being, as we all know them to be, thus inherited. Indeed, we may sometimes observe a mere tendency to produce a rudiment inherited: for instance, in the common snapdragon (Antirrhinum) a rudiment of a fifth stamen so often appears, that this plant must have an inherited tendency to produce it.

As all the species of the same genus are supposed, on my theory, to have descended from a common parent, it might be expected that they would occasionally vary in an analogous manner; so that a variety of one species would resemble in some of its characters another species; this other species being on my view only a well-marked and permanent variety. But characters thus gained would probably be of an unimportant nature, for the presence of all import-ant characters will be governed by natural selection, in accordance with the diverse habits of the species, and will not be left to the mutual action of the conditions of life and of a similar inherited constitution. It might further be expected that the species of the same genus would occasionally exhibit reversions to lost ancestral characters. As, how-ever, we never know the exact character of the common ancestor of a group, we could not distinguish these two cases: if, for instance, we did not know that the rock-pigeon was not feather-footed or turn-crowned, we could not have told, whether these characters in our domestic breeds were reversions or only analogous variations; but we might have inferred that the blueness was a case of reversion, from the number of the markings, which are correlated with the blue tint, and which it does not appear probable would all appear together from simple variation. More especially we might have inferred this, from the blue colour and marks so often appearing when distinct breeds of diverse colours are crossed. Hence, though under nature it must generally be left doubtful, what cases are reversions to an anciently existing character, and what are new but analogous vari-ations, yet we ought, on my theory, sometimes to find the varying offspring of a species assuming characters (either from reversion or from analogous variation) which already occur in some other members of the same group. And this undoubtedly is the case in nature.

A considerable part of the difficulty in recognizing a variable species in our systematic works, is due to its varieties mocking, as it were, some of the other species of the same genus. A considerable catalogue, also, could be given of forms intermediate between two other forms, which themselves must be doubtfully ranked as either varieties or species; and this shows, unless all these forms be considered as independently created species, that the one in varying has assumed some of the characters of the other, so as to produce the intermediate form. But the best evidence is afforded by parts or organs of an important and uniform nature occasionally varying so as to acquire, in some degree, the character of the same part or organ in an allied species. I have collected a long list of such cases; but here, as before, I lie under a great disadvantage in not being able to give them. I can only repeat that such cases certainly do occur, and seem to me very remarkable.

I will, however, give one curious and complex case, not indeed as affecting any important character, but from occurring in several species of the same genus, partly under domestication and partly under nature. It is a case apparently of reversion. The ass not rarely has very distinct transverse bars on its legs, like those on the legs of the zebra: it has been asserted that these are plainest in the foal, and from inquiries which I have made, I believe this to be true. It has also been asserted that the stripe on each shoulder is sometimes double. The shoulder-stripe is certainly very variable in length and outline. A white ass, but *not* an albino, has been described without either spinal or shoulder stripe; and these stripes are sometimes very obscure, or actually quite lost, in dark-coloured asses. The koulan of Pallas is said to have been seen with a double shoulder-stripe. The hemionus has no shoulder-stripe; but traces of it, as stated by Mr. Blyth and others, occasionally appear: and I have been informed by Colonel Poole that the foals of this species are generally striped on the legs, and faintly on the shoulder. The quagga, though so plainly barred like a zebra over the body, is without bars on the legs;

but Dr. Gray has figured one specimen with very distinct zebra-like bars on the hocks.

With respect to the horse, I have collected cases in England of the spinal stripe in horses of the most distinct breeds, and of *all* colours; transverse bars on the legs are not rare in duns, mouse-duns, and in one instance in a chestnut: a faint shoulder-stripe may sometimes be seen in duns, and I have seen a trace in a bay horse. My son made a careful examination and sketch for me of a dun Belgian cart-horse with a double stripe on each shoulder and with leg-stripes; and a man, whom I can implicitly trust, has examined for me a small dun Welch pony with *three* short parallel stripes on each shoulder.

In the north-west part of India the Kattywar breed of horses is so generally striped, that, as I hear from Colonel Poole, who examined the breed for the Indian Government, a horse without stripes is not considered as purely-bred. The spine is always striped; the legs are generally barred; and the shoulder-stripe, which is sometimes double and sometimes treble, is common; the side of the face, moreover, is sometimes striped. The stripes are plainest in the foal; and sometimes quite disappear in old horses. Colonel Poole has seen both grey and bay Kattywar horses striped when first foaled. I have, also, reason to suspect, from information given me by Mr. W. W. Edwards, that with the English race-horse the spinal stripe is much commoner in the foal than in the full-grown animal. Without here entering on further details, I may state that I have collected cases of leg and shoulder-stripes in horses of very different breeds, in various countries from Britain to Eastern China; and from Norway in the north to the Malay Archipelago in the south. In all parts of the world these stripes occur far oftenest in duns and mouse-duns; by the term dun a large range of colour is included, from one between brown and black to a close approach to cream-colour.

I am aware that Colonel Hamilton Smith, who has written on this subject, believes that the several breeds of the horse have descended from several aboriginal species – one of which, the dun, was striped; and that the above-described appearances are all due to ancient crosses with the dun stock. But I am not at all satisfied with this theory, and should be loth to apply it to breeds so distinct as the heavy

Belgian cart-horse, Welch ponies, cobs, the lanky Kattywar race, &c., inhabiting the most distant parts of the world.

Now let us turn to the effects of crossing the several species of the horse-genus. Rollin asserts, that the common mule from the ass and horse is particularly apt to have bars on its legs. I once saw a mule with its legs so much striped that any one at first would have thought that it must have been the product of a zebra; and Mr. W. C. Martin, in his excellent treatise on the horse, has given a figure of a similar mule. In four coloured drawings, which I have seen, of hybrids between the ass and zebra, the legs were much more plainly barred than the rest of the body; and in one of them there was a double shoulder-stripe. In Lord Moreton's famous hybrid from a chestnut mare and male quagga, the hybrid, and even the pure offspring subsequently produced from the mare by a black Arabian sire, were much more plainly barred across the legs than is even the pure quagga. Lastly, and this is another most remarkable case, a hybrid has been figured by Dr. Gray (and he informs me that he knows of a second case) from the ass and the hemionus; and this hybrid, though the ass seldom has stripes on its legs and the hemionus has none and has not even a shoulder-stripe, nevertheless had all four legs barred, and had three short shoulder-stripes, like those on the dun Welch pony, and even had some zebra-like stripes on the sides of its face. With respect to this last fact, I was so convinced that not even a stripe of colour appears from what would commonly be called an accident, that I was led solely from the occurrence of the face-stripes on this hybrid from the ass and hemionus, to ask Colonel Poole whether such face-stripes ever occur in the eminently striped Kattywar breed of horses, and was, as we have seen, answered in the affirmative.

What now are we to say to these several facts? We see several very distinct species of the horse-genus becoming, by simple variation, striped on the legs like a zebra, or striped on the shoulders like an ass. In the horse we see this tendency strong whenever a dun tint appears – a tint which approaches to that of the general colouring of the other species of the genus. The appearance of the stripes is not accompanied by any change of form or by any other new character. We see this tendency to become striped most strongly displayed in hybrids from between several of the most distinct species. Now observe the case of the several breeds of pigeons: they are descended from a pigeon (including two or three sub-species or geographical races) of a bluish colour, with certain bars and other marks; and when any breed assumes by simple variation a bluish tint, these bars and other marks invariably reappear; but without any other change of form or character. When the oldest and truest breeds of various colours are crossed, we see a strong tendency for the blue tint and bars and marks to reappear in the mongrels. I have stated that the most probable hypothesis to account for the reappearance of very ancient characters, is – that there is a *tendency* in the young of each successive generation to produce the long-lost character, and that this tendency, from unknown causes, sometimes prevails. And we have just seen that in several species of the horse-genus the stripes are either plainer or appear more commonly in the young than in the old. Call the breeds of pigeons, some of which have bred true for centuries, species; and how exactly parallel is the case with that of the species of the horse-genus! For myself, I venture confidently to look back thousands on thousands of generations, and I see an animal striped like a zebra, but perhaps otherwise very differently constructed, the common parent of our domestic horse, whether or not it be descended from one or more wild stocks, of the ass, the hemionus, quagga, and zebra.

He who believes that each equine species was independently created, will, I presume, assert that each species has been created with a tendency to vary, both under nature and under domestication, in this particular manner, so as often to become striped like other species of the genus; and that each has been created with a strong tendency, when crossed with species inhabiting distant quarters of the world, to produce hybrids resembling in their stripes, not their own parents, but other species of the genus. To admit this view is, as it seems to me, to reject a real for an unreal, or at least for an unknown, cause. It makes the works of God a mere mockery and deception; I would almost as soon believe with the old and ignorant cosmogonists, that fossil shells had never lived, but had been created in stone so as to mock the shells now living on the sea-shore.

Summary. Our ignorance of the laws of variation is profound. Not in one case out of a hundred can we pretend to assign any reason why this or that part differs, more or less, from the same part in the parents. But whenever we have the means of instituting a comparison, the same laws appear to have acted in producing the lesser differences between varieties of the same species, and the greater differences between species of the same genus. The external conditions of life, as climate and food, &c., seem to have induced some slight modifications. Habit in producing constitutional differences, and use in strengthening, and disuse in weakening and diminishing organs, seem to have been more potent in their effects. Homologous parts tend to vary in the same way, and homologous parts tend to cohere. Modifications in hard parts and in external parts sometimes affect softer and internal parts. When one part is largely developed, perhaps it tends to draw nourishment from the adjoining parts; and every part of the structure which can be saved without detriment to the individual, will be saved. Changes of structure at an early age will generally affect parts subsequently developed; and there are very many other correlations of growth, the nature of which we are utterly unable to understand. Multiple parts are variable in number and in structure, perhaps arising from such parts not having been closely specialized to any particular function, so that their modifications have not been closely checked by natural selection. It is probably from this same cause that organic beings low in the scale of nature are more variable than those which have their whole organization more specialized, and are higher in the scale. Rudimentary organs, from being useless, will be disregarded by natural selection, and hence probably are variable. Specific characters – that is, the characters which have come to differ since the several species of the same genus branched off from a common parent – are more variable than generic characters, or those which have long been inherited, and have not differed within this same period. In these remarks we have referred to special parts or organs being still variable, because they have recently varied and thus come to differ; but we have also seen in the second Chapter that the same principle applies to the whole individual; for in a district where many species of any genus are found – that is, where there has been much former variation and differentiation, or where the manufactory of new specific forms has been actively at work – there, on an average, we now find most varieties or incipient species. Secondary sexual characters are highly variable, and such characters differ much in the species of the same group. Variability in the same parts of the organization has generally been taken advantage of in giving secondary sexual differences to the sexes of the same species, and specific differences to the several species of the same genus. Any part or organ developed to an extraordinary size or in an extraordinary manner, in comparison with the same part or organ in the allied species, must have gone through an extraordinary amount of modification since the genus arose; and thus we can understand why it should often still be variable in a much higher degree than other parts; for variation is a long-continued and slow process, and natural selection will in such cases not as yet have had time to overcome the tendency to further variability and to reversion to a less modified state. But when a species with any extraordinarily-developed organ has become the parent of many modified descendants – which on my view must be a very slow process, requiring a long lapse of time – in this case, natural selection may readily have succeeded in giving a fixed character to the organ, in however extraordinary a manner it may be developed. Species inheriting nearly the same constitution from a common parent and exposed to similar influences will naturally tend to present analogous variations, and these same species may occasionally revert to some of the characters of their ancient progenitors. Although new and important modifications may not arise from reversion and analogous variation, such modifications will add to the beautiful and harmonious diversity of nature.

Whatever the cause may be of each slight difference in the offspring from their parents – and a cause for each must exist – it is the steady accumulation, through natural selection, of such differences, when beneficial to the individual, that gives rise to all the more important modifications of structure, by which the innumerable beings on the face of this earth are enabled to struggle with each other, and the best adapted to survive.

CHAPTER VI

DIFFICULTIES ON THEORY

Difficulties on the theory of descent with modification – Transitions – Absence or rarity of transitional varieties – Transitions in habits of life – Diversified habits in the same species – Species with habits widely different from those of their allies – Organs of extreme perfection – Means of transition – Cases of difficulty – Natura non facit saltum – Organs of small importance – Organs not in all cases absolutely perfect – The law of Unity of Type and of the Conditions of Existence embraced by the theory of Natural Selection

LONG before having arrived at this part of my work, a crowd of difficulties will have occurred to the reader. Some of them are so grave that to this day I can never reflect on them without being staggered; but, to the best of my judgement, the greater number are only apparent, and those that are real are not, I think, fatal to my theory.

These difficulties and objections may be classed under the following heads: – Firstly, why, if species have descended from other species by insensibly fine gradations, do we not everywhere see innumerable transitional forms? Why is not all nature in confusion instead of the species being, as we see them, well defined?

Secondly, is it possible that an animal having, for instance, the structure and habits of a bat, could have been formed by the modification of some animal with wholly different habits? Can we believe that natural selection could produce, on the one hand, organs of trifling importance, such as the tail of a giraffe, which serves as a fly-flapper, and, on the other hand, organs of such wonderful structure, as the eye, of which we hardly as yet fully understand the inimitable perfection?

Thirdly, can instincts be acquired and modified through natural selection? What shall we say to so marvellous an instinct as that which leads the bee to make cells, which have practically anticipated the discoveries of profound mathematicians?

Fourthly, how can we account for species, when crossed, being sterile and producing sterile offspring, whereas, when varieties are crossed, their fertility is unimpaired?

The two first heads shall be here discussed – Instinct and Hybridism in separate chapters.

On the absence or rarity of transitional varieties. As natural selection acts solely by the preservation of profitable modifications, each new form will tend in a fully-stocked country to take the place of, and finally to exterminate, its own less improved parent or other less-favoured forms with which it comes into competition. Thus extinction and natural selection will, as we have seen, go hand in hand. Hence, if we look at each species as descended from some other unknown form, both the parent and all the transitional varieties will generally have been exterminated by the very process of formation and perfection of the new form.

But, as by this theory innumerable transitional forms must have existed, why do we not find them embedded in countless numbers in the crust of the earth? It will be much more convenient to discuss this question in the chapter on the Imperfection of the geological record; and I will here only state that I believe the answer mainly lies in the record being incomparably less perfect than is generally supposed; the imperfection of the record being chiefly due to organic beings not inhabiting profound depths of the sea, and to their remains being embedded and preserved to a future age only in masses of sediment sufficiently thick and extensive to withstand an enormous amount of future degradation; and such fossiliferous masses can be accumulated only where much sediment is deposited on the shallow bed of the sea, whilst it slowly subsides. These contingencies will concur only rarely, and after enormously long intervals. Whilst the bed of the sea is stationary or is rising, or when very little sediment is being deposited, there will be blanks in our geological history. The crust of the earth is a vast museum; but the natural collections have been made only at intervals of time immensely remote.

But it may be urged that when several closely-allied species inhabit the same territory we surely ought to find at the present time many transitional forms. Let us take a simple case: in travelling from north to south over a continent, we generally meet at successive intervals with closely allied or representative species, evidently filling nearly the same place in the natural economy of the land. These representative species often meet and interlock; and as the one becomes rarer and rarer, the other becomes more and more frequent, till the one replaces the other. But if we compare these species where they intermingle, they are generally as absolutely distinct from each other in every detail of structure as are specimens taken from the metropolis inhabited by each. By my theory these allied species have descended from a common parent; and during the process of modification, each has become adapted to the conditions of life of its own region, and has supplanted and exterminated its original parent and all the transitional varieties between its past and present states. Hence we ought not to expect at the present time to meet with numerous transitional varieties in each region, though they must have existed there, and may be embedded there in a fossil condition. But in the intermediate region, having intermediate conditions of life, why do we not now find closely-linking intermediate varieties? This difficulty for a long time quite confounded me. But I think it can be in large part explained.

In the first place we should be extremely cautious in inferring, because an area is now continuous, that it has been continuous during a long period. Geology would lead us to believe that almost every continent has been broken up into islands even during the later tertiary periods; and in such islands distinct species might have been separately formed without the possibility of intermediate varieties existing in the intermediate zones. By changes in the form of the land and of climate, marine areas now continuous must often have existed within recent times in a far less continuous and uniform condition than at present. But I will pass over this way of escaping from the difficulty; for I believe that many perfectly defined species have been formed on strictly continuous areas; though I do not doubt that the formerly broken condition of areas now continuous has played an important part in the formation of new species, more especially with freely-crossing and wandering animals.

In looking at species as they are now distributed over a wide area, we generally find them tolerably numerous over a large territory, then becoming somewhat abruptly rarer and rarer on the confines, and finally disappearing. Hence the neutral territory between two representative species is generally narrow in comparison with the territory proper to each. We see the same fact in ascending mountains, and sometimes it is quite remarkable how abruptly, as Alph. De Candolle has observed, a common alpine species disappears. The same fact has been noticed by Forbes in sounding the depths of the sea with the dredge. To those who look at climate and the physical conditions of life as the all-important elements of distribution, these facts ought to cause surprise, as climate and height or depth graduate away insensibly. But when we bear in mind that almost every species, even in its metropolis, would increase immensely in numbers, were it not for other competing species; that nearly all either prey on or serve as prey for others; in short, that each organic being is either directly or indirectly related in the most important manner to other organic beings, we must see that the range of the inhabitants of any country by no means exclusively depends on insensibly changing physical conditions, but in large part on the presence of other species, on which it depends, or by which it is destroyed, or with which it comes into competition; and as these species are already defined objects (however they may have become so), not blending one into another by insensible gradations, the range of any one species, depending as it does on the range of others, will tend to be sharply defined. Moreover, each species on the confines of its range, where it exists in lessened numbers, will, during fluctuations in the number of its enemies or of its prey, or in the seasons, be extremely liable to utter extermination; and thus its geographical range will come to be still more sharply defined.

If I am right in believing that allied or representative species, when inhabiting a continuous area, are generally so distributed that each has a wide range, with a comparatively narrow neutral territory between them, in which they become rather suddenly rarer and rarer; then, as varieties do not essentially differ from species, the same rule will probably apply to both; and if we in imagination adapt a varying species to a very large area, we shall have to adapt two varieties to two large areas, and a third variety to a narrow intermediate zone. The intermediate variety, consequently, will exist in lesser numbers from inhabiting a narrow and lesser area; and practically, as far as I can make out, this rule holds good with varieties in a state of nature. I have met with striking instances of the rule in the case of varieties intermediate between well-marked varieties in the genus Balanus. And it would appear from information given me by Mr. Watson, Dr. Asa Gray, and Mr. Wollaston, that generally when varieties intermediate between two other forms occur, they are much rarer numerically than the forms which they connect. Now, if we may trust these facts and inferences, and therefore conclude that varieties linking two other varieties together have generally existed in lesser numbers than the forms which they connect, then, I think, we can understand why intermediate varieties should not endure for very long periods; – why as a general rule they should be exterminated and disappear, sooner than the forms which they originally linked together.

For any form existing in lesser numbers would, as already remarked, run a greater chance of being exterminated than one existing in large numbers; and in this particular case the intermediate form would be eminently liable to the inroads of closely allied forms existing on both sides of it. But a far more important consideration, as I believe, is that, during the process of further modification, by which two varieties are supposed on my theory to be converted and perfected into two distinct species, the two which exist in larger numbers from inhabiting larger areas, will have a great advantage over the intermediate variety, which exists in smaller numbers in a narrow and intermediate zone. For forms existing in larger numbers will always have a better chance, within any given period, of presenting further favourable variations for natural selection to seize on, than will the rarer forms which exist in lesser numbers. Hence, the more common forms, in the race for life, will tend to beat and supplant the less common forms, for these will be more slowly modified and improved. It is the same principle which, as I believe, accounts for the common species in each country, as shown in the second chapter, presenting on an average a greater number of well-marked varieties than do the rarer species. I may illustrate what I mean by supposing three varieties of sheep to be kept, one adapted to an extensive mountainous region; a second to a comparatively narrow, hilly tract; and a third to wide plains at the base; and that the inhabitants are all trying with equal steadiness and skill to improve their stocks by selection; the chances in this case will be strongly in favour of the great holders on the mountains or on the plains improving their breeds more quickly than the small holders on the intermediate narrow, hilly tract; and consequently the improved mountain or plain breed will soon take the place of the less improved hill breed; and thus the two breeds, which originally existed in greater numbers, will come into close contact with each other, without the interposition of the supplanted, intermediate hill-variety.

To sum up, I believe that species come to be tolerably well-defined objects, and do not at any one period present an inextricable chaos of varying and intermediate links: firstly, because new varieties are very slowly formed, for variation is a very slow process, and natural selection can do nothing until favourable variations chance to occur, and until a place in the natural polity of the country can be better filled by some modification of some one or more of its inhabitants. And such new places will depend on slow changes of climate, or on the occasional immigration of new inhabitants, and, probably, in a still more important degree, on some of the old inhabitants becoming slowly modified, with the new forms thus produced and the old ones acting and reacting on each other. So that, in any one region and at any one time, we ought only to see a few species presenting slight modifications of structure in some degree permanent: and this assuredly we do see.

Secondly, areas now continuous must often have existed within the recent period in isolated portions, in which many forms, more especially amongst the classes which unite for each birth and wander much, may have separately been rendered sufficiently distinct to rank as representative species. In this case, intermediate varieties between the several representative species and their common parent, must formerly have existed in each broken portion of the land, but these links will have been supplanted and exterminated during the process of natural selection, so that they will no longer exist in a living state.

Thirdly, when two or more varieties have been formed in different portions of a strictly continuous area, intermediate varieties will, it is probable, at first have been formed in the intermediate zones, but they will generally have had a short duration. For these intermediate varieties will, from reasons already assigned (namely from what we know of the actual distribution of closely allied or representative species, and likewise of acknowledged varieties), exist in the intermediate zones in lesser numbers than the varieties which they tend to connect. From this cause alone the intermediate varieties will be liable to accidental extermination; and during the process of further modification through natural selection, they will almost certainly be beaten and supplanted by the forms which they connect; for these from existing in greater numbers will, in the aggregate, present more variation, and thus be further improved through natural selection and gain further advantages.

Lastly, looking not to any one time, but to all time, if my theory be true, numberless intermediate varieties, linking most closely all the species of the same group together, must assuredly have existed; but the very process of natural selection constantly tends, as has been so often remarked, to exterminate the parent-forms and the intermediate links. Consequently evidence of their former existence could be found only amongst fossil remains, which are preserved, as we shall in a future chapter attempt to show, in an extremely imperfect and intermittent record.

On the origin and transitions of organic beings with peculiar habits and structure. It has been asked by the opponents of such views as I hold, how, for instance, a land carnivorous animal could have been converted into one with aquatic habits; for how could the animal in its transitional state have subsisted? It would be easy to show that within the same group carnivorous animals exist having every intermediate grade between truly aquatic and strictly terrestrial habits; and as each exists by a struggle for life, it is clear that each is well adapted in its habits to its place in nature. Look at the Mustela vison of North America, which has webbed feet and which resembles an otter in its fur, short legs, and form of tail; during summer this animal dives for and preys on fish, but during the long winter it leaves the frozen waters, and preys like other polecats on mice and land animals. If a different case had been taken, and it had been asked how an insectivorous quadruped could possibly have been converted into a flying bat, the question would have been far more difficult, and I could have given no answer. Yet I think such difficulties have very little weight.

Here, as on other occasions, I lie under a heavy disadvantage, for out of the many striking cases which I have collected, I can give only one or two instances of transitional habits and structures in closely allied species of the same genus; and of diversified habits, either constant or occasional, in the same species. And it seems to me that nothing less than a long list of such cases is sufficient to lessen the difficulty in any particular case like that of the bat.

Look at the family of squirrels; here we have the finest gradation from animals with their tails only slightly flattened, and from others, as Sir J. Richardson has remarked, with the posterior part of their bodies rather wide and with the skin on their flanks rather full, to the so-called flying squirrels; and flying squirrels have their limbs and even the base of the tail united by a broad expanse of skin, which serves as a parachute and allows them to glide through the air to an astonishing distance from tree to tree. We cannot doubt that each structure is of use to each kind of squirrel in its own country, by enabling it to escape birds or beasts of prey, or to collect food more quickly, or, as there is reason to believe, by lessening the danger from occasional falls. But it does not follow from this fact that the structure of each squirrel is the best that it is possible to conceive under all natural conditions. Let the climate and vegetation change, let other competing rodents or new beasts of prey immigrate, or old ones become modified, and all analogy would lead us to believe that some at least of the squirrels would decrease in numbers or become exterminated, unless they also became modified and improved in structure in a corresponding manner. Therefore, I can see no difficulty, more especially under changing conditions of life, in the continued preservation of individuals with fuller and fuller flank-membranes, each modification being useful, each being propagated, until by the accumulated effects of this process of natural selection, a perfect so-called flying squirrel was produced.

Now look at the Galeopithecus or flying lemur, which formerly was falsely ranked amongst bats. It has an extremely wide flank-membrane, stretching from the corners of the jaw to the tail, and including the limbs and the elongated fingers: the flank-membrane is, also, furnished with an extensor muscle. Although no graduated links of structure, fitted for gliding through the air, now connect the Galeopithecus with the other Lemuridae, yet I can see no difficulty in supposing that such links formerly existed, and that each had been formed by the same steps as in the case of the less perfectly gliding squirrels; and that each grade of structure had been useful to its possessor. Nor can I see any insuperable difficulty in further believing it possible that the membrane-connected fingers and fore-arm of the Galeopithecus might be greatly lengthened by natural selection; and this, as far as the organs of flight are concerned, would convert it into a bat. In bats which have the wing-membrane extended from the top of the shoulder to the tail, including the hind-legs, we perhaps see traces of an apparatus originally constructed for gliding through the air rather than for flight.

If about a dozen genera of birds had become extinct or were unknown, who would have ventured to have surmised that birds might have existed which used their wings solely as flappers, like the logger-headed duck (Micropterus of Eyton); as fins in the water and front legs on the land, like the penguin; as sails, like the ostrich; and functionally for no purpose, like the Apteryx. Yet the structure of each of these birds is good for it, under the conditions of life to which it is exposed, for each has to live by a struggle; but it is not necessarily the best possible under all possible conditions. It must not be inferred from these remarks that any of the grades of wing-structure here alluded to, which perhaps may all have resulted from disuse, indicate the natural steps by which birds have acquired their perfect power of flight; but they serve, at least, to show what diversified means of transition are possible.

Seeing that a few members of such water-breathing classes as the Crustacea and Mollusca are adapted to live on the land, and seeing that we have flying birds and mammals, flying insects of the most diversified types, and formerly had flying reptiles, it is conceivable that flying-fish, which now glide far through the air, slightly rising and turning by the aid of their fluttering fins, might have been modified into perfectly winged animals. If this had been effected, who would have ever imagined that in an early transitional state they had been inhabitants of the open ocean, and had used their incipient organs of flight exclusively, as far as we know, to escape being devoured by other fish?

When we see any structure highly perfected for any particular habit, as the wings of a bird for flight, we should bear in mind that animals displaying early transitional grades of the structure will seldom continue to exist to the present day, for they will have been supplanted by the very process of perfection through natural selection. Furthermore, we may conclude that transitional grades between structures fitted for very different habits of life will rarely have been developed at an early period in great numbers and under many subordinate forms. Thus, to return to our imaginary illustration of the flying-fish, it does not seem probable that fishes capable of true flight would have been developed under many subordinate forms, for taking prey of many kinds in many ways, on the land and in the water, until their organs of flight had come to a high stage of perfection, so as to have given them a decided advantage over other animals in the battle for life. Hence the chance of discovering species with transitional grades of structure in a fossil condition will always be less, from their having existed in lesser numbers, than in the case of species with fully developed structures.

I will now give two or three instances of diversified and of changed habits in the individuals of the same species. When either case occurs, it would be easy for natural selection to fit the animal, by some modification of its structure, for its changed habits, or exclusively for one of its several different habits. But it is difficult to tell, and immaterial for us, whether habits generally change first and structure afterwards; or whether slight modifications of structure lead to changed habits; both probably often change almost simultaneously. Of cases of changed habits it will suffice merely to allude to that of the many British insects which now feed on exotic plants, or exclusively on artificial substances. Of diversified habits innumerable instances could be given: I have often watched a tyrant flycatcher (Saurophagus sulphuratus) in South America, hovering over one spot and then proceeding to another, like a kestrel, and at other times standing stationary on the margin of water, and then dashing like a kingfisher at a fish. In our own country the larger titmouse (Parus major) may be seen climbing branches, almost like a creeper; it often, like a shrike, kills small birds by blows on the head; and I have many times seen and heard it hammering the seeds of the yew on a branch, and thus breaking them like a nuthatch. In North America the black bear was seen by Hearne swimming for hours with widely open mouth, thus catching, like a whale, insects in the water. Even in so extreme a case as this, if the supply of insects were constant, and if better adapted competitors did not already exist in the country, I can see no difficulty in a race of bears being rendered, by natural selection, more and more aquatic in their structure and habits, with larger and larger mouths, till a creature was produced as monstrous as a whale.

As we sometimes see individuals of a species following habits widely different from those both of their own species and of the other species of the same genus, we might expect, on my theory, that such individuals would occasionally have given rise to new species, having anomalous habits, and with their structure either slightly or considerably modified from that of their proper type. And such instances do occur in nature. Can a more striking instance of adaptation be given than that of a woodpecker for climbing trees and for seizing insects in the chinks of the bark? Yet in North America there are woodpeckers which feed largely on fruit, and others with elongated wings which chase insects on the wing; and on the plains of La Plata, where not a tree grows, there is a woodpecker, which in every essential part of its organization, even in its colouring, in the harsh tone of its voice, and undulatory flight, told me plainly of its close blood-relationship to our common species; yet it is a woodpecker which never climbs a tree!

Petrels are the most aërial and oceanic of birds, yet in the quiet Sounds of Tierra del Fuego, the Puffinuria berardi, in its general habits, in its astonishing power of diving, its manner of swimming, and of flying when unwillingly it takes flight, would be mistaken by any one for an auk or grebe; nevertheless, it is essentially a petrel, but with many parts of its organization profoundly modified. On the other hand, the acutest observer by examining the dead body of the water-ouzel would never have suspected its sub-aquatic habits; yet this anomalous member of the strictly terrestrial thrush family wholly subsists by diving, – grasping the stones with its feet and using its wings under water.

He who believes that each being has been created as we now see it, must occasionally have felt surprise when he has met with an animal having habits and structure not at all in agreement. What can be plainer than that the webbed feet of ducks and geese are formed for swimming; yet there are upland geese with webbed feet which rarely or never go near the water; and no one except Audubon has seen the frigate-bird, which has all its four toes webbed, alight on the surface of the sea. On the other hand, grebes and coots are eminently aquatic, although their toes are only bordered by membrane. What seems plainer than that the long toes of grallatores are formed for walking over swamps and floating plants, yet the water-hen is nearly as aquatic as the coot; and the landrail nearly as terrestrial as the quail or partridge. In such cases, and many others could be given, habits have changed without a corresponding change of structure. The webbed feet of the upland goose may be said to have become rudimentary in function, though not in structure. In the frigate-bird, the deeply-scooped membrane between the toes shows that structure has begun to change.

He who believes in separate and innumerable acts of creation will say, that in these cases it has pleased the Creator to cause a being of one type to take the place of one of another type; but this seems to me only restating the fact in dignified language. He who believes in the struggle for existence and in the principle of natural selection, will acknowledge that every organic being is constantly endeavouring to increase in numbers; and that if any one being vary ever so little, either in habits or structure, and thus gain an advantage over some other inhabitant of the country, it will seize on the place of that inhabitant, however different it may be from its own place. Hence it will cause him no surprise that there should be geese and frigate-birds with webbed feet, either living on the dry land or most rarely alighting on the water; that there should be long-toed corncrakes living in meadows instead of in swamps; that there should be wood-peckers where not a tree grows; that there should be diving thrushes, and petrels with the habits of auks.

Organs of extreme perfection and complication. To suppose that the eye, with all its inimitable contrivances for adjusting the focus to different distances, for admitting different amounts of light, and for the correction of spherical and chromatic aberration, could have been formed by natural selection, seems, I freely confess, absurd in the highest possible degree. Yet reason tells me, that if numerous gradations from a perfect and complex eye to one very imperfect and simple, each grade being useful to its possessor, can be shown to exist; if further, the eye does vary ever so slightly, and the variations be inherited, which is certainly the case; and if any variation or modification in the organ be ever useful to an animal under changing conditions of life, then the difficulty of believing that a perfect and complex eye could be formed by natural selection, though insuper-able by our imagination, can hardly be considered real. How a nerve comes to be sensitive to light, hardly concerns us more than how life itself first originated; but I may remark that several facts make me suspect that any sensitive nerve may be rendered sensitive to light, and likewise to those coarser vibrations of the air which produce sound.

In looking for the gradations by which an organ in any species has been perfected, we ought to look exclusively to its lineal ancestors; but this is scarcely ever possible, and we are forced in each case to look to species of the same group, that is to the collateral descendants from the same original parent-form, in order to see what gradations are possible, and for the chance of some gradations having been trans-mitted from the earlier stages of descent, in an unaltered or little altered condition. Amongst existing Vertebrata, we find but a small amount of gradation in the structure of the eye, and from fossil species we can learn nothing on this head. In this great class we should probably have to descend far beneath the lowest known fossiliferous stratum to discover the earlier stages, by which the eye has been perfected.

In the Articulata we can commence a series with an optic nerve merely coated with pigment, and without any other mechanism; and from this low stage, numerous gradations of structure, branching off in two fundamentally different lines, can be shown to exist, until we reach a moderately high stage of perfection. In certain crustaceans, for instance, there is a double cornea, the inner one divided into facets, within each of which there is a lens-shaped swelling. In other crustaceans the transparent cones which are coated by pigment, and which properly act only by excluding lateral pencils of light, are convex at their upper ends and must act by convergence; and at their lower ends there seems to be an imperfect vitreous substance. With these facts, here far too briefly and imperfectly given, which show that there is much graduated diversity in the eyes of living crustaceans, and bearing in mind how small the number of living animals is in proportion to those which have become extinct, I can see no very great difficulty (not more than in the case of many other structures) in believing that natural selection has converted the simple apparatus of an optic nerve merely coated with pigment and invested by transparent membrane, into an optical instrument as perfect as is possessed by any member of the great Articulate class.

He who will go thus far, if he find on finishing this treatise that large bodies of facts, otherwise inexplicable, can be explained by the theory of descent, ought not to hesitate to go further, and to admit that a structure even as perfect as the eye of an eagle might be formed by natural selection, although in this case he does not know any of the transitional grades. His reason ought to conquer his imagination; though I have felt the difficulty far too keenly to be surprised at any degree of hesitation in extending the principle of natural selection to such startling lengths.

It is scarcely possible to avoid comparing the eye to a telescope. We know that this instrument has been perfected by the long-continued efforts of the highest human intellects; and we naturally infer that the eye has been formed by a somewhat analogous process. But may not this inference be presumptuous? Have we any right to assume that the Creator works by intellectual powers like those of man? If we must compare the eye to an optical instrument, we ought in imagination to take a thick layer of transparent tissue, with a nerve sensitive to light beneath, and then suppose every part of this layer to be continually changing slowly in density, so as to separate into layers of different densities and thicknesses, placed at different distances from each other, and with the surfaces of each layer slowly changing in form. Further we must suppose that there is a power always intently watching each slight accidental alteration in the transparent layers; and carefully selecting each alteration which, under varied circumstances, may in any way, or in any degree, tend to produce a distincter image. We must suppose each new state of the instrument to be multiplied by the million; and each to be preserved till a better be produced, and then the old ones to be destroyed. In living bodies, variation will cause the slight alterations, generation will multiply them almost infinitely, and natural selection will pick out with unerring skill each improvement. Let this process go on for millions on millions of years; and during each year on millions of individuals of many kinds; and may we not believe that a living optical instrument might thus be formed as superior to one of glass, as the works of the Creator are to those of man?

If it could be demonstrated that any complex organ existed, which could not possibly have been formed by numerous, successive, slight modifications, my theory would absolutely break down. But I can find out no such case. No doubt many organs exist of which we do not know the transitional grades, more especially if we look to much-isolated species, round which, according to my theory, there has been much extinction. Or again, if we look to an organ common to all the members of a large class, for in this latter case the organ must have been first formed at an extremely remote period, since which all the many members of the class have been developed; and in order to discover the early transitional grades through which the organ has passed, we should have to look to very ancient ancestral forms, long since become extinct.

We should be extremely cautious in concluding that an organ could not have been formed by transitional gradations of some kind. Numerous cases could be given amongst the lower animals of the same organ performing at the same time wholly distinct functions; thus the alimentary canal respires, digests, and excretes in the larva of the dragon-fly and in the fish Cobites. In the Hydra, the animal may be turned inside out, and the exterior surface will then digest and the stomach respire. In such cases natural selection might easily specialize, if any advantage were thus gained, a part or organ, which had performed two functions, for one function alone, and thus wholly change its nature by insensible steps. Two distinct organs sometimes perform simultaneously the same function in the same individual; to give one instance, there are fish with gills or branchiae that breathe the air dissolved in the water, at the same time that they breathe free air in their swimbladders, this latter organ having a ductus pneumaticus for its supply, and being divided by highly vascular partitions. In these cases, one of the two organs might with ease be modified and perfected so as to perform all the work by itself, being aided during the process of modification by the other organ; and then this other organ might be modified for some other and quite distinct purpose, or be quite obliterated.

The illustration of the swimbladder in fishes is a good one, because it shows us clearly the highly important fact that an organ originally constructed for one purpose, namely flotation, may be converted into one for a wholly different purpose, namely respiration. The swimbladder has, also, been worked in as an accessory to the auditory organs of certain fish, or, for I do not know which view is now generally held, a part of the auditory apparatus has been worked in as a complement to the swimbladder. All physiologists admit that the swimbladder is homologous, or 'ideally similar,' in position and structure with the lungs of the higher vertebrate animals: hence there seems to me to be no great difficulty in believing that natural selection has actually converted a swimbladder into a lung, or organ used exclusively for respiration.

I can, indeed, hardly doubt that all vertebrate animals having true lungs have descended by ordinary generation from an ancient prototype, of which we know nothing, furnished with a floating apparatus or swimbladder. We can thus, as I infer from Professor Owen's interesting description of these parts, understand the strange fact that every particle of food and drink which we swallow has to pass over the orifice of the trachea, with some risk of falling into the lungs, notwithstanding the beautiful contrivance by which the glottis is closed. In the higher Vertebrata the branchiae have wholly disappeared – the slits on the sides of the neck and the loop-like course of the arteries still marking in the embryo their former position. But it is conceivable that the now utterly lost branchiae might have been gradually worked in by natural selection for some quite distinct purpose: in the same manner as, on the view entertained by some naturalists that the branchiae and dorsal scales of Annelids are homologous with the wings and wing-covers of insects, it is probable that organs which at a very ancient period served for respiration have been actually converted into organs of flight.

In considering transitions of organs, it is so important to bear in mind the probability of conversion from one function to another, that I will give one more instance. Pedunculated cirripedes have two minute folds of skin, called by me the ovigerous frena, which serve, through the means of a sticky secretion, to retain the eggs until they are hatched within the sack. These cirripedes have no branchiae, the whole surface of the body and sack, including the small frena, serving for respiration. The Balanidae or sessile cirripedes, on the other hand, have no ovigerous frena, the eggs lying loose at the bottom of the sack, in the well-enclosed shell; but they have large folded branchiae. Now I think no one will dispute that the ovigerous frena in the one family are strictly homologous with the branchiae of the other family; indeed, they graduate into each other. Therefore I do not doubt that little folds of skin, which originally served as ovigerous frena, but which, likewise, very slightly aided the act of respiration, have been gradually converted by natural selection into branchiae, simply through an increase in their size and the obliteration of their adhesive glands. If all pedunculated cirripedes had become extinct, and they have already suffered far more extinction than have sessile cirripedes, who would ever have imagined that the branchiae in this latter family had originally existed as organs for preventing the ova from being washed out of the sack?

Although we must be extremely cautious in concluding that any organ could not possibly have been produced by successive transitional gradations, yet, undoubtedly, grave cases of difficulty occur, some of which will be discussed in my future work.

One of the gravest is that of neuter insects, which are often very differently constructed from either the males or fertile females; but this case will be treated of in the next chapter. The electric organs of fishes offer another case of special difficulty; it is impossible to conceive by what steps these wondrous organs have been produced; but, as Owen and others have remarked, their intimate structure closely resembles that of common muscle; and as it has lately been shown that Rays have an organ closely analogous to the electric apparatus, and yet do not, as Matteuchi asserts, discharge any electricity, we must own that we are far too ignorant to argue that no transition of any kind is possible.

The electric organs offer another and even more serious difficulty; for they occur in only about a dozen fishes, of which several are widely remote in their affinities. Generally when the same organ appears in several members of the same class, especially if in members having very different habits of life, we may attribute its presence to inheritance from a common ancestor; and its absence in some of the members to its loss through disuse or natural selection. But if the electric organs had been inherited from one ancient progenitor thus provided, we might have expected that all electric fishes would have been specially related to each other. Nor does geology at all lead to the belief that formerly most fishes had electric organs, which most of their modified descendants have lost. The presence of luminous organs in a few insects, belonging to different families and orders, offers a parallel case of difficulty. Other cases could be given; for instance in plants, the very curious contrivance of a mass of pollen-grains, borne on a foot-stalk with a sticky gland at the end, is the same in Orchis and Asclepias, – genera almost as remote as possible amongst flowering plants. In all these cases of two very distinct species furnished with apparently the same anomalous organ, it should be observed that, although the general appearance and function of the organ may be the same, yet some fundamental difference can generally be detected. I am inclined to believe that in nearly the same way as two men have sometimes independently hit on the very same invention, so natural selection, working for the good of each being and taking advantage of analogous variations, has sometimes modified in very nearly the same manner two parts in two organic beings, which owe but little of their structure in common to inheritance from the same ancestor.

Although in many cases it is most difficult to conjecture by what transitions an organ could have arrived at its present state; yet, considering that the proportion of living and known forms to the extinct and unknown is very small, I have been astonished how rarely an organ can be named, towards which no transitional grade is known to lead. The truth of this remark is indeed shown by that old canon in natural history of 'Natura non facit saltum.' We meet with this admission in the writings of almost every experienced naturalist; or, as Milne Edwards has well expressed it, nature is prodigal in variety, but niggard in innovation. Why, on the theory of Creation, should this be so? Why should all the parts and organs of many independent beings, each supposed to have been separately created for its proper place in nature, be so invariably linked together by graduated steps? Why should not Nature have taken a leap from structure to structure? On the theory of natural selection, we can clearly understand why she should not; for natural selection can act only by taking advantage of slight successive variations; she can never take a leap, but must advance by the shortest and slowest steps.

Organs of little apparent importance. As natural selection acts by life and death, – by the preservation of individuals with any favourable variation, and by the destruction of those with any unfavourable deviation of structure, – I have sometimes felt much difficulty in understanding the origin of simple parts, of which the importance does not seem sufficient to cause the preservation of successively varying individuals. I have sometimes felt as much difficulty, though of a very different kind, on this head, as in the case of an organ as perfect and complex as the eye.

In the first place, we are much too ignorant in regard to the whole economy of any one organic being, to say what slight modifications would be of importance or not. In a former chapter I have given instances of most trifling characters, such as the down on fruit and the colour of the flesh, which, from determining the attacks of insects or from being correlated with constitutional differences, might assuredly be acted on by natural selection. The tail of the giraffe looks like an artificially constructed fly-flapper; and it seems at first incredible that this could have been adapted for its present purpose by successive slight modifications, each better and better, for so trifling an object as driving away flies; yet we should pause before being too positive even in this case, for we know that the distribution and existence of cattle and other animals in South America absolutely depends on their power of resisting the attacks of insects: so that individuals which could by any means defend themselves from these small enemies, would be able to range into new pastures and thus gain a great advantage. It is not that the larger quadrupeds are actually destroyed (except in some rare cases) by the flies, but they are incessantly harassed and their strength reduced, so that they are more subject to disease, or not so well enabled in a coming dearth to search for food, or to escape from beasts of prey.

Organs now of trifling importance have probably in some cases been of high importance to an early progenitor, and, after having been slowly perfected at a former period, have been transmitted in nearly the same state, although now become of very slight use; and any actually injurious deviations in their structure will always have been checked by natural selection. Seeing how important an organ of locomotion the tail is in most aquatic animals, its general presence and use for many purposes in so many land animals, which in their lungs or modified swimbladders betray their aquatic origin, may perhaps be thus accounted for. A well-developed tail having been formed in an aquatic animal, it might subsequently come to be worked in for all sorts of purposes, as a fly-flapper, an organ of prehension, or as an aid in turning, as with the dog, though the aid must be slight, for the hare, with hardly any tail, can double quickly enough.

In the second place, we may sometimes attribute importance to characters which are really of very little importance, and which have originated from quite secondary causes, independently of natural selection. We should remember that climate, food, &c., probably have some little direct influence on the organization; that characters reappear from the law of reversion; that correlation of growth will have had a most important influence in modifying various structures; and finally, that sexual selection will often have largely modified the external characters of animals having a will, to give one male an advantage in fighting with another or in charming the females. Moreover when a modification of structure has primarily arisen from the above or other unknown causes, it may at first have been of no advantage to the species, but may subsequently have been taken advantage of by the descendants of the species under new conditions of life and with newly acquired habits.

To give a few instances to illustrate these latter remarks. If green woodpeckers alone had existed, and we did not know that there were many black and pied kinds, I dare say that we should have thought that the green colour was a beautiful adaptation to hide this tree-frequenting bird from its enemies; and consequently that it was a character of importance and might have been acquired through natural selection; as it is, I have no doubt that the colour is due to some quite distinct cause, probably to sexual selection. A trailing bamboo in the Malay Archipelago climbs the loftiest trees by the aid of exquisitely constructed hooks clustered around the ends of the branches, and this contrivance, no doubt, is of the highest service to the plant; but as we see nearly similar hooks on many trees which are not climbers, the hooks on the bamboo may have arisen from unknown laws of growth, and have been subsequently taken advantage of by the plant undergoing further modification and becoming a climber. The naked skin on the head of a vulture is generally looked at as a direct adaptation for wallowing in putridity; and so it may be, or it may possibly be due to the direct action of putrid matter; but we should be very cautious in drawing any such inference, when we see that the skin on the head of the clean-feeding male turkey is likewise naked. The sutures in the skulls of young mammals have been advanced as a beautiful adaptation for aiding parturition, and no doubt they facilitate, or may be indispensable for this act; but as sutures occur in the skulls of young birds and reptiles, which have only to escape from a broken egg, we may infer that this structure has arisen from the laws of growth, and has been taken advantage of in the parturition of the higher animals.

We are profoundly ignorant of the causes producing slight and unimportant variations; and we are immediately made concious of this by reflecting on the differences in the breeds of our domesticated animals in different countries, – more especially in the less civilized countries where there has been but little artificial selection. Careful observers are convinced that a damp climate affects the growth of the hair, and that with the hair the horns are correlated. Mountain breeds always differ from lowland breeds; and a mountainous country would probably affect the hind limbs from exercising them more, and possibly even the form of the pelvis; and then by the law of homologous variation, the front limbs and even the head would probably be affected. The shape, also, of the pelvis might affect by pressure the shape of the head of the young in the womb. The laborious breathing necessary in high regions would, we have some reason to believe, increase the size of the chest; and again correlation would come into play. Animals kept by savages in different countries often have to struggle for their own subsistence, and would be exposed to a certain extent to natural selection, and individuals with slightly different constitutions would succeed best under different climates; and there is reason to believe that constitution and colour are correlated. A good observer, also, states that in cattle susceptibility to the attacks of flies is correlated with colour, as is the liability to be poisoned by certain plants; so that colour would be thus subjected to the action of natural selection. But we are far too ignorant to speculate on the relative importance of the several known and unknown laws of variation; and I have here alluded to them only to show that, if we are unable to account for the characteristic differences of our domestic breeds, which nevertheless we generally admit to have arisen through ordinary generation, we ought not to lay too much stress on our ignorance of the precise cause of the slight analogous differences between species. I might have adduced for this same purpose the differences between the races of man, which are so strongly marked; I may add that some little light can apparently be thrown on the origin of these differences, chiefly through sexual selection of a particular kind, but without here entering on copious details my reasoning would appear frivolous.

The foregoing remarks lead me to say a few words on the protest lately made by some naturalists, against the utilitarian doctrine that every detail of structure has been produced for the good of its possessor. They believe that very many structures have been created for beauty in the eyes of man, or for mere variety. This doctrine, if true, would be absolutely fatal to my theory. Yet I fully admit that many structures are of no direct use to their possessors. Physical conditions probably have had some little effect on structure, quite independently of any good thus gained. Correlation of growth has no doubt played a most important part, and a useful modification of one part will often have entailed on other parts diversified changes of no direct use. So again characters which formerly were useful, or which formerly had arisen from correlation of growth, or from other un-known cause, may reappear from the law of reversion, though now of no direct use. The effects of sexual selection, when displayed in beauty to charm the females, can be called useful only in rather a forced sense. But by far the most important consideration is that the chief part of the organization of every being is simply due to inherit-ance; and consequently, though each being assuredly is well fitted for its place in nature, many structures now have no direct relation to the habits of life of each species. Thus, we can hardly believe that the webbed feet of the upland goose or of the frigate-bird are of special use to these birds; we cannot believe that the same bones in the arm of the monkey, in the fore leg of the horse, in the wing of the bat, and in the flipper of the seal, are of special use to these animals. We may safely attribute these structures to inheritance. But to the progenitor of the upland goose and of the frigate-bird, webbed feet no doubt were as useful as they now are to the most aquatic of existing birds. So we may believe that the progenitor of the seal had not a flipper, but a foot with five toes fitted for walking or grasping; and we may further venture to believe that the several bones in the limbs of the monkey, horse, and bat, which have been inherited from a common progeni-tor, were formerly of more special use to that progenitor, or its progenitors, than they now are to these animals having such widely diversified habits. Therefore we may infer that these several bones might have been acquired through natural selection, subjected for-merly, as now, to the several laws of inheritance, reversion, correl-ation of growth, &c. Hence every detail of structure in every living creature (making some little allowance for the direct action of phys-ical conditions) may be viewed, either as having been of special use to some ancestral form, or as being now of special use to the descendants of this form – either directly, or indirectly through the complex laws of growth.

Natural selection cannot possibly produce any modification in any one species exclusively for the good of another species; though throughout nature one species incessantly takes advantage of, and profits by, the structure of another. But natural selection can and does often produce structures for the direct injury of other species, as we see in the fang of the adder, and in the ovipositor of the ichneumon, by which its eggs are deposited in the living bodies of other insects. If it could be proved that any part of the structure of any one species had been formed for the exclusive good of another species, it would annihilate my theory, for such could not have been produced through natural selection. Although many statements may be found in works on natural history to this effect, I cannot find even one which seems to me of any weight. It is admitted that the rattlesnake has a poison-fang for its own defence and for the destruction of its prey; but some authors suppose that at the same time this snake is furnished with a rattle for its own injury, namely, to warn its prey to escape. I would almost as soon believe that the cat curls the end of its tail when preparing to spring, in order to warn the doomed mouse. But I have not space here to enter on this and other such cases.

Natural selection will never produce in a being anything injurious to itself, for natural selection acts solely by and for the good of each. No organ will be formed, as Paley has remarked, for the purpose of causing pain or for doing an injury to its possessor. If a fair balance be struck between the good and evil caused by each part, each will be found on the whole advantageous. After the lapse of time, under changing conditions of life, if any part comes to be injurious, it will be modified; or if it be not so, the being will become extinct, as myriads have become extinct.

Natural selection tends only to make each organic being as perfect as, or slightly more perfect than, the other inhabitants of the same country with which it has to struggle for existence. And we see that this is the degree of perfection attained under nature. The endemic productions of New Zealand, for instance, are perfect one compared with another; but they are now rapidly yielding before the advancing legions of plants and animals introduced from Europe. Natural selection will not produce absolute perfection, nor do we always meet, as far as we can judge, with this high standard under nature. The correction for the aberration of light is said, on high authority, not to be perfect even in that most perfect organ, the eye. If our reason leads us to admire with enthusiasm a multitude of inimitable contrivances in nature, this same reason tells us, though we may easily err on both sides, that some other contrivances are less perfect. Can we consider the sting of the wasp or of the bee as perfect, which, when used against many attacking animals, cannot be withdrawn, owing to the backward serratures, and so inevitably causes the death of the insect by tearing out its viscera?

If we look at the sting of the bee, as having originally existed in a remote progenitor as a boring and serrated instrument, like that in so many members of the same great order, and which has been modified but not perfected for its present purpose, with the poison originally adapted to cause galls subsequently intensified, we can perhaps understand how it is that the use of the sting should so often cause the insect's own death: for if on the whole the power of stinging be useful to the community, it will fulfil all the requirements of natural selection, though it may cause the death of some few members. If we admire the truly wonderful power of scent by which the males of many insects find their females, can we admire the production for this single purpose of thousands of drones, which are utterly useless to the community for any other end, and which are ultimately slaughtered by their industrious and sterile sisters? It may be difficult, but we ought to admire the savage instinctive hatred of the queen-bee, which urges her instantly to destroy the young queens her daughters as soon as born, or to perish herself in the combat; for undoubtedly this is for the good of the community; and maternal love or maternal hatred, though the latter fortunately is most rare, is all the same to the inexorable principle of natural selection. If we admire the several ingenious contrivances, by which the flowers of the orchis and of many other plants are fertilized through insect agency, can we consider as equally perfect the elaboration by our fir-trees of dense clouds of pollen, in order that a few granules may be wafted by a chance breeze on to the ovules?

Summary of Chapter. We have in this chapter discussed some of the difficulties and objections which may be urged against my theory. Many of them are very grave; but I think that in the discussion light has been thrown on several facts, which on the theory of independent acts of creation are utterly obscure. We have seen that species at any one period are not indefinitely variable, and are not linked together by a multitude of intermediate gradations, partly because the process of natural selection will always be very slow, and will act, at any one time, only on a very few forms; and partly because the very process of natural selection almost implies the continual supplanting and extinction of preceding and intermediate gradations. Closely allied species, now living on a continuous area, must often have been formed when the area was not continuous, and when the conditions of life did not insensibly graduate away from one part to another. When two varieties are formed in two districts of a continuous area, an intermediate variety will often be formed, fitted for an intermediate zone; but from reasons assigned, the intermediate variety will usually exist in lesser numbers than the two forms which it connects; consequently the two latter, during the course of further modification, from existing in greater numbers, will have a great advantage over the less numerous intermediate variety, and will thus generally succeed in supplanting and exterminating it.

We have seen in this chapter how cautious we should be in concluding that the most different habits of life could not graduate into each other; that a bat, for instance, could not have been formed by natural selection from an animal which at first could only glide through the air.

We have seen that a species may under new conditions of life change its habits, or have diversified habits, with some habits very unlike those of its nearest congeners. Hence we can understand bearing in mind that each organic being is trying to live wherever it can live, how it has arisen that there are upland geese with webbed feet, ground woodpeckers, diving thrushes, and petrels with the habits of auks.

Although the belief that an organ so perfect as the eye could have been formed by natural selection, is more than enough to stagger any one; yet in the case of any organ, if we know of a long series of gradations in complexity, each good for its possessor, then, under changing conditions of life, there is no logical impossibility in the acquirement of any conceivable degree of perfection through natural selection. In the cases in which we know of no intermediate or transitional states, we should be very cautious in concluding that none could have existed, for the homologies of many organs and their intermediate states show that wonderful metamorphoses in function are at least possible. For instance, a swimbladder has apparently been converted into an air-breathing lung. The same organ having performed simultaneously very different functions, and then having been specialized for one function; and two very distinct organs having performed at the same time the same function, the one having been perfected whilst aided by the other, must often have largely facilitated transitions.

We are far too ignorant, in almost every case, to be enabled to assert that any part or organ is so unimportant for the welfare of a species, that modifications in its structure could not have been slowly accumulated by means of natural selection. But we may confidently believe that many modifications, wholly due to the laws of growth, and at first in no way advantageous to a species, have been subsequently taken advantage of by the still further modified descendants of this species. We may, also, believe that a part formerly of high importance has often been retained (as the tail of an aquatic animal by its terrestrial descendants), though it has become of such small importance that it could not, in its present state, have been acquired by natural selection, – a power which acts solely by the preservation of profitable variations in the struggle for life.

Natural selection will produce nothing in one species for the exclusive good or injury of another; though it may well produce parts, organs, and excretions highly useful or even indispensable, or highly injurious to another species, but in all cases at the same time useful to the owner. Natural selection in each well-stocked country, must act chiefly through the competition of the inhabitants one with another, and consequently will produce perfection, or strength in the battle for life, only according to the standard of that country. Hence the inhabitants of one country, generally the smaller one, will often yield, as we see they do yield, to the inhabitants of another and generally larger country. For in the larger country there will have existed more individuals, and more diversified forms, and the competition will have been severer, and thus the standard of perfection will have been rendered higher. Natural selection will not necessarily produce absolute perfection; nor, as far as we can judge by our limited faculties, can absolute perfection be everywhere found.

On the theory of natural selection we can clearly understand the full meaning of that old canon in natural history, 'Natura non facit saltum.' This canon, if we look only to the present inhabitants of the world, is not strictly correct, but if we include all those of past times, it must by my theory be strictly true.

It is generally acknowledged that all organic beings have been formed on two great laws – Unity of Type, and the Conditions of Existence. By unity of type is meant that fundamental agreement in structure, which we see in organic beings of the same class, and which is quite independent of their habits of life. On my theory, unity of type is explained by unity of descent. The expression of conditions of existence, so often insisted on by the illustrious Cuvier, is fully embraced by the principle of natural selection. For natural selection acts by either now adapting the varying parts of each being to its organic and inorganic conditions of life; or by having adapted them during long-past periods of time: the adaptations being aided in some cases by use and disuse, being slightly affected by the direct action of the external conditions of life, and being in all cases subjected to the several laws of growth. Hence, in fact, the law of the Conditions of Existence is the higher law; as it includes, through the inheritance of former adaptations, that of Unity of Type.

CHAPTER VII

INSTINCT

Instincts comparable with habits, but different in their origin – Instincts graduated – Aphides and ants – Instincts variable – Domestic instincts, their origin – Natural instincts of the cuckoo, ostrich, and parasitic bees – Slave-making ants – Hive-bee, its cell-making instinct – Difficulties on the theory of the Natural Selection of instincts – Neuter or sterile insects – Summary

THE subject of instinct might have been worked into the previous chapters; but I have thought that it would be more convenient to treat the subject separately, especially as so wonderful an instinct as that of the hive-bee making its cells will probably have occurred to many readers, as a difficulty sufficient to overthrow my whole theory. I must premise, that I have nothing to do with the origin of the primary mental powers, any more than I have with that of life itself. We are concerned only with the diversities of instinct and of the other mental qualities of animals within the same class.

I will not attempt any definition of instinct. It would be easy to show that several distinct mental actions are commonly embraced by this term; but every one understands what is meant, when it is said that instinct impels the cuckoo to migrate and to lay her eggs in other birds' nests. An action, which we ourselves should require experience to enable us to perform, when performed by an animal, more especially by a very young one, without any experience, and when performed by many individuals in the same way, without their knowing for what purpose it is performed, is usually said to be instinctive. But I could show that none of these characters of instinct are universal. A little dose, as Pierre Huber expresses it, of judgement or reason, often comes into play, even in animals very low in the scale of nature.

Frederick Cuvier and several of the older metaphysicians have compared instinct with habit. This comparison gives, I think, a remarkably accurate notion of the frame of mind under which an instinctive action is performed, but not of its origin. How unconsciously many habitual actions are performed, indeed not rarely in direct opposition to our conscious will! yet they may be modified by the will or reason. Habits easily become associated with other habits, and with certain periods of time and states of the body. When once acquired, they often remain constant throughout life. Several other points of resemblance between instincts and habits could be pointed out. As in repeating a well-known song, so in instincts, one action follows another by a sort of rhythm; if a person be interrupted in a song, or in repeating anything by rote, he is generally forced to go back to recover the habitual train of thought: so P. Huber found it was with a caterpillar, which makes a very complicated hammock; for if he took a caterpillar which had completed its hammock up to, say, the sixth stage of construction, and put it into a hammock completed up only to the third stage, the caterpillar simply re-performed the fourth, fifth, and sixth stages of construction. If, however, a caterpillar were taken out of a hammock made up, for instance, to the third stage, and were put into one finished up to the sixth stage, so that much of its work was already done for it, far from feeling the benefit of this, it was much embarrassed, and, in order to complete its hammock, seemed forced to start from the third stage, where it had left off, and thus tried to complete the already finished work.

If we suppose any habitual action to become inherited – and I think it can be shown that this does sometimes happen – then the resemblance between what originally was a habit and an instinct becomes so close as not to be distinguished. If Mozart, instead of playing the pianoforte at three years old with wonderfully little practice, had played a tune with no practice at all, he might truly be said to have done so instinctively. But it would be the most serious error to suppose that the greater number of instincts have been acquired by habit in one generation, and then transmitted by inheritance to succeeding generations. It can be clearly shown that the most wonderful instincts with which we are acquainted, namely, those of the hive-bee and of many ants, could not possibly have been thus acquired.

It will be universally admitted that instincts are as important as corporeal structure for the welfare of each species, under its present conditions of life. Under changed conditions of life, it is at least possible that slight modifications of instinct might be profitable to a species; and if it can be shown that instincts do vary ever so little, then I can see no difficulty in natural selection preserving and continually accumulating variations of instinct to any extent that may be profitable. It is thus, as I believe, that all the most complex and wonderful instincts have originated. As modifications of corporeal structure arise from, and are increased by, use or habit, and are diminished or lost by disuse, so I do not doubt it has been with instincts. But I believe that the effects of habit are of quite subordinate importance to the effects of the natural selection of what may be called accidental variations of instincts; – that is of variations produced by the same unknown causes which produce slight deviations of bodily structure.

No complex instinct can possibly be produced through natural selection, except by the slow and gradual accumulation of numerous, slight, yet profitable, variations. Hence, as in the case of corporeal structures, we ought to find in nature, not the actual transitional gradations by which each complex instinct has been acquired – for these could be found only in the lineal ancestors of each species – but we ought to find in the collateral lines of descent some evidence of such gradations; or we ought at least to be able to show that gradations of some kind are possible; and this we certainly can do. I have been surprised to find, making allowance for the instincts of animals having been but little observed except in Europe and North America, and for no instinct being known amongst extinct species, how very generally gradations, leading to the most complex instincts, can be discovered. The canon of 'Natura non facit saltum' applies with almost equal force to instincts as to bodily organs. Changes of instinct may sometimes be facilitated by the same species having different instincts at different periods of life, or at different seasons of the year, or when placed under different circumstances, &c.; in which case either one or the other instinct might be preserved by natural selection. And such instances of diversity of instinct in the same species can be shown to occur in nature.

Again as in the case of corporeal structure, and conformably with my theory, the instinct of each species is good for itself, but has never, as far as we can judge, been produced for the exclusive good of others. One of the strongest instances of an animal apparently performing an action for the sole good of another, with which I am acquainted, is that of aphides voluntarily yielding their sweet excretion to ants: that they do so voluntarily, the following facts show. I removed all the ants from a group of about a dozen aphides on a dock-plant, and prevented their attendance during several hours. After this interval, I felt sure that the aphides would want to excrete. I watched them for some time through a lens, but not one excreted; I then tickled and stroked them with a hair in the same manner, as well as I could, as the ants do with their antennae; but not one excreted. Afterwards I allowed an ant to visit them, and it immediately seemed, by its eager way of running about, to be well aware what a rich flock it had discovered; it then began to play with its antennae on the abdomen first of one aphis and then of another; and each aphis, as soon as it felt the antennae, immediately lifted up its abdomen and excreted a limpid drop of sweet juice, which was eagerly devoured by the ant. Even the quite young aphides behaved in this manner, showing that the action was instinctive, and not the result of experience. But as the execretion is extremely viscid, it is probably a convenience to the aphides to have it removed; and therefore probably the aphides do not instinctively excrete for the sole good of the ants. Although I do not believe that any animal in the world performs an action for the exclusive good of another of a distinct species, yet each species tries to take advantage of the instincts of others, as each takes advantage of the weaker bodily structure of others. So again, in some few cases, certain instincts cannot be considered as absolutely perfect; but as details on this and other such points are not indispensable, they may be here passed over.

As some degree of variation in instincts under a state of nature, and the inheritance of such variations, are indispensable for the action of natural selection, as many instances as possible ought to have been here given; but want of space prevents me. I can only assert, that instincts certainly do vary – for instance, the migratory instinct, both in extent and direction, and in its total loss. So it is with the nests of birds, which vary partly in dependence on the situations chosen, and on the nature and temperature of the country inhabited, but often from causes wholly unknown to us: Audubon has given several remarkable cases of differences in nests of the same species in the northern and southern United States. Fear of any particular enemy is certainly an instinctive quality, as may be seen in nestling birds, though it is strengthened by experience, and by the sight of fear of the same enemy in other animals. But fear of man is slowly acquired, as I have elsewhere shown, by various animals inhabiting desert islands; and we may see an instance of this, even in England, in the greater wildness of all our large birds than of our small birds; for the large birds have been most persecuted by man. We may safely attribute the greater wildness of our large birds to this cause; for in uninhabited islands large birds are not more fearful than small; and the magpie, so wary in England, is tame in Norway, as is the hooded crow in Egypt.

That the general disposition of individuals of the same species, born in a state of nature, is extremely diversified, can be shown by a multitude of facts. Several cases also, could be given, of occasional and strange habits in certain species, which might, if advantageous to the species, give rise, through natural selection, to quite new instincts. But I am well aware that these general statements, without facts given in detail, can produce but a feeble effect on the reader's mind. I can only repeat my assurance, that I do not speak without good evidence.

The possibility, or even probability, of inherited variations of instinct in a state of nature will be strengthened by briefly considering a few cases under domestication. We shall thus also be enabled to see the respective parts which habit and the selection of so-called acci-dental variations have played in modifying the mental qualities of our domestic animals. A number of curious and authentic instances could be given of the inheritance of all shades of disposition and tastes, and likewise of the oddest tricks, associated with certain frames of mind or periods of time. But let us look to the familiar case of the several breeds of dogs: it cannot be doubted that young pointers (I have myself seen a striking instance) will sometimes point and even back other dogs the very first time that they are taken out; retrieving is certainly in some degree inherited by retrievers; and a tendency to run round, instead of at, a flock of sheep, by shepherd-dogs. I cannot see that these actions, performed without experience by the young, and in nearly the same manner by each individual, performed with eager delight by each breed, and without the end being known, – for the young pointer can no more know that he points to aid his master, than the white butterfly knows why she lays her eggs on the leaf of the cabbage, – I cannot see that these actions differ essentially from true instincts. If we were to see one kind of wolf, when young and without any training, as soon as it scented its prey, stand motionless like a statue, and then slowly crawl forward with a peculiar gait; and another kind of wolf rushing round, instead of at, a herd of deer, and driving them to a distant point, we should assuredly call these actions instinctive. Domestic instincts, as they may be called, are certainly far less fixed or invariable than natural instincts; but they have been acted on by far less rigorous selection, and have been transmitted for an incomparably shorter period, under less fixed conditions of life.

How strongly these domestic instincts, habits, and dispositions are inherited, and how curiously they become mingled, is well shown when different breeds of dogs are crossed. Thus it is known that a cross with a bull-dog has affected for many generations the courage and obstinacy of greyhounds; and a cross with a greyhound has given to a whole family of shepherd-dogs a tendency to hunt hares. These domestic instincts, when thus tested by crossing, resemble natural instincts, which in a like manner become curiously blended together, and for a long period exhibit traces of the instincts of either parent: for example, Le Roy describes a dog, whose great-grandfather was a wolf, and this dog showed a trace of its wild parentage only in one way, by not coming in a straight line to his master when called.

Domestic instincts are sometimes spoken of as actions which have become inherited solely from long-continued and compulsory habit, but this, I think, is not true. No one would ever have thought of teaching, or probably could have taught, the tumbler-pigeon to tumble, – an action which, as I have witnessed, is performed by young birds, that have never seen a pigeon tumble. We may believe that some one pigeon showed a slight tendency to this strange habit, and that the long-continued selection of the best individuals in successive generations made tumblers what they now are; and near Glasgow there are house-tumblers, as I hear from Mr. Brent, which cannot fly eighteen inches high without going head over heels. It may be doubted whether any one would have thought of training a dog to point, had not some one dog naturally shown a tendency in this line; and this is known occasionally to happen, as I once saw in a pure terrier. When the first tendency was once displayed, methodical selection and the inherited effects of compulsory training in each successive generation would soon complete the work; and unconscious selection is still at work, as each man tries to procure, without intending to improve the breed, dogs which will stand and hunt best. On the other hand, habit alone in some cases has sufficed; no animal is more difficult to tame than the young of the wild rabbit; scarcely any animal is tamer than the young of the tame rabbit; but I do not suppose that domestic rabbits have ever been selected for tameness; and I presume that we must attribute the whole of the inherited change from extreme wildness to extreme tameness, simply to habit and long-continued close confinement.

Natural instincts are lost under domestication: a remarkable instance of this is seen in those breeds of fowls which very rarely or never become 'broody,' that is, never wish to sit on their eggs. Familiarity alone prevents our seeing how universally and largely the minds of our domestic animals have been modified by domestication. It is scarcely possible to doubt that the love of man has become instinctive in the dog. All wolves, foxes, jackals, and species of the cat genus, when kept tame, are most eager to attack poultry, sheep, and pigs; and this tendency has been found incurable in dogs which have been brought home as puppies from countries, such as Tierra del Fuego and Australia, where the savages do not keep these domestic animals. How rarely, on the other hand, do our civilized dogs, even when quite young, require to be taught not to attack poultry, sheep, and pigs! No doubt they occasionally do make an attack, and are then beaten; and if not cured, they are destroyed; so that habit, with some degree of selection, has probably concurred in civilizing by inheritance our dogs. On the other hand, young chickens have lost, wholly by habit, that fear of the dog and cat which no doubt was originally instinctive in them, in the same way as it is so plainly instinctive in young pheasants, though reared under a hen. It is not that chickens have lost all fear, but fear only of dogs and cats, for if the hen gives the danger-chuckle, they will run (more especially young turkeys) from under her, and conceal themselves in the surrounding grass or thickets; and this is evidently done for the instinctive purpose of allowing, as we see in wild ground-birds, their mother to fly away. But this instinct retained by our chickens has become useless under domestication, for the mother-hen has almost lost by disuse the power of flight.

Hence, we may conclude, that domestic instincts have been acquired and natural instincts have been lost partly by habit, and partly by man selecting and accumulating during successive gener-ations, peculiar mental habits and actions, which at first appeared from what we must in our ignorance call an accident. In some cases compulsory habit alone has sufficed to produce such inherited mental changes; in other cases compulsory habit has done nothing, and all has been the result of selection, pursued both methodically and uncon-sciously; but in most cases, probably, habit and selection have acted together.

We shall, perhaps, best understand how instincts in a state of nature have become modified by selection, by considering a few cases. I will select only three, out of the several which I shall have to discuss in my future work, – namely, the instinct which leads the cuckoo to lay her eggs in other birds' nests; the slave-making instinct of certain ants; and the comb-making power of the hive-bee: these two latter instincts have generally, and most justly, been ranked by naturalists as the most wonderful of all known instincts.

It is now commonly admitted that the more immediate and final cause of the cuckoo's instinct is, that she lays her eggs, not daily, but at intervals of two or three days; so that, if she were to make her own nest and sit on her own eggs, those first laid would have to be left for some time unincubated, or there would be eggs and young birds of different ages in the same nest. If this were the case, the process of laying and hatching might be inconveniently long, more especially as she has to migrate at a very early period; and the first hatched young would probably have to be fed by the male alone. But the American cuckoo is in this predicament; for she makes her own nest and has eggs and young successively hatched, all at the same time. It has been asserted that the American cuckoo occasionally lays her eggs in other birds' nests; but I hear on the high authority of Dr. Brewer, that this is a mistake. Nevertheless, I could give several instances of various birds which have been known occasionally to lay their eggs in other birds' nests. Now let us suppose that the ancient progenitor of our European cuckoo had the habits of the American cuckoo; but that occasionally she laid an egg in another bird's nest. If the old bird profited by this occasional habit, or if the young were made more vigorous by advantage having been taken of the mistaken maternal instinct of another bird, than by their own mother's care, encumbered as she can hardly fail to be by having eggs and young of different ages at the same time; then the old birds or the fostered young would gain an advantage. And analogy would lead me to believe, that the young thus reared would be apt to follow by inheritance the occasional and aberrant habit of their mother, and in their turn would be apt to lay their eggs in other birds' nests, and thus be successful in rearing their young. By a continued process of this nature, I believe that the strange instinct of our cuckoo could be, and has been, generated. I may add that, according to Dr. Gray and to some other observers, the European cuckoo has not utterly lost all maternal love and care for her own offspring.

The occasional habit of birds laying their eggs in other birds' nests, either of the same or of a distinct species, is not very uncommon with the Gallinaceae; and this perhaps explains the origin of a singular instinct in the allied group of ostriches. For several hen ostriches, at least in the case of the American species, unite and lay first a few eggs in one nest and then in another; and these are hatched by the males. This instinct may probably be accounted for by the fact of the hens laying a large number of eggs; but, as in the case of the cuckoo, at intervals of two or three days. This instinct, however, of the American ostrich has not as yet been perfected; for a surprising number of eggs lie strewed over the plains, so that in one day's hunting I picked up no less than twenty lost and wasted eggs.

Many bees are parasitic, and always lay their eggs in the nests of bees of other kinds. This case is more remarkable than that of the cuckoo; for these bees have not only their instincts but their structure modified in accordance with their parasitic habits; for they do not possess the pollen-collecting apparatus which would be necessary if they had to store food for their own young. Some species, likewise, of Sphegidae (wasp-like insects) are parasitic on other species; and M. Fabre has lately shown good reason for believing that although the Tachytes nigra generally makes its own burrow and stores it with paralysed prey for its own larvae to feed on, yet that when this insect finds a burrow already made and stored by another sphex, it takes advantage of the prize, and becomes for the occasion parasitic. In this case, as with the supposed case of the cuckoo, I can see no difficulty in natural selection making an occasional habit permanent, if of advantage to the species, and if the insect whose nest and stored food are thus feloniously appropriated, be not thus exterminated.

Slave-making instinct. This remarkable instinct was first discovered in the Formica (Polyerges) rufescens by Pierre Huber, a better observer even than his celebrated father. This ant is absolutely dependent on its slaves; without their aid, the species would certainly become extinct in a single year. The males and fertile females do no work. The workers or sterile females, though most energetic and courageous in capturing slaves, do no other work. They are incapable of making their own nests, or of feeding their own larvae. When the old nest is found inconvenient, and they have to migrate, it is the slaves which determine the migration, and actually carry their masters in their jaws. So utterly helpless are the masters, that when Huber shut up thirty of them without a slave, but with plenty of the food which they like best, and with their larvae and pupae to stimulate them to work, they did nothing; they could not even feed themselves, and many perished of hunger. Huber then introduced a single slave (F. fusca), and she instantly set to work, fed and saved the survivors; made some cells and tended the larvae, and put all to rights. What can be more extraordinary than these well-ascertained facts? If we had not known of any other slave-making ant, it would have been hopeless to have speculated how so wonderful an instinct could have been perfected.

Formica sanguinea was likewise first discovered by P. Huber to be a slave-making ant. This species is found in the southern parts of England, and its habits have been attended to by Mr. F. Smith, of the British Museum, to whom I am much indebted for information on this and other subjects. Although fully trusting to the statements of Huber and Mr. Smith, I tried to approach the subject in a sceptical frame of mind, as any one may well be excused for doubting the truth of so extraordinary and odious an instinct as that of making slaves. Hence I will give the observations which I have myself made, in some little detail. I opened fourteen nests of F. sanguinea, and found a few slaves in all. Males and fertile females of the slave-species are found only in their own proper communities, and have never been observed in the nests of F. sanguinea. The slaves are black and not above half the size of their red masters, so that the contrast in their appearance is very great. When the nest is slightly disturbed, the slaves occasionally come out, and like their masters are much agitated and defend the nest: when the nest is much disturbed and the larvae and pupae are exposed, the slaves work energetically with their masters in carrying them away to a place of safety. Hence, it is clear, that the slaves feel quite at home. During the months of June and July, on three successive years, I have watched for many hours several nests in Surrey and Sussex, and never saw a slave either leave or enter a nest. As, during these months, the slaves are very few in number, I thought that they might behave differently when more numerous; but Mr. Smith informs me that he has watched the nests at various hours during May, June and August, both in Surrey and Hampshire, and has never seen the slaves, though present in large numbers in August, either leave or enter the nest. Hence he considers them as strictly household slaves. The masters, on the other hand, may be constantly seen bringing in materials for the nest, and food of all kinds. During the present year, however, in the month of July, I came across a community with an unusually large stock of slaves, and I observed a few slaves mingled with their masters leaving the nest, and marching along the same road to a tall Scotch-fir-tree, twenty-five yards distant, which they ascended together, probably in search of aphides or cocci. According to Huber, who had ample opportunities for observation, in Switzerland the slaves habitually work with their masters in making the nest, and they alone open and close the doors in the morning and evening; and, as Huber expressly states, their principal office is to search for aphides. This difference in the usual habits of the masters and slaves in the two countries, probably depends merely on the slaves being captured in greater numbers in Switzerland than in England.

One day I fortunately chanced to witness a migration from one nest to another, and it was a most interesting spectacle to behold the masters carefully carrying, as Huber has described, their slaves in their jaws. Another day my attention was struck by about a score of the slave-makers haunting the same spot, and evidently not in search of food; they approached and were vigorously repulsed by an independent community of the slave species (F. fusca); sometimes as many as three of these ants clinging to the legs of the slave-making F. sanguinea. The latter ruthlessly killed their small opponents, and carried their dead bodies as food to their nest, twenty-nine yards distant; but they were prevented from getting any pupae to rear as slaves. I then dug up a small parcel of the pupae of F. fusca from another nest, and put them down on a bare spot near the place of combat; they were eagerly seized, and carried off by the tyrants, who perhaps fancied that, after all, they had been victorious in their late combat.

At the same time I laid on the same place a small parcel of the pupae of another species, F. flava, with a few of these little yellow ants still clinging to the fragments of the nest. This species is sometimes, though rarely, made into slaves, as has been described by Mr. Smith. Although so small a species, it is very courageous, and I have seen it ferociously attack other ants. In one instance I found to my surprise an independent community of F. flava under a stone beneath a nest of the slave-making F. sanguinea; and when I had accidentally disturbed both nests, the little ants attacked their big neighbours with surprising courage. Now I was curious to ascertain whether F. sanguinea could distinguish the pupae of F. fusca, which they habitually make into slaves, from those of the little and furious F. flava, which they rarely capture, and it was evident that they did at once distinguish them: for we have seen that they eagerly and instantly seized the pupae of F. fusca, whereas they were much terrified when they came across the pupae, or even the earth from the nest of F. flava, and quickly ran away; but in about a quarter of an hour, shortly after all the little yellow ants had crawled away, they took heart and carried off the pupae.

One evening I visited another community of F. sanguinea, and found a number of these ants entering their nest, carrying the dead bodies of F. fusca (showing that it was not a migration) and numerous pupae. I traced the returning file burthened with booty, for about forty yards, to a very thick clump of heath. Whence I saw the last individual of F. sanguinea emerge, carrying a pupa; but I was not able to find the desolated nest in the thick heath. The nest, however, must have been close at hand, for two or three individuals of F. fusca were rushing about in the greatest agitation, and one was perched motionless with its own pupa in its mouth on the top of a spray of heath over its ravaged home.

Such are the facts, though they did not need confirmation by me, in regard to the wonderful instinct of making slaves. Let it be observed what a contrast the instinctive habits of F. sanguinea present with those of the F. rufescens. The latter does not build its own nest, does not determine its own migrations, does not collect food for itself or its young, and cannot even feed itself: it is absolutely dependent on its numerous slaves. Formica sanguinea, on the other hand, possesses much fewer slaves, and in the early part of the summer extremely few. The masters determine when and where a new nest shall be formed, and when they migrate, the masters carry the slaves. Both in Switzerland and England the slaves seem to have the exclusive care of the larvae, and the masters alone go on slave-making expeditions. In Switzerland the slaves and masters work together, making and bringing materials for the nest: both, but chiefly the slaves, tend, and milk as it may be called, their aphides; and thus both collect food for the community. In England the masters alone usually leave the nest to collect building materials and food for themselves, their slaves and larvae. So that the masters in this country receive much less service from their slaves than they do in Switzerland.

By what steps the instinct of F. sanguinea originated I will not pretend to conjecture. But as ants, which are not slave-makers, will, as I have seen, carry off pupae of other species, if scattered near their nests, it is possible that pupae originally stored as food might become developed; and the ants thus unintentionally reared would then follow their proper instincts, and do what work they could. If their presence proved useful to the species which had seized them – if it were more advantageous to this species to capture workers than to procreate them – the habit of collecting pupae originally for food might by natural selection be strengthened and rendered permanent for the very different purpose of raising slaves. When the instinct was once acquired, if carried out to a much less extent even than in our British F. sanguinea, which, as we have seen, is less aided by its slaves than the same species in Switzerland, I can see no difficulty in natural selection increasing and modifying the instinct – always supposing each modification to be of use to the species – until an ant was formed as abjectly dependent on its slaves as is the Formica rufescens.

Cell-making instinct of the Hive-Bee. I will not here enter on minute details on this subject, but will merely give an outline of the conclusions at which I have arrived. He must be a dull man who can examine the exquisite structure of a comb, so beautifully adapted to its end, without enthusiastic admiration. We hear from mathematicians that bees have practically solved a recondite problem, and have made their cells of the proper shape to hold the greatest possible amount of honey, with the least possible consumption of precious wax in their construction. It has been remarked that a skilful workman, with fitting tools and measures, would find it very difficult to make cells of wax of the true form, though this is perfectly effected by a crowd of bees working in a dark hive. Grant whatever instincts you please, and it seems at first quite inconceivable how they can make all the necessary angles and planes, or even perceive when they are correctly made. But the difficulty is not nearly so great as it at first appears: all this beautiful work can be shown, I think, to follow from a few very simple instincts.

I was led to investigate this subject by Mr. Waterhouse, who has shown that the form of the cell stands in close relation to the presence of adjoining cells; and the following view may, perhaps, be considered only as a modification of his theory. Let us look to the great principle of gradation, and see whether Nature does not reveal to us her method of work. At one end of a short series we have humblebees, which use their old cocoons to hold honey, sometimes adding to them short tubes of wax, and likewise making separate and very irregular rounded cells of wax. At the other end of the series we have the cells of the hive-bee, placed in a double layer: each cell, as is well known, is an hexagonal prism, with the basal edges of its six sides bevelled so as to join on to a pyramid, formed of three rhombs. These rhombs have certain angles, and the three which form the pyramidal base of a single cell on one side of the comb, enter into the composition of the bases of three adjoining cells on the opposite side. In the series between the extreme perfection of the cells of the hive-bee and the simplicity of those of the humble-bee, we have the cells of the Mexican Melipona domestica, carefully described and figured by Pierre Huber. The Melipona itself is intermediate in structure between the hive and humble-bee, but more nearly related to the latter: it forms a nearly regular waxen comb of cylindrical cells, in which the young are hatched, and, in addition, some large cells of wax for holding honey. These latter cells are nearly spherical and of nearly equal sizes, and are aggregated into an irregular mass. But the important point to notice, is that these cells are always made at that degree of nearness to each other, that they would have intersected or broken into each other, if the spheres had been completed; but this is never permitted, the bees building perfectly flat walls of wax between the spheres which thus tend to intersect. Hence each cell consists of an outer spherical portion and of two, three, or more perfectly flat surfaces, according as the cell adjoins two, three or more other cells. When one cell comes into contact with three other cells, which, from the spheres being nearly of the same size, is very frequently and necessarily the case, the three flat surfaces are united into a pyramid; and this pyramid, as Huber has remarked, is manifestly a gross imitation of the three-sided pyramidal basis of the cell of the hive-bee. As in the cells of the hive-bee, so here, the three plane surfaces in any one cell necessarily enter into the construction of three adjoining cells. It is obvious that the Melipona saves wax by this manner of building; for the flat walls between the adjoining cells are not double, but are of the same thickness as the outer spherical portions, and yet each flat portion forms a part of two cells.

Reflecting on this case, it occurred to me that if the Melipona had made its spheres at some given distance from each other, and had made them of equal sizes and had arranged them symmetrically in a double layer, the resulting structure would probably have been as perfect as the comb of the hive-bee. Accordingly I wrote to Professor Miller, of Cambridge, and this geometer has kindly read over the following statement, drawn up from his information, and tells me that it is strictly correct: —

If a number of equal spheres be described with their centres placed in two parallel layers; with the centre of each sphere at the distance of radius × $\sqrt{2}$, or radius × 1.41421 (or at some lesser distance), from the centres of the six surrounding spheres in the same layer; and at the same distance from the centres of the adjoining spheres in the other and parallel layer; then, if planes of intersection between the several spheres in both layers be formed, there will result a double layer of hexagonal prisms united together by pyramidal bases formed of three rhombs; and the rhombs and the sides of the hexagonal prisms will have every angle identically the same with the best measurements which have been made of the cells of the hive-bee.

Hence we may safely conclude that if we could slightly modify the instincts already possessed by the Melipona, and in themselves not very wonderful, this bee would make a structure as wonderfully perfect as that of the hive-bee. We must suppose the Melipona to make her cells truly spherical, and of equal sizes; and this would not be very surprising, seeing that she already does so to a certain extent, and seeing what perfectly cylindrical burrows in wood many insects can make, apparently by turning round on a fixed point. We must suppose the Melipona to arrange her cells in level layers, as she already does her cylindrical cells; and we must further suppose, and this is the greatest difficulty, that she can somehow judge accurately at what distance to stand from her fellow-labourers when several are making their spheres; but she is already so far enabled to judge of distance, that she always describes her spheres so as to intersect largely; and then she unites the points of intersection by perfectly flat surfaces. We have further to suppose, but this is no difficulty, that after hexagonal prisms have been formed by the intersection of adjoining spheres in the same layer, she can prolong the hexagon to any length requisite to hold the stock of honey; in the same way as the rude humble-bee adds cylinders of wax to the circular mouths of her old cocoons. By such modifications of instincts in themselves not very wonderful, — hardly more wonderful than those which guide a bird to make its nest, – I believe that the hive-bee has acquired, through natural selection, her inimitable architectural powers.

But this theory can be tested by experiment. Following the example of Mr. Tegetmeier, I separated two combs, and put between them a long, thick, square strip of wax: the bees instantly began to excavate minute circular pits in it; and as they deepened these little pits, they made them wider and wider until they were converted into shallow basins, appearing to the eye perfectly true or parts of a sphere, and of about the diameter of a cell. It was most interesting to me to observe that wherever several bees had begun to excavate these basins near together, they had begun their work at such a distance from each other, that by the time the basins had acquired the above stated width (*i.e.* about the width of an ordinary cell), and were in depth about one sixth of the diameter of the sphere of which they formed a part, the rims of the basins intersected or broke into each other. As soon as this occurred, the bees ceased to excavate, and began to build up flat walls of wax on the lines of intersection between the basins, so that each hexagonal prism was built upon the festooned edge of a smooth basin, instead of on the straight edges of a three-sided pyramid as in the case of ordinary cells.

I then put into the hive, instead of a thick, square piece of wax, a thin and narrow, knife-edged ridge, coloured with vermilion. The bees instantly began on both sides to excavate little basins near to each other, in the same way as before; but the ridge of wax was so thin, that the bottoms of the basins, if they had been excavated to the same depth as in the former experiment, would have broken into each other from the opposite sides. The bees, however, did not suffer this to happen, and they stopped their excavations in due time; so that the basins, as soon as they had been a little deepened, came to have flat bottoms; and these flat bottoms, formed by thin little plates of the vermilion wax having been left ungnawed, were situated, as far as the eye could judge, exactly along the planes of imaginary inter-section between the basins on the opposite sides of the ridge of wax. In parts, only little bits, in other parts, large portions of a rhombic plate had been left between the opposed basins, but the work, from the unnatural state of things, had not been neatly performed. The bees must have worked at very nearly the same rate on the opposite side of the ridge of vermilion wax, as they circularly gnawed away and deepened the basins on both sides, in order to have succeeded in thus leaving flat plates between the basins, by stopping work along the intermediate planes or planes of intersection.

Considering how flexible thin wax is, I do not see that there is any difficulty in the bees, whilst at work on the two sides of a strip of wax, perceiving when they have gnawed the wax away to the proper thinness, and then stopping their work. In ordinary combs it has appeared to me that the bees do not always succeed in working at exactly the same rate from the opposite sides; for I have noticed half-completed rhombs at the base of a just-commenced cell, which were slightly concave on one side, where I suppose that the bees had excavated too quickly, and convex on the opposed side, where the bees had worked less quickly. In one well-marked instance, I put the comb back into the hive and allowed the bees to go on working for a short time and again examined the cell, and I found that the rhombic plate had been completed, and had become *perfectly flat*: it was absolutely impossible, from the extreme thinness of the little rhombic plate, that they could have affected this by gnawing away the convex side; and I suspect that the bees in such cases stand in the opposed cells and push and bend the ductile and warm wax (which as I have tried is easily done) into its proper intermediate plane, and thus flatten it.

From the experiment of the ridge of vermilion wax, we can clearly see that if the bees were to build for themselves a thin wall of wax, they could make their cells of the proper shape, by standing at the proper distance from each other, by excavating at the same rate, and by endeavouring to make equal spherical hollows, but never allowing the spheres to break into each other. Now bees, as may be clearly seen by examining the edge of a growing comb, do make a rough, circumferential wall or rim all round the comb; and they gnaw into this from the opposite sides, always working circularly as they deepen each cell. They do not make the whole three-sided pyramidal base of any one cell at the same time, but only the one rhombic plate which stands on the extreme growing margin, or the two plates, as the case may be; and they never complete the upper edges of the rhombic plates, until the hexagonal walls are commenced. Some of these statements differ from those made by the justly celebrated elder Huber, but I am convinced of their accuracy; and if I had space, I could show that they are conformable with my theory.

Huber's statement that the very first cell is excavated out of a little parallel-sided wall of wax, is not, as far as I have seen, strictly correct; the first commencement having always been a little hood of wax; but I will not here enter on these details. We see how important a part excavation plays in the construction of the cells; but it would be a great error to suppose that the bees cannot build up a rough wall of wax in the proper position – that is, along the plane of intersection between two adjoining spheres. I have several specimens showing clearly that they can do this. Even in the rude circumferential rim or wall of wax round a growing comb, flexures may sometimes be observed, corresponding in position to the planes of the rhombic basal plates of future cells. But the rough wall of wax has in every case to be finished off, by being largely gnawed away on both sides. The manner in which the bees build is curious; they always make the first rough wall from ten to twenty times thicker than the excessively thin finished wall of the cell, which will ultimately be left. We shall understand how they work, by supposing masons first to pile up a broad ridge of cement, and then to begin cutting it away equally on both sides near the ground, till a smooth, very thin wall is left in the middle; the masons always piling up the cut-away cement, and adding fresh cement, on the summit of the ridge. We shall thus have a thin wall steadily growing upward; but always crowned by a gigantic coping. From all the cells, both those just commenced and those completed, being thus crowned by a strong coping of wax, the bees can cluster and crawl over the comb without injuring the delicate hexagonal walls, which are only about one four-hundredth of an inch in thickness; the plates of the pyramidal basis being about twice as thick. By this singular manner of building, strength is continually given to the comb, with the utmost ultimate economy of wax.

It seems at first to add to the difficulty of understanding how the cells are made, that a multitude of bees all work together; one bee after working a short time at one cell going to another, so that, as Huber has stated, a score of individuals work even at the commencement of the first cell. I was able practically to show this fact, by covering the edges of the hexagonal walls of a single cell, or the extreme margin of the circumferential rim of a growing comb, with an extremely thin layer of melted vermilion wax; and I invariably found that the colour was most delicately diffused by the bees – as delicately as a painter could have done with his brush – by atoms of the coloured wax having been taken from the spot on which it had been placed, and worked into the growing edges of the cells all round. The work of construction seems to be a sort of balance struck between many bees, all instinctively standing at the same relative distance from each other, all trying to sweep equal spheres, and then building up, or leaving ungnawed, the planes of intersection between these spheres. It was really curious to note in cases of difficulty, as when two pieces of comb met at an angle, how often the bees would entirely pull down and rebuild in different ways the same cell, sometimes recurring to a shape which they had at first rejected.

When bees have a place on which they can stand in their proper positions for working, – for instance, on a slip of wood, placed directly under the middle of a comb growing downwards so that the comb has to be built over one face of the slip – in this case the bees can lay the foundations of one wall of a new hexagon, in its strictly proper place, projecting beyond the other completed cells. It suffices that the bees should be enabled to stand at their proper relative distances from each other and from the walls of the last completed cells, and then, by striking imaginary spheres, they can build up a wall intermediate between two adjoining spheres; but, as far as I have seen, they never gnaw away and finish off the angles of a cell till a large part both of that cell and of the adjoining cells has been built. This capacity in bees of laying down under certain circumstances a rough wall in its proper place between two just-commenced cells, is important, as it bears on a fact, which seems at first quite subversive of the foregoing theory; namely, that the cells on the extreme margin of wasp-combs are sometimes strictly hexagonal; but I have not space here to enter on this subject. Nor does there seem to me any great difficulty in a single insect (as in the case of a queen-wasp) making hexagonal cells, if she work alternately on the inside and outside of two or three cells commenced at the same time, always standing at the proper relative distance from the parts of the cells just begun, sweeping spheres or cylinders, and building up intermediate planes. It is even conceivable that an insect might, by fixing on a point at which to commence a cell, and then moving outside, first to one point, and then to five other points, at the proper relative distances from the central point and from each other, strike the planes of intersection, and so make an isolated hexagon: but I am not aware that any such case has been observed; nor would any good be derived from a single hexagon being built, as in its construction more materials would be required than for a cylinder.

As natural selection acts only by the accumulation of slight modifications of structure or instinct, each profitable to the individual under its conditions of life, it may reasonably be asked, how a long and graduated succession of modified architectural instincts, all tending towards the present perfect plan of construction, could have profited the progenitors of the hive-bee? I think the answer is not difficult: it is known that bees are often hard pressed to get sufficient nectar; and I am informed by Mr. Tegetmeier that it has been experimentally found that no less than from twelve to fifteen pounds of dry sugar are consumed by a hive of bees for the secretion of each pound of wax; so that a prodigious quantity of fluid nectar must be collected and consumed by the bees in a hive for the secretion of the wax necessary for the construction of their combs. Moreover, many bees have to remain idle for many days during the process of secretion. A large store of honey is indispensable to support a large stock of bees during the winter; and the security of the hive is known mainly to depend on a large number of bees being supported. Hence the saving of wax by largely saving honey must be a most important element of success in any family of bees. Of course the success of any species of bee may be dependent on the number of its parasites or other enemies, or on quite distinct causes, and so be altogether independent of the quantity of honey which the bees could collect. But let us suppose that this latter circumstance determined, as it probably often does determine, the numbers of a humble-bee which could exist in a country; and let us further suppose that the community lived throughout the winter, and consequently required a store of honey: there can in this case be no doubt that it would be an advantage to our humble-bee, if a slight modification of her instinct led her to make her waxen cells near together, so as to intersect a little; for a wall in common even to two adjoining cells, would save some little wax. Hence it would continually be more and more advantageous to our humble-bee, if she were to make her cells more and more regular, nearer together, and aggregated into a mass, like the cells of the Melipona; for in this case a large part of the bounding surface of each cell would serve to bound other cells, and much wax would be saved. Again, from the same cause, it would be advantageous to the Melipona, if she were to make her cells closer together, and more regular in every way than at present; for then, as we have seen, the spherical surfaces would wholly disappear, and would all be replaced by plane surfaces; and the Melipona would make a comb as perfect as that of the hive-bee. Beyond this stage of perfection in architecture, natural selection could not lead; for the comb of the hive-bee, as far as we can see, is absolutely perfect in economizing wax.

Thus, as I believe, the most wonderful of all known instincts, that of the hive-bee, can be explained by natural selection having taken advantage of numerous, successive, slight modifications of simpler instincts; natural selection having by slow degrees, more and more perfectly, led the bees to sweep equal spheres at a given distance from each other in a double layer, and to build up and excavate the wax along the planes of intersection. The bees, of course, no more knowing that they swept their spheres at one particular distance from each other, than they know what are the several angles of the hexagonal prisms and of the basal rhombic plates. The motive power of the process of natural selection having been economy of wax; that individual swarm which wasted least honey in the secretion of wax, having succeeded best, and having transmitted by inheritance its newly acquired economical instinct to new swarms, which in their turn will have had the best chance of succeeding in the struggle for existence.

No doubt many instincts of very difficult explanation could be opposed to the theory of natural selection, – cases, in which we cannot see how an instinct could possibly have originated; cases, in which no intermediate gradations are known to exist; cases of instinct of apparently such trifling importance, that they could hardly have been acted on by natural selection; cases of instincts almost identically the same in animals so remote in the scale of nature, that we cannot account for their similarity by inheritance from a common parent, and must therefore believe that they have been acquired by independent acts of natural selection. I will not here enter on these several cases, but will confine myself to one special difficulty, which at first appeared to me insuperable, and actually fatal to my whole theory. I allude to the neuters or sterile females in insect-communities: for these neuters often differ widely in instinct and in structure from both the males and fertile females, and yet, from being sterile, they cannot propagate their kind.

The subject well deserves to be discussed at great length, but I will here take only a single case, that of working or sterile ants. How the workers have been rendered sterile is a difficulty; but not much greater than that of any other striking modification of structure; for it can be shown that some insects and other articulate animals in a state of nature occasionally become sterile; and if such insects had been social, and it had been profitable to the community that a number should have been annually born capable of work, but incapable of procreation, I can see no very great difficulty in this being effected by natural selection. But I must pass over this preliminary difficulty. The great difficulty lies in the working ants differing widely from both the males and the fertile females in structure, as in the shape of the thorax and in being destitute of wings and sometimes of eyes, and in instinct. As far as instinct alone is concerned, the prodigious difference in this respect between the workers and the perfect females, would have been far better exemplified by the hive-bee. If a working ant or other neuter insect had been an animal in the ordinary state, I should have unhesitatingly assumed that all its characters had been slowly acquired through natural selection; namely, by an individual having been born with some slight profitable modification of structure, this being inherited by its offspring, which again varied and were again selected, and so onwards. But with the working ant we have an insect differing greatly from its parents, yet absolutely sterile; so that it could never have transmitted successively acquired modifications of structure or instinct to its progeny. It may well be asked how is it possible to reconcile this case with the theory of natural selection?

First, let it be remembered that we have innumerable instances, both in our domestic productions and in those in a state of nature, of all sorts of differences of structure which have become correlated to certain ages, and to either sex. We have differences correlated not only to one sex, but to that short period alone when the reproductive system is active, as in the nuptial plumage of many birds, and in the hooked jaws of the male salmon. We have even slight differences in the horns of different breeds of cattle in relation to an artificially imperfect state of the male sex; for oxen of certain breeds have longer horns than in other breeds, in comparison with the horns of the bulls or cows of these same breeds. Hence I can see no real difficulty in any character having become correlated with the sterile condition of certain members of insect-communities: the difficulty lies in understanding how such correlated modifications of structure could have been slowly accumulated by natural selection.

This difficulty, though appearing insuperable, is lessened, or, as I believe, disappears, when it is remembered that selection may be applied to the family, as well as to the individual, and may thus gain the desired end. Thus, a well-flavoured vegetable is cooked, and the individual is destroyed; but the horticulturist sows seeds of the same stock, and confidently expects to get nearly the same variety; breeders of cattle wish the flesh and fat to be well marbled together; the animal has been slaughtered, but the breeder goes with confidence to the same family. I have such faith in the powers of selection, that I do not doubt that a breed of cattle, always yielding oxen with extraordinarily long horns, could be slowly formed by carefully watching which individual bulls and cows, when matched, produced oxen with the longest horns; and yet no one ox could ever have propagated its kind. Thus I believe it has been with social insects: a slight modification of structure, or instinct, correlated with the sterile condition of certain members of the community, has been advantageous to the community: consequently the fertile males and females of the same community flourished, and transmitted to their fertile offspring a tendency to produce sterile members having the same modification. And I believe that this process has been repeated, until that prodigious amount of difference between the fertile and sterile females of the same species has been produced, which we see in many social insects.

But we have not as yet touched on the climax of the difficulty; namely, the fact that the neuters of several ants differ, not only from the fertile females and males, but from each other, sometimes to an almost incredible degree, and are thus divided into two or even three castes. The castes, moreover, do not generally graduate into each other, but are perfectly well defined; being as distinct from each other, as are any two species of the same genus, or rather as any two genera of the same family. Thus in Eciton, there are working and soldier neuters, with jaws and instincts extraordinarily different: in Cryptocerus, the workers of one caste alone carry a wonderful sort of shield on their heads, the use of which is quite unknown: in the Mexican Myrmecocystus, the workers of one caste never leave the nest; they are fed by the workers of another caste, and they have an enormously developed abdomen which secretes a sort of honey, supplying the place of that excreted by the aphides, or the domestic cattle as they may be called, which our European ants guard or imprison.

It will indeed be thought that I have an overweening confidence in the principle of natural selection, when I do not admit that such wonderful and well-established facts at once annihilate my theory. In the simpler case of neuter insects all of one caste or of the same kind, which have been rendered by natural selection, as I believe to be quite possible, different from the fertile males and females, – in this case, we may safely conclude from the analogy of ordinary variations, that each successive, slight, profitable modification did not probably at first appear in all the individual neuters in the same nest, but in a few alone; and that by the long-continued selection of the fertile parents which produced most neuters with the profitable modification, all the neuters ultimately came to have the desired character. On this view we ought occasionally to find neuter-insects of the same species, in the same nest, presenting gradations of structure; and this we do find, even often, considering how few neuter-insects out of Europe have been carefully examined. Mr. F. Smith has shown how surprisingly the neuters of several British ants differ from each other in size and sometimes in colour; and that the extreme forms can sometimes be perfectly linked together by individuals taken out of the same nest: I have myself compared perfect gradations of this kind. It often happens that the larger or the smaller sized workers are the most numerous; or that both large and small are numerous, with those of an intermediate size scanty in numbers. Formica flava has larger and smaller workers, with some of intermediate size; and, in this species, as Mr. F. Smith has observed, the larger workers have simple eyes (ocelli), which though small can be plainly distinguished, whereas the smaller workers have their ocelli rudimentary. Having carefully dissected several specimens of these workers, I can affirm that the eyes are far more rudimentary in the smaller workers than can be accounted for merely by their proportionally lesser size; and I fully believe, though I dare not assert so positively, that the workers of intermediate size have their ocelli in an exactly intermediate condition. So that we here have two bodies of sterile workers in the same nest, differing not only in size, but in their organs of vision, yet connected by some few members in an intermediate condition. I may digress by adding, that if the smaller workers had been the most useful to the community, and those males and females had been continually selected, which produced more and more of the smaller workers, until all the workers had come to be in this condition; we should then have had a species of ant with neuters very nearly in the same condition with those of Myrmica. For the workers of Myrmica have not even rudiments of ocelli, though the male and female ants of this genus have well-developed ocelli.

I may give one other case: so confidently did I expect to find gradations in important points of structure between the different castes of neuters in the same species, that I gladly availed myself of Mr. F. Smith's offer of numerous specimens from the same nest of the driver ant (Anomma) of West Africa. The reader will perhaps best appreciate the amount of difference in these workers, by my giving not the actual measurements, but a strictly accurate illustration: the difference was the same as if we were to see a set of workmen building a house of whom many were five feet four inches high, and many sixteen feet high; but we must suppose that the larger workmen had heads four instead of three times as big as those of the smaller men, and jaws nearly five times as big. The jaws, moreover, of the working ants of the several sizes differed wonderfully in shape, and in the form and number of the teeth. But the important fact for us is, that though the

workers can be grouped into castes of different sizes, yet they graduate insensibly into each other, as does the widely-different structure
of their jaws. I speak confidently on this latter point, as Mr. Lubbock
made drawings for me with the camera lucida of the jaws which I had
dissected from the workers of the several sizes.

With these facts before me, I believe that natural selection, by
acting on the fertile parents, could form a species which should
regularly produce neuters, either all of large size with one form of
jaw, or all of small size with jaws having a widely different structure;
or lastly, and this is our climax of difficulty, one set of workers of one
size and structure, and simultaneously another set of workers of a
different size and structure; – a graduated series having been first
formed, as in the case of the driver ant, and then the extreme forms,
from being the most useful to the community, having been produced
in greater and greater numbers through the natural selection of the
parents which generated them; until none with an intermediate
structure were produced.

Thus, as I believe, the wonderful fact of two distinctly defined
castes of sterile workers existing in the same nest, both widely different from each other and from their parents, has originated. We can see
how useful their production may have been to a social community of
insects, on the same principle that the division of labour is useful to
civilized man. As ants work by inherited instincts and by inherited
tools or weapons, and not by acquired knowledge and manufactured
instruments, a perfect division of labour could be effected with them
only by the workers being sterile; for had they been fertile, they
would have intercrossed, and their instincts and structure would
have become blended. And nature has, as I believe, effected this
admirable division of labour in the communities of ants, by the
means of natural selection. But I am bound to confess, that, with all
my faith in this principle, I should never have anticipated that natural
selection could have been efficient in so high a degree, had not the
case of these neuter insects convinced me of the fact. I have, therefore, discussed this case, at some little but wholly insufficient length,
in order to show the power of natural selection, and likewise because
this is by far the most serious special difficulty, which my theory has
encountered. The case, also, is very interesting, as it proves that with

animals, as with plants, any amount of modification in structure can be effected by the accumulation of numerous, slight, and as we must call them accidental, variations, which are in any manner profitable, without exercise, or habit having come into play. For no amount of exercise, or habit, or volition, in the utterly sterile members of a community could possibly have affected the structure or instincts of the fertile members, which alone leave descendants. I am surprised that no one has advanced this demonstrative case of neuter insects, against the well-known doctrine of Lamarck.

Summary. I have endeavoured briefly in this chapter to show that the mental qualities of our domestic animals vary, and that the variations are inherited. Still more briefly I have attempted to show that instincts vary slightly in a state of nature. No one will dispute that instincts are of the highest importance to each animal. Therefore I can see no difficulty, under changing conditions of life, in natural selection accumulating slight modifications of instinct to any extent, in any useful direction. In some cases habit or use and disuse have probably come into play. I do not pretend that the facts given in this chapter strengthen in any great degree my theory; but none of the cases of difficulty, to the best of my judgement, annihilate it. On the other hand, the fact that instincts are not always absolutely perfect and are liable to mistakes; – that no instinct has been produced for the exclusive good of other animals, but that each animal takes advantage of the instincts of others; – that the canon in natural history, of 'natura non facit saltum' is applicable to instincts as well as to corporeal structure, and is plainly explicable on the foregoing views, but is otherwise inexplicable, – all tend to corroborate the theory of natural selection.

This theory is, also, strengthened by some few other facts in regard to instincts; as by that common case of closely allied, but certainly distinct, species, when inhabiting distant parts of the world and living under considerably different conditions of life, yet often retaining nearly the same instincts. For instance, we can understand on the principle of inheritance, how it is that the thrush of South America lines its nest with mud, in the same peculiar manner as does our British thrush: how it is that the male wrens (Troglodytes) of North

America, build 'cock-nests,' to roost in, like the males of our distinct Kitty-wrens, – a habit wholly unlike that of any other known bird. Finally, it may not be a logical deduction, but to my imagination it is far more satisfactory to look at such instincts as the young cuckoo ejecting its foster-brothers, – ants making slaves, – the larvae of ichneumonidae feeding within the live bodies of caterpillars, – not as specially endowed or created instincts, but as small consequences of one general law, leading to the advancement of all organic beings, namely, multiply, vary, let the strongest live and the weakest die.

CHAPTER VIII

HYBRIDISM

Distinction between the sterility of first crosses and of hybrids – Sterility various in degree, not universal, affected by close interbreeding, removed by domestication – Laws governing the sterility of hybrids – Sterility not a special endowment, but incidental on other differences – Causes of the sterility of first crosses and of hybrids – Parallelism between the effects of changed conditions of life and crossing – Fertility of varieties when crossed and of their mongrel offspring not universal – Hybrids and mongrels compared independently of their fertility – Summary

THE view generally entertained by naturalists is that species, when intercrossed, have been specially endowed with the quality of sterility, in order to prevent the confusion of all organic forms. This view certainly seems at first probable, for species within the same country could hardly have kept distinct had they been capable of crossing freely. The importance of the fact that hybrids are very generally sterile, has, I think, been much underrated by some late writers. On the theory of natural selection the case is especially important, inasmuch as the sterility of hybrids could not possibly be of any advantage to them, and therefore could not have been acquired by the continued preservation of successive profitable degrees of sterility. I hope, however, to be able to show that sterility is not a specially acquired or endowed quality, but is incidental on other acquired differences.

In treating this subject, two classes of facts, to a large extent fundamentally different, have generally been confounded together; namely, the sterility of two species when first crossed, and the sterility of the hybrids produced from them.

Pure species have of course their organs of reproduction in a perfect condition, yet when intercrossed they produce either few or no offspring. Hybrids, on the other hand, have their reproductive organs functionally impotent, as may be clearly seen in the state of the male element in both plants and animals; though the organs themselves are perfect in structure, as far as the microscope reveals. In the first case the two sexual elements which go to form the embryo are perfect; in the second case they are either not at all developed, or are imperfectly developed. This distinction is important, when the cause of the sterility, which is common to the two cases, has to be considered. The distinction has probably been slurred over, owing to the sterility in both cases being looked on as a special endowment, beyond the province of our reasoning powers.

The fertility of varieties, that is of the forms known or believed to have descended from common parents, when intercrossed, and likewise the fertility of their mongrel offspring, is, on my theory, of equal importance with the sterility of species; for it seems to make a broad and clear distinction between varieties and species.

First, for the sterility of species when crossed and of their hybrid offspring. It is impossible to study the several memoirs and works of those two conscientious and admirable observers, Kölreuter and Gärtner, who almost devoted their lives to this subject, without being deeply impressed with the high generality of some degree of sterility. Kölreuter makes the rule universal; but then he cuts the knot, for in ten cases in which he found two forms, considered by most authors as distinct species, quite fertile together, he unhesitatingly ranks them as varieties. Gärtner, also, makes the rule equally universal; and he disputes the entire fertility of Kölreuter's ten cases. But in these and in many other cases, Gärtner is obliged carefully to count the seeds, in order to show that there is any degree of sterility. He always compares the maximum number of seeds produced by two species when crossed and by their hybrid offspring, with the average number produced by both pure parent-species in a state of nature. But a serious cause of error seems to me to be here introduced: a plant to be hybridized must be castrated, and, what is often more important, must be secluded in order to prevent pollen being brought to it by insects from other plants. Nearly all the plants experimentized on by Gärtner were potted, and apparently were kept in a chamber in his house. That these processes are often injurious to the fertility of a plant cannot be doubted; for Gärtner gives in his table about a score of cases of plants which he castrated, and artificially fertilized with their own pollen, and (excluding all cases such as the Leguminosae, in which there is an acknowledged difficulty in the manipulation) half of these twenty plants had their fertility in some degree impaired. Moreover, as Gärtner during several years repeatedly crossed the primrose and cowslip, which we have such good reason to believe to be varieties, and only once or twice succeeded in getting fertile seed; as he found the common red and blue pimpernels (Anagallis arvensis and coerulea), which the best botanists rank as varieties, absolutely sterile together; and as he came to the same conclusion in several other analogous cases; it seems to me that we may well be permitted to doubt whether many other species are really so sterile, when intercrossed, as Gärtner believes.

It is certain, on the one hand, that the sterility of various species when crossed is so different in degree and graduates away so insensibly, and, on the other hand, that the fertility of pure species is so easily affected by various circumstances, that for all practical purposes it is most difficult to say where perfect fertility ends and sterility begins. I think no better evidence of this can be required than that the two most experienced observers who have ever lived, namely, Kölreuter and Gärtner, should have arrived at diametrically opposite conclusions in regard to the very same species. It is also most instructive to compare – but I have not space here to enter on details – the evidence advanced by our best botanists on the question whether certain doubtful forms should be ranked as species or varieties, with the evidence from fertility adduced by different hybridizers, or by the same author, from experiments made during different years. It can thus be shown that neither sterility nor fertility affords any clear distinction between species and varieties; but that the evidence from this source graduates away, and is doubtful in the same degree as is the evidence derived from other constitutional and structural differences.

In regard to the sterility of hybrids in successive generations; though Gärtner was enabled to rear some hybrids, carefully guarding them from a cross with either pure parent, for six or seven, and in one case for ten generations, yet he asserts positively that their fertility never increased, but generally greatly decreased. I do not doubt that this is usually the case, and that the fertility often suddenly decreases in the first few generations. Nevertheless I believe that in all these experiments the fertility has been diminished by an independent cause, namely, from close interbreeding. I have collected so large a body of facts, showing that close interbreeding lessens fertility, and, on the other hand, that an occasional cross with a distinct individual or variety increases fertility, that I cannot doubt the correctness of this almost universal belief amongst breeders. Hybrids are seldom raised by experimentalists in great numbers; and as the parent-species, or other allied hybrids, generally grow in the same garden, the visits of insects must be carefully prevented during the flowering season: hence hybrids will generally be fertilized during each generation by their own individual pollen; and I am convinced that this would be injurious to their fertility, already lessened by their hybrid origin. I am strengthened in this conviction by a remarkable statement repeatedly made by Gärtner, namely, that if even the less fertile hybrids be artificially fertilized with hybrid pollen of the same kind, their fertility, notwithstanding the frequent ill effects of manipulation, sometimes decidedly increases, and goes on increasing. Now, in artificial fertilization pollen is as often taken by chance (as I know from my own experience) from the anothers of another flower, as from the anthers of the flower itself which is to be fertilized; so that a cross between two flowers, though probably on the same plant, would be thus effected. Moreover, whenever complicated experiments are in progress, so careful an observer as Gärtner would have castrated his hybrids, and this would have insured in each generation a cross with the pollen from a distinct flower, either from the same plant or from another plant of the same hybrid nature. And thus, the strange fact of the increase of fertility in the successive generations of *artificially fertilized* hybrids may, I believe, be accounted for by close interbreeding having been avoided.

Now let us turn to the results arrived at by the third most experienced hybridizer, namely, the Hon. and Rev. W. Herbert. He is as emphatic in his conclusion that some hybrids are perfectly fertile – as fertile as the pure parent-species – as are Kölreuter and Gärtner that some degree of sterility between distinct species is a universal law of nature. He experimentized on some of the very same species as did Gärtner. The difference in their results may, I think, be in part accounted for by Herbert's great horticultural skill, and by his having hothouses at his command. Of his many important statements I will here give only a single one as an example, namely, that 'every ovule in a pod of Crinum capense fertilized by C. revolutum produced a plant, which (he says) I never saw to occur in a case of its natural fecundation.' So that we here have perfect, or even more than commonly perfect, fertility in a first cross between two distinct species.

This case of the Crinum leads me to refer to a most singular fact, namely, that there are individual plants, as with certain species of Lobelia, and with all the species of the genus Hippeastrum, which can be far more easily fertilized by the pollen of another and distinct species, than by their own pollen. For these plants have been found to yield seed to the pollen of a distinct species, though quite sterile with their own pollen, notwithstanding that their own pollen was found to be perfectly good, for it fertilized distinct species. So that certain individual plants and all the individuals of certain species can actually be hybridized much more readily than they can be self-fertilized! For instance, a bulb of Hippeastrum aulicum produced four flowers; three were fertilized by Herbert with their own pollen, and the fourth was subsequently fertilized by the pollen of a compound hybrid descended from three other and distinct species: the result was that 'the ovaries of the three first flowers soon ceased to grow, and after a few days perished entirely, whereas the pod impregnated by the pollen of the hybrid made vigorous growth and rapid progress to maturity, and bore good seed, which vegetated freely.' In a letter to me, in 1839, Mr. Herbert told me that he had then tried the experiment during five years, and he continued to try it during several subsequent years, and always with the same result. This result has, also, been confirmed by other observers in the case of Hippeastrum with its sub-genera, and in the case of some other genera, as Lobelia, Passiflora and Verbascum. Although the plants in these experiments appeared perfectly healthy, and although both the ovules and pollen of the same flower were perfectly good with respect to other species, yet as they were functionally imperfect in their mutual self-action, we must infer that the plants were in an unnatural state. Nevertheless these facts show on what slight and mysterious causes the lesser or greater fertility of species when crossed, in comparison with the same species when self-fertilized, sometimes depends.

The practical experiments of horticulturists, though not made with scientific precision, deserve some notice. It is notorious in how complicated a manner the species of Pelargonium, Fuchsia, Calceolaria, Petunia, Rhododendron, &c., have been crossed, yet many of these hybrids seed freely. For instance, Herbert asserts that a hybrid from Calceolaria integrifolia and plantaginea, species most widely dissimilar in general habit, 'reproduced itself as perfectly as if it had been a natural species from the mountains of Chile.' I have taken some pains to ascertain the degree of fertility of some of the complex crosses of Rhododendrons, and I am assured that many of them are perfectly fertile. Mr. C. Noble, for instance, informs me that he raises stocks for grafting from a hybrid between Rhod. Ponticum and Catawbiense, and that this hybrid 'seeds as freely as it is possible to imagine.' Had hybrids, when fairly treated, gone on decreasing in fertility in each successive generation, as Gärtner believes to be the case, the fact would have been notorious to nurserymen. Horticulturists raise large beds of the same hybrids, and such alone are fairly treated, for by insect agency the several individuals of the same hybrid variety are allowed to freely cross with each other, and the injurious influence of close interbreeding is thus prevented. Any one may readily convince himself of the efficiency of insect-agency by examining the flowers of the more sterile kinds of hybrid rhododendrons, which produce no pollen, for he will find on their stigmas plenty of pollen brought from other flowers.

In regard to animals, much fewer experiments have been carefully tried than with plants. If our systematic arrangements can be trusted, that is if the genera of animals are as distinct from each other, as are the genera of plants, then we may infer that animals more widely separated in the scale of nature can be more easily crossed than in the case of plants; but the hybrids themselves are, I think, more sterile. I doubt whether any case of a perfectly fertile hybrid animal can be considered as thoroughly well authenticated. It should, however, be borne in mind that, owing to few animals breeding freely under confinement, few experiments have been fairly tried: for instance, the canary-bird has been crossed with nine other finches, but as not one of these nine species breeds freely in confinement, we have no right to expect that the first crosses between them and the canary, or that their hybrids, should be perfectly fertile. Again, with respect to the fertility in successive generations of the more fertile hybrid animals, I hardly know of an instance in which two families of the same hybrid have been raised at the same time from different parents, so as to avoid the ill effects of close interbreeding. On the contrary, brothers and sisters have usually been crossed in each successive generation, in opposition to the constantly repeated admonition of every breeder. And in this case, it is not at all surprising that the inherent sterility in the hybrids should have gone on increasing. If we were to act thus, and pair brothers and sisters in the case of any pure animal, which from any cause had the least tendency to sterility, the breed would assuredly be lost in a very few generations.

Although I do not know of any thoroughly well-authenticated cases of perfectly fertile hybrid animals, I have some reason to believe that the hybrids from Cervulus vaginalis and Reevesii, and from Phasianus colchicus with P. torquatus and with P. versicolor are perfectly fertile. The hybrids from the common and Chinese geese (A. cygnoides), species which are so different that they are generally ranked in distinct genera, have often bred in this country with either pure parent, and in one single instance they have bred *inter se*. This was effected by Mr. Eyton, who raised two hybrids from the same parents but from different hatches; and from these two birds he raised no less than eight hybrids (grandchildren of the pure geese) from one nest. In India, however, these cross-bred geese must be far more fertile; for I am assured by two eminently capable judges, namely Mr. Blyth and Capt. Hutton, that whole flocks of these crossed geese are kept in various parts of the country; and as they are kept for profit, where neither pure parent-species exists, they must certainly be highly fertile.

A doctrine which originated with Pallas, has been largely accepted by modern naturalists; namely, that most of our domestic animals have descended from two or more aboriginal species, since com-mingled by intercrossing. On this view, the aboriginal species must either at first have produced quite fertile hybrids, or the hybrids must have become in subsequent generations quite fertile under domestication. This latter alternative seems to me the most probable, and I am inclined to believe in its truth, although its rests on no direct evidence. I believe, for instance, that our dogs have descended from several wild stocks; yet, with perhaps the exception of certain indigenous domestic dogs of South America, all are quite fertile together; and analogy makes me greatly doubt, whether the several aboriginal species would at first have freely bred together and have produced quite fertile hybrids. So again there is reason to believe that our European and the humped Indian cattle are quite fertile together; but from facts communicated to me by Mr. Blyth, I think they must be considered as distinct species. On this view of the origin of many of our domestic animals, we must either give up the belief of the almost universal sterility of distinct species of animals when crossed; or we must look at sterility, not as an indelible characteristic, but as one capable of being removed by domestication.

Finally, looking to all the ascertained facts on the intercrossing of plants and animals, it may be concluded that some degree of sterility, both in first crosses and in hybrids, is an extremely general result; but that it cannot, under our present state of knowledge, be considered as absolutely universal.

Laws governing the Sterility of first Crosses and of Hybrids. We will now consider a little more in detail the circumstances and rules governing the sterility of first crosses and of hybrids. Our chief object will be to see whether or not the rules indicate that species have specially been endowed with this quality, in order to prevent their crossing and blending together in utter confusion. The following rules and conclusions are chiefly drawn up from Gärtner's admirable work on the hybridization of plants. I have taken much pains to ascertain how far the rules apply to animals, and considering how scanty our knowledge is in regard to hybrid animals, I have been surprised to find how generally the same rules apply to both kingdoms.

It has been already remarked, that the degree of fertility, both of first crosses and of hybrids, graduates from zero to perfect fertility. It is surprising in how many curious ways this gradation can be shown to exist; but only the barest outline of the facts can here be given. When pollen from a plant of one family is placed on the stigma of a plant of a distinct family, it exerts no more influence than so much inorganic dust. From this absolute zero of fertility, the pollen of different species of the same genus applied to the stigma of some one species, yields a perfect gradation in the number of seeds produced, up to nearly complete or even quite complete fertility; and, as we have seen, in certain abnormal cases, even to an excess of fertility, beyond that which the plant's own pollen will produce. So in hybrids themselves, there are some which never have produced, and probably never would produce, even with the pollen of either pure parent, a single fertile seed: but in some of these cases a first trace of fertility may be detected, by the pollen of one of the pure parent-species causing the flower of the hybrid to wither earlier than it otherwise would have done; and the early withering of the flower is well known to be a sign of incipient fertilization. From this extreme degree of sterility we have self-fertilized hybrids producing a greater and greater number of seeds up to perfect fertility.

Hybrids from two species which are very difficult to cross, and which rarely produce any offspring, are generally very sterile; but the parallelism between the difficulty of making a first cross, and the sterility of the hybrids thus produced – two classes of facts which are generally confounded together – is by no means strict. There are many cases, in which two pure species can be united with unusual facility, and produce numerous hybrid-offspring, yet these hybrids are remarkably sterile. On the other hand, there are species which can be crossed very rarely, or with extreme difficulty, but the hybrids, when at last produced, are very fertile. Even within the limits of the same genus, for instance in Dianthus, these two opposite cases occur.

The fertility, both of first crosses and of hybrids, is more easily affected by unfavourable conditions, than is the fertility of pure species. But the degree of fertility is likewise innately variable; for it is not always the same when the same two species are crossed under the same circumstances, but depends in part upon the constitution of the individuals which happen to have been chosen for the experiment. So it is with hybrids, for their degree of fertility is often found to differ greatly in the several individuals raised from seed out of the same capsule and exposed to exactly the same conditions.

By the term systematic affinity is meant, the resemblance between species in structure and in constitution, more especially in the structure of parts which are of high physiological importance and which differ little in the allied species. Now the fertility of first crosses between species, and of the hybrids produced from them, is largely governed by their systematic affinity. This is clearly shown by hybrids never having been raised between species ranked by systematists in distinct families; and on the other hand, by very closely allied species generally uniting with facility. But the correspondence between systematic affinity and the facility of crossing is by no means strict. A multitude of cases could be given of very closely allied species which will not unite, or only with extreme difficulty; and on the other hand of very distinct species which unite with the utmost facility. In the same family there may be a genus, as Dianthus, in which very many species can most readily be crossed; and another genus, as Silene, in which the most persevering efforts have failed to produce between extremely close species a single hybrid. Even within the limits of the same genus, we meet with this same difference; for instance, the many species of Nicotiana have been more largely crossed than the species of almost any other genus; but Gärtner found that N. acuminata, which is not a particularly distinct species, obstinately failed to fertilize, or to be fertilized by, no less than eight other species of Nicotiana. Very many analogous facts could be given.

No one has been able to point out what kind, or what amount, of difference in any recognizable character is sufficient to prevent two species crossing. It can be shown that plants most widely different in habit and general appearance, and having strongly marked differences in every part of the flower, even in the pollen, in the fruit, and in the cotyledons, can be crossed. Annual and perennial plants, deciduous and evergreen trees, plants inhabiting different stations and fitted for extremely different climates, can often be crossed with ease.

By a reciprocal cross between two species, I mean the case, for instance, of a stallion-horse being first crossed with a female-ass, and then a male-ass with a mare: these two species may then be said to have been reciprocally crossed. There is often the widest possible difference in the facility of making reciprocal crosses. Such cases are highly important, for they prove that the capacity in any two species to cross is often completely independent of their systematic affinity, or of any recognizable difference in their whole organization. On the other hand, these cases clearly show that the capacity for crossing is connected with constitutional differences imperceptible by us, and confined to the reproductive system. This difference in the result of reciprocal crosses between the same two species was long ago observed by Kölreuter. To give an instance: Mirabilis jalappa can easily be fertilized by the pollen of M. longiflora, and the hybrids thus produced are sufficiently fertile; but Kölreuter tried more than two hundred times, during eight following years, to fertilize reciprocally M. longiflora with the pollen of M. jalappa, and utterly failed. Several other equally striking cases could be given. Thuret has observed the same fact with certain sea-weeds or Fuci. Gärtner, moreover, found that this difference of facility in making reciprocal crosses is extremely common in a lesser degree. He has observed it even between forms so closely related (as Matthiola annua and glabra) that many botanists rank them only as varieties. It is also a remarkable fact, that hybrids raised from reciprocal crosses, though of course compounded of the very same two species, the one species having first been used as the father and then as the mother, generally differ in fertility in a small, and occasionally in a high degree.

Several other singular rules could be given from Gärtner: for instance, some species have a remarkable power of crossing with other species; other species of the same genus have a remarkable power of impressing their likeness on their hybrid offspring; but these two powers do not at all necessarily go together. There are certain hybrids which instead of having, as is usual, an intermediate character between their two parents, always closely resemble one of them; and such hybrids, though externally so like one of their pure parent-species, are with rare exceptions extremely sterile. So again amongst hybrids which are usually intermediate in structure between their parents, exceptional and abnormal individuals sometimes are born, which closely resemble one of their pure parents; and these hybrids are almost always utterly sterile, even when the other hybrids raised from seed from the same capsule have a considerable degree of fertility. These facts show how completely fertility in the hybrid is independent of its external resemblance to either pure parent.

Considering the several rules now given, which govern the fertility of first crosses and of hybrids, we see that when forms, which must be considered as good and distinct species, are united, their fertility graduates from zero to perfect fertility, or even to fertility under certain conditions in excess. That their fertility, besides being eminently susceptible to favourable and unfavourable conditions, is innately variable. That it is by no means always the same in degree in the first cross and in the hybrids produced from this cross. That the fertility of hybrids is not related to the degree in which they resemble in external appearance either parent. And lastly, that the facility of making a first cross between any two species is not always governed by their systematic affinity or degree of resemblance to each other. This latter statement is clearly proved by reciprocal crosses between the same two species, for according as the one species or the other is used as the father or the mother, there is generally some difference, and occasionally the widest possible difference, in the facility of effecting an union. The hybrids, moreover, produced from reciprocal crosses often differ in fertility.

Now do these complex and singular rules indicate that species have been endowed with sterility simply to prevent their becoming confounded in nature? I think not. For why should the sterility be so extremely different in degree, when various species are crossed, all of which we must suppose it would be equally important to keep from blending together? Why should the degree of sterility be innately variable in the individuals of the same species? Why should some species cross with facility, and yet produce very sterile hybrids; and other species cross with extreme difficulty, and yet produce fairly fertile hybrids? Why should there often be so great a difference in the result of a reciprocal cross between the same two species? Why, it may even be asked, has the production of hybrids been permitted? To grant to species the special power of producing hybrids, and then to stop their further propagation by different degrees of sterility, not strictly related to the facility of the first union between their parents, seems to be a strange arrangement.

The foregoing rules and facts, on the other hand, appear to me clearly to indicate that the sterility both of first crosses and of hybrids is simply incidental or dependent on unknown differences, chiefly in the reproductive systems, of the species which are crossed. The differences being of so peculiar and limited a nature, that, in reciprocal crosses between two species the male sexual element of the one will often freely act on the female sexual element of the other, but not in a reversed direction. It will be advisable to explain a little more fully by an example what I mean by sterility being incidental on other differences, and not a specially endowed quality. As the capacity of one plant to be grafted or budded on another is so entirely unimportant for its welfare in a state of nature, I presume that no one will suppose that this capacity is a *specially* endowed quality, but will admit that it is incidental on differences in the laws of growth of the two plants. We can sometimes see the reason why one tree will not take on another, from differences in their rate of growth, in the hardness of their wood, in the period of the flow or nature of their sap, &c.; but in a multitude of cases we can assign no reason whatever. Great diversity in the size of two plants, one being woody and the other herbaceous, one being evergreen and the other deciduous, and adaptation to widely different climates, does not always prevent the two grafting together. As in hybridization, so with grafting, the capacity is limited by systematic affinity, for no one has been able to graft trees together belonging to quite distinct families; and, on the other hand, closely allied species, and varieties of the same species, can usually, but not invariably, be grafted with ease. But this capacity, as in hybridization, is by no means absolutely governed by systematic affinity. Although many distinct genera within the same family have been grafted together, in other cases species of the same genus will not take on each other. The pear can be grafted far more readily on the quince, which is ranked as a distinct genus, than on the apple, which is a member of the same genus. Even different varieties of the pear take with different degrees of facility on the quince; so do different varieties of the apricot and peach on certain varieties of the plum.

As Gärtner found that there was sometimes an innate difference in different *individuals* of the same two species in crossing; so Sagaret believes this to be the case with different individuals of the same two species in being grafted together. As in reciprocal crosses, the facility of effecting an union is often very far from equal, so it sometimes is in grafting; the common gooseberry, for instance, cannot be grafted on the currant, whereas the currant will take, though with difficulty, on the gooseberry.

We have seen that the sterility of hybrids, which have their reproductive organs in an imperfect condition, is a very different case from the difficulty of uniting two pure species, which have their reproductive organs perfect; yet these two distinct cases run to a certain extent parallel. Something analogous occurs in grafting; for Thouin found that three species of Robinia, which seeded freely on their own roots, and which could be grafted with no great difficulty on another species, when thus grafted were rendered barren. On the other hand, certain species of Sorbus, when grafted on other species, yielded twice as much fruit as when on their own roots. We are reminded by this latter fact of the extraordinary case of Hippeastrum, Lobelia, &c., which seeded much more freely when fertilized with the pollen of distinct species, than when self-fertilized with their own pollen.

We thus see, that although there is a clear and fundamental difference between the mere adhesion of grafted stocks, and the union of the male and female elements in the act of reproduction, yet that there is a rude degree of parallelism in the results of grafting and of crossing distinct species. And as we must look at the curious and complex laws governing the facility with which trees can be grafted on each other as incidental on unknown differences in their vegetative systems, so I believe that the still more complex laws governing the facility of first crosses, are incidental on unknown differences, chiefly in their reproductive systems. These differences, in both cases, follow to a certain extent, as might have been expected, systematic affinity, by which every kind of resemblance and dissimilarity between organic beings is attempted to be expressed. The facts by no means seem to me to indicate that the greater or lesser difficulty of either grafting or crossing together various species has been a special endowment; although in the case of crossing, the difficulty is as important for the endurance and stability of specific forms, as in the case of grafting it is unimportant for their welfare.

Causes of the Sterility of first Crosses and of Hybrids. We may now look a little closer at the probable causes of the sterility of first crosses and of hybrids. These two cases are fundamentally different, for, as just remarked, in the union of two pure species the male and female sexual elements are perfect, whereas in hybrids they are imperfect. Even in first crosses, the greater or lesser difficulty in effecting a union apparently depends on several distinct causes. There must sometimes be a physical impossibility in the male element reaching the ovule, as would be the case with a plant having a pistil too long for the pollen-tubes to reach the ovarium. It has also been observed that when pollen of one species is placed on the stigma of a distantly allied species, though the pollen-tubes protrude, they do not penetrate the stigmatic surface. Again, the male element may reach the female element, but be incapable of causing an embryo to be developed, as seems to have been the case with some of Thuret's experiments on Fuci. No explanation can be given of these facts, any more than why certain trees cannot be grafted on others. Lastly, an embryo may be developed, and then perish at an early period. This latter alternative has not been sufficiently attended to; but I believe, from observations communicated to me by Mr. Hewitt, who has had great experience in hybridizing gallinaceous birds, that the early death of the embryo is a very frequent cause of sterility in first crosses. I was at first very unwilling to believe in this view; as hybrids, when once born, are generally healthy and long-lived, as we see in the case of the common mule. Hybrids, however, are differently circumstanced before and after birth: when born and living in a country where their two parents can live, they are generally placed under suitable conditions of life. But a hybrid partakes of only half of the nature and constitution of its mother, and therefore before birth, as long as it is nourished within its mother's womb or within the egg or seed produced by the mother, it may be exposed to conditions in some degree unsuitable, and consequently be liable to perish at an early period; more especially as all very young beings seem eminently sensitive to injurious or unnatural conditions of life.

In regard to the sterility of hybrids, in which the sexual elements are imperfectly developed, the case is very different. I have more than once alluded to a large body of facts, which I have collected, showing that when animals and plants are removed from their natural conditions, they are extremely liable to have their reproductive systems seriously affected. This, in fact, is the great bar to the domestication of animals. Between the sterility thus superinduced and that of hybrids, there are many points of similarity. In both cases the sterility is independent of general health, and is often accompanied by excess of size or great luxuriance. In both cases, the sterility occurs in various degrees; in both, the male element is the most liable to be affected; but sometimes the female more than the male. In both, the tendency goes to a certain extent with systematic affinity, or whole groups of animals and plants are rendered impotent by the same unnatural conditions; and whole groups of species tend to produce sterile hybrids. On the other hand, one species in a group will sometimes resist great changes of conditions with unimpaired fertility; and certain species in a group will produce unusually fertile hybrids. No one can tell, till he tries, whether any particular animal will breed under confinement or any plant seed freely under culture; nor can he tell, till he tries, whether any two species of a genus will produce more or less sterile hybrids. Lastly, when organic beings are placed during several generations under conditions not natural to them, they are extremely liable to vary, which is due, as I believe, to their reproductive systems having been specially affected, though in a lesser degree than when sterility ensues. So it is with hybrids, for hybrids in successive generations are eminently liable to vary, as every experimentalist has observed.

Thus we see that when organic beings are placed under new and unnatural conditions, and when hybrids are produced by the unnatural crossing of two species, the reproductive system, independently of the general state of health, is affected by sterility in a very similar manner. In the one case, the conditions of life have been disturbed, though often in so slight a degree as to be inappreciable by us; in the other case, or that of hybrids, the external conditions have remained the same, but the organization has been disturbed by two different structures and constitutions having been blended into one. For it is scarcely possible that two organizations should be compounded into one, without some disturbance occurring in the development, or periodical action, or mutual relation of the different parts and organs one to another, or to the conditions of life. When hybrids are able to breed *inter se*, they transmit to their offspring from generation to generation the same compounded organization, and hence we need not be surprised that their sterility, though in some degree variable, rarely diminishes.

It must, however, be confessed that we cannot understand, excepting on vague hypotheses, several facts with respect to the sterility of hybrids; for instance, the unequal fertility of hybrids produced from reciprocal crosses; or the increased sterility in those hybrids which occasionally and exceptionally resemble closely either pure parent. Nor do I pretend that the foregoing remarks go to the root of the matter: no explanation is offered why an organism, when placed under unnatural conditions, is rendered sterile. All that I have attempted to show, is that in two cases, in some respects allied, sterility is the common result, − in the one case from the conditions of life having been disturbed, in the other case from the organization having been disturbed by two organizations having been compounded into one.

It may seem fanciful, but I suspect that a similar parallelism extends to an allied yet very different class of facts. It is an old and almost universal belief, founded, I think, on a considerable body of evidence, that slight changes in the conditions of life are beneficial to all living things. We see this acted on by farmers and gardeners in their frequent exchanges of seed, tubers, &c., from one soil or climate to another, and back again. During the convalescence of animals, we plainly see that great benefit is derived from almost any change in the habits of life. Again, both with plants and animals, there is abundant evidence, that a cross between very distinct individuals of the same species, that is between members of different strains or sub-breeds, gives vigour and fertility to the offspring. I believe, indeed, from the facts alluded to in our fourth chapter, that a certain amount of crossing is indispensable even with hermaphrodites; and that close interbreeding continued during several generations between the nearest relations, especially if these be kept under the same conditions of life, always induces weakness and sterility in the progeny.

Hence it seems that, on the one hand, slight changes in the conditions of life benefit all organic beings, and on the other hand, that slight crosses, that is crosses between the males and females of the same species which have varied and become slightly different, give vigour and fertility to the offspring. But we have seen that greater changes, or changes of a particular nature, often render organic beings in some degree sterile; and that greater crosses, that is crosses between males and females which have become widely or specifically different, produce hybrids which are generally sterile in some degree. I cannot persuade myself that this parallelism is an accident or an illusion. Both series of facts seem to be connected together by some common but unknown bond, which is essentially related to the principle of life.

Fertility of Varieties when crossed, and of their Mongrel offspring. It may be urged, as a most forcible argument, that there must be some essential distinction between species and varieties, and that there must be some error in all the foregoing remarks, inasmuch as varieties, however much they may differ from each other in external appearance, cross with perfect facility, and yield perfectly fertile offspring. I fully admit that this is almost invariably the case. But if we look to varieties produced under nature, we are immediately involved in hopeless difficulties; for if two hitherto reputed varieties be found in any degree sterile together, they are at once ranked by most naturalists as species. For instance, the blue and red pimpernel, the primrose and cowslip, which are considered by many of our best botanists as varieties, are said by Gärtner not to be quite fertile when crossed, and he consequently ranks them as undoubted species. If we thus argue in a circle, the fertility of all varieties produced under nature will assuredly have to be granted.

If we turn to varieties, produced, or supposed to have been produced, under domestication, we are still involved in doubt. For when it is stated, for instance, that the German Spitz dog unites more easily than other dogs with foxes, or that certain South American indigenous domestic dogs do not readily cross with European dogs, the explanation which will occur to everyone, and probably the true one, is that these dogs have descended from several aboriginally distinct species. Nevertheless the perfect fertility of so many domestic varieties, differing widely from each other in appearance, for instance of the pigeon or of the cabbage, is a remarkable fact; more especially when we reflect how many species there are, which, though resembling each other most closely, are utterly sterile when intercrossed. Several considerations, however, render the fertility of domestic varieties less remarkable than at first appears. It can, in the first place, be clearly shown that mere external dissimilarity between two species does not determine their greater or lesser degree of sterility when crossed; and we may apply the same rule to domestic varieties. In the second place, some eminent naturalists believe that a long course of domestication tends to eliminate sterility in the successive generations of hybrids, which were at first only slightly sterile; and if this be so, we surely ought not to expect to find sterility both appearing and disappearing under nearly the same conditions of life. Lastly, and this seems to me by far the most important consideration, new races of animals and plants are produced under domestication by man's methodical and unconscious power of selection, for his own use and pleasure: he neither wishes to select, nor could select, slight differences in the reproductive system, or other constitutional difference correlated with the reproductive system. He supplies his several varieties with the same food; treats them in nearly the same manner, and does not wish to alter their general habits of life. Nature acts uniformly and slowly during vast periods of time on the whole organization, in any way which may be for each creature's own good; and thus she may, either directly, or more probably indirectly, through correlation, modify the reproductive system in the several descendants from any one species. Seeing this difference in the process of selection, as carried on by man and nature, we need not be surprised at some difference in the result.

I have as yet spoken as if the varieties of the same species were invariably fertile when intercrossed. But it seems to me impossible to resist the evidence of the existence of a certain amount of sterility in the few following cases, which I will briefly abstract. The evidence is at least as good as that from which we believe in the sterility of a multitude of species. The evidence is, also, derived from hostile witnesses, who in all other cases consider fertility and sterility as safe criterions of specific distinction. Gärtner kept during several years a dwarf kind of maize with yellow seeds, and a tall variety with red seeds, growing near each other in his garden; and although these plants have separated sexes, they never naturally crossed. He then fertilized thirteen flowers of the one with the pollen of the other; but only a single head produced any seed, and this one head produced only five grains. Manipulation in this case could not have been injurious, as the plants have separated sexes. No one, I believe, has suspected that these varieties of maize are distinct species; and it is important to notice that the hybrid plants thus raised were themselves *perfectly* fertile; so that even Gärtner did not venture to consider the two varieties as specifically distinct.

Girou de Buzareingues crossed three varieties of gourd, which like the maize has separated sexes, and he asserts that their mutual fertilization is by so much the less easy as their differences are greater. How far these experiments may be trusted, I know not; but the forms experimentized on, are ranked by Sagaret, who mainly founds his classification by the test of infertility, as varieties.

The following case is far more remarkable, and seems at first quite incredible; but it is the result of an astonishing number of experiments made during many years on nine species of Verbascum, by so good an observer and so hostile a witness, as Gärtner: namely, that yellow and white varieties of the same species of Verbascum when intercrossed produce less seed, than do either coloured varieties when fertilized with pollen from their own coloured flowers. Moreover, he asserts that when yellow and white varieties of one species are crossed with yellow and white varieties of a *distinct* species, more seed is produced by the crosses between the same coloured flowers, than between those which are differently coloured. Yet these varieties of Verbascum present no other difference besides the mere colour of the flower; and one variety can sometimes be raised from the seed of the other.

From observations which I have made on certain varieties of hollyhock, I am inclined to suspect that they present analogous facts.

Kölreuter, whose accuracy has been confirmed by every subsequent observer, has proved the remarkable fact, that one variety of the common tobacco is more fertile, when crossed with a widely distinct species, than are the other varieties. He experimentized on five forms, which are commonly reputed to be varieties, and which he tested by the severest trial, namely, by reciprocal crosses, and he found their mongrel offspring perfectly fertile. But one of these five varieties, when used either as father or mother, and crossed with the Nicotiana glutinosa, always yielded hybrids not so sterile as those which were produced from the four other varieties when crossed with N. gluti-nosa. Hence the reproductive system of this one variety must have been in some manner and in some degree modified.

From these facts; from the great difficulty of ascertaining the infertility of varieties in a state of nature, for a supposed variety if infertile in any degree would generally be ranked as species; from man selecting only external characters in the production of the most distinct domestic varieties, and from not wishing or being able to produce recondite and functional differences in the reproductive system; from these several considerations and facts, I do not think that the very general fertility of varieties can be proved to be of universal occurrence, or to form a fundamental distinction between varieties and species. The general fertility of varieties does not seem to me sufficient to overthrow the view which I have taken with respect to the very general, but not invariable, sterility of first crosses and of hybrids, namely, that it is not a special endowment, but is incidental on slowly acquired modifications, more especially in the reproduct-ive systems of the forms which are crossed.

Hybrids and Mongrels compared, independently of their fertility. Inde-pendently of the question of fertility, the offspring of species when crossed and of varieties when crossed may be compared in several other respects. Gärtner, whose strong wish was to draw a marked line of distinction between species and varieties, could find very few and, as it seems to me, quite unimportant differences between the so-called hybrid offspring of species, and the so-called mongrel offspring of varieties. And, on the other hand, they agree most closely in very many important respects.

I shall here discuss this subject with extreme brevity. The most important distinction is, that in the first generation mongrels are more variable than hybrids; but Gärtner admits that hybrids from species which have long been cultivated are often variable in the first gener-ation; and I have myself seen striking instances of this fact. Gärtner further admits that hybrids between very closely allied species are more variable than those from very distinct species; and this shows that the difference in the degree of variability graduates away.

When mongrels and the more fertile hybrids are propagated for several generations an extreme amount of variability in their offspring is notorious; but some few cases both of hybrids and mongrels long retaining uniformity of character could be given. The variability, however, in the successive generations of mongrels is, perhaps, greater than in hybrids.

This greater variability of mongrels than of hybrids does not seem to me at all surprising. For the parents of mongrels are varieties, and mostly domestic varieties (very few experiments having been tried on natural varieties), and this implies in most cases that there has been recent variability; and therefore we might expect that such variability would often continue and be super-added to that arising from the mere act of crossing. The slight degree of variability in hybrids from the first cross or in the first generation, in contrast with their extreme variability in the succeeding generations, is a curious fact and deserves attention. For it bears on and corroborates the view which I have taken on the cause of ordinary variability; namely, that it is due to the reproductive system being eminently sensitive to any change in the conditions of life, being thus often rendered either impotent or at least incapable of its proper function of producing offspring identical with the parent-form. Now hybrids in the first generation are descended from species (excluding those long cultivated) which have not had their reproductive systems in any way affected, and they are not variable; but hybrids themselves have their reproductive systems seriously affected, and their descendants are highly variable.

But to return to our comparison of mongrels and hybrids: Gärtner further insists that when any two species, although most closely allied to each other, are crossed with a third species, the hybrids are widely different from each other; whereas if two very distinct varieties of one species are crossed with another species, the hybrids do not differ much. But this conclusion, as far as I can make out, is founded on a single experiment; and seems directly opposed to the results of several experiments made by Kölreuter.

These alone are the unimportant differences, which Gärtner is able to point out, between hybrid and mongrel plants. On the other hand, the resemblance in mongrels and in hybrids to their respective parents, more especially in hybrids produced from nearly related species, follows according to Gärtner the same laws. When two species are crossed, one has sometimes a prepotent power of impressing its likeness on the hybrid; and so I believe it to be with varieties of plants. With animals one variety certainly often has this prepotent power over another variety. Hybrid plants produced from a reciprocal cross, generally resemble each other closely; and so it is with mongrels from a reciprocal cross. Both hybrids and mongrels can be reduced to either pure parent-form, by repeated crosses in successive generations with either parent.

These several remarks are apparently applicable to animals; but the subject is here excessively complicated, partly owing to the existence of secondary sexual characters; but more especially owing to prepotency in transmitting likeness running more strongly in one sex than in the other, both when one species is crossed with another, and when one variety is crossed with another variety. For instance, I think these authors are right, who maintain that the ass has a prepotent power over the horse, so that both the mule and the hinny more resemble the ass than the horse; but that the prepotency runs more strongly in the male-ass than in the female, so that the mule, which is the offspring of the male-ass and mare, is more like an ass, than is the hinny, which is the offspring of the female-ass and stallion.

Much stress has been laid by some authors on the supposed fact, that mongrel animals alone are born closely like one of their parents; but it can be shown that this does sometimes occur with hybrids; yet I grant much less frequently with hybrids than with mongrels. Looking to the cases which I have collected of cross-bred animals closely resembling one parent, the resemblances seem chiefly confined to characters almost monstrous in their nature, and which have suddenly appeared – such as albinism, melanism, deficiency of tail or horns, or additional fingers and toes; and do not relate to characters which have been slowly acquired by selection. Consequently, sudden reversions to the perfect character of either parent would be more likely to occur with mongrels, which are descended from varieties often suddenly produced and semi-monstrous in character, than with hybrids, which are descended from species slowly and naturally produced. On the whole I entirely agree with Dr. Prosper Lucas, who, after arranging an enormous body of facts with respect to animals, comes to the conclusion, that the laws of resemblance of the child to its parents are the same, whether the two parents differ much or little from each other, namely in the union of individuals of the same variety, or of different varieties, or of distinct species.

Laying aside the question of fertility and sterility, in all other respects there seems to be a general and close similarity in the offspring of crossed species, and of crossed varieties. If we look at species as having been specially created, and at varieties as having been produced by secondary laws, this similarity would be an astonishing fact. But it harmonizes perfectly with the view that there is no essential distinction between species and varieties.

Summary of Chapter. First crosses between forms sufficiently distinct to be ranked as species, and their hybrids, are very generally, but not universally, sterile. The sterility is of all degrees, and is often so slight that the two most careful experimentalists who have ever lived, have come to diametrically opposite conclusions in ranking forms by this test. The sterility is innately variable in individuals of the same species, and is eminently susceptible of favourable and unfavourable conditions. The degree of sterility does not strictly follow systematic affinity, but is governed by several curious and complex laws. It is generally different, and sometimes widely different, in reciprocal crosses between the same two species. It is not always equal in degree in a first cross and in the hybrid produced from this cross.

In the same manner as in grafting trees, the capacity of one species or variety to take on another, is incidental on generally unknown differences in their vegetative systems, so in crossing, the greater or less facility of one species to unite with another, is incidental on unknown differences in their reproductive systems. There is no more reason to think that species have been specially endowed with various degrees of sterility to prevent them crossing and blending in nature, than to think that trees have been specially endowed with various and somewhat analogous degrees of difficulty in being grafted together in order to prevent them becoming inarched in our forests.

The sterility of first crosses between pure species, which have their reproductive systems perfect, seems to depend on several circumstances; in some cases largely on the early death of the embryo. The sterility of hybrids, which have their reproductive systems imperfect, and which have had this system and their whole organization disturbed by being compounded of two distinct species, seems closely allied to that sterility which so frequently affects pure species, when their natural conditions of life have been disturbed. This view is supported by a parallelism of another kind; – namely, that the crossing of forms only slightly different is favourable to the vigour and fertility of their offspring; and that slight changes in the conditions of life are apparently favourable to the vigour and fertility of all organic beings. It is not surprising that the degree of difficulty in uniting two species, and the degree of sterility of their hybrid-offspring should generally correspond, though due to distinct causes; for both depend on the amount of difference of some kind between the species which are crossed. Nor is it surprising that the facility of effecting a first cross, the fertility of the hybrids produced, and the capacity of being grafted together – though this latter capacity evidently depends on widely different circumstances – should all run, to a certain extent, parallel with the systematic affinity of the forms which are subjected to experiment; for systematic affinity attempts to express all kinds of resemblance between all species.

First crosses between forms known to be varieties, or sufficiently alike to be considered as varieties, and their mongrel offspring, are very generally, but not quite universally, fertile. Nor is this nearly general and perfect fertility surprising, when we remember how liable we are to argue in a circle with respect to varieties in a state of nature; and when we remember that the greater number of varieties have been produced under domestication by the selection of mere external differences, and not of differences in the reproductive system. In all other respects, excluding fertility, there is a close general resemblance between hybrids and mongrels. Finally, then, the facts briefly given in this chapter do not seem to me opposed to, but even rather to support the view, that there is no fundamental distinction between species and varieties.

CHAPTER IX

ON THE IMPERFECTION OF
THE GEOLOGICAL RECORD

On the absence of intermediate varieties at the present day
– On the nature of extinct intermediate varieties; on their
number – On the vast lapse of time, as inferred from the
rate of deposition and of denudation – On the poorness of
our palaeontological collections – On the intermittence of
geological formations – On the absence of intermediate
varieties in any one formation – On the sudden appearance
of groups of species – On their sudden appearance in the
lowest known fossiliferous strata

IN the sixth chapter I enumerated the chief objections which might
be justly urged against the views maintained in this volume. Most
of them have now been discussed. One, namely the distinctness of
specific forms, and their not being blended together by innumerable
transitional links, is a very obvious difficulty. I assigned reasons why
such links do not commonly occur at the present day, under the
circumstances apparently most favourable for their presence, namely
on an extensive and continuous area with graduated physical conditions. I endeavoured to show, that the life of each species depends in a
more important manner on the presence of other already defined
organic forms, than on climate; and, therefore, that the really
governing conditions of life do not graduate away quite insensibly
like heat or moisture. I endeavoured, also, to show that intermediate
varieties, from existing in lesser numbers than the forms which they
connect, will generally be beaten out and exterminated during
the course of further modification and improvement. The main
cause, however, of innumerable intermediate links not now occurring everywhere throughout nature depends on the very process
of natural selection, through which new varieties continually take
the places of and exterminate their parent-forms. But just in proportion as this process of extermination has acted on an enormous

scale, so must the number of intermediate varieties, which have formerly existed on the earth, be truly enormous. Why then is not every geological formation and every stratum full of such intermediate links? Geology assuredly does not reveal any such finely graduated organic chain; and this, perhaps, is the most obvious and gravest objection which can be urged against my theory. The explanation lies, as I believe, in the extreme imperfection of the geological record.

In the first place it should always be borne in mind what sort of intermediate forms must, on my theory, have formerly existed. I have found it difficult, when looking at any two species, to avoid picturing to myself, forms *directly* intermediate between them. But this is a wholly false view; we should always look for forms intermediate between each species and a common but unknown progenitor; and the progenitor will generally have differed in some respects from all its modified descendants. To give a simple illustration: the fantail and pouter pigeons have both descended from the rock-pigeon; if we possessed all the intermediate varieties which have ever existed, we should have an extremely close series between both and the rock-pigeon; but we should have no varieties directly intermediate between the fantail and pouter; none, for instance, combining a tail somewhat expanded with a crop somewhat enlarged, the characteristic features of these two breeds. These two breeds, moreover, have become so much modified, that if we had no historical or indirect evidence regarding their origin, it would not have been possible to have determined from a mere comparison of their structure with that of the rock-pigeon, whether they had descended from this species or from some other allied species, such as C. oenas.

So with natural species, if we look to forms very distinct, for instance to the horse and tapir, we have no reason to suppose that links ever existed directly intermediate between them, but between each and an unknown common parent. The common parent will have had in its whole organization much general resemblance to the tapir and to the horse; but in some points of structure may have differed considerably from both, even perhaps more than they differ from each other. Hence in all such cases, we should be unable to recognize the parent-form of any two or more species, even if we closely compared the structure of the parent with that of its modified descendants, unless at the same time we had a nearly perfect chain of the intermediate links.

It is just possible by my theory, that one of two living forms might have descended from the other; for instance, a horse from a tapir; and in this case *direct* intermediate links will have existed between them. But such a case would imply that one form had remained for a very long period unaltered, whilst its descendants had undergone a vast amount of change; and the principle of competition between organism and organism, between child and parent, will render this a very rare event; for in all cases the new and improved forms of life will tend to supplant the old and unimproved forms.

By the theory of natural selection all living species have been connected with the parent-species of each genus, by differences not greater than we see between the varieties of the same species at the present day; and these parent-species, now generally extinct, have in their turn been similarly connected with more ancient species; and so on backwards, always converging to the common ancestor of each great class. So that the number of intermediate and transitional links, between all living and extinct species, must have been inconceivably great. But assuredly, if this theory be true, such have lived upon this earth.

On the lapse of Time. Independently of our not finding fossil remains of such infinitely numerous connecting links, it may be objected, that time will not have sufficed for so great an amount of organic change, all changes having been effected very slowly through natural selection. It is hardly possible for me even to recall to the reader, who may not be a practical geologist, the facts leading the mind feebly to comprehend the lapse of time. He who can read Sir Charles Lyell's grand work on the Principles of Geology, which the future historian will recognize as having produced a revolution in natural science, yet does not admit how incomprehensibly vast have been the past periods of time, may at once close this volume. Not that it suffices to study the Principles of Geology, or to read special treatises by different observers on separate formations, and to mark how each author attempts to give an inadequate idea of the duration of each formation or even each stratum. A man must for years examine for himself great piles of superimposed strata, and watch the sea at work grinding down old rocks and making fresh sediment, before he can hope to comprehend anything of the lapse of time, the monuments of which we see around us.

It is good to wander along lines of sea-coast, when formed of moderately hard rocks, and mark the process of degradation. The tides in most cases reach the cliffs only for a short time twice a day, and the waves eat into them only when they are charged with sand or pebbles; for there is reason to believe that pure water can effect little or nothing in wearing away rock. At last the base of the cliff is undermined, huge fragments fall down, and these remaining fixed, have to be worn away, atom by atom, until reduced in size they can be rolled about by the waves, and then are more quickly ground into pebbles, sand, or mud. But how often do we see along the bases of retreating cliffs rounded boulders, all thickly clothed by marine productions, showing how little they are abraded and how seldom they are rolled about! Moreover, if we follow for a few miles any line of rocky cliff, which is undergoing degradation, we find that it is only here and there, along a short length or round a promontory, that the cliffs are at the present time suffering. The appearance of the surface and the vegetation show that elsewhere years have elapsed since the waters washed their base.

He who most closely studies the action of the sea on our shores, will, I believe, be most deeply impressed with the slowness with which rocky coasts are worn away. The observations on this head by Hugh Miller, and by that excellent observer Mr. Smith of Jordan Hill, are most impressive. With the mind thus impressed, let any one examine beds of conglomerate many thousand feet in thickness, which, though probably formed at a quicker rate than many other deposits, yet, from being formed of worn and rounded pebbles, each of which bears the stamp of time, are good to show how slowly the mass has been accumulated. Let him remember Lyell's profound remark, that the thickness and extent of sedimentary formations are the result and measure of the degradation which the earth's crust has elsewhere suffered. And what an amount of degradation is implied by the sedimentary deposits of many countries! Professor Ramsay has given me the maximum thickness, in most cases from actual measurement, in a few cases from estimate, of each formation in different parts of Great Britain; and this is the result:

	Feet
Palaeozoic strata (not including igneous beds)	57,154
Secondary strata	13,190
Tertiary strata	2,240

– making altogether 72,584 feet; that is, very nearly thirteen and three-quarters British miles. Some of these formations, which are represented in England by thin beds, are thousands of feet in thickness on the Continent. Moreover, between each successive formation, we have, in the opinion of most geologists, enormously long blank periods. So that the lofty pile of sedimentary rocks in Britain, gives but an inadequate idea of the time which has elapsed during their accumulation; yet what time this must have consumed! Good observers have estimated that sediment is deposited by the great Mississippi river at the rate of only 600 feet in a hundred thousand years. This estimate may be quite erroneous; yet, considering over what wide spaces very fine sediment is transported by the currents of the sea, the process of accumulation in any one area must be extremely slow.

But the amount of denudation which the strata have in many places suffered, independently of the rate of accumulation of the degraded matter, probably offers the best evidence of the lapse of time. I remember having been much struck with the evidence of denudation, when viewing volcanic islands, which have been worn by the waves and pared all round into perpendicular cliffs of one or two thousand feet in height; for the gentle slope of the lava-streams, due to their formerly liquid state, showed at a glance how far the hard, rocky beds had once extended into the open ocean. The same story is still more plainly told by faults, – those great cracks along which the strata have been upheaved on one side, or thrown down on the other, to the height or depth of thousands of feet; for since the crust cracked, the surface of the land has been so completely planed down by the action of the sea, that no trace of these vast dislocations is externally visible.

The Craven fault, for instance, extends for upwards of 30 miles, and along this line the vertical displacement of the strata has varied from 600 to 3000 feet. Prof. Ramsay has published an account of a downthrow in Anglesea of 2300 feet; and he informs me that he fully believes there is one in Merionethshire of 12,000 feet; yet in these cases there is nothing on the surface to show such prodigious movements; the pile of rocks on the one or other side having been smoothly swept away. The consideration of these facts impresses my mind almost in the same manner as does the vain endeavour to grapple with the idea of eternity.

I am tempted to give one other case, the well-known one of the denudation of the Weald. Though it must be admitted that the denudation of the Weald has been a mere trifle, in comparison with that which has removed masses of our palaeozoic strata, in parts ten thousand feet in thickness, as shown in Prof. Ramsay's masterly memoir on this subject. Yet it is an admirable lesson to stand on the North Downs and to look at the distant South Downs; for, remembering that at no great distance to the west the northern and southern escarpments meet and close, one can safely picture to oneself the great dome of rocks which must have covered up the Weald within so limited a period as since the latter part of the Chalk formation. The distance from the northern to the southern Downs is about 22 miles, and the thickness of the several formations is on an average about 1100 feet, as I am informed by Prof. Ramsay. But if, as some geologists suppose, a range of older rocks underlies the Weald, on the flanks of which the overlying sedimentary deposits might have accumulated in thinner masses than elsewhere, the above estimate would be erroneous; but this source of doubt probably would not greatly affect the estimate as applied to the western extremity of the district. If, then, we knew the rate at which the sea commonly wears away a line of cliff of any given height, we could measure the time requisite to have denuded the Weald. This, of course, cannot be done; but we may, in order to form some crude notion on the subject, assume that the sea would eat into cliffs 500 feet in height at the rate of one inch in a century. This will at first appear much too small an allowance; but it is the same as if we were to assume a cliff one yard in height to be eaten back along a whole line of coast at the rate of one yard in nearly every twenty-two years. I doubt whether any rock, even as soft as chalk, would yield at this rate excepting on the most exposed coasts;

though no doubt the degradation of a lofty cliff would be more rapid from the breakage of the fallen fragments. On the other hand, I do not believe that any line of coast, ten or twenty miles in length, ever suffers degradation at the same time along its whole indented length; and we must remember that almost all strata contain harder layers or nodules, which from long resisting attrition form a breakwater at the base. Hence, under ordinary circumstances, I conclude that for a cliff 500 feet in height, a denudation of one inch per century for the whole length would be an ample allowance. At this rate, on the above data, the denudation of the Weald must have required 306,662,400 years; or say three hundred million years.

The action of fresh water on the gently inclined Wealden district, when upraised, could hardly have been great, but it would somewhat reduce the above estimate. On the other hand, during oscillations of level, which we know this area has undergone, the surface may have existed for millions of years as land, and thus have escaped the action of the sea: when deeply submerged for perhaps equally long periods, it would, likewise, have escaped the action of the coast-waves. So that in all probability a far longer period than 300 million years has elapsed since the latter part of the Secondary period.

I have made these few remarks because it is highly important for us to gain some notion, however imperfect, of the lapse of years. During each of these years, over the whole world, the land and the water has been peopled by hosts of living forms. What an infinite number of generations, which the mind cannot grasp, must have succeeded each other in the long roll of years! Now turn to our richest geological museums, and what a paltry display we behold!

On the poorness of our Palaeontological collections. That our palaeontological collections are very imperfect, is admitted by every one. The remark of that admirable palaeontologist, the late Edward Forbes, should not be forgotten, namely, that numbers of our fossil species are known and named from single and often broken specimens, or from a few specimens collected on some one spot. Only a small portion of the surface of the earth has been geologically explored, and no part with sufficient care, as the important discoveries made every year in Europe prove. No organism wholly soft can be preserved. Shells and bones will decay and disappear when left on the bottom of the sea, where sediment is not accumulating. I believe we are continually taking a most erroneous view, when we tacitly admit to ourselves that sediment is being deposited over nearly the whole bed of the sea, at a rate sufficiently quick to embed and preserve fossil remains. Throughout an enormously large proportion of the ocean, the bright blue tint of the water bespeaks its purity. The many cases on record of a formation conformably covered, after an enormous interval of time, by another and later formation, without the underlying bed having suffered in the interval any wear and tear, seem explicable only on the view of the bottom of the sea not rarely lying for ages in an unaltered condition. The remains which do become embedded, if in sand or gravel, will when the beds are upraised generally be dissolved by the percolation of rain-water. I suspect that but few of the very many animals which live on the beach between high and low water-mark are preserved. For instance, the several species of the Chthamalinae (a sub-family of sessile cirri-pedes) coat the rocks all over the world in infinite numbers: they are all strictly littoral, with the exception of a single Mediterranean species, which inhabits deep water and has been found fossil in Sicily, whereas not one other species has hitherto been found in any tertiary formation: yet it is now known that the genus Chthamalus existed during the chalk period. The molluscan genus Chiton offers a par-tially analogous case.

With respect to the terrestrial productions which lived during the Secondary and Palaeozoic periods, it is superfluous to state that our evidence from fossil remains is fragmentary in an extreme degree. For instance, not a land shell is known belonging to either of these vast periods, with one exception discovered by Sir C. Lyell in the carboniferous strata of North America. In regard to mammiferous remains, a single glance at the historical table published in the Supplement to Lyell's Manual, will bring home the truth, how accidental and rare is their preservation, far better than pages of detail. Nor is their rarity surprising, when we remember how large a proportion of the bones of tertiary mammals have been discovered either in caves or in lacustrine deposits; and that not a cave or true lacustrine bed is known belonging to the age of our secondary or palaeozoic formations.

But the imperfection in the geological record mainly results from another and more important cause than any of the foregoing; namely, from the several formations being separated from each other by wide intervals of time. When we see the formations tabulated in written works, or when we follow them in nature, it is difficult to avoid believing that they are closely consecutive. But we know, for instance, from Sir R. Murchison's great work on Russia, what wide gaps there are in that country between the superimposed formations; so it is in North America, and in many other parts of the world. The most skilful geologist, if his attention had been exclusively confined to these large territories, would never have suspected that during the periods which were blank and barren in his own country, great piles of sediment, charged with new and peculiar forms of life, had elsewhere been accumulated. And if in each separate territory, hardly any idea can be formed of the length of time which has elapsed between the consecutive formations, we may infer that this could nowhere be ascertained. The frequent and great changes in the mineralogical composition of consecutive formations, generally implying great changes in the geography of the surrounding lands, whence the sediment has been derived, accords with the belief of vast intervals of time having elapsed between each formation.

But we can, I think, see why the geological formations of each region are almost invariably intermittent; that is, have not followed each other in close sequence. Scarcely any fact struck me more when examining many hundred miles of the South American coasts, which have been upraised several hundred feet within the recent period, than the absence of any recent deposits sufficiently extensive to last for even a short geological period. Along the whole west coast, which is inhabited by a peculiar marine fauna, tertiary beds are so scantily developed, that no record of several successive and peculiar marine faunas will probably be preserved to a distant age. A little reflection will explain why along the rising coast of the western side of South America, no extensive formations with recent or tertiary remains can anywhere be found, though the supply of sediment must for ages have been great, from the enormous degradation of the coast-rocks and from muddy streams entering the sea. The explanation, no doubt, is, that the littoral and sub-littoral deposits are continually worn away, as soon as they are brought up by the slow and gradual rising of the land within the grinding action of the coast-waves.

We may, I think, safely conclude that sediment must be accumulated in extremely thick, solid, or extensive masses, in order to withstand the incessant action of the waves, when first upraised and during subsequent oscillations of level. Such thick and extensive accumulations of sediment may be formed in two ways; either, in profound depths of the sea, in which case, judging from the researches of E. Forbes, we may conclude that the bottom will be inhabited by extremely few animals, and the mass when upraised will give a most imperfect record of the forms of life which then existed; or, sediment may be accumulated to any thickness and extent over a shallow bottom, if it continue slowly to subside. In this latter case, as long as the rate of subsidence and supply of sediment nearly balance each other, the sea will remain shallow and favourable for life, and thus a fossiliferous formation thick enough, when upraised, to resist any amount of degradation, may be formed.

I am convinced that all our ancient formations, which are rich in fossils, have thus been formed during subsidence. Since publishing my views on this subject in 1845, I have watched the progress of Geology, and have been surprised to note how author after author, in treating of this or that great formation, has come to the conclusion that it was accumulated during subsidence. I may add, that the only ancient tertiary formation on the west coast of South America, which has been bulky enough to resist such degradation as it has as yet suffered, but which will hardly last to a distant geological age, was certainly deposited during a downward oscillation of level, and thus gained considerable thickness.

All geological facts tell us plainly that each area has undergone numerous slow oscillations of level, and apparently these oscillations have affected wide spaces. Consequently formations rich in fossils and sufficiently thick and extensive to resist subsequent degradation, may have been formed over wide spaces during periods of subsidence, but only where the supply of sediment was sufficient to keep the sea shallow and to embed and preserve the remains before they had time to decay. On the other hand, as long as the bed of the sea remained stationary, *thick* deposits could not have been accumulated in the shallow parts, which are the most favourable to life. Still less could this have happened during the alternate periods of elevation; or, to speak more accurately, the beds which were then accumulated will have been destroyed by being upraised and brought within the limits of the coast-action.

Thus the geological record will almost necessarily be rendered intermittent. I feel much confidence in the truth of these views, for they are in strict accordance with the general principles inculcated by Sir C. Lyell; and E. Forbes independently arrived at a similar conclusion.

One remark is here worth a passing notice. During periods of elevation the area of the land and of the adjoining shoal parts of the sea will be increased, and new stations will often be formed; – all circumstances most favourable, as previously explained, for the formation of new varieties and species; but during such periods there will generally be a blank in the geological record. On the other hand, during subsidence, the inhabited area and number of inhabitants will decrease (excepting the productions on the shores of a continent when first broken up into an archipelago), and consequently during subsidence, though there will be much extinction, fewer new varieties or species will be formed; and it is during these very periods of subsidence, that our great deposits rich in fossils have been accumulated. Nature may almost be said to have guarded against the frequent discovery of her transitional or linking forms.

From the foregoing considerations it cannot be doubted that the geological record, viewed as a whole, is extremely imperfect; but if we confine our attention to any one formation, it becomes more difficult to understand, why we do not therein find closely graduated varieties between the allied species which lived at its commencement and at its close. Some cases are on record of the same species presenting distinct varieties in the upper and lower parts of the same formation, but, as they are rare, they may be here passed over. Although each formation has indisputably required a vast number of years for its deposition, I can see several reasons why each should not include a graduated series of links between the species which then lived; but I can by no means pretend to assign due proportional wieght to the following considerations.

Although each formation may mark a very long lapse of years, each perhaps is short compared with the period requisite to change one species into another. I am aware that two palaeontologists, whose opinions are worthy of much deference, namely Bronn and Wood-ward, have concluded that the average duration of each formation is twice or thrice as long as the average duration of specific forms. But insuperable difficulties, as it seems to me, prevent us coming to any just conclusion on this head. When we see a species first appearing in the middle of any formation, it would be rash in the extreme to infer that it had not elsewhere previously existed. So again when we find a species disappearing before the uppermost layers have been de-posited, it would be equally rash to suppose that it then became wholly extinct. We forget how small the area of Europe is compared with the rest of the world; nor have the several stages of the same formation throughout Europe been correlated with perfect accuracy.

With marine animals of all kinds, we may safely infer a large amount of migration during climatal and other changes; and when we see a species first appearing in any formation, the probability is that it only then first immigrated into that area. It is well known, for instance, that several species appeared somewhat earlier in the palaeo-zoic beds of North America than in those of Europe; time having apparently been required for their migration from the American to the European seas. In examining the latest deposits of various quarters of the world, it has everywhere been noted, that some few still existing species are common in the deposit, but have become extinct in the immediately surrounding sea; or, conversely, that some are now abundant in the neighbouring sea, but are rare or absent in this particular deposit. It is an excellent lesson to reflect on the ascertained amount of migration of the inhabitants of Europe during the Glacial period, which forms only a part of one whole geological period; and likewise to reflect on the great changes of level, on the inordinately great change of climate, on the prodigious lapse of time, all included within this same glacial period. Yet it may be doubted whether in any quarter of the world, sedimentary deposits, *including fossil remains*, have gone on accumulating within the same area during the whole of this period. It is not, for instance, probable that sediment was deposited during the whole of the glacial period near the mouth of the Mississippi, within that limit of depth at which marine animals can flourish; for we know what vast geographical changes occurred in other parts of America during this space of time. When such beds as were deposited in shallow water near the mouth of the Mississippi during some part of the glacial period shall have been upraised, organic remains will probably first appear and disappear at different levels, owing to the migration of species and to geographical changes. And in the distant future, a geologist examining these beds, might be tempted to conclude that the average duration of life of the embedded fossils had been less than that of the glacial period, instead of having been really far greater, that is extending from before the glacial epoch to the present day.

In order to get a perfect gradation between two forms in the upper and lower parts of the same formation, the deposit must have gone on accumulating for a very long period, in order to have given sufficient time for the slow process of variation; hence the deposit will generally have to be a very thick one; and the species undergoing modification will have had to live on the same area throughout this whole time. But we have seen that a thick fossiliferous formation can only be accumulated during a period of subsidence; and to keep the depth approximately the same, which is necessary in order to enable the same species to live on the same space, the supply of sediment must nearly have counterbalanced the amount of subsidence. But this same movement of subsidence will often tend to sink the area whence the sediment is derived, and thus diminish the supply whilst the downward movement continues. In fact, this nearly exact balancing between the supply of sediment and the amount of subsidence is probably a rare contingency; for it has been observed by more than one palaeontologist, that very thick deposits are usually barren of organic remains, except near their upper or lower limits.

It would seem that each separate formation, like the whole pile of formations in any country, has generally been intermittent in its accumulation. When we see, as is so often the case, a formation composed of beds of different mineralogical composition, we may reasonably suspect that the process of deposition has been much interrupted, as a change in the currents of the sea and a supply of sediment of a different nature will generally have been due to geo-graphical changes requiring much time. Nor will the closest inspection of a formation give any idea of the time which its deposition has consumed. Many instances could be given of beds only a few feet in thickness, representing formations, elsewhere thousands of feet in thickness, and which must have required an enormous period for their accumulation; yet no one ignorant of this fact would have suspected the vast lapse of time represented by the thinner formation. Many cases could be given of the lower beds of a formation having been upraised, denuded, submerged, and then re-covered by the upper beds of the same formation, – facts, showing what wide, yet easily overlooked, intervals have occurred in its accumulation. In other cases we have the plainest evidence in great fossilized trees, still standing upright as they grew, of many long intervals of time and changes of level during the process of deposition, which would never even have been suspected, had not the trees chanced to have been preserved: thus, Messrs. Lyell and Dawson found carboniferous beds 1400 feet thick in Nova Scotia, with ancient root-bearing strata, one above the other, at no less than sixty-eight different levels. Hence, when the same species occur at the bottom, middle, and top of a formation, the probability is that they have not lived on the same spot during the whole period of deposition, but have disappeared and reappeared, perhaps many times, during the same geological period. So that if such species were to undergo a considerable amount of modification during any one geological period, a section would not probably include all the fine intermediate gradations which must on my theory have existed between them, but abrupt, though perhaps very slight, changes of form.

It is all-important to remember that naturalists have no golden rule by which to distinguish species and varieties; they grant some little variability to each species, but when they meet with a somewhat greater amount of difference between any two forms, they rank both as species, unless they are enabled to connect them together by close intermediate gradations. And this from the reasons just assigned we can seldom hope to effect in any one geological section. Supposing B and C to be two species, and a third, A, to be found in an underlying bed; even if A were strictly intermediate between B and

C, it would simply be ranked as a third and distinct species, unless at the same time it could be most closely connected with either one or both forms by intermediate varieties. Nor should it be forgotten, as before explained, that A might be the actual progenitor of B and C, and yet might not at all necessarily be strictly intermediate between them in all points of structure. So that we might obtain the parent-species and its several modified descendants from the lower and upper beds of a formation, and unless we obtained numerous transitional gradations, we should not recognize their relationship, and should consequently be compelled to rank them all as distinct species.

It is notorious on what excessively slight differences many palae-ontologists have founded their species; and they do this the more readily if the specimens come from different sub-stages of the same formation. Some experienced conchologists are now sinking many of the very fine species of D'Orbigny and others into the rank of varieties; and on this view we do find the kind of evidence of change which on my theory we ought to find. Moreover, if we look to rather wider intervals, namely, to distinct but consecutive stages of the same great formation, we find that the embedded fossils, though almost universally ranked as specifically different, yet are far more closely allied to each other than are the species found in more widely separated formations; but to this subject I shall have to return in the following chapter.

One other consideration is worth notice: with animals and plants that can propagate rapidly and are not highly locomotive, there is reason to suspect, as we have formerly seen, that their varieties are generally at first local; and that such local varieties do not spread widely and supplant their parent-forms until they have been modified and perfected in some considerable degree. According to this view, the chance of discovering in a formation in any one country all the early stages of transition between any two forms, is small, for the successive changes are supposed to have been local or confined to some one spot. Most marine animals have a wide range; and we have seen that with plants it is those which have the widest range, that oftenest present varieties; so that with shells and other marine animals, it is probably those which have had the widest range, far exceeding the limits of the known geological formations of Europe, which have oftenest given rise, first to local varieties and ultimately to new species; and this again would greatly lessen the chance of our being able to trace the stages of transition in any one geological formation.

It should not be forgotten, that at the present day, with perfect specimens for examination, two forms can seldom be connected by intermediate varieties and thus proved to be the same species, until many specimens have been collected from many places; and in the case of fossil species this could rarely be effected by palaeontologists. We shall, perhaps, best perceive the improbability of our being enabled to connect species by numerous, fine, intermediate, fossil links, by asking ourselves whether, for instance, geologists at some future period will be able to prove, that our different breeds of cattle, sheep, horses, and dogs have descended from a single stock or from several aboriginal stocks; or, again, whether certain sea-shells inhabiting the shores of North America, which are ranked by some conchologists as distinct species from their European representatives, and by other conchologists as only varieties, are really varieties or are, as it is called, specifically distinct. This could be effected only by the future geologist discovering in a fossil state numerous intermediate gradations; and such success seems to me improbable in the highest degree.

Geological research, though it has added numerous species to existing and extinct genera, and has made the intervals between some few groups less wide than they otherwise would have been, yet has done scarcely anything in breaking down the distinction between species, by connecting them together by numerous, fine, intermediate varieties; and this not having been effected, is probably the gravest and most obvious of all the many objections which may be urged against my views. Hence it will be worth while to sum up the foregoing remarks, under an imaginary illustration. The Malay Archipelago is of about the size of Europe from the North Cape to the Mediterranean, and from Britain to Russia; and therefore equals all the geological formations which have been examined with any accuracy, excepting those of the United States of America. I fully agree with Mr. Godwin-Austen, that the present condition of the Malay Archipelago, with its numerous large islands separated by wide and shallow seas, probably represents the former state of Europe, when most of our formations were accumulating.

The Malay Archipelago is one of the richest regions of the whole world in organic beings; yet if all the species were to be collected which have ever lived there, how imperfectly would they represent the natural history of the world!

But we have every reason to believe that the terrestrial productions of the archipelago would be preserved in an excessively imperfect manner in the formations which we suppose to be there accumulating. I suspect that not many of the strictly littoral animals, or of those which lived on naked submarine rocks, would be embedded; and those embedded in gravel or sand, would not endure to a distant epoch. Wherever sediment did not accumulate on the bed of the sea, or where it did not accumulate at a sufficient rate to protect organic bodies from decay, no remains could be preserved.

In our archipelago, I believe that fossiliferous formations could be formed of sufficient thickness to last to an age, as distant in futurity as the secondary formations lie in the past, only during periods of subsidence. These periods of subsidence would be separated from each other by enormous intervals, during which the area would be either stationary or rising; whilst rising, each fossiliferous formation would be destroyed, almost as soon as accumulated, by the incessant coast-action, as we now see on the shores of South America. During the periods of subsidence there would probably be much extinction of life; during the periods of elevation, there would be much variation, but the geological record would then be least perfect.

It may be doubted whether the duration of any one great period of subsidence over the whole or part of the archipelago, together with a contemporaneous accumulation of sediment, would *exceed* the average duration of the same specific forms; and these contingencies are indispensable for the preservation of all the transitional gradations between any two or more species. If such gradations were not fully preserved, transitional varieties would merely appear as so many distinct species. It is, also, probable that each great period of subsidence would be interrupted by oscillations of level, and that slight climatal changes would intervene during such lengthy periods; and in these cases the inhabitants of the archipelago would have to migrate, and no closely consecutive record of their modifications could be preserved in any one formation.

Very many of the marine inhabitants of the archipelago now range thousands of miles beyond its confines; and analogy leads me to believe that it would be chiefly these far-ranging species which would oftenest produce new varieties; and the varieties would at first generally be local or confined to one place, but if possessed of any decided advantage, or when further modified and improved, they would slowly spread and supplant their parent-forms. When such varieties returned to their ancient homes, as they would differ from their former state, in a nearly uniform, though perhaps extremely slight degree, they would, according to the principles followed by many palaeontologists, be ranked as new and distinct species.

If then, there be some degree of truth in these remarks, we have no right to expect to find in our geological formations, an infinite number of those fine transitional forms, which on my theory assuredly have connected all the past and present species of the same group into one long and branching chain of life. We ought only to look for a few links, some more closely, some more distantly related to each other; and these links, let them be ever so close, if found in different stages of the same formation, would, by most palaeontologists, be ranked as distinct species. But I do not pretend that I should ever have suspected how poor a record of the mutations of life, the best preserved geological section presented, had not the difficulty of our not discovering innumerable transitional links between the species which appeared at the commencement and close of each formation, pressed so hardly on my theory.

On the sudden appearance of whole groups of Allied Species. The abrupt manner in which whole groups of species suddenly appear in certain formations, has been urged by several palaeontologists, for instance, by Agassiz, Pictet, and by none more forcibly than by Professor Sedgwick, as a fatal objection to the belief in the transmutation of species. If numerous species, belonging to the same genera or families, have really started into life all at once, the fact would be fatal to the theory of descent with slow modification through natural selection. For the development of a group of forms, all of which have descended from some one progenitor, must have been an extremely slow process; and the progenitors must have lived long ages before their modified descendants. But we continually over-rate the perfection of the geological record, and falsely infer, because certain genera or families have not been found beneath a certain stage, that they did not exist before that stage. We continually forget how large the world is, compared with the area over which our geological formations have been carefully examined; we forget that groups of species may elsewhere have long existed and have slowly multiplied before they invaded the ancient archipelagoes of Europe and of the United States. We do not make due allowance for the enormous intervals of time, which have probably elapsed between our consecutive formations, – longer perhaps in some cases than the time required for the accumulation of each formation. These intervals will have given time for the multiplication of species from some one or some few parent-forms; and in the succeeding formation such species will appear as if suddenly created.

I may here recall a remark formerly made, namely that it might require a long succession of ages to adapt an organism to some new and peculiar line of life, for instance to fly through the air; but that when this had been effected, and a few species had thus acquired a great advantage over other organisms, a comparatively short time would be necessary to produce many divergent forms, which would be able to spread rapidly and widely throughout the world.

I will now give a few examples to illustrate these remarks; and to show how liable we are to error in supposing that whole groups of species have suddenly been produced. I may recall the well-known fact that in geological treatises, published not many years ago, the great class of mammals was always spoken of as having abruptly come in at the commencement of the tertiary series. And now one of the richest known accumulations of fossil mammals belongs to the middle of the secondary series; and one true mammal has been discovered in the new red sandstone at nearly the commencement of this great series. Cuvier used to urge that no monkey occurred in any tertiary stratum; but now extinct species have been discovered in India, South America, and in Europe even as far back as the eocene stage. The most striking case, however, is that of the Whale family; as these animals have huge bones, are marine, and range over the world, the fact of not a single bone of a whale having been discovered in any secondary formation, seemed fully to justify the belief that this great and distinct order had been suddenly produced in the interval between the latest secondary and earliest tertiary formation. But now we may read in the Supplement to Lyell's 'Manual,' published in 1858, clear evidence of the existence of whales in the upper greensand, some time before the close of the secondary period.

I may give another instance, which from having passed under my own eyes has much struck me. In a memoir on Fossil Sessile Cirripedes, I have stated that, from the number of existing and extinct tertiary species; from the extraordinary abundance of the individuals of many species all over the world, from the Arctic regions to the equator, inhabiting various zones of depths from the upper tidal limits to 50 fathoms; from the perfect manner in which specimens are preserved in the oldest tertiary beds; from the ease with which even a fragment of a valve can be recognized; from all these circumstances, I inferred that had sessile cirripedes existed during the secondary periods, they would certainly have been preserved and discovered; and as not one species had been discovered in beds of this age, I concluded that this great group had been suddenly developed at the commencement of the tertiary series. This was a sore trouble to me, adding as I thought one more instance of the abrupt appearance of a great group of species. But my work had hardly been published, when a skilful palaeontologist, M. Bosquet, sent me a drawing of a perfect specimen of an unmistakeable sessile cirripede, which he had himself extracted from the chalk of Belgium. And, as if to make the case as striking as possible, this sessile cirripede was a Chthamalus, a very common, large, and ubiquitous genus, of which not one specimen has as yet been found even in any tertiary stratum. Hence we now positively know that sessile cirripedes existed during the secondary period; and these cirripedes might have been the progenitors of our many tertiary and existing species.

The case most frequently insisted on by palaeontologists of the apparently sudden appearance of a whole group of species, is that of the teleostean fishes, low down in the Chalk period. This group includes the large majority of existing species. Lately, Professor Pictet has carried their existence one sub-stage further back; and some palaeontologists believe that certain much older fishes, of which the affinities are as yet imperfectly known, are really teleostean. Assuming, however, that the whole of them did appear, as Agassiz believes, at the commencement of the chalk formation, the fact would certainly be highly remarkable; but I cannot see that it would be an insuperable difficulty on my theory, unless it could likewise be shown that the species of this group appeared suddenly and simultaneously throughout the world at this same period. It is almost superfluous to remark that hardly any fossil-fish are known from south of the equator; and by running through Pictet's Palaeontology it will be seen that very few species are known from several formations in Europe. Some few families of fish now have a confined range; the teleostean fish might formerly have had a similarly confined range, and after having been largely developed in some one sea, might have spread widely. Nor have we any right to suppose that the seas of the world have always been so freely open from south to north as they are at present. Even at this day, if the Malay Archipelago were converted into land, the tropical parts of the Indian Ocean would form a large and perfectly enclosed basin, in which any great group of marine animals might be multiplied; and here they would remain confined, until some of the species became adapted to a cooler climate, and were enabled to double the southern capes of Africa or Australia, and thus reach other and distant seas.

From these and similar considerations, but chiefly from our ignorance of the geology of other countries beyond the confines of Europe and the United States; and from the revolution in our palaeontological ideas on many points, which the discoveries of even the last dozen years have effected, it seems to me to be about as rash in us to dogmatize on the succession of organic beings throughout the world, as it would be for a naturalist to land for five minutes on some one barren point in Australia, and then to discuss the number and range of its productions.

On the sudden appearance of groups of Allied Species in the lowest known fossiliferous strata. There is another and allied difficulty, which is much graver. I allude to the manner in which numbers of species of the same group, suddenly appear in the lowest known fossiliferous rocks. Most of the arguments which have convinced me that all the existing species of the same group have descended from one progenitor, apply with nearly equal force to the earliest known species. For instance, I cannot doubt that all the Silurian trilobites have descended from some one crustacean, which must have lived long before the Silurian age, and which probably differed greatly from any known animal. Some of the most ancient Silurian animals, as the Nautilus, Lingula, &c., do not differ much from living species; and it cannot on my theory be supposed, that these old species were the progenitors of all the species of the orders to which they belong, for they do not present characters in any degree intermediate between them. If, moreover, they had been the progenitors of these orders, they would almost certainly have been long ago supplanted and exterminated by their numerous and improved descendants.

Consequently, if my theory be true, it is indisputable that before the lowest Silurian stratum was deposited, long periods elapsed, as long as, or probably far longer than, the whole interval from the Silurian age to the present day; and that during these vast, yet quite unknown, periods of time, the world swarmed with living creatures.

To the question why we do not find records of these vast primordial periods, I can give no satisfactory answer. Several of the most eminent geologists, with Sir R. Murchison at their head, are convinced that we see in the organic remains of the lowest Silurian stratum the dawn of life on this planet. Other highly competent judges, as Lyell and the late E. Forbes, dispute this conclusion. We should not forget that only a small portion of the world is known with accuracy. M. Barrande has lately added another and lower stage to the Silurian system, abounding with new and peculiar species. Traces of life have been detected in the Longmynd beds beneath Barrande's so-called primordial zone. The presence of phosphatic nodules and bituminous matter in some of the lowest azoic rocks, probably indicates the former existence of life at these periods. But the difficulty of understanding the absence of vast piles of fossiliferous strata, which on my theory no doubt were somewhere accumulated before the Silurian epoch, is very great. If these most ancient beds had been wholly worn away by denudation, or obliterated by metamorphic action, we ought to find only small remnants of the formations next succeeding them in age, and these ought to be very generally in a metamorphosed condition. But the descriptions which we now possess of the Silurian deposits over immense territories in Russia and in North America, do not support the view, that the older a formation is, the more it has suffered the extremity of denudation and metamorphism.

The case at present must remain inexplicable; and may be truly urged as a valid argument against the views here entertained. To show that it may hereafter receive some explanation, I will give the following hypothesis. From the nature of the organic remains, which do not appear to have inhabited profound depths, in the several formations of Europe and of the United States; and from the amount of sediment, miles in thickness, of which the formations are composed, we may infer that from first to last large islands or tracts of land, whence the sediment was derived, occurred in the neighbourhood of the existing continents of Europe and North America. But we do not know what was the state of things in the intervals between the successive formations; whether Europe and the United States during these intervals existed as dry land, or as a submarine surface near land, on which sediment was not deposited, or again as the bed of an open and unfathomable sea.

Looking to the existing oceans, which are thrice as extensive as the land, we see them studded with many islands; but not one oceanic island is as yet known to afford even a remnant of any palaeozoic or secondary formation. Hence we may perhaps infer, that during the palaeozoic and secondary periods, neither continents nor continental islands existed where our oceans now extend; for had they existed there, palaeozoic and secondary formations would in all probability have been accumulated from sediment derived from their wear and tear; and would have been at least partially upheaved by the oscillations of level, which we may fairly conclude must have intervened during these enormously long periods. If then we may infer anything from these facts, we may infer that where our oceans now extend, oceans have extended from the remotest period of which we have any record; and on the other hand, that where continents now exist, large tracts of land have existed, subjected no doubt to great oscillations of level, since the earliest silurian period. The coloured map appended to my volume on Coral Reefs, led me to conclude that the great oceans are still mainly areas of subsidence, the great archipelagoes still areas of oscillations of level, and the continents areas of elevation. But have we any right to assume that things have thus remained from eternity? Our continents seem to have been formed by a preponderance, during many oscillations of level, of the force of elevation; but may not the areas of preponderant movement have changed in the lapse of ages? At a period immeasurably antecedent to the silurian epoch, continents may have existed where oceans are now spread out; and clear and open oceans may have existed where our continents now stand. Nor should we be justified in assuming that if, for instance, the bed of the Pacific Ocean were now converted into a continent, we should there find formations older than the silurian strata, supposing such to have been formerly deposited; for it might well happen that strata which had subsided some miles nearer to the centre of the earth, and which had been pressed on by an enormous weight of superincumbent water, might have undergone far more metamorphic action than strata which have always remained nearer to the surface. The immense areas in some parts of the world, for instance in South America, of bare metamorphic rocks, which must have been heated under great pressure, have always seemed to me to require some special explanation; and we may perhaps believe that we see in these large areas, the many formations long anterior to the silurian epoch in a completely metamorphosed condition.

The several difficulties here discussed, namely our not finding in the successive formations infinitely numerous transitional links between the many species which now exist or have existed; the sudden manner in which whole groups of species appear in our European formations; the almost entire absence, as at present known, of fossiliferous formations beneath the Silurian strata, are all undoubtedly of the gravest nature. We see this in the plainest manner by the fact that all the most eminent palaeontologists, namely

Cuvier, Owen, Agassiz, Barrande, Falconer, E. Forbes, &c., and all our greatest geologists, as Lyell, Murchison, Sedgwick, &c., have unanimously, often vehemently, maintained the immutability of species. But I have reason to believe that one great authority, Sir Charles Lyell, from further reflexion entertains grave doubts on this subject. I feel how rash it is to differ from these great authorities, to whom, with others, we owe all our knowledge. Those who think the natural geological record in any degree perfect, and who do not attach much weight to the facts and arguments of other kinds given in this volume, will undoubtedly at once reject my theory. For my part, following out Lyell's metaphor, I look at the natural geological record, as a history of the world imperfectly kept, and written in a changing dialect; of this history we possess the last volume alone, relating only to two or three countries. Of this volume, only here and there a short chapter has been preserved; and of each page, only here and there a few lines. Each word of the slowly-changing language, in which the history is supposed to be written, being more or less different in the interrupted succession of chapters, may represent the apparently abruptly changed forms of life, entombed in our consecutive, but widely separated formations. On this view, the difficulties above discussed are greatly diminished, or even disappear.

CHAPTER X

ON THE GEOLOGICAL SUCCESSION OF ORGANIC BEINGS

On the slow and successive appearance of new species – On their different rates of change – Species once lost do not reappear – Groups of species follow the same general rules in their appearance and disappearance as do single species – On Extinction – On simultaneous changes in the forms of life throughout the world – On the affinities of extinct species to each other and to living species – On the state of development of ancient forms – On the succession of the same types within the same areas – Summary of preceding and present chapters

LET us now see whether the several facts and rules relating to the geological succession of organic beings, better accord with the common view of the immutability of species, or with that of their slow and gradual modification, through descent and natural selection.

New species have appeared very slowly, one after another, both on the land and in the waters. Lyell has shown that it is hardly possible to resist the evidence on this head in the case of the several tertiary stages; and every year tends to fill up the blanks between them, and to make the percentage system of lost and new forms more gradual. In some of the most recent beds, though undoubtedly of high antiquity if measured by years, only one or two species are lost forms, and only one or two are new forms, having here appeared for the first time, either locally, or, as far as we know, on the face of the earth. If we may trust the observations of Philippi in Sicily, the successive changes in the marine inhabitants of that island have been many and most gradual. The secondary formations are more broken; but, as Bronn has remarked, neither the appearance nor disappearance of their many now extinct species has been simultaneous in each separate formation.

Species of different genera and classes have not changed at the same rate, or in the same degree. In the oldest tertiary beds a few living shells may still be found in the midst of a multitude of extinct forms. Falconer has given a striking instance of a similar fact, in an existing crocodile associated with many strange and lost mammals and reptiles in the sub-Himalayan deposits. The Silurian Lingula differs but little from the living species of this genus; whereas most of the other Silurian Molluscs and all the Crustaceans have changed greatly. The productions of the land seem to change at a quicker rate than those of the sea, of which a striking instance has lately been observed in Switzerland. There is some reason to believe that organisms, considered high in the scale of nature, change more quickly than those that are low: though there are exceptions to this rule. The amount of organic change, as Pictet has remarked, does not strictly correspond with the succession of our geological formations; so that between each two consecutive formations, the forms of life have seldom changed in exactly the same degree. Yet if we compare any but the most closely related formations, all the species will be found to have undergone some change. When a species has once disappeared from the face of the earth, we have reason to believe that the same identical form never reappears. The strongest apparent exception to this latter rule, is that of the so-called 'colonies' of M. Barrande, which intrude for a period in the midst of an older formation, and then allow the pre-existing fauna to reappear; but Lyell's explanation, namely, that it is a case of temporary migration from a distinct geographical province, seems to me satisfactory.

These several facts accord well with my theory. I believe in no fixed law of development, causing all the inhabitants of a country to change abruptly, or simultaneously, or to an equal degree. The process of modification must be extremely slow. The variability of each species is quite independent of that of all others. Whether such variability be taken advantage of by natural selection, and whether the variations be accumulated to a greater or lesser amount, thus causing a greater or lesser amount of modification in the varying species, depends on many complex contingencies, – on the variability being of a beneficial nature, on the power of intercrossing, on the rate of breeding, on the slowly changing physical conditions of the country, and more especially on the nature of the other inhabitants with which the varying species comes into competition. Hence it is by no means surprising that one species should retain the same identical form much longer than others; or, if changing, that it should change less. We see the same fact in geographical distribution; for instance, in the land-shells and coleopterous insects of Madeira having come to differ considerably from their nearest allies on the continent of Europe, whereas the marine shells and birds have remained unaltered. We can perhaps understand the apparently quicker rate of change in terrestrial and in more highly organized productions compared with marine and lower productions, by the more complex relations of the higher beings to their organic and inorganic conditions of life, as explained in a former chapter. When many of the inhabitants of a country have become modified and improved, we can understand, on the principle of competition, and on that of the many all-important relations of organism to organism, that any form which does not become in some degree modified and improved, will be liable to be exterminated. Hence we can see why all the species in the same region do at last, if we look to wide enough intervals of time, become modified; for those which do not change will become extinct.

In members of the same class the average amount of change, during long and equal periods of time, may, perhaps, be nearly the same; but as the accumulation of long-enduring fossiliferous formations depends on great masses of sediment having been deposited on areas whilst subsiding, our formations have been almost necessarily accumulated at wide and irregularly intermittent intervals; consequently the amount of organic change exhibited by the fossils embedded in consecutive formations is not equal. Each formation, on this view, does not mark a new and complete act of creation, but only an occasional scene, taken almost at hazard, in a slowly changing drama.

We can clearly understand why a species when once lost should never reappear, even if the very same conditions of life, organic and inorganic, should recur. For though the offspring of one species might be adapted (and no doubt this has occurred in innumerable instances) to fill the exact place of another species in the economy of nature, and thus supplant it; yet the two forms – the old and the new – would not be identically the same; for both would almost certainly inherit different characters from their distinct progenitors. For instance, it is just possible, if our fantail-pigeons were all destroyed, that fanciers, by striving during long ages for the same object, might make a new breed hardly distinguishable from our present fantail; but if the parent rock-pigeon were also destroyed, and in nature we have every reason to believe that the parent-form will generally be supplanted and exterminated by its improved offspring, it is quite incredible that a fantail, identical with the existing breed, could be raised from any other species of pigeon, or even from the other well-established races of the domestic pigeon, for the newly-formed fantail would be almost sure to inherit from its new progenitor some slight characteristic differences.

Groups of species, that is, genera and families, follow the same general rules in their appearance and disappearance as do single species, changing more or less quickly, and in a greater or lesser degree. A group does not reappear after it has once disappeared; or its existence, as long as it lasts, is continuous. I am aware that there are some apparent exceptions to this rule, but the exceptions are surprisingly few, so few, that E. Forbes, Pictet, and Woodward (though all strongly opposed to such views as I maintain) admit its truth; and the rule strictly accords with my theory. For as all the species of the same group have descended from some one species, it is clear that as long as any species of the group have appeared in the long succession of ages, so long must its members have continuously existed, in order to have generated either new and modified or the same old and unmodified forms. Species of the genus Lingula, for instance, must have continuously existed by an unbroken succession of generations, from the lowest Silurian stratum to the present day.

We have seen in the last chapter that the species of a group sometimes falsely appear to have come in abruptly; and I have attempted to give an explanation of this fact, which if true would have been fatal to my views. But such cases are certainly exceptional; the general rule being a gradual increase in number, till the group reaches its maximum, and then, sooner or later, it gradually decreases. If the number of the species of a genus, or the number of the genera of a family, be represented by a vertical line of varying thickness, crossing the successive geological formations in which the species are found, the line will sometimes falsely appear to begin at its lower end, not in a sharp point, but abruptly; it then gradually thickens upwards, sometimes keeping for a space of equal thickness, and ultimately thins out in the upper beds, marking the decrease and final extinction of the species. This gradual increase in number of the species of a group is strictly conformable with my theory; as the species of the same genus, and the genera of the same family, can increase only slowly and progressively; for the process of modification and the production of a number of allied forms must be slow and gradual, – one species giving rise first to two or three varieties, these being slowly converted into species, which in their turn produce by equally slow steps other species, and so on, like the branching of a great tree from a single stem, till the group becomes large.

On Extinction. We have as yet spoken only incidentally of the disappearance of species and of groups of species. On the theory of natural selection the extinction of old forms and the production of new and improved forms are intimately connected together. The old notion of all the inhabitants of the earth having been swept away at successive periods by catastrophes, is very generally given up, even by those geologists, as Elie de Beaumont, Murchison, Barrande, &c., whose general views would naturally lead them to this conclusion. On the contrary, we have every reason to believe, from the study of the tertiary formations, that species and groups of species gradually disappear, one after another, first from one spot, then from another, and finally from the world. Both single species and whole groups of species last for very unequal periods; some groups, as we have seen, having endured from the earliest known dawn of life to the present day; some having disappeared before the close of the palaeozoic period. No fixed law seems to determine the length of time during which any single species or any single genus endures. There is reason to believe that the complete extinction of the species of a group is generally a slower process than their production: if the appearance and disappearance of a group of species be represented, as before, by a vertical line of varying thickness, the line is found to taper more gradually at its upper end, which marks the progress of extermination, than at its lower end, which marks the first appearance and increase in numbers of the species. In some cases, however, the extermination of whole groups of beings, as of ammonites towards the close of the secondary period, has been wonderfully sudden.

The whole subject of the extinction of species has been involved in the most gratuitous mystery. Some authors have even supposed that as the individual has a definite length of life, so have species a definite duration. No one I think can have marvelled more at the extinction of species, than I have done. When I found in La Plata the tooth of a horse embedded with the remains of Mastodon, Megatherium, Toxodon, and other extinct monsters, which all co-existed with still living shells at a very late geological period, I was filled with astonishment; for seeing that the horse, since its introduc-tion by the Spaniards into South America, has run wild over the whole country and has increased in numbers at an unparalleled rate, I asked myself what could so recently have exterminated the former horse under conditions of life apparently so favourable. But how utterly groundless was my astonishment! Professor Owen soon per-ceived that the tooth, though so like that of the existing horse, belonged to an extinct species. Had this horse been still living, but in some degree rare, no naturalist would have felt the least surprise at its rarity; for rarity is the attribute of a vast number of species of all classes, in all countries. If we ask ourselves why this or that species is rare, we answer that something is unfavourable in its conditions of life; but what that something is, we can hardly ever tell. On the supposition of the fossil horse still existing as a rare species, we might have felt certain from the analogy of all other mammals, even of the slow-breeding elephant, and from the history of the naturaliza-tion of the domestic horse in South America, that under more favourable conditions it would in a very few years have stocked the whole continent. But we could not have told what the unfavourable conditions were which checked its increase, whether some one or several contingencies, and at what period of the horse's life, and in what degree, they severally acted. If the conditions had gone on, however slowly, becoming less and less favourable, we assuredly should not have perceived the fact, yet the fossil horse would certainly have become rarer and rarer, and finally extinct; – its place being seized on by some more successful competitor.

It is most difficult always to remember that the increase of every living being is constantly being checked by unperceived injurious agencies; and that these same unperceived agencies are amply sufficient to cause rarity, and finally extinction. We see in many cases in the more recent tertiary formations, that rarity precedes extinction; and we know that this has been the progress of events with those animals which have been exterminated, either locally or wholly, through man's agency. I may repeat what I published in 1845, namely, that to admit that species generally become rare before they become extinct – to feel no surprise at the rarity of a species, and yet to marvel greatly when it ceases to exist, is much the same as to admit that sickness in the individual is the forerunner of death – to feel no surprise at sickness, but when the sick man dies, to wonder and to suspect that he died by some unknown deed of violence.

The theory of natural selection is grounded on the belief that each new variety, and ultimately each new species, is produced and maintained by having some advantage over those with which it comes into competition; and the consequent extinction of less-favoured forms almost inevitably follows. It is the same with our domestic productions: when a new and slightly improved variety has been raised, it at first supplants the less improved varieties in the same neighbourhood; when much improved it is transported far and near, like our short-horn cattle, and takes the place of other breeds in other countries. Thus the appearance of new forms and the disappearance of old forms, both natural and artificial, are bound together. In certain flourishing groups, the number of new specific forms which have been produced within a given time is probably greater than that of the old forms which have been exterminated; but we know that the number of species has not gone on indefinitely increasing, at least during the later geological periods, so that looking to later times we may believe that the production of new forms has caused the extinction of about the same number of old forms.

The competition will generally be most severe, as formerly explained and illustrated by examples, between the forms which are most like each other in all respects. Hence the improved and modified descendants of a species will generally cause the extermination of the parent-species; and if many new forms have been developed from any one species, the nearest allies of that species, *i.e.* the species of the same genus, will be the most liable to extermination. Thus, as I believe, a number of new species descended from one species, that is a new genus, comes to supplant an old genus, belonging to the same family. But it must often have happened that a new species belonging to some one group will have seized on the place occupied by a species belonging to a distinct group, and thus caused its extermination; and if many allied forms be developed from the successful intruder, many will have to yield their places; and it will generally be allied forms, which will suffer from some inherited inferiority in common. But whether it be species belonging to the same or to a distinct class, which yield their places to other species which have been modified and improved, a few of the sufferers may often long be preserved, from being fitted to some peculiar line of life, or from inhabiting some distant and isolated station, where they have escaped severe competition. For instance, a single species of Trigonia, a great genus of shells in the secondary formations, survives in the Australian seas; and a few members of the great and almost extinct group of Ganoid fishes still inhabit our fresh waters. Therefore the utter extinction of a group is generally, as we have seen, a slower process than its production.

With respect to the apparently sudden extermination of whole families or orders, as of Trilobites at the close of the palaeozoic period and of Ammonites at the close of the secondary period, we must remember what has been already said on the probable wide intervals of time between our consecutive formations; and in these intervals there may have been much slow extermination. More-over, when by sudden immigration or by unusually rapid development, many species of a new group have taken possession of a new area, they will have exterminated in a correspondingly rapid manner many of the old inhabitants; and the forms which thus yield their places will commonly be allied, for they will partake of some inferiority in common.

Thus, as it seems to me, the manner in which single species and whole groups of species become extinct, accords well with the theory of natural selection. We need not marvel at extinction; if we must marvel, let it be at our presumption in imagining for a moment that we understand the many complex contingencies, on which the existence of each species depends. If we forget for an instant, that each species tends to increase inordinately, and that some check is always in action, yet seldom perceived by us, the whole economy of nature will be utterly obscured. Whenever we can precisely say why this species is more abundant in individuals than that; why this species and not another can be naturalized in a given country; then, and not till then, we may justly feel surprise why we cannot account for the extinction of this particular species or group of species.

On the Forms of Life changing almost simultaneously throughout the World. Scarcely any palaeontological discovery is more striking than the fact, that the forms of life change almost simultaneously throughout the world. Thus our European Chalk formation can be recognized in many distant parts of the world, under the most different climates, where not a fragment of the mineral chalk itself can be found; namely, in North America, in equatorial South America, in Tierra del Fuego, at the Cape of Good Hope, and in the peninsula of India. For at these distant points, the organic remains in certain beds present an unmistakable degree of resemblance to those of the Chalk. It is not that the same species are met with; for in some cases not one species is identically the same, but they belong to the same families, genera, and sections of genera, and sometimes are similarly characterized in such trifling points as mere superficial sculpture. Moreover other forms, which are not found in the Chalk of Europe, but which occur in the formations either above or below, are similarly absent at these distant points of the world. In the several successive palaeozoic formations of Russia, Western Europe and North America, a similar parallelism in the forms of life has been observed by several authors: so it is, according to Lyell, with the several European and North American tertiary deposits. Even if the few fossil species which are common to the Old and New Worlds be kept wholly out of view, the general parallelism in the successive forms of life, in the stages of the widely separated palaeozoic and tertiary periods, would still be manifest, and the several formations could be easily correlated.

These observations, however, relate to the marine inhabitants of distant parts of the world: we have not sufficient data to judge whether the productions of the land and of fresh water change at distant points in the same parallel manner. We may doubt whether they have thus changed: if the Megatherium, Mylodon, Macrauchenia, and Toxodon had been brought to Europe from La Plata, without any information in regard to their geological position, no one would have suspected that they had coexisted with still living sea-shells; but as these anomalous monsters coexisted with the Mastodon and Horse, it might at least have been inferred that they had lived during one of the latter tertiary stages.

When the marine forms of life are spoken of as having changed simultaneously throughout the world, it must not be supposed that this expression relates to the same thousandth or hundred-thousandth year, or even that it has a very strict geological sense; for if all the marine animals which live at the present day in Europe, and all those that lived in Europe during the pleistocene period (an enormously remote period as measured by years, including the whole glacial epoch), were to be compared with those now living in South America or in Australia, the most skilful naturalist would hardly be able to say whether the existing or the pleistocene inhabitants of Europe resembled most closely those of the southern hemisphere. So, again, several highly competent observers believe that the existing productions of the United States are more closely related to those which lived in Europe during certain later tertiary stages, than to those which now live here; and if this be so, it is evident that fossiliferous beds deposited at the present day on the shores of North America would hereafter be liable to be classed with somewhat older European beds. Nevertheless, looking to a remotely future epoch, there can, I think, be little doubt that all the more modern *marine* formations, namely, the upper pliocene, the pleistocene and strictly modern beds, of Europe, North and South America, and Australia, from containing fossil remains in some degree allied, and from not including those forms which are only found in the older underlying deposits, would be correctly ranked as simultaneous in a geological sense.

The fact of the forms of life changing simultaneously, in the above large sense, at distant parts of the world, has greatly struck those admirable observers, MM. de Verneuil and d'Archiac. After referring to the parallelism of the palaeozoic forms of life in various parts of Europe, they add, 'If struck by this strange sequence, we turn our attention to North America, and there discover a series of analogous phenomena, it will appear certain that all these modifications of species, their extinction, and the introduction of new ones, cannot be owing to mere changes in marine currents or other causes more or less local and temporary, but depend on general laws which govern the whole animal kingdom.' M. Barrande has made forcible remarks to precisely the same effect. It is, indeed, quite futile to look to changes of currents, climate, or other physical conditions, as the cause of these great mutations in the forms of life throughout the world, under the most different climates. We must, as Barrande has remarked, look to some special law. We shall see this more clearly when we treat of the present distribution of organic beings, and find how slight is the relation between the physical conditions of various countries, and the nature of their inhabitants.

This great fact of the parallel succession of the forms of life throughout the world, is explicable on the theory of natural selection. New species are formed by new varieties arising, which have some advantage over older forms; and those forms, which are already dominant, or have some advantage over the other forms in their own country, would naturally oftenest give rise to new varieties or incipient species; for these latter must be victorious in a still higher degree in order to be preserved and to survive. We have distinct evidence on this head, in the plants which are dominant, that is, which are commonest in their own homes, and are most widely diffused, having produced the greatest number of new varieties. It is also natural that the dominant, varying, and far-spreading species, which already have invaded to a certain extent the territories of other species, should be those which would have the best chance of spreading still further, and of giving rise in new countries to new varieties and species. The process of diffusion may often be very slow, being dependent on climatal and geographical changes, or on strange accidents, but in the long run the dominant forms will generally succeed in spreading. The diffusion would, it is probable, be slower with the terrestrial inhabitants of distinct continents than with the marine inhabitants of the continuous sea. We might therefore expect to find, as we apparently do find, a less strict degree of parallel succession in the productions of the land than of the sea.

Dominant species spreading from any region might encounter still more dominant species, and then their triumphant course, or even their existence, would cease. We know not at all precisely what are all the conditions most favourable for the multiplication of new and dominant species; but we can, I think, clearly see that a number of individuals, from giving a better chance of the appearance of favourable variations, and that severe competition with many already existing forms, would be highly favourable, as would be the power of spreading into new territories. A certain amount of isolation, recurring at long intervals of time, would probably be also favourable, as before explained. One quarter of the world may have been most favourable for the production of new and dominant species on the land, and another for those in the waters of the sea. If two great regions had been for a long period favourably circumstanced in an equal degree, whenever their inhabitants met, the battle would be prolonged and severe; and some from one birthplace and some from the other might be victorious. But in the course of time, the forms dominant in the highest degree, wherever produced, would tend everywhere to prevail. As they prevailed, they would cause the extinction of other and inferior forms; and as these inferior forms would be allied in groups by inheritance, whole groups would tend slowly to disappear; though here and there a single member might long be enabled to survive.

Thus, as it seems to me, the parallel, and, taken in a large sense, simultaneous, succession of the same forms of life throughout the world, accords well with the principle of new species having been formed by dominant species spreading widely and varying; the new species thus produced being themselves dominant owing to inheritance, and to having already had some advantage over their parents or over other species; these again spreading, varying, and producing new species. The forms which are beaten and which yield their places to the new and victorious forms, will generally be allied in groups, from inheriting some inferiority in common; and therefore as new and improved groups spread throughout the world, old groups will disappear from the world; and the succession of forms in both ways will everywhere tend to correspond.

There is one other remark connected with this subject worth making. I have given my reasons for believing that all our greater fossiliferous formations were deposited during periods of subsidence; and that blank intervals of vast duration occurred during the periods when the bed of the sea was either stationary or rising, and likewise when sediment was not thrown down quickly enough to embed and preserve organic remains. During these long and blank intervals I suppose that the inhabitants of each region underwent a considerable amount of modification and extinction, and that there was much migration from other parts of the world. As we have reason to believe that large areas are affected by the same movement, it is probable that strictly contemporaneous formations have often been accumulated over very wide spaces in the same quarter of the world; but we are far from having any right to conclude that this has invariably been the case, and that large areas have invariably been affected by the same movements. When two formations have been deposited in two regions during nearly, but not exactly the same period, we should find in both, from the causes explained in the foregoing paragraphs, the same general succession in the forms of life; but the species would not exactly correspond; for there will have been a little more time in the one region than in the other for modification, extinction, and immigration.

I suspect that cases of this nature have occurred in Europe. Mr. Prestwich, in his admirable Memoirs on the eocene deposits of England and France, is able to draw a close general parallelism between the successive stages in the two countries; but when he compares certain stages in England with those in France, although he finds in both a curious accordance in the numbers of the species belonging to the same genera, yet the species themselves differ in a manner very difficult to account for, considering the proximity of the two areas, – unless, indeed, it be assumed that an isthmus separated two seas inhabited by distinct, but contemporaneous, faunas. Lyell has made similar observations on some of the later tertiary formations. Barrande, also, shows that there is a striking general parallelism in the successive Silurian deposits of Bohemia and Scandinavia; nevertheless he finds a surprising amount of difference in the species. If the several formations in these regions have not been deposited during the same exact periods, – a formation in one region often corresponding with a blank interval in the other, – and if in both regions the species have gone on slowly changing during the accumulation of the several formations and during the long intervals of time between them; in this case, the several formations in the two regions could be arranged in the same order, in accordance with the general succession of the form of life, and the order would falsely appear to be strictly parallel; nevertheless the species would not all be the same in the apparently corresponding stages in the two regions.

On the Affinities of extinct Species to each other, and to living forms. Let us now look to the mutual affinities of extinct and living species. They all fall into one grand natural system; and this fact is at once explained on the principle of descent. The more ancient any form is, the more, as a general rule, it differs from living forms. But, as Buckland long ago remarked, all fossils can be classed either in still existing groups, or between them. That the extinct forms of life help to fill up the wide intervals between existing genera, families, and orders, cannot be disputed. For if we confine our attention either to the living or to the extinct alone, the series is far less perfect than if we combine both into one general system. With respect to the Vertebrata, whole pages could be filled with striking illustrations from our great palaeontologist, Owen, showing how extinct animals fall in between existing groups. Cuvier ranked the Ruminants and Pachyderms, as the two most distinct orders of mammals; but Owen has discovered so many fossil links, that he has had to alter the whole classification of these two orders; and has placed certain pachyderms in the same sub-order with ruminants: for example, he dissolves by fine gradations the apparently wide difference between the pig and the camel. In regard to the Invertebrata, Barrande, and a higher authority could not be named, asserts that he is every day taught that palaeozoic animals, though belonging to the same orders, families, or genera with those living at the present day, were not at this early epoch limited in such distinct groups as they now are.

Some writers have objected to any extinct species or group of species being considered as intermediate between living species or groups. If by this term it is meant that an extinct form is directly intermediate in all its characters between two living forms, the objection is probably valid. But I apprehend that in a perfectly natural classification many fossil species would have to stand between living species, and some extinct genera between living genera, even between genera belonging to distinct families. The most common case, especially with respect to very distinct groups, such as fish and reptiles, seems to be, that supposing them to be distinguished at the present day from each other by a dozen characters, the ancient members of the same two groups would be distinguished by a somewhat lesser number of characters, so that the two groups, though formerly quite distinct, at that period made some small approach to each other.

It is a common belief that the more ancient a form is, by so much the more it tends to connect by some of its characters groups now widely separated from each other. This remark no doubt must be restricted to those groups which have undergone much change in the course of geological ages; and it would be difficult to prove the truth of the proposition, for every now and then even a living animal, as the Lepidosiren, is discovered having affinities directed towards very distinct groups. Yet if we compare the older Reptiles and Batrachians, the older Fish, the older Cephalopods, and the eocene Mammals, with the more recent members of the same classes, we must admit that there is some truth in the remark.

Let us see how far these several facts and inferences accord with the theory of descent with modification. As the subject is somewhat complex, I must request the reader to turn to the diagram in the fourth chapter. We may suppose that the numbered letters represent genera, and the dotted lines diverging from them the species in each genus. The diagram is much too simple, too few genera and too few species being given, but this is unimportant for us. The horizontal lines may represent successive geological formations, and all the forms beneath the uppermost line may be considered as extinct. The three existing genera, a^{14}, q^{14}, p^{14}, will form a small family; b^{14} and f^{14} a closely allied family or sub-family; and o^{14}, e^{14}, m^{14}, a third family.

These three families, together with the many extinct genera on the several lines of descent diverging from the parent-form A, will form an order; for all will have inherited something in common from their ancient and common progenitor. On the principle of the continued tendency to divergence of character, which was formerly illustrated by this diagram, the more recent any form is, the more it will generally differ from its ancient progenitor. Hence we can understand the rule that the most ancient fossils differ most from existing forms. We must not, however, assume that divergence of character is a necessary contingency; it depends solely on the descendants from a species being thus enabled to seize on many and different places in the economy of nature. Therefore it is quite possible, as we have seen in the case of some Silurian forms, that a species might go on being slightly modified in relation to its slightly altered conditions of life, and yet retain throughout a vast period the same general characteristics. This is represented in the diagram by the letter F ^{14}.

All the many forms, extinct and recent, descended from A, make, as before remarked, one order; and this order, from the continued effects of extinction and divergence of character, has become divided into several sub-families and families, some of which are supposed to have perished at different periods, and some to have endured to the present day.

By looking at the diagram we can see that if many of the extinct forms, supposed to be embedded in the successive formations, were discovered at several points low down in the series, the three existing families on the uppermost line would be rendered less distinct from each other. If, for instance, the genera a^1, a^5, a^{10}, f^8, m^3, m^6, m^9, were disinterred, these three families would be so closely linked together that they probably would have to be united into one great family, in nearly the same manner as has occurred with ruminants and pachyderms. Yet he who objected to call the extinct genera, which thus linked the living genera of three families together, intermediate in character, would be justified, as they are intermediate, not directly, but only by a long and circuitous course through many widely different forms. If many extinct forms were to be discovered above one of the middle horizontal lines or geological formations – for instance, above No. VI. – but none from beneath this line, then only the two families on the left hand (namely, a^{14}, &c., and b^{14}, &c.) would have to be united into one family; and the two other families (namely, a^{14} to f^{14} now including five genera, and o^{14} to m^{14}) would yet remain distinct. These two families, however, would be less distinct from each other than they were before the discovery of the fossils. If, for instance, we suppose the existing genera of the two families to differ from each other by a dozen characters, in this case the genera, at the early period marked VI., would differ by a lesser number of characters; for at this early stage of descent they have not diverged in character from the common progenitor of the order, nearly so much as they subsequently diverged. Thus it comes that ancient and extinct genera are often in some slight degree intermediate in character between their modified descendants, or between their collateral relations.

In nature the case will be far more complicated than is represented in the diagram; for the groups will have been more numerous, they will have endured for extremely unequal lengths of time, and will have been modified in various degrees. As we possess only the last volume of the geological record, and that in a very broken condition, we have no right to expect, expect in very rare cases, to fill up wide intervals in the natural system, and thus unite distinct families or orders. All that we have a right to expect, is that those groups, which have within known geological periods undergone much modification, should in the older formations make some slight approach to each other; so that the older members should differ less from each other in some of their characters than do the existing members of the same groups; and this by the concurrent evidence of our best palaeontologists seems frequently to be the case.

Thus, on the theory of descent with modification, the main facts with respect to the mutual affinities of the extinct forms of life to each other and to living forms, seem to me explained in a satisfactory manner. And they are wholly inexplicable on any other view.

On this same theory, it is evident that the fauna of any great period in the earth's history will be intermediate in general character between that which preceded and that which succeeded it. Thus, the species which lived at the sixth great stage of descent in the diagram are the modified offspring of those which lived at the fifth stage, and are the parents of those which became still more modified at the seventh stage; hence they could hardly fail to be nearly intermediate in character between the forms of life above and below. We must, however, allow for the entire extinction of some preceding forms, and for the coming in of quite new forms by immigration, and for a large amount of modification, during the long and blank intervals between the successive formations. Subject to these allowances, the fauna of each geological period undoubtedly is intermediate in character, between the preceding and succeeding faunas. I need give only one instance, namely, the manner in which the fossils of the Devonian system, when this system was first discovered, were at once recognized by palaeontologists as intermediate in character between those of the overlying carboniferous, and underlying Silurian system. But each fauna is not necessarily exactly intermediate, as unequal intervals of time have elapsed between consecutive formations.

It is no real objection to the truth of the statement, that the fauna of each period as a whole is nearly intermediate in character between the preceding and succeeding faunas, that certain genera offer exceptions to the rule. For instance, mastodons and elephants, when arranged by Dr. Falconer in two series, first according to their mutual affinities and then according to their periods of existence, do not accord in arrangement. The species extreme in character are not the oldest, or the most recent; nor are those which are intermediate in character, intermediate in age. But supposing for an instant, in this and other such cases, that the record of the first appearance and disappearance of the species was perfect, we have no reason to believe that forms successively produced necessarily endure for corresponding lengths of time: a very ancient form might occasionally last much longer than a form elsewhere subsequently produced, especially in the case of terrestrial productions inhabiting separated districts. To compare small things with great: if the principal living and extinct races of the domestic pigeon were arranged as well as they could be in serial affinity, this arrangement would not closely accord with the order in time of their production, and still less with the order of their disappearance; for the parent rock-pigeon now lives; and many varieties between the rock-pigeon and the carrier have become extinct; and carriers which are extreme in the important character of length of beak originated earlier than short-beaked tumblers, which are at the opposite end of the series in this same respect.

Closely connected with the statement, that the organic remains from an intermediate formation are in some degree intermediate in character, is the fact, insisted on by all palaeontologists, that fossils from two consecutive formations are far more closely related to each other, than are the fossils from two remote formations. Pictet gives as a well-known instance, the general resemblance of the organic remains from the several stages of the chalk formation, though the species are distinct in each stage. This fact alone, from its generality, seems to have shaken Professor Pictet in his firm belief in the immutability of species. He who is acquainted with the distribution of existing species over the globe, will not attempt to account for the close resemblance of the distinct species in closely consecutive formations, by the physical conditions of the ancient areas having remained nearly the same. Let it be remembered that the forms of life, at least those inhabiting the sea, have changed almost simultaneously throughout the world, and therefore under the most different climates and conditions. Consider the prodigious vicissitudes of climate during the pleistocene period, which includes the whole glacial period, and note how little the specific forms of the inhabitants of the sea have been affected.

On the theory of descent, the full meaning of the fact of fossil remains from closely consecutive formations, though ranked as distinct species, being closely related, is obvious. As the accumulation of each formation has often been interrupted, and as long blank intervals have intervened between successive formations, we ought not to expect to find, as I attempted to show in the last chapter, in any one or two formations all the intermediate varieties between the species which appeared at the commencement and close of these periods; but we ought to find after intervals, very long as measured by years, but only moderately long as measured geologically, closely allied forms, or, as they have been called by some authors, representative species; and these we assuredly do find. We find, in short, such evidence of the slow and scarcely sensible mutation of specific forms, as we have a just right to expect to find.

On the state of Development of Ancient Forms. There has been much discussion whether recent forms are more highly developed than ancient. I will not here enter on this subject, for naturalists have not as yet defined to each other's satisfaction what is meant by high and low forms. But in one particular sense the more recent forms must, on my theory, be higher than the more ancient; for each new species is formed by having had some advantage in the struggle for life over other and preceding forms. If under a nearly similar climate, the eocene inhabitants of one quarter of the world were put into competition with the existing inhabitants of the same or some other quarter, the eocene fauna or flora would certainly be beaten and exterminated; as would a secondary fauna by an eocene, and a palaeozoic fauna by a secondary fauna. I do not doubt that this process of improvement has affected in a marked and sensible manner the organization of the more recent and victorious forms of life, in comparison with the ancient and beaten forms; but I can see no way of testing this sort of progress. Crustaceans, for instance, not the highest in their own class, may have beaten the highest molluscs. From the extraordinary manner in which European productions have recently spread over New Zealand, and have seized on places which must have been previously occupied, we may believe, if all the animals and plants of Great Britain were set free in New Zealand, that in the course of time a multitude of British forms would become thoroughly naturalized there, and would exterminate many of the natives. On the other hand, from what we see now occurring in New Zealand, and from hardly a single inhabitant of the southern hemisphere having become wild in any part of Europe, we may doubt, if all the productions of New Zealand were set free in Great Britain, whether any considerable number would be enabled to seize on places now occupied by our native plants and animals. Under this point of view, the productions of Great Britain, may be said to be higher than those of New Zealand. Yet the most skilful naturalist from an examination of the species of the two countries could not have foreseen this result.

Agassiz insists that ancient animals resemble to a certain extent the embryos of recent animals of the same classes; or that the geological succession of extinct forms is in some degree parallel to the embryological development of recent forms. I must follow Pictet and Huxley in thinking that the truth of this doctrine is very far from proved. Yet I fully expect to see it hereafter confirmed, at least in regard to subordinate groups, which have branched off from each other within comparatively recent times. For this doctrine of Agassiz accords well with the theory of natural selection. In a future chapter I shall attempt to show that the adult differs from its embryo, owing to variations supervening at a not early age, and being inherited at a corresponding age. This process, whilst it leaves the embryo almost unaltered, continually adds, in the course of successive generations, more and more difference to the adult.

Thus the embryo comes to be left as a sort of picture, preserved by nature, of the ancient and less modified condition of each animal. This view may be true, and yet it may never be capable of full proof. Seeing, for instance, that the oldest known mammals, reptiles, and fish strictly belong to their own proper classes, though some of these old forms are in a slight degree less distinct from each other than are the typical members of the same groups at the present day, it would be vain to look for animals having the common embryological character of the Vertebrata, until beds far beneath the lowest Silurian strata are discovered – a discovery of which the chance is very small.

On the succession of the same Types within the same areas, during the later tertiary periods. Mr. Clift many years ago showed that the fossil mammals from the Australian caves were closely allied to the living marsupials of that continent. In South America, a similar relationship is manifest, even to an uneducated eye, in the gigantic pieces of armour like those of the armadillo, found in several parts of La Plata; and Professor Owen has shown in the most striking manner that most of the fossil mammals, buried there in such numbers, are related to South American types. This relationship is even more clearly seen in the wonderful collection of fossil bones made by MM. Lund and Clausen in the caves of Brazil. I was so much impressed with these facts that I strongly insisted, in 1839 and 1845, on this 'law of the succession of types,' – on 'this wonderful relationship in the same continent between the dead and the living.' Professor Owen has subsequently extended the same generalization to the mammals of the Old World. We see the same law in this author's restorations of the extinct and gigantic birds of New Zealand. We see it also in the birds of the caves of Brazil. Mr. Woodward has shown that the same law holds good with sea-shells, but from the wide distribution of most genera of molluscs, it is not well displayed by them. Other cases could be added, as the relation between the extinct and living land-shells of Madeira; and between the extinct and living brackish-water shells of the Aralo-Caspian Sea.

Now what does this remarkable law of the succession of the same types within the same areas mean? He would be a bold man, who after comparing the present climate of Australia and of parts of South America under the same latitude, would attempt to account, on the one hand, by dissimilar physical conditions for the dissimilarity of the inhabitants of these two continents, and, on the other hand, by similarity of conditions, for the uniformity of the same types in each during the later tertiary periods. Nor can it be pretended that it is an immutable law that marsupials should have been chiefly or solely produced in Australia; or that Edentata and other American types should have been solely produced in South America. For we know that Europe in ancient times was peopled by numerous marsupials; and I have shown in the publications above alluded to, that in America the law of distribution of terrestrial mammals was formerly different from what it now is. North America formerly partook strongly of the present character of the southern half of the continent; and the southern half was formerly more closely allied, than it is at present, to the northern half. In a similar manner we know from Falconer and Cautley's discoveries, that northern India was formerly more closely related in its mammals to Africa than it is at the present time. Analogous facts could be given in relation to the distribution of marine animals.

On the theory of descent with modification, the great law of the long enduring, but not immutable, succession of the same types within the same areas, is at once explained; for the inhabitants of each quarter of the world will obviously tend to leave in that quarter, during the next succeeding period of time, closely allied though in some degree modified descendants. If the inhabitants of one continent formerly differed greatly from those of another continent, so will their modified descendants still differ in nearly the same manner and degree. But after very long intervals of time and after great geographical changes, permitting much inter-migration, the feebler will yield to the more dominant forms, and there will be nothing immutable in the laws of past and present distribution.

It may be asked in ridicule, whether I suppose that the megatherium and other allied huge monsters have left behind them in South America the sloth, armadillo, and anteater, as their degenerate descendants. This cannot for an instant be admitted. These huge animals have become wholly extinct, and have left no progeny. But in the caves of Brazil, there are many extinct species which are closely allied in size and in other characters to the species still living in South America; and some of these fossils may be the actual progenitors of living species. It must not be forgotten that, on my theory, all the species of the same genus have descended from some one species; so that if six genera, each having eight species, be found in one geological formation, and in the next succeeding formation there be six other allied or representative genera with the same number of species, then we may conclude that only one species of each of the six older genera has left modified descendants, constituting the six new genera. The other seven species of the old genera have all died out and have left no progeny. Or, which would probably be a far commoner case, two or three species of two or three alone of the six older genera will have been the parents of the six new genera; the other old species and the other whole genera having become utterly extinct. In failing orders, with the genera and species decreasing in numbers, as apparently is the case of the Edentata of South America, still fewer genera and species will have left modified blood-descendants.

Summary of the preceding and present Chapters. I have attempted to show that the geological record is extremely imperfect; that only a small portion of the globe has been geologically explored with care; that only certain classes of organic beings have been largely preserved in a fossil state; that the number both of specimens and of species, preserved in our museums, is absolutely as nothing compared with the incalculable number of generations which must have passed away even during a single formation; that, owing to subsidence being necessary for the accumulation of fossiliferous deposits thick enough to resist future degradation, enormous intervals of time have elapsed between the successive formations; that there has probably been more extinction during the periods of subsidence, and more variation during the periods of elevation, and during the latter the record will have been least perfectly kept; that each single formation has not been continuously deposited; that the duration of each formation is, perhaps, short compared with the average duration of specific forms; that migration has played an important part in the first appearance of new forms in any one area and formation; that widely ranging species are those which have varied most, and have oftenest given rise to new species; and that varieties have at first often been local. All these causes taken conjointly, must have tended to make the geological record extremely imperfect, and will to a large extent explain why we do not find interminable varieties, connecting together all the extinct and existing forms of life by the finest graduated steps.

He who rejects these views on the nature of the geological record, will rightly reject my whole theory. For he may ask in vain where are the numberless transitional links which must formerly have connected the closely allied or representative species, found in the several stages of the same great formation. He may disbelieve in the enormous intervals of time which have elapsed between our consecutive formations; he may overlook how important a part migration must have played, when the formations of any one great region alone, as that of Europe, are considered; he may urge the apparent, but often falsely apparent, sudden coming in of whole groups of species. He may ask where are the remains of those infinitely numerous organisms which must have existed long before the first bed of the Silurian system was deposited: I can answer this latter question only hypothetically, by saying that as far as we can see, where our oceans now extend they have for an enormous period extended, and where our oscillating continents now stand they have stood ever since the Silurian epoch; but that long before that period, the world may have presented a wholly different aspect; and that the older continents, formed of formations older than any known to us, may now all be in a metamorphosed condition, or may lie buried under the ocean.

Passing from these difficulties, all the other great leading facts in palaeontology seem to me simply to follow on the theory of descent with modification through natural selection. We can thus understand how it is that new species come in slowly and successively; how species of different classes do not necessarily change together, or at the same rate, or in the same degree; yet in the long run that all undergo modification to some extent. The extinction of old forms is the almost inevitable consequence of the production of new forms. We can understand why when a species has once disappeared it never reappears. Groups of species increase in numbers slowly, and endure for unequal periods of time; for the process of modification is necessarily slow, and depends on many complex contingencies. The dominant species of the larger dominant groups tend to leave many modified descendants, and thus new sub-groups and groups are formed. As these are formed, the species of the less vigorous groups, from their inferiority inherited from a common progenitor, tend to become extinct together, and to leave no modified offspring on the face of the earth. But the utter extinction of a whole group of species may often be a very slow process, from the survival of a few descendants, lingering in protected and isolated situations. When a group has once wholly disappeared, it does not reappear; for the link of generation has been broken.

We can understand how the spreading of the dominant forms of life, which are those that oftenest vary, will in the long run tend to people the world with allied, but modified, descendants; and these will generally succeed in taking the places of those groups of species which are their inferiors in the struggle for existence. Hence, after long intervals of time, the productions of the world will appear to have changed simultaneously.

We can understand how it is that all the forms of life, ancient and recent, make together one grand system; for all are connected by generation. We can understand, from the continued tendency to divergence of character, why the more ancient a form is, the more it generally differs from those now living. Why ancient and extinct forms often tend to fill up gaps between existing forms, sometimes blending two groups previously classed as distinct into one; but more commonly only bringing them a little closer together. The more ancient a form is, the more often, apparently, it displays characters in some degree intermediate between groups now distinct;

for the more ancient a form is, the more nearly it will be related to, and consequently resemble, the common progenitor of groups, since become widely divergent. Extinct forms are seldom directly intermediate between existing forms; but are intermediate only by a long and circuitous course through many extinct and very different forms. We can clearly see why the organic remains of closely consecutive formations are more closely allied to each other, than are those of remote formations; for the forms are more closely linked together by generation: we can clearly see why the remains of an intermediate formation are intermediate in character.

The inhabitants of each successive period in the world's history have beaten their predecessors in the race for life, and are, in so far, higher in the scale of nature; and this may account for that vague yet ill-defined sentiment, felt by many palaeontologists, that organization on the whole has progressed. If it should hereafter be proved that ancient animals resemble to a certain extent the embryos of more recent animals of the same class, the fact will be intelligible. The succession of the same types of structure within the same areas during the later geological periods ceases to be mysterious, and is simply explained by inheritance.

If then the geological record be as imperfect as I believe it to be, and it may at least be asserted that the record cannot be proved to be much more perfect, the main objections to the theory of natural selection are greatly diminished or disappear. On the other hand, all the chief laws of palaeontology plainly proclaim, as it seems to me, that species have been produced by ordinary generation: old forms having been supplanted by new and improved forms of life, produced by the laws of variation still acting round us, and preserved by Natural Selection.

CHAPTER XI

GEOGRAPHICAL DISTRIBUTION

Present distribution cannot be accounted for by differences
in physical conditions – Importance of barriers – Affinity of
the productions of the same continent – Centres of creation
– Means of dispersal, by changes of climate and of the level
of the land, and by occasional means – Dispersal during the
Glacial period co-extensive with the world

IN considering the distribution of organic beings over the face of the
globe, the first great fact which strikes us is, that neither the similarity
nor the dissimilarity of the inhabitants of various regions can be
accounted for by their climatal and other physical conditions. Of
late, almost every author who has studied the subject has come to this
conclusion. The case of America alone would almost suffice to prove
its truth: for if we exclude the northern parts where the circumpolar
land is almost continuous, all authors agree that one of the most
fundamental divisions in geographical distribution is that between
the New and Old Worlds; yet if we travel over the vast American
continent, from the central parts of the United States to its extreme
southern point, we meet with the most diversified conditions; the
most humid districts, arid deserts, lofty mountains, grassy plains,
forests, marshes, lakes, and great rivers, under almost every temperature. There is hardly a climate or condition in the Old World which
cannot be paralleled in the New – at least as closely as the same species
generally require; for it is a most rare case to find a group of organisms
confined to any small spot, having conditions peculiar in only a slight
degree; for instance, small areas in the Old World could be pointed
out hotter than any in the New World, yet these are not inhabited by
a peculiar fauna or flora. Notwithstanding this parallelism in the
conditions of the Old and New Worlds, how widely different are
their living productions!

In the southern hemisphere, if we compare large tracts of land
in Australia, South Africa, and western South America, between

latitudes 25° and 35°, we shall find parts extremely similar in all their conditions, yet it would not be possible to point out three faunas and floras more utterly dissimilar. Or again we may compare the productions of South America south of lat. 35° with those north of 25°, which consequently inhabit a considerably different climate, and they will be found incomparably more closely related to each other, than they are to the productions of Australia or Africa under nearly the same climate. Analogous facts could be given with respect to the inhabitants of the sea.

A second great fact which strikes us in our general review is, that barriers of any kind, or obstacles to free migration, are related in a close and important manner to the differences between the productions of various regions. We see this in the great difference of nearly all the terrestrial productions of the New and Old Worlds, excepting in the northern parts, where the land almost joins, and where, under a slightly different climate, there might have been free migration for the northern temperate forms, as there now is for the strictly arctic productions. We see the same fact in the great difference between the inhabitants of Australia, Africa, and South America under the same latitude: for these countries are almost as much isolated from each other as is possible. On each continent, also, we see the same fact; for on the opposite sides of lofty and continuous mountain-ranges, and of great deserts, and sometimes even of large rivers, we find different productions; though as mountain-chains, deserts, &c., are not as impassable, or likely to have endured so long as the oceans separating continents, the differences are very inferior in degree to those characteristic of distinct continents.

Turning to the sea, we find the same law. No two marine faunas are more distinct, with hardly a fish, shell, or crab in common, than those of the eastern and western shores of South and Central America; yet these great faunas are separated only by the narrow, but impassable, isthmus of Panama. Westward of the shores of America, a wide space of open ocean extends, with not an island as a halting-place for emigrants; here we have a barrier of another kind, and as soon as this is passed we meet in the eastern islands of the Pacific, with another and totally distinct fauna. So that here three marine faunas range far northward and southward, in parallel lines not far from each other, under corresponding climates; but from being separated from each other by impassable barriers, either of land or open sea, they are wholly distinct. On the other hand, proceeding still further westward from the eastern islands of the tropical parts of the Pacific, we encounter no impassable barriers, and we have innumerable islands as halting-places, until after travelling over a hemisphere we come to the shores of Africa; and over this vast space we meet with no well-defined and distinct marine faunas. Although hardly one shell, crab or fish is common to the above-named three approximate faunas of Eastern and Western America and the eastern Pacific islands, yet many fish range from the Pacific into the Indian Ocean, and many shells are common to the eastern islands of the Pacific and the eastern shores of Africa, on almost exactly opposite meridians of longitude.

A third great fact, partly included in the foregoing statements, is the affinity of the productions of the same continent or sea, though the species themselves are distinct at different points and stations. It is a law of the widest generality, and every continent offers innumerable instances. Nevertheless the naturalist in travelling, for instance, from north to south never fails to be struck by the manner in which successive groups of beings, specifically distinct, yet clearly related, replace each other. He hears from closely allied, yet distinct kinds of birds, notes nearly similar, and sees their nests similarly constructed, but not quite alike, with eggs coloured in nearly the same manner. The plains near the Straits of Magellan are inhabited by one species of Rhea (American ostrich), and northward the plains of La Plata by another species of the same genus; and not by a true ostrich or emeu, like those found in Africa and Australia under the same latitude. On these same plains of La Plata, we see the agouti and bizcacha, animals having nearly the same habits as our hares and rabbits and belonging to the same order of Rodents, but they plainly display an American type of structure. We ascend the lofty peaks of the Cordillera and we find an alpine species of bizcacha; we look to the waters, and we do not find the beaver or musk-rat, but the coypu and capybara, rodents of the American type. Innumerable other instances could be given. If we look to the islands off the American shore, however much they may differ in geological structure, the inhabitants, though they may be all peculiar species, are essentially American. We may look back to past ages, as shown in the last chapter, and we find American types then prevalent on the American continent and in the American seas. We see in these facts some deep organic bond, prevailing throughout space and time, over the same areas of land and water, and independent of their physical conditions. The naturalist must feel little curiosity, who is not led to inquire what this bond is.

This bond, on my theory, is simply inheritance, that cause which alone, as far as we positively know, produces organisms quite like, or, as we see in the case of varieties nearly like each other. The dissimilarity of the inhabitants of different regions may be attributed to modification through natural selection, and in a quite subordinate degree to the direct influence of different physical conditions. The degree of dissimilarity will depend on the migration of the more dominant forms of life from one region into another having been effected with more or less ease, at periods more or less remote; – on the nature and number of the former immigrants; – and on their action and reaction, in their mutual struggles for life; – the relation of organism to organism being, as I have already often remarked, the most important of all relations. Thus the high importance of barriers comes into play by checking migration; as does time for the slow process of modification through natural selection. Widely-ranging species, abounding in individuals, which have already triumphed over many competitors in their own widely-extended homes will have the best chance of seizing on new places, when they spread into new countries. In their new homes they will be exposed to new conditions, and will frequently undergo further modification and improvement; and thus they will become still further victorious, and will produce groups of modified descendants. On this principle of inheritance with modification, we can understand how it is that sections of genera, whole genera, and even families are confined to the same areas, as is so commonly and notoriously the case.

I believe, as was remarked in the last chapter, in no law of necessary development. As the variability of each species is an independent property, and will be taken advantage of by natural selection, only so far as it profits the individual in its complex struggle for life, so the degree of modification in different species will be no uniform quantity. If, for instance, a number of species, which stand in direct competition with each other, migrate in a body into a new and afterwards isolated country, they will be little liable to modification; for neither migration nor isolation in themselves can do anything. These principles come into play only by bringing organisms into new relations with each other, and in a lesser degree with the surrounding physical conditions. As we have seen in the last chapter that some forms have retained nearly the same character from an enormously remote geological period, so certain species have migrated over vast spaces, and have not become greatly modified.

On these views, it is obvious, that the several species of the same genus, though inhabiting the most distant quarters of the world, must originally have proceeded from the same source, as they have descended from the same progenitor. In the case of those species, which have undergone during whole geological periods but little modification, there is not much difficulty in believing that they may have migrated from the same region; for during the vast geographical and climatal changes which will have supervened since ancient times, almost any amount of migration is possible. But in many other cases, in which we have reason to believe that the species of a genus have been produced within comparatively recent times, there is great difficulty on this head. It is also obvious that the individuals of the same species, though now inhabiting distant and isolated regions, must have proceeded from one spot, where their parents were first produced: for, as explained in the last chapter, it is incredible that individuals identically the same should ever have been produced through natural selection from parents specifically distinct.

We are thus brought to the question which has been largely discussed by naturalists, namely, whether species have been created at one or more points of the earth's surface. Undoubtedly there are very many cases of extreme difficulty, in understanding how the same species could possibly have migrated from some one point to the several distant and isolated points, where now found. Nevertheless the simplicity of the view that each species was first produced within a single region captivates the mind. He who rejects it, rejects the *vera causa* of ordinary generation with subsequent migration, and calls in the agency of a miracle. It is universally admitted, that in most cases the area inhabited by a species is continuous; and when a plant or animal inhabits two points so distant from each other, or with an interval of such a nature, that the space could not be easily passed over by migration, the fact is given as something remarkable and exceptional. The capacity of migrating across the sea is more distinctly limited in terrestrial mammals, than perhaps in any other organic beings; and, accordingly, we find no inexplicable cases of the same mammal inhabiting distant points of the world. No geologist will feel any difficulty in such cases as Great Britain having been formerly united to Europe, and consequently possessing the same quadrupeds. But if the same species can be produced at two separate points, why do we not find a single mammal common to Europe and Australia or South America? The conditions of life are nearly the same, so that a multitude of European animals and plants have become naturalized in America and Australia; and some of the aboriginal plants are identically the same at these distant points of the northern and southern hemispheres. The answer, as I believe, is, that mammals have not been able to migrate, whereas some plants, from their varied means of dispersal, have migrated across the vast and broken interspace. The great and striking influence which barriers of every kind have had on distribution, is intelligible only on the view that the great majority of species have been produced on one side alone, and have not been able to migrate to the other side. Some few families, many sub-families, very many genera, and a still greater number of sections of genera are confined to a single region; and it has been observed by several naturalists, that the most natural genera, or those genera in which the species are most closely related to each other, are generally local, or confined to one area. What a strange anomaly it would be, if, when coming one step lower in the series, to the individuals of the same species, a directly opposite rule prevailed; and species were not local, but had been produced in two or more distinct areas!

Hence it seems to me, as it has to many other naturalists, that the view of each species having been produced in one area alone, and having subsequently migrated from that area as far as its powers of migration and subsistence under past and present conditions permitted, is the most probable. Undoubtedly many cases occur, in which we cannot explain how the same species could have passed from one point to the other. But the geographical and climatal changes, which have certainly occurred within recent geological times, must have interrupted or rendered discontinuous the formerly continuous range of many species. So that we are reduced to consider whether the exceptions to continuity of range are so numerous and of so grave a nature, that we ought to give up the belief, rendered probable by general considerations, that each species has been produced within one area, and has migrated thence as far as it could. It would be hopelessly tedious to discuss all the exceptional cases of the same species, now living at distant and separated points; nor do I for a moment pretend that any explanation could be offered of many such cases. But after some preliminary remarks, I will discuss a few of the most striking classes of facts; namely, the existence of the same species on the summits of distant mountain-ranges, and at distant points in the arctic and antarctic regions; and secondly (in the following chapter), the wide distribution of freshwater productions; and thirdly, the occurrence of the same terrestrial species on islands and on the mainland, though separated by hundreds of miles of open sea. If the existence of the same species at distant and isolated points of the earth's surface, can in many instances be explained on the view of each species having migrated from a single birthplace; then, considering our ignorance with respect to former climatal and geographical changes and various occasional means of transport, the belief that this has been the universal law, seems to me incomparably the safest.

In discussing this subject, we shall be enabled at the same time to consider a point equally important for us, namely, whether the several distinct species of a genus, which on my theory have all descended from a common progenitor, can have migrated (undergoing modification during some part of their migration) from the area inhabited by their progenitor. If it can be shown to be almost invariably the case, that a region, of which most of its inhabitants are closely related to, or belong to the same genera with the species of a second region, has probably received at some former period immigrants from this other region, my theory will be strengthened; for we can clearly understand, on the principle of modification, why the inhabitants of a region should be related to those of another region, whence it has been stocked. A volcanic island, for instance, upheaved and formed at the distance of a few hundreds of miles from a continent, would probably receive from it in the course of time a few colonists, and their descendants, though modified, would still be plainly related by inheritance to the inhabitants of the continent. Cases of this nature are common, and are, as we shall hereafter more fully see, inexplicable on the theory of independent creation. This view of the relation of species in one region to those in another, does not differ much (by substituting the word variety for species) from that lately advanced in an ingenious paper by Mr. Wallace, in which he concludes, that 'every species has come into existence coincident both in space and time with a pre-existing closely allied species.' And I now know from correspondence, that this coincidence he attributes to generation with modification.

The previous remarks on 'single and multiple centres of creation' do not directly bear on another allied question, – namely whether all the individuals of the same species have descended from a single pair, or single hermaphrodite, or whether, as some authors suppose, from many individuals simultaneously created. With those organic beings which never intercross (if such exist), the species, on my theory, must have descended from a succession of improved varieties, which will never have blended with other individuals or varieties, but will have supplanted each other; so that, at each successive stage of modification and improvement, all the individuals of each variety will have descended from a single parent. But in the majority of cases, namely, with all organisms which habitually unite for each birth, or which often intercross, I believe that during the slow process of modification the individuals of the species will have been kept nearly uniform by intercrossing; so that many individuals will have gone on simultaneously changing, and the whole amount of modification will not have been due, at each stage, to descent from a single parent. To illustrate what I mean: our English race-horses differ slightly from the horses of every other breed; but they do not owe their difference and superiority to descent from any single pair, but to continued care in selecting and training many individuals during many generations.

Before discussing the three classes of facts, which I have selected as presenting the greatest amount of difficulty on the theory of 'single centres of creation,' I must say a few words on the means of dispersal.

Means of Dispersal. Sir C. Lyell and other authors have ably treated this subject. I can give here only the briefest abstract of the more important facts. Change of climate must have had a powerful influence on migration: a region when its climate was different may have been a high road for migration, but now be impassable; I shall, however, presently have to discuss this branch of the subject in some detail. Changes of level in the land must also have been highly influential: a narrow isthmus now separates two marine faunas; submerge it, or let it formerly have been submerged, and the two faunas will now blend or may formerly have blended: where the sea now extends, land may at a former period have connected islands or possibly even continents together, and thus have allowed terrestrial productions to pass from one to the other. No geologist will dispute that great mutations of level have occurred within the period of existing organisms. Edward Forbes insisted that all the islands in the Atlantic must recently have been connected with Europe or Africa, and Europe likewise with America. Other authors have thus hypothetically bridged over every ocean, and have united almost every island to some mainland. If indeed the arguments used by Forbes are to be trusted, it must be admitted that scarcely a single island exists which has not recently been united to some continent. This view cuts the Gordian knot of the dispersal of the same species to the most distant points, and removes many a difficulty: but to the best of my judgement we are not authorized in admitting such enormous geographical changes within the period of existing species. It seems to me that we have abundant evidence of great oscillations of level in our continents; but not of such vast changes in their position and extension, as to have united them within the recent period to each other and to the several intervening oceanic islands. I freely admit the former existence of many islands, now buried beneath the sea, which may have served as halting places for plants and for many animals during their migration. In the coral-producing oceans such sunken islands are now marked, as I believe, by rings of coral or atolls standing over them. Whenever it is fully admitted, as I believe it will some day be, that each species has proceeded from a single birthplace, and when in the course of time we know something definite about the means of distribution, we shall be enabled to speculate with security on the former extension of the land. But I do not believe that it will ever be proved that within the recent period continents which are now quite separate, have been continuously, or almost continuously, united with each other, and with the many existing oceanic islands. Several facts in distribution, – such as the great difference in the marine faunas on the opposite sides of almost every continent, – the close relation of the tertiary inhabitants of several lands and even seas to their present inhabitants, – a certain degree of relation (as we shall hereafter see) between the distribution of mammals and the depth of the sea, – these and other such facts seem to me opposed to the admission of such prodigious geographical revolutions within the recent period, as are necessitated in the view advanced by Forbes and admitted by his many followers. The nature and relative proportions of the inhabitants of oceanic islands likewise seem to me opposed to the belief of their former continuity with continents. Nor does their almost universally volcanic composition favour the admission that they are the wrecks of sunken continents; – if they had originally existed as mountain-ranges on the land, some at least of the islands would have been formed, like other mountain-summits, of granite, metamorphic schists, old fossiliferous or other such rocks, instead of consisting of mere piles of volcanic matter.

I must now say a few words on what are called accidental means, but which more properly might be called occasional means of distribution. I shall here confine myself to plants. In botanical works, this or that plant is stated to be ill adapted for wide dissemination; but for transport across the sea, the greater or less facilities may be said to be almost wholly unknown. Until I tried, with Mr. Berkeley's aid, a few experiments, it was not even known how far seeds could resist the injurious action of sea-water. To my surprise I found that out of 87 kinds, 64 germinated after an immersion of 28 days, and a few survived an immersion of 137 days. For convenience sake I chiefly tried small seeds, without the capsule or fruit; and as all of these sank in a few days, they could not be floated across wide spaces of the sea, whether or not they were injured by the salt-water. Afterwards I tried some larger fruits, capsules, &c., and some of these floated for a long time. It is well known what a difference there is in the buoyancy of green and seasoned timber; and it occurred to me that floods might wash down plants or branches, and that these might be dried on the banks, and then by a fresh rise in the stream be washed into the sea. Hence I was led to dry stems and branches of 94 plants with ripe fruit, and to place them on sea water. The majority sank quickly, but some which whilst green floated for a very short time, when dried floated much longer; for instance, ripe hazel-nuts sank immediately, but when dried, they floated for 90 days and afterwards when planted they germinated; an asparagus plant with ripe berries floated for 23 days, when dried it floated for 85 days, and the seeds afterwards germinated: the ripe seeds of Helosciadium sank in two days, when dried they floated for above 90 days, and afterwards germinated. Altogether out of the 94 dried plants, 18 floated for above 28 days, and some of the 18 floated for a very much longer period. So that as $\frac{64}{87}$ seeds germinated after an immersion of 28 days; and as $\frac{18}{94}$ plants with ripe fruit (but not all the same species as in the foregoing experiment) floated, after being dried, for above 28 days, as far as we may infer anything from these scanty facts, we may conclude that the seeds of $\frac{14}{100}$ plants of any country might be floated by sea-currents during 28 days, and would retain their power of germination. In Johnston's Physical Atlas, the average rate of the several Atlantic currents is 33 miles per diem (some currents running at the rate of 60 miles per diem); on this average, the seeds of $\frac{14}{100}$ plants belonging to one country might be floated across 924 miles of sea to another country; and when stranded, if blown to a favourable spot by an inland gale, they would germinate.

Subsequently to my experiments, M. Martens tried similar ones, but in a much better manner, for he placed the seeds in a box in the actual sea, so that they were alternately wet and exposed to the air like really floating plants. He tried 98 seeds, mostly different from mine; but he chose many large fruits and likewise seeds from plants which live near the sea; and this would have favoured the average length of their flotation and of their resistance to the injurious action of the salt-water. On the other hand he did not previously dry the plants or branches with the fruit; and this, as we have seen, would have caused some of them to have floated much longer. The result was that $\frac{18}{98}$ of his seeds floated for 42 days, and were then capable of germination. But I do not doubt that plants exposed to the waves would float for a less time than those protected from violent movement as in our experiments. Therefore it would perhaps be safer to assume that the seeds of about $\frac{10}{100}$ plants of a flora, after having been dried, could be floated across a space of sea 900 miles in width, and would then germinate. The fact of the larger fruits often floating longer than the small, is interesting; as plants with large seeds or fruit could hardly be transported by any other means; and Alph. de Candolle has shown that such plants generally have restricted ranges.

But seeds may be occasionally transported in another manner. Drift timber is thrown up on most islands, even on those in the midst of the widest oceans; and the natives of the coral-islands in the Pacific, procure stones for their tools, solely from the roots of drifted trees, these stones being a valuable royal tax. I find on examination, that when irregularly shaped stones are embedded in the roots of trees, small parcels of earth are very frequently enclosed in their interstices and behind them, – so perfectly that not a particle could be washed away in the longest transport: out of one small portion of earth thus *completely* enclosed by wood in an oak about 50 years old, three dicotyledonous plants germinated: I am certain of the accuracy of this observation. Again, I can show that the carcasses of birds, when floating on the sea, sometimes escape being immediately devoured; and seeds of many kinds in the crops of floating birds long retain their vitality: peas and vetches, for instance, are killed by even a few days' immersion in sea-water; but of some taken out of the crop of a pigeon, which had floated on artificial salt-water for 30 days, to my surprise nearly all germinated.

Living birds can hardly fail to be highly effective agents in the transportation of seeds. I could give many facts showing how frequently birds of many kinds are blown by gales to vast distances across the ocean. We may I think safely assume that under such circumstances their rate of flight would often be 35 miles an hour; and some authors have given a far higher estimate. I have never seen an instance of nutritious seeds passing through the intestines of a bird; but hard seeds of fruit will pass uninjured through even the digestive organs of a turkey. In the course of two months, I picked up in my garden 12 kinds of seeds, out of the excrement of small birds, and these seemed perfect, and some of them, which I tried, germinated. But the following fact is more important: the crops of birds do not secrete gastric juice, and do not in the least injure, as I know by trial, the germination of seeds; now after a bird has found and devoured a large supply of food, it is positively asserted that all the grains do not pass into the gizzard for 12 or even 18 hours. A bird in this interval might easily be blown to the distance of 500 miles, and hawks are known to look out for tired birds, and the contents of their torn crops might thus readily get scattered. Mr. Brent informs me that a friend of his had to give up flying carrier-pigeons from France to England, as the hawks on the English coast destroyed so many on their arrival. Some hawks and owls bolt their prey whole, and after an interval of from twelve to twenty hours, disgorge pellets, which, as I know from experiments made in the Zoological Gardens, include seeds capable of germination. Some seeds of the oat, wheat, millet, canary, hemp, clover, and beet germinated after having been from twelve to twenty-one hours in the stomachs of different birds of prey; and two seeds of beet grew after having been thus retained for two days and fourteen hours. Freshwater fish, I find, eat seeds of many land and water plants: fish are frequently devoured by birds, and thus the seeds might be transported from place to place. I forced many kinds of seeds into the stomachs of dead fish, and then gave their bodies to fishing-eagles, storks, and pelicans; these birds after an interval of many hours, either rejected the seeds in pellets or passed them in their excrement; and several of these seeds retained their power of germination. Certain seeds, however, were always killed by this process.

Although the beaks and feet of birds are generally quite clean, I can show that earth sometimes adheres to them: in one instance I removed twenty-two grains of dry argillaceous earth from one foot of a partridge, and in this earth there was a pebble quite as large as the seed of a vetch. Thus seeds might occasionally be transported to great distances; for many facts could be given showing that soil almost everywhere is charged with seeds. Reflect for a moment on the millions of quails which annually cross the Mediterranean; and can we doubt that the earth adhering to their feet would sometimes include a few minute seeds? But I shall presently have to recur to this subject.

As icebergs are known to be sometimes loaded with earth and stones, and have even carried brushwood, bones, and the nest of a land-bird, I can hardly doubt that they must occasionally have transported seeds from one part to another of the arctic and antarctic regions, as suggested by Lyell; and during the Glacial period from one part of the now temperate regions to another. In the Azores, from the large number of the species of plants common to Europe, in comparison with the plants of other oceanic islands nearer to the mainland, and (as remarked by Mr. H. C. Watson) from the somewhat northern character of the flora in comparison with the latitude, I suspected that these islands had been partly stocked by ice-borne seeds, during the Glacial epoch. At my request Sir C. Lyell wrote to M. Hartung to inquire whether he had observed erratic boulders on these islands, and he answered that he had found large fragments of granite and other rocks, which do not occur in the archipelago. Hence we may safely infer that icebergs formerly landed their rocky burthens on the shores of these mid-ocean islands, and it is at least possible that they may have brought thither the seeds of northern plants.

Considering that the several above means of transport, and that several other means, which without doubt remain to be discovered, have been in action year after year, for centuries and tens of thousands of years, it would I think be a marvellous fact if many plants had not thus become widely transported. These means of transport are sometimes called accidental, but this is not strictly correct: the currents of the sea are not accidental, nor is the direction of prevalent gales of wind. It should be observed that scarcely any means of transport would carry seeds for very great distances; for seeds do not retain their vitality when exposed for a great length of time to the action of seawater; nor could they be long carried in the crops or intestines of birds. These means, however, would suffice for occasional transport across tracts of sea some hundred miles in breadth, or from island to island, or from a continent to a neighbouring island, but not from one distant continent to another. The floras of distant continents would not by such means become mingled in any great degree; but would remain as distinct as we now see them to be. The currents, from their course, would never bring seeds from North America to

Britain, though they might and do bring seeds from the West Indies to our western shores, where, if not killed by so long an immersion in salt-water, they could not endure our climate. Almost every year, one or two land-birds are blown across the whole Atlantic Ocean, from North America to the western shores of Ireland and England; but seeds could be transported by these wanderers only by one means, namely, in dirt sticking to their feet, which is in itself a rare accident. Even in this case, how small would the chance be of a seed falling on favourable soil, and coming to maturity! But it would be a great error to argue that because a well-stocked island, like Great Britain, has not, as far as is known (and it would be very difficult to prove this), received within the last few centuries, through occasional means of transport, immigrants from Europe or any other continent, that a poorly-stocked island, though standing more remote from the mainland, would not receive colonists by similar means. I do not doubt that out of twenty seeds or animals transported to an island, even if far less well-stocked than Britain, scarcely more than one would be so well fitted to its new home, as to become naturalized. But this, as it seems to me, is no valid argument against what would be effected by occasional means of transport, during the long lapse of geological time, whilst an island was being upheaved and formed, and before it had become fully stocked with inhabitants. On almost bare land, with few or no destructive insects or birds living there, nearly every seed, which chanced to arrive, would be sure to germinate and survive.

Dispersal during the Glacial period. The identity of many plants and animals, on mountain-summits, separated from each other by hundreds of miles of lowlands, where the Alpine species could not possibly exist, is one of the most striking cases known of the same species living at distant points, without the apparent possibility of their having migrated from one to the other. It is indeed a remarkable fact to see so many of the same plants living on the snowy regions of the Alps or Pyrenees, and in the extreme northern parts of Europe; but it is far more remarkable, that the plants on the White Mountains, in the United States of America, are all the same with those of Labrador, and nearly all the same, as we hear from Asa Gray, with those on the loftiest mountains of Europe. Even as long ago as 1747, such facts led Gmelin to conclude that the same species must have been independently created at several distinct points; and we might have remained in this same belief, had not Agassiz and others called vivid attention to the Glacial period, which, as we shall immediately see, affords a simple explanation of these facts. We have evidence of almost every conceivable kind, organic and inorganic, that within a very recent geological period, central Europe and North America suffered under an Arctic climate. The ruins of a house burnt by fire do not tell their tale more plainly, than do the mountains of Scotland and Wales, with their scored flanks, polished surfaces, and perched boulders, of the icy streams with which their valleys were lately filled. So greatly has the climate of Europe changed, that in Northern Italy, gigantic moraines, left by old glaciers, are now clothed by the vine and maize. Throughout a large part of the United States, erratic boulders, and rocks scored by drifted icebergs and coast-ice, plainly reveal a former cold period.

The former influence of the glacial climate on the distribution of the inhabitants of Europe, as explained with remarkable clearness by Edward Forbes, is substantially as follows. But we shall follow the changes more readily, by supposing a new glacial period to come slowly on, and then pass away, as formerly occurred. As the cold came on, and as each more southern zone became fitted for arctic beings and ill-fitted for their former more temperate inhabitants, the latter would be supplanted and arctic productions would take their places. The inhabitants of the more temperate regions would at the same time travel southward, unless they were stopped by barriers, in which case they would perish. The mountains would become covered with snow and ice, and their former Alpine inhabitants would descend to the plains. By the time that the cold had reached its maximum, we should have a uniform arctic fauna and flora, covering the central parts of Europe, as far south as the Alps and Pyrenees, and even stretching into Spain. The now temperate regions of the United States would likewise be covered by arctic plants and animals, and these would be nearly the same with those of Europe; for the present circumpolar inhabitants, which we suppose to have everywhere travelled southward, are remarkably uniform round the world. We may suppose that the Glacial period came on a little earlier or later in

North America than in Europe, so will the southern migration there have been a little earlier or later; but this will make no difference in the final result.

As the warmth returned, the arctic forms would retreat north-ward, closely followed up in their retreat by the productions of the more temperate regions. And as the snow melted from the bases of the mountains, the arctic forms would seize on the cleared and thawed ground, always ascending higher and higher, as the warmth increased, whilst their brethren were pursuing their northern jour-ney. Hence, when the warmth had fully returned, the same arctic species, which had lately lived in a body together on the lowlands of the Old and New Worlds, would be left isolated on distant mountain-summits (having been exterminated on all lesser heights) and in the arctic regions of both hemispheres.

Thus we can understand the identity of many plants at points so immensely remote as on the mountains of the United States and of Europe. We can thus also understand the fact that the Alpine plants of each mountain-range are more especially related to the arctic forms living due north or nearly due north of them: for the migration as the cold came on, and the re-migration on the returning warmth, will generally have been due south and north. The Alpine plants, for example, of Scotland, as remarked by Mr. H. C. Watson, and those of the Pyrenees, as remarked by Ramond, are more especially allied to the plants of northern Scandinavia; those of the United States to Labrador; those of the mountains of Siberia to the arctic regions of that country. These views, grounded as they are on the perfectly well-ascertained occurrence of a former Glacial period, seem to me to explain in so satisfactory a manner the present distribution of the Alpine and Arctic productions of Europe and America, that when in other regions we find the same species on distant mountain-summits, we may almost conclude without other evidence, that a colder climate permitted their former migration across the low intervening tracts, since become too warm for their existence.

If the climate, since the Glacial period, has ever been in any degree warmer than at present (as some geologists in the United States believe to have been the case, chiefly from the distribution of the fossil Gnathodon), then the arctic and temperate productions will at a very late period have marched a little further north, and subsequently have retreated to their present homes; but I have met with no satisfactory evidence with respect to this intercalated slightly warmer period, since the Glacial period.

The arctic forms, during their long southern migration and re-migration northward, will have been exposed to nearly the same climate, and, as is especially to be noticed, they will have kept in a body together; consequently their mutual relations will not have been much disturbed, and, in accordance with the principles inculcated in this volume, they will not have been liable to much modification. But with our Alpine productions, left isolated from the moment of the returning warmth, first at the bases and ultimately on the summits of the mountains, the case will have been somewhat different; for it is not likely that all the same arctic species will have been left on mountain ranges distant from each other, and have survived there ever since; they will, also, in all probability have become mingled with ancient Alpine species, which must have existed on the mountains before the commencement of the Glacial epoch, and which during its coldest period will have been temporarily driven down to the plains; they will, also, have been exposed to somewhat different climatal influences. Their mutual relations will thus have been in some degree disturbed; consequently they will have been liable to modification; and this we find has been the case; for if we compare the present Alpine plants and animals of the several great European mountain-ranges, though very many of the species are identically the same, some present varieties, some are ranked as doubtful forms, and some few are distinct yet closely allied or representative species.

In illustrating what, as I believe, actually took place during the Glacial period, I assumed that at its commencement the arctic productions were as uniform round the polar regions as they are at the present day. But the foregoing remarks on distribution apply not only to strictly arctic forms, but also to many subarctic and to some few northern temperate forms, for some of these are the same on the lower mountains and on the plains of North America and Europe; and it may be reasonably asked how I account for the necessary degree of uniformity of the sub-arctic and northern temperate forms round the world, at the commencement of the Glacial period. At the present day, the sub-arctic and northern temperate productions of the Old and New Worlds are separated from each other by the Atlantic Ocean and by the extreme northern part of the Pacific. During the Glacial period, when the inhabitants of the Old and New Worlds lived further southwards than at present, they must have been still more completely separated by wider spaces of ocean. I believe the above difficulty may be surmounted by looking to still earlier changes of climate of an opposite nature. We have good reason to believe that during the newer Pliocene period, before the Glacial epoch, and whilst the majority of the inhabitants of the world were specifically the same as now, the climate was warmer than at the present day. Hence we may suppose that the organisms now living under the climate of latitude 60°, during the Pliocene period lived further north under the Polar Circle, in latitude 66°–67°; and that the strictly arctic productions then lived on the broken land still nearer to the pole. Now if we look at a globe, we shall see that under the Polar Circle there is almost continuous land from western Europe, through Siberia, to eastern America. And to this continuity of the circumpolar land, and to the consequent freedom for inter-migration under a more favourable climate, I attribute the necessary amount of uniformity in the sub-arctic and northern temperate productions of the Old and New Worlds, at a period anterior to the Glacial epoch.

Believing, from reasons before alluded to, that our continents have long remained in nearly the same relative position, though subjected to large, but partial oscillations of level, I am strongly inclined to extend the above view, and to infer that during some earlier and still warmer period, such as the older Pliocene period, a large number of the same plants and animals inhabited the almost continuous circumpolar land; and that these plants and animals, both in the Old and New Worlds, began slowly to migrate southwards as the climate became less warm, long before the commencement of the Glacial period. We now see, as I believe, their descendants, mostly in a modified condition, in the central parts of Europe and the United States. On this view we can understand the relationship, with very little identity, between the productions of North America and Europe, – a relationship which is most remarkable, considering the distance of the two areas, and their separation by the Atlantic Ocean. We can further understand the singular fact remarked on by several observers, that the productions of Europe and America during the later tertiary stages were more closely related to each other than they are at the present time; for during these warmer periods the northern parts of the Old and New Worlds will have been almost continuously united by land, serving as a bridge, since rendered impassable by cold, for the inter-migration of their inhabitants.

During the slowly decreasing warmth of the Pliocene period, as soon as the species in common, which inhabited the New and Old Worlds, migrated south of the Polar Circle, they must have been completely cut off from each other. This separation, as far as the more temperate productions are concerned, took place long ages ago. And as the plants and animals migrated southward, they will have become mingled in the one great region with the native American productions, and have had to compete with them; and in the other great region, with those of the Old World. Consequently we have here everything favourable for much modification, – for far more modification than with the Alpine productions, left isolated, within a much more recent period, on the several mountain-ranges and on the arctic lands of the two Worlds. Hence it has come, that when we compare the now living productions of the temperate regions of the New and Old Worlds, we find very few identical species (though Asa Gray has lately shown that more plants are identical than was formerly supposed), but we find in every great class many forms, which some naturalists rank as geographical races, and others as distinct species; and a host of closely allied or representative forms which are ranked by all naturalists as specifically distinct.

As on the land, so in the waters of the sea, a slow southern migration of a marine fauna, which during the Pliocene or even a somewhat earlier period, was nearly uniform along the continuous shores of the Polar Circle, will account, on the theory of modification, for many closely allied forms now living in areas completely sundered. Thus, I think, we can understand the presence of many existing and tertiary representative forms on the eastern and western shores of temperate North America; and the still more striking case of many closely allied crustaceans (as described in Dana's admirable work), of some fish and other marine animals, in the Mediterranean and in the seas of Japan, – areas now separated by a continent and by nearly a hemisphere of equatorial ocean.

These cases of relationship, without identity, of the inhabitants of seas now disjoined, and likewise of the past and present inhabitants of the temperate lands of North America and Europe, are inexplicable on the theory of creation. We cannot say that they have been created alike, in correspondence with the nearly similar physical conditions of the areas; for if we compare, for instance, certain parts of South America with the southern continents of the Old World, we see countries closely corresponding in all their physical conditions, but with their inhabitants utterly dissimilar.

But we must return to our more immediate subject, the Glacial period. I am convinced that Forbes's view may be largely extended. In Europe we have the plainest evidence of the cold period, from the western shores of Britain to the Oural range, and southward to the Pyrenees. We may infer, from the frozen mammals and nature of the mountain vegetation, that Siberia was similarly affected. Along the Himalaya, at points 900 miles apart, glaciers have left the marks of their former low descent; and in Sikkim, Dr. Hooker saw maize growing on gigantic ancient moraines. South of the equator, we have some direct evidence of former glacial action in New Zealand; and the same plants, found on widely separated mountains in this island, tell the same story. If one account which has been published can be trusted, we have direct evidence of glacial action in the southeastern corner of Australia.

Looking to America; in the northern half, ice-borne fragments of rock have been observed on the eastern side as far south as lat. 36°–37°, and on the shores of the Pacific, where the climate is now so different, as far south as lat. 46°; erratic boulders have, also, been noticed on the Rocky Mountains. In the Cordillera of Equatorial South America, glaciers once extended far below their present level. In central Chile I was astonished at the structure of a vast mound of detritus, about 800 feet in height, crossing a valley of the Andes; and this I now feel convinced was a gigantic moraine, left far below any existing glacier. Further south on both sides of the continent, from lat. 41° to the southernmost extremity, we have the clearest evidence of former glacial action, in huge boulders transported far from their parent source.

We do not know that the Glacial epoch was strictly simultaneous at these several far distant points on opposite sides of the world. But we have good evidence in almost every case, that the epoch was included within the latest geological period. We have, also, excellent evidence, that it endured for an enormous time, as measured by years, at each point. The cold may have come on, or have ceased, earlier at one point of the globe than at another, but seeing that it endured for long at each, and that it was contemporaneous in a geological sense, it seems to me probable that it was, during a part at least of the period, actually simultaneous throughout the world. Without some distinct evidence to the contrary, we may at least admit as probable that the glacial action was simultaneous on the eastern and western sides of North America, in the Cordillera under the equator and under the warmer temperate zones, and on both sides of the southern extremity of the continent. If this be admitted, it is difficult to avoid believing that the temperature of the whole world was at this period simultaneously cooler. But it would suffice for my purpose, if the temperature was at the same time lower along certain broad belts of longitude.

On this view of the whole world, or at least of broad longitudinal belts, having been simultaneously colder from pole to pole, much light can be thrown on the present distribution of identical and allied species. In America, Dr. Hooker has shown that between forty and fifty of the flowering plants of Tierra del Fuego, forming no inconsiderable part of its scanty flora, are common to Europe, enormously remote as these two points are; and there are many closely allied species. On the lofty mountains of equatorial America a host of peculiar species belonging to European genera occur. On the highest mountains of Brazil, some few European genera were found by Gardner, which do not exist in the wide intervening hot countries. So on the Silla of Caraccas the illustrious Humboldt long ago found species belonging to genera characteristic of the Cordillera. On the mountains of Abyssinia, several European forms and some few representatives of the peculiar flora of the Cape of Good Hope occur. At the Cape of Good Hope a very few European species, believed not to have been introduced by man, and on the mountains, some few representative European forms are found, which have not been discovered in the intertropical parts of Africa. On the Himalaya, and on the isolated mountain-ranges of the peninsula of India, on the heights of Ceylon, and on the volcanic cones of Java, many plants occur, either identically the same or representing each other, and at the same time representing plants of Europe, not found in the intervening hot lowlands. A list of the genera collected on the loftier peaks of Java raises a picture of a collection made on a hill in Europe! Still more striking is the fact that southern Australian forms are clearly represented by plants growing on the summits of the mountains of Borneo. Some of these Australian forms, as I hear from Dr. Hooker, extend along the heights of the peninsula of Malacca, and are thinly scattered, on the one hand over India and on the other as far as Japan.

On the southern mountains of Australia, Dr. F. Müller has discovered several European species; other species, not introduced by man, occur on the lowlands; and a long list can be given, as I am informed by Dr. Hooker, of European genera, found in Australia, but not in the intermediate torrid regions. In the admirable 'Introduction to the Flora of New Zealand,' by Dr. Hooker, analogous and striking facts are given in regard to the plants of that large island. Hence we see that throughout the world, the plants growing on the more lofty mountains, and on the temperate lowlands of the northern and southern hemispheres, are sometimes identically the same; but they are much oftener specifically distinct, though related to each other in a most remarkable manner.

This brief abstract applies to plants alone: some strictly analogous facts could be given on the distribution of terrestrial animals. In marine productions, similar cases occur; as an example, I may quote a remark by the highest authority, Prof. Dana, that 'it is certainly a wonderful fact that New Zealand should have a closer resemblance in its crustacea to Great Britain, its antipode, than to any other part of the world.' Sir J. Richardson, also, speaks of the reappearance on the shores of New Zealand, Tasmania, &c., of northern forms of fish. Dr. Hooker informs me that twenty-five species of Algae are common to New Zealand and to Europe, but have not been found in the intermediate tropical seas.

It should be observed that the northern species and forms found in the southern parts of the southern hemisphere, and on the mountain-ranges of the intertropical regions, are not arctic, but belong to the northern temperate zones. As Mr. H. C. Watson has recently remarked, 'In receding from polar towards equatorial latitudes, the Alpine or mountain floras really become less and less arctic.' Many of the forms living on the mountains of the warmer regions of the earth and in the southern hemisphere are of doubtful value, being ranked by some naturalists as specifically distinct, by others as varieties; but some are certainly identical, and many, though closely related to northern forms, must be ranked as distinct species.

Now let us see what light can be thrown on the foregoing facts, on the belief, supported as it is by a large body of geological evidence, that the whole world, or a large part of it, was during the Glacial period simultaneously much colder than at present. The Glacial period, as measured by years, must have been very long; and when we remember over what vast spaces some naturalized plants and animals have spread within a few centuries, this period will have been ample for any amount of migration. As the cold came slowly on, all the tropical plants and other productions will have retreated from both sides towards the equator, followed in the rear by the temperate productions, and these by the arctic; but with the latter we are not now concerned. The tropical plants probably suffered much extinction; how much no one can say; perhaps formerly the tropics supported as many species as we see at the present day crowded together at the Cape of Good Hope, and in parts of temperate Australia. As we know that many tropical plants and animals can withstand a considerable amount of cold, many might have escaped extermination during a moderate fall of temperature, more especially by escaping into the warmest spots. But the great fact to bear in mind is, that all tropical productions will have suffered to a certain extent. On the other hand, the temperate productions, after migrating nearer to the equator, though they will have been placed under somewhat new conditions, will have suffered less. And it is certain that many temperate plants, if protected from the inroads of competitors, can withstand a much warmer climate than their own. Hence, it seems to me possible, bearing in mind that the tropical productions were in a suffering state and could not have presented a firm front against intruders, that a certain number of the more vigorous and dominant temperate forms might have penetrated the native ranks and have reached or even crossed the equator. The invasion would, of course, have been greatly favoured by high land, and perhaps by a dry climate; for Dr. Falconer informs me that it is the damp with the heat of the tropics which is so destructive to perennial plants from a temperate climate. On the other hand, the most humid and hottest districts will have afforded an asylum to the tropical natives. The mountain-ranges north-west of the Himalaya, and the long line of the Cordillera, seem to have afforded two great lines of invasion: and it is a striking fact, lately communicated to me by Dr. Hooker, that all the flowering plants, about forty-six in number, common to Tierra del Fuego and to Europe still exist in North America, which must have lain on the line of march. But I do not doubt that some temperate productions entered and crossed even the *lowlands* of the tropics at the period when the cold was most intense, – when arctic forms had migrated some twenty-five degrees of latitude from their native country and covered the land at the foot of the Pyrenees. At this period of extreme cold, I believe that the climate under the equator at the level of the sea was about the same with that now felt there at the height of six or seven thousand feet. During this the coldest period, I suppose that large spaces of the tropical lowlands were clothed with a mingled tropical and temperate vegetation, like that now growing with strange luxuriance at the base of the Himalaya, as graphically described by Hooker.

Thus, as I believe, a considerable number of plants, a few terrestrial animals, and some marine productions, migrated during the Glacial period from the northern and southern temperate zones into the intertropical regions, and some even crossed the equator. As the warmth returned, these temperate forms would naturally ascend the higher mountains, being exterminated on the lowlands; those which had not reached the equator, would re-migrate northward or southward towards their former homes; but the forms, chiefly northern, which had crossed the equator, would travel still further from their homes into the more temperate latitudes of the opposite hemisphere. Although we have reason to believe from geological evidence that the whole body of arctic shells underwent scarcely any modification during their long southern migration and re-migration northward, the case may have been wholly different with those intruding forms which settled themselves on the inter-tropical mountains, and in the southern hemisphere. These being surrounded by strangers will have had to compete with many new forms of life; and it is probable that selected modifications in their structure, habits, and constitutions will have profited them. Thus many of these wanderers, though still plainly related by inheritance to their brethren of the northern or southern hemispheres, now exist in their new homes as well-marked varieties or as distinct species.

It is a remarkable fact, strongly insisted on by Hooker in regard to America, and by Alph. de Candolle in regard to Australia, that many more identical plants and allied forms have apparently migrated from the north to the south, than in a reversed direction. We see, however, a few southern vegetable forms on the mountains of Borneo and Abyssinia. I suspect that this preponderant migration from north to south is due to the greater extent of land in the north, and to the northern forms having existed in their own homes in greater numbers, and having consequently been advanced through natural selection and competition to a higher stage of perfection or dominat-ing power, than the southern forms. And thus, when they became commingled during the Glacial period, the northern forms were enabled to beat the less powerful southern forms. Just in the same manner as we see at the present day, that very many European productions cover the ground in La Plata, and in a lesser degree in Australia, and have to a certain extent beaten the natives; whereas extremely few southern forms have become naturalized in any part of Europe, though hides, wool, and other objects likely to carry seeds have been largely imported into Europe during the last two or three centuries from La Plata, and during the last thirty or forty years from Australia. Something of the same kind must have occurred on the intertropical mountains: no doubt before the Glacial period they were stocked with endemic Alpine forms; but these have almost everywhere largely yielded to the more dominant forms, generated in the larger areas and more efficient workshops of the north. In many islands the native productions are nearly equalled or even outnumbered by the naturalized; and if the natives have not been actually exterminated, their numbers have been greatly reduced, and this is the first stage towards extinction. A mountain is an island on the land; and the intertropical mountains before the Glacial period must have been completely isolated; and I believe that the productions of these islands on the land yielded to those produced within the larger areas of the north, just in the same way as the productions of real islands have everywhere lately yielded to continental forms, naturalized by man's agency.

I am far from supposing that all difficulties are removed on the view here given in regard to the range and affinities of the allied species which live in the northern and southern temperate zones and on the mountains of the intertropical regions. Very many difficulties remain to be solved. I do not pretend to indicate the exact lines and means of migration, or the reason why certain species and not others have migrated; why certain species have been modified and have given rise to new groups of forms, and others have remained unaltered. We cannot hope to explain such facts, until we can say why one species and not another becomes naturalized by man's agency in a foreign land; why one ranges twice or thrice as far, and is twice or thrice as common, as another species within their own homes.

I have said that many difficulties remain to be solved: some of the most remarkable are stated with admirable clearness by Dr. Hooker in his botanical works on the antarctic regions. These cannot be here discussed. I will only say that as far as regards the occurrence of identical species at points so enormously remote as Kerguelen Land, New Zealand, and Fuegia, I believe that towards the close of the Glacial period, icebergs, as suggested by Lyell, have been largely concerned in their dispersal. But the existence of several quite distinct species, belonging to genera exclusively confined to the south, at these and other distant points of the southern hemisphere, is, on my theory of descent with modification, a far more remarkable case of difficulty. For some of these species are so distinct, that we cannot suppose that there has been time since the commencement of the Glacial period for their migration, and for their subsequent modification to the necessary degree. The facts seem to me to indicate that peculiar and very distinct species have migrated in radiating lines from some common centre; and I am inclined to look in the southern, as in the northern hemisphere, to a former and warmer period, before the commencement of the Glacial period, when the antarctic lands, now covered with ice, supported a highly peculiar and isolated flora. I suspect that before this flora was exterminated by the Glacial epoch, a few forms were widely dispersed to various points of the southern hemisphere by occasional means of transport, and by the aid, as halting-places, of existing and now sunken islands, and perhaps at the commencement of the Glacial period, by icebergs. By these means, as I believe, the southern shores of America, Australia, New Zealand have become slightly tinted by the same peculiar forms of vegetable life.

Sir C. Lyell in a striking passage has speculated, in language almost identical with mine, on the effects of great alternations of climate on geographical distribution. I believe that the world has recently felt one of his great cycles of change; and that on this view, combined with modification through natural selection, a multitude of facts in the present distribution both of the same and of allied forms of life can be explained. The living waters may be said to have flowed during one short period from the north and from the south, and to have crossed at the equator; but to have flowed with greater force from the north so as to have freely inundated the south. As the tide leaves its drift in horizontal lines, though rising higher on the shores where the tide rises highest, so have the living waters left their living drift on our mountain-summits, in a line gently rising from the arctic lowlands to a great height under the equator. The various beings thus left stranded may be compared with savage races of man, driven up and surviving in the mountain-fastnesses of almost every land, which serve as a record, full of interest to us, of the former inhabitants of the surrounding lowlands.

CHAPTER XII

GEOGRAPHICAL DISTRIBUTION –
continued

Distribution of fresh-water productions – On the inhabitants of oceanic islands – Absence of Batrachians and of
terrestrial Mammals – On the relation of the inhabitants
of islands to those of the nearest mainland – On colonization from the nearest source with subsequent modification
– Summary of the last and present chapters

As lakes and river-systems are separated from each other by barriers
of land, it might have been thought that fresh-water productions
would not have ranged widely within the same country, and as the sea
is apparently a still more impassable barrier, that they never would
have extended to distant countries. But the case is exactly the reverse.
Not only have many fresh-water species, belonging to quite different
classes, an enormous range, but allied species prevail in a remarkable
manner throughout the world. I well remember, when first collecting
in the fresh waters of Brazil, feeling much surprise at the similarity of
the fresh-water insects, shells, &c., and at the dissimilarity of the
surrounding terrestrial beings, compared with those of Britain.

But this power in fresh-water productions of ranging widely,
though so unexpected, can, I think, in most cases be explained by
their having become fitted, in a manner highly useful to them, for
short and frequent migrations from pond to pond, or from stream to
stream; and liability to wide dispersal would follow from this capacity
as an almost necessary consequence. We can here consider only a few
cases. In regard to fish, I believe that the same species never occur in
the fresh waters of distant continents. But on the same continent the
species often range widely and almost capriciously; for two river-
systems will have some fish in common and some different. A few
facts seem to favour the possibility of their occasional transport by
accidental means; like that of the live fish not rarely dropped
by whirlwinds in India, and the vitality of their ova when removed

from the water. But I am inclined to attribute the dispersal of fresh-water fish mainly to slight changes within the recent period in the level of the land, having caused rivers to flow into each other. Instances, also, could be given of this having occurred during floods, without any change of level. We have evidence in the loess of the Rhine of considerable changes of level in the land within a very recent geological period, and when the surface was peopled by existing land and fresh-water shells. The wide difference of the fish on opposite sides of continuous mountain-ranges, which from an early period must have parted river-systems and completely prevented their inosculation, seems to lead to this same conclusion. With respect to allied fresh-water fish occurring at very distant points of the world, no doubt there are many cases which cannot at present be explained: but some fresh-water fish belong to very ancient forms, and in such cases there will have been ample time for great geographical changes, and consequently time and means for much migration. In the second place, salt-water fish can with care be slowly accustomed to live in fresh water; and, according to Valenciennes, there is hardly a single group of fishes confined exclusively to fresh water, so that we may imagine that a marine member of a fresh-water group might travel far along the shores of the sea, and subsequently become modified and adapted to the fresh waters of a distant land.

Some species of fresh-water shells have a very wide range, and allied species, which, on my theory, are descended from a common parent and must have proceeded from a single source, prevail throughout the world. Their distribution at first perplexed me much, as their ova are not likely to be transported by birds, and they are immediately killed by sea water, as are the adults. I could not even understand how some naturalized species have rapidly spread throughout the same country. But two facts, which I have observed – and no doubt many others remain to be observed – throw some light on this subject. When a duck suddenly emerges from a pond covered with duck-weed, I have twice seen these little plants adhering to its back; and it has happened to me, in removing a little duck-weed from one aquarium to another, that I have quite unintentionally stocked the one with fresh-water shells from the other. But another agency is perhaps more effectual: I suspended a duck's feet, which might represent those of a bird sleeping in a natural pond, in an aquarium, where many ova of fresh-water shells were hatching; and I found that numbers of the extremely minute and just hatched shells crawled on the feet, and clung to them so firmly that when taken out of the water they could not be jarred off, though at a somewhat more advanced age they would voluntarily drop off. These just hatched molluscs, though aquatic in their nature, survived on the duck's feet, in damp air, from twelve to twenty hours; and in this length of time a duck or heron might fly at least six or seven hundred miles, and would be sure to alight on a pool or rivulet, if blown across sea to an oceanic island or to any other distant point. Sir Charles Lyell also informs me that a Dyticus has been caught with an Ancylus (a fresh-water shell like a limpet) firmly adhering to it; and a water-beetle of the same family, a Colymbetes, once flew on board the *Beagle*, when forty-five miles distant from the nearest land: how much farther it might have flown with a favouring gale no one can tell.

With respect to plants, it has long been known what enormous ranges many fresh-water and even marsh-species have, both over continents and to the most remote oceanic islands. This is strikingly shown, as remarked by Alph. de Candolle, in large groups of terrestrial plants, which have only a very few aquatic members; for these latter seem immediately to acquire, as if in consequence, a very wide range. I think favourable means of dispersal explain this fact. I have before mentioned that earth occasionally, though rarely, adheres in some quantity to the feet and beaks of birds. Wading birds, which frequent the muddy edges of ponds, if suddenly flushed, would be the most likely to have muddy feet. Birds of this order I can show are the greatest wanderers, and are occasionally found on the most remote and barren islands in the open ocean; they would not be likely to alight on the surface of the sea, so that the dirt would not be washed off their feet; when making land, they would be sure to fly to their natural fresh-water haunts. I do not believe that botanists are aware how charged the mud of ponds is with seeds: I have tried several little experiments, but will here give only the most striking case: I took in February three table-spoonfuls of mud from three different points, beneath water, on the edge of a little pond; this mud when dry weighed only $6\frac{3}{4}$ ounces; I kept it covered up in my study for six months, pulling up and counting each plant as it grew; the plants were of many kinds, and were altogether 537 in number; and yet the viscid mud was all contained in a breakfast cup! Considering these facts, I think it would be an inexplicable circumstance if water-birds did not transport the seeds of fresh-water plants to vast distances, and if consequently the range of these plants was not very great. The same agency may have come into play with the eggs of some of the smaller fresh-water animals.

Other and unknown agencies probably have also played a part. I have stated that fresh-water fish eat some kinds of seeds, though they reject many other kinds after having swallowed them; even small fish swallow seeds of moderate size, as of the yellow water-lily and Potamogeton. Herons and other birds, century after century, have gone on daily devouring fish; they then take flight and go to other waters, or are blown across the sea; and we have seen that seeds retain their power of germination, when rejected in pellets or in excrement, many hours afterwards. When I saw the great size of the seeds of that fine water-lily, the Nelumbium, and remembered Alph. de Candolle's remarks on this plant, I thought that its distribution must remain quite inexplicable; but Audubon states that he found the seeds of the great southern water-lily (probably, according to Dr. Hooker, the Nelumbium luteum) in a heron's stomach; although I do not know the fact, yet analogy makes me believe that a heron flying to another pond and getting a hearty meal of fish, would probably reject from its stomach a pellet containing the seeds of the Nelumbium undigested; or the seeds might be dropped by the bird whilst feeding its young, in the same way as fish are known sometimes to be dropped.

In considering these several means of distribution, it should be remembered that when a pond or stream is first formed, for instance, on a rising islet, it will be unoccupied; and a single seed or egg will have a good chance of succeeding. Although there will always be a struggle for life between the individuals of the species, however few, already occupying any pond, yet as the number of kinds is small, compared with those on the land, the competition will probably be less severe between aquatic than between terrestrial species; consequently an intruder from the waters of a foreign country, would have a better chance of seizing on a place, than in the case of terrestrial colonists. We should, also, remember that some, perhaps many, fresh-water productions are low in the scale of nature, and that we have reason to believe that such low beings change or become modified less quickly than the high; and this will give longer time than the average for the migration of the same aquatic species. We should not forget the probability of many species having formerly ranged as continuously as fresh-water productions ever can range, over immense areas, and having subsequently become extinct in intermediate regions. But the wide distribution of fresh-water plants and of the lower animals, whether retaining the same identical form or in some degree modified, I believe mainly depends on the wide dispersal of their seeds and eggs by animals, more especially by fresh-water birds, which have large powers of flight, and naturally travel from one to another and often distant piece of water. Nature, like a careful gardener, thus takes her seeds from a bed of a particular nature, and drops them in another equally well fitted for them.

On the Inhabitants of Oceanic Islands. We now come to the last of the three classes of facts, which I have selected as presenting the greatest amount of difficulty, on the view that all the individuals both of the same and of allied species have descended from a single parent; and therefore have all proceeded from a common birthplace, notwithstanding that in the course of time they have come to inhabit distant points of the globe. I have already stated that I cannot honestly admit Forbes's view on continental extensions, which, if legitimately followed out, would lead to the belief that within the recent period all existing islands have been nearly or quite joined to some continent. This view would remove many difficulties, but it would not, I think, explain all the facts in regard to insular productions. In the following remarks I shall not confine myself to the mere question of dispersal; but shall consider some other facts, which bear on the truth of the two theories of independent creation and of descent with modification.

The species of all kinds which inhabit oceanic islands are few in number compared with those on equal continental areas: Alph. de Candolle admits this for plants, and Wollaston for insects. If we look to the large size and varied stations of New Zealand, extending over 780 miles of latitude, and compare its flowering plants, only 750 in number, with those on an equal area at the Cape of Good Hope or in Australia, we must, I think, admit that something quite independently of any difference in physical conditions has caused so great a difference in number. Even the uniform county of Cambridge has 847 plants, and the little island of Anglesea 764, but a few ferns and a few introduced plants are included in these numbers, and the comparison in some other respects is not quite fair. We have evidence that the barren island of Ascension aboriginally possessed under half-a-dozen flowering plants; yet many have become naturalized on it, as they have on New Zealand and on every other oceanic island which can be named. In St. Helena there is reason to believe that the naturalized plants and animals have nearly or quite exterminated many native productions. He who admits the doctrine of the creation of each separate species, will have to admit, that a sufficient number of the best adapted plants and animals have not been created on oceanic islands; for man has unintentionally stocked them from various sources far more fully and perfectly than has nature.

Although in oceanic islands the number of kinds of inhabitants is scanty, the proportion of endemic species (*i.e.* those found nowhere else in the world) is often extremely large. If we compare, for instance, the number of the endemic land-shells in Madeira, or of the endemic birds in the Galapagos Archipelago, with the number found on any continent, and then compare the area of the islands with that of the continent, we shall see that this is true. This fact might have been expected on my theory, for, as already explained, species occasionally arriving after long intervals in a new and isolated district, and having to compete with new associates, will be eminently liable to modification, and will often produce groups of modified descendants. But it by no means follows, that, because in an island nearly all the species of one class are peculiar, those of another class, or of another section of the same class, are peculiar; and this difference seems to depend on the species which do not become modified having immigrated with facility and in a body, so that their mutual relations have not been much disturbed. Thus in the Galapagos Islands nearly every land-bird, but only two out of the eleven marine birds, are peculiar;

and it is obvious that marine birds could arrive at these islands more easily than land-birds. Bermuda, on the other hand, which lies at about the same distance from North America as the Galapagos Islands do from South America, and which has a very peculiar soil, does not possess one endemic land-bird; and we know from Mr. J. M. Jones's admirable account of Bermuda, that very many North American birds, during their great annual migrations, visit either periodically or occasionally this island. Madeira does not possess one peculiar bird, and many European and African birds are almost every year blown there, as I am informed by Mr. E. V. Harcourt. So that these two islands of Bermuda and Madeira have been stocked by birds, which for long ages have struggled together in their former homes, and have become mutually adapted to each other; and when settled in their new homes, each kind will have been kept by the others to their proper places and habits, and will consequently have been little liable to modification. Madeira, again, is inhabited by a wonderful number of peculiar land-shells, whereas not one species of sea-shell is confined to its shores: now, though we do not know how sea-shells are dispersed, yet we can see that their eggs or larvae, perhaps attached to seaweed or floating timber, or to the feet of wading-birds, might be transported far more easily than land-shells, across three or four hundred miles of open sea. The different orders of insects in Madeira apparently present analogous facts.

Oceanic islands are sometimes deficient in certain classes, and their places are apparently occupied by the other inhabitants; in the Galapagos Islands reptiles, and in New Zealand gigantic wingless birds, take the place of mammals. In the plants of the Galapagos Islands, Dr. Hooker has shown that the proportional numbers of the different orders are very different from what they are elsewhere. Such cases are generally accounted for by the physical conditions of the islands; but this explanation seems to me not a little doubtful. Facility of immigration, I believe, has been at least as important as the nature of the conditions.

Many remarkable little facts could be given with respect to the inhabitants of remote islands. For instance, in certain islands not tenanted by mammals, some of the endemic plants have beautifully hooked seeds; yet few relations are more striking than the adaptation of hooked seeds for transportal by the wool and fur of quadrupeds. This case presents no difficulty on my view, for a hooked seed might be transported to an island by some other means; and the plant then becoming slightly modified, but still retaining its hooked seeds, would form an endemic species, having as useless an appendage as any rudimentary organ, – for instance, as the shrivelled wings under the soldered elytra of many insular beetles. Again, islands often possess trees or bushes belonging to orders which elsewhere include only herbaceous species; now trees, as Alph. de Candolle has shown, generally have, whatever the cause may be, confined ranges. Hence trees would be little likely to reach distant oceanic islands; and an herbaceous plant, though it would have no chance of successfully competing in stature with a fully developed tree, when established on an island and having to compete with herbaceous plants alone, might readily gain an advantage by growing taller and taller and overtopping the other plants. If so, natural selection would often tend to add to the stature of herbaceous plants when growing on an island, to whatever order they belonged, and thus convert them first into bushes and ultimately into trees.

With respect to the absence of whole orders on oceanic islands, Bory St. Vincent long ago remarked that Batrachians (frogs, toads, newts) have never been found on any of the many islands with which the great oceans are studded. I have taken pains to verify this assertion, and I have found it strictly true. I have, however, been assured that a frog exists on the mountains of the great island of New Zealand; but I suspect that this exception (if the information be correct) may be explained through glacial agency. This general absence of frogs, toads, and newts on so many oceanic islands cannot be accounted for by their physical conditions; indeed it seems that islands are peculiarly well fitted for these animals; for frogs have been introduced into Madeira, the Azores, and Mauritius, and have multiplied so as to become a nuisance. But as these animals and their spawn are known to be immediately killed by sea-water, on my view we can see that there would be great difficulty in their transportal across the sea, and therefore why they do not exist on any oceanic island. But why, on the theory of creation, they should not have been created there, it would be very difficult to explain.

Mammals offer another and similar case. I have carefully searched the oldest voyages, but have not finished my search; as yet I have not found a single instance, free from doubt, of a terrestrial mammal (excluding domesticated animals kept by the natives) inhabiting an island situated above 300 miles from a continent or great continental island; and many islands situated at a much less distance are equally barren. The Falkland Islands, which are inhabited by a wolf-like fox, come nearest to an exception; but this group cannot be considered as oceanic, as it lies on a bank connected with the mainland; moreover, icebergs formerly brought boulders to its western shores, and they may have formerly transported foxes, as so frequently now happens in the arctic regions. Yet it cannot be said that small islands will not support small mammals, for they occur in many parts of the world on very small islands, if close to a continent; and hardly an island can be named on which our smaller quadrupeds have not become natural- ized and greatly multiplied. It cannot be said, on the ordinary view of creation, that there has not been time for the creation of mammals; many volcanic islands are sufficiently ancient, as shown by the stu- pendous degradation which they have suffered and by their tertiary strata: there has also been time for the production of endemic species belonging to other classes; and on continents it is thought that mammals appear and disappear at a quicker rate than other and lower animals. Though terrestrial mammals do not occur on oceanic islands, aërial mammals do occur on almost every island. New Zea- land possesses two bats found nowhere else in the world: Norfolk Island, the Viti Archipelago, the Bonin Islands, the Caroline and Marianne Archipelagoes, and Mauritius, all possess their peculiar bats. Why, it may be asked, has the supposed creative force produced bats and no other mammals on remote islands? On my view this question can easily be answered; for no terrestrial mammal can be transported across a wide space of sea, but bats can fly across. Bats have been seen wandering by day far over the Atlantic Ocean; and two North American species either regularly or occasionally visit Bermuda, at the distance of 600 miles from the mainland. I hear from Mr. Tomes, who has specially studied this family, that many of the same species have enormous ranges, and are found on continents and on far distant islands. Hence we have only to suppose that such wandering species have been modified through natural selection in their new homes in relation to their new position, and we can understand the presence of endemic bats on islands, with the absence of all terrestrial mammals.

Besides the absence of terrestrial mammals in relation to the remoteness of islands from continents, there is also a relation, to a certain extent independent of distance, between the depth of the sea separating an island from the neighbouring mainland, and the presence in both of the same mammiferous species or of allied species in a more or less modified condition. Mr. Windsor Earl has made some striking observations on this head in regard to the great Malay Archipelago, which is traversed near Celebes by a space of deep ocean; and this space separates two widely distinct mammalian faunas. On either side the islands are situated on moderately deep submarine banks, and they are inhabited by closely allied or identical quadrupeds. No doubt some few anomalies occur in this great archipelago, and there is much difficulty in forming a judgement in some cases owing to the probable naturalization of certain mammals through man's agency; but we shall soon have much light thrown on the natural history of this archipelago by the admirable zeal and researches of Mr. Wallace. I have not as yet had time to follow up this subject in all other quarters of the world; but as far as I have gone, the relation generally holds good. We see Britain separated by a shallow channel from Europe, and the mammals are the same on both sides; we meet with analogous facts on many islands separated by similar channels from Australia. The West Indian Islands stand on a deeply submerged bank, nearly 1000 fathoms in depth, and here we find American forms, but the species and even the genera are distinct. As the amount of modification in all cases depends to a certain degree on the lapse of time, and as during changes of level it is obvious that islands separated by shallow channels are more likely to have been continuously united within a recent period to the mainland than islands separated by deeper channels, we can understand the frequent relation between the depth of the sea and the degree of affinity of the mammalian inhabitants of islands with those of a neighbouring continent, – an explicable relation on the view of independent acts of creation.

All the foregoing remarks on the inhabitants of oceanic islands, – namely, the scarcity of kinds – the richness in endemic forms in particular classes or sections of classes, – the absence of whole groups, as of batrachians, and of terrestrial mammals notwithstanding the presence of aërial bats, – the singular proportions of certain orders of plants, – herbaceous forms having been developed into trees, &c., – seem to me to accord better with the view of occasional means of transport having been largely efficient in the long course of time, than with the view of all our oceanic islands having been formerly connected by continuous land with the nearest continent; for on this latter view the migration would probably have been more complete; and if modification be admitted, all the forms of life would have been more equally modified, in accordance with the paramount importance of the relation of organism to organism.

I do not deny that there are many and grave difficulties in understanding how several of the inhabitants of the more remote islands, whether still retaining the same specific form or modified since their arrival, could have reached their present homes. But the probability of many islands having existed as halting-places, of which not a wreck now remains, must not be overlooked. I will here give a single instance of one of the cases of difficulty. Almost all oceanic islands, even the most isolated and smallest, are inhabited by land-shells, generally by endemic species, but sometimes by species found elsewhere. Dr. Aug. A. Gould has given several interesting cases in regard to the land-shells of the islands of the Pacific. Now it is notorious that land-shells are very easily killed by salt; their eggs, at least such as I have tried, sink in sea-water and are killed by it. Yet there must be, on my view, some unknown, but highly efficient means for their transportal. Would the just-hatched young occasionally crawl on and adhere to the feet of birds roosting on the ground, and thus get transported? It occurred to me that land-shells, when hybernating and having a membranous diaphragm over the mouth of the shell, might be floated in chinks of drifted timber across moderately wide arms of the sea. And I found that several species did in this state withstand uninjured an immersion in sea-water during seven days: one of these shells was the Helix pomatia, and after it had again hybernated I put it in sea-water for twenty days, and it perfectly recovered. As this species has a thick calcareous operculum, I removed it, and when it had formed a new membranous one, I immersed it for fourteen days in sea-water, and it recovered and crawled away: but more experiments are wanted on this head.

The most striking and important fact for us in regard to the inhabitants of islands, is their affinity to those of the nearest mainland, without being actually the same species. Numerous instances could be given of this fact. I will give only one, that of the Galapagos Archipelago, situated under the equator, between 500 and 600 miles from the shores of South America. Here almost every product of the land and water bears the unmistakeable stamp of the American continent. There are twenty-six land-birds, and twenty-five of those are ranked by Mr. Gould as distinct species, supposed to have been created here; yet the close affinity of most of these birds to American species in every character, in their habits, gestures, and tones of voice, was manifest. So it is with the other animals, and with nearly all the plants, as shown by Dr. Hooker in his admirable memoir on the Flora of this archipelago. The naturalist, looking at the inhabitants of these volcanic islands in the Pacific, distant several hundred miles from the continent, yet feels that he is standing on American land. Why should this be so? why should the species which are supposed to have been created in the Galapagos Archipelago, and nowhere else, bear so plain a stamp of affinity to those created in America? There is nothing in the conditions of life, in the geological nature of the islands, in their height or climate, or in the proportions in which the several classes are associated together, which resembles closely the conditions of the South American coast: in fact there is a considerable dissimilarity in all these respects. On the other hand, there is a considerable degree of resemblance in the volcanic nature of the soil, in climate, height, and size of the islands, between the Galapagos and Cape de Verde Archipelagos: but what an entire and absolute difference in their inhabitants! The inhabitants of the Cape de Verde Islands are related to those of Africa, like those of the Galapagos to America. I believe this grand fact can receive no sort of explanation on the ordinary view of independent creation; whereas on the view here maintained, it is obvious that the Galapagos Islands would be likely to receive colonists, whether by occasional means of transport or by formerly continuous land, from America; and the Cape de Verde Islands from Africa; and that such colonists would be liable to modifications; – the principle of inheritance still betraying their original birthplace.

Many analogous facts could be given: indeed it is an almost universal rule that the endemic productions of islands are related to those of the nearest continent, or of other near islands. The exceptions are few, and most of them can be explained. Thus the plants of Kerguelen Land, though standing nearer to Africa than to America, are related, and that very closely, as we know from Dr. Hooker's account, to those of America: but on the view that this island has been mainly stocked by seeds brought with earth and stones on icebergs, drifted by the prevailing currents, this anomaly disappears. New Zealand in its endemic plants is much more closely related to Australia, the nearest mainland, than to any other region: and this is what might have been expected; but it is also plainly related to South America, which, although the next nearest continent, is so enormously remote, that the fact becomes an anomaly. But this difficulty almost disappears on the view that both New Zealand, South America, and other southern lands were long ago partially stocked from a nearly intermediate though distant point, namely from the antarctic islands, when they were clothed with vegetation, before the commencement of the Glacial period. The affinity, which, though feeble, I am assured by Dr. Hooker is real, between the flora of the southwestern corner of Australia and of the Cape of Good Hope, is a far more remarkable case, and is at present inexplicable: but this affinity is confined to the plants, and will, I do not doubt, be some day explained.

The law which causes the inhabitants of an archipelago, though specifically distinct, to be closely allied to those of the nearest continent, we sometimes see displayed on a small scale, yet in a most interesting manner, within the limits of the same archipelago. Thus the several islands of the Galapagos Archipelago are tenanted, as I have elsewhere shown, in a quite marvellous manner, by very closely related species; so that the inhabitants of each separate island, though mostly distinct, are related in an incomparably closer degree to each other than to the inhabitants of any other part of the world. And this is just what might have been expected on my view, for the islands are situated so near each other that they would almost certainly receive immigrants from the same original source, or from each other. But this dissimilarity between the endemic inhabitants of the islands may be used as an argument against my views; for it may be asked, how has it happened in the several islands situated within sight of each other, having the same geological nature, the same height, climate, &c., that many of the immigrants should have been differently modified, though only in a small degree. This long appeared to me a great difficulty: but it arises in chief part from the deeply-seated error of considering the physical conditions of a country as the most import-ant for its inhabitants; whereas it cannot, I think, be disputed that the nature of the other inhabitants, with which each has to compete, is at least as important, and generally a far more important element of success. Now if we look to those inhabitants of the Galapagos Archipelago which are found in other parts of the world (laying on one side for the moment the endemic species, which cannot be here fairly included, as we are considering how they have come to be modified since their arrival), we find a considerable amount of differ-ence in the several islands. This difference might indeed have been expected on the view of the islands having been stocked by occasional means of transport – a seed, for instance, of one plant having been brought to one island, and that of another plant to another island. Hence when in former times an immigrant settled on any one or more of the islands, or when it subsequently spread from one island to another, it would undoubtedly be exposed to different conditions of life in the different islands, for it would have to compete with different sets of organisms: a plant, for instance, would find the best-fitted ground more perfectly occupied by distinct plants in one island than in another, and it would be exposed to the attacks of somewhat different enemies. If then it varied, natural selection would probably favour different varieties in the different islands. Some species, however, might spread and yet retain the same character throughout the group, just as we see on continents some species spreading widely and remaining the same.

The really surprising fact in this case of the Galapagos Archipelago, and in a lesser degree in some analogous instances, is that the new species formed in the separate islands have not quickly spread to the other islands. But the islands, though in sight of each other, are separated by deep arms of the sea, in most cases wider than the British Channel, and there is no reason to suppose that they have at any former period been continuously united. The currents of the sea are rapid and sweep across the archipelago, and gales of wind are extraordinarily rare; so that the islands are far more effectually separated from each other than they appear to be on a map. Nevertheless a good many species, both those found in other parts of the world and those confined to the archipelago, are common to the several islands, and we may infer from certain facts that these have probably spread from some one island to the others. But we often take, I think, an erroneous view of the probability of closely allied species invading each other's territory, when put into free intercommunication. Undoubtedly if one species has any advantage whatever over another, it will in a very brief time wholly or in part supplant it; but if both are equally well fitted for their own places in nature, both probably will hold their own places and keep separate for almost any length of time. Being familiar with the fact that many species, naturalized through man's agency, have spread with astonishing rapidity over new countries, we are apt to infer that most species would thus spread; but we should remember that the forms which become naturalized in new countries are not generally closely allied to the aboriginal inhabitants, but are very distinct species, belonging in a large proportion of cases, as shown by Alph. de Candolle, to distinct genera. In the Galapagos Archipelago, many even of the birds, though so well adapted for flying from island to island, are distinct on each; thus there are three closely-allied species of mocking-thrush of Chatham Island to be blown to Charles Island, which has its own mocking-thrush: why should it succeed in establishing itself there? We may safely infer that Charles Island is well stocked with its own species, for annually more eggs are laid there than can possibly be reared; and we may infer that the mocking-thrush peculiar to Charles Island is at least as well fitted for its home as is the species peculiar to Chatham Island. Sir C. Lyell and Mr. Wollaston have communicated to me a remarkable fact bearing on this subject; namely, that Madeira and the adjoining islet of Porto Santo possess many distinct but representative land-shells, some of which live in crevices of stone; and although large quantities of stone are annually transported from Porto Santo to Madeira, yet this latter island has not become colonized by the Porto Santo species: nevertheless both islands have been colonized by some European land-shells, which no doubt had some advantage over the indigenous species. From these considerations I think we need not greatly marvel at the endemic and representative species, which inhabit the several islands of the Galapagos Archipelago, not having universally spread from island to island. In many other instances, as in the several districts of the same continent, pre-occupation has probably played an important part in checking the commingling of species under the same conditions of life. Thus, the south-east and south-west corners of Australia have nearly the same physical conditions, and are united by continuous land, yet they are inhabited by a vast number of distinct mammals, birds, and plants.

The principle which determines the general character of the fauna and flora of oceanic islands, namely, that the inhabitants, when not identically the same, yet are plainly related to the inhabitants of that region whence colonists could most readily have been derived, – the colonists having been subsequently modified and better fitted to their new homes, – is of the widest application throughout nature. We see this on every mountain, in every lake and marsh. For Alpine species, excepting in so far as the same forms, chiefly of plants, have spread widely throughout the world during the recent Glacial epoch, are related to those of the surrounding lowlands; – thus we have in South America, Alpine humming-birds, Alpine rodents, Alpine plants, &c., all of strictly American forms, and it is obvious that a mountain, as it became slowly upheaved, would naturally be colonized from the surrounding lowlands. So it is with the inhabitants of lakes and marshes, excepting in so far as great facility of transport has given the same general forms to the whole world. We see this same principle in the blind animals inhabiting the caves of America and of Europe. Other analogous facts could be given. And it will, I believe, be universally found to be true, that wherever in two regions, let them be ever so distant, many closely allied or representative species occur, there will likewise be found some identical species, showing, in accordance with the foregoing view, that at some former period there has been intercommunication or migration between the two regions. And wherever many closely-allied species occur, there will be found many forms which some naturalists rank as distinct species, and some as varieties; these doubtful forms showing us the steps in the process of modification.

This relation between the power and extent of migration of a species, either at the present time or at some former period under different physical conditions, and the existence at remote points of the world of other species allied to it, is shown in another and more general way. Mr. Gould remarked to me long ago, that in those genera of birds which range over the world, many of the species have very wide ranges. I can hardly doubt that this rule is generally true, though it would be difficult to prove it. Amongst mammals, we see it strikingly displayed in Bats, and in a lesser degree in the Felidae and Canidae. We see it, if we compare the distribution of butterflies and beetles. So it is with most fresh-water productions, in which so many genera range over the world, and many individual species have enormous ranges. It is not meant that in world-ranging genera all the species have a wide range, or even that they have on an *average* a wide range; but only that some of the species range very widely; for the facility with which widely-ranging species vary and give rise to new forms will largely determine their average range. For instance, two varieties of the same species inhabit America and Europe, and the species thus has an immense range; but, if the variation had been a little greater, the two varieties would have been ranked as distinct species, and the common range would have been greatly reduced. Still less is it meant, that a species which apparently has the capacity of crossing barriers and ranging widely, as in the case of certain power-fully-winged birds, will necessarily range widely; for we should never forget that to range widely implies not only the power of crossing barriers, but the more important power of being victorious in distant lands in the struggle for life with foreign associates. But on the view of all the species of a genus having descended from a single parent, though now distributed to the most remote points of the world, we ought to find, and I believe as a general rule we do find, that some at least of the species range very widely; for it is necessary that the unmodified parent should range widely, undergoing modification during its diffusion, and should place itself under diverse conditions favourable for the conversion of its offspring, firstly into new varieties and ultimately into new species.

In considering the wide distribution of certain genera, we should bear in mind that some are extremely ancient, and must have branched off from a common parent at a remote epoch; so that in such cases there will have been ample time for great climatal and geographical changes and for accidents of transport; and consequently for the migration of some of the species into all quarters of the world, where they may have become slightly modified in relation to their new conditions. There is, also, some reason to believe from geological evidence that organisms low in the scale within each great class, generally change at a slower rate than the higher forms; and consequently the lower forms will have had a better chance of ranging widely and of still retaining the same specific character. This fact, together with the seeds and eggs of many low forms being very minute and better fitted for distant transportation, probably accounts for a law which has long been observed, and which has lately been admirably discussed by Alph. de Candolle in regard to plants, namely, that the lower any group of organisms is, the more widely it is apt to range.

The relations just discussed, – namely, low and slowly-changing organisms ranging more widely than the high, – some of the species of widely-ranging genera themselves ranging widely, – such facts, as alpine, lacustrine, and marsh productions being related (with the exceptions before specified) to those on the surrounding low lands and dry lands, though these stations are so different – the very close relation of the distinct species which inhabit the islets of the same archipelago, – and especially the striking relation of the inhabitants of each whole archipelago or island to those of the nearest mainland, – are, I think, utterly inexplicable on the ordinary view of the independent creation of each species, but are explicable on the view of colonization from the nearest and readiest source, together with the subsequent modification and better adaptation of the colonists to their new homes.

Summary of last and present Chapters. In these chapters I have endeavoured to show, that if we make due allowance for our ignorance of the full effects of all the changes of climate and of the level of the land, which have certainly occurred within the recent period, and of other similar changes which may have occurred within the same period; if we remember how profoundly ignorant we are with respect to the many and curious means of occasional transport, – a subject which has hardly ever been properly experimentized on; if we bear in mind how often a species may have ranged continuously over a wide area, and then have become extinct in the intermediate tracts, I think the difficulties in believing that all the individuals of the same species, wherever located, have descended from the same parents, are not insuperable. And we are led to this conclusion, which has been arrived at by many naturalists under the designation of single centres of creation, by some general considerations, more especially from the importance of barriers and from the analogical distribution of sub-genera, genera, and families.

With respect to the distinct species of the same genus, which on my theory must have spread from one parent-source; if we make the same allowances as before for our ignorance, and remember that some forms of life change most slowly, enormous periods of time being thus granted for their migration, I do not think that the difficulties are insuperable; though they often are in this case, and in that of the individuals of the same species, extremely grave.

As exemplifying the effects of climatal changes on distribution, I have attempted to show how important has been the influence of the modern Glacial period, which I am fully convinced simultaneously affected the whole world, or at least great meridional belts. As showing how diversified are the means of occasional transport, I have discussed at some little length the means of dispersal of fresh-water productions.

If the difficulties be not insuperable in admitting that in the long course of time the individuals of the same species, and likewise of allied species, have proceeded from some one source; then I think all the grand leading facts of geographical distribution are explicable on the theory of migration (generally of the more dominant forms of life), together with subsequent modification and the multiplication of new forms. We can thus understand the high importance of barriers, whether of land or water, which separate our several zoological and botanical provinces. We can thus understand the localization of sub-genera, genera, and families; and how it is that under different latitudes, for instance in South America, the inhabitants of the plains and mountains, of the forests, marshes, and deserts, are in so mysterious a manner linked together by affinity, and are likewise linked to the extinct beings which formerly inhabited the same continent. Bearing in mind that the mutual relations of organism to organism are of the highest importance, we can see why two areas having nearly the same physical conditions should often be inhabited by very different forms of life; for according to the length of time which has elapsed since new inhabitants entered one region; according to the nature of the communication which allowed certain forms and not others to enter, either in greater or lesser numbers; according or not, as those which entered happened to come in more or less direct competition with each other and with the aborigines; and according as the immigrants were capable of varying more or less rapidly, there would ensue in different regions, independently of their physical conditions, infinitely diversified conditions of life, – there would be an almost endless amount of organic action and reaction, – and we should find, as we do find, some groups of beings greatly, and some only slightly modified, – some developed in great force, some existing in scanty numbers – in the different great geographical provinces of the world.

On these same principles, we can understand, as I have endeavoured to show, why oceanic islands should have few inhabitants, but of these a great number should be endemic or peculiar; and why, in relation to the means of migration, one group of beings, even within the same class, should have all its species endemic, and another group should have all its species common to other quarters of the world. We can see why whole groups of organisms, as batrachians and terrestrial mammals, should be absent from oceanic islands, whilst the most isolated islands possess their own peculiar species of aërial mammals or bats. We can see why there should be some relation between the presence of mammals, in a more or less modified condition, and the depth of the sea between an island and the mainland. We can clearly see why all the inhabitants of an archipelago, though specifically distinct on the several islets, should be closely related to each other, and likewise be related, but less closely, to those of the nearest continent or other source whence immigrants were probably derived. We can see why in two areas, however distant from each other, there should be a correlation, in the presence of identical species, of varieties, of doubtful species, and of distinct but representative species.

As the late Edward Forbes often insisted, there is a striking parallelism in the laws of life throughout time and space: the laws governing the succession of forms in past times being nearly the same with those governing at the present time the differences in different areas. We see this in many facts. The endurance of each species and group of species is continuous in time; for the exceptions to the rule are so few, that they may fairly be attributed to our not having as yet discovered in an intermediate deposit the forms which are therein absent, but which occur above and below: so in space, it certainly is the general rule that the area inhabited by a single species, or by a group of species, is continuous; and the exceptions, which are not rare, may, as I have attempted to show, be accounted for by migration at some former period under different conditions or by occasional means of transport, and by the species having become extinct in the intermediate tracts. Both in time and space, species and groups of species have their points of maximum development. Groups of species, belonging either to a certain period of time, or to a certain area, are often characterized by trifling characters in common, as of sculpture or colour. In looking to the long succession of ages, as in now looking to distant provinces throughout the world, we find that some organisms differ little, whilst others belonging to a different class, or to a different order, or even only to a different family of the same order, differ greatly. In both time and space the lower members of each class generally change less than the higher; but there are in both cases marked exceptions to the rule. On my theory these several relations throughout time and space are intelligible; for whether we look to the forms of life which have changed during successive ages within the same quarter of the world, or to those which have changed after having migrated into distant quarters, in both cases the forms within each class have been connected by the same bond of ordinary generation; and the more nearly any two forms are related in blood, the nearer they will generally stand to each other in time and space; in both cases the laws of variation have been the same, and modifications have been accumulated by the same power of natural selection.

CHAPTER XIII

MUTUAL AFFINITIES OF ORGANIC BEINGS: MORPHOLOGY: EMBRYOLOGY: RUDIMENTARY ORGANS

CLASSIFICATION, groups subordinate to groups – Natural system – Rules and difficulties in classification, explained on the theory of descent with modification – Classification of varieties – Descent always used in classification – Analogical or adaptive characters – Affinities, general, complex and radiating – Extinction separates and defines groups – MORPHOLOGY, between members of the same class, between parts of the same individual – EMBRYOLOGY, laws of, explained by variations not supervening at an early age, and being inherited at a corresponding age – RUDIMENTARY ORGANS; their origin explained – Summary

FROM the first dawn of life, all organic beings are found to resemble each other in descending degrees, so that they can be classed in groups under groups. This classification is evidently not arbitrary like the grouping of the stars in constellations. The existence of groups would have been of simple signification, if one group had been exclusively fitted to inhabit the land, and another the water; one to feed on flesh, another on vegetable matter, and so on; but the case is widely different in nature; for it is notorious how commonly members of even the same sub-group have different habits. In our second and fourth chapters, on Variation and on Natural Selection, I have attempted to show that it is the widely ranging, the much diffused and common, that is the dominant species belonging to the larger genera, which vary most. The varieties, or incipient species, thus produced ultimately become converted, as I believe, into new and distinct species; and these, on the principle of inheritance, tend to produce other new and dominant species. Consequently the groups which are now large, and which generally include many dominant species, tend to go on increasing indefinitely in size. I further attempted to show that from the varying descendants of each species trying to occupy as many and as different places as possible in the economy of nature, there is a constant tendency in their characters to diverge. This conclusion was supported by looking at the great diversity of the forms of life which, in any small area, come into the closest competition, and by looking to certain facts in naturalization.

I attempted also to show that there is a constant tendency in the forms which are increasing in number and diverging in character, to supplant and exterminate the less divergent, the less improved, and preceding forms. I request the reader to turn to the diagram illustrating the action, as formerly explained, of these several principles; and he will see that the inevitable result is that the modified descendants proceeding from one progenitor become broken up into groups subordinate to groups. In the diagram each letter on the uppermost line may represent a genus including several species; and all the genera on this line form together one class, for all have descended from one ancient but unseen parent, and, consequently, have inherited something in common. But the three genera on the left hand have, on this same principle, much in common, and form a sub-family, distinct from that including the next two genera on the right hand, which diverged from a common parent at the fifth stage of descent. These five genera have also much, though less, in common; and they form a family distinct from that including the three genera still further to the right hand, which diverged at a still earlier period. And all these genera, descended from (A), form an order distinct from the genera descended from (I). So that we here have many species descended from a single progenitor grouped into genera; and the genera are included in, or subordinate to, sub-families, families, and orders, all united into one class. Thus, the grand fact in natural history of the subordination of group under group, which, from its familiarity, does not always sufficiently strike us, is in my judgement fully explained.

Naturalists try to arrange the species, genera, and families in each class, on what is called the Natural System. But what is meant by this system? Some authors look at it merely as a scheme for arranging together those living objects which are most alike, and for separating those which are most unlike; or as an artificial means for enunciating, as briefly as possible, general propositions, – that is, by one sentence to give the characters common, for instance, to all mammals, by another those common to all carnivora, by another those common to the dog-genus, and then by adding a single sentence, a full description is given of each kind of dog. The ingenuity and utility of this system are indisputable. But many naturalists think that something more is meant by the Natural System; they believe that it reveals the plan of the Creator; but unless it be specified whether order in time or space, or what else is meant by the plan of the Creator, it seems to me that nothing is thus added to our knowledge. Such expressions as that famous one of Linnaeus, and which we often meet with in a more or less concealed form, that the characters do not make the genus, but that the genus gives the characters, seem to imply that something more is included in our classification, than mere resemblance. I believe that something more is included; and that propinquity of descent, – the only known cause of the similarity of organic beings, – is the bond, hidden as it is by various degrees of modification, which is partially revealed to us by our classifications.

Let us now consider the rules followed in classification, and the difficulties which are encountered on the view that classification either gives some unknown plan of creation, or is simply a scheme for enunciating general propositions and of placing together the forms most like each other. It might have been thought (and was in ancient times thought) that those parts of the structure which determined the habits of life, and the general place of each being in the economy of nature, would be of very high importance in classification. Nothing can be more false. No one regards the external similarity of a mouse to a shrew, of a dugong to a whale, of a whale to a fish, as of any importance. These resemblances, though so intimately connected with the whole life of the being, are ranked as merely 'adaptive or analogical characters;' but to the consideration of these resemblances we shall have to recur. It may even be given as a general rule, that the less any part of the organization is concerned with special habits, the more important it becomes for classification. As an instance: Owen, in speaking of the dugong, says, 'The generative organs being those which are most remotely related to the habits and food of an animal, I have always regarded as affording very clear indications of its true affinities. We are least likely in the modifications of these organs to mistake a merely adaptive for an essential character.' So with plants, how remarkable it is that the organs of vegetation, on which their whole life depends, are of little signification, excepting in the first main divisions; whereas the organs of reproduction, with their product the seed, are of paramount importance!

We must not, therefore, in classifying, trust to resemblances in parts of the organization, however important they may be for the welfare of the being in relation to the outer world. Perhaps from this cause it has partly arisen, that almost all naturalists lay the greatest stress on resemblances in organs of high vital or physiological importance. No doubt this view of the classificatory importance of organs which are important is generally, but by no means always, true. But their importance for classification, I believe, depends on their greater constancy throughout large groups of species; and this constancy depends on such organs having generally been subjected to less change in the adaptation of the species to their conditions of life. That the mere physiological importance of an organ does not determine the classificatory value, is almost shown by the one fact, that in allied groups, in which the same organ, as we have every reason to suppose, has nearly the same physiological value, its classificatory value is widely different. No naturalist can have worked at any group without being struck with this fact; and it has been most fully acknowledged in the writings of almost every author. It will suffice to quote the highest authority, Robert Brown, who in speaking of certain organs in the Proteaceae, says their generic importance, 'like that of all their parts, not only in this but, as I apprehend, in every natural family, is very unequal, and in some cases seems to be entirely lost.' Again in another work he says, the genera of the Connaraceae 'differ in having one or more ovaria, in the existence or absence of albumen, in the imbricate or valvular aestivation. Any one of these characters singly is frequently of more than generic importance, though here even when all taken together they appear insufficient to separate Cnestis from Connarus.' To give an example amongst insects, in one great division of the Hymenoptera, the antennae, as Westwood has remarked, are most constant in structure; in another division they differ much, and the differences are of quite subordinate value in classification; yet no one probably will say that the antennae in these two divisions of the same order are of unequal physiological importance. Any number of instances could be given of the varying importance for classification of the same important organ within the same group of beings.

Again, no one will say that rudimentary or atrophied organs are of high physiological or vital importance; yet, undoubtedly, organs in this condition are often of high value in classification. No one will dispute that the rudimentary teeth in the upper jaws of young ruminants, and certain rudimentary bones of the leg, are highly serviceable in exhibiting the close affinity between Ruminants and Pachyderms. Robert Brown has strongly insisted on the fact that the rudimentary florets are of the highest importance in the classification of the Grasses.

Numerous instances could be given of characters derived from parts which must be considered of very trifling physiological importance, but which are universally admitted as highly serviceable in the definition of whole groups. For instance, whether or not there is an open passage from the nostrils to the mouth, the only character, according to Owen, which absolutely distinguishes fishes and reptiles – the inflection of the angle of the jaws in Marsupials – the manner in which the wings of insects are folded – mere colour in certain Algae – mere pubescence on parts of the flower in grasses – the nature of the dermal covering, as hair or feathers, in the Vertebrata. If the Ornithorhynchus had been covered with feathers instead of hair, this external and trifling character would, I think, have been considered by naturalists as important an aid in determining the degree of affinity of this strange creature to birds and reptiles, as an approach in structure in any one internal and important organ.

The importance, for classification, of trifling characters, mainly depends on their being correlated with several other characters of more or less importance. The value indeed of an aggregate of characters is very evident in natural history. Hence, as has often been remarked, a species may depart from its allies in several characters, both of high physiological importance and of almost universal prevalence, and yet leave us in no doubt where it should be ranked.

Hence, also, it has been found, that a classification founded on any single character, however important that may be, has always failed; for no part of the organization is universally constant. The importance of an aggregate of characters, even when none are important, alone explains, I think, that saying of Linnaeus, that the characters do not give the genus, but the genus gives the characters; for this saying seems founded on an appreciation of many trifling points of resemblance, too slight to be defined. Certain plants, belonging to the Malpighiaceae, bear perfect and degraded flowers; in the latter, as A. de Jussieu has remarked, 'the greater number of the characters proper to the species, to the genus, to the family, to the class, disappear, and thus laugh at our classification.' But when Aspicarpa produced in France, during several years, only degraded flowers, departing so wonderfully in a number of the most important points of structure from the proper type of the order, yet M. Richard sagaciously saw, as Jussieu observes, that this genus should still be retained amongst the Malpighiaceae. This case seems to me well to illustrate the spirit with which our classifications are sometimes necessarily founded.

Practically when naturalists are at work, they do not trouble themselves about the physiological value of the characters which they use in defining a group, or in allocating any particular species. If they find a character nearly uniform, and common to a great number of forms, and not common to others, they use it as one of high value; if common to some lesser number, they use it as of subordinate value. This principle has been broadly confessed by some naturalists to be the true one; and by none more clearly than by that excellent botanist, Aug. St. Hilaire. If certain characters are always found correlated with others, though no apparent bond of connexion can be discovered between them, especial value is set on them. As in most groups of animals, important organs, such as those for propelling the blood, or for aërating it, or those for propagating the race, are found nearly uniform, they are considered as highly serviceable in classification; but in some groups of animals all these, the most important vital organs, are found to offer characters of quite subordinate value.

We can see why characters derived from the embryo should be of equal importance with those derived from the adult, for our classifications of course include all ages of each species. But it is by no means obvious, on the ordinary view, why the structure of the embryo should be more important for this purpose than that of the adult, which alone plays its full part in the economy of nature. Yet it has been strongly urged by those great naturalists, Milne Edwards and Agassiz, that embryonic characters are the most important of any in the classification of animals; and this doctrine has very generally been admitted as true. The same fact holds good with flowering plants, of which the two main divisions have been founded on characters derived from the embryo, – on the number and position of the embryonic leaves or cotyledons, and on the mode of development of the plumule and radicle. In our discussion on embryology, we shall see why such characters are so valuable, on the view of classification tacitly including the idea of descent.

Our classifications are often plainly influenced by chains of affinities. Nothing can be easier than to define a number of characters common to all birds; but in the case of crustaceans, such definition has hitherto been found impossible. There are crustaceans at the opposite ends of the series, which have hardly a character in common; yet the species at both ends, from being plainly allied to others, and these to others, and so onwards, can be recognized as unequivocally belonging to this, and to no other class of the Articulata.

Geographical distribution has often been used, though perhaps not quite logically, in classification, more especially in very large groups of closely allied forms. Temminck insists on the utility or even necessity of this practice in certain groups of birds; and it has been followed by several entomologists and botanists.

Finally, with respect to the comparative value of the various groups of species, such as orders, sub-orders, families, sub-families, and genera, they seem to be, at least at present, almost arbitrary. Several of the best botanists, such as Mr. Bentham and others, have strongly insisted on their arbitrary value. Instances could be given amongst plants and insects, of a group of forms, first ranked by practised naturalists as only a genus, and then raised to the rank of a sub-family or family; and this has been done, not because further research has detected important structural differences, at first overlooked, but because numerous allied species, with slightly different grades of difference, have been subsequently discovered.

All the foregoing rules and aids and difficulties in classification are explained, if I do not greatly deceive myself, on the view that the natural system is founded on descent with modification; that the characters which naturalists consider as showing true affinity between any two or more species, are those which have been inherited from a common parent, and, in so far, all true classification is genealogical; that community of descent is the hidden bond which naturalists have been unconsciously seeking, and not some unknown plan of creation, or the enunciation of general propositions, and the mere putting together and separating objects more or less alike.

But I must explain my meaning more fully. I believe that the *arrangement* of the groups within each class, in due subordination and relation to the other groups, must be strictly genealogical in order to be natural; but that the *amount* of difference in the several branches or groups, though allied in the same degree in blood to their common progenitor, may differ greatly, being due to the different degrees of modification which they have undergone; and this is expressed by the forms being ranked under different genera, families, sections, or orders. The reader will best understand what is meant, if he will take the trouble to refer to the diagram in the fourth chapter. We will suppose the letters A to L to represent allied genera, which lived during the Silurian epoch, and these have descended from a species which existed at an unknown anterior period. Species of three of these genera (A, F, and I) have transmitted modified descendants to the present day, represented by the fifteen genera (a^{14} to z^{14}) on the uppermost horizontal line. Now all these modified descendants from a single species, are represented as related in blood or descent to the same degree; they may metaphorically be called cousins to the same millionth degree; yet they differ widely and in different degrees from each other. The forms descended from A, now broken up into two or three families, constitute a distinct order from those descended from I, also broken up into two families. Nor can the existing species, descended from A, be ranked in the same genus with the parent A; or those from I, with the parent I. But the existing genus F^{14} may be supposed to have been but slightly modified; and it will then rank with the parent-genus F; just as some few still living organic beings belong to Silurian genera. So that the amount or value of the differences between organic beings all related to each other in the same degree in blood, has come to be widely different. Nevertheless their genealogical *arrangement* remains strictly true, not only at the present time, but at each successive period of descent. All the modified descendants from A will have inherited something in common from their common parent, as will all the descendants from I; so will it be with each subordinate branch of descendants, at each successive period. If, however, we choose to suppose that any of the descendants of A or of I have been so much modified as to have more or less completely lost traces of their parentage, in this case, their places in a natural classification will have been more or less completely lost, – as sometimes seems to have occurred with existing organisms. All the descendants of the genus F, along its whole line of descent, are supposed to have been but little modified, and they yet form a single genus. But this genus, though much isolated, will still occupy its proper intermediate position; for F originally was intermediate in character between A and I, and the several genera descended from these two genera will have inherited to a certain extent their characters. This natural arrangement is shown, as far as is possible on paper, in the diagram, but in much too simple a manner. If a branching diagram had not been used, and only the names of the groups had been written in a linear series, it would have been still less possible to have given a natural arrangement; and it is notoriously not possible to represent in a series, on a flat surface, the affinities which we discover in nature amongst the beings of the same group. Thus, on the view which I hold, the natural system is genealogical in its arrangement, like a pedigree; but the degrees of modification which the different groups have undergone, have to be expressed by ranking them under different so-called genera, sub-families, families, sections, orders, and classes.

It may be worth while to illustrate this view of classification, by taking the case of languages. If we possessed a perfect pedigree of mankind, a genealogical arrangement of the races of man would afford the best classification of the various languages now spoken throughout the world; and if all extinct languages, and all intermediate and slowly changing dialects, had to be included, such an arrangement would, I think, be the only possible one. Yet it might be that some very ancient language had altered little, and had given rise to few new languages, whilst others (owing to the spreading and subsequent isolation and states of civilization of the several races, descended from a common race) had altered much, and had given rise to many new languages and dialects. The various degrees of difference in the languages from the same stock, would have to be expressed by groups subordinate to groups; but the proper or even only possible arrangement would still be genealogical; and this would be strictly natural, as it would connect together all languages, extinct and modern, by the closest affinities, and would give the filiation and origin of each tongue.

In confirmation of this view, let us glance at the classification of varieties, which are believed or known to have descended from one species. These are grouped under species, with sub-varieties under varieties; and with our domestic productions, several other grades of difference are requisite, as we have seen with pigeons. The origin of the existence of groups subordinate to groups, is the same with varieties as with species, namely, closeness of descent with various degrees of modification. Nearly the same rules are followed in classifying varieties, as with species. Authors have insisted on the necessity of classing varieties on a natural instead of an artificial system; we are cautioned, for instance, not to class two varieties of the pine-apple together, merely because their fruit, though the most important part, happens to be nearly identical; no one puts the swedish and common turnips together, though the esculent and thickened stems are so similar. Whatever part is found to be most constant, is used in classing varieties: thus the great agriculturist Marshall says the horns are very useful for this purpose with cattle, because they are less variable than the shape or colour of the body, &c.; whereas with sheep the horns are much less serviceable, because less constant. In classing varieties, I apprehend if we had a real pedigree, a genealogical classification would be universally preferred; and it has been attempted by some authors. For we might feel sure, whether there had been more or less modification, the principle of inheritance would keep the forms together which were allied in the greatest number of points. In tumbler pigeons, though some sub-varieties differ from the others in the important character of having a longer beak, yet all are kept together from having the common habit of tumbling; but the short-faced breed has nearly or quite lost this habit; nevertheless, without any reasoning or thinking on the subject, these tumblers are kept in the same group, because allied in blood and alike in some other respects. If it could be proved that the Hottentot had descended from the Negro, I think he would be classed under the Negro group, however much he might differ in colour and other important characters from negroes.

With species in a state of nature, every naturalist has in fact brought descent into his classification; for he includes in his lowest grade, or that of a species, the two sexes; and how enormously these sometimes differ in the most important characters, is known to every naturalist: scarcely a single fact can be predicated in common of the males and hermaphrodites of certain cirripedes, when adult, and yet no one dreams of separating them. The naturalist includes as one species the several larval stages of the same individual, however much they may differ from each other and from the adult; as he likewise includes the so-called alternate generations of Steenstrup, which can only in a technical sense be considered as the same individual. He includes monsters; he includes varieties, not solely because they closely resemble the parent-form, but because they are descended from it. He who believes that the cowslip is descended from the primrose, or conversely, ranks them together as a single species, and gives a single definition. As soon as three Orchidean forms (Monochanthus, Myanthus, and Catasetum), which had previously been ranked as three distinct genera, were known to be sometimes produced on the same spike, they were immediately included as a single species. But it may be asked, what ought we to do, if it could be proved that one species of kangaroo had been produced, by a long course of modification, from a bear? Ought we to rank this one species with bears, and what should we do with the other species? The supposition is of course preposterous; and I might answer by the *argumentum ad hominem*, and ask what should be done if a perfect kangaroo were seen to come out of the womb of a bear? According to all analogy, it would be ranked with bears; but then assuredly all the other species of the kangaroo family would have to be classed under the bear genus. The whole case is preposterous; for where there has been close descent in common, there will certainly be close resemblance or affinity.

As descent has universally been used in classing together the individuals of the same species, though the males and females and larvae are sometimes extremely different; and as it has been used in classing varieties which have undergone a certain, and sometimes a considerable amount of modification, may not this same element of descent have been unconsciously used in grouping species under genera, and genera under higher groups, though in these cases the modification has been greater in degree, and has taken a longer time to complete? I believe it has thus been unconsciously used; and only thus can I understand the several rules and guides which have been followed by our best systematists. We have no written pedigrees; we have to make out community of descent by resemblances of any kind. Therefore we choose those characters which, as far as we can judge, are the least likely to have been modified in relation to the conditions of life to which each species has been recently exposed. Rudimentary structures on this view are as good as, or even sometimes better than, other parts of the organization. We care not how trifling a character may be – let it be the mere inflection of the angle of the jaw, the manner in which an insect's wing is folded, whether the skin be covered by hair or feathers – if it prevail throughout many and different species, especially those having very different habits of life, it assumes high value; for we can account for its presence in so many forms with such different habits, only by its inheritance from a common parent. We may err in this respect in regard to single points of structure, but when several characters, let them be ever so trifling, occur together throughout a large group of beings having different habits, we may feel almost sure, on the theory of descent, that these characters have been inherited from a common ancestor. And we know that such correlated or aggregated characters have especial value in classification.

We can understand why a species or a group of species may depart, in several of its most important characteristics, from its allies, and yet be safely classed with them. This may be safely done, and is often done, as long as a sufficient number of characters, let them be ever so unimportant, betrays the hidden bond of community of descent. Let two forms have not a single character in common, yet if these extreme forms are connected together by a chain of intermediate groups, we may at once infer their community of descent, and we put them all into the same class. As we find organs of high physiological importance – those which serve to preserve life under the most diverse conditions of existence – are generally the most constant, we attach especial value to them; but if these same organs, in another group or section of a group, are found to differ much, we at once value them less in our classification. We shall hereafter, I think, clearly see why embryological characters are of such high classificatory importance. Geographical distribution may sometimes be brought usefully into play in classing large and widely-distributed genera, because all the species of the same genus, inhabiting any distinct and isolated region, have in all probability descended from the same parents.

We can understand, on these views, the very important distinction between real affinities and analogical or adaptive resemblances. Lamarck first called attention to this distinction, and he has been ably followed by Macleay and others. The resemblance, in the shape of the body and in the fin-like anterior limbs, between the dugong, which is a pachydermatous animal, and the whale, and between both these mammals and fishes, is analogical. Amongst insects there are innumerable instances: thus Linnaeus, misled by external appearances, actually classed an homopterous insect as a moth. We see something of the same kind even in our domestic varieties, as in the thickened stems of the common and swedish turnip. The resemblance of the greyhound and race-horse is hardly more fanciful than the analogies which have been drawn by some authors between very distinct animals. On my view of characters being of real importance for classification, only in so far as they reveal descent, we can clearly understand why analogical or adaptive character, although of the utmost importance to the welfare of the being, are almost valueless to the systematist. For animals, belonging to two most distinct lines of descent, may readily become adapted to similar conditions, and thus assume a close external resemblance; but such resemblances will not reveal – will rather tend to conceal their blood-relationship to their proper lines of descent. We can also understand the apparant paradox, that the very same characters are analogical when one class or order is compared with another, but give true affinities when the members of the same class or order are compared one with another: thus the shape of the body and fin-like limbs are only analogical when whales are compared with fishes, being adaptations in both classes for swimming through the water; but the shape of the body and fin-like limbs serve as characters exhibiting true affinity between the several members of the whale family; for these cetaceans agree in so many characters, great and small, that we cannot doubt that they have inherited their general shape of body and structure of limbs from a common ancestor. So it is with fishes.

As members of distinct classes have often been adapted by successive slight modifications to live under nearly similar circumstances, – to inhabit for instance the three elements of land, air, and water, – we can perhaps understand how it is that a numerical parallelism has sometimes been observed between the sub-groups in distinct classes. A naturalist, struck by a parallelism of this nature in any one class, by arbitrarily raising or sinking the value of the groups in other classes (and all our experience shows that this valuation has hitherto been arbitrary), could easily extend the parallelism over a wide range; and thus the septenary, quinary, quaternary, and ternary classifications have probably arisen.

As the modified descendants of dominant species, belonging to the larger genera, tend to inherit the advantages, which made the groups to which they belong large and their parents dominant, they are almost sure to spread widely, and to seize on more and more places in the economy of nature. The larger and more dominant groups thus tend to go on increasing in size; and they consequently supplant many smaller and feebler groups. Thus we can account for the fact that all organisms, recent and extinct, are included under a few great orders, under still fewer classes, and all in one great natural system. As showing how few the higher groups are in number, and how widely spread they are throughout the world, the fact is striking, that the discovery of Australia has not added a single insect belonging to a new order; and that in the vegetable kingdom, as I learn from Dr. Hooker, it has added only two or three orders of small size.

In the chapter on geological succession I attempted to show, on the principle of each group having generally diverged much in character during the long-continued process of modification, how it is that the more ancient forms of life often present characters in some slight degree intermediate between existing groups. A few old and intermediate parent-forms having occasionally transmitted to the present day descendants but little modified, will give to us our so-called osculant or aberrant groups. The more aberrant any form is, the greater must be the number of connecting forms which on my theory have been exterminated and utterly lost. And we have some evidence of aberrant forms having suffered severely from extinction, for they are generally represented by extremely few species; and such species as do occur are generally very distinct from each other, which again implies extinction. The genera Ornithorhynchus and Lepido-siren, for example, would not have been less aberrant had each been represented by a dozen species instead of by a single one; but such richness in species, as I find after some investigation, does not commonly fall to the lot of aberrant genera. We can, I think, account for this fact only by looking at aberrant forms as failing groups conquered by more successful competitors, with a few members preserved by some unusual coincidence of favourable circumstances.

Mr. Waterhouse has remarked that, when a member belonging to one group of animals exhibits an affinity to a quite distinct group, this affinity in most cases is general and not special: thus, according to Mr. Waterhouse, of all Rodents, the bizcacha is most nearly related to Marsupials; but in the points in which it approaches this order, its relations are general, and not to any one marsupial species more than to another. As the points of affinity of the bizcacha to Marsupials are believed to be real and not merely adaptive, they are due on my theory to inheritance in common. Therefore we must suppose either that all Rodents, including the bizcacha, branched off from some very ancient Marsupial, which will have had a character in some degree intermediate with respect to all existing Marsupials; or that both Rodents and Marsupials branched off from a common progenitor, and that both groups have since undergone much modification in divergent directions. On either view we may suppose that the bizca-cha has retained, by inheritance, more of the character of its ancient progenitor than have other Rodents; and therefore it will not be specially related to any one existing Marsupial, but indirectly to all or nearly all Marsupials, from having partially retained the character of their common progenitor, or of an early member of the group. On the other hand, of all Marsupials, as Mr. Waterhouse has remarked, the phascolomys resembles most nearly, not any one species, but the general order of Rodents. In this case, however, it may be strongly suspected that the resemblance is only analogical, owing to the phascolomys having become adapted to habits like those of a Rodent. The elder De Candolle has made nearly similar observations on the general nature of the affinities of distinct orders of plants.

On the principle of the multiplication and gradual divergence in character of the species descended from a common parent, together with their retention by inheritance of some characters in common, we can understand the excessively complex and radiating affinities by which all the members of the same family or higher group are connected together. For the common parent of a whole family of species, now broken up by extinction into distinct groups and sub-groups, will have transmitted some of its characters, modified in various ways and degrees, to all; and the several species will conse-quently be related to each other by circuitous lines of affinity of various lengths (as may be seen in the diagram so often referred to), mounting up through many predecessors. As it is difficult to show the blood-relationship between the numerous kindred of any ancient and noble family, even by the aid of a genealogical tree, and almost impossible to do this without this aid, we can understand the extra-ordinary difficulty which naturalists have experienced in describing, without the aid of a diagram, the various affinities which they per-ceive between the many living and extinct members of the same great natural class.

Extinction, as we have seen in the fourth chapter, has played an important part in defining and widening the intervals between the several groups in each class. We may thus account even for the distinctness of whole classes from each other – for instance, of birds from all other vertebrate animals – by the belief that many ancient forms of life have been utterly lost, through which the early progeni-tors of birds were formerly connected with the early progenitors of the other vertebrate classes. There has been less entire extinction of the forms of life which once connected fishes with batrachians.

There has been still less in some other classes, as in that of the Crustacea, for here the most wonderfully diverse forms are still tied together by a long, but broken, chain of affinities. Extinction has only separated groups: it has by no means made them; for if every form which has ever lived on this earth were suddenly to reappear, though it would be quite impossible to give definitions by which each group could be distinguished from other groups, as all would blend together by steps as fine as those between the finest existing varieties, nevertheless a natural classification, or at least a natural arrangement, would be possible. We shall see this by turning to the diagram: the letters, A to L, may represent eleven Silurian genera, some of which have produced large groups of modified descendants. Every intermediate link between these eleven genera and their primordial parent, and every intermediate link in each branch and sub-branch of their descendants, may be supposed to be still alive; and the links to be as fine as those between the finest varieties. In this case it would be quite impossible to give any definition by which the several members of the several groups could be distinguished from their more immediate parents; or these parents from their ancient and unknown progenitor. Yet the natural arrangement in the diagram would still hold good; and, on the principle of inheritance, all the forms descended from A, or from I, would have something in common. In a tree we can specify this or that branch, though at the actual fork the two unite and blend together. We could not, as I have said, define the several groups; but we could pick out types, or forms, representing most of the characters of each group, whether large or small, and thus give a general idea of the value of the differences between them. This is what we should be driven to, if we were ever to succeed in collecting all the forms in any class which have lived throughout all time and space. We shall certainly never succeed in making so perfect a collection: nevertheless, in certain classes, we are tending in this direction; and Milne Edwards has lately insisted, in an able paper, on the high importance of looking to types, whether or not we can separate and define the groups to which such types belong.

Finally, we have seen that natural selection, which results from the struggle for existence, and which almost inevitably induces extinction and divergence of character in the many descendants from one dominant parent-species, explains that great and universal feature in the affinities of all organic beings, namely, their subordination in group under group. We use the element of descent in classing the individuals of both sexes and of all ages, although having few characters in common, under one species; we use descent in classing acknowledged varieties, however different they may be from their parent; and I believe this element of descent is the hidden bond of connexion which naturalists have sought under the term of the Natural System. On this idea of the natural system being, in so far as it has been perfected, genealogical in its arrangement, with the grades of difference between the descendants from a common parent, expressed by the terms genera, families, orders, &c., we can understand the rules which we are compelled to follow in our classification. We can understand why we value certain resemblances far more than others; why we are permitted to use rudimentary and useless organs, or others of trifling physiological importance; why, in comparing one group with a distinct group, we summarily reject analogical or adaptive characters, and yet use these same characters within the limits of the same group. We can clearly see how it is that all living and extinct forms can be grouped together in one great system; and how the several members of each class are connected together by the most complex and radiating lines of affinities. We shall never, probably, disentangle the inextricable web of affinities between the members of any one class; but when we have a distinct object in view, and do not look to some unknown plan of creation, we may hope to make sure but slow progress.

Morphology. We have seen that the members of the same class, independently of their habits of life, resemble each other in the general plan of their organization. This resemblance is often expressed by the term 'unity of type;' or by saying that the several parts and organs in the different species of the class are homologous. The whole subject is included under the general name of Morphology. This is the most interesting department of natural history, and may be said to be its very soul. What can be more curious than that the hand of a man, formed for grasping, that of a mole for digging, the leg of the horse, the paddle of the porpoise, and the wing of the bat, should all be constructed on the same pattern, and should include the same bones, in the same relative positions? Geoffroy St. Hilaire has insisted strongly on the high importance of relative connexion in homologous organs: the parts may change to almost any extent in form and size, and yet they always remain connected together in the same order. We never find, for instance, the bones of the arm and forearm, or of the thigh and leg, transposed. Hence the same names can be given to the homologous bones in widely different animals. We see the same great law in the construction of the mouths of insects: what can be more different than the immensely long spiral proboscis of a sphinx-moth, the curious folded one of a bee or bug, and the great jaws of a beetle? – yet all these organs, serving for such different purposes, are formed by infinitely numerous modifications of an upper lip, mandibles, and two pairs of maxillae. Analogous laws govern the construction of the mouths and limbs of crustaceans. So it is with the flowers of plants.

Nothing can be more hopeless than to attempt to explain this similarity of pattern in members of the same class, by utility or by the doctrine of final causes. The hopelessness of the attempt has been expressly admitted by Owen in his most interesting work on the 'Nature of Limbs.' On the ordinary view of the independent creation of each being, we can only say that so it is; – that it has so pleased the Creator to construct each animal and plant.

The explanation is manifest on the theory of the natural selection of successive slight modifications, – each modification being profitable in some way to the modified form, but often affecting by correlation of growth other parts of the organization. In changes of this nature, there will be little or no tendency to modify the original pattern, or to transpose parts. The bones of a limb might be shortened and widened to any extent, and become gradually enveloped in thick membrane, so as to serve as a fin; or a webbed foot might have all its bones, or certain bones, lengthened to any extent, and the membrane connecting them increased to any extent, so as to serve as a wing: yet in all this great amount of modification there will be no tendency to alter the framework of bones or the relative connexion of the several parts. If we suppose that the ancient progenitor, the archetype as it may be called, of all mammals, had its limbs constructed on the existing general pattern, for whatever purpose they served, we can at once perceive the plain signification of the homologous construction of the limbs throughout the whole class. So with the mouths of insects, we have only to suppose that their common progenitor had an upper lip, mandibles, and two pair of maxillae, these parts being perhaps very simple in form; and then natural selection will account for the infinite diversity in structure and function of the mouths of insects. Nevertheless, it is conceivable that the general pattern of an organ might become so much obscured as to be finally lost, by the atrophy and ultimately by the complete abortion of certain parts, by the soldering together of other parts, and by the doubling or multiplication of others, – variations which we know to be within the limits of possibility. In the paddles of the extinct gigantic sea-lizards, and in the mouths of certain suctorial crustaceans, the general pattern seems to have been thus to a certain extent obscured.

There is another and equally curious branch of the present subject; namely, the comparison not of the same part in different members of a class, but of the different parts or organs in the same individual. Most physiologists believe that the bones of the skull are homologous with – that is correspond in number and in relative connexion with – the elemental parts of a certain number of vertebrae. The anterior and posterior limbs in each member of the vertebrate and articulate classes are plainly homologous. We see the same law in comparing the wonderfully complex jaws and legs in crustaceans. It is familiar to almost every one, that in a flower the relative position of the sepals, petals, stamens, and pistils, as well as their intimate structure, are intelligible in the view that they consist of metamorphosed leaves, arranged in a spire. In monstrous plants, we often get direct evidence of the possibility of one organ being transformed into another; and we can actually see in embryonic crustaceans and in many other animals, and in flowers, that organs, which when mature become extremely different, are at an early stage of growth exactly alike.

How inexplicable are these facts on the ordinary view of creation! Why should the brain be enclosed in a box composed of such numerous and such extraordinarily shaped pieces of bone? As Owen has remarked, the benefit derived from the yielding of the separate pieces in the act of parturition of mammals, will by no means explain the same construction in the skulls of birds. Why should similar bones have been created in the formation of the wing and leg of a bat, used as they are for such totally different purposes? Why should one crustacean, which has an extremely complex mouth formed of many parts, consequently always have fewer legs; or conversely, those with many legs have simpler mouths? Why should the sepals, petals, stamens, and pistils in any individual flower, though fitted for such widely different purposes, be all constructed on the same pattern?

On the theory of natural selection, we can satisfactorily answer these questions. In the vertebrata, we see a series of internal vertebrae bearing certain processes and appendages; in the articulata, we see the body divided into a series of segments, bearing external appendages; and in flowering plants, we see a series of successive spiral whorls of leaves. An indefinite repetition of the same part or organ is the common characteristic (as Owen has observed) of all low or little-modified forms; therefore we may readily believe that the unknown progenitor of the vertebrata possessed many vertebrae; the unknown progenitor of the articulata, many segments; and the unknown progenitor of flowering plants, many spiral whorls of leaves. We have formerly seen that parts many times repeated are eminently liable to vary in number and structure; consequently it is quite probable that natural selection, during a long-continued course of modification, should have seized on a certain number of the primordially similar elements, many times repeated, and have adapted them to the most diverse purposes. And as the whole amount of modification will have been effected by slight successive steps, we need not wonder at discovering in such parts or organs, a certain degree of fundamental resemblance, retained by the strong principle of inheritance.

In the great class of molluscs, though we can homologize the parts of one species with those of another and distinct species, we can indicate but few serial homologies; that is, we are seldom enabled to say that one part or organ is homologous with another in the same individual. And we can understand this fact; for in molluscs, even in the lowest members of the class, we do not find nearly so much indefinite repetition of any one part, as we find in the other great classes of the animal and vegetable kingdoms.

Naturalists frequently speak of the skull as formed of metamorphosed vertebrae: the jaws of crabs as metamorphosed legs; the stamens and pistils of flowers as metamorphosed leaves; but it would in these cases probably be more correct, as Professor Huxley has remarked, to speak of both skull and vertebrae, both jaws and legs, &c., – as having been metamorphosed, not one from the other, but from some common element. Naturalists, however, use such language only in a metaphorical sense: they are far from meaning that during a long course of descent, primordial organs of any kind – vertebrae in the one case and legs in the other – have actually been modified into skulls or jaws. Yet so strong is the appearance of a modification of this nature having occurred, that naturalists can hardly avoid employing language having this plain signification. On my view these terms may be used literally; and the wonderful fact of the jaws, for instance, of a crab retaining numerous characters, which they would probably have retained through inheritance, if they had really been metamorphosed during a long course of descent from true legs, or from some simple appendage, is explained.

Embryology. It has already been casually remarked that certain organs in the individual, which when mature become widely different and serve for different purposes, are in the embryo exactly alike. The embryos, also, of distinct animals within the same class are often strikingly similar: a better proof of this cannot be given, than a circumstance mentioned by Agassiz, namely, that having forgotten to ticket the embryo of some vertebrate animal, he cannot now tell whether it be that of a mammal, bird, or reptile. The vermiform larvae of moths, flies, beetles, &c., resemble each other much more closely than do the mature insects; but in the case of larvae, the embryos are active, and have been adapted for special lines of life. A trace of the law of embryonic resemblance, sometimes lasts till a rather late age: thus birds of the same genus, and of closely allied genera, often resemble each other in their first and second plumage; as we see in the spotted feathers in the thrush group. In the cat tribe, most of the species are striped or spotted in lines; and stripes can be plainly distinguished in the whelp of the lion. We occasionally though rarely see something of this kind in plants: thus the embryonic leaves of the ulex or furze, and the first leaves of the phyllodineous acaceas, are pinnate or divided like the ordinary leaves of the leguminosae.

The points of structure, in which the embryos of widely different animals of the same class resemble each other, often have no direct relation to their conditions of existence. We cannot, for instance, suppose that in the embryos of the vertebrata the peculiar loop-like course of the arteries near the branchial slits are related to similar conditions, – in the young mammal which is nourished in the womb of its mother, in the egg of the bird which is hatched in a nest, and in the spawn of a frog under water. We have no more reason to believe in such a relation, than we have to believe that the same bones in the hand of a man, wing of a bat, and fin of a porpoise, are related to similar conditions of life. No one will suppose that the stripes on the whelp of a lion, or the spots on the young blackbird, are of any use to these animals, or are related to the conditions to which they are exposed.

The case, however, is different when an animal during any part of its embryonic career is active, and has to provide for itself. The period of activity may come on earlier or later in life; but whenever it comes on, the adaptation of the larva to its conditions of life is just as perfect and as beautiful as in the adult animal. From such special adaptations, the similarity of the larvae or active embryos of allied animals is sometimes much obscured; and cases could be given of the larvae of two species, or of two groups of species, differing quite as much, or even more, from each other than do their adult parents. In most cases, however, the larvae, though active, still obey more or less closely the law of common embryonic resemblance. Cirripedes afford a good instance of this: even the illustrious Cuvier did not perceive that a barnacle was, as it certainly is, a crustacean; but a glance at the larva shows this to be the case in an unmistakeable manner. So again the two main divisions of cirripedes, the pedunculated and sessile, which differ widely in external appearance, have larvae in all their several stages barely distinguishable.

The embryo in the course of development generally rises in organization: I use this expression, though I am aware that it is hardly possible to define clearly what is meant by the organization being higher or lower. But no one probably will dispute that the butterfly is higher than the caterpillar. In some cases, however, the mature animal is generally considered as lower in the scale than the larva, as with certain parasitic crustaceans. To refer once again to cirripedes: the larvae in the first stage have three pairs of legs, a very simple single eye, and a prosciformed mouth, with which they feed largely, for they increase much in size. In the second stage, answering to the chrysalis stage of butterflies, they have six pairs of beautifully constructed natatory legs, a pair of magnificent compound eyes, and extremely complex antennae; but they have a closed and imperfect mouth, and cannot feed: their function at this stage is, to search by their well-developed organs of sense, and to reach by their active powers of swimming, a proper place on which to become attached and to undergo their final metamorphosis. When this is completed they are fixed for life: their legs are now converted into prehensile organs; they again obtain a well-constructed mouth; but they have no antennae, and their two eyes are now reconverted into a minute, single, and very simple eye-spot. In this last and complete state, cirripedes may be considered as either more highly or more lowly organized than they were in the larval condition. But in some genera the larvae become developed either into hermaphrodites having the ordinary structure, or into what I have called complemental males: and in the latter, the development has assuredly been retrograde; for the male is a mere sack, which lives for a short time, and is destitute of mouth, stomach, or other organ of importance, excepting for reproduction.

We are so much accustomed to see differences in structure between the embryo and the adult, and likewise a close similarity in the embryos of widely different animals within the same class, that we might be led to look at these facts as necessarily contingent in some manner on growth. But there is no obvious reason why, for instance, the wing of a bat, or the fin of a porpoise, should not have been sketched out with all the parts in proper proportion, as soon as any structure became visible in the embryo. And in some whole groups of animals and in certain members of other groups, the embryo does not at any period differ widely from the adult: thus

Owen has remarked in regard to cuttle-fish, 'there is no metamorphosis; the cephalopodic character is manifested long before the parts of the embryo are completed;' and again in spiders, 'there is nothing worthy to be called a metamorphosis.' The larvae of insects, whether adapted to the most diverse and active habits, or quite inactive, being fed by their parents or placed in the midst of proper nutriment, yet nearly all pass through a similar worm-like stage of development; but in some few cases, as in that of Aphis, if we look to the admirable drawings by Professor Huxley of the development of this insect, we see no trace of the vermiform stage.

How, then, can we explain these several facts in embryology, – namely the very general, but not universal difference in structure between the embryo and the adult; – of parts in the same individual embryo, which ultimately become very unlike and serve for diverse purposes, being at this early period of growth alike; – of embryos of different species within the same class, generally, but not universally, resembling each other; – of the structure of the embryo not being closely related to its conditions of existence, except when the embryo becomes at any period of life active and has to provide for itself; – of the embryo apparently having sometimes a higher organization than the mature animal, into which it is developed. I believe that all these facts can be explained, as follows, on the view of descent with modification.

It is commonly assumed, perhaps from monstrosities often affecting the embryo at a very early period, that slight variations necessarily appear at an equally early period. But we have little evidence on this head – indeed the evidence rather points the other way; for it is notorious that breeders of cattle, horses, and various fancy animals, cannot positively tell, until some time after the animal has been born, what its merits or form will ultimately turn out. We see this plainly in our own children; we cannot always tell whether the child will be tall or short, or what its precise features will be. The question is not, at what period of life any variation has been caused, but at what period it is fully displayed. The cause may have acted, and I believe generally has acted, even before the embryo is formed; and the variation may be due to the male and female sexual elements having been affected by the conditions to which either parent, or their ancestors, have been exposed. Nevertheless an effect thus caused at a very early period, even before the formation of the embryo, may appear late in life; as when an hereditary disease, which appears in old age alone, has been communicated to the offspring from the reproductive element of one parent. Or again, as when the horns of cross-bred cattle have been affected by the shape of the horns of either parent. For the welfare of a very young animal, as long as it remains in its mother's womb, or in the egg, or as long as it is nourished and protected by its parent, it must be quite unimportant whether most of its characters are fully acquired a little earlier or later in life. It would not signify, for instance, to a bird which obtained its food best by having a long beak, whether or not it assumed a beak of this particular length, as long as it was fed by its parents. Hence, I conclude, that it is quite possible, that each of the many successive modifications, by which each species has acquired its present structure, may have supervened at a not very early period of life; and some direct evidence from our domestic animals supports this view. But in other cases it is quite possible that each successive modification, or most of them, may have appeared at an extremely early period.

I have stated in the first chapter, that there is some evidence to render it probable, that at whatever age any variation first appears in the parent, it tends to reappear at a corresponding age in the offspring. Certain variations can only appear at corresponding ages, for instance, peculiarities in the caterpillar, cocoon, or imago states of the silk-moth; or, again, in the horns of almost full-grown cattle. But further than this, variations which, for all that we can see, might have appeared earlier or later in life, tend to appear at a corresponding age in the offspring and parent. I am far from meaning that this is invariably the case; and I could give a good many cases of variations (taking the word in the largest sense) which have supervened at an earlier age in the child than in the parent.

These two principles, if their truth be admitted, will, I believe, explain all the above specified leading facts in embryology. But first let us look at a few analogous cases in domestic varieties. Some authors who have written on Dogs, maintain that the greyhound and bulldog, though appearing so different, are really varieties most closely allied, and have probably descended from the same wild stock;

hence I was curious to see how far their puppies differed from each other: I was told by breeders that they differed just as much as their parents, and this, judging by the eye, seemed almost to be the case; but on actually measuring the old dogs and their six-days old puppies, I found that the puppies had not nearly acquired their full amount of proportional difference. So, again, I was told that the foals of cart and race-horses differed as much as the full-grown animals; and this surprised me greatly, as I think it probable that the difference between these two breeds has been wholly caused by selection under domestication; but having had careful measurements made of the dam and of a three-days old colt of a race and heavy cart-horse, I find that the colts have by no means acquired their full amount of proportional difference.

As the evidence appears to me conclusive, that the several domestic breeds of Pigeon have descended from one wild species, I compared young pigeons of various breeds, within twelve hours after being hatched; I carefully measured the proportions (but will not here give details) of the beak, width of mouth, length of nostril and of eyelid, size of feet and length of leg, in the wild stock, in pouters, fantails, runts, barbs, dragons, carriers, and tumblers. Now some of these birds, when mature, differ so extraordinarily in length and form of beak, that they would, I cannot doubt, be ranked in distinct genera, had they been natural productions. But when the nestling birds of these several breeds were placed in a row, though most of them could be distinguished from each other, yet their proportional differences in the above specified several points were incomparably less than in the full-grown birds. Some characteristic points of difference – for instance, that of the width of mouth – could hardly be detected in the young. But there was one remarkable exception to this rule, for the young of the short-faced tumbler differed from the young of the wild rock-pigeon and of the other breeds, in all its proportions, almost exactly as much as in the adult state.

The two principles above given seem to me to explain these facts in regard to the later embryonic stages of our domestic varieties. Fanciers select their horses, dogs, and pigeons, for breeding, when they are nearly grown up: they are indifferent whether the desired qualities and structures have been acquired earlier or later in life, if the full-grown animal possesses them. And the cases just given, more especially that of pigeons, seem to show that the characteristic differences which give value to each breed, and which have been accumulated by man's selection, have not generally first appeared at an early period of life, and have been inherited by the offspring at a corresponding not early period. But the case of the short-faced tumbler, which when twelve hours old had acquired its proper proportions, proves that this is not the universal rule; for here the characteristic differences must either have appeared at an earlier period than usual, or, if not so, the differences must have been inherited, not at the corresponding, but at an earlier age.

Now let us apply these facts and the above two principles – which latter, though not proved true, can be shown to be in some degree probable – to species in a state of nature. Let us take a genus of birds, descended on my theory from some one parent-species, and of which the several new species have become modified through natural selection in accordance with their diverse habits. Then, from the many slight successive steps of variation having supervened at a rather late age, and having been inherited at a corresponding age, the young of the new species of our supposed genus will manifestly tend to resemble each other much more closely than do the adults, just as we have seen in the case of pigeons. We may extend this view to whole families or even classes. The fore-limbs, for instance, which served as legs in the parent-species, may become, by a long course of modification, adapted in one descendant to act as hands, in another as paddles, in another as wings; and on the above two principles – namely of each successive modification supervening at a rather late age, and being inherited at a corresponding late age – the fore-limbs in the embryos of the several descendants of the parent-species will still resemble each other closely, for they will not have been modified. But in each individual new species, the embryonic fore-limbs will differ greatly from the fore-limbs in the mature animal; the limbs in the latter having undergone much modification at a rather late period of life, and having thus been converted into hands, or paddles, or wings. Whatever influence long-continued exercise or use on the one hand, and disuse on the other, may have in modifying an organ, such influence will mainly affect the mature animal, which has come to its full powers of activity and has to gain its own living; and the effects thus produced will be inherited at a corresponding mature age. Whereas the young will remain unmodified, or be modified in a lesser degree, by the effects of use and disuse.

In certain cases the successive steps of variation might supervene, from causes of which we are wholly ignorant, at a very early period of life, or each step might be inherited at an earlier period than that at which it first appeared. In either case (as with the short-faced tumbler) the young or embryo would closely resemble the mature parent-form. We have seen that this is the rule of development in certain whole groups of animals, as with cuttle-fish and spiders, and with a few members of the great class of insects, as with Aphis. With respect to the final cause of the young in these cases not undergoing any metamorphosis, or closely resembling their parents from their earliest age, we can see that this would result from the two following contingencies; firstly, from the young, during a course of modification carried on for many generations, having to provide for their own wants at a very early stage of development, and secondly, from their following exactly the same habits of life with their parents; for in this case, it would be indispensable for the existence of the species, that the child should be modified at a very early age in the same manner with its parents, in accordance with their similar habits. Some further explanation, however, of the embryo not undergoing any metamorphosis is perhaps requisite. If, on the other hand, it profited the young to follow habits of life in any degree different from those of their parent, and consequently to be constructed in a slightly different manner, then, on the principle of inheritance at corresponding ages, the active young or larvae might easily be rendered by natural selection different to any conceivable extent from their parents. Such differences might, also, become correlated with successive stages of development; so that the larvae, in the first stage, might differ greatly from the larvae in the second stage, as we have seen to be the case with cirripedes. The adult might become fitted for sites or habits, in which organs of locomotion or of the senses, &c., would be useless; and in this case the final metamorphosis would be said to be retrograde.

As all the organic beings, extinct and recent, which have ever lived on this earth have to be classed together, and as all have been connected by the finest gradations, the best, or indeed, if our collections were nearly perfect, the only possible arrangement, would be genealogical. Descent being on my view the hidden bond of connexion which naturalists have been seeking under the term of the natural system. On this view we can understand how it is that, in the eyes of most naturalists, the structure of the embryo is even more important for classification than that of the adult. For the embryo is the animal in its less modified state; and in so far it reveals the structure of its progenitor. In two groups of animals, however much they may at present differ from each other in structure and habits, if they pass through the same or similar embryonic stages, we may feel assured that they have both descended from the same or nearly similar parents, and are therefore in that degree closely related. Thus, community in embryonic structure reveals community of descent. It will reveal this community of descent, however much the structure of the adult may have been modified and obscured; we have seen, for instance, that cirripedes can at once be recognized by their larvae as belonging to the great class of crustaceans. As the embryonic state of each species and group of species partially shows us the structure of their less modified ancient progenitors, we can clearly see why ancient and extinct forms of life should resemble the embryos of their descendants, – our existing species. Agassiz believes this to be a law of nature; but I am bound to confess that I only hope to see the law hereafter proved true. It can be proved true in those cases alone in which the ancient state, now supposed to be represented in many embryos, has not been obliterated, either by the successive variations in a long course of modification having supervened at a very early age, or by the variations having been inherited at an earlier period than that at which they first appeared. It should also be borne in mind, that the supposed law of resemblance of ancient forms of life to the embryonic stages of recent forms, may be true, but yet, owing to the geological record not extending far enough back in time, may remain for a long period, or for ever, incapable of demonstration.

Thus, as it seems to me, the leading facts in embryology, which are second in importance to none in natural history, are explained on the principle of slight modifications not appearing, in the many descendants from some one ancient progenitor, at a very early period in the life of each, though perhaps caused at the earliest, and being inherited at a corresponding not early period. Embryology rises greatly in interest, when we thus look at the embryo as a picture, more or less obscured, of the common parent-form of each great class of animals.

Rudimentary, atrophied, or aborted organs. Organs or parts in this strange condition, bearing the stamp of inutility, are extremely common throughout nature. For instance, rudimentary mammae are very general in the males of mammals: I presume that the 'bastard-wing' in birds may be safely considered as a digit in a rudimentary state: in very many snakes one lobe of the lungs is rudimentary; in other snakes there are rudiments of the pelvis and hind limbs. Some of the cases of rudimentary organs are extremely curious; for instance, the presence of teeth in foetal whales, which when grown up have not a tooth in their heads; and the presence of teeth, which never cut through the gums, in the upper jaws of our unborn calves. It has even been stated on good authority that rudiments of teeth can be detected in the beaks of certain embryonic birds. Nothing can be plainer than that wings are formed for flight, yet in how many insects do we see wings so reduced in size as to be utterly incapable of flight, and not rarely lying under wing-cases, firmly soldered together!

The meaning of rudimentary organs is often quite unmistakeable: for instance there are beetles of the same genus (and even of the same species) resembling each other most closely in all respects, one of which will have full-sized wings, and another mere rudiments of membrane; and here it is impossible to doubt, that the rudiments represent wings. Rudimentary organs sometimes retain their potentiality, and are merely not developed: this seems to be the case with the mammae of male mammals, for many instances are on record of these organs having become well developed in full-grown males, and having secreted milk. So again there are normally four developed and two rudimentary teats in the udders of the genus Bos, but in our domestic cows the two sometimes become developed and give milk. In individual plants of the same species the petals sometimes occur as mere rudiments, and sometimes in a well-developed state. In plants with separated sexes, the male flowers often have a rudiment of a pistil; and Kölreuter found that by crossing such male plants with an hermaphrodite species, the rudiment of the pistil in the hybrid offspring was much increased in size; and this shows that the rudiment and the perfect pistil are essentially alike in nature.

An organ serving for two purposes, may become rudimentary or utterly aborted for one, even the more important purpose; and remain perfectly efficient for the other. Thus in plants, the office of the pistil is to allow the pollen-tubes to reach the ovules protected in the ovarium at its base. The pistil consists of a stigma supported on the style; but in some Compositae, the male florets, which of course cannot be fecundated, have a pistil, which is in a rudimentary state, for it is not crowned with a stigma; but the style remains well developed, and is clothed with hairs as in other compositae, for the purpose of brushing the pollen out of the surrounding anthers. Again, an organ may become rudimentary for its proper purpose, and be used for a distinct object: in certain fish the swimbladder seems to be rudimentary for its proper function of giving buoyancy, but has become converted into a nascent breathing organ or lung. Other similar instances could be given.

Rudimentary organs in the individuals of the same species are very liable to vary in degree of development and in other respects. Moreover, in closely allied species, the degree to which the same organ has been rendered rudimentary occasionally differs much. This latter fact is well exemplified in the state of the wings of the female moths in certain groups. Rudimentary organs may be utterly aborted; and this implies, that we find in an animal or plant no trace of an organ, which analogy would lead us to expect to find, and which is occasionally found in monstrous individuals of the species. Thus in the snapdragon (antirrhinum) we generally do not find a rudiment of a fifth stamen; but this may sometimes be seen. In tracing the homologies of the same part in different members of a class, nothing is more common, or more necessary, than the use and discovery of rudiments. This is well shown in the drawings given by Owen of the bones of the leg of the horse, ox, and rhinoceros.

It is an important fact that rudimentary organs, such as teeth in the upper jaws of whales and ruminants, can often be detected in the embryo, but afterwards wholly disappear. It is also, I believe, a universal rule, that a rudimentary part or organ is of greater size relatively to the adjoining parts in the embryo, than in the adult; so that the organ at this early age is less rudimentary, or even cannot be said to be in any degree rudimentary. Hence, also, a rudimentary organ in the adult, is often said to have retained its embryonic condition.

I have now given the leading facts with respect to rudimentary organs. In reflecting on them, every one must be struck with astonishment: for the same reasoning power which tells us plainly that most parts and organs are exquisitely adapted for certain purposes, tells us with equal plainness that these rudimentary or atrophied organs, are imperfect and useless. In works on natural history rudimentary organs are generally said to have been created 'for the sake of symmetry,' or in order 'to complete the scheme of nature;' but this seems to me no explanation, merely a restatement of the fact. Would it be thought sufficient to say that because planets revolve in elliptic courses round the sun, satellites follow the same course round the planets, for the sake of symmetry, and to complete the scheme of nature? An eminent physiologist accounts for the presence of rudimentary organs, by supposing that they serve to excrete matter in excess, or injurious to the system; but can we suppose that the minute papilla, which often represents the pistil in male flowers, and which is formed merely of cellular tissue, can thus act? Can we suppose that the formation of rudimentary teeth which are subsequently absorbed, can be of any service to the rapidly growing embryonic calf by the excretion of precious phosphate of lime? When a man's fingers have been amputated, imperfect nails sometimes appear on the stumps: I could as soon believe that these vestiges of nails have appeared, not from unknown laws of growth, but in order to excrete horny matter, as that the rudimentary nails on the fin of the manatee were formed for this purpose.

On my view of descent with modification, the origin of rudimentary organs is simple. We have plenty of cases of rudimentary organs in our domestic productions, – as the stump of a tail in tailless breeds, – the vestige of an ear in earless breeds, – the reappearance of minute dangling horns in hornless breeds of cattle, more especially, according to Youatt, in young animals, – and the state of the whole flower in the cauliflower. We often see rudiments of various parts in monsters.

But I doubt whether any of these cases throw light on the origin of rudimentary organs in a state of nature, further than by showing that rudiments can be produced; for I doubt whether species under nature ever undergo abrupt changes. I believe that disuse has been the main agency; that it has led in successive generations to the gradual reduction of various organs, until they have become rudimentary, – as in the case of the eyes of animals inhabiting dark caverns, and of the wings of birds inhabiting oceanic islands, which have seldom been forced to take flight, and have ultimately lost the power of flying. Again, an organ useful under certain conditions, might become injurious under others, as with the wings of beetles living on small and exposed islands; and in this case natural selection would continue slowly to reduce the organ, until it was rendered harmless and rudimentary.

Any change in function, which can be effected by insensibly small steps, is within the power of natural selection; so that an organ rendered, during changed habits of life, useless or injurious for one purpose, might easily be modified and used for another purpose. Or an organ might be retained for one alone of its former functions. An organ, when rendered useless, may well be variable, for its variations cannot be checked by natural selection. At whatever period of life disuse or selection reduces an organ, and this will generally be when the being has come to maturity and to its full powers of action, the principle of inheritance at corresponding ages will reproduce the organ in its reduced state at the same age, and consequently will seldom affect or reduce it in the embryo. Thus we can understand the greater relative size of rudimentary organs in the embryo, and their lesser relative size in the adult. But if each step of the process of reduction were to be inherited, not at the corresponding age, but at an extremely early period of life (as we have good reason to believe to be possible) the rudimentary part would tend to be wholly lost, and we should have a case of complete abortion. The principle, also, of economy, explained in a former chapter, by which the materials forming any part or structure, if not useful to the possessor, will be saved as far as is possible, will probably often come into play; and this will tend to cause the entire obliteration of a rudimentary organ.

As the presence of rudimentary organs is thus due to the tendency in every part of the organization, which has long existed, to be inherited – we can understand, on the genealogical view of classification, how it is that systematists have found rudimentary parts as useful as, or even sometimes more useful than, parts of high physiological importance. Rudimentary organs may be compared with the letters in a word, still retained in the spelling, but become useless in the pronunciation, but which serve as a clue in seeking for its derivation. On the view of descent with modification, we may conclude that the existence of organs in a rudimentary, imperfect, and useless condition, or quite aborted, far from presenting a strange difficulty, as they assuredly do on the ordinary doctrine of creation, might even have been anticipated, and can be accounted for by the laws of inheritance.

Summary. In this chapter I have attempted to show, that the subordination of group to group in all organisms throughout all time; that the nature of the relationship, by which all living and extinct beings are united by complex, radiating, and circuitous lines of affinities into one grand system; the rules followed and the difficulties encountered by naturalists in their classifications; the value set upon characters, if constant and prevalent, whether of high vital importance, or of the most trifling importance, or, as in rudimentary organs, of no importance; the wide opposition in value between analogical or adaptive characters, and characters of true affinity; and other such rules; – all naturally follow on the view of the common parentage of those forms which are considered by naturalists as allied, together with their modification through natural selection, with its contingencies of extinction and divergence of character. In considering this view of classification, it should be borne in mind that the element of descent has been universally used in ranking together the sexes, ages, and acknowledged varieties of the same species, however different they may be in structure. If we extend the use of this element of descent, – the only certainly known cause of similarity in organic beings, – we shall understand what is meant by the natural system: it is genealogical in its attempted arrangement, with the grades of acquired difference marked by the terms varieties, species, genera, families, orders, and classes.

On this same view of descent with modification, all the great facts in Morphology become intelligible, – whether we look to the same pattern displayed in the homologous organs, to whatever purpose applied, of the different species of a class; or to the homologous parts constructed on the same pattern in each individual animal and plant.

On the principle of successive slight variations, not necessarily or generally supervening at a very early period of life, and being inherited at a corresponding period, we can understand the great leading facts in Embryology; namely, the resemblance in an individual embryo of the homologous parts, which when matured will become widely different from each other in structure and function; and the resemblance in different species of a class of the homologous parts or organs, though fitted in the adult members for purposes as different as possible. Larvae are active embryos, which have become specially modified in relation to their habits of life, through the principle of modifications being inherited at corresponding ages. On this same principle – and bearing in mind, that when organs are reduced in size, either from disuse or selection, it will generally be at that period of life when the being has to provide for its own wants, and bearing in mind how strong is the principle of inheritance – the occurrence of rudimentary organs and their final abortion, present to us no inexplicable difficulties; on the contrary, their presence might have been even anticipated. The importance of embryological characters and of rudimentary organs in classification is intelligible, on the view that an arrangement is only so far natural as it is genealogical.

Finally, the several classes of facts which have been considered in this chapter, seem to me to proclaim so plainly, that the innumerable species, genera, and families of organic beings, with which this world is peopled, have all descended, each within its own class or group, from common parents, and have all been modified in the course of descent, that I should without hesitation adopt this view, even if it were unsupported by other facts or arguments.

CHAPTER XIV

RECAPITULATION AND CONCLUSION

Recapitulation of the difficulties on the theory of Natural Selection – Recapitulation of the general and special circumstances in its favour – Causes of the general belief in the immutability of species – How far the theory of natural selection may be extended – Effects of its adoption on the study of Natural history – Concluding remarks

AS this whole volume is one long argument, it may be convenient to the reader to have the leading facts and inferences briefly recapitulated.

That many and grave objections may be advanced against the theory of descent with modification through natural selection, I do not deny. I have endeavoured to give to them their full force. Nothing at first can appear more difficult to believe than that the more complex organs and instincts should have been perfected, not by means superior to, though analogous with, human reason, but by the accumulation of innumerable slight variations, each good for the individual possessor. Nevertheless, this difficulty, though appearing to our imagination insuperably great, cannot be considered real if we admit the following propositions, namely, – that gradations in the perfection of any organ or instinct, which we may consider, either do now exist or could have existed, each good of its kind, – that all organs and instincts are, in ever so slight a degree, variable, – and, lastly, that there is a struggle for existence leading to the preservation of each profitable deviation of structure or instinct. The truth of these propositions cannot, I think, be disputed.

It is, no doubt, extremely difficult even to conjecture by what gradations many structures have been perfected, more especially amongst broken and failing groups of organic beings; but we see so many strange gradations in nature, as is proclaimed by the canon, 'Natura non facit saltum,' that we ought to be extremely cautious in saying that any organ or instinct, or any whole being, could not have arrived at its present state by many graduated steps. There are, it must be admitted, cases of special difficulty on the theory of natural selection; and one of the most curious of these is the existence of two or three defined castes of workers or sterile females in the same community of ants; but I have attempted to show how this difficulty can be mastered.

With respect to the almost universal sterility of species when first crossed, which forms so remarkable a contrast with the almost universal fertility of varieties when crossed, I must refer the reader to the recapitulation of the facts given at the end of the eighth chapter, which seem to me conclusively to show that this sterility is no more a special endowment than is the incapacity of two trees to be grafted together; but that it is incidental on constitutional differences in the reproductive systems of the intercrossed species. We see the truth of this conclusion in the vast difference in the result, when the same two species are crossed reciprocally; that is, when one species is first used as the father and then as the mother.

The fertility of varieties when intercrossed and of their mongrel offspring cannot be considered as universal; nor is their very general fertility surprising when we remember that it is not likely that either their constitutions or their reproductive systems should have been profoundly modified. Moreover, most of the varieties which have been experimentized on have been produced under domestication; and as domestication apparently tends to eliminate sterility, we ought not to expect it also to produce sterility.

The sterility of hybrids is a very different case from that of first crosses, for their reproductive organs are more or less functionally impotent; whereas in first crosses the organs on both sides are in a perfect condition. As we continually see that organisms of all kinds are rendered in some degree sterile from their constitutions having been disturbed by slightly different and new conditions of life, we need not feel surprise at hybrids being in some degree sterile, for their constitutions can hardly fail to have been disturbed from being compounded of two distinct organizations. This parallelism is supported by another parallel, but directly opposite, class of facts; namely, that the vigour and fertility of all organic beings are increased by slight changes in their conditions of life, and that the offspring of slightly modified forms or varieties acquire from being crossed increased vigour and fertility. So that, on the one hand, considerable changes in the conditions of life and crosses between greatly modified forms, lessen fertility; and on the other hand, lesser changes in the conditions of life and crosses between less modified forms, increase fertility.

Turning to geographical distribution, the difficulties encountered on the theory of descent with modification are grave enough. All the individuals of the same species, and all the species of the same genus, or even higher group, must have descended from common parents; and therefore, in however distant and isolated parts of the world they are now found, they must in the course of successive generations have passed from some one part to the others. We are often wholly unable even to conjecture how this could have been effected. Yet, as we have reason to believe that some species have retained the same specific form for very long periods, enormously long as measured by years, too much stress ought not to be laid on the occasional wide diffusion of the same species; for during very long periods of time there will always be a good chance for wide migration by many means. A broken or interrupted range may often be accounted for by the extinction of the species in the intermediate regions. It cannot be denied that we are as yet very ignorant of the full extent of the various climatal and geographical changes which have affected the earth during modern periods; and such changes will obviously have greatly facilitated migration. As an example, I have attempted to show how potent has been the influence of the Glacial period on the distribution both of the same and of representative species throughout the world. We are as yet profoundly ignorant of the many occasional means of transport. With respect to distinct species of the same genus inhabiting very distant and isolated regions, as the process of modification has necessarily been slow, all the means of migration will have been possible during a very long period; and consequently the difficulty of the wide diffusion of species of the same genus is in some degree lessened.

As on the theory of natural selection an interminable number of intermediate forms must have existed, linking together all the species in each group by gradations as fine as our present varieties, it may be asked, Why do we not see these linking forms all around us? Why are not all organic beings blended together in an inextricable chaos? With respect to existing forms, we should remember that we have no right to expect (excepting in rare cases) to discover *directly* connecting links between them, but only between each and some extinct and supplanted form. Even on a wide area, which has during a long period remained continuous, and of which the climate and other conditions of life change insensibly in going from a district occupied by one species into another district occupied by a closely allied species, we have no just right to expect often to find intermediate varieties in the intermediate zone. For we have reason to believe that only a few species are undergoing change at any one period; and all changes are slowly effected. I have also shown that the intermediate varieties which will at first probably exist in the intermediate zones, will be liable to be supplanted by the allied forms on either hand; and the latter, from existing in greater numbers, will generally be modified and improved at a quicker rate than the intermediate varieties, which exist in lesser numbers; so that the intermediate varieties will, in the long run, be supplanted and exterminated.

On this doctrine of the extermination of an infinitude of connecting links, between the living and extinct inhabitants of the world, and at each successive period between the extinct and still older species, why is not every geological formation charged with such links? Why does not every collection of fossil remains afford plain evidence of the gradation and mutation of the forms of life? We meet with no such evidence, and this is the most obvious and forcible of the many objections which may be urged against my theory. Why, again, do whole groups of allied species appear, though certainly they often falsely appear, to have come in suddenly on the several geological stages? Why do we not find great piles of strata beneath the Silurian system, stored with the remains of the progenitors of the Silurian groups of fossils? For certainly on my theory such strata must somewhere have been deposited at these ancient and utterly unknown epochs in the world's history.

I can answer these questions and grave objections only on the supposition that the geological record is far more imperfect than most geologists believe. It cannot be objected that there has not been time sufficient for any amount of organic change; for the lapse of time has been so great as to be utterly inappreciable by the human intellect. The number of specimens in all our museums is absolutely as nothing compared with the countless generations of countless species which certainly have existed. We should not be able to recognize a species as the parent of any one or more species if we were to examine them ever so closely, unless we likewise possessed many of the intermediate links between their past or parent and present states; and these many links we could hardly ever expect to discover, owing to the imperfection of the geological record. Numerous existing doubtful forms could be named which are probably varieties; but who will pretend that in future ages so many fossil links will be discovered, that naturalists will be able to decide, on the common view, whether or not these doubtful forms are varieties? As long as most of the links between any two species are unknown, if any one link or intermediate variety be discovered, it will simply be classed as another and distinct species. Only a small portion of the world has been geologically explored. Only organic beings of certain classes can be preserved in a fossil condition, at least in any great number. Widely ranging species vary most, and varieties are often at first local, – both causes rendering the discovery of intermediate links less likely. Local varieties will not spread into other and distant regions until they are considerably modified and improved; and when they do spread, if discovered in a geological formation, they will appear as if suddenly created there, and will be simply classed as new species. Most formations have been intermittent in their accumulation; and their duration, I am inclined to believe, has been shorter than the average duration of specific forms. Successive formations are separated from each other by enormous blank intervals of time; for fossiliferous formations, thick enough to resist future degradation, can be accumulated only where much sediment is deposited on the subsiding bed of the sea. During the alternate periods of elevation and of stationary level the record will be blank. During these latter periods there will probably be more variability in the forms of life; during periods of subsidence, more extinction.

With respect to the absence of fossiliferous formations beneath the lowest Silurian strata, I can only recur to the hypothesis given in the ninth chapter. That the geological record is imperfect all will admit; but that it is imperfect to the degree which I require, few will be inclined to admit. If we look to long enough intervals of time, geology plainly declares that all species have changed; and they have changed in the manner which my theory requires, for they have changed slowly and in a graduated manner. We clearly see this in the fossil remains from consecutive formations invariably being much more closely related to each other, than are the fossils from formations distant from each other in time.

Such is the sum of the several chief objections and difficulties which may justly be urged against my theory; and I have now briefly recapitulated the answers and explanations which can be given to them. I have felt these difficulties far too heavily during many years to doubt their weight. But it deserves especial notice that the more important objections relate to questions on which we are confessedly ignorant; nor do we know how ignorant we are. We do not know all the possible transitional gradations between the simplest and the most perfect organs; it cannot be pretended that we know all the varied means of Distribution during the long lapse of years, or that we know how imperfect the Geological Record is. Grave as these several difficulties are, in my judgement they do not overthrow the theory of descent with modification.

Now let us turn to the other side of the argument. Under domestication we see much variability. This seems to be mainly due to the reproductive system being eminently susceptible to changes in the conditions of life; so that this system, when not rendered impotent, fails to reproduce offspring exactly like the parent-form. Variability is governed by many complex laws, – by correlation of growth, by use and disuse, and by the direct action of the physical conditions of life. There is much difficulty in ascertaining how much modification our domestic productions have undergone; but we may safely infer that the amount has been large, and that modifications can be inherited for long periods. As long as the conditions of life remain the same, we have reason to believe that a modification, which has already been inherited for many generations, may continue to be inherited for an almost infinite number of generations. On the other hand we have evidence that variability, when it has once come into play, does not wholly cease; for new varieties are still occasionally produced by our most anciently domesticated productions.

Man does not actually produce variability; he only unintentionally exposes organic beings to new conditions of life, and then nature acts on the organization, and causes variability. But man can and does select the variations given to him by nature, and thus accumulate them in any desired manner. He thus adapts animals and plants for his own benefit or pleasure. He may do this methodically, or he may do it unconsciously by preserving the individuals most useful to him at the time, without any thought of altering the breed. It is certain that he can largely influence the character of a breed by selecting, in each successive generation, individual differences so slight as to be quite inappreciable by an uneducated eye. This process of selection has been the great agency in the production of the most distinct and useful domestic breeds. That many of the breeds produced by man have to a large extent the character of natural species, is shown by the inextricable doubts whether very many of them are varieties or aboriginal species.

There is no obvious reason why the principles which have acted so efficiently under domestication should not have acted under nature. In the preservation of favoured individuals and races, during the constantly-recurrent Struggle for Existence, we see the most powerful and ever-acting means of selection. The struggle for exist-ence inevitably follows from the high geometrical ratio of increase which is common to all organic beings. This high rate of increase is proved by calculation, by the effects of a succession of peculiar seasons, and by the results of naturalization, as explained in the third chapter. More individuals are born than can possibly survive. A grain in the balance will determine which individual shall live and which shall die, – which variety or species shall increase in number, and which shall decrease, or finally become extinct. As the individuals of the same species come in all respects into the closest competition with each other, the struggle will generally be most severe between them; it will be almost equally severe between the varieties of the same species, and next in severity between the species of the same genus. But the struggle will often be very severe between beings most remote in the scale of nature. The slightest advantage in one being, at any age or during any season, over those with which it comes into competition, or better adaptation in however slight a degree to the surrounding physical conditions, will turn the balance.

With animals having separated sexes there will in most cases be a struggle between the males for possession of the females. The most vigorous individuals, or those which have most successfully struggled with their conditions of life, will generally leave most progeny. But success will often depend on having special weapons or means of defence, or on the charms of the males; and the slightest advantage will lead to victory.

As geology plainly proclaims that each land has undergone great physical changes, we might have expected that organic beings would have varied under nature, in the same way as they generally have varied under the changed conditions of domestication. And if there be any variability under nature, it would be an unaccountable fact if natural selection had not come into play. It has often been asserted, but the assertion is quite incapable of proof, that the amount of variation under nature is a strictly limited quantity. Man, though acting on external characters alone and often capriciously, can produce within a short period a great result by adding up mere individual differences in his domestic productions; and every one admits that there are at least individual differences in species under nature. But, besides such differences, all naturalists have admitted the existence of varieties, which they think sufficiently distinct to be worthy of record in systematic works. No one can draw any clear distinction between individual differences and slight varieties; or between more plainly marked varieties and sub-species, and species. Let it be observed how naturalists differ in the rank which they assign to the many representative forms in Europe and North America.

If then we have under nature variability and a powerful agent always ready to act and select, why should we doubt that variations in any way useful to beings, under their excessively complex relations of life, would be preserved, accumulated, and inherited? Why, if man can by patience select variations most useful to himself, should nature fail in selecting variations useful, under changing conditions of life, to her living products? What limit can be put to this power, acting during long ages and rigidly scrutinizing the whole constitution, structure, and habits of each creature, – favouring the good and rejecting the bad? I can see no limit to this power, in slowly and beautifully adapting each form to the most complex relations of life. The theory of natural selection, even if we looked no further than this, seems to me to be in itself probable. I have already recapitulated, as fairly as I could, the opposed difficulties and objections: now let us turn to the special facts and arguments in favour of the theory.

On the view that species are only strongly marked and permanent varieties, and that each species first existed as a variety, we can see why it is that no line of demarcation can be drawn between species, commonly supposed to have been produced by special acts of creation, and varieties which are acknowledged to have been produced by secondary laws. On this same view we can understand how it is that in each region where many species of a genus have been produced, and where they now flourish, these same species should present many varieties; for where the manufactory of species has been active, we might expect, as a general rule, to find it still in action; and this is the case if varieties be incipient species. Moreover, the species of the large genera, which afford the greater number of varieties or incipient species, retain to a certain degree the character of varieties; for they differ from each other by a less amount of difference than do the species of smaller genera. The closely allied species also of the larger genera apparently have restricted ranges, and they are clustered in little groups round other species – in which respects they resemble varieties. These are strange relations on the view of each species having been independently created, but are intelligible if all species first existed as varieties.

As each species tends by its geometrical ratio of reproduction to increase inordinately in number; and as the modified descendants of each species will be enabled to increase by so much the more as they become more diversified in habits and structure, so as to be enabled to seize on many and widely different places in the economy of nature, there will be a constant tendency in natural selection to preserve the most divergent offspring of any one species. Hence during a long-continued course of modification, the slight differences, characteristic of varieties of the same species, tend to be augmented into the greater differences characteristic of species of the same genus.

New and improved varieties will inevitably supplant and exterminate the older, less improved and intermediate varieties; and thus species are rendered to a large extent defined and distinct objects. Dominant species belonging to the larger groups tend to give birth to new and dominant forms; so that each large group tends to become still larger, and at the same time more divergent in character. But as all groups cannot thus succeed in increasing in size, for the world would not hold them, the more dominant groups beat the less dominant. This tendency in the large groups to go on increasing in size and diverging in character, together with the almost inevitable contingency of much extinction, explains the arrangement of all the forms of life, in groups subordinate to groups, all within a few great classes, which we now see everywhere around us, and which has prevailed throughout all time. This grand fact of the grouping of all organic beings seems to me utterly inexplicable on the theory of creation.

As natural selection acts solely by accumulating slight, successive, favourable variations, it can produce no great or sudden modification; it can act only by very short and slow steps. Hence the canon of 'Natura non facit saltum,' which every fresh addition to our knowledge tends to make more strictly correct, is on this theory simply intelligible. We can plainly see why nature is prodigal in variety, though niggard in innovation. But why this should be a law of nature if each species has been independently created, no man can explain.

Many other facts are, as it seems to me, explicable on this theory. How strange it is that a bird, under the form of woodpecker, should have been created to prey on insects on the ground; that upland geese, which never or rarely swim, should have been created with webbed feet; that a thrush should have been created to dive and feed on sub-aquatic insects; and that a petrel should have been created with habits and structure fitting it for the life of an auk or grebe! and so on in endless other cases. But on the view of each species constantly trying to increase in number, with natural selection always ready to adapt the slowly varying descendants of each to any unoccupied or ill-occupied place in nature, these facts cease to be strange, or perhaps might even have been anticipated.

As natural selection acts by competition, it adapts the inhabitants of each country only in relation to the degree of perfection of their associates; so that we need feel no surprise at the inhabitants of any one country, although on the ordinary view supposed to have been specially created and adapted for that country, being beaten and supplanted by the naturalized productions from another land. Nor ought we to marvel if all the contrivances in nature be not, as far as we can judge, absolutely perfect; and if some of them be abhorrent to our ideas of fitness. We need not marvel at the sting of the bee causing the bee's own death; at drones being produced in such vast numbers for one single act, and being then slaughtered by their sterile sisters; at the astonishing waste of pollen by our fir-trees; at the instinctive hatred of the queen bee for her own fertile daughters; at ichneumonidae feeding within the live bodies of caterpillars; and at other such cases. The wonder indeed is, on the theory of natural selection, that more cases of the want of absolute perfection have not been observed.

The complex and little known laws governing variation are the same, as far as we can see, with the laws which have governed the production of so-called specific forms. In both cases physical conditions seem to have produced but little direct effect; yet when varieties enter any zone, they occasionally assume some of the characters of the species proper to that zone. In both varieties and species, use and disuse seem to have produced some effect; for it is difficult to resist this conclusion when we look, for instance, at the logger-headed duck, which has wings incapable of flight, in nearly the same condition as in the domestic duck; or when we look at the burrowing tucutucu, which is occasionally blind, and then at certain moles, which are habitually blind and have their eyes covered with skin; or when we look at the blind animals inhabiting the dark caves of America and Europe. In both varieties and species correction of growth seems to have played a most important part, so that when one part has been modified other parts are necessarily modified. In both varieties and species reversions to long-lost characters occur. How inexplicable on the theory of creation is the occasional appearance of stripes on the shoulder and legs of the several species of the horse-genus and in their hybrids! How simply is this fact explained if we believe that these species have descended from a striped progenitor, in the same manner as the several domestic breeds of pigeon have descended from the blue and barred rock-pigeon!

On the ordinary view of each species having been independently created, why should the specific characters, or those by which the species of the same genus differ from each other, be more variable than the generic characters in which they all agree? Why, for instance, should the colour of a flower be more likely to vary in any one species of a genus, if the other species, supposed to have been created independently, have differently coloured flowers, than if all the species of the genus have the same coloured flowers? If species are only well-marked varieties, of which the characters have become in a high degree permanent, we can understand this fact; for they have already varied since they branched off from a common progenitor in certain characters, by which they have come to be specifically distinct from each other; and therefore these same characters would be more likely still to be variable than the generic characters which have been inherited without change for an enormous period. It is inexplicable on the theory of creation why a part developed in a very unusual manner in any one species of a genus, and therefore, as we may naturally infer, of great importance to the species, should be eminently liable to variation; but, on my view, this part has undergone, since the several species branched off from a common progenitor, an unusual amount of variability and modification, and therefore we might expect this part generally to be still variable. But a part may be developed in the most unusual manner, like the wing of a bat, and yet not be more variable than any other structure, if the part be common to many subordinate forms, that is, if it has been inherited for a very long period; for in this case it will have been rendered constant by long-continued natural selection.

Glancing at instincts, marvellous as some are, they offer no greater difficulty than does corporeal structure on the theory of the natural selection of successive, slight, but profitable modifications. We can thus understand why nature moves by graduated steps in endowing different animals of the same class with their several instincts. I have attempted to show how much light the principle of gradation throws on the admirable architectural powers of the hive-bee. Habit no doubt sometimes comes into play in modifying instincts; but it certainly is not indispensable, as we see, in the case of neuter insects, which leave no progeny to inherit the effects of long-continued habit. On the view of all the species of the same genus having descended from a common parent, and having inherited much in common, we can understand how it is that allied species, when placed under considerably different conditions of life, yet should follow nearly the same instincts; why the thrush of South America, for instance, lines her nest with mud like our British species. On the view of instincts having been slowly acquired through natural selection we need not marvel at some instincts being apparently not perfect and liable to mistakes, and at many instincts causing other animals to suffer.

If species be only well-marked and permanent varieties, we can at once see why their crossed offspring should follow the same complex laws in their degrees and kinds of resemblance to their parents, – in being absorbed into each other by successive crosses, and in other such points, – as do the crossed offspring of acknowledged varieties. On the other hand, these would be strange facts if species have been independently created, and varieties have been produced by secondary laws.

If we admit that the geological record is imperfect in an extreme degree, then such facts as the record gives, support the theory of descent with modification. New species have come on the stage slowly and at successive intervals; and the amount of change, after equal intervals of time, is widely different in different groups. The extinction of species and of whole groups of species, which has played so conspicuous a part in the history of the organic world, almost inevitably follows on the principle of natural selection; for old forms will be supplanted by new and improved forms. Neither single species nor groups of species reappear when the chain of ordinary generation has once been broken. The gradual diffusion of dominant forms, with the slow modification of their descendants, causes the forms of life, after long intervals of time, to appear as if they had changed simultaneously throughout the world. The fact of the fossil remains of each formation being in some degree intermediate in character between the fossils in the formations above and below, is simply explained by their intermediate position in the chain of descent. The grand fact that all extinct organic beings belong to the same system with recent beings, falling either into the same or into intermediate groups, follows from the living and the extinct being the offspring of common parents. As the groups which have descended from an ancient progenitor have generally diverged in character, the progenitor with its early descendants will often be intermediate in character in comparison with its later descendants; and thus we can see why the more ancient a fossil is, the oftener it stands in some degree intermediate between existing and allied groups. Recent forms are generally looked at as being, in some vague sense, higher than ancient and extinct forms; and they are in so far higher as the later and more improved forms have conquered the older and less improved organic beings in the struggle for life. Lastly, the law of the long endurance of allied forms on the same continent, – of marsupials in Australia, of edentata in America, and other such cases, – is intelligible, for within a confined country, the recent and the extinct will naturally be allied by descent.

Looking to geographical distribution, if we admit that there has been during the long course of ages much migration from one part of the world to another, owing to former climatal and geographical changes and to the many occasional and unknown means of dispersal, then we can understand, on the theory of descent with modification, most of the great leading facts in Distribution. We can see why there should be so striking a parallelism in the distribution of organic beings throughout space, and in their geological succession throughout time; for in both cases the beings have been connected by the bond of ordinary generation, and the means of modification have been the same. We see the full meaning of the wonderful fact, which must have struck every traveller, namely, that on the same continent, under the most diverse conditions, under heat and cold, on mountain and lowland, on deserts and marshes, most of the inhabitants within each great class are plainly related; for they will generally be descendants of the same progenitors and early colonists. On this same principle of former migration, combined in most cases with modification, we can understand, by the aid of the Glacial period, the identity of some few plants, and the close alliance of many others, on the most distant mountains, under the most different climates; and likewise the close alliance of some of the inhabitants of the sea in the northern and southern temperate zones, though separated by the whole intertropical ocean. Although two areas may present the same physical conditions of life, we need feel no surprise at their inhabitants being widely different, if they have been for a long period completely separated from each other; for as the relation of organism to organism is the most important of all relations, and as the two areas will have received colonists from some third source or from each other, at various periods and in different proportions, the course of modification in the two areas will inevitably be different.

On this view of migration, with subsequent modification, we can see why oceanic islands should be inhabited by few species, but of these, that many should be peculiar. We can clearly see why those animals which cannot cross wide spaces of ocean, as frogs and terrestrial mammals, should not inhabit oceanic islands; and why, on the other hand, new and peculiar species of bats, which can traverse the ocean, should so often be found on islands far distant from any continent. Such facts as the presence of peculiar species of bats, and the absence of all other mammals, on oceanic islands, are utterly inexplicable on the theory of independent acts of creation.

The existence of closely allied or representative species in any two areas, implies, on the theory of descent with modification, that the same parents formerly inhabited both areas; and we almost invariably find that wherever many closely allied species inhabit two areas, some identical species common to both still exist. Wherever many closely allied yet distinct species occur, many doubtful forms and varieties of the same species likewise occur. It is a rule of high generality that the inhabitants of each area are related to the inhabitants of the nearest source whence immigrants might have been derived. We see this in nearly all the plants and animals of the Galapagos archipelago, of Juan Fernandez, and of the other American islands being related in the most striking manner to the plants and animals of the neighbouring American mainland; and those of the Cape de Verde archipelago and other African islands to the African mainland. It must be admitted that these facts receive no explanation on the theory of creation.

The fact, as we have seen, that all past and present organic beings constitute one grand natural system, with group subordinate to group, and with extinct groups often falling in between recent groups, is intelligible on the theory of natural selection with its contingencies of extinction and divergence of character. On these same principles we see how it is, that the mutual affinities of the species and genera within each class are so complex and circuitous. We see why certain characters are far more serviceable than others for classification; − why adaptive characters, though of paramount importance to the being, are of hardly any importance in classification; why characters derived from rudimentary parts, though of no service to the being, are often of high classificatory value; and why embryological characters are the most valuable of all. The real affinities of all organic beings are due to inheritance or community of descent. The natural system is a genealogical arrangement, in which we have to discover the lines of descent by the most permanent characters, however slight their vital importance may be.

The framework of bones being the same in the hand of a man, wing of a bat, fin of the porpoise, and leg of the horse, − the same number of vertebrae forming the neck of the giraffe and of the elephant, − and innumerable other such facts, at once explain themselves on the theory of descent with slow and slight successive modifications. The similarity of pattern in the wing and leg of a bat, though used for such different purposes, − in the jaws and legs of a crab, − in the petals, stamens, and pistils of a flower, is likewise intelligible on the view of the gradual modification of parts or organs, which were alike in the early progenitor of each class. On the principle of successive variations not always supervening at an early age, and being inherited at a corresponding not early period of life, we can clearly see why the embryos of mammals, birds, reptiles, and fishes should be so closely alike, and should be so unlike the adult forms. We may cease marvelling at the embryo of an air-breathing mammal or bird having branchial slits and arteries running in loops, like those in a fish which has to breathe the air dissolved in water, by the aid of well-developed branchiae.

Disuse, aided sometimes by natural selection, will often tend to reduce an organ, when it has become useless by changed habits or under changed conditions of life; and we can clearly understand on this view the meaning of rudimentary organs. But disuse and selection will generally act on each creature, when it has come to maturity and has to play its full part in the struggle for existence, and will thus have little power of acting on an organ during early life; hence the organ will not be much reduced or rendered rudimentary at this early age. The calf, for instance, has inherited teeth, which never cut through the gums of the upper jaw, from an early progenitor having well-developed teeth; and we may believe, that the teeth in the mature animal were reduced, during successive generations, by disuse or by the tongue and palate having been fitted by natural selection to browse without their aid; whereas in the calf, the teeth have been left untouched by selection or disuse, and on the principle of inheritance at corresponding ages have been inherited from a remote period to the present day. On the view of each organic being and each separate organ having been specially created, how utterly inexplicable it is that parts, like the teeth in the embryonic calf or like the shrivelled wings under the soldered wing-covers of some beetles, should thus so frequently bear the plain stamp of inutility! Nature may be said to have taken pains to reveal, by rudimentary organs and by homologous structures, her scheme of modification, which it seems that we wilfully will not understand.

I have now recapitulated the chief facts and considerations which have thoroughly convinced me that species have changed, and are still slowly changing by the preservation and accumulation of successive slight favourable variations. Why, it may be asked, have all the most eminent living naturalists and geologists rejected this view of the mutability of species? It cannot be asserted that organic beings in a state of nature are subject to no variation; it cannot be proved that the amount of variation in the course of long ages is a limited quantity; no clear distinction has been, or can be, drawn between species and well-marked varieties. It cannot be maintained that species when inter-crossed are invariably sterile, and varieties invariably fertile; or that sterility is a special endowment and sign of creation. The belief that species were immutable productions was almost unavoidable as long as the history of the world was thought to be of short duration; and now that we have acquired some idea of the lapse of time, we are too apt to assume, without proof, that the geological record is so perfect that it would have afforded us plain evidence of the mutation of species, if they had undergone mutation.

But the chief cause of our natural unwillingness to admit that one species has given birth to other and distinct species, is that we are always slow in admitting any great change of which we do not see the intermediate steps. The difficulty is the same as that felt by so many geologists, when Lyell first insisted that long lines of inland cliffs had been formed, and great valleys excavated, by the slow action of the coast-waves. The mind cannot possibly grasp the full meaning of the term of a hundred million years; it cannot add up and perceive the full effects of many slight variations, accumulated during an almost infinite number of generations.

Although I am fully convinced of the truth of the views given in this volume under the form of an abstract, I by no means expect to convince experienced naturalists whose minds are stocked with a multitude of facts all viewed, during a long course of years, from a point of view directly opposite to mine. It is so easy to hide our ignorance under such expressions as the 'plan of creation,' 'unity of design,' &c., and to think that we give an explanation when we only restate a fact. Any one whose disposition leads him to attach more weight to unexplained difficulties than to the explanation of a certain number of facts will certainly reject my theory. A few naturalists, endowed with much flexibility of mind, and who have already begun to doubt on the immutability of species, may be influenced by this volume; but I look with confidence to the future, to young and rising naturalists, who will be able to view both sides of the question with impartiality. Whoever is led to believe that species are mutable will do good service by conscientiously expressing his conviction; for only thus can the load of prejudice by which this subject is overwhelmed be removed.

Several eminent naturalists have of late published their belief that a multitude of reputed species in each genus are not real species; but that other species are real, that is, have been independently created. This seems to me a strange conclusion to arrive at. They admit that a multitude of forms, which till lately they themselves thought were special creations, and which are still thus looked at by the majority of naturalists, and which consequently have every external characteristic feature of true species, – they admit that these have been produced by variation, but they refuse to extend the same view to other and very slightly different forms. Nevertheless they do not pretend that they can define, or even conjecture, which are the created forms of life, and which are those produced by secondary laws. They admit variation as a *vera causa* in one case, they arbitrarily reject it in another, without assigning any distinction in the two cases. The day will come when this will be given as a curious illustration of the blindness of preconceived opinion. These authors seem no more startled at a miraculous act of creation than at an ordinary birth. But do they really believe that at innumerable periods in the earth's history certain elemental atoms have been commanded suddenly to flash into living tissues? Do they believe that at each supposed act of creation one individual or many were produced? Were all the infinitely numerous kinds of animals and plants created as eggs or seed, or as full grown? and in the case of mammals, were they created bearing the false marks of nourishment from the mother's womb? Although naturalists very properly demand a full explanation of every difficulty from those who believe in the mutability of species, on their own side they ignore the whole subject of the first appearance of species in what they consider reverent silence.

It may be asked how far I extend the doctrine of the modification of species. The question is difficult to answer, because the more distinct the forms are which we may consider, by so much the arguments fall away in force. But some arguments of the greatest weight extend very far. All the members of whole classes can be connected together by chains of affinities, and all can be classified on the same principle, in groups subordinate to groups. Fossil remains sometimes tend to fill up very wide intervals between existing orders. Organs in a rudimentary condition plainly show that an early progenitor had the organ in a fully developed state; and this in some instances necessarily implies an enormous amount of modi-fication in the descendants. Throughout whole classes various struc-tures are formed on the same pattern, and at an embryonic age the species closely resemble each other. Therefore I cannot doubt that the theory of descent with modification embraces all the members of the same class. I believe that animals have descended from at most only four or five progenitors, and plants from an equal or lesser number.

Analogy would lead me one step further, namely, to the belief that all animals and plants have descended from some one prototype. But analogy may be a deceitful guide. Nevertheless all living things have much in common, in their chemical composition, their germinal vesicles, their cellular structure, and their laws of growth and reproduction. We see this even in so trifling a circumstance as that the same poison often similarly affects plants and animals; or that the poison secreted by the gall-fly produces monstrous growths on the wild rose or oak-tree. Therefore I should infer from analogy that probably all the organic beings which have ever lived on this earth have descended from some one primordial form, into which life was first breathed.

When the views entertained in this volume on the origin of species, or when analogous views are generally admitted, we can dimly foresee that there will be a considerable revolution in natural history. Systematists will be able to pursue their labours as at present; but they will not be incessantly haunted by the shadowy doubt whether this or that form be in essence a species. This I feel sure, and I speak after experience, will be no slight relief. The endless disputes whether or not some fifty species of British brambles are true species will cease. Systematists will have only to decide (not that this will be easy) whether any form be sufficiently constant and distinct from other forms, to be capable of definition; and if definable, whether the differences be sufficiently important to deserve a specific name. This latter point will become a far more essential consideration than it is at present; for differences, however slight, between any two forms, if not blended by intermediate gradations, are looked at by most naturalists as sufficient to raise both forms to the rank of species. Hereafter we shall be compelled to acknowledge that the only distinction between species and well-marked varieties is, that the latter are known, or believed, to be connected at the present day by intermediate gradations, whereas species were formerly thus connected. Hence, without quite rejecting the consideration of the present existence of intermediate gradations between any two forms, we shall be led to weigh more carefully and to value higher the actual amount of difference between them. It is quite possible that forms now generally acknowledged to be merely varieties may here-after be thought worthy of specific names, as with the primrose and cowslip; and in this case scientific and common language will come into accordance. In short, we shall have to treat species in the same manner as those naturalists treat genera, who admit that genera are merely artificial combinations made for convenience. This may not be a cheering prospect; but we shall at least be freed from the vain search for the undiscovered and undiscoverable essence of the term species.

The other and more general departments of natural history will rise greatly in interest. The terms used by naturalists of affinity, relationship, community of type, paternity, morphology, adaptive characters, rudimentary and aborted organs, &c., will cease to be metaphorical, and will have a plain signification. When we no longer look at an organic being as a savage looks at a ship, as at something wholly beyond his comprehension; when we regard every produc-tion of nature as one which has had a history; when we contemplate every complex structure and instinct as the summing up of many contrivances, each useful to the possessor, nearly in the same way as when we look at any great mechanical invention as the summing up of the labour, the experience, the reason, and even the blunders of numerous workmen; when we thus view each organic being, how far more interesting, I speak from experience, will the study of natural history become!

A grand and almost untrodden field of inquiry will be opened, on the causes and laws of variation, on correlation of growth, on the effects of use and disuse, on the direct action of external conditions, and so forth. The study of domestic productions will rise immensely in value. A new variety raised by man will be a far more important and interesting subject for study than one more species added to the infinitude of already recorded species. Our classifications will come to be, as far as they can be so made, genealogies; and will then truly give what may be called the plan of creation. The rules for classifying will no doubt become simpler when we have a definite object in view. We possess no pedigrees or armorial bearings; and we have to discover and trace the many diverging lines of descent in our natural genealogies, by characters of any kind which have long been inherited. Rudimentary organs will speak infallibly with respect to the nature of long-lost structures. Species and groups of species, which are called aberrant, and which may fancifully be called living fossils, will aid us in forming a picture of the ancient forms of life. Embryology will reveal to us the structure, in some degree obscured, of the prototypes of each great class.

When we can feel assured that all the individuals of the same species, and all the closely allied species of most genera, have within a not very remote period descended from one parent, and have migrated from some one birthplace; and when we better know the many means of migration, then, by the light which geology now throws, and will continue to throw, on former changes of climate and of the level of the land, we shall surely be enabled to trace in an admirable manner the former migrations of the inhabitants of the whole world. Even at present, by comparing the differences of the inhabitants of the sea on the opposite sides of a continent, and the nature of the various inhabitants of that continent in relation to their apparent means of immigration, some light can be thrown on ancient geography.

The noble science of Geology loses glory from the extreme imperfection of the record. The crust of the earth with its embedded remains must not be looked at as a well-filled museum, but as a poor collection made at hazard and at rare intervals. The accumulation of each great fossiliferous formation will be recognized as having depended on an unusual concurrence of circumstances, and the blank intervals between the successive stages as having been of vast duration. But we shall be able to gauge with some security the duration of these intervals by a comparison of the preceding and succeeding organic forms. We must be cautious in attempting to correlate as strictly contemporaneous two formations, which include few identical species, by the general succession of their forms of life. As species are produced and exterminated by slowly acting and still existing causes, and not by miraculous acts of creation and by catastrophes; and as the most important of all causes of organic change is one which is almost independent of altered and perhaps suddenly altered physical conditions, namely, the mutual relation of organism to organism, – the improvement of one being entailing the improvement or the extermination of others; it follows, that the amount of organic change in the fossils of consecutive formations probably serves as a fair measure of the lapse of actual time. A number of species, however, keeping in a body might remain for a long period unchanged, whilst within this same period, several of these species, by migrating into new countries and coming into competition with foreign associates, might become modified; so that we must not overrate the accuracy of organic change as a measure of time. During early periods of the earth's history, when the forms of life were probably fewer and simpler, the rate of change was probably slower; and at the first dawn of life, when very few forms of the simplest structure existed, the rate of change may have been slow in an extreme degree. The whole history of the world, as at present known, although of a length quite incomprehensible by us, will hereafter be recognized as a mere fragment of time, compared with the ages which have elapsed since the first creature, the progenitor of innumerable extinct and living descendants, was created.

In the distant future I see open fields for far more important researches. Psychology will be based on a new foundation, that of the necessary acquirement of each mental power and capacity by gradation. Light will be thrown on the origin of man and his history.

Authors of the highest eminence seem to be fully satisfied with the view that each species has been independently created. To my mind it accords better with what we know of the laws impressed on matter by the Creator, that the production and extinction of the past and present inhabitants of the world should have been due to secondary causes, like those determining the birth and death of the individual. When I view all beings not as special creations, but as the lineal descendants of some few beings which lived long before the first bed of the Silurian system was deposited, they seem to me to become ennobled. Judging from the past, we may safely infer that not one living species will transmit its unaltered likeness to a distant futurity. And of the species now living very few will transmit progeny of any kind to a far distant futurity; for the manner in which all organic beings are grouped, shows that the greater number of species of each genus, and all the species of many genera, have left no descendants, but have become utterly extinct. We can so far take a prophetic glance into futurity as to foretell that it will be the common and widely-spread species, belonging to the larger and dominant groups, which will ultimately prevail and procreate new and dominant species. As all the living forms of life are the lineal descendants of those which lived long before the Silurian epoch, we may feel certain that the ordinary succession by generation has never once been broken, and that no cataclysm has desolated the whole world. Hence we may look with some confidence to a secure future of equally inappreciable length. And as natural selection works solely by and for the good of each being, all corporeal and mental endowments will tend to progress towards perfection.

It is interesting to contemplate an entangled bank, clothed with many plants of many kinds, with birds singing on the bushes, with various insects flitting about, and with worms crawling through the damp earth, and to reflect that these elaborately constructed forms, so different from each other, and dependent on each other in so complex a manner, have all been produced by laws acting around us. These laws, taken in the largest sense, being Growth with Reproduction; Inheritance which is almost implied by reproduction; Variability from the indirect and direct action of the external conditions of life, and from use and disuse; a Ratio of Increase so high as to lead to a Struggle for Life, and as a consequence to Natural Selection, entailing Divergence of Character and the Extinction of less-improved forms. Thus, from the war of nature, from famine and death, the most exalted object which we are capable of conceiving, namely, the production of the higher animals, directly follows. There is grandeur in this view of life, with its several powers, having been originally breathed into a few forms or into one; and that, whilst this planet has gone cycling on according to the fixed law of gravity, from so simple a beginning endless forms most beautiful and most wonderful have been, and are being, evolved.

GLOSSARY

OF THE PRINCIPAL SCIENTIFIC TERMS
USED IN THE PRESENT VOLUME*

ABERRANT: Forms or groups of animals or plants which deviate in important characters from their nearest allies, so as not to be easily included in the same group with them, are said to be aberrant.

ABERRATION (IN OPTICS): In the refraction of light by a convex lens the rays passing through different parts of the lens are brought to a focus at slightly different distances, – this is called *spherical aberration*; at the same time the coloured rays are separated by the prismatic action of the lens and likewise brought to a focus at different distances, – this is *chromatic aberration*.

ABNORMAL: Contrary to the general rule.

ABORTED: An organ is said to be aborted, when its development has been arrested at a very early stage.

ALBINISM: Albinos are animals in which the usual colouring matters characteristic of the species have not been produced in the skin and its appendages. Albinism is the state of being an albino.

ALGAE: A class of plants including the ordinary sea-weeds and the filamentous fresh-water weeds.

ALTERNATION OF GENERATIONS: This term is applied to a peculiar mode of reproduction which prevails among many of the lower animals, in which the egg produces a living form quite different from its parent, but from which the parent-form is

*I am indebted to the kindness of Mr. W. S. Dallas for this Glossary, which has been given because several readers have complained to me that some of the terms used were unintelligible to them. Mr. Dallas has endeavoured to give the explanations of the terms in as popular a form as possible.

reproduced by a process of budding, or by the division of the substance of the first product of the egg.

AMMONITES: A group of fossil, spiral, chambered shells, allied to the existing pearly Nautilus, but having the partitions between the chambers waved in complicated patterns at their junction with the outer wall of the shell.

ANALOGY: The resemblance of structures which depends upon similarity of function, as in the wings of insects and birds. Such structures are said to be *analogous*, and to be *analogues* of each other.

ANIMALCULE: A minute animal: generally applied to those visible only by the microscope.

ANNELIDS: A class of worms in which the surface of the body exhibits a more or less distinct division into rings or segments, generally provided with appendages for locomotion and with gills. It includes the ordinary marine worms, the earthworms, and the leeches.

ANTENNAE: Jointed organs appended to the head in Insects, Crustacea and Centipedes, and not belonging to the mouth.

ANTHERS: The summits of the stamens of flowers, in which the pollen or fertilizing dust is produced.

APLACENTALIA, APLACENTATA OR APLACENTAL MAMMALS. See *Mammalia*.

ARCHETYPAL: Of or belonging to the Archetype, or ideal primitive form upon which all the beings of a group seem to be organized.

ARTICULATA: A great division of the Animal Kingdom characterized generally by having the surface of the body divided into rings called segments, a greater or less number of which are furnished with jointed legs (such as Insects, Crustaceans and Centipedes).

ASYMMETRICAL: Having the two sides unlike.

ATROPHIED: Arrested in development at a very early stage.

BALANUS: The genus including the common Acorn-shells which live in abundance on the rocks of the sea-coast.

BATRACHIANS: A class of animals allied to the Reptiles, but undergoing a peculiar metamorphosis in which the young animal is generally aquatic and breathes by gills. (*Examples*, Frogs, Toads, and Newts.)

BOULDERS: Large transported blocks of stone generally embedded in clays or gravels.

BRACHIOPODA: A class of marine Mollusca, or soft-bodied animals, furnished with a bivalve shell, attached to submarine objects by a stalk which passes through an aperture in one of the valves, and furnished with fringed arms, by the action of which food is carried to the mouth.

BRANCHIAE: Gills or organs for respiration in water.

BRANCHIAL: Pertaining to gills or branchiae.

CAMBRIAN SYSTEM: A series of very ancient alaeozoic rocks, between the Laurentian and the Silurian. Until recently these were regarded as the oldest fossiliferous rocks.

CANIDAE: The Dog-family, including the Dog, Wolf, Fox, Jackal, &c.

CARAPACE: The shell enveloping the anterior part of the body in Crustaceans generally; applied also to the hard shelly pieces of the Cirripedes.

CARBONIFEROUS: This term is applied to the great formation which includes among other rocks, the coal-measures. It belongs to the oldest, or Palaeozoic, system of formations.

CAUDAL: Of or belonging to the tail.

CEPHALOPODS: The highest class of the Mollusca, or soft-bodied animals, characterized by having the mouth surrounded by a greater or less number of fleshy arms or tentacles, which, in most living species, are furnished with sucking-cups. (*Examples*, Cuttle-fish, Nautilus.)

CETACEA: An order of Mammalia, including the Whales, Dolphins, &c., having the form of the body fish-like, the skin naked and only the fore-limbs developed.

CHELONIA: An order of Reptiles including the Turtles, Tortoises, &c.

CIRRIPEDES: An order of Crustaceans including the Barnacles and Acorn-shells. Their young resemble those of many other Crustaceans in form; but when mature they are always attached to other objects, either directly or by means of a stalk, and their bodies are enclosed by a calcareous shell composed of several pieces, two of which can open to give issue to a bunch of curled, jointed tentacles, which represent the limbs.

COCCUS: The genus of Insects including the Cochineal. In these the male is a minute, winged fly, and the female generally a motionless, berry-like mass.

COCOON: A case usually of silky material, in which insects are frequently enveloped during the second or resting-stage (pupa) of their existence. The term 'cocoon-stage' is here used as equivalent to 'pupa-stage.'

COELOSPERMOUS: A term applied to those fruits of the Umbelliferae which have the seed hollowed on the inner face.

COLEOPTERA: Beetles, an order of Insects, having a biting mouth and the first pair of wings more or less horny, forming sheaths for the second pair, and usually meeting in a straight line down the middle of the back.

COLUMN: A peculiar organ in the flowers of Orchids, in which the stamens, style and stigma (or the reproductive parts) are united.

COMPOSITAE OR COMPOSITOUS PLANTS: Plants in which the inflorescence consists of numerous small flowers (florets) brought together into a dense head, the base of which is enclosed by a common envelope. (*Examples*, the Daisy, Dandelion, &c.)

CONFERVAE: The filamentous weeds of fresh water.

CONGLOMERATE: A rock made up of fragments of rock or pebbles, cemented together by some other material.

COROLLA: The second envelope of a flower usually composed of coloured, leaf-like organs (petals), which may be united by their edges either in the basal part or throughout.

CORRELATION: The normal coincidence of one phenomenon, character, &c., with another.

CORYMB: A bunch of flowers in which those springing from the lower part of the flower stalk are supported on long stalks so as to be nearly on a level with the upper ones.

COTYLEDONS: The first or seed-leaves of plants.

CRUSTACEANS: A class of articulated animals, having the skin of the body generally more or less hardened by the deposition of calcareous matter, breathing by means of gills. (*Examples*, Crab, Lobster, Shrimp, &c.)

CURCULIO: The old generic term for the Beetles known as Weevils, characterized by their four-jointed feet, and by the head being produced into a sort of beak, upon the sides of which the antennae are inserted.

CUTANEOUS: Of or belonging to the skin.

DEGRADATION: The wearing down of land by the action of the sea or of meteoric agencies.

DENUDATION: The wearing away of the surface of the land by water.

DEVONIAN SYSTEM OR FORMATION: A series of Palaeozoic rocks, including the Old Red Sandstone.

DICOTYLEDONS OR DICOTYLEDONOUS PLANTS: A class of plants characterized by having two seed-leaves, by the formation of new wood between the bark and the old wood (oxogenous growth) and by the reticulation of the veins of the leaves. The parts of the flowers are generally in multiples of five.

DIFFERENTIATION: The separation or discrimination of parts or organs which in simpler forms of life are more or less united.

DIMORPHIC: Having two distinct forms. – Dimorphism is the condition of the appearance of the same species under two dissimilar forms.

DIOECIOUS: Having the organs of the sexes upon distinct individuals.

DIORITE: A peculiar form of Greenstone.

DORSAL: Of or belonging to the back.

EDENTATA: A peculiar order of Quadrupeds, characterized by the absence of at least the middle incisor (front) teeth in both jaws. (*Examples*, the Sloths and Armadillos.)

ELYTRA: The hardened fore-wings of Beetles, serving as sheaths for the membranous hind-wings, which constitute the true organs of flight.

EMBRYO: The young animal undergoing development within the egg or womb.

EMBRYOLOGY: The study of the development of the embryo.

ENDEMIC: Peculiar to a given locality.

ENTOMOSTRACA: A division of the class Crustacea, having all the segments of the body usually distinct, gills attached to the feet or organs of the mouth, and the feet fringed with fine hairs. They are generally of small size.

EOCENE: The earliest of the three divisions of the Tertiary epoch of geologists. Rocks of this age contain a small proportion of shells identical with species now living.

EPHEMEROUS INSECTS: Insects allied to the May-fly.

FAUNA: The totality of the animals naturally inhabiting a certain country or region, or which have lived during a given geological period.

FELIDAE: The Cat-family.

FERAL: Having become wild from a state of cultivation or domestication.

FLORA: The totality of the plants growing naturally in a country, or during a given geological period.

FLORETS: Flowers imperfectly developed in some respects, and collected into a dense spike or head, as in the Grasses, the Dandelion, &c.

FOETAL: Of or belonging to the foetus, or embryo in course of development.

FORAMINIFERA: A class of animals of very low organization, and generally of small size, having a jelly-like body, from the surface of which delicate filaments can be given off and retracted for the prehension of external objects, and having a calcareous or sandy shell, usually divided into chambers, and perforated with small apertures.

FOSSILIFEROUS: Containing fossils.

FOSSORIAL: Having a faculty of digging. The Fossorial Hymenoptera are a group of Wasp-like Insects, which burrow in sandy soil to make nests for their young.

FRENUM (PL. FRENA): A small band or fold of skin.

FUNGI (SING. FUNGUS): A class of cellular plants, of which Mushrooms, Toadstools, and Moulds, are familiar examples.

FURCULA: The forked bone formed by the union of the collar-bones in many birds, such as the common Fowl.

GALLINACEOUS BIRDS: An order of Birds of which the common Fowl, Turkey and Pheasant, are well-known examples.

GALLUS: The genus of birds which includes the common Fowl.

GANGLION: A swelling or knot from which nerves are given off as from a centre.

GANOID FISHES: Fishes covered with peculiar enamelled bony scales. Most of them are extinct.

GERMINAL VESICLE: A minute vesicle in the eggs of animals, from which the development of the embryo proceeds.

GLACIAL PERIOD: A period of great cold and of enormous extension of ice upon the surface of the earth. It is believed that glacial periods have occurred repeatedly during the geological history of the earth, but the term is generally applied to the close of the Tertiary epoch, when nearly the whole of Europe was subjected to an arctic climate.

GLAND: An organ which secretes or separates some peculiar product from the blood or sap of animals or plants.

GLOTTIS: The opening of the windpipe into the oesophagus or gullet.

GNEISS: A rock approaching granite in composition, but more or less laminated, and really produced by the alteration of a sedimentary deposit after its consolidation.

GRALLATORES: The so-called Wading-birds (Storks, Cranes, Snipes, &c.), which are generally furnished with long legs, bare of feathers above the heel, and have no membranes between the toes.

GRANITE: A rock consisting essentially of crystals of felspar and mica in a mass of quartz.

HABITAT: The locality in which a plant or animal naturally lives.

HEMIPTERA: An order or sub-order of Insects, characterized by the possession of a jointed beak or rostrum, and by having the fore-wings horny in the basal portion and membranous at the extremity, where they cross each other. This group includes the various species of Bugs.

HERMAPHRODITE: Possessing the organs of both sexes.

HOMOLOGY: That relation between parts which results from their development from corresponding embryonic parts, either in different animals, as in the case of the arm of man, the fore-leg of a quadruped, and the wing of a bird; or in the same individual, as in the case of the fore and hind legs in quadrupeds, and the segments or rings and their appendages of which the body of a worm, a centipede, &c., is composed. The latter is called *serial homology*. The parts which stand in such a relation to each other are said to be *homologous*, and one such part or organ is called the *homologue* of the other. In different plants the parts of the flower are homologous, and in general these parts are regarded as homologous with leaves.

HOMOPTERA: An order or sub-order of Insects having (like the Hemiptera) a jointed beak, but in which the fore-wings are either wholly membranous or wholly leathery. The *Cicadoe*, Frog-hoppers, and *Aphides*, are well-known examples.

HYBRID: The offspring of the union of two distinct species.

HYMENOPTERA: An order of Insects possessing biting jaws and usually four membranous wings in which there are a few veins. Bees and Wasps are familiar examples of this group.

HYPERTROPHIED: Excessively developed.

ICHNEUMONIDAE: A family of Hymenopterous insects, the members of which lay their eggs in the bodies or eggs of other insects.

IMAGO: The perfect (generally winged) reproductive state of an insect.

INDIGENS: The aboriginal animal or vegetable inhabitants of a country or region.

INFLORESCENCE: The mode of arrangement of the flowers of plants.

INFUSORIA: A class of microscopic Animalcules, so called from their having originally been observed in infusions of vegetable matters. They consist of a gelatinous material enclosed in a delicate membrane, the whole or part of which is furnished with short vibrating hairs (called cilia), by means of which the animalcules swim through the water or convey the minute particles of their food to the orifice of the mouth.

INSECTIVOROUS: Feeding on Insects.

INVERTEBRATA, OR INVERTEBRATE ANIMALS: Those animals which do not possess a backbone or spinal column.

LACUNAE: Spaces left among the tissues in some of the lower animals, and serving in place of vessels for the circulation of the fluids of the body.

LAMELLATED: Furnished with lamellae or little plates.

LARVA (PL. LARVAE): The first condition of an insect at its issuing from the egg, when it is usually in the form of a grub, caterpillar, or maggot.

LARYNX: The upper part of the windpipe opening into the gullet.

LAURENTIAN: A group of greatly altered and very ancient rocks, which is greatly developed along the course of the St. Laurence, whence the name. It is in these that the earliest known traces of organic bodies have been found.

LEGUMINOSAE: An order of plants represented by the common Peas and Beans, having an irregular flower in which one petal stands up like a wing, and the stamens and pistil are enclosed in a sheath formed by two other petals. The fruit is a pod (or legume).

LEMURIDAE: A group of four-handed animals, distinct from the Monkeys and approaching the Insectivorous Quadrupeds in some of their characters and habits. Its members have the nostrils curved or twisted, and a claw instead of a nail upon the first finger of the hind hands.

LEPIDOPTERA: An order of Insects, characterized by the possession of a spiral proboscis, and of four large more or less scaly wings. It includes the well-known Butterflies and Moths.

LITTORAL: Inhabiting the seashore.

LOESS: A marly deposit of recent (Post-Tertiary) date, which occupies a great part of the valley of the Rhine.

MALACOSTRACA: The higher division of the Crustacea, including the ordinary Crabs, Lobsters, Shrimps, &c., together with the Woodlice and Sand-hoppers.

MAMMALIA: The highest class of animals, including the ordinary hairy quadrupeds, the Whales, and Man, and characterized by the production of living young which are nourished after birth by milk from the teats (*Mammoe, Mammary glands*) of the mother. A striking difference in embryonic development has led to the division of this class into two great groups, in one of these, when the embryo has attained a certain stage, a vascular connection, called the *placenta*, is formed between the embryo and the mother; in the other this is wanting, and the young are produced in a very incomplete state. The former, including the greater part of the class, are called *Placental mammals*; the latter, or *Aplacental mammals*, include the Marsupials and Monotremes (*Ornithorhynchus*).

MAMMIFEROUS: Having mammae or teats (See *Mammalia*).

MANDIBLES, IN INSECTS: The first or uppermost pair of jaws, which are generally solid, horny, biting organs. In Birds the term is applied to both jaws with their horny coverings. In Quadrupeds the mandible is properly the lower jaw.

MARSUPIALS: An order of Mammalia in which the young are born in a very incomplete state of development, and carried by the mother, while sucking, in a ventral pouch (marsupium), such as the Kangaroos, Opossums, &c. (See *Mammalia*).

MAXILLAE, IN INSECTS: The second or lower pair of jaws, which are composed of several joints and furnished with peculiar jointed appendages called palpi, or feelers.

MELANISM: The opposite of albinism; an undue development of colouring material in the skin and its appendages.

METAMORPHIC ROCKS: Sedimentary rocks which have undergone alteration, generally by the action of heat, subsequently to their deposition and consolidation.

MOLLUSCA: One of the great divisions of the Animal Kingdom, including those animals which have a soft body, usually furnished with a shell, and in which the nervous ganglia, or centres, present no definite general arrangement. They are generally known under the denomination of 'shell-fish'; the cuttle-fish, and the common snails, whelks, oysters, mussels, and cockles, may serve as examples of them.

MONOCOTYLEDONS, OR MONOCOTYLEDONOUS PLANTS: Plants in which the seed sends up only a single seed-leaf (or cotyledon), characterized by the absence of consecutive layers of wood in the stem (endogenous growth), by the veins of the leaves being generally straight, and by the parts of the flowers being generally in multiples of three. (*Examples*, Grasses, Lilies, Orchids, Palms, &c.)

MORAINES: The accumulations of fragments of rock brought down by glaciers.

MORPHOLOGY: The law of form or structure independent of function.

MYSIS-STAGE: A stage in the development of certain Crustaceans (Prawns), in which they closely resemble the adults of a genus (*Mysis*) belonging to a slightly lower group.

NASCENT: Commencing development.

NATATORY: Adapted for the purpose of swimming.

NAUPLIUS-FORM: The earliest stage in the development of many Crustacea, especially belonging to the lower groups. In this stage the animal has a short body, with indistinct indications of a division into segments, and three pairs of fringed limbs. This form of the common fresh-water *Cyclops* was described as a distinct genus under the name of *Nauplius*.

NEURATION: The arrangement of the veins or nervures in the wings of Insects.

NEUTERS: Imperfectly developed females of certain social insects (such as Ants and Bees), which perform all the labours of the community. Hence they are also called *workers*.

NICTITATING MEMBRANE: A semi-transparent membrane, which can be drawn across the eye in Birds and Reptiles, either to moderate the effects of a strong light or to sweep particles of dust, &c., from the surface of the eye.

OCELLI: The simple eyes or stemmata of Insects, usually situated on the crown of the head between the great compound eyes.

OESOPHAGUS: The gullet.

OOLITIC: A great series of secondary rocks, so called from the texture of some of its members, which appear to be made up of a mass of small *egg-like* calcareous bodies.

OPERCULUM: A calcareous plate employed by many Mollusca to close the aperture of their shell. The *opercular valves* of Cirripedes are those which close the aperture of the shell.

ORBIT: The bony cavity for the reception of the eye.

ORGANISM: An organized being, whether plant or animal.

ORTHOSPHERMOUS: A term applied to those fruits of the Umbelliferae which have the seed straight.

OSCULANT: Forms or groups apparently intermediate between and connecting other groups are said to be osculant.

OVA: Eggs.

OVARIUM OR OVARY (IN PLANTS): The lower part of the pistil or female organ of the flower, containing the ovules or incipient seeds; by growth after the other organs of the flower have fallen, it usually becomes converted into the fruit.

OVIGEROUS: Egg-bearing.

OVULES (OF PLANTS): The seeds in the earliest condition.

PACHYDERMS: A group of Mammalia, so called from their thick skins, and including the Elephant, Rhinoceros, Hippopotamus, &c.

PALAEOZOIC: The oldest system of fossiliferous rocks.

PALPI: Jointed appendages to some of the organs of the mouth in Insects and Crustacea.

PAPILIONACEAE: An order of Plants (see *Leguminosae*). The flowers of these plants are called *papilionaceous*, or butterfly-like, from the fancied resemblance of the expanded superior petals to the wings of a butterfly.

PARASITE: An animal or plant living upon or in, and at the expense of, another organism.

PARTHENOGENESIS: The production of living organisms from unimpregnated eggs or seeds.

PEDUNCULATED: Supported upon a stem or stalk. The pedunculated oak has its acorns borne upon a footstool.

PELORIA OR *PELORISM*: The appearance of regularity of structure in the flowers of plants which normally bear irregular flowers.

PELVIS: The bony arch to which the hind limbs of vertebrate animals are articulated.

PETALS: The leaves of the corolla, or second circle of organs in a flower. They are usually of delicate texture and brightly coloured.

PHYLLODINEOUS: Having flattened, leaf-like twigs or leafstalks instead of true leaves.

PIGMENT: The colouring material produced generally in the superficial parts of animals. The cells secreting it are called *pigment-cells*.

PINNATE: Bearing leaflets on each side of a central stalk.

PISTILS: The female organs of a flower, which occupy a position in the centre of the other floral organs. The pistil is generally divisible into the ovary or germen, the style and the stigma.

PLACENTALIA, PLACENTATA, OR PLACENTAL MAMMALS: See *Mammalia*.

PLANTIGRADES: Quadrupeds which walk upon the whole sole of the foot, like the Bears.

PLASTIC: Readily capable of change.

PLEISTOCENE PERIOD: The latest portion of the Tertiary epoch.

PLUMULE (IN PLANTS): The minute bud between the seed-leaves of newly germinated plants.

PLUTONIC ROCKS: Rocks supposed to have been produced by igneous action in the depths of the earth.

POLLEN: The male element in flowering plants; usually a fine dust produced by the anthers, which, by contact with the stigma, effects the fecundation of the seeds. This impregnation is brought about by means of tubes (*pollen-tubes*) which issue from the pollen-grains adhering to the stigma, and penetrate through the tissues until they reach the ovary.

POLYANDROUS (FLOWERS): Flowers having many stamens.

POLYGAMOUS PLANTS: Plants in which some flowers are unisexual and others hermaphrodite. The unisexual (male and female) flowers, may be on the same or on different plants.

POLYMORPHIC: Presenting many forms.

POLYZOARY: The common structure formed by the cells of the Polyzoa, such as the well-known Sea-mats.

PREHENSILE: Capable of grasping.

PREPOTENT: Having a superiority of power.

PRIMARIES: The feathers forming the tip of the wing of a bird, and inserted upon that part which represents the hand of man.

PROCESSES: Projecting portions of bones, usually for the attachment of muscles, ligaments, &c.

PROPOLIS: A resinous material collected by the Hive-Bees from the opening buds of various trees.

PROTEAN: Exceedingly variable.

PROTOZOA: The lowest great division of the Animal Kingdom. These animals are composed of a gelatinous material, and show scarcely any trace of distinct organs. The Infusoria, Foraminifera, and Sponges, with some other forms, belong to this division.

PUPA (PL. PUPAE): The second stage in the development of an Insect, from which it emerges in the perfect (winged) reproductive form. In most insects the *pupal stage* is passed in perfect repose. The *chrysalis* is the pupal state of butterflies.

RADICLE: The minute root of an embryo plant.

RAMUS: One half of the lower jaw in the Mammalia. The portion which rises to articulate with the skull is called the *ascending ramus*.

RANGE: The extent of country over which a plant or animal is naturally spread. *Range in time* expresses the distribution of a species or group through the fossiliferous beds of the earth's crust.

RETINA: The delicate inner coat of the eye, formed by nervous filaments spreading from the optic nerve, and serving for the perception of the impressions produced by light.

RETROGRESSION: Backward development. When an animal, as it approaches maturity, becomes less perfectly organized than might be expected from its early stages and known relationships, it is said to undergo a *retrograde development* or *metamorphosis*.

RHIZOPODS: A class of lowly organized animals (Protozoa), having a gelatinous body, the surface of which can be protruded in the form of root-like processes or filaments, which serve for locomotion and the prehension of food. The most important order is that of the Foraminifera.

RODENTS: The gnawing Mammalia, such as the Rats, Rabbits, and Squirrels. They are especially characterized by the possession of a single pair of chisel-like cutting teeth in each jaw, between which and the grinding teeth there is a great gap.

RUBUS: The Bramble Genus.

RUDIMENTARY: Very imperfectly developed.

RUMINANTS: The group of Quadrupeds which ruminate or chew the cud, such as oxen, sheep, and deer. They have divided hoofs, and are destitute of front teeth in the upper jaw.

SACRAL: Belonging to the sacrum, or the bone composed usually of two or more united vertebrae to which the sides of the pelvis in vertebrate animals are attached.

SARCODE: The gelatinous material of which the bodies of the lowest animals (Protozoa) are composed.

SCUTELLAE: The horny plates with which the feet of birds are generally more or less covered, especially in front.

SEDIMENTARY FORMATIONS: Rocks deposited as sediments from water.

SEGMENTS: The transverse rings of which the body of an articulate animal or Annelid is composed.

SEPALS: The leaves or segments of the calyx, or outermost envelope of an ordinary flower. They are usually green, but sometimes brightly coloured.

SERRATURES: Teeth like those of a saw.

SESSILE: Not supported on a stem or footstalk.

SILURIAN SYSTEM: A very ancient system of fossiliferous rocks belonging to the earlier parts of the Palaeozoic series.

SPECIALIZATION: The setting apart of a particular organ for the performance of a particular function.

SPINAL CORD: The central portion of the nervous system in the Vertebrata, which descends from the brain through the arches of the vertebrae, and gives off nearly all the nerves to the various organs of the body.

STAMENS: The male organs of flowering plants, standing in a circle within the petals. They usually consist of a filament and an anther, the anther being the essential part in which the pollen, or fecundating dust, is formed.

STERNUM: The breast-bone.

STIGMA: The apical portion of the pistil in flowering plants.

STIPULES: Small leafy organs placed at the base of the footstalks of the leaves in many plants.

STYLE: The middle portion of the perfect pistil, which rises like a column from the ovary and supports the stigma at its summit.

SUBCUTANEOUS: Situated beneath the skin.

SUCTORIAL: Adapted for sucking.

SUTURES (IN THE SKULL): The lines of junction of the bones of which the skull is composed.

TARSUS (PL. TARSI): The jointed feet of articulate animals, such as Insects.

TELEOSTEAN FISHES: Fishes of the kind familiar to us in the present day, having the skeleton usually completely ossified and the scales horny.

TENTACULA OR TENTACLES: Delicate fleshy organs of prehension or touch possessed by many of the lower animals.

TERTIARY: The latest geological epoch, immediately preceding the establishment of the present order of things.

TRACHEA: The windpipe or passage for the admission of air to the lungs.

TRIDACTYLE: Three-fingered, or composed of three movable parts attached to a common base.

TRILOBITES: A peculiar group of extinct Crustaceans, somewhat resembling the Woodlice in external form, and, like some of them, capable of rolling themselves up into a ball. Their remains are found only in the Palaeozoic rocks, and most abundantly in those of Silurian age.

TRIMORPHIC: Presenting three distinct forms.

UMBELLIFERAE: An order of plants in which the flowers, which contain five stamens and a pistil with two styles, are supported upon footstalks which spring from the top of the flower stem and spread out like the wires of an umbrella, so as to bring all the flowers in the same head (*umbel*) nearly to the same level. (*Examples*, Parsley and Carrot.)

UNGULATA: Hoofed quadrupeds.

UNICELLULAR: Consisting of a single cell.

VASCULAR: Containing blood-vessels.

VERMIFORM: Like a worm.

VERTEBRATA; OR VERTEBRATE ANIMALS: The highest division of the animal kingdom, so called from the presence in most cases of a backbone composed of numerous joints or *vertebrae*, which constitutes the centre of the skeleton and at the same time supports the central parts of the nervous system.

WHORLS: The circles or spiral lines in which the parts of plants are arranged upon the axis of growth.

WORKERS: See Neuters.

ZOËA-STAGE: The earliest stage in the development of many of the higher Crustacea, so called from the name of *Zoëa* applied to these young animals when they were supposed to constitute a peculiar genus.

ZOOIDS: In many of the lower animals (such as the Corals, Medusae, &c.) reproduction takes place in two ways, namely, by means of eggs and by a process of budding with or without separation from the parent of the product of the latter, which is often very different from that of the egg. The individuality of the species is represented by the whole of the form produced between two sexual reproductions; and these forms, which are apparently individual animals, have been called *zooids*.

INDEXES

INDEX TO *THE VOYAGE OF THE BEAGLE*

CHINUA ACHEBE
Things Fall Apart

THE ARABIAN NIGHTS
(2 vols, tr. Husain Haddawy)

AUGUSTINE
The Confessions

JANE AUSTEN
Emma
Mansfield Park
Northanger Abbey
Persuasion
Pride and Prejudice
Sanditon and Other Stories
Sense and Sensibility

HONORÉ DE BALZAC
Cousin Bette
Eugénie Grandet
Old Goriot

SIMONE DE BEAUVOIR
The Second Sex

SAMUEL BECKETT
Molloy, Malone Dies,
The Unnamable
(US only)

SAUL BELLOW
The Adventures of Augie March

HECTOR BERLIOZ
The Memoirs of Hector Berlioz

WILLIAM BLAKE
Poems and Prophecies

JORGE LUIS BORGES
Ficciones

JAMES BOSWELL
The Life of Samuel Johnson
The Journal of a Tour to
the Hebrides

CHARLOTTE BRONTË
Jane Eyre
Villette

EMILY BRONTË
Wuthering Heights

MIKHAIL BULGAKOV
The Master and Margarita

SAMUEL BUTLER
The Way of all Flesh

JAMES M. CAIN
The Postman Always Rings Twice
Double Indemnity
Mildred Pierce
Selected Stories
(in 1 vol.)

ITALO CALVINO
If on a winter's night a traveler

ALBERT CAMUS
The Outsider (UK)
The Stranger (US)

WILLA CATHER
Death Comes for the Archbishop
My Ántonia
(US only)

MIGUEL DE CERVANTES
Don Quixote

RAYMOND CHANDLER
The novels (2 vols)
Collected Stories

GEOFFREY CHAUCER
Canterbury Tales

ANTON CHEKHOV
My Life and Other Stories
The Steppe and Other Stories

KATE CHOPIN
The Awakening

CARL VON CLAUSEWITZ
On War

S. T. COLERIDGE
Poems

WILKIE COLLINS
The Moonstone
The Woman in White

CONFUCIUS
The Analects

JOSEPH CONRAD
Heart of Darkness
Lord Jim
Nostromo
The Secret Agent
Typhoon and Other Stories
Under Western Eyes
Victory

This book is set in BEMBO which was cut
by the punch-cutter Francesco Griffo
for the Venetian printer-publisher
Aldus Manutius in early 1495
and first used in a pamphlet
by a young scholar
named Pietro
Bembo.